TABLE 3.4 Common Fourier transform pairs

	Time	Frequency
1	$\delta(t)$	1
2	$e^{-at}u_h(t),\ a>0$	$\dfrac{1}{a+j\omega}$
3	$p_\tau^h(t)$	$\tau\,\mathrm{sinc}\!\left(\dfrac{\omega\tau}{2\pi}\right)=\tau\,\dfrac{\sin(\omega\tau/2)}{\omega\tau/2}$
4	$\Delta_\tau(t)$	$\dfrac{\tau}{2}\,\mathrm{sinc}^2\!\left(\dfrac{\omega\tau}{4}\right)$
5	1	$2\pi\delta(\omega)$
6	const	const $\times\,2\pi\delta(\omega)$
7	$\cos(\omega_0 t)$	$\pi[\delta(\omega+\omega_0)+\delta(\omega-\omega_0)]$
8	$\sin(\omega_0 t)$	$j\pi[\delta(\omega+\omega_0)-\delta(\omega-\omega_0)]$
9	$\mathrm{sgn}(t)$	$\dfrac{2}{j\omega}$
10	$u_h(t)$	$\dfrac{1}{j\omega}+\pi\delta(\omega)$
11	$e^{-\alpha|t|},\ \alpha>0$	$\dfrac{2\alpha}{\alpha^2+\omega^2}$
12	$e^{j\omega_0 t}$	$2\pi\delta(\omega-\omega_0)$
13	$\displaystyle\sum_{n=-\infty}^{\infty} X_n e^{jn\omega_0 t},\ \ \omega_0=\dfrac{2\pi}{T}$	$2\pi\displaystyle\sum_{n=-\infty}^{\infty} X_n\delta(\omega-n\omega_0)$
14	$\displaystyle\sum_{n=-\infty}^{\infty}\delta(t-nT_0)$	$\omega_0\displaystyle\sum_{n=-\infty}^{\infty}\delta(\omega-n\omega_0),\ \ \omega_0=\dfrac{2\pi}{T_0}$

Linear Dynamic Systems
and Signals

Linear Dynamic Systems and Signals

Zoran Gajić

Pearson Education, Inc.
Upper Saddle River, New Jersey 07458

Library of Congress Cataloging-in-Publication Data

Gajic, Zoran.
    Linear dynamic systems and signals / Zoran Gajic.
      p. cm.
    Includes bibliographical references and index.
    ISBN 0-201-61854-0
    1. Signal processing—Mathematics. 2. Linear systems. I. Title.

    TK5102.9 .G355 2002   621.382$'$2—dc21   2002072593

Vice President and Editorial Director, ECS: *Marcia J. Horton*
Publisher: *Tom Robbins*
Associate Editor: *Alice Dworkin*
Editorial Assistant: *Jody McDonnell*
Vice President and Director of Production and Manufacturing, ESM: *David W. Riccardi*
Executive Managing Editor: *Vince O'Brien*
Managing Editor: *David A. George*
Production Editor: *Kathy Ewing*
Director of Creative Services: *Paul Belfanti*
Creative Director: *Carole Anson*
Art Director: *Jayne Conte*
Art Editor: *Greg Dulles*
Manufacturing Manager: *Trudy Pisciotti*
Manufacturing Buyer: *Lynda Castillo*
Marketing Manager: *Holly Stark*

  © 2003 by Pearson Education, Inc.
  Pearson Education, Inc.
  Upper Saddle River, NJ 07458

All rights reserved. No part of this book may be reproduced in any form or by any means, without permission in writing from the publisher.

The author and publisher of this book have used their best efforts in preparing this book. These efforts include the development, research, and testing of the theories and programs to determine their effectiveness. The author and publisher make no warranty of any kind, expressed or implied, with regard to these programs or the documentation contained in this book. The author and publisher shall not be liable in any event for incidental or consequential damages in connection with, or arising out of, the furnishing, performance, or use of these programs.

MATLAB and Simulink are registered trademarks of The MathWorks, Inc., 3 Apple Hill Drive, Natick, MA 01760-2098

Printed in the United States of America
10 9 8 7 6 5 4 3 2 1

ISBN 0-201-61854-0

Pearson Education Ltd., *London*
Pearson Education Australia Pty., Ltd., *Sydney*
Pearson Education Singapore, Pte. Ltd
Pearson Education North Asia Ltd, *Hong Kong*
Pearson Education Canada, Inc., *Toronto*
Pearson Educación de Mexico, S.A. de C.V.
Pearson Education—Japan, *Tokyo*
Pearson Education Malaysia, Pte. Ltd.
Pearson Education, Inc., *Upper Saddle River, New Jersey*

To my son Zoran-Zoki Gajić and his desire to learn

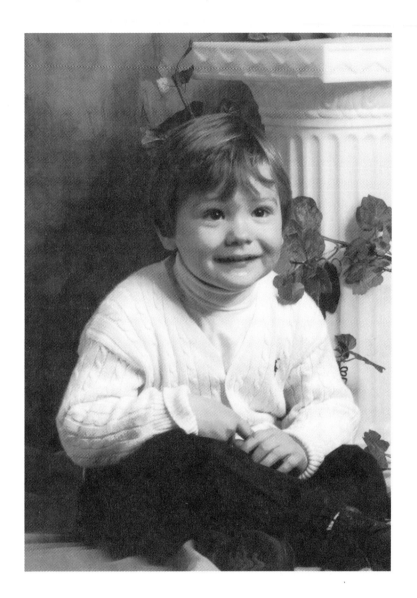

Contents

	Preface	xv
	About the Author	xxi
1	**Introduction to Linear Systems**	**1**
	1.1 Continuous and Discrete Signals and Linear Systems	2
	1.1.1 Continuous and Discrete Signals and Sampling	3
	1.1.2 Continuous- and Discrete-Time Systems	5
	1.2 System Linearity and Time Invariance	13
	1.2.1 System Linearity	13
	1.2.2 Linear System Time Invariance	16
	1.3 Mathematical Modeling of Systems	18
	1.4 System Classification	24
	1.5 MATLAB System Analysis and Design	26
	1.6 Organization of the Text	27
	1.7 Summary	28
	1.8 References	29
	1.9 Problems	30
2	**Introduction to Signals**	**33**
	2.1 Common Signals in Linear Systems	33
	2.1.1 Impulse Delta Signal	42
	2.2 Signal Operations	51
	2.3 Signal Classification	56
	2.4 MATLAB Laboratory Experiment on Signals	58
	2.5 Summary	59
	2.6 References	62
	2.7 Problems	63
I	**Frequency Domain Techniques**	**71**
3	**Fourier Series and Fourier Transform**	**73**
	3.1 Fourier Series	74
	3.1.1 From Fourier Series to Fourier Transform	84
	3.2 Fourier Transform and Its Properties	86
	3.2.1 Properties of the Fourier Transform	88
	3.2.2 Inverse Fourier Transform	104

3.3		Fourier Transform in System Analysis	106
	3.3.1	System Transfer Function and System Response	107
	3.3.2	Frequency Spectra	110
3.4		Fourier Series in System Analysis	116
	3.4.1	System Response to Periodic Inputs	116
	3.4.2	System Response to Sinusoidal Inputs	121
3.5		From Fourier Transform to Laplace Transform	124
3.6		Fourier Analysis MATLAB Laboratory Experiment	127
3.7		Summary	128
3.8		References	131
3.9		Problems	132

4 Laplace Transform — 143

4.1		Laplace Transform and Its Properties	144
	4.1.1	Definitions and Existence Condition	144
	4.1.2	Properties of the Laplace Transform	146
4.2		Inverse Laplace Transform	157
4.3		Laplace Transform in Linear System Analysis	165
	4.3.1	System Transfer Function and Impulse Response	166
	4.3.2	System Zero-State Response	170
	4.3.3	Unit Step and Ramp Responses	173
	4.3.4	Complete System Response	175
	4.3.5	Case Studies	179
4.4		Block Diagrams	183
4.5		From Laplace Transform to $\mathcal{Z}$-Transform	189
4.6		MATLAB Laboratory Experiment	191
4.7		Summary	192
4.8		References	195
4.9		Problems	196

5 $\mathcal{Z}$-Transform — 208

5.1		$\mathcal{Z}$-Transform and Its Properties	209
5.2		Inverse of the $\mathcal{Z}$-Transform	222
5.3		$\mathcal{Z}$-Transform in Linear System Analysis	228
	5.3.1	Two Formulations of Discrete-Time Linear Systems	228
	5.3.2	System Response Using the Integral Formulation	232
	5.3.3	System Response Using the Derivative Formulation	242
	5.3.4	Case Study: An ATM Computer Network Switch	249
5.4		Block Diagrams	254
5.5		Discrete-Time Frequency Spectra	258
	5.5.1	System Response to Sinusoidal Inputs	260
5.6		$\mathcal{Z}$-Transform MATLAB Laboratory Experiment	262
5.7		Summary	263

| | 5.8 | References . . . . | 266 |
| | 5.9 | Problems . . . . | 267 |

II Time Domain Techniques   277

6 Convolution   279
    6.1 Convolution of Continuous-Time Signals 280
           6.1.1 Graphical Convolution 283
    6.2 Convolution for Linear Continuous-Time Systems 292
    6.3 Convolution of Discrete-Time Signals 296
           6.3.1 Sliding Tape Method 298
    6.4 Convolution for Linear Discrete-Time Systems 303
    6.5 Numerical Convolution Using MATLAB 305
    6.6 MATLAB Laboratory Experiments on Convolution 309
           6.6.1 Convolution of Signals 309
           6.6.2 Convolution for Linear Dynamic Systems 310
    6.7 Summary 311
    6.8 Reference 312
    6.9 Problems 312

7 System Response in the Time Domain   318
    7.1 Solving Linear Differential Equations 318
    7.2 Solving Linear Difference Equations 327
    7.3 Discrete-Time System Impulse Response 335
           7.3.1 Direct Method for Finding the Impulse Response 335
           7.3.2 Impulse Response by Linearity and Time Invariance . . . 338
    7.4 Continuous-Time System Impulse Response 343
    7.5 Complete Continuous-Time System Response 348
    7.6 Complete Discrete-Time System Response 349
    7.7 Stability of Continuous-Time Linear Systems 352
           7.7.1 Internal Stability of Continuous-Time Linear Systems . . 352
           7.7.2 Routh–Hurwitz Stability Criterion 355
           7.7.3 Continuous-Time Linear System BIBO Stability 363
    7.8 Stability of Discrete-Time Linear Systems 364
           7.8.1 Internal Stability of Discrete-Time Linear Systems . . . 365
           7.8.2 Algebraic Stability Tests for Discrete Systems 367
           7.8.3 Discrete-Time Linear System BIBO Stability 369
    7.9 MATLAB Experiment on Continuous-Time Systems 371
    7.10 MATLAB Experiment on Discrete-Time Systems 373
    7.11 Summary 373
    7.12 References 379
    7.13 Problems 379

8 State Space Approach 388
- 8.1 State Space Models 391
- 8.2 Time Response from the State Space Equation 396
  - 8.2.1 Time Domain Solution 396
  - 8.2.2 Solution Using the Laplace Transform 401
  - 8.2.3 State Space Model and Transfer Function 404
  - 8.2.4 Impulse and Step Responses 405
- 8.3 Discrete-Time Models 407
  - 8.3.1 Difference Equations and State Space Form 407
  - 8.3.2 Discretization of Continuous-Time Systems 408
  - 8.3.3 Solution of the Discrete-Time State Space Equation 411
  - 8.3.4 Solution Using the $\mathcal{Z}$-transform 413
  - 8.3.5 Discrete-Time Impulse and Step Responses 416
- 8.4 System Characteristic Equation and Eigenvalues 418
- 8.5 Cayley–Hamilton Theorem 421
- 8.6 Linearization of Nonlinear Systems 426
- 8.7 State Space MATLAB Laboratory Experiments 433
  - 8.7.1 Experiment 1–The Inverted Pendulum 433
  - 8.7.2 Experiment 2–Response of Continuous Systems 435
  - 8.7.3 Experiment 3–Response of Discrete Systems 437
- 8.8 Summary 438
- 8.9 References 442
- 8.10 Problems 443

III Linear Systems In Electrical Engineering 457

9 Signals in Digital Signal Processing 459
- 9.1 Sampling Theorem 459
  - 9.1.1 Sampling with an Ideal Sampler and the DTFT 465
  - 9.1.2 Sampling with a Physically Realizable Sampler 467
- 9.2 Discrete-Time Fourier Transform (DTFT) 469
  - 9.2.1 DTFT in Linear Systems 476
  - 9.2.2 From DTFT to the Double-Sided $\mathcal{Z}$-Transform 478
- 9.3 Double-Sided $\mathcal{Z}$-Transform 480
  - 9.3.1 Double-Sided $\mathcal{Z}$-Transform in Linear Systems 484
- 9.4 Discrete Fourier Transform (DFT) 485
  - 9.4.1 Fast Fourier Transform (FFT) 490
- 9.5 Discrete-Time Fourier Series (DFS) 491
- 9.6 Correlation of Discrete-Time Signals 493
- 9.7 IIR and FIR Filters 495
- 9.8 Laboratory Experiment on Signal Processing 497
- 9.9 Summary 498

9.10	References	502
9.11	Problems	502

10 Signals in Communication Systems 507
10.1	Signal Transmission in Communication Systems	508
10.2	Signal Correlation, Energy, and Power Spectra	512
10.3	Hilbert Transform	517
10.4	Ideal Filters	521
10.5	Modulation and Demodulation	522
10.6	Digital Communication Systems	529
10.7	Communication Systems Laboratory Experiment	530
10.8	Summary	532
10.9	References	535
10.10	Problems	535

11 Linear Electrical Circuits 538
11.1	Basic Relations	539
	11.1.1 Equivalence Between Voltage and Current Sources	542
11.2	First-Order Linear Electrical Circuits	543
	11.2.1 RC Electrical Circuits	544
	11.2.2 RL Electrical Circuits	549
11.3	Second-Order Linear Electrical Circuits	554
	11.3.1 Cascade LC Circuit Driven by a Voltage Source	554
	11.3.2 Series Connection of R, L, and C Elements	557
	11.3.3 Parallel Connection of R, L, and C Elements	559
11.4	Higher-Order Linear Electrical Circuits	559
11.5	MATLAB Laboratory Experiment	563
11.6	Summary	563
11.7	References	564
11.8	Problems	564

12 Linear Control Systems 569
12.1	The Essence of Feedback	570
12.2	Transient Response of Second-Order Systems	575
	12.2.1 Transient Response of Higher-Order Systems	579
12.3	Feedback System Steady State Errors	581
12.4	Feedback System Frequency Characteristics	585
12.5	Bode Diagrams	587
12.6	Common Dynamic Controllers: PD, PI, and PID	599
12.7	Laboratory Experiment on Control Systems	601
12.8	Summary	603
12.9	References	605
12.10	Problems	606

A Linear Algebra — 611

- Square Matrices and Matrix Trace — 611
- Identity and Diagonal Matrices — 611
- Matrix Addition (Subtraction) — 612
- Matrix Product — 612
- Multiplication of a Matrix by a Scalar — 613
- Matrix Transpose — 613
- Matrix Determinant — 614
  - Determinant of Matrix Product — 614
- Matrix Inversion — 615
  - Determinant of Matrix Inversion — 615
  - Matrix Product Inversion — 615
- Systems of Linear Algebraic Equations — 616
  - Independent Vectors — 616
  - Matrix Rank — 616
  - Matrix Null Space — 616
  - Matrix Range Space — 616
- Eigenvalues and Eigenvectors — 616
- Matrix Integration and Differentiation — 617
- Matrix Exponential — 617
- Integral of a Matrix Exponential — 617
- Proof of The Cayley-Hamilton Theorem — 617
- References — 618

B Some Results from Calculus — 619

- Standard Trigonometric Formulas — 619
- Basics of Complex Numbers — 619
- Summation Formulas — 619
- Results About Integrals — 620
  - Leibniz Formula — 620
- Derivative Result — 621
- Cauchy-Schwarz Inequalities — 621
- References — 621

C Introduction to MATLAB — 622

- Basic MATLAB Functions — 622
  - The Help Facility — 622
  - Numbers and Arithmetic Expressions — 624
- Vectors and Matrices — 624
- Matrix Operations — 625
  - Transpose — 625
  - Matrix Addition and Subtraction — 626
  - Matrix Multiplication — 626
  - Matrix Powers — 626

Contents   xiii

    Eigenvalues and Eigenvectors 626
    Characteristic Polynomial 627
 Polynomials 628
 Plots 629
 Loops 629
 Script Files 630
 Some MATLAB Functions used in Linear Systems 630
 References 632

D Introduction to Simulink 633
 Basic Rules for Building Block Diagrams 633
 Reference 635

Index 637

Preface

This textbook is intended for college sophomores and juniors whose electrical engineering programs include linear systems and signals courses. It can be also used by other engineering students interested in linear dynamic systems and signals—especially biomedical, aerospace, mechanical, and industrial engineering students. The text is based on the author's sixteen years of teaching linear systems and signals to undergraduate and graduate students in the Department of Electrical and Computer Engineering, Rutgers University.

In the electrical engineering curriculum, a course in linear dynamic systems and signals is a prerequisite for courses in control systems, communication systems, and digital signal processing. In addition, many problems in wireless communications, networking, signal processing, electronics, photonics, and robotics are now studied from the dynamic system point of view. The book presents both continuous- and discrete-time linear systems and signals. Historically, teachers and students have first studied continuous-time phenomena, and then applied the results to the discrete-time domain. However, with the development of modern computers, discrete-time analysis is increasingly dominant, especially from the application and computation points of view. Furthermore, some phenomena are much easier to explain and understand in the discrete-time domain than in the continuous-time domain, since discrete-time linear systems are represented by simple recursive formulas.

It should be emphasized that this textbook reflects the most recent changes in electrical and computer engineering curricula at U.S. universities. Due to the increased number of computer engineering courses over the last 15 years, and due to newly introduced senior level courses in wireless communications, networking, photonics, and signal processing, some courses in classical electrical engineering areas have had to be modified, condensed, combined, or even eliminated. These changes primarily affected courses in the principles of electrical engineering, electrical circuits, systems and signals, control systems, electromagnetic fields, communication systems, electronics, power systems, and related courses. The modification of linear systems and signals courses has gone in two directions: (a) teaching it at the sophomore level, as the course on signals and time-frequency transforms, with little emphasis on system dynamics (in general, sophomore students do not have sufficient knowledge of differential equations); (b) teaching it as a junior (or even senior) level course with emphasis on system dynamics, and including some topics from electrical circuits, feedback systems, communications, and signal processing. This textbook has taken a twofold approach: Chapters 2–6 and 9–10 have been written in the direction of (a), and Chapters 1, 7–8, and 11–12 have been written in the direction of (b).

The book is divided into three major parts: 1) the frequency domain approach to linear dynamic systems; 2) the time domain approach to linear dynamic systems; and 3) the linear system approach to electrical engineering. An introduction to continuous- and discrete-time signals is presented in Chapter 2. Signal transforms (Fourier, Laplace, and $\mathcal{Z}$-transform) are presented in Chapters 3–5. The Fourier series and Fourier transform

TABLE 1 Suggested Topics for a One-Semester Course with Emphasis on Linear Dynamic Systems

One-semester course with emphasis on dynamic systems

Chapter 1: Section 1.3 may be skipped
Chapter 2
Chapter 3: Fourier series and transform may be less emphasized
Chapter 4
Chapter 5
Chapter 6: Convolution (can be taught after Chapter 2)
Chapter 7: Selected topics
Chapter 8: State space (Section 8.6 may be omitted)

TABLE 2 Suggested Topics for a One-Semester Course with Emphasis on Signals

One-semester course with emphasis on signals

Chapter 1: Introduction and Section 1.1.1 only
Chapter 2
Chapter 3
Chapter 4
Chapter 5
Chapter 6: Convolution (can be taught after Chapter 2)
Chapter 9: DFDT, DFT, FFT, and DFS
Chapter 10: Sections 10.1–4

are presented in Chapter 3 in the continuous-time domain. Chapter 9 gives a full coverage of the discrete-time Fourier transform and its variants. The time domain approach to linear systems presents continuous- and discrete-time convolution (Chapter 6), methods for solving differential and difference equations (Chapter 7), and continuous- and discrete-time state space approaches (Chapter 8). Additional topics on signals in digital signal processing and communication systems are presented in Chapters 9 and 10. Instructors intending to teach a linear *dynamic* systems course should be able to cover most of Chapters 1–8 in one semester (see Table 1). Instructors more interested in signals than in systems should cover in detail Chapters 2–6 and 9, and selected parts of Chapter 10 (see Table 2).

The complete book can be thoroughly covered in a two-semester course. Chapters from the third part of the book (Chapters 9–12) explain how to approach other linear system areas of electrical engineering. Since many systems in electrical engineering are linear, the reader will find these chapters extremely useful in combination with other advanced undergraduate courses in electrical engineering (such as control systems, robotics, signal processing, communications, neural networks, computer/communication networks, power systems, and electronics). In that case, the reader will conclude that the linear system course (area) is not just another electrical engineering course, but the *core course around which several courses (areas) of electrical engineering evolve*.

TABLE 3 Suggested Topics for a Two-Semester Course Independently Covering Continuous- and Discrete-time Signals and Systems

First-semester course on continuous-time signals and systems	*Second-semester course on discrete-time signals and systems*
Chapter 1: Continuous part	Chapter 1: Discrete part
Chapter 2: Continuous part	Chapter 2: Discrete part
Chapter 3	Chapter 5
Chapter 4	Chapter 6: Discrete part
Chapter 6: Continuous part	Chapter 7: Discrete part
Chapter 7: Continuous part	Chapter 8: Discrete part
Chapter 8: Continuous part	Chapter 9
Chapters 10–12: Selected topics	Chapter 10: Section 10.6

A one-semester junior-level course with emphasis on *linear dynamic systems* can cover the topics presented in Table 1. At Rutgers University, we also cover the introductory topics from Chapter 12 on linear feedback systems, since an undergraduate control course is not required in our curriculum. A one-semester sophomore/junior-level course *with emphasis on signals* can be taught according to Table 2. A two-semester sophomore/junior-level course with a distinction between continuous- and discrete-time signals/systems (with the first semester on continuous-time signals and systems, and the second semester on discrete-time signals and systems) can be taught by organizing material according to Table 3.

The main goal in linear dynamic system analysis is to find the system response due to any excitation (input). It is shown that the most systematic way to achieve that goal is to first find the system impulse response. Finding the system impulse response directly in the continuous-time domain is a very tedious task. From the author's teaching experience, students are able to find the system impulse response in the frequency domain rather easily, but they have difficulty in the time domain. To avoid this problem, *all important system concepts (including the system impulse response) are introduced first in the frequency domain*. It is very simple to define the system transfer function in the frequency domain. Having defined (and obtained) the transfer function, it is very simple to define (and obtain) the system impulse response in an inverse procedure, in which signals are mapped from the frequency domain to the time domain. In this book, the continuous-time system impulse response is obtained as the inverse Laplace transform of the system transfer function. From the impulse response, we derive the most important result of linear dynamic systems theory, which states that for a system at rest (with no initial energy stored in the system), *the system response to an arbitrary input is the convolution of the system impulse response and the given input signal*. The convolution operation is studied thoroughly in Chapter 6.

After all of the important linear system concepts are introduced in the frequency domain, we interpret these concepts in the time domain and develop the time domain techniques for finding the response of continuous- and discrete-time linear systems. This leads to the state space technique as a highlight of the time domain approach for studying linear systems. Due to the rapid development of electrical engineering and other engineering disciplines in the last two decades, and the frequent use of modern computer packages (such as MATLAB®) for system analysis, it is imperative that modern courses in linear system analysis give extensive coverage of the state space technique. This book achieves that goal by requiring only elementary knowledge of linear algebra and differential equations. Using this mathematical background, the main state space concepts are slowly and thoroughly developed, and new notions are fully explained. In general, it is well known and accepted that the frequency domain gives a better understanding of considered physical phenomena, but the time domain is more efficient from the computational point of view. In the last part of this book, we demonstrate how to use linear system theory concepts to solve problems in other fields of electrical engineering: signal processing, control systems, communication systems, and electrical circuits.

The material presented in this book has been class-tested for several years in the required junior-level course in linear systems and signals at Rutgers University. The book includes many examples and problems. Most are of analytical nature; some, especially those referring to higher-order systems, are performed (or ought to be performed) using the MATLAB package. The real-world linear system examples are given in terms of differential/difference equations or in the state space form (system, input, and output matrices), without explaining the physics of linear systems and the development of corresponding mathematical models. The author believes that most sophomore/junior students will have difficulty grasping all of the modeling issues from diverse linear system disciplines (electrical, mechanical, chemical, and biomedical engineering). Consequently, real physical systems are represented by the numerical entries in corresponding differential/difference equations and matrices (for the state space description). At some points, to show the relationship between mathematical models, state space forms, and system transfer functions, mathematical modeling is fully explained.

The undergraduate linear systems and signals course at Rutgers University is associated with the linear systems and signals laboratory, which over the years has been slowly evolving from a hardware oriented laboratory to a fully software oriented laboratory based on MATLAB. This computer-aided system design package is used quite often in this book for solving problems and designing laboratory experiments. MATLAB is a good learning tool, which helps students to get a better understanding of main linear systems and signal concepts. It is especially useful for studying higher-order linear dynamic systems.

This book provides concurrently a laboratory manual for linear systems and signals. Since undergraduate laboratories at major universities, in the United States and abroad, are more and more software oriented, we use the MATLAB package to design corresponding laboratory experiments. After each important topic, a laboratory experiment is presented. However, the inclusion of MATLAB only helps to deepen the understanding and practical working knowledge of the main linear system theory concepts. *By no means is MATLAB essential for the material presented in this textbook.*

The book is supplemented with a teacher's solution manual for problems and laboratory experiments, available only to instructors who have adopted the text for classroom use. The MATLAB programs and numerical data used in this book may be obtained at the homepage of the book, http://www.ece.rutgers.edu/~gajic/systems.html. The same web site also contains additional problems and their complete or partial solutions, laboratory experiments, sample exams, and list of corrections (if any). The author is in the process of developing transparencies to help instructors teach the corresponding course(s) on linear systems and signals.

The author approached the writing of this book with the desire to present linear systems theory essentials in sufficient detail, yet explain these essentials in a such way that every student in electrical engineering will be able to use it as a self-study guide. After taking this course, engineering students should be well equipped to cope with all types of linear *dynamic* system problems, especially those encountered in related courses (such as communications, signal processing, controls, networking, robotics, power systems, electrical circuits, and electronics). Hence, the main purpose of this course is to develop unified techniques for recognizing and solving linear *dynamic* system problems regardless of their origin.

Many people made valuable contributions to this book. The author is particularly thankful for support and contributions from his former doctoral students (many presently university professors) M. Lim (Korea University), M. Qureshi (University of Western Sydney), D. Skatarić (University of Belgrade), X. Shen (University of Waterloo), W-C. Su (Taiwan's National Chung-Hsing University), and researchers I. Borno from AT&T Bell Labs and L. Qian from Lucent Bell Labs. Former teaching assistants M. Sheth and M. Wehle have corrected earlier versions of the text and helped in the development of MATLAB laboratory experiments. Former undergraduate student T. Carpenter-Alvarez (presently a professor at New Jersey Institute of Technology) helped to develop laboratory experiments, and M. Funk and G. Topalović read and corrected some parts of the manuscript. I am thankful to my colleagues Professors G. Hung and J. Li (Rutgers University), V. Kecman (University of New Zealand), M. Lelić (University of Tuzla, presently a researcher at Corning Incorporated), and S. Orfanidis (Rutgers University) for providing useful suggestions and interesting application examples.

The author is indebted to the people who made this book possible, editor Paul Becker and Associate Editor Alice Dworkin, of Prentice Hall. They took an interest in the book, spent a great deal of time with the author discussing it, solicited numerous reviews for the manuscript, gave valuable suggestions, and encouraged the author to finalize this endless project. Finally, the author is thankful to numerous reviewers for their suggestions and recommendations. The following reviewers provided especially useful comments: Maruthi R. Akella of University of Texas at Austin, Er-Wei Bai of University of Iowa, Robert A. Paz of New Mexico State University, Rodney Roberts of Florida State University, and Gang Tao of University of Virginia.

Special thanks go to my wife Verica Radisavljević-Gajić for drawing all the figures in the textbook and solutions manual.

ZORAN GAJIĆ
Piscataway, New Jersey

About the Author

Professor Zoran Gajić has been teaching linear systems, controls, and networking courses in the Electrical and Computer Engineering Department at Rutgers University, New Jersey since 1984. He is the author or coauthor of more than fifty journal papers, primarily published in *IEEE Transactions on Automatic Control* and *IFAC Automatica* journals, and seven books in the fields of linear and bilinear control systems published by Academic Press, Prentice Hall International, Marcel Dekker, and Springer Verlag. Professor Gajić has delivered two plenary lectures at international conferences and presented almost 100 conference papers. He serves on the editorial board of the journal *Dynamics of Continuous, Discrete, and Impulsive Systems,* and has been a guest editor of a special issue of that journal, on control systems technology. Professor Gajić received the B.S. and M.S. degrees in Electrical Engineering from the University of Belgrade, and the M.S. degree in Applied Mathematics and the Ph.D. degree in Systems Science Engineering from Michigan State University. He is a life master of the U.S. Chess Federation and a master of the World Chess Federation.

CHAPTER 1

Introduction to Linear Systems

This book presents the theory of continuous- and discrete-time, linear, time-invariant, dynamic systems. At the same time, the text presents, in sufficient detail (Chapters 2–6, 9–10), the theory of continuous- and discrete-time signals, which is an integral part of the theory of continuous- and discrete-time linear dynamic systems. At points, the text considers applications of the presented theory to natural and artificial (man-made) linear dynamic systems.

The theory of linear dynamic systems represents a foundation for several electrical engineering disciplines, including electrical circuits, control systems, communication systems, signal processing systems, power systems, networking, and electronics. Computers connected into networks can also be studied as dynamic systems. Wireless communication networks have long been recognized as dynamic systems. Traffic highways of the future are already the subject of broad research as dynamic systems, and an intensive search for the most efficient networks of highways is underway. Robotics, aerospace, photonics, chemical, and automotive industries are producing new and challenging models of dynamic systems that must be analyzed, optimized, and utilized. Many processes in economics, and in the biological and social sciences, are analyzed using dynamic systems theory.

The rapid scientific development that started in the 1960s brought a tremendous number of new scientific results. The successes of the U.S. space program (landing a man on the moon in 1969, and an autonomous moving vehicle-laboratory on Mars in 1997) and the aircraft industry were heavily based on the power of dynamic systems theory. We have witnessed the real explosion of the computer industry since the middle of the 1980s, and the rapid development of signal processing, parallel computing, neural networks, and wireless and optical communications since the beginning of the 1990s. In years to come, many scientific areas will evolve around vastly enhanced modern computers with the ability to solve, by sheer brute force, very complex problems. As pointed out in [1], the already established "information superhighway" may just be a synonym for numerous possibilities for "informational breakthroughs" in almost all scientific and engineering areas, with the use of modern computers. The power of computers and the speed of information exchange will increase continually due to the tremendous bandwidth that optical fiber (replacing copper wire) can provide. In 2001, commercial computer information exchange using optical fiber transmission lines is performed at the speed of 10 Gb/s (gigabits per second). In the near future, computer/communication networking speeds of several Tb/s (terabits per second) will be expected.

The theory of linear dynamic systems has been used in engineering and science to study real physical dynamic systems. This theory is at least as old as the mathematical theory of differential equations, since differential equations describe the dynamic behavior of real physical systems. "As an imperative of the modern time, linear system theory must preserve its classic values and incorporate them into modern scientific trends, which are based on the already developed fast and reliable software packages for numerical computations, dynamic simulation, symbolic computations, and computer graphics." [1] One of these packages, MATLAB, has already gained broad acceptance in the scientific community, academia, and industry. It represents an expert system for solving many engineering and scientific problems, including many linear dynamic system problems. We will use MATLAB throughout this book to solve many linear system theory problems, and to facilitate a deeper understanding and analysis of problems under consideration.

Linear dynamic systems can be studied in either the *frequency domain* or the *time domain*. The frequency domain technique for the analysis of linear dynamic systems is based on the existence of corresponding time-frequency transformations that lead to the concept of the *system transfer function*. This technique will be presented in Chapters 3–5 for both continuous- and discrete-time linear dynamic systems. The time domain technique is based on differential and difference equations, where the difference equations are discrete-time counterparts to the differential equations. The time domain technique for the analysis of linear systems is computationally more powerful than the frequency domain technique, especially in the case of higher-order linear dynamic systems. However, the frequency domain technique very often gives a better understanding of the phenomena under consideration. The highlight of the time domain analysis technique is the *state space approach*. The modern approach to linear systems requires an extensive study of this approach, which is undertaken in Chapter 8. (The state space approach requires some basic knowledge of linear algebra.) The state space time domain method is more convenient for describing and studying higher-order linear dynamic systems than the frequency domain method. Scientific problems to be addressed in the future will often be of very high dimensions.

This introductory chapter is organized as follows: In the first section, we introduce continuous- and discrete-time, time-invariant, linear systems, represented respectively by constant coefficient differential and difference equations. Analysis of these two classes of linear dynamic systems is the main subject of this text. In that respect, the notions of continuous- and discrete-time scales (domains) are introduced, and the sampling operation that relates continuous and discrete signals is defined. In Section 1.2 we discuss the concepts of system linearity and time invariance. Modeling of dynamic systems is presented in Section 1.3. A system classification is given in Section 1.4. In the concluding sections, we outline the book's structure and organization, and indicate the use of MATLAB and Simulink toolboxes as teaching tools in computer oriented linear system analysis.

1.1 CONTINUOUS AND DISCRETE SIGNALS AND LINEAR SYSTEMS

In this section we introduce standard terminology used in the analysis of continuous- and discrete-time linear dynamic systems. We define the notions of continuous- and discrete-time scales and relate them through the sampling operation. In that direction, we formally introduce continuous- and discrete-time signals.

1.1.1 Continuous and Discrete Signals and Sampling

Everyday life abounds in signals. All of us are familiar with telephone (voice) and video signals that carry useful information. Astronomers use telescopes to observe signals coming from the universe, doctors use stethoscopes to observe signals coming from the human body, and engineering students use oscilloscopes to monitor and record signals coming from electrical and electronic circuits. Typically, common signals have very complex forms, which can be synthesized from very simple elementary signals that can be represented by very simple mathematical functions (for example, sine and cosine functions). In Chapter 2 we will thoroughly study elementary signals that can be used to describe more complex signals. Computer and communication signals used to transmit data, voice, and video are composed of streams of zeros and ones, where zeros (in fiber optics communications) are characterized by absence of light and ones are characterized by presence of light. In general, signals are either continuous or discrete in nature.

A *continuous signal* is a mathematical function of an independent variable $t \in \Re$, where $\Re$ represents the set of real numbers. In addition, it is required that signals are *uniquely* defined in t except for a finite number of points. For example, the function $f(t) = \sqrt{t}$ does not qualify as a signal, even for $t > 0$, since the square root of t has two values for any positive t. A continuous signal is represented in Figure 1.1. Very often, especially in the study of dynamic systems, the independent variable t represents time. In such cases, $f(t)$ is a time function.

Note that signals are real mathematical functions, but some transforms applied to signals can produce complex signals that have both real and imaginary parts. For example, in the analysis of alternating current electrical circuits we use phasors, rotating vectors in the complex plane, $I(j\omega) = |I(j\omega)| \angle I(j\omega)$, where ω represents the angular frequency of rotation, $\angle I(j\omega)$ denotes phase, and $|I(j\omega)|$ is the amplitude of the alternating current. Complex plane representation is useful in simplifying circuit analysis; however, the complex signal just defined represents in fact a real sinusoidal signal, oscillating with the corresponding amplitude, frequency, and phase represented by $|I(j\omega)| \sin(\omega t + \angle I(j\omega))$. Complex signal representation of real signals will be encountered in this text in many application examples. In addition, in several chapters on signal transforms (Fourier, Laplace, $\mathcal{Z}$-transform), we will present complex domain equivalents of real signals.

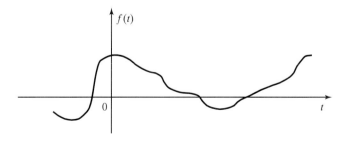

FIGURE 1.1: A continuous signal

4 Chapter 1 Introduction to Linear Systems

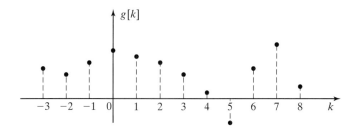

FIGURE 1.2: A discrete signal

A *discrete signal* is a uniquely defined mathematical function (single-valued function) of an independent variable $k \in Z$, where Z denotes the set of integers. Such a signal is represented in Figure 1.2. In order to clearly distinguish between continuous and discrete signals, we will use parentheses for arguments of continuous signals and square brackets for arguments of discrete signals, as demonstrated in Figures 1.1 and 1.2. If k represents discrete time (counted in the number of seconds, minutes, hours, days, and so on), then $g[k]$ defines a discrete-time signal.

Sampling

Continuous and discrete signals can be related through a sampling operation, in the sense that a discrete signal can be obtained by performing sampling on a continuous-time signal with the uniform sampling period T, as presented in Figure 1.3. Since T is a given quantity, we will use $f(kT) \triangleq f[k]$ in order to simplify notation.

It is obvious from Figure 1.3 that when the sampling period is small, the discrete signal approximates very well the corresponding continuous signal. Since modern computers are digital, they can easily handle discrete signals, and the only way that digital computers can deal with continuous signals is to have them discretized. In the discretization procedure, the sampling period T must be small enough to get a good approximation for the corresponding continuous signal. However, the smaller the sampling period, the more data required to represent the corresponding continuous signal, which makes computer calculations with that signal more complex and slower. Hence, in the sampling procedure we must take T small enough to satisfy two conflicting requirements: a good approximation with minimal data. The complete answer to this problem leads to the famous Shannon sampling theorem, one of the most important theorems in signal processing and digital communications. This theorem will be presented in Chapter 9.

Note that digital computers process *digital signals*. Digital signals are discrete signals that are also discretized with respect to signal magnitude (signal quantization). Detailed studies of signal sampling and quantization are done in signal processing and communications courses. For the purpose of this course it is only sufficient to understand the notions of signal sampling and signal quantization.

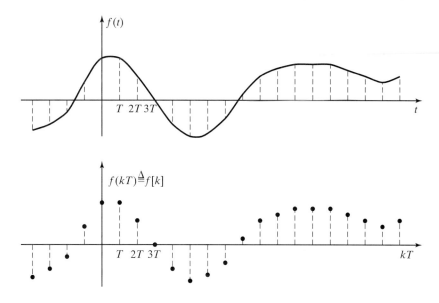

FIGURE 1.3: Sampling of a continuous signal

1.1.2 Continuous- and Discrete-Time Systems

Everything that changes in time and space can be considered a dynamic system. Aircraft, cars, trains, bicycles, humans, animals, robots, are all examples of moving systems. Also, currents and voltages in electrical and electronic systems, bits in computers systems, photons in light technology systems, chemical processes, the stock market, the growth of a national economy, and biological processes all change in time and space and can be considered as typical examples of dynamic systems. Of course, it is difficult to describe the dynamics of moving objects (systems) using equations, and it is equally difficult to solve the corresponding equations since, in general, they can have very complex forms. To obtain such equations (mathematical models) would be very beneficial, since one would be able to analyze corresponding systems using their mathematical models and perform countless analytical and computer simulation studies. In this introductory course to linear dynamic systems we must limit our attention to the classes of dynamic systems that we can mathematically describe and analyze. Two main classes of dynamic systems are continuous- and discrete-time systems with concentrated or lumped parameters. In this course, we will study only *linear* continuous- and discrete-time systems with *constant* concentrated parameters. We will find that a large number of practical engineering systems belong to these two categories. In Section 1.4, we will give a much broader classification of dynamic systems.

Continuous-time systems are represented by *differential equations,* and discrete-time systems are represented by *difference equations*. It is well known from basic physics and engineering courses that many electrical and mechanical dynamic systems are described by differential equations. Undergraduate junior students are more or less familiar with analytical methods for solving first-, second-, and higher-order linear differential equations with constant coefficients. In general, differential equations of very high order can be solved numerically by using digital computers.

Difference equations are studied in engineering for the first time in linear systems and digital signal processing courses. In order to solve differential equations numerically, we must provide digital computers with corresponding recursive formulas. Recursive formulas are in fact difference equations. Difference equations are used to describe dynamic systems that evolve in discrete time (for example, bank accounts, the growth of a national economy, the evolution of a stock market, and so on). Differential equations can also be discretized and replaced by difference equations (this will be demonstrated in Chapter 8). It can be said in general that difference equations are discrete-time counterparts to differential equations.

Continuous- and discrete-time, *linear, time-invariant, dynamic systems* are described, respectively, by *linear* differential and difference equations with *constant coefficients*. Mathematical models of such systems that have one input and one output are given by

$$\frac{d^n y(t)}{dt^n} + a_{n-1}\frac{d^{n-1} y(t)}{dt^{n-1}} + \cdots + a_1\frac{dy(t)}{dt} + a_0 y(t) = f(t) \quad (1.1)$$

and

$$y[k+n] + a_{n-1}y[k+n-1] + \cdots + a_1 y[k+1] + a_0 y[k] = f[k] \quad (1.2)$$

where n is the order of the system, y is the *system output*, and f is the external forcing function representing the *system input*. In this text, we study only *time-invariant* continuous and discrete linear systems for which the *coefficients a_i, $i = 0, 1, \ldots, n-1$, are constants*. Linear time-varying systems, whose coefficients vary in time, are usually studied in a graduate course on linear systems.

Initial Conditions

In addition to the *external forcing function,* the system is also driven by its *internal forces* coming from the *system initial conditions* (accumulated system energy at the given initial time). It is well known from elementary differential equations that to find the solution for (1.1), a solution of a differential equation of order n, the set of n initial conditions must be specified as follows:

$$y(t_0), \frac{dy(t_0)}{dt}, \ldots, \frac{d^{n-1}y(t_0)}{dt^{n-1}} \quad (1.3)$$

where t_0 denotes the initial time. In the discrete-time domain, for a difference equation of order n, the set of n initial conditions must be specified. For (1.2), the initial conditions are given by

$$y[k_0], y[k_0 + 1], \ldots, y[k_0 + n - 1] \quad (1.4)$$

It is interesting to note that *in the discrete-time domain the initial conditions carry information about the evolution of the system output* in time, from some initial time k_0 to $k_0 + n - 1$. Those values are the system output past values, and they must be used to determine the system output current value, that is, $y[k_0 + n]$. In contrast, *for continuous-time systems all initial conditions are defined at the initial time t_0.*

System Response

The main goal in the analysis of dynamic systems is to find the system response (system output) due to external (system inputs) and internal (system initial conditions) forces. It is known from the elementary theory of differential equations that the solution of a linear differential equation has two additive components: the *homogenous and particular solutions*. The homogenous solution is contributed by the initial conditions and the particular solution comes from the forcing function. In engineering, the homogenous solution is sometimes called the *system natural response,* and the particular solution is called the *system forced response.* Hence, we have

$$y(t) = y_h(t) + y_p(t) \tag{1.5}$$

The same result holds for linear difference equations, that is

$$y[k] = y_h[k] + y_p[k] \tag{1.6}$$

Since undergraduate students are usually introduced to difference equations in signal processing and linear systems courses, we will give detailed coverage of methods for solving difference (and differential) equations in Chapter 7.

Homogeneous and particular solutions of differential equations correspond, respectively (not identically, in general), to the so-called zero-input and zero-state responses of dynamic systems. In the following, we give the formal definitions of the zero-state and zero-input responses.

DEFINITION 1.1: The continuous-time (discrete-time) linear system response solely contributed by the system initial conditions is called the *system zero-input response (forcing function is set to zero).* It is denoted by y_{zi}.

DEFINITION 1.2: The continuous-time (discrete-time) linear system response solely contributed by the system forcing function is called the *system zero-state response (system initial conditions are set to zero).* It is denoted by y_{zs}.

In view of Definitions 1.1 and 1.2, it also follows that the linear system response has two components: one component contributed by the system initial conditions, y_{zi}, and another component contributed by the system forcing function (input), y_{zs}; that is, for continuous-time linear systems, we have

$$y(t) = y_{zi}(t) + y_{zs}(t) \tag{1.7}$$

and for discrete-time linear systems, we have

$$y[k] = y_{zi}[k] + y_{zs}[k] \tag{1.8}$$

In the following example, we demonstrate a method for finding the zero-input and zero-state responses of a simple second-order linear dynamic system. In addition, we demonstrate differences and similarities between the homogenous solution of the linear differential equation with constant coefficients, and the zero-input response of the corresponding linear time-invariant system, as well as indicating differences and similarities between the particular solution and the system zero-state response.

EXAMPLE 1.1

Consider the following second-order system:

$$\frac{d^2y(t)}{dt^2} + 4\frac{dy(t)}{dt} + 3y(t) = e^{-2t}, \quad t \geq 0, \quad y(0) = 1, \quad \frac{dy(0)}{dt} = 1$$

Using elementary knowledge from differential equations, we have

$$y(t) = y_h(t) + y_p(t) = H_1 e^{-t} + H_2 e^{-3t} + \alpha e^{-2t}, \quad t \geq 0$$

Note that a particular solution is any solution that satisfies the differential equation. Usually, a particular solution has the form of the forcing function and its derivatives. Since a particular solution satisfies

$$\frac{d^2 y_p(t)}{dt^2} + 4\frac{dy_p(t)}{dt} + 3y_p(t) = e^{-2t}, \quad t \geq 0, \text{ (no initial conditions imposed)}$$

we can guess its form as $y_p(t) = \alpha e^{-2t}$, which when substituted into the preceding differential equation easily produces $\alpha = -1$. Hence

$$y_p(t) = -e^{-2t}, \quad t \geq 0$$

The constants for $y_h(t)$ are determined such that

$$y(0) = 1 = y_h(0) + y_p(0) = H_1 + H_2 - 1$$

$$\frac{dy(0)}{dt} = 1 = \frac{dy_h(0)}{dt} + \frac{dy_p(0)}{dt} = -H_1 - 3H_2 + 2$$

which implies $H_1 = 5/2, H_2 = -1/2$. Hence

$$y_h(t) = \frac{5}{2}e^{-t} - \frac{1}{2}e^{-3t}, \quad t \geq 0,$$

so that the sought solution is given by

$$y(t) = y_h(t) + y_p(t) = \frac{5}{2}e^{-t} - \frac{1}{2}e^{-3t} - e^{-2t}, \quad t \geq 0$$

The zero-input response is obtained from the original differential equation (system) with no forcing function (no input), that is,

$$\frac{d^2 y_{zi}(t)}{dt^2} + 4\frac{dy_{zi}(t)}{dt} + 3y_{zi}(t) = 0, \quad t \geq 0, \quad y_{zi}(0) = 1, \quad \frac{dy_{zi}(0)}{dt} = 1$$

which produces

$$y_{zi}(t) = K_1 e^{-t} + K_2 e^{-3t}, \quad t \geq 0$$

Using the values for the corresponding initial conditions, we get

$$y_{zi}(t) = 2e^{-t} - e^{-3t}, \quad t \geq 0$$

The zero-state response is obtained from the system differential equation with all initial conditions set to zero, that is,

$$\frac{d^2 y_{zs}(t)}{dt^2} + 4\frac{dy_{zs}(t)}{dt} + 3y_{zs}(t) = e^{-2t}, \quad t \geq 0, \quad y_{zs}(0) = 0, \quad \frac{dy_{zs}(0)}{dt} = 0$$

The solution of this differential equation is given by

$$y_{zs}(t) = C_1 e^{-t} + C_2 e^{-3t} - e^{-2t}, \quad t \geq 0$$

Since both initial conditions for the zero-state response are equal to zero, we obtain $C_1 = C_2 = 0.5$, which implies that

$$y_{zs}(t) = \frac{1}{2}e^{-t} + \frac{1}{2}e^{-3t} - e^{-2t}, \quad t \geq 0$$

It is easy to check that

$$y(t) = y_{zi}(t) + y_{zs}(t) = y_h(t) + y_p(t) = \frac{5}{2}e^{-t} - \frac{1}{2}e^{-3t} - e^{-2t}, \quad t \geq 0$$

It can also be seen that, for this example,

$$y_h(t) = \frac{5}{2}e^{-t} - \frac{1}{2}e^{-3t} \neq y_{zi}(t) = 2e^{-t} - e^{-3t}$$

and

$$y_p(t) = -e^{-2t} \neq y_{zs}(t) = \frac{1}{2}e^{-t} + \frac{1}{2}e^{-3t} - e^{-2t}$$

However, in Chapter 6, we will see that if we use the convolution technique to find $y_p(t)$ then, $y_p(t) = y_{zs}(t)$ and $y_h(t) = y_{zi}(t)$.

A discrete-time example corresponding to Example 1.1 will not be presented here, due to the fact that undergraduate junior students are not familiar with general methods for solving difference equations. That issue will be considered in Chapters 5 and 7.

Sometimes, in the linear system literature, the zero-state response is superficially called the system *steady state response,* and the zero-input response is called the system *transient response.* More precisely, the *transient response represents the system response in the time interval immediately after the initial time, say from* $t_0 = 0$ *to* t_1, *contributed by both the system input and the system initial conditions.* The *system steady state response stands for the system response in the long run after some* $t > t_1$. This distinction between the transient and steady state responses is demonstrated in Figure 1.4. The component of the system transient response, contributed by the system initial conditions, in most cases decays quickly to zero. Hence, after a certain time interval, the system response is most

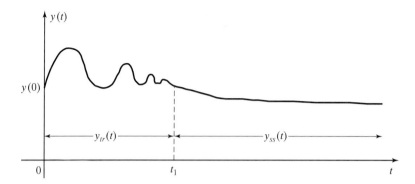

FIGURE 1.4: Transient and steady state responses

likely determined by the forcing function only. Note that the steady state is not necessarily constant in time, as demonstrated in Example 1.2.

In addition to the definition of the system transient and steady state responses given in Figure 1.4, the following definitions are used in electrical circuits and control systems [1]:

$$y(t) = \bar{y}_{tr}(t) + \bar{y}_{ss}(t)$$

with the system steady state response often evaluated as

$$\bar{y}_{ss} = \lim_{t \to \infty} \{y(t)\}$$

Note that in this definition of the steady state response, $\bar{y}_{ss}$ in fact represents the final value of the system response, $y(\infty)$, assuming that such a value exists.

EXAMPLE 1.2

Consider the same system as in Example 1.1, with the same initial conditions, but take the forcing function as $f(t) = \sin(t)$, that is,

$$\frac{d^2 y(t)}{dt^2} + 4\frac{dy(t)}{dt} + 3y(t) = \sin(t), \quad t \geq 0, \quad y(0) = 1, \quad \frac{dy(0)}{dt} = 1$$

The complete response of this system (the solution of the differential equation) is given by

$$y(t) = y_h(t) + y_p(t) = H_1 e^{-t} + H_2 e^{-3t} + \alpha \sin(t) + \beta \cos(t), \quad t \geq 0$$

The constants α and β can be determined by plugging $y_p(t)$ into the differential equation and equating coefficients that multiply sine and cosine functions, which leads to a system of two linear algebraic equations with two unknowns whose solutions are $\alpha = 0.1$ and $\beta = -0.2$. The constants H_1 and H_2 are obtained from the initial conditions, following the procedure of Example 1.1, as $H_1 = 9/4$ and $H_2 = -21/20$. Hence, the sought solution is given by

$$y(t) = \frac{9}{4} e^{-t} - \frac{21}{20} e^{-3t} + \frac{1}{10} \sin(t) - \frac{2}{10} \cos(t), \quad t \geq 0$$

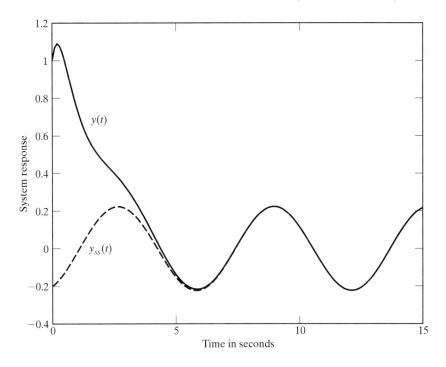

FIGURE 1.5: System complete response (solid line) and steady state response (dashed line) for Example 1.2

It is easy to see that the system response exponential functions decay to zero rapidly, so that the system steady state response is determined by

$$y_{ss}(t) \approx \frac{1}{10}\sin(t) - \frac{2}{10}\cos(t), \quad t \geq t_1$$

The plots of $y(t)$ and $y_{ss}(t)$ are given in Figure 1.5.

It can be seen from the figure that the transient ends roughly at $t_1 = 6$ s, hence after that time the system is in its steady state.

From the preceding presentation, we have seen that linear dynamic systems process input signals in order to produce output signals. The processing rule is given in the form of differential/difference equations. Sometimes, linear dynamic systems are called linear signal processors. A block diagram representation of a linear system, processing one input and producing one output, corresponding either to (1.1) or (1.2), is given in Figure 1.6.

In general, the *system input signal can be differentiated by the system,* so that a description of time-invariant linear continuous-time systems more general than the one

FIGURE 1.6: Input-output block diagram of a system

given in (1.1) is

$$\frac{d^n y(t)}{dt^n} + a_{n-1}\frac{d^{n-1} y(t)}{dt^{n-1}} + \cdots + a_1 \frac{dy(t)}{dt} + a_0 y(t)$$
$$= b_m \frac{d^m f(t)}{dt^m} + b_{m-1}\frac{d^{m-1} f(t)}{dt^{m-1}} + \cdots + b_1 \frac{df(t)}{dt} + b_0 f(t) \quad (1.9)$$

This system differentiation of input signals leads to some interesting system properties (discussed in Chapters 3–4). The corresponding form of time-invariant linear discrete-time systems (more general than the one given in (1.2)) is

$$y[k+n] + a_{n-1} y[k+n-1] + \cdots + a_1 y[k+1] + a_0 y[k]$$
$$= b_m f[k+m] + b_{m-1} f[k+m-1] + \cdots + b_1 f[k+1] + b_0 f[k] \quad (1.10)$$

The coefficients a_i, $i = 0, 1, 2, \ldots, n-1$, and b_j, $j = 0, 1, \ldots, m$, in (1.9) and (1.10) are constants. Note that for real physical systems $n \geq m$.

The problem of finding the system response for the given input signal $f(t)$ or $f[k]$ is the central problem in the analysis of linear systems. It is basically the problem of solving the corresponding linear differential or difference equation, (1.9) or (1.10). This problem can be solved either by using knowledge from the mathematical theory of linear differential and/or difference equations, or by the engineering frequency domain approach—based on the concept of the system transfer function. This leads to the conclusion that linear systems can be studied either in the time domain (to be generalized in Chapter 8 to the state space approach) or in the frequency domain (the transfer function approach). Chapters 3–5 of this text will be dedicated to the frequency domain techniques, and Chapters 6–8 will study time domain techniques for the analysis of continuous- and discrete-time linear time-invariant systems.

It should be emphasized that in order to solve (1.9) we need to broaden our knowledge beyond a basic course in differential equations. Namely, *due to the derivatives of the input signal present on the right-hand side of Equation (1.9), impulses that instantly change system initial conditions are generated at the initial time*. These impulses are called the impulse delta functions (signals). The impulse delta signal and its role in the derivative operation will be studied in detail in Chapter 2. We will learn that one method for solving equation (1.9) will be based on the Laplace transform. That method is fine when the order of the differential equation is not too high, and when we can easily find the Laplace transform of the function $f(t)$. The Laplace transform will be presented in Chapter 4. Another method for solving (1.9) requires using $y(t) = y_h(t) + y_p(t)$ with *the particular solution*

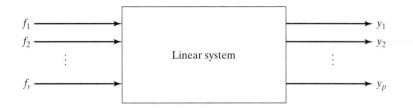

FIGURE 1.7: Block diagram of a multi-input multioutput system

of (1.9) being obtained through the convolution operation. The convolution operation will be introduced in Chapter 2 and used in Chapters 3 and 4 for the analysis of linear time-invariant systems. The convolution operation will be studied in detail in Chapter 6 and its use for solving (1.9) will be fully explained in Sections 7.1 and 8.2. (The reader is referred to Problem 1.13 to observe difficulties encountered while solving equation (1.9) using only elementary knowledge from differential equations.)

The system represented in Figure 1.6, and given in formulas (1.9–10), has only one input f and one output y. Such systems are known as *single-input single-output systems*. In general, systems may have several inputs and several outputs, say r inputs $f_1, f_2, \ldots, f_r$, and p outputs, $y_1, y_2, \ldots, y_p$. These systems are known as *multi-input multioutput systems*. They are also called *multivariable systems*. A block diagram for a multi-input multi-output system is represented in Figure 1.7.

The problem of obtaining differential (difference) equations that describe dynamics of real physical systems is known as *mathematical modeling*. In Section 1.3, this problem will be addressed and mathematical models for several real physical systems will be derived. Note that mathematical models of many other natural and artificial systems are available in the system literature (see, for example, [2–8]).

1.2 SYSTEM LINEARITY AND TIME INVARIANCE

In Section 1.1, the concept of system linearity is tacitly introduced by stating that linear dynamic systems are described by linear differential/difference equations. We have also stated that the concept of time invariance is related to differential/difference equations with constant coefficients. In this section we discuss the concepts of system linearity and time invariance in more detail.

1.2.1 System Linearity

The concept of system linearity is presented for continuous-time systems. Similar derivations and explanations are valid for presentation of the linearity concept of discrete-time linear dynamic systems. Before we derive and state the linearity property of continuous-time linear dynamic systems, we need the following definition.

DEFINITION 1.3: A *system at rest* is a system that has no initial internal energy; that is, all its initial conditions are equal to zero.

Chapter 1 Introduction to Linear Systems

It is very well known from basic differential equations courses that with any differential equation of order n, a set of n initial conditions of the form given in (1.3) is associated. It follows from Definition 1.3 that for a system at rest, the initial conditions are set to zero; that is,

$$y(t_0) = 0, \quad \frac{dy(t_0)}{dt} = 0, \quad \ldots, \quad \frac{d^{n-1}y(t_0)}{dt^{n-1}} = 0 \quad (1.11)$$

Systems at rest are also called systems with zero initial conditions.

The linearity property of continuous-time linear dynamic systems is the consequence of the linearity property of mathematical derivatives, that is,

$$\frac{d^i}{dt^i}(y_1(t)) + \frac{d^i}{dt^i}(y_2(t)) = \frac{d^i}{dt^i}(y_1(t) + y_2(t)), \quad i = 1, 2, \ldots$$

Consider the general nth order continuous-time linear differential equation that describes the behavior of an nth order linear dynamic system given by (1.9). Assume that the *system is at rest*, and that it is driven either by $f_1(t)$ or $f_2(t)$, which respectively produce the system outputs $y_1(t)$ and $y_2(t)$; that is,

$$\frac{d^n y_1(t)}{dt^n} + a_{n-1}\frac{d^{n-1}y_1(t)}{dt^{n-1}} + \cdots + a_1\frac{dy_1(t)}{dt} + a_0 y_1(t)$$

$$= b_m \frac{d^m f_1(t)}{dt^m} + b_{m-1}\frac{d^{m-1}f_1(t)}{dt^{m-1}} + \cdots + b_1\frac{df_1(t)}{dt} + b_0 f_1(t) \quad (1.12)$$

and

$$\frac{d^n y_2(t)}{dt^n} + a_{n-1}\frac{d^{n-1}y_2(t)}{dt^{n-1}} + \cdots + a_1\frac{dy_2(t)}{dt} + a_0 y_2(t)$$

$$= b_m \frac{d^m f_2(t)}{dt^m} + b_{m-1}\frac{d^{m-1}f_2(t)}{dt^{m-1}} + \cdots + b_1\frac{df_2(t)}{dt} + b_0 f_2(t) \quad (1.13)$$

Assume now that the *same system at rest* is driven by a linear combination $\alpha f_1(t) + \beta f_2(t)$, where α and β are known constants. Multiplying (1.12) by α and multiplying (1.13) by β and adding the two differential equations, we obtain the following differential equation:

$$\frac{d^n(\alpha y_1 + \beta y_2)}{dt^n} + a_{n-1}\frac{d^{n-1}(\alpha y_1 + \beta y_2)}{dt^{n-1}} + \cdots + a_0(\alpha y_1 + \beta y_2)$$

$$= b_m \frac{d^m(\alpha f_1 + \beta f_2)}{dt^m} + b_{m-1}\frac{d^{m-1}(\alpha f_1 + \beta f_2)}{dt^{m-1}} + \cdots + b_0(\alpha f_1 + \beta f_2) \quad (1.14)$$

where for simplicity of notation we have dropped the time dependence of y_1, y_2, f_1, f_2 on t. It is easy to conclude from (1.14) that the output of the system at rest (the solution of the corresponding differential equation) due to a linear combination of system inputs $\alpha f_1(t) + \beta f_2(t)$ is equal to the corresponding linear combination of the system outputs, that is $\alpha y_1(t) + \beta y_2(t)$. This, basically, is the *linearity principle*. Note that the linearity principle is valid under the assumption that the system initial conditions are zero (system at rest). The linearity principle is, in fact, the *superposition principle,* the well-known principle of elementary electrical circuit theory.

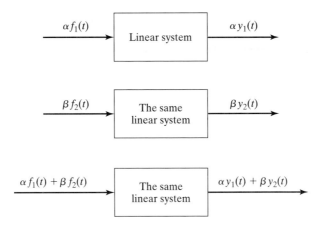

FIGURE 1.8: Graphical representation of the linearity principle for a system at rest

The linearity principle can be put into a formal mathematical framework as follows. If we introduce symbolic notation, the solutions of equations (1.12) and (1.13) can be recorded as

$$y_1(t) = \mathbf{L}\{f_1(t)\}, \quad y_2(t) = \mathbf{L}\{f_2(t)\} \tag{1.15}$$

where $\mathbf{L}$ stands for a linear integral type operator.[†] Note that equations (1.12) and (1.13) can be multiplied by some constants, say α and β, producing (using terminology introduced in (1.15))

$$\alpha y_1(t) = \mathbf{L}\{\alpha f_1(t)\} = \alpha \mathbf{L}\{f_1(t)\}$$
$$\beta y_2(t) = \mathbf{L}\{\beta f_2(t)\} = \beta \mathbf{L}\{f_2(t)\} \tag{1.16}$$

Adding equations in (1.16) leads to the conclusion that

$$\alpha y_1(t) + \beta y_2(t) = \mathbf{L}\{\alpha f_1(t)\} + \mathbf{L}\{\beta f_2(t)\} \tag{1.17}$$

It follows from (1.15–17) that the linearity principle can be mathematically stated as follows:

$$\mathbf{L}\{\alpha f_1(t) + \beta f_2(t)\} = \alpha \mathbf{L}\{f_1(t)\} + \beta \mathbf{L}\{f_2(t)\} \tag{1.18}$$

Using a similar reasoning, we can state the linearity principle for an arbitrary number of inputs; that is,

$$\mathbf{L}\{\alpha_1 f_1(t) + \alpha_2 f_2(t) + \cdots + \alpha_N f_N(t)\} = \alpha_1 \mathbf{L}\{f_1(t)\} + \alpha_2 \mathbf{L}\{f_2(t)\} + \cdots$$
$$+ \alpha_N \mathbf{L}\{f_N(t)\} \tag{1.19}$$

where α_i, $i = 1, 2, \ldots, N$, are constants.

The linearity principle stated analytically in (1.19) is demonstrated graphically in Figure 1.8.

[†] In order to get a solution of an nth order differential equation, the corresponding differential equation must be integrated n times. That is why linear dynamic systems can be modelled as integrators.

EXAMPLE 1.3

Let the response of a linear system at rest due to the system input $f_1(t) = 1, t \geq 0$, be given by $y_1(t) = 0.5 - e^{-t} + 0.5e^{-2t}, t > 0$, and let the response of the same system at rest due to another system input $f_2(t) = \sin(t), t > 0$, be $y_2(t) = 0.5e^{-t} - 0.2e^{-2t} + (\sqrt{10}/10)\cos(t + \theta), t > 0$; then the response of the same system at rest due to the input given by $2f_1(t) + 3f_2(t) = 2 + 3\sin(t), t > 0$, is simply obtained as

$$y(t) = 2y_1(t) + 3y_2(t) = 1 - 0.5e^{-t} + 0.4e^{-2t} + \frac{3\sqrt{10}}{10}\cos(t+\theta), \quad t > 0 \quad \clubsuit$$

EXAMPLE 1.4

The solution of the differential equation

$$\frac{d^2 y(t)}{dt^2} + 2\frac{dy(t)}{dt} + y(t) = 3e^{-t}\sin(2t) + 5\ln(t), \quad \frac{dy(0)}{dt} = 0, \quad y(0) = 0$$

can be obtained by using the linearity principle, in terms of solutions of two simpler problems, as $y(t) = 3y_1(t) + 5y_2(t)$, where

$$\frac{d^2 y_1(t)}{dt^2} + 2\frac{dy_1(t)}{dt} + y_1(t) = e^{-t}\sin(2t), \quad \frac{dy_1(0)}{dt} = 0, \quad y_1(0) = 0$$

$$\frac{d^2 y_2(t)}{dt^2} + 2\frac{dy_2(t)}{dt} + y_2(t) = \ln(t), \quad \frac{dy_2(0)}{dt} = 0, \quad y_2(0) = 0 \quad \clubsuit$$

It is straightforward to verify, by using dual arguments to (1.15–19), that the linear difference equation (1.10) also obeys the linearity principle. Hence, in the discrete-time domain, the linearity principle can be analytically recorded as

$$\mathbf{L}\{\alpha_1 f_1[k] + \alpha_2 f_2[k] + \cdots + \alpha_N f_N[k]\} = \alpha_1 \mathbf{L}\{f_1[k]\} + \alpha_2 \mathbf{L}\{f_2[k]\} + \cdots + \alpha_N \mathbf{L}\{f_N[k]\} \quad (1.20)$$

where $\alpha_i, i = 1, 2, \ldots, N$, are constants.

Note that the linearity (superposition) principle in discrete time can be extended to include a countable infinite number of system inputs. In that case, however additional conditions that guarantee convergence of infinite sums must be imposed [9].

1.2.2 Linear System Time Invariance

It was indicated in Section 1.1 that for a general nth order linear dynamic system, represented by

$$\frac{d^n y(t)}{dt^n} + a_{n-1}\frac{d^{n-1} y(t)}{dt^{n-1}} + a_{n-2}\frac{d^{n-2} y(t)}{dt^{n-2}} + \cdots + a_1 \frac{dy(t)}{dt} + a_0 y(t)$$

$$= b_m \frac{d^m f(t)}{dt^m} + b_{m-1}\frac{d^{m-1} f(t)}{dt^{m-1}} + \cdots + b_1 \frac{df(t)}{dt} + b_0 f(t) \quad (1.21)$$

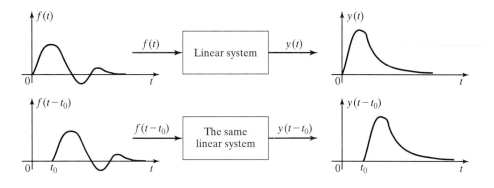

FIGURE 1.9: Graphical representation of system time invariance

the coefficients $a_i, i = 0, 1, 2, \ldots, n-1$, and $b_i, i = 1, 2, \ldots, m$, are assumed to be constant, which indicates that the given system is *time invariant*.

Here, we give an additional clarification of system time invariance. Consider a system at rest. The system time invariance is manifested by the constant shape in time (waveform) of the system output response due to the given input. The output response of a system at rest is invariant regardless of the initial time of the input. If the system input is shifted in time, the system output response due to the same input will be shifted in time by the same amount and, in addition, it will preserve the same waveform. The corresponding graphical interpretation of the time invariance principle is shown in Figure 1.9.

The same arguments presented for the time invariance of continuous-time linear dynamic systems, described by differential equation (1.9), hold for the time invariance of discrete-time linear dynamic systems, described by difference equation (1.10).

The next two examples demonstrate the use of the time invariance principle to simplify analysis of linear time-invariant systems.

EXAMPLE 1.5

Let the response of a time-invariant linear system at rest due to $f_1(t), t > 0$, be given by $y_1(t) = 3e^{-t} - 4e^{-2t}, t > 0$. Then, the system response due to the shifted system input defined by

$$f_2(t) = \begin{cases} f_1(t-6), & t > 6 \\ 0, & t < 6 \end{cases}$$

is

$$y_2(t) = \begin{cases} 3e^{-(t-6)} - 4e^{-2(t-6)}, & t > 6 \\ 0, & t < 6 \end{cases}$$

EXAMPLE 1.6

The solution of the differential equation

$$\frac{d^2 y(t)}{dt^2} + 2\frac{dy(t)}{dt} + 3y(t) = e^{-(t-5)} \sin(t-5), \quad \frac{dy(0)}{dt} = 0, \quad y(0) = 0$$

can be obtained using the time invariance principle by first solving a simpler differential equation,

$$\frac{d^2y_1(t)}{dt^2} + 2\frac{dy_1(t)}{dt} + 3y_1(t) = e^{-t}\sin t, \quad \frac{dy_1(0)}{dt} = 0, \quad y_1(0) = 0$$

and then shifting its solution by five time units, which leads to the result $y(t) = y_1(t-5)$, $t > 5$, and $y(t) = 0$ for $t < 5$.

The system linearity and time invariance principles will be extensively used in subsequent chapters, first of all to simplify the solution of the main linear system theory problem: finding the system response due to arbitrary input signals.

1.3 MATHEMATICAL MODELING OF SYSTEMS[†]

Mathematical modeling of physical systems is based on the application of known physical laws to the given systems, which leads to mathematical equations describing the behavior of the systems under consideration. In addition, systems (processes) in economic, biological, chemical, social, and other sciences are also governed by the corresponding laws, whose mathematical models can be put in the form of either differential or difference equations, as demonstrated in several examples in this section. We will limit our attention to mathematical models of linear time-invariant systems in continuous- and discrete-time domains. In that direction, we either derive or simply present differential/difference equations that describe the dynamics of an electrical circuit, a mechanical system, national income, an amortization process, the heartbeat, eye movement, vehicle lateral motion, a computer switch, and a commercial aircraft.

Electrical Circuit

Consider a simple RLC electrical circuit from [1], presented in Figure 1.10. Applying the basic circuit laws for voltages and currents (see, for example, [10]), we obtain

$$e_i(t) = L\frac{di_1(t)}{dt} + R_1 i_1(t) + e_0(t) \tag{1.22}$$

and

$$e_0(t) = R_2 i_2(t) = \frac{1}{C}\int_0^t i_3(\tau)d\tau + v_C(0) \Rightarrow i_3 = C\frac{de_0(t)}{dt} \tag{1.23}$$

$$i_1(t) = i_2(t) + i_3(t) \tag{1.24}$$

Using (1.23) in (1.24) produces

$$i_1(t) = \frac{1}{R_2}e_0(t) + C\frac{de_0(t)}{dt} \tag{1.25}$$

Taking the derivative of (1.25) and combining (1.22) and (1.25), we obtain the desired second-order differential equation, which relates the input and output of the system, and

---

[†]This section may be skipped without loss of continuity. Most of the models considered in this section will be used in subsequent chapters to demonstrate presented theory.

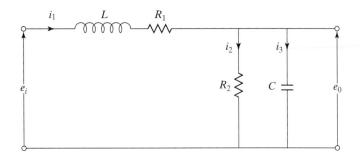

FIGURE 1.10: An RLC network

represents a mathematical model of the circuit given in Figure 1.10:

$$\frac{d^2 e_0(t)}{dt^2} + \left(\frac{L + R_1 R_2 C}{R_2 LC}\right) \frac{de_0(t)}{dt} + \left(\frac{R_1 + R_2}{R_2 LC}\right) e_0(t) = \frac{1}{LC} e_i(t) \quad (1.26)$$

In order to solve this differential equation for $e_0(t)$, the initial conditions $e_0(0)$ and $de_0(0)/dt$ must be known (determined). For electrical circuits, the initial conditions are usually specified in terms of capacitor voltages and inductor currents. Hence, in this example, $e_0(0)$ and $de_0(0)/dt$ should be expressed in terms of $v_C(0)$ and $i_1(0)$.

Note that in this mathematical model $e_i(t)$ represents the system input and $e_0(t)$ is the system output. However, any of the currents and any of the voltages can play the roles of either input or output variables.

Mechanical System

A translational mechanical system (considered in [1]) is represented in Figure 1.11. Using the basic laws of dynamics (see, for example, [11]), we obtain

$$F_1 = m_1 \frac{d^2 y_1(t)}{dt^2} + B_1 \left(\frac{dy_1(t)}{dt} - \frac{dy_2(t)}{dt}\right) + k_1 (y_1(t) - y_2(t)) \quad (1.27)$$

and

$$F_2 = m_2 \frac{d^2 y_2(t)}{dt^2} + B_2 \frac{dy_2(t)}{dt} + k_2 y_2(t) - B_1 \left(\frac{dy_1(t)}{dt} - \frac{dy_2(t)}{dt}\right) - k_1 (y_1(t) - y_2(t))$$

$$(1.28)$$

This system has two inputs, F_1 and F_2, and two outputs, $y_1(t)$ and $y_2(t)$. Equations (1.27) and (1.28) can be rewritten as

$$m_1 \frac{d^2 y_1(t)}{dt^2} + B_1 \frac{dy_1(t)}{dt} + k_1 y_1(t) - B_1 \frac{dy_2(t)}{dt} - k_1 y_2(t) = F_1 \quad (1.29)$$

and

$$-B_1 \frac{dy_1(t)}{dt} - k_1 y_1(t) + m_2 \frac{d^2 y_2(t)}{dt^2} + (B_1 + B_2) \frac{dy_2(t)}{dt} + (k_1 + k_2) y_2(t) = F_2 \quad (1.30)$$

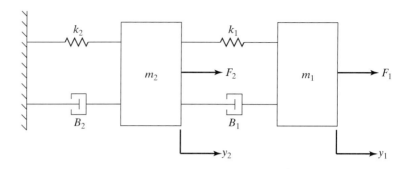

FIGURE 1.11: A translational mechanical system

Mathematical models of many other interesting physical systems can be found in [2–8]. Sometimes, due to the complexity of dynamic systems, it is impossible to establish mathematical relations describing the dynamic behavior of the systems under consideration. In those cases, we perform experiments to obtain data that can be used in establishing some mathematical relations caused (induced) by system dynamics. Techniques for experimentally obtaining mathematical models of dynamic systems are studied within the scientific area called *system identification*. (For a classic textbook on system identification, see reference [12].)

National Income
The national income is governed by the following set of difference equations [6–7]:

$$y[k] = c[k] + i[k] + f[k]$$
$$c[k] = \alpha y[k-1] \quad (1.31)$$
$$i[k] = \beta(c[k] - c[k-1])$$

where α, β are positive constants, $y[k]$ is the national income, $c[k]$ represents consumer expenditures, $i[k]$ is induced private investment, and $f[k]$ represents government expenditures.

From equations (1.31), by simple algebra, we obtain the difference equation that represents the mathematical model for the national income

$$y[k+2] - \alpha(1+\beta)y[k+1] + \alpha\beta y[k] = f[k+2] \quad (1.32)$$

This model is known as the Samuelson model [6–7]. The system input is the "government" and the system output is the national income. The unit for discrete-time k is usually taken as either one year or three months (quarterly calculations).

Assuming that the initial conditions $y[0]$ and $y[1]$ are known, and that the input signal represented by $f[2], f[3], \ldots$, is known, we can iterate (1.32) and recursively find its solution for any k. For example, for $k = 0$, we have

$$y[2] - \alpha(1+\beta)y[1] + \alpha\beta y[0] = f[2]$$

which implies the value for $y[2]$ in terms of $y[1]$, $y[0]$, and $f[2]$. Having found $y[2]$, we can obtain $y[3]$ as follows

$$y[3] - \alpha(1+\beta)y[2] + \alpha\beta y[1] = f[3]$$

and so on; thus, we can obtain $y[k]$ for an arbitrary discrete-time k. Iterating is computationally involved; in order to find, for example $y[100]$, we must evaluate all previous values of $y[k]$, that is, $y[99]$, $y[98]$, ... In Chapters 5 and 7 we will study more convenient and better methods for solving linear difference equations.

Amortization

If we purchase a house, or a car, and take a loan of d dollars with a fixed interest rate of R percent per year ($r = R/12$ per month), then the loan is paid back through the process known in economics as amortization. Using simple logic, it is not hard to conclude that the outstanding principal, $y[k]$, at the $k+1$ discrete-time instant (month), is given by the recursive formula (difference equation)

$$y[k+1] = y[k] + ry[k] - f[k+1] = (1+r)y[k] - f[k+1] \quad (1.33)$$

where $f[k+1]$ stands for the payment made in the $(k+1)$st discrete-time instant (month).

Let us assume that the monthly loan payment is constant, say $f[k] = p$. The question that we wish to answer is: What is the monthly loan payment needed to pay back the entire loan of d dollars within N months? (Typically, $N = 48$ for the purchase of a car, and $N = 240$ or $N = 360$ for the purchase of a house.) The answer to this question can be easily obtained by iterating (1.33) and producing the corresponding recursive formula. Since $y[0] = d$ and $f[1] = f[2] = \cdots = f[N] = p$ are known, we have from (1.33), for $k = 1$,

$$y[1] = (1+r)y[0] - f[1] = (1+r)d - p$$

For $k = 2$, we obtain

$$y[2] = (1+r)y[1] - f[2] = (1+r)\{(1+r)d - p\} - p = (1+r)^2 d - (1+r)p - p$$

Continuing this procedure for $k = 3, \ldots, N$, we can recognize the pattern and conclude that

$$y[3] = (1+r)^3 d - (1+r)^2 p - (1+r)p - p$$

and, in general,

$$y[N] = (1+r)^N d - (1+r)^{N-1} p - \cdots - (1+r)^2 p - (1+r)p - p$$

$$= (1+r)^N d - p \sum_{i=0}^{N-1} (1+r)^i \quad (1.34)$$

The recursive formula thus obtained represents the solution of difference equation (1.33). The formula can be simplified even further using the known summation formula

$$\sum_{i=0}^{n} q^i = \frac{q^{n+1} - 1}{q - 1}, \quad q \neq 1 \quad (1.35)$$

Applying this formula to the second term in (1.34), we obtain

$$y[N] = (1+r)^N d - p \frac{(1+r)^N - 1}{(1+r) - 1} = (1+r)^N d - p \frac{(1+r)^N - 1}{r} \quad (1.36)$$

From the last expression, we conclude that the loan is paid back when $y[N] \stackrel{\scriptscriptstyle\triangle}{=} 0$, which implies that the formula for the required monthly payment is

$$0 = (1+r)^N d - p \frac{(1+r)^N - 1}{r} \Rightarrow p = \frac{r(1+r)^N}{(1+r)^N - 1} d \quad (1.37)$$

EXAMPLE 1.7

Assume that a student purchases a car and takes a loan of $10,000 with an interest rate of $R = 5\%$ per year, or equivalently $r = (5/12)\%$ per month, for the period of 4 years = 48 months. According to formula (1.37), the monthly payment required to pay back the loan within 48 months is given by

$$p = \frac{\frac{0.05}{12}\left(1 + \frac{0.05}{12}\right)^{48}}{\left(1 + \frac{0.05}{12}\right)^{48} - 1} 10000 = \frac{0.00417 \times 1.221}{1.221 - 1} 10000 = 0.023 \times 10000 = 230$$

Note that the total amount that the student must pay during the 48 months is equal to $230 \times 48 = \$11,040$.

Heartbeat Dynamics

The dynamics of a heartbeat (where the diastole is the relaxed state and the systole the contracted state of the heart) can be approximately described by the following set of linear differential equations:

$$\begin{aligned} \dot{x}_1(t) &= -\frac{2}{\epsilon} x_1(t) - \frac{1}{\epsilon} x_2(t) - \frac{1}{\epsilon} x_3(t) \\ \dot{x}_2(t) &= -2x_1(t) - 2x_2(t) \\ \dot{x}_3(t) &= -x_2(t) \end{aligned} \quad (1.38)$$

where $x_1(t)$ is the length of the muscle fiber, $x_2(t)$ represents the tension in the fiber caused by blood pressure, $x_3(t)$ represents the dynamics of an electrochemical process that governs the heart beat, and ϵ is a small positive parameter [8]. The system presented in (1.38) is driven by the initial condition that characterizes the heart's diastolic state, whose normalized value, in this model, is equal to $(x_1(0), x_2(0), x_3(0)) = (1, -1, 0)$. The presented model of a heartbeat is known in the literature as Zeeman's model [8, 13].

Since, in the above model, ϵ is a small positive number, we conclude that $dx_1(t)/dt$ is large, which implies that the cardiac muscle shortens quickly from its full length (normalized at $x_1(0) = 1$) to some steady state value at the end of the systolic period, at which time it rapidly goes to the diastolic state characterized by the given initial condition (heart's equilibrium point). For 60 beats/min or 1 beat/s, the systolic period is $T_{sis} = 0.4\,s$ and the diastolic period is $T_{dia} = 0.6$ beat/s. The rate or velocity of shortening of the cardiac muscle during the systolic period is proportional to ϵ.

Eye Movement (Oculomotor Dynamics)

The dynamics of eye movement (muscles, eye, and orbit) can be modeled by the second-order system represented by

$$\frac{d^2 y(t)}{dt^2} + \left(\frac{1}{\tau_1} + \frac{1}{\tau_2}\right)\frac{dy(t)}{dt} + \frac{1}{\tau_1 \tau_2} y(t) = \frac{1}{\tau_1 \tau_2} f(t) \tag{1.39}$$

where $\tau_1 = 13$ ms and $\tau_2 = 224$ ms are respectively the minor and major eye time constants [14, 15]. $y(t)$ is the eye position in degrees and $f(t)$ is the eye stimulus force in degrees (reference eye position, target position). Several other mathematical models for eye movement exist in the biomedical engineering literature, including a simple first-order model [16], and a more complex model of order six [17]. The eye movement mathematical model of order six will be presented in Chapter 8, Problem 8.46.

Vehicle Lateral Error

In a recent study [18], a mathematical model was derived for the dynamics of vehicle lateral error:

$$\frac{d^4 e(t)}{dt^4} + 24.32\frac{d^3 e(t)}{dt^3} + 151.92\frac{d^2 e(t)}{dt^2} = 114.26\frac{d^2 \theta(t)}{dt^2} + 1535.49\frac{d\theta(t)}{dt} + 3592.09\,\theta(t) \tag{1.40}$$

where $e(t)$ stands for the corresponding lateral error and $\theta(t)$ represents the steering angle.

Computer Communication Network Switch

Data messages sent over modern computer communication networks are divided into much smaller units called packets, usually of fixed length. Such packets travel over network links. The links are connected by nodes (switches) whose roles are to direct the network traffic to the desired destination, and to store packets awaiting transfer due to network congestion. One of the main techniques for transmission of data packets in a computer communication network that has many nodes and many users is the asynchronous transfer mode (ATM) technique. The data packet queue length, $q[k]$, and the data packet arrival rate, $y[k]$, in a computer communication network ATM switch that stores and/or routes data packets, can be modeled by the set of difference equations

$$q[k+1] = q[k] + y[k+1-d] - f[k]$$

$$y[k+1] = y[k] - \sum_{j=0}^{1} \alpha_j (q[k-j] - q^0) - \sum_{i=0}^{d} \beta_i y[k-i] \tag{1.41}$$

where $f[k]$ is the service rate of the buffer (which represents switch memory of a rather limited size), q^0 is the desired buffer's steady state queue length, d is the round trip delay between the information source and the switch expressed in discrete-time units, and α_j and

β_i are gains to be chosen by a network engineer to assure network stability and eliminate packet flow congestion [19]. In addition, it is known that $\sum_{i=0}^{d} \beta_i = 0$ and $\sum_{j=0}^{-1} \alpha_j > 0$. It is often assumed that the service rate $f[k] = \mu$ is constant, which can be provided only when $q[k] \geq \mu$. If $q[k] < \mu$, then the service rate is equal to $f[k] = q[k]$.

In Chapters 5 and 8 we will study the dynamics of the preceding system of linear difference equations. We will show how to get only one difference equation either for the queue length or the packet arrival rate. Using realistic numerical data, it will be shown in Chapter 5 that $d = 12$, and in Chapter 8 that the system of difference equations (1.41) determines a linear time-invariant discrete-time system of order 14 (see Problem 8.47).

Commercial Aircraft

The linearized equations governing the motion of a commercial aircraft are given by

$$\frac{d\alpha(t)}{dt} = -0.313\alpha(t) + 56.7q(t) + 0.232 f_e(t)$$

$$\frac{dq(t)}{dt} = -0.0139\alpha(t) - 0.426q(t) + 0.0203 f_e(t) \qquad (1.42)$$

$$\frac{d\theta(t)}{dt} = 56.7q(t)$$

where $\alpha(t)$ is the aircraft angle of attack, $q(t)$ is the pitch rate, and $\theta(t)$ represents the pitch angle [20]. The driving force $f_e(t)$ stands for the elevator deflection angle. In practice, the aircraft pitch rate is controlled by an appropriate choice of the input signal $f_e(t)$. Differentiating this system of three first-order linear differential equations, it can be replaced by one third-order linear differential equation that gives direct dependence of $\theta(t)$ on $f_e(t)$; that is,

$$\frac{d^3\theta(t)}{dt^3} + 0.739 \frac{d^2\theta(t)}{dt^2} + 0.921 \frac{d\theta(t)}{dt} = 1.151 \frac{df_e(t)}{dt} + 0.1774 f_e(t) \qquad (1.43)$$

This model will be further analyzed in subsequent chapters.

1.4 SYSTEM CLASSIFICATION

Real-world systems are either static or dynamic. Static systems are represented by algebraic equations—for example, describing electrical circuits with resistors and constant voltage sources, or in statics indicating that at equilibrium the sums of all forces are equal to zero. Studying static systems is simple; hence, static systems will not be the subject of this book.

Dynamic systems are, in general, described either by differential/difference equations (also known as *systems with concentrated* or *lumped parameters*) or by partial differential equations (known as *systems with distributed parameters*). For example, electric power transmissions, wave propagation, behavior of antennas, propagation of light through optical fiber, and heat conduction represent dynamic systems described by partial differential equations. For example, one-dimensional electromagnetic wave propagation is described by the partial differential equation

$$\frac{\partial^2 E(t, x)}{\partial t^2} + c^2 \frac{\partial^2 E(t, x)}{\partial x^2} = 0$$

where $E(t, x)$ is the electric field, t represents time, x is the spatial coordinate, and c is the constant that characterizes the medium. Systems with distributed parameters are also known as *infinite dimensional systems,* in contrast to systems with concentrated parameters, which are known as *finite dimensional systems* (represented by differential/difference equations of finite orders, $n < \infty$). Distributed parameter systems are very difficult to study, since their analysis is based on advanced mathematics. We will pay attention to concentrated parameter systems—that is, dynamic systems described by differential/difference equations. It is important to point out that many real physical systems belong to the category of concentrated (lumped) parameter systems, and a large number of them will be encountered in this book.

Dynamic systems with lumped parameters can be either linear or nonlinear. *Linear dynamic systems* are described by linear differential/difference equations, and obey the linearity principle; *nonlinear dynamic systems* are described by nonlinear differential/difference equations. For example, the motion of a simple pendulum, a familiar system studied in basic physics courses, is described by the nonlinear differential equation

$$\frac{d^2\theta(t)}{dt^2} + \frac{g}{m} \sin(\theta(t)) = 0 \tag{1.44}$$

where $\theta(t)$ is the pendulum angle with respect to the vertical axis, $g = 9.8 \, \text{m/s}^2$ is the gravitational constant, and m is the pendulum mass. Note that $\sin(\theta(t))$ is a nonlinear function.

We can also distinguish between *time-invariant systems* (systems with constant coefficients) obeying the time invariance principle, and *time-varying systems* whose parameters change in time. For example, the linear time-varying model of the Erbium-doped optical fiber amplifier (which revolutionized optical communications in the middle of the 1990s), derived in [21], is given by

$$\frac{dN(t)}{dt} + \frac{1}{\tau_l(t)} N(t) = b_p(t) P_p + \sum_{i=1}^{n} b_i(t) P_{si} \tag{1.45}$$

where $N(t)$ represents deviation from the nominal value of the average level of the normalized number of Erbium atoms in the upper excited state, $\tau_l(t)$ is the time-varying time constant, P_p and P_{si} are respectively laser pump and optical signal power deviations from their nominal values, and $b_p(t)$ and $b_i(t)$ are corresponding time-varying coefficients. The difficulty in dealing with linear time-varying systems is that we can solve analytically only the first-order time-varying linear differential equation. For higher-order (two and above) linear time-varying differential equations, *analytical solutions do not exist* in general.

Some system parameters and variables can vary randomly (for example, the generated power of a solar cell, house humidity, and temperature). Also, system inputs may be random signals (for example, aircraft under wind disturbances, electric current under electron thermal noise). Systems that have random parameters and/or process random signals are called *stochastic systems*. Stochastic systems can be either linear or nonlinear, time invariant or time varying, continuous or discrete. In contrast to stochastic systems, we have *deterministic systems*, whose parameters and input signals are deterministic quantities.

Real-world physical systems are known as *nonanticipatory systems* or *causal systems*. Real physical systems cannot predict the future. In other words, they cannot produce an output before the actual input is applied to the system. Let us be more specific: Let the input $f(t_1)$ be applied to a system at time t_1. Then the real physical system can only produce the system output at time t_1, that is, $y(t_1)$. The real physical system cannot, at time t_1, produce information about $y(t)$ for $t > t_1$. That is, the system is unable to predict the future input values and produce the future system response $y(t)$, based on the information that the system has at time t_1. The system causality can also be defined by the statement that the system input $f(t_2)$ has no impact on the system output $y(t_1)$ for $t_2 > t_1$. In contrast to nonanticipatory (causal) systems, we have *anticipatory* or *noncausal systems*. Anticipatory systems are encountered in digital signal processing—they are artificial systems.

Dynamic systems are also *systems with memory*. Namely, the system output at time t_1 depends not only on the system input at time t_1, but also on all previous values of the system input. Let $y(t) = \phi(f(t))$ be the solution of the corresponding differential equation representing a dynamic system. Then the fact that the dynamic system possesses memory can be formally recorded as $y(t_1) = \phi(f(t)), t_0 \leq t \leq t_1$. In contrast, static systems have no memory. If the relationship $y(t) = \phi(f(t))$ came from a static system, then we would have $y(t_1) = \phi(f(t_1))$. That is, for static systems the output at time t_1 depends only on the input at time t_1. Static systems are known as *memoryless systems* or *instantaneous systems*. For example, an electric resistor is a static system since its voltage (system output) is an instantaneous function of its current (system input), so that $y(t) = v(t) = Ri(t) = Rf(t)$ for any t.

Analog systems deal with continuous-time signals, which can take a continuum of values with respect to the signal magnitude. *Digital systems* process digital signals, whose magnitudes can take only a finite number of values. In digital systems, signals are discretized with respect to both time and magnitude (signal sampling and quantization).

1.5 MATLAB SYSTEM ANALYSIS AND DESIGN

MATLAB is a very advanced and reliable computer package, which can be used for computer-aided system analysis and design. In addition to handling standard linear algebra problems, it has several specialized toolboxes, such as Control, Signal Processing, Communications, Neural Networks, Optimization, Identification, and Symbolic. We will also refer to the Simulink package, which is basically another MATLAB tool-box. Simulink is very convenient for simulation (finding system responses due to given inputs) of linear and nonlinear dynamic systems.

MATLAB is user friendly. It takes only a few hours to master most of the functions needed for the purpose of this course. MATLAB will help students to obtain a deeper understanding of the main system concepts and techniques, and to study higher-order real physical linear systems, which would otherwise be impossible. Many MATLAB problems, laboratory experiments, and case studies will be encountered in this book. More about MATLAB and its functions relevant to system analysis can be found in Appendix C; a brief introduction to Simulink is presented in Appendix D.

1.6 ORGANIZATION OF THE TEXT

This book is divided into three parts: Part I, *Frequency Domain Techniques,* Chapters 3–5; Part II, *Time Domain Techniques,* Chapters 6–8; and Part III, *Linear Systems in Electrical Engineering,* Chapters 9–12.

Chapters 1 and 2 serve as an introduction, in which we define continuous- and discrete-time linear time-invariant dynamic systems and explain the principles of linearity and time invariance, and present a general theory of signals used in linear systems. (Some signals, their properties, and corresponding signal operations are presented in Chapters 9 and 10, within the system approach to signal processing and communications.)

In the first part of the book, we study Fourier series and Fourier transform (Chapter 3), Laplace transform (Chapter 4), and $\mathcal{Z}$–transform (Chapter 5). *All important linear system theory concepts are first introduced in the frequency domain.* It is natural to define the system transfer function in the frequency domain. We introduce the system impulse response, and derive the most important result of linear system theory, which states: *the system zero-state response due to an arbitrary input is given by the signal convolution.*

The signal convolution is thoroughly studied in Chapter 6. After all of the important linear system concepts are introduced in the frequency domain, we interpret these concepts in the time domain and develop the time domain techniques for finding the response of continuous- and discrete-time linear systems (Chapters 6–8). This leads to the state space technique as a highlight of the time domain approach for studying linear systems. Due to the rapid development of electrical engineering in the preceding two decades and the frequent use of modern computer packages (like MATLAB) for system analysis, it is imperative for modern courses on linear system analysis to give an extensive coverage of the state space technique. Using elementary knowledge of linear algebra and differential equations, the main state space concepts are slowly and thoroughly developed, and new notions are fully explained in Chapter 8.

Chapters 1–8 are essential for an undergraduate linear systems course. These chapters can and should be covered at least in parts during a one-semester course on linear systems and signals. In the third part of the book, we demonstrate how to use linear system theory results to solve problems in other fields of electrical engineering: digital signal processing, communication systems, electrical circuits, and control systems. Some additional signal and system concepts specific to these fields are presented in Chapters 9 and 10. Instructors more interested in signals than in dynamic systems should also cover selected topics from Chapters 9 and 10; instructors more interested in dynamic systems than in signals may skip Chapter 3 entirely, and cover during one semester Chapters 1–2, 4–8, and selected parts of Chapters 11 and 12 (on electrical circuits and linear feedback systems).

1.7 SUMMARY

Linear time-invariant dynamic systems, which are the main subject of this book, are described by linear differential (continuous-time) and difference (discrete-time) equations with constant coefficients given, respectively, by

$$\frac{d^n y(t)}{dt^n} + a_{n-1}\frac{d^{n-1} y(t)}{dt^{n-1}} + \cdots + a_1 \frac{dy(t)}{dt} + a_0 y(t)$$
$$= b_m \frac{d^m f(t)}{dt^m} + b_{m-1}\frac{d^{m-1} f(t)}{dt^{m-1}} + \cdots + b_1 \frac{df(t)}{dt} + b_0 f(t)$$

and

$$y[k+n] + a_{n-1} y[k+n-1] + \cdots + a_1 y[k+1] + a_0 y[k]$$
$$= b_m f[k+m] + b_{m-1} f[k+m-1] + \cdots + b_1 f[k+1] + b_0 f[k]$$

where f represents the system input, y is the system output, and a_i, $i = 0, 1, 2, \ldots, n-1$, and b_j, $j = 0, 1, \ldots, m$, are constants. *The central problem that we are faced with in this course is finding solutions (system responses) to the preceding equations for given initial conditions and due to arbitrary forcing functions (system inputs).* This problem can be solved either in the time domain by using the results of the mathematical theory of differential and difference equations, or in the frequency domain by using the corresponding time-frequency domain transforms. Both approaches will be considered.

For engineering and science students, the importance of mastering the mapping from the system input to the system output (the solution to the problem of finding the system response) can be symbolically compared to a chess game goal of converting (mapping) a pawn into a queen, as illustrated in Figure 1.12.

In this chapter, we have introduced two main concepts that characterize time-invariant linear systems: time invariance and system linearity. *Time invariance basically says: if we shift the system input by a certain time interval, the system output will be shifted by the same amount of time and preserve the same waveform as the original system output.* System linearity is basically the familiar concept of superposition known to electrical engineering students from basic courses on electrical circuits. *The linearity concept states*

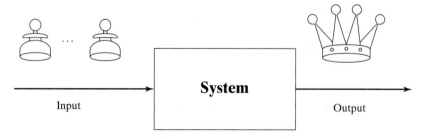

FIGURE 1.12: A symbolic representation of the mapping from the system input to the system output

that if $y_1(t)$ is the system output caused by the system input $f_1(t)$ and $y_2(t)$ is the system output due to the system input $f_2(t)$, then the system response due to $\alpha f_1(t) + \beta f_2(t)$ is equal to $\alpha y_1(t) + \beta y_2(t)$, where α and β are arbitrary real constants. Students should use system linearity whenever possible since it simplifies calculations. The linearity concept can be generalized to several system inputs.

The material presented in this book uses the *universal language of mathematics* to study real systems described by linear differential and/or difference equations with constant coefficients, without taking into account the nature of the systems under consideration. In a rapidly developing technological world, the knowledge that students get from this course will help them *to cope with modern linear dynamic systems* coming from such diverse disciplines as electrical, mechanical, biomedical, aerospace, industrial, and chemical engineering, and physics, chemistry, biology, economics, sociology, and other sciences. In addition, the course in linear systems and signals is an *important and excellent prerequisite for undergraduate courses in controls, communications, signal processing, electronics, robotics, and power systems*.

1.8 REFERENCES

[1] Z. Gajić and M. Lelić, *Modern Control Systems Engineering,* Prentice Hall International, London, 1996.

[2] P. Wellstead, *Physical System Modelling,* Academic Press, London, 1979.

[3] V. Kecman, *State Space Models of Lumped and Distributed Systems,* Springer-Verlag, Berlin, 1988.

[4] L. Ljung, *Mathematical Modeling,* Prentice Hall, Englewood Cliffs, NJ, 1994.

[5] R. Dorf, *Modern Control Systems,* Addison Wesley, Reading, MA, 1998.

[6] J. Cadzow, *Discrete-Time Systems,* Prentice Hall, Englewood Cliffs, NJ, 1973.

[7] S. Elaydi, *An Introduction to Difference Equations,* Springer-Verlag, NY, 1996.

[8] P. Tu, *Dynamical Systems: An Introduction with Applications in Economics and Biology,* Springer-Verlag, Berlin, 1994.

[9] I. Sandberg, "The superposition scandal," *Circuits, Systems, and Signal Processing,* 17:733–35, 1998.

[10] R. Thomas and A. Rosa, *Analysis and Design of Linear Circuits,* Prentice Hall, Englewood Cliffs, NJ, 1994.

[11] D. Greenwood, *Principles of Dynamics,* Prentice Hall, Englewood Cliffs, NJ, 1988.

[12] L. Ljung, *System Identification: Theory for the User,* Prentice Hall, Englewood Cliffs, NJ, 1987.

[13] E. Zeeman, "Differential equations for the heart and nerve impulse," in *Dynamical Systems,* M. Peixoto (ed.), Academic Press, New York, 1973.

[14] J. Horng, J. Semmlow, G. Hung, and K. Ciuffreda, "Initial component control in disparity vergence: A model-based study," *IEEE Transactions on Biomedical Engineering,* 45:249–57, 1998.

30   Chapter 1   Introduction to Linear Systems

[15] T. Alvarez, J. Semmlow, W. Yuan, and P. Munoz, "Dynamic details of disparity convergence eye movements," *Annals of Biomedical Engineering,* 27:380–90, 1999.

[16] P. Gamlin, and L. Mays, "Dynamic properties of medial rectus motoneurons during vergence eye movements," *Journal of Neurophysiology,* 67:64–74, 1992.

[17] F. Szidarovszky and A. Bahill, *Linear Systems Theory,* CRC Press, Boca Raton, FL, 1992.

[18] R. Byrne, C. Abdallah, and P. Dorato, "Experimental results in robust lateral control of highway vehicles," *IEEE Control Systems Magazine,* 18:70–76, 1998.

[19] L. Benmohamed and S. Meerkov, "Feedback control of congestion in packet switching networks: The case of a single congested node," *IEEE/ACM Transactions on Networking,* 1:693–708, 1993.

[20] W. Messner and D. Tilbury, *Control Tutorials for MATLAB and SIMULINK: A Web-Based Approach,* Addison Wesley Longman, Reading, MA, 1998.

[21] Z. Gajić and M. Lelić, "Linear time varying model of erbium-doped fiber amplifier dynamics and its stability," Report R14343, Corning Inc., Corning, NY, 1999.

1.9 PROBLEMS

1.1. Which of the following continuous-time mathematical functions qualify as continuous-time signals? (For simplicity, assume $t \geq 0$.)

(a) $e^{-2t} \sin(t^2 + 0.1\pi)$

(b) $\sqrt[3]{t} \ln(t)$

(c) $\cos\left(\sqrt{t} - \dfrac{\pi}{3}\right)$

(d) $\dfrac{1}{1-t^2} e^{3t}$

(e) $(1 + t^4) \tan\left(\dfrac{t}{1-t}\right)$

(f) $\cos(\sqrt{2}\, t)$

Hint: Uniqueness.

1.2. Consider the first-order linear time-invariant system represented by the differential equation

$$\frac{dy(t)}{dt^2} + y(t) = f(t) = 1, \quad y(0) = 2, \quad t \geq 0$$

Find the system response and its zero-state and zero-input components. What are the response steady state and transient components?

Answer: $y(t) = 1 + e^{-t}$, $y_{zi}(t) = 2e^{-t}$, $y_{zs}(t) = 1 - e^{-t}$, $\bar{y}_{ss} = 1$, $\bar{y}_{tr}(t) = e^{-t}$

1.3. Repeat Problem 1.2 for the system input equal to $f(t) = \sin(t), t \geq 0$.

1.4. Consider the second-order linear time-invariant system

$$\frac{d^2y(t)}{dt} + 2\frac{dy(t)}{dt} + y(t) = e^{-t}, \quad y(0) = 0, \quad \frac{dy(0)}{dt} = -1, \quad t \geq 0$$

Find the system response and its zero-state and zero-input components. What are the response steady state and transient components?

1.5. Repeat Problem 1.4 for the system input equal to $f(t) = 1 + \cos(2t)$, $t \geq 0$.

1.6. Find the zero-state, zero-input, transient, and steady state responses for the output voltage of the electric circuit presented in Figure 1.10, whose differential equation is given in (1.26). Assume that the circuit initial conditions are $e_0(0) = 1$, $e_0(0)/dt = 2$ and that the input signal is equal to $e_i(t) = 6\,\text{V}$. The circuit parameters have the following values: $R_1 = R_2 = 10\,\text{k}\Omega$, $L = 10\,\text{mH}$, and $C = 1\,\mu\text{F}$.

1.7. Find the zero-state, zero-input, transient, and steady state responses for eye movement dynamics whose differential equation is given in (1.39). Assume that the initial conditions are $y(0) = 10°$, $dy(0)/dt = 0$ and that the stimulus force is equal to $f(t) = 30°$.

1.8. Assume that the responses of a continuous-time linear time-invariant system at rest due to inputs $f_1(t)$, $f_2(t)$, and $f_3(t)$ are respectively given by $y_1(t)$, $y_2(t)$, and $y_3(t)$. What is the zero-state response of the same system due to the input signal given by $2f_1(t) - 5f_2(t) - f_3(t)$.
Answer: $2y_1(t) - 5y_2(t) - y_3(t)$

1.9. Using the linearity principle, find the response of a linear time-invariant system at rest defined by

$$\frac{dy(t)}{dt} + 3y(t) = 1 + 3e^{-t} - 5\cos(t)$$

1.10. Using the time invariance principle, find the response of the system represented by

$$\frac{d^2y(t)}{dt^2} + 4\frac{dy(t)}{dt} + 3y(t) = f(t), \quad f(t) = \begin{cases} \sin(t-2), & t \geq 2 \\ 0, & t < 0 \end{cases}$$

Assume that the system initial conditions are zero.

1.11. Consider a discrete-time linear time-invariant system at rest. Let its response to $(-1)^k$, $k \geq 0$, be $(0.1)^k + 2(-0.5)^k - 3(-1)^k$. What is the system response due to the input signal equal to $3(-1)^k$, $k \geq 0$? What is the system response due to the input signal defined by $f[k] = 3(-1)^{k-3}$, $k \geq 3$, and $f[k] = 0$, $k < 3$?

1.12. Consider a time-invariant linear system at rest and assume that the input $f(t)$ produces the output response $y(t) = e^{-2t}\cos(3t + \pi/4)$. What is the response of this system due to the input signal defined by

$$f_1(t) = \begin{cases} -2f(t-4), & t \geq 4 \\ 0, & t < 4 \end{cases}$$

Answer: $y_1(t) = -2e^{-(t-4)}\cos(3(t-4) + \pi/4)$, $t > 4$, and $y_1(t) = 0$, $t < 4$

1.13. Can you solve, using the classical method for solving differential equations, the second-order differential equation that represents the dynamics of a second-order linear time-invariant system that differentiates the system input? For example, consider a system at rest, and try to find its zero-state response using the classical method:

$$\frac{d^2y(t)}{dt^2} + 3\frac{dy(t)}{dt} + 2y(t) = \frac{df(t)}{dt} + 3f(t), \quad f(t) = e^{-5t}, \quad t \geq 0$$

Hint and Answer: You will make a mistake if you use ordinary calculus and evaluate $df(t)/dt + 3f(t) = -2e^{-5t}$, $t \geq 0$. The system response is given by $y(t) = \frac{1}{2}e^{-t} - \frac{1}{3}e^{-2t} - \frac{1}{6}e^{-3t}$, $t \geq 0$. This answer can be obtained either by using the Laplace transform, to be studied in Chapter 4, or by introducing a new function called the impulse delta function, $\delta(t)$ (to be presented in Chapter 2), and correctly finding the derivative of the right-hand side of the given differential equation using the generalized calculus (to be introduced in Chapter 2) by identifying a missing term, that is, $df(t)/dt + 3f(t) = -2e^{-5t} + \delta(t)$.

1.14. Assume that you would like to buy a car and that you can get a loan with an interest rate of $R = 5\%$ per year. You want to limit your expenditure such that your monthly payment does not exceed $300 during the loan period of 48 months. How much can you spend? How much will you pay in interest during the period of 48 months?

1.15. Assume that you want to purchase a house, and you take a loan of $200,000 with an interest rate of $R = 8\%$ per year, or equivalently $r = (8/12)\%$ per month, for the period of 20 years $= 240$ months. Use formula (1.37) to find the monthly payment required to pay back the loan within 240 months. Find the total dollar amount that you pay for the house, and find the interest paid during 240 months.

1.16. Solve Problem 1.15 assuming that the loan is taken over a 30–year period and compare the results with the corresponding results obtained in Problem 1.15.

1.17. Is an electrical diode a static or dynamic system? Is it a linear or nonlinear system?

1.18. Is computer memory a static or dynamic system? Is it a linear or nonlinear system? Is it a time-varying or time-invariant system?

1.19. Identify the following systems as linear or nonlinear, and time varying or time invariant

(a) $\dfrac{d^3y(t)}{dt^3} + \dfrac{d^2y(t)}{dt^2} + ty(t) = f(t)$

(b) $\dfrac{d^2y(t)}{dt^2} + 2y(t)\dfrac{dy(t)}{dt} + y(t) = \dfrac{df(t)}{dt} + 3f(t)$

(c) $\dfrac{d^5y(t)}{dt^5} - 2\dfrac{d^2y(t)}{dt^2} + 5\dfrac{dy(t)}{dt} - y(t) = \dfrac{d^3f(t)}{dt^3} + \dfrac{d^2f(t)}{dt^2} - 2f(t)$

(d) $t^2\dfrac{d^2y(t)}{dt^2} + y^2(t) = f(t)$

Identify terms that are specific for particular systems.
Answer:
(a) linear and time-varying system;
(b) nonlinear time-invariant system;
(c) linear time-invariant system;
(d) nonlinear time-varying system.

CHAPTER 2

Introduction to Signals

In many areas of engineering and the sciences we are very often faced with signals. Signals are mathematical functions that describe the behavior of corresponding natural and artificial (man-made) physical phenomena. Since in nature everything is unique, we require that the functions that represent signals are uniquely determined. Dealing with signals is the same as dealing with mathematical functions. We can define operations on signals, such as signal addition, subtraction, multiplication, differentiation, integration, time shifting, time reversal, time scaling, and frequency scaling, in the same way as we are used to doing in calculus with functions. In a MATLAB laboratory experiment on signals, given at the end of this chapter, students will be asked to perform most of these operations on time signals commonly used in electrical engineering. (The purpose of the experiment is for students to become familiar with MATLAB while reviewing some basic operations with mathematical functions (signals).)

Signals can be represented either in continuous- or discrete-time domains. In this chapter we treat both continuous- and discrete-time signals in parallel. Section 2.1 presents the basic signals common in engineering and the sciences. This section also introduces the impulse delta signal and its properties. The impulse delta signal is particularly important for linear dynamic systems: it will help us to derive a general formula for finding the system response due to any input signal. In Section 2.2, the concepts of continuous- and discrete-time convolutions are defined and the notion of the "discrete-time derivative" is discussed. The concept of convolution is extremely important for linear system theory, such that an entire chapter will be devoted to it (Chapter 6). In Section 2.3, we give a signal classification and introduce several definitions related to signals.

2.1 COMMON SIGNALS IN LINEAR SYSTEMS

In this section we present common signals used in linear system analysis: step, sign (signum), ramp, triangular pulse, rectangular pulse, parabolic, sine, sinc, and impulse delta signals. These signals are presented simultaneously in both continuous- and discrete-time domains. Except for the impulse delta signal, the signals studied in this section are described by *ordinary functions,* the functions that we are familiar with from basic calculus courses. The impulse delta signal belongs to the class of *distribution (singular)* functions. Due to its particular importance for linear system theory, the impulse delta signal will be studied in an independent subsection.

Three slightly different definitions of the *unit step* signal can be found in engineering literature on linear systems and signals. Two commonly used definitions in undergraduate signals and systems books [1–11] are

$$u(t) = \begin{cases} 1, & t \geq 0 \\ 0, & t < 0 \end{cases} \quad (2.1a)$$

and ([12–24])

$$u_1(t) = \begin{cases} 1, & t > 0 \\ 0, & t < 0 \end{cases} \quad (2.1b)$$

Note that in the second case the value of $u_1(0)$ is not defined.

The third definition of the unit step signal (also known as Heaviside's unit step signal, in honor of the famous nineteenth-century electrical engineer) is given [25–27] by

$$u_h(t) = \begin{cases} 1, & t > 0 \\ 0.5, & t = 0 \\ 0, & t < 0 \end{cases} \quad (2.2)$$

Definitions (2.1) are simpler and easier to work with than definition (2.2). From the linear dynamic system perspective (linear dynamic systems are integrators[†]) the exact values of the preceding signals at zero are irrelevant. It will be shown in Chapter 3 (on Fourier analysis) that all three unit step signals have the same Fourier transform, but the inverse procedure of recovering the continuous-time signal from the corresponding Fourier transform produces Heaviside's unit step signal—that is, the one whose value at $t = 0$ is 0.5. Note that all three definitions of the unit step signal have a jump discontinuity at zero ($f(0^+) \neq f(0^-)$). The key point is that the inverse Fourier transform produces, at the point of jump discontinuities the average value of the signal at the given point. In the case of the unit step signals, that means $(1 + 0)/2 = 0.5$ at $t = 0$. The need for the Heaviside unit step signal in the procedure of finding the corresponding Fourier transform is also discussed in [28, p. 179].

Bearing in mind that the primary concern of this book is the study of linear dynamic systems, and for simplicity, we will use the definition of the unit step signal as given in (2.1a), that is $u(t)$, except where explicitly indicated that the presentation holds for the Heaviside unit step signal $u_h(t)$. In Chapter 3 on Fourier analysis, the signal $u_h(t)$ will be used exclusively.

The graphical presentation of the unit step signal $u(t)$ is given in Figure 2.1.

The Heaviside unit step signal can be expressed in terms of the sign function. The sign function, well known in mathematics, is defined by

$$\text{sgn}(t) = \begin{cases} 1, & t > 0 \\ 0, & t = 0 \\ -1, & t < 0 \end{cases} \quad (2.3)$$

[†]Continuous-time linear dynamic systems are described by differential equations, hence in order to find their responses (solutions of differential equations) integration must be performed.

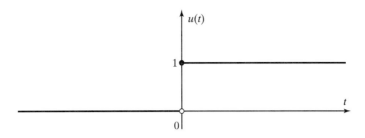

FIGURE 2.1: Continuous-time unit step signal $u(t)$

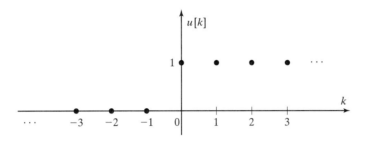

FIGURE 2.2: Discrete-time unit step signal $u[k]$

The sign function is also known as the signum function. The formula of interest, which will be used in the next chapter on the Fourier transform, relates signals defined in (2.2) and (2.3) as

$$u_h(t) = \frac{1}{2} + \frac{1}{2}\operatorname{sgn}(t) \tag{2.4}$$

Discrete-time representation of the unit step signal can be obtained by sampling $u(t)$ with the sampling period T. The corresponding *discrete unit step signal* is defined by

$$u(kT) \triangleq u[k] = \begin{cases} 1, & k \geq 0 \\ 0, & k < 0 \end{cases} \tag{2.5}$$

where kT stands for discrete time. Since T is a fixed positive constant, we will use throughout this book only k to indicate discrete-time instant kT, where k is any integer. That is, in our notation, unless explicitly indicated otherwise, for any discrete signal the following holds: $f(kT) \triangleq f[k]$. This convention will simplify notation. The discrete-time unit step signal is represented in Figure 2.2.

The *unit ramp signal* is defined in the continuous-time domain by

$$r(t) = \begin{cases} t, & t \geq 0 \\ 0, & t < 0 \end{cases} \tag{2.6}$$

The unit ramp signal has slope equal to 1 for $t > 0$. We can also introduce the ramp signal that has arbitrary slope α for $t > 0$ as $r_\alpha(t) = \alpha r(t)$.

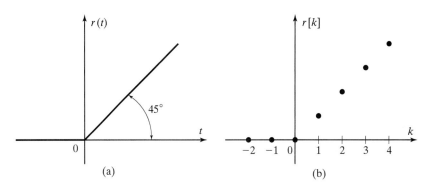

FIGURE 2.3: (a) Continuous-time and (b) discrete-time unit ramp signals

Sampling the unit ramp signal we get its discrete counterpart as

$$r(kT) \triangleq r[k] = \begin{cases} k, & k \geq 0 \\ 0, & k < 0 \end{cases} \tag{2.7}$$

The corresponding graphical representations are given in Figure 2.3.

It is easy to observe that in continuous time we have the following relationships between the unit step and unit ramp signals:

$$\begin{aligned} u(t) &= \frac{dr(t)}{dt}, \quad t \neq 0 \\ r(t) &= \int_{-\infty}^{t} u(\tau)\,d\tau = \begin{cases} t, & t \geq 0 \\ 0, & t < 0 \end{cases} \end{aligned} \tag{2.8}$$

Note that the ramp signal is not differentiable at $t = 0$. The corresponding discrete-time relations will be discussed later, after we introduce the notion of the discrete-time derivative.

Similarly, we can introduce the *parabolic signal* as

$$f_p(t) = \begin{cases} t^2, & t \geq 0 \\ 0, & t < 0 \end{cases}$$

and, in general, a family of signals of the form

$$f_n(t) = \begin{cases} t^n, & t \geq 0 \\ 0, & t < 0 \end{cases}, \quad n = 3, 4, 5, \ldots$$

These signals appear in some signal processing and control system applications, but they will not be used in this course. The corresponding discrete-time equivalents can also be defined as

$$f_p[k] = \begin{cases} k^2, & k \geq 0 \\ 0, & k < 0 \end{cases}$$

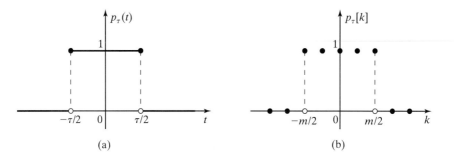

FIGURE 2.4: Rectangular pulses: (a) continuous-time and (b) discrete-time

and
$$f_n[k] = \begin{cases} k^n, & k \geq 0 \\ 0, & k < 0 \end{cases}, \quad n = 3, 4, 5, \ldots$$

The *rectangular pulse* is mathematically defined by

$$p_\tau(t) = \begin{cases} 1, & -\tau/2 \leq t \leq \tau/2 \\ 0, & \text{elsewhere} \end{cases} \tag{2.9}$$

This signal is presented in Figure 2.4(a).

In the linear systems literature, one can find another definition of the rectangular pulse as

$$p_\tau^h(t) = \begin{cases} 1, & -\tau/2 < t < \tau/2 \\ 0.5, & t = \pm\tau/2 \\ 0, & \text{elsewhere} \end{cases} \tag{2.10}$$

This definition will be used in Chapter 3, since the inverse Fourier transformation of a rectangular pulse signal produces exactly the signal given in (2.10). In addition, this definition of the rectangular pulse is consistent with the definition of the Heaviside unit step signal, since

$$p_\tau^h(t) = u_h\left(t + \frac{\tau}{2}\right) - u_h\left(t - \frac{\tau}{2}\right) \tag{2.11}$$

Note that in view of definitions (2.1a) and (2.9), the dual relationship is not valid at $t = \tau/2$, since at that point the corresponding formula implies $p_\tau(\tau/2) = 0$, that is,

$$p_\tau(t) = u\left(t + \frac{\tau}{2}\right) - u\left(t - \frac{\tau}{2}\right), \quad \forall t \text{ except for } t = \frac{\tau}{2} \tag{2.12}$$

However, from definition (2.9) we have $p_\tau(\tau/2) = 1$. We have pointed out that linear dynamic systems are integrators (they integrate input signals), so that the input signal value at a given point has no impact on the system response; hence, in the following we will not be picky about a specific value of a signal at a point unless explicitly indicated. Relationships

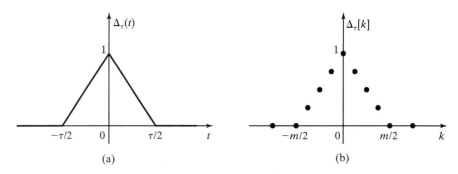

FIGURE 2.5: Triangular pulses: continuous-time (a) and discrete-time (b)

that are valid everywhere except at a selected number of points—like (2.12)—will be considered correct (from the linear system theory point of view).

The *triangular pulse* is defined by

$$\Delta_\tau(t) = \begin{cases} 0, & t \leq -\tau/2 \\ 1 + 2t/\tau, & -\tau/2 \leq t \leq 0 \\ 1 - 2t/\tau, & 0 \leq t \leq \tau/2 \\ 0, & \tau/2 \leq t \end{cases} \quad (2.13)$$

The graphical presentation of this signal is given in Figure 2.5(a).

Discrete versions of the rectangular and triangular pulses can be easily obtained by sampling. Two examples of these signals, which are self explanatory, are given in Figures 2.4(b) and 2.5(b). In these figures, m is defined by $m = 2[\tau/2T]$, where $[\cdot]$ stands for the integer part operation and T represents the sampling period. The discrete-time rectangular pulse is analytically defined as

$$p_m[k] = \begin{cases} 1, & -m/2 \leq k \leq m/2 \\ 0, & \text{elsewhere} \end{cases} \quad (2.14)$$

Similarly, one can define analytically the discrete-time triangular pulse.

EXAMPLE 2.1

The triangular pulse (2.13) can be represented in terms of unit step and/or ramp signals, as demonstrated in the following three cases:

(a) $\Delta_\tau(t) = \dfrac{2}{\tau}\left[r\left(t + \dfrac{\tau}{2}\right) - 2r(t) + r\left(t - \dfrac{\tau}{2}\right)\right]$

(b) $\Delta_\tau(t) = \dfrac{2}{\tau}\left[r\left(t + \dfrac{\tau}{2}\right) - 2r(t)\right]\left[u\left(t + \dfrac{\tau}{2}\right) - u\left(t - \dfrac{\tau}{2}\right)\right]$

(c) $\Delta_\tau(t) = \dfrac{2}{\tau}r\left(t + \dfrac{\tau}{2}\right)\left[u\left(t + \dfrac{\tau}{2}\right) - u(t)\right] + \left[1 - \dfrac{2}{\tau}r(t)\right]\left[u(t) - u\left(t - \dfrac{\tau}{2}\right)\right]$

Note that the third expression produces a "hole" at $t=0$, since at that point formula (c) implies $\Delta_\tau(0) = 0$. However, from the linear dynamic system theory point of view, that hole is irrelevant, and the signal defined in (c) can be used to represent the triangular pulse.

Sine and cosine signals are commonly used in engineering to represent oscillatory physical phenomena, such as vibrations of mechanical systems, alternating electrical currents, and voltages. Those functions (signals) are very well known to all college students from basic high school courses. Here, we just indicate that the sine and cosine signals basically represent the same signal except for a phase angle difference, due to the simple trigonometric formulas

$$\cos(\theta t) = \sin\left(\theta t + \frac{\pi}{2}\right), \quad \sin(\theta t) = \cos\left(\theta t - \frac{\pi}{2}\right) \tag{2.15}$$

The *sinc signal* plays a very important role in Fourier analysis, communication systems, and signal processing. It is defined by

$$\operatorname{sinc}(t) = \frac{\sin(\pi t)}{\pi t} \tag{2.16}$$

At $t=0$, by the very well-known trigonometric limit $\sin(0)/0=1$, it follows that $\operatorname{sinc}(0)=1$. For $t \neq 0$, the zeros of the sinc signal are at $t = \pm n$, $n = 1, 2, 3, \ldots$. When $t \to \pm\infty$, the sinc signal tends to zero. It can be easily concluded that the sinc signal is even, since $\operatorname{sinc}(-t) = \operatorname{sinc}(t)$. From this information, we can draw the graph of this signal, as presented in Figure 2.6.

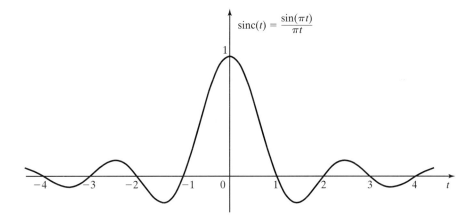

FIGURE 2.6: Continuous-time sinc signal

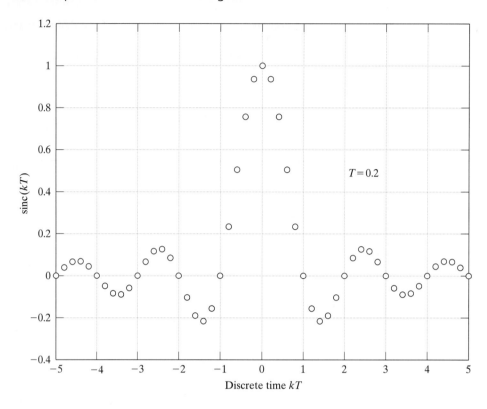

FIGURE 2.7: Discrete-time sinc signal obtained using MATLAB

The discrete-time sinc signal can be obtained by discretizing the corresponding continuous-time sinc signal. The discrete-time sinc signal plot, obtained using MATLAB, is presented in Figure 2.7.

The following MATLAB script is used to draw this signal:

```
T=0.2                       % defines the sampling period
kT=-5:T:5                   % discretizes the time axis
y=sinc(kT)                  % finds values for the sinc signal
plot(kT,y,'o')              % 'o' small letter o for plotting
ylabel('sinc(kT)')
xlabel('Discrete time kT')
axis([-5 5 -0.4 1.2])       % set ups the scale for x and y axis
grid
text(2,0.5,'T=0.2')         % places text 'T=0.2' at coordinates x, y
print -dps fig2.7.ps        % forms figure's postscript file
```

Section 2.1 Common Signals in Linear Systems 41

The formal definition of the discrete-time sinc signal is given by

$$\text{sinc}(kT) = \frac{\sin(\pi kT)}{\pi kT} \triangleq \frac{\sin \pi[k]}{\pi[k]} \triangleq \text{sinc}[k] \tag{2.17}$$

In subsequent chapters, we will show that it is easy to find the response of a linear system due to basic input signals such as step and ramp signals. More complex signals can often be represented as linear combinations of step and ramp signals. This signal representation in terms of basic signals together with the superposition principle (the linearity property introduced in Chapter 1) will help to easily obtain the linear system response due to input signals that have complex waveform (shape).

EXAMPLE 2.2

Consider the signal given in Figure 2.8.

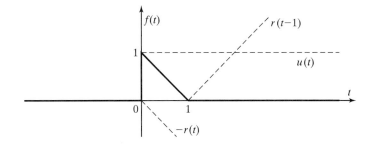

FIGURE 2.8: A simple signal

It is easy to observe that the signal presented by the solid lines in Figure 2.8 can be represented in terms of unit step and unit ramp signals as follows:

$$f(t) = u(t) - r(t) + r(t-1)$$

The elementary signals $u(t)$, $-r(t)$, and $r(t-1)$ are represented in the same figure using dashed lines. Note that this signal representation is not unique; hence, other representations are possible, for example

$$f(t) = r(-t+1) - r(-t) - u(-t)$$

EXAMPLE 2.3

Consider the complex shape signal represented in Figure 2.9.
This signal can be represented in terms of step and ramp signals (as demonstrated in Figure 2.10), leading to the following result:

$$f(t) = r(t+1) - r(t) - 2u(t) + r(t-1) - r(t-2)$$

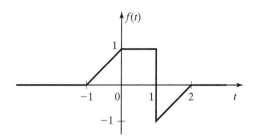

FIGURE 2.9: A complex shape signal

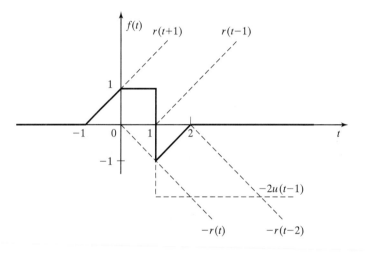

FIGURE 2.10: Signal representation in terms of step and ramp signals

Note that we first draw the signal $r(t+1)$ and then subtract the signal $r(t)$, which produces the desired signal in the time interval from $-\infty$ to 1 and the signal equal to 1 for $t > 1$. Then, we subtract the signal $2u(t-1)$, which produces the signal equal to -1 in the time interval $t > 1$. By adding the ramp signal $r(t-1)$, we get the desired signal in the time interval $1 < t < 2$. Since the desired signal is equal to zero for $t > 2$, we subtract $r(t-2)$ to level off the impact of the signal $r(t-1)$.

2.1.1 Impulse Delta Signal

For many centuries, mathematicians and scientists have worked with ordinary mathematical functions. The need for a new class of functions was first observed at the end of the nineteenth century by O. Heaviside while analyzing electrical circuits (see Problems 2.18 and 2.19 and formulas (11.20) and (11.30) in Chapter 11). In the 1920s, P. Dirac came to the same conclusion while studying some problems in relativistic mechanics [29]. The new class of functions—the so-called *distributions* or *singular functions* that, together

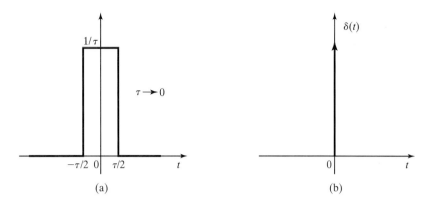

FIGURE 2.11: (a) Approximation of the impulse delta signal and (b) symbolic representation for $\delta(t)$

with ordinary functions, form the set of *generalized functions*—was formally introduced in mathematics by the end of the 1930s by S. Sobolev. The field of generalized functions attained maturity during the 1940s in the papers by S. Sobolev and L. Schwartz [30–31]. The generalized functions play a very important role in the analysis of linear dynamic systems, and they are used in almost all engineering and scientific disciplines.

The impulse delta function (signal) is extremely important for linear system theory. Loosely speaking, this "strange" function has no time structure. It is equal to zero everywhere except at zero, where it is equal to ∞.[†] However, its integral is well behaved, and is defined by

$$\int_{-\infty}^{\infty} \delta(t)\,dt = \int_{0^-}^{0^+} \delta(t)\,dt = 1 \qquad (2.18)$$

The impulse delta signal can be visualized as a mathematical artifice of the rectangular pulse represented in Figure 2.11(a), in the limit when the width of the pulse tends to zero. We also give the symbolic notation for $\delta(t)$ in Figure 2.11(b).

Notice that the area of the rectangular pulse is always equal to 1. The impulse delta signal is obtained in the limit when $\tau \to 0$, that is,

$$\delta(t) = \lim_{\tau \to 0} \left\{ \frac{1}{\tau} p_\tau(t) \right\} \qquad (2.19)$$

In the literature, the impulse delta signal is also called the Dirac impulse function, in honor of the great physicist and mathematician P. Dirac.

[†]In a more rigorous definition, the impulse delta function is undefined at zero and it could be equal to ∞ at zero [32].

EXAMPLE 2.4

Note that from the definition of the impulse delta signal it follows that

$$\int_{-5}^{3} \delta(t-4)\,dt = 0, \quad \int_{-5}^{4^+} \delta(t-4)\,dt = 1$$

In the first case, the impulse delta signal is located outside of the integration limits, whereas in the second case it is within the integration limits. ∫

The shifted impulse delta signal is defined by

$$\delta(t-t_0) = \begin{cases} \infty, & t = t_0 \\ 0, & t \neq t_0 \end{cases} \quad \text{and} \quad \int_{-\infty}^{\infty} \delta(t-t_0)\,dt = \int_{t_0^-}^{t_0^+} \delta(t-t_0)\,dt = 1 \quad (2.20)$$

This signal is represented in Figure 2.12.

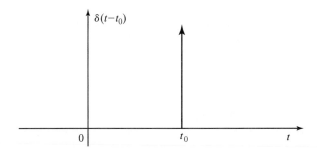

FIGURE 2.12: Shifted continuous-time impulse delta signal

The impulse delta signal (function) in *mathematics* is defined by the integral

$$\int_{-\infty}^{\infty} f(t)\delta(t-t_0)\,dt = f(t_0) \qquad (2.21)$$

where $f(t)$ is an ordinary function continuous at $t = t_0$. In engineering, we prefer to call (2.21) *the sifting property* of the impulse delta function, since the effect of the impulse delta function in this integral is to take out (sift) a particular value of the function $f(t)$ at $t = t_0$.

EXAMPLE 2.5

By using the sifting property of the impulse delta signal, the following integral can be calculated:

$$\int_{-\infty}^{\infty} \{[e^{-5t}\cos(2t) + t^2]\delta(t) + (2t+1)\delta(t-2)\}\,dt = [1+0] + (4+1) = 6 \quad ∫$$

What can be said about the integral

$$\int_{-\infty}^{t_0} f(t)\delta(t-t_0)\,dt = ?$$

with $f(t)$ being an ordinary function continuous at $t = t_0$? This integral sometimes appears in actual derivations (such as in the well-known paper by M. Athans and E. Tse [33]). The following result is used in [33] to derive one of the classic results of linear control theory:

$$\int_{-\infty}^{t_0} f(t)\delta(t-t_0)\,dt = \frac{1}{2}f(t_0) \tag{2.22}$$

Hence

$$\int_{0^-}^{0} \delta(t)\,dt = \frac{1}{2}, \quad \int_{-1}^{1} \delta(t-1)f(t)\,dt = \frac{1}{2}f(1)$$

Another visualization of the impulse delta signal can be obtained by considering the triangular pulse in Figure 2.13(a) in the limit when $\tau \to 0$. It is obvious from this figure that

$$\delta(t) = \lim_{\tau \to 0}\left\{\frac{1}{\tau}\Delta_{2\tau}(t)\right\} \tag{2.23}$$

This representation of the impulse delta signal helps to visualize the *derivative of the impulse delta signal* as the limit of the signal represented in Figure 2.13(b) when $\tau \to 0$, that is,

$$\frac{d\delta(t)}{dt} = \lim_{\tau \to 0}\left\{\frac{1}{\tau^2}p_\tau\left(t+\frac{\tau}{2}\right) - \frac{1}{\tau^2}p_\tau\left(t-\frac{\tau}{2}\right)\right\} \tag{2.24}$$

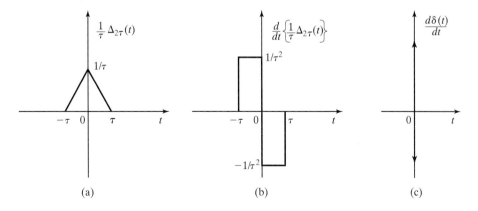

FIGURE 2.13: Approximations of (a) the impulse delta signal, (b) its derivative, and (c) the symbolic representation of the derivative of the impulse delta signal

46 Chapter 2 Introduction to Signals

The corresponding derivative is symbolically represented in Figure 2.13(c) with two impulses of width zero that tend to plus and minus infinity. The derivative of the unit impulse delta signal is also known in the literature as the unit-doublet. Note that the unit-doublet $d\delta(t)/dt$ evaluated at $t = 0^+$ and $t = 0^-$ produces zero values like the impulse delta signal.

Mathematically, we can define the derivative of the delta impulse signal using its integral representation, introduced in (2.21), as

$$\int_{-\infty}^{\infty} f(t) \frac{d\delta(t - t_0)}{dt} dt \tag{2.25}$$

Similarly, we can define the second- and higher-order derivatives of the impulse delta signal by the integrals

$$\int_{-\infty}^{\infty} f(t) \frac{d^i \delta(t - t_0)}{dt^i} dt, \quad i = 1, 2, \ldots \tag{2.26}$$

The second derivative of $\delta(t)$ is called the unit-triplet. The $(n - 1)$th derivative of $\delta(t)$ is called the unit-n-tuplet. The unit-triplet and unit-n-tuplets are also graphically represented using exactly the same plot as the one given in Figure 2.13(c) for the derivative of the unit impulse delta signal. Note that all $d^i\delta(t)/dt^i$, $i = 1, 2, \ldots$, evaluated at $t = 0^+$ and $t = 0^-$ produce zero values.

The formulas for the derivatives of the impulse delta signal, (2.25) and (2.26), will be further clarified and simplified in the subsequent development of the properties of the impulse delta signal.

In addition to the sifting property given by (2.21), the following properties of the impulse delta signal in continuous time can be established by using either basic integral properties or integration by parts.

Time Scaling Property

$$\int_{-\infty}^{\infty} f(t)\delta(at - t_0) dt = \frac{1}{|a|} f\left(\frac{t_0}{a}\right) \Rightarrow \delta(at - t_0) = \frac{1}{|a|} \delta\left(t - \frac{t_0}{a}\right) \tag{2.27}$$

Proof. Consider first the case when $a > 0$, and introduce the change of variables

$$at - t_0 = \sigma \Rightarrow t = \frac{\sigma + t_0}{a}, \quad dt = \frac{d\sigma}{a}$$

which implies

$$\int_{t=-\infty}^{t=\infty} f(t)\delta(at - t_0) dt = \frac{1}{a} \int_{\sigma=-\infty}^{\sigma=\infty} f\left(\frac{\sigma + t_0}{a}\right) \delta(\sigma) d\sigma = \frac{1}{a} f\left(\frac{t_0}{a}\right)$$

For $a < 0$, the same change of variables implies that the integral defined on the left-hand side in formula (2.27) has the form

$$\frac{1}{a}\int_{\sigma=\infty}^{\sigma=-\infty} f\left(\frac{\sigma+t_0}{a}\right)\delta(\sigma)\,d\sigma = -\frac{1}{a}\int_{\sigma=-\infty}^{\sigma=\infty} f\left(\frac{\sigma+t_0}{a}\right)\delta(\sigma)\,d\sigma = -\frac{1}{a}f\left(\frac{t_0}{a}\right)$$

Note that the integration limits are now changed and we must factor -1 to switch the integration limits. Putting together both cases $a > 0$ and $a < 0$, we obtain the time scaling result stated in formula (2.27). □

Derivative Property Let $\delta^{(1)}(t)$ denote $d\delta(t)/dt$; then

$$\int_{-\infty}^{\infty} f(t)\delta^{(1)}(t-t_0)\,dt = -f^{(1)}(t_0) \tag{2.28}$$

where $f^{(1)}(t_0)$ stands for $df(t)/dt$ evaluated at $t = t_0$. In general, it can also be shown that for the nth derivative we have

$$\int_{-\infty}^{\infty} f(t)\delta^{(n)}(t-t_0)\,dt = (-1)^n f^{(n)}(t_0) \tag{2.29}$$

Proof. Integrating by parts the integral defined in (2.28), we obtain

$$\int_{-\infty}^{\infty} f(t)\delta^{(1)}(t-t_0)\,dt = f(\infty)\delta(\infty) - f(-\infty)\delta(-\infty)$$

$$- \int_{-\infty}^{\infty} f^{(1)}(t)\delta(t-t_0)\,dt = 0 - 0 - f^{(1)}(t_0)$$

Similarly, integrating n times by parts, we get the general formula (2.29) for the nth derivative. □

From formula (2.21), we can get the following properties of the impulse delta signal:

$$\begin{aligned} f(t)\delta(t) &= f(0)\delta(t) \\ f(t)\delta(t-t_0) &= f(t_0)\delta(t-t_0) \end{aligned} \tag{2.30}$$

and from (2.21) and (2.29), the following holds:

$$\delta^{(2n)}(-t) = \delta^{(2n)}(t), \quad \delta^{(2n+1)}(-t) = -\delta^{(2n+1)}(t), \quad n = 0, 1, 2, \ldots \tag{2.31}$$

It follows for $n = 0$ that $\delta(-t) = \delta(t)$, hence $\delta(t)$ is an even function.

Proof. The property stated in (2.30) can be proved as follows: Let $\varphi(t)$ be another ordinary function continuous at $t = 0$. Then, by using (2.21), we have

$$\int_{-\infty}^{\infty} f(t)\varphi(t)\delta(t)\,dt = f(0)\varphi(0) = f(0)\int_{-\infty}^{\infty} \varphi(t)\delta(t)\,dt$$

$$= \int_{-\infty}^{\infty} f(0)\varphi(t)\delta(t)\,dt$$

Comparing the first and the last integral in the preceding expression, the property given in (2.30) follows. It is left to students as a homework problem (see Problem 2.13) to prove property (2.31). Using the preceding proof, we can also establish an interesting result that holds in the generalized calculus—that is, $t^n \delta(t) = 0$, where n is any positive real number. □

There are several other properties of the impulse delta signal, but for the purpose of this text, it is sufficient to know only those presented.

EXAMPLE 2.6

Using properties (2.27–29) of the impulse delta signal we can evaluate the following integrals as

$$\int_{-\infty}^{\infty} \delta(2t-1)e^{-3t}\sin(\pi t)\,dt = \frac{1}{2}e^{-\frac{3}{2}}\sin\left(\frac{\pi}{2}\right) = \frac{1}{2}e^{-\frac{3}{2}}$$

$$\int_{-\infty}^{3}(t^3 + 2\sin(t) - 2)\delta^{(1)}(t-1)\,dt = (-1)\frac{d}{dt}(t^3 + 2\sin(\pi t) - 2)_{|t=1}$$

$$= -(3 + 2\pi\cos(\pi)) = -3 + 2\pi$$

At this point we leave the continuous-time impulse delta signal and introduce its discrete-time counterpart. Applications of the continuous-time impulse delta signal in linear system analysis will be considered in Chapters 3, 4, and 7.

Discrete-Time Impulse Delta Signal

Note that the sampling technique makes no sense in an attempt to obtain the discrete-time impulse delta signal from the continuous-time impulse delta signal, since the continuous-time impulse delta signal has no time structure. In the discrete-time domain, the impulse delta signal is defined as a very nice signal that is equal to 1 at $k = 0$ and 0 everywhere else, that is,

$$\delta[k] = \begin{cases} 1, & k = 0 \\ 0, & k \neq 0 \end{cases} \tag{2.32}$$

This form for the impulse delta signal in the discrete-time domain can be justified by using a discrete version of the integral (2.21):

$$\sum_{k=-\infty}^{k=\infty} f[k]\delta[k-k_0] = f[k_0] \qquad (2.33)$$

Since this infinite sum must produce only $f[k_0]$ (sifting property), apparently the discrete-time impulse delta signal must be zero everywhere except at k_0, where it must be equal to 1. Thus, the shifted version of the discrete-time impulse delta signal is defined by

$$\delta[k-k_0] = \begin{cases} 1, & k = k_0 \\ 0, & k \neq k_0 \end{cases} \qquad (2.34)$$

The shifted discrete-time impulse delta signal is also called the Kronecker delta function.

Note that the discrete-time impulse delta signal satisfies the following property, which corresponds to the (2.30) property of the continuous-time impulse delta signal:

$$f[k]\delta[k] = f[0]\delta[k] \quad \text{or} \quad f[k]\delta[k-k_0] = f[k_0]\delta[k-k_0] \qquad (2.35)$$

The discrete-time impulse delta signal and its shifted version are graphically presented in Figure 2.14.

Generalized Derivative

The importance of the impulse delta function in mathematics lies in the fact that we can define the *generalized derivative* in terms of the impulse delta function. It is well known that at the point of jump discontinuity ($f(t_1^+) \neq f(t_1^-)$) the function $f(t)$ has no derivative in the ordinary sense. However, from the geometric point of view—since the derivative stands for a slope of the tangent at the given point, t_1—we can say that the derivative at the point of jump discontinuity is equal to infinity. Since the shifted impulse delta signal $\delta(t-t_1)$ is equal to infinity at $t = t_1$, we can use the impulse delta function in order to define the generalized derivative of functions that have jump discontinuities.

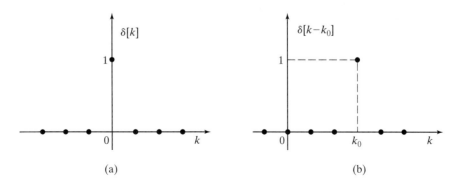

FIGURE 2.14: (a) Discrete-time impulse delta signal and (b) its shifted version

DEFINITION 2.1: Consider a function $f(t)$ that has jump discontinuities at the points $t_1, t_2, \ldots, t_j$. The generalized derivative of $f(t)$ is defined by

$$\frac{Df(t)}{Dt} = \sum_{i=1}^{j} (f(t_i^+) - f(t_i^-))\delta(t - t_i) + \frac{df(t)}{dt}\bigg|_{t \neq t_1, t_2, \ldots, t_j}$$

where $Df(t)/Dt$ stands for the generalized derivative, and df/dt is the ordinary derivative at the points where it exists.

EXAMPLE 2.7

Find the generalized derivative of the signal shown in Figure 2.15(a).

In this example, we have two jump discontinuities (at -1 and 2), so that the generalized derivative according to Definition 2.1 is obtained as

$$\frac{Df(t)}{Dt} = [f(-1^+) - f(-1^-)]\delta(t+1) + [f(2^+) - f(2^-)]\delta(t-2)$$
$$+ \frac{df(t)}{dt}[u(t+1) - u(t-2)] = \delta(t+1) + \delta(t-2) - \frac{2}{3}[u(t+1) - u(t-2)]$$

The generalized derivative is plotted in Figure 2.15(b).

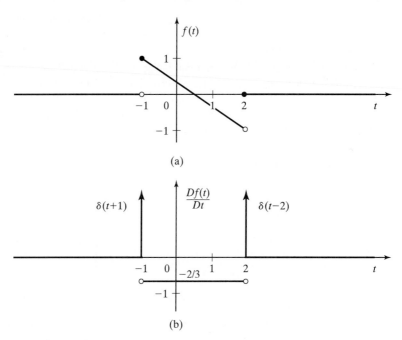

FIGURE 2.15: (a) Continuous-time signal and (b) its generalized derivative

EXAMPLE 2.8

Using the notion of the generalized derivative, the following relationship between the impulse delta signal and the unit step signals, $u(t)$ and $u_h(t)$, can be established:

$$\delta(t) = \frac{Du(t)}{Dt} = \frac{Du_h(t)}{Dt} \qquad (2.36)$$

Using in addition the result established in (2.22), the following holds:

$$u_h(t) = \int_{-\infty}^{t} \delta(\tau)\,d\tau = \begin{cases} 1, & t > 0 \\ 0.5, & t = 0 \\ 0, & t < 0 \end{cases} \qquad (2.37)$$

⚑

2.2 SIGNAL OPERATIONS

In this section we introduce some signal operations that are of interest for linear dynamic systems. Additional signal operations used in signal processing and communication systems will be presented in Chapters 9 and 10.

Signals are mathematical functions, hence all known mathematical operations with functions are applicable to both continuous- and discrete-time signals. We can add, subtract, and multiply signals in exactly the same way that we add, subtract, and multiply mathematical functions. We can also perform signal differentiation and integration in the continuous-time domain. In this section we define the operation that plays the role of discrete-time differentiation, and introduce definitions of continuous- and discrete-time signal convolutions. We also demonstrate how to plot some continuous-time signals using corresponding MATLAB functions.

One extremely important operation that we can perform on signals is convolution. The convolution plays a fundamental role in linear system theory, and will be used very often in this book. Here, we give only the definitions for both continuous- and discrete-time convolutions. These definitions will be needed in subsequent chapters, particularly in Chapter 3, where we introduce the main concepts of linear dynamic system theory.

DEFINITION 2.2: CONTINUOUS-TIME CONVOLUTION Given continuous-time signals $g(t)$ and $v(t)$, the continuous-time convolution is defined by

$$\begin{aligned} g(t) * v(t) &= \int_{-\infty}^{\infty} v(\tau)g(t-\tau)\,d\tau \\ &= \int_{-\infty}^{\infty} v(t-\tau)g(\tau)\,d\tau = v(t) * g(t), \quad -\infty \le t \le \infty \end{aligned} \qquad (2.38)$$

where the asterisk denotes the convolution operator.

The equality of the two formulas in (2.38) can be easily established by using a simple change of variables. Note that in the convolution integral t is a parameter and τ is a dummy variable of integration.

This formula also states the commutativity property of the convolution. Other properties of the continuous-time convolution follow from the properties of integrals; they will be studied in detail in Chapter 6. The use of the convolution concept in the analysis of linear time-invariant systems will be considered in Chapters 3–6 and 8.

DEFINITION 2.3: DISCRETE-TIME CONVOLUTION Given discrete-time signals $g[k]$ and $v[k]$, the discrete-time convolution is defined by

$$g[k] * v[k] = \sum_{m=-\infty}^{m=\infty} g[m]v[k-m]$$
$$= \sum_{m=-\infty}^{m=\infty} g[k-m]v[m] = v[k] * g[k], \quad -\infty \leq k \leq \infty \quad (2.39)$$

where k is a parameter and m is a dummy variable of summation.

It will be shown in Chapters 3 and 5 that the main result of linear system theory (which states that for a linear dynamic system at rest (zero initial conditions), the system output response is the convolution of the system input excitation and the system impulse response) is just an interpretation of formulas (2.38) and (2.39).

Note that in Chapters 9 and 10 we will introduce an operation on signals that is important for both digital signal processing and communication systems—the *signal correlation*. The signal correlation is similar in form to the signal convolution in both continuous- and discrete-time domains, but it has a completely different physical meaning. The signal correlation will be used to determine the energy distribution in the signal. Continuous- and discrete-time signal correlations are respectively defined by

$$R_{vg}(t) = \int_{-\infty}^{\infty} v(\tau)g(\tau+t)\,d\tau, \quad -\infty \leq t \leq \infty \quad (2.40)$$

and

$$R_{vg}[k] = \sum_{m=-\infty}^{m=\infty} g[m]v[m+k], \quad -\infty \leq k \leq \infty \quad (2.41)$$

DEFINITION 2.4: FORWARD DIFFERENCE (DISCRETE-TIME "DERIVATIVE")
The discrete-time counterpart of the derivative is the forward difference. Consider a discrete-time signal $f[k]$ defined in some discrete-time interval $k \in [k_1 \; k_2]$. Then, the forward difference (discrete-time "derivative") of $f[k]$ in the given interval is defined by

$$\Delta f[k] = f[k+1] - f[k] \quad (2.42)$$

This definition can be justified with the following reasoning. The continuous-time derivative geometrically represents the slope of the tangent at the given point, hence the derivative can be approximated as

$$\frac{df(t)}{dt} \approx \frac{f(t+\Delta t) - f(t)}{\Delta t}$$

Taking $t = k\Delta t$, where k is an integer, we have

$$\Delta t \frac{df(k\Delta t)}{dt} \approx f((k+1)\Delta t) - f(k\Delta t)$$

It can be seen that the approximation of the continuous-time derivative is proportional to the forward difference. In this text, we will call the forward difference the discrete-time derivative. For notational convenience we omit Δt, and formula (2.42) follows.

EXAMPLE 2.9

Using definitions of the discrete-time unit step and ramp signals given in (2.5) and (2.7), it can be observed that

$$\Delta r[k] = r[k+1] - r[k] = u[k] \qquad (2.43)$$

that is, the discrete-time unit step signal is the discrete-time derivative of the discrete-time unit ramp signal.

EXAMPLE 2.10

It is easy to see that

$$\Delta u[k-1] = u[k] - u[k-1] = \begin{cases} 1, k=0 \\ 0, k \neq 0 \end{cases} = \delta[k] \qquad (2.44)$$

Thus, the impulse delta signal is the derivative of the unit step signal for both continuous-time (see (2.36)) and discrete-time domains.

An attempt to define the discrete-time integral is meaningless, since an integral, in general, represents an area. In the discrete-time domain all integrals are zero since functions are defined at discrete points, each having zero interval. However, we know from calculus that an integral can be interpreted as an infinite sum. That gives us an idea of how to interpret formula (2.37) in the discrete-time domain. By formally replacing the corresponding integral in (2.37) by an infinite sum, we obtain

$$u[k] = \sum_{m=-\infty}^{k} \delta[m] \qquad (2.45)$$

where m is a dummy variable of summation. Since $\delta[m]$ is equal to 1 only for $m=0$ and equal to zero for all other m, we see that if $k<0$ the infinite summation defined in (2.45) does not include the delta impulse at zero, thus the sum is equal to zero. If $k \geq 0$ the summation (2.45) will always be equal to 1, since the delta impulse at zero (the only signal equal to 1) is included within the limits of summation. The obtained result is consistent with the definition of the discrete-time unit step signal. Hence, we can conclude that the formula given in (2.45) is correct.

Using the same logic, we can try to interpret in the discrete-time domain the integral in (2.8) (which states that the continuous-time ramp signal is the integral of the unit step signal) as

$$r[k] = \sum_{m=-\infty}^{k} u[m] \quad (?) \qquad (2.46)$$

Is this formula correct? It can be seen for $k < 0$ that $r[k] = 0$, but for $k = 0$ we have $r[0] = u[0] = 1$. $k = 1$ implies that $r[1] = 0 + \cdots + 0 + u[0] + u[1] = 2$ which, based on the definition of the discrete-time ramp signal (2.7), indicates that formula (2.46) is not correct. Let us slightly modify the formula:

$$r[k] = \sum_{m=-\infty}^{k-1} u[m] \qquad (2.47)$$

Using the same argument as before, we can establish that $r[0] = 0$, $r[1] = u[0] = 1$, $r[2] = u[0] + u[1] = 2, \ldots, r[k] = u[0] + u[1] + \cdots + u[k-1] = k$. Thus, formula (2.47) is correct. From (2.46–47), we can conclude that we must be careful when converting continuous-time integrals into discrete-time sums. In most cases, the analogies can be established (with or without minor modifications) between continuous-time integrals and infinite discrete-time sums.

EXAMPLE 2.11

In this example we demonstrate how to use MATLAB to plot some continuous-time signals. The program plots the following signals: $u(t)$, $p_6(t)$, $r(t)$, $\Delta_4(t)$. The MATLAB code is as follows:

```
% continuous-time unit step
t1=-10:0; t2=0:10; t=[t1 t2];
% semicolon suppresses printing to screen/paper
u=[zeros(1,11) ones(1,11)]
subplot(221); plot(t,u); axis([-10 10 -0.5 1.5]); grid
title('Unit step'); xlabel('Time t'); ylabel('u(t)')
% continuous-time rectangular pulse
t1=-10:-3; t2=-3:3; t3=3:10; t=[t1 t2 t3];
p6=[zeros(1,8) ones(1,7) zeros(1,8)]
subplot(222); plot(t,p6); axis([-10 10 -0.5 1.5]); grid
title('Rectangular pulse'); xlabel('Time t');
ylabel('p6(t)')
% continuous-time ramp signal
t1=-10:-1; t2=0:10; t=[t1 t2]
r=[zeros(1,10) t2]
subplot(223); plot(t,r); grid
title('Unit ramp'); xlabel('Time t'); ylabel('r(t)')
```

Section 2.2 Signal Operations 55

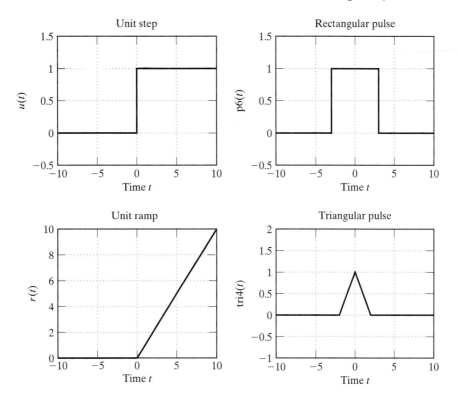

FIGURE 2.16: Common signals plotted by MATLAB

```
% continuous-time triangular pulse
t1=-10:-3; t2=-2:0; t3=0:2; t4=3:10; t=[t1 t2 t3 t4]
tri4=[zeros(1,8) 1+0.5*t2 1-0.5*t3 zeros(1,8)]
subplot(224); plot(t,tri4); axis([-10 10 -1 2]); grid
title('Triangular pulse'); xlabel('Time t');
ylabel('tri4(t)')
print -dps fig2.16.ps
```

The corresponding plots are presented in Figure 2.16.

EXAMPLE 2.12

Note that MATLAB has a built-in function for generating (and plotting) the unit step signal. It is used as follows: stepfun(t,t0), where t defines the time interval during which the step signal must be generated and t0 is the time instant at which the jump from 0 to 1 occurs. For example, t=-3:0.01:5; t0=1; u=stepfun(t,t0); plot(t,u), calculates and plots $u(t-1)$. It should be pointed out that this presentation requires a very small time increment for a correct plot. Repeat the preceding program with `t=-3:0.1:5` and `t=-3:0.001:5` and compare the plots obtained. This problem, however, is not present in calculating the discrete-time step function. For example, `k=-3:1:5; k0=2; du=stepfun(k,k0); plot(k,du,'*')` produces the plot for the discrete-time unit step function shifted to the right by two discrete-time units, $u[k-2]$. The MATLAB function `stepfun` is very convenient when plotting more complex signals.

2.3 SIGNAL CLASSIFICATION

Since signals represent mathematical functions, the obvious signal classification is the one known from the classification of mathematical functions. Like mathematical functions, signals can be classified as periodic or aperiodic (nonperiodic), even or odd, real or complex, continuous-time or discrete-time, deterministic or stochastic (random), sinusoidal, exponential, and so on.

Let us review here that a *periodic signal* satisfies

$$f(t) = f(t + T_p), \quad T_p < \infty \tag{2.48}$$

for all t and some T_p, where T_p is the time period after which the signal repeats itself. For example, for sine and cosine functions, $T_p = 2\pi$.

Even signals are symmetrical with respect to the vertical axis; that is,

$$f(-t) = f(t) \tag{2.49}$$

For example, $\cos(t)$ and $\text{sinc}(t)$ are even signals.

Odd signals are symmetrical with respect to the origin, hence they satisfy

$$f(-t) = -f(t) \tag{2.50}$$

For example, $\sin(t)$ is an odd signal.

In addition to the classification known from calculus, in linear systems and signals we distinguish between energy and power signals. Let us first introduce definitions of the *signal energy* and *power*.

Recall from elementary electrical engineering courses that the electrical energy developed on a resistor is proportional either to the square of the constant current through the resistor or to the square of the constant voltage on the resistor. Similarly, from basic dynamics we know that a mass moving at a constant velocity has kinetic energy that is proportional to the square of the velocity. Hence, the square of the signal serves as a measure of signal energy. In the case in which the signal changes in time, we must integrate (or sum, in the case of discrete-time signals) over a given time period of interest. For example, in the case of a time-varying current $i(t)$, the energy developed (dissipated as heat) on the resistor during the time interval from t_1 to t_2 is given by

$$E_{[t_1,t_2]} = R \int_{t_1}^{t_2} i^2(t)\, dt$$

Now we can define the signal energy as follows.

DEFINITION 2.5: The *continuous-time signal energy* over the time interval $[t_1, t_2]$ of length $L = t_2 - t_1$ is defined by

$$E_L = \int_{t_1}^{t_2} |f(t)|^2 \, dt \qquad (2.51)$$

The *total continuous-time signal energy* is given by

$$E_\infty = \int_{-\infty}^{\infty} |f(t)|^2 \, dt \qquad (2.52)$$

Note that the definitions given in (2.51) and (2.52) are general and hold even for complex signals, in which case $f(t)f^*(t) = |f(t)|^2$, where $f^*(t)$ denotes the complex conjugate signal. In the case of real signals, the absolute values in (2.51) and (2.52) can be removed.

Recall from elementary physics that the power is work (energy) over time (speed of work). In order to get the expression for the average signal power, we must divide the corresponding expression for energy by the length of the time interval, so that we have the following definition.

DEFINITION 2.6: The *continuous-time signal average power* is defined by

$$P_\infty = \lim_{L \to \infty} \frac{1}{L} \int_{-L/2}^{L/2} |f(t)|^2 \, dt \qquad (2.53)$$

We can also define the signal energy and power in the discrete-time domain.

DEFINITION 2.7: The *discrete-time signal energy* over the time interval $[k_1, k_2]$ of length $M = k_2 - k_1$ is defined by

$$E_M = \sum_{k=k_1}^{k=k_2} |f[k]|^2 \qquad (2.54)$$

The *total discrete-time signal energy* is given by

$$E_\infty = \sum_{k=-\infty}^{k=\infty} |f[k]|^2 \qquad (2.55)$$

DEFINITION 2.8: The *discrete-time signal average power* is defined by

$$P = \lim_{M \to \infty} \frac{1}{2M+1} \sum_{k=-M}^{k=M} |f[k]|^2 \qquad (2.56)$$

58 Chapter 2 Introduction to Signals

Based on their energy and power, signals are classified as follows:

1. *Energy signals* have finite total energy, $E_\infty < \infty$, and zero average power, $P_\infty = 0$.
2. *Power signals* have infinite total energy and finite average power, that is, $E_\infty = \infty$, $P_\infty < \infty$.

For example, the rectangular and triangular pulses are energy signals. Periodic signals have infinite energy and very often finite average power; thus, in most cases, periodic signals are power signals.

Finally, signals are classified as causal and anticausal.

Causal signals satisfy $f(t) = 0$ for all $t < 0$. If a signal is not causal, that is, if $f(t) \neq 0$ for some $t < 0$, the signal is *anticausal*. Similarly, discrete-time signals $f[k] = 0$ for $k < 0$ are causal, otherwise they are anticausal. Anticausal signals are common in signal processing; signals encountered in real-world dynamic systems are causal.

2.4 MATLAB LABORATORY EXPERIMENT ON SIGNALS

Purpose: This experiment introduces the graphical representation of common signals used in linear systems. Time shifting, time scaling, signal addition, and signal multiplication will also be demonstrated. It is important to emphasize that signals are mathematical functions—thus, the signal operations given in the following are known from calculus.

Part 1. Use MATLAB to plot the following continuous-time signals in the time interval $t \in [-10 \ 10]$.

1. $u(t)$ (unit step signal), $u(t-3)$, $u(t-5)$.
2. $p_6(t)$ (unit rectangular pulse), $p_6(t-3)$, $p_6(t+5)$.
3. $r(t)$ (unit ramp signal), $r(t-3)$, $r(t+5)$.
4. $\Delta_4(t)$ (unit triangular pulse), $\Delta_4(t-3)$, $\Delta_4(t+5)$.

Part 2. Plot approximations of the impulse delta signal and the sinc signal.

5. Plot an approximation for $\delta(t)$ (impulse delta signal). *Hint:* $\delta(t)$ can be approximated by a rectangular pulse of width τ and amplitude $1/\tau$ when $\tau \to 0$. Take $\tau = 0.3, 0.2, 0.1$.
6. Use $\text{sinc}(t) = \sin(\pi t)/\pi t$ with `t=-5:0.1:5; t=-15:0.1:15; t=-30:0.1:30`. The sinc signal can be obtained as $\text{sinc}(t) = \sin(\pi t)./\pi t$. (Note that the operation ./ stands for pointwise division.) MATLAB also has the built-in function sinc. To get information about any MATLAB function, type `help function name`; in this case type `help sinc`.

Part 3. In this part, we demonstrate time scaling and time shifting operations. Plot the signals given in the following. Take `t=0:0.1:6.28`.

7. $y_1(t) = \sin(t)$, $y_2(t) = \sin(2t)$, $y_3(t) = \sin(5t)$. Plot all three signals in the same figure. Use `plot(t,y1,'o',t,y2,'-',t,y3)`.

8. $\sin(4(t-1))$, $\sin(2t-3)$. Explain the figures obtained in (7) and (8).
9. $e^{-at}\sin(at)$ for $a = 0.5, 1, 5$. Use .* as pointwise multiplication. Comment on the effect of time scaling.

Part 4. Some signal operations. Plot the following signals.

10. $u(t) + r(t)$.
11. $p_2(t) + \Delta_3(t)$.
12. $\cos(5t + \sin(2t))$. Expand the time axis such that it includes one signal period.

Part 5. Calculate and plot the following discrete-time signals.

13. $u[k-1], r[k+2]$.
14. $r[-k-1] * u[k-2]$.
15. $(-0.5)^k u[k-2] * u[-k+10]$.

Submit a report composed of fifteen figures for fifteen problems and, where required, comment on the results obtained.

2.5 SUMMARY

Study Guide for Chapter Two: Students must know geometrical shapes and analytical expressions for common (elementary) signals presented in Section 2.1. From the linear system theory point of view, it is not crucial to define exact signal values at discontinuity points (linear systems are integrators). The definition and properties of the impulse delta signal are very important for all classes of linear systems, including communication and signal processing systems. Finding generalized derivatives of signals that have jump discontinuities is a simple procedure that demonstrates the use of the impulse delta signal in mathematics and engineering. Standard problems: (1) express a given signal in terms of elementary signals; (2) plot graphs of signals represented by elementary signals; (3) use the definition and properties of the impulse delta signal to simplify integrals and analytical expressions; (4) find generalized derivatives of signals that have jump discontinuities; (5) define analytically and plot some elementary signals.

Unit Step Signals:

$$u(t) = \begin{cases} 1, & t \geq 0 \\ 0, & t < 0 \end{cases}, \quad u[k] = \begin{cases} 1, & k \geq 0 \\ 0, & k < 0 \end{cases}, \quad u_h(t) = \begin{cases} 1, & t > 0 \\ 0.5, & t = 0 \\ 0, & t < 0 \end{cases}$$

Unit Ramp Signals:

$$r(t) = \begin{cases} t, & t \geq 0 \\ 0, & t < 0 \end{cases}, \quad r[k] = \begin{cases} k, & k \geq 0 \\ 0, & k < 0 \end{cases}$$

Relations Between Unit Step and Unit Ramp Signals:

$$u(t) = \frac{dr(t)}{dt}, t \neq 0, \quad r(t) = \int_{-\infty}^{t} u(\sigma)\, d\sigma$$

Chapter 2 Introduction to Signals

Triangular and Rectangular Pulses:

$$\Delta_\tau(t) = \begin{cases} 0, & t \leq -\tau/2 \\ 1 + 2t/\tau, & -\tau/2 \leq t \leq 0 \\ 1 - 2t/\tau, & 0 \leq t \leq \tau/2 \\ 0, & \tau/2 \leq t \end{cases}$$

$$p_\tau(t) = \begin{cases} 1, & -\tau/2 \leq t \leq \tau/2 \\ 0, & \text{elsewhere} \end{cases}, \quad p_m[k] = \begin{cases} 1, & -m/2 \leq k \leq m/2 \\ 0, & \text{elsewhere} \end{cases}$$

Sinc Signals:

$$\text{sinc}(t) = \frac{\sin(\pi t)}{\pi t}, \quad \text{sinc}(kT) = \frac{\sin(\pi kT)}{\pi kT} \triangleq \text{sinc}[k] = \frac{\sin \pi [k]}{\pi [k]}$$

Impulse Delta Signal and Its Properties:

$$\delta(t - t_0) = \begin{cases} \infty, & t = t_0 \\ 0, & t \neq t_0 \end{cases} \quad \text{and} \quad \int_{-\infty}^{\infty} \delta(t - t_0)\, dt = \int_{t_0^-}^{t_0^+} \delta(t - t_0)\, dt = 1$$

(i) $\displaystyle\int_{-\infty}^{\infty} f(t)\delta(t - t_0)\, dt = f(t_0)$

(ii) $\displaystyle\int_{-\infty}^{\infty} f(t)\delta(at - t_0)\, dt = \frac{1}{|a|} f\left(\frac{t_0}{a}\right)$

(iii) $\displaystyle\int_{-\infty}^{\infty} f(t)\delta^{(n)}(t - t_0)\, dt = (-1)^n f^{(n)}(t_0), \quad n = 1, 2, \ldots$

(iv) $f(t)\delta(t - t_0) = f(t_0)\delta(t - t_0)$

Discrete-Time Impulse Delta Signal and Its Properties:

$$\delta[k - k_0] = \begin{cases} 1, & k = k_0 \\ 0, & k \neq k_0 \end{cases}, \quad \sum_{k=-\infty}^{k=\infty} f[k]\delta[k - k_0] = f[k_0]$$

Generalized Derivative:

$$\frac{Df(t)}{Dt} = \sum_{i=1}^{j} (f(t_i^+) - f(t_i^-))\delta(t - t_i) + \frac{df(t)}{dt}\bigg|_{t \neq t_1, t_2, \ldots, t_j}$$

Relationships Between Continuous-Time Unit Step and Impulse Delta Signals:

$$u_h(t) = \int_{-\infty}^{t} \delta(\tau)\, d\tau, \quad \delta(t) = \frac{Du(t)}{Dt} = \frac{Du_h(t)}{Dt}$$

Continuous-Time Convolution:
$$g(t) * v(t) = \int_{-\infty}^{\infty} v(\tau)g(t-\tau)\,d\tau$$
$$= \int_{-\infty}^{\infty} v(t-\tau)g(\tau)\,d\tau = v(t) * g(t), \quad -\infty \le t \le \infty$$

Discrete-Time Convolution:
$$g[k] * v[k] = \sum_{m=-\infty}^{m=\infty} g[m]v[k-m]$$
$$= \sum_{m=-\infty}^{m=\infty} g[k-m]v[m] = v[k] * g[k], \quad -\infty \le k \le \infty$$

Discrete-Time Derivative:
$$\Delta f[k] = f[k+1] - f[k]$$

Relationships Among Discrete-Time Unit Step, Unit Ramp, and Impulse Delta Signals:
$$\delta[k] = u[k] - u[k-1] = \begin{cases} 1, & k=0 \\ 0, & k\ne 0 \end{cases}, \quad u[k] = \sum_{m=-\infty}^{k} \delta[m]$$

$$u[k] = r[k+1] - r[k], \quad r[k] = \sum_{m=-\infty}^{k-1} u[m]$$

Continuous- and Discrete-Time Signal Energy and Power:
$$E_L = \int_{t_1}^{t_2} |f(t)|^2\,dt, \quad L = t_2 - t_1, \quad E_\infty = \int_{-\infty}^{\infty} |f(t)|^2\,dt$$

$$P_\infty = \lim_{L \to \infty} \frac{1}{L} \int_{-L/2}^{L/2} |f(t)|^2\,dt$$

$$E_M = \sum_{k=k_1}^{k=k_2} |f[k]|^2, \quad M = k_2 - k_1, \quad E_\infty = \sum_{k=-\infty}^{k=\infty} |f[k]|^2$$

$$P_\infty = \lim_{M \to \infty} \frac{1}{2M+1} \sum_{k=-M}^{k=M} |f[k]|^2$$

Causal Signals:
$$f(t) = 0, \quad t < 0, \quad f[k] = 0, \quad k < 0$$

2.6 REFERENCES

[1] R. Gabel and R. Roberts, *Signals and Linear Systems,* Wiley, New York, 1980.

[2] W. Siebert, *Circuits, Signals, and Systems,* McGraw-Hill/MIT Press, Cambridge, MA, 1986.

[3] C. Chen, *System and Signal Analysis,* Saunders College Publishing, New York, 1989.

[4] C. McGillem and G. Cooper, *Continuous and Discrete Signal and System Analysis,* Saunders College Publishing, Philadelphia, 1991.

[5] H. Kwakernaak and R. Sivan, *Modern Signals and Systems,* Prentice Hall, Englewood Cliffs, NJ, 1991.

[6] B. Lathi, *Linear Systems and Signals,* Berkeley-Cambridge Press, Carmichael, CA, 1992.

[7] F. Taylor, *Principles of Signals and Systems,* McGraw-Hill, New York, 1994.

[8] R. Sturm and D. Kirk, *Contemporary Linear Systems,* PWS Publishing Company, Boston, MA, 1996.

[9] E. Kamen and B. Heck, *Fundamentals of Signals and Systems: Using MATLAB,* Prentice Hall, Upper Saddle River, NJ, 1997.

[10] D. Lindner, *Signals and Systems,* WCB/McGraw-Hill, Boston, MA, 1999.

[11] S. Haykin and B. Van Veen, *Signals and Systems,* Wiley, New York, 1999.

[12] H. Neff, *Continuous and Discrete Linear Systems,* Harper & Row, New York, 1984.

[13] J. Cadzow and H. Van Landingham, *Signals, Systems, and Transforms*, Prentice Hall, Englewood Cliffs, NJ, 1985.

[14] M. O'Flynn and E. Moriarty, *Linear Systems: Time Domain and Frequency Analysis,* Wiley, New York, 1987.

[15] R. Ziemer, W. Tranter, and D. Fannin, *Signals and Systems: Continuous and Discrete,* Macmillan Publishing Company, New York, 1989.

[16] S. Soliman and M. Srinath, *Continuous and Discrete Signals and Systems,* Prentice Hall, Englewood Cliffs, NJ, 1990.

[17] N. Sinha, *Linear Systems,* Wiley, New York, 1991.

[18] L. Balmer, *Signals and Systems: An Introduction,* Prentice Hall International, London, 1991.

[19] R. Houts and O. Alkin, *Signal Analysis in Linear Systems,* Saunders College Publishing, Philadelphia, 1991.

[20] G. Carlson, *Signal and Linear System Analysis,* Houghton Mifflin, Boston, MA, 1992.

[21] P. Kraniauskas, *Transforms in Signals and Systems,* Addison Wesley, Wokingham, England, 1992.

[22] H. Hsu, *Signals and Systems,* McGraw-Hill, New York, 1995.

[23] C. Phillips and J. Parr, *Signals, Systems, and Transforms,* Prentice Hall, Englewood Cliffs, NJ, 1995.

[24] A. Oppenheim and A. Willsky with H. Nawab, *Signals & Systems,* Prentice Hall, Upper Saddle River, NJ, 1997.

[25] A. Papoulis, *The Fourier Integral and Its Applications,* McGraw-Hill, New York, 1961.

[26] R. Bracewell, *The Fourier Transform and Its Applications*, McGraw-Hill, New York, 1965.

[27] R. Gray and J. Goodman, *Fourier Transforms: An Introduction for Engineers,* Kluwer Academic Publishers, Boston, MA, 1995.

[28] S. Karni and W. Byatt, *Mathematical Methods in Continuous and Discrete Systems,* Holt, Rinehart and Winston, New York, 1982.

[29] P. Dirac, "The physical interpretation of the quantum mechanics," *Proc. Roy. Soc.,* Sec. A113, 134–48, London, 1926/27.

[30] I. Gelfand and G. Shilov, *Generalized Functions,* Academic Press, New York, 1964.

[31] A. Zemanian, *Distribution Theory and Transform Analysis,* McGraw-Hill, New York, 1965.

[32] T. Kailath, *Linear Systems*, Prentice Hall, Englewood Cliffs, NJ, 1980.

[33] M. Athans and E. Tse, "A direct derivation of the optimal linear filter using the maximum principle," *IEEE Trans. Automatic Control,* AC-12:690–98, 1967.

2.7 PROBLEMS

2.1. Plot the graphs of the following signals:

(a) $u(t) - \frac{1}{2}\text{sgn}(t)$

(b) $u_1(t) - \frac{1}{2}\text{sgn}(t)$

(c) $u_h(t) - \frac{1}{2}\text{sgn}(t)$

2.2. Verify graphically that $p_\tau(t) = u(t + \tau/2)u(-t + \tau/2)$.

2.3. Plot the graph of the following signal:

$$f(t) = \text{sgn}(t) + p_4(t)u(t) - 2\Delta_2(t)$$

2.4. Represent graphically the following continuous-time signals:

(a) $u(-t + 2)$

(b) $u(-t - 2)$

(c) $r(-t + 3)$

(d) $r(-t - 1)$

2.5. Plot the graphs of the following signals:

(a) $f_1(t) = u(t - 3) - r(-t + 1)$

(b) $f_2(t) = r(t - 2) - u(t - 4)$

(c) $f_3(t) = r(t) + u(t - 2)$

(d) $f_4(t) = -r(-t - 2) - u(t - 5)$

2.6. Plot the graph of the following signal:

$$f(t) = u(t+1) + u(t-1) - r(t-1) + r(t-3)$$

2.7. Represent signals given in Figure 2.17 in terms of step and ramp signals.

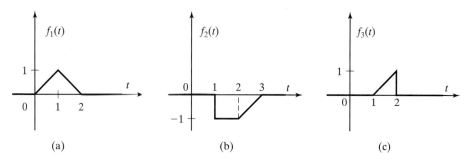

FIGURE 2.17: Continuous-time signals

2.8. Represent the signal given in Figure 2.18 in terms of step and ramp signals.

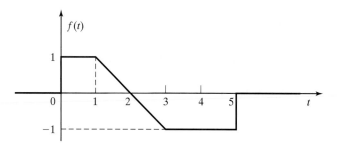

FIGURE 2.18: A continuous-time signal

Answer: $f(t) = u(t) - r(t-1) + r(t-3) + u(t-5)$.

2.9. Plot the graph of the discrete-time signal

$$f[k] = r[k-2] - r[k]u[-k+2] + p_4[k-1]$$

2.10. Represent graphically the discrete-time signal

$$f[k] = \Delta_4[k]u[-k+1] + p_2[k-1]u[k-1]$$

2.11. Find the value $f[3]$ of the discrete-time signal represented by

$$f[k] = \text{sinc}[k] - r[-k+5] + \Delta_6[k]p_4[k]$$

2.12. Give an analytical expression for the discrete-time triangular pulse.

2.13. Prove the impulse delta signal property given in (2.31) for $n = 1, 2$, and draw the conclusion that (2.31) holds for any positive integer n.

Hint: Introduce the change of variables $\gamma = -t$.

2.14. Derive the derivative property of the impulse delta signal that also includes time scaling; that is, evaluate the following integrals:

$$\int_{-\infty}^{\infty} f(t)\delta^{(1)}(at - t_0)\,dt, \quad \int_{-\infty}^{\infty} f(t)\delta^{(1)}(a(t - t_0))\,dt$$

2.15. Generalize the results of Problem 2.14 to the nth derivative with time scaling; that is, find formulas for evaluating the following integrals:

$$\int_{-\infty}^{\infty} f(t)\delta^{(n)}(at - t_0)\,dt, \quad \int_{-\infty}^{\infty} f(t)\delta^{(n)}(a(t - t_0))\,dt$$

2.16. Use the formula obtained in Problem 2.15 to evaluate the following integral:

$$\int_{-\infty}^{\infty} \left[e^{-3t}\delta^{(2)}(3t + 5) + 4\delta^{(1)}(-t + 2) \right] dt$$

2.17. Establish the relationship between $\delta(\omega)$ and $\delta(f)$ for $\omega = 2\pi f$.

2.18. Consider an RC electrical circuit as given in Figure 2.19. Assume that no initial charge exists on the capacitor at time $t = 0^-$ when the switch is closed. Find the current in this circuit assuming that $R \to 0$ and $E =$ constant.

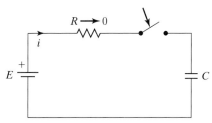

FIGURE 2.19: An RC circuit

2.19. Consider a simple electrical circuit as given in Figure 2.20. Assume that no initial charge exists on the capacitor at time $t = 0^-$ when the switch is closed. Find the current in this circuit assuming that $\varepsilon \to 0$.

2.20. Using the properties of the impulse delta signal, evaluate the integral

$$\int_{-\infty}^{\infty} (t^2 + \sin(t) + e^{-2t})\left[\delta(t) + 3\delta(2t - 1) + 4\delta^{(1)}(t - 2)\right] dt$$

Answer: $1 + \frac{3}{2}\left(\frac{1}{4} + \sin\left(\frac{1}{2}\right) + e^{-1}\right) - 4(4 + \cos(2) - 2e^{-4})$.

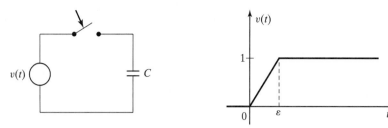

FIGURE 2.20: A simple circuit

2.21. Using the properties of the impulse delta signal, evaluate the integral

$$\int_{-\infty}^{\infty} (t\sin(t) + e^{-t})\left[\delta^{(3)}(t) - \delta^{(2)}(t-2)\right] dt$$

Answer: $1 - 2\cos(2) + 2\sin(2) - e^{-2}$.

2.22. Evaluate the following integrals:

(a) $\int_{-3}^{5} e^{-4t} u(t)\delta(t-4)\, dt$

(b) $\int_{-3}^{5} e^{-t}\delta(t-6)\, dt$

(c) $\int_{-3}^{5} e^{-2t} \sin(t-3)\delta(t-5)\, dt$

(d) $\int_{-3}^{5} e^{-t}\delta(-t+3)\, dt$

2.23. Use the properties of the impulse delta signal to simplify the following expression:

$$2\cos(\pi t)\delta(t-1) + \int_{-\infty}^{\infty} e^{-3t}\left[\delta^{(2)}(t-1) + \delta(2t-3)\right] dt$$
$$+ \int_{-2}^{+2} \tan(2t)\delta(t-3)\, dt + \int_{-3}^{4} 5\delta(t+2)\, dt$$

Answer: $-2\delta(t-1) + 9e^{-3} + 0.5e^{-4.5} + 5$.

2.24. Use the properties of the impulse delta signal to simplify the following expression:

$$e^{-2t}\delta(2t-1) + \int_{-\infty}^{\infty} \sin(\pi(t-1))\left[\delta^{(2)}(t-1) + \delta(2t-2)\right]dt$$

$$+ \int_{-2}^{+2} \tan(2t)\delta(2t-4)\,dt + \int_{-3}^{4} 5\delta(t+2)\,dt$$

2.25. Simplify the following expressions:
  (a) $e^{-5t}\sin(2t+4)\delta(t-1)$
  (b) $(-2)^k \cos[2k-1]\delta[k-4]$
  Answer: (a) $e^{-5}\sin(6)\delta(t-1)$, (b) $(-2)^4 \cos[7]\delta[k-4]$.

2.26. Examine the impact of time scaling and time shifting operations on the following signals:
  (a) $p_2(3t)$, $p_2(3t-2)$, $p_4(4(t-5))$
  (b) $u(2t-3)$, $u(-3t+2)$
  by plotting corresponding graphs.

2.27. Plot the following time scaled and time shifted signals:
  (a) $r(-4(t+2))$
  (b) $u(-3t-1)$
  (c) $p_2(-2t-4)$

2.28. Find the generalized derivatives for the signals given in Problems 2.5 and 2.6.

2.29. Find the generalized derivative of the signal

$$f(t) = (t+2)u(t-2) + \cos(t)u(t-3) + e^{-4t}\sin(t)$$

Hint: Use the derivative product rule.
Answer: $u(t-2) + 4\delta(t-2) - \sin(t)u(t-3) + \cos(3)\delta(t-3) + e^{-4t}(\cos(t) - 4\sin(t))$.

2.30. Draw graphs of the following signals determined by elementary signals:
  (a) $f_1(t) = r(t-2) - r(t)u(-t+2) + p_4(t)$
  (b) $f_2(t) = u(t+1) + r(-t-2) - u(-t+2) - r(t-4)$
  and find their generalized derivatives.

2.31. Distinguish between causal and anticausal signals in the following:
  (a) $e^{-2t}\sin(3t)$
  (b) $\cos(2t-3)u(t-1)$
  (c) $p_4(t-3)$
  (d) $p_4(t-4)u(t+2)$
  (e) $\Delta_2(t)r(t)$

(f) $r(t+1)u(t)$
(g) $p_2[k]\cos[(k-2)\pi]$
(h) $(-2)^{k-2}\sin[k\pi]u[k-1]$

2.32. Find the generalized derivatives of the signals presented in Figure 2.21.

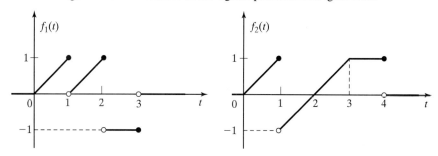

FIGURE 2.21: Continuous-time signals

2.33. Plot the derivative of the signal presented in Figure 2.22 and conclude that signal differentiation can amplify small signals. Very often small signals represent system noise; hence, signal differentiation should be considered with special care since it amplifies noise.

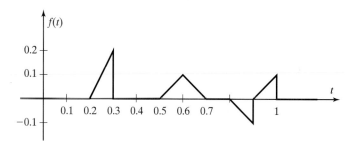

FIGURE 2.22: A small amplitude signal (noise)

2.34. Find the integral of the signal represented in Figure 2.22 and plot the signal obtained. Conclude that signal integration has a smoothing effect.

2.35. Find the first four derivatives (generalized) of a continuous signal $\sin(t)u(t)$; that is, evaluate

$$\frac{d^i}{dt^i}\{\sin(t)u(t)\}, \quad i=1,2,3,4$$

and conclude that differentiation introduces impulses (delta impulse signal and its derivatives) even for continuous signals.

Answer: $d^4\{\sin(t)u(t)\}/dt^4 = \sin(t)u(t) - \delta(t) + \delta^{(2)}(t)$

2.36. Find the first four derivatives (generalized) of the signal $tu(t)$.

2.37. Repeat Problem 2.36 for the signal $u(t)\cos(t)$.

2.38. Derive the general formula for the nth derivative of the signal $e^{\alpha t}u(t)$, $\alpha > 0$.
Answer: $\alpha^n e^{\alpha t}u(t) + \alpha^{n-1}\delta(t) + \alpha^{n-2}\delta^{(1)}(t) + \cdots + \alpha\delta^{(n-2)}(t) + \delta^{(n-1)}(t)$

2.39. Represent the *saturation signal,* defined by

$$\text{sat}(t) = \begin{cases} \alpha t, & 0 \leq t \leq t_0 \\ \beta, & t_0 \leq t \leq \infty \\ 0, & t < 0 \end{cases}, \quad \alpha, \beta = \text{constants}$$

in terms of step and ramp signals.

2.40. Using MATLAB plot the signal defined in Problem 2.39 for $\alpha = 1$ and $\beta = 2$.

2.41. Use MATLAB and its logical functions == and >= to represent, respectively, the discrete-time impulse delta signal $\delta[k-3]$ and the unit step signal $u[k-1]$ in the time interval $[0, 10]$.

2.42. Use MATLAB to plot the *damped sinusoidal* signal $e^{\lambda t}\sin(2t)$, $\text{Re}\{\lambda\} < 0$ in the time interval $t \in [0, 1]$. Consider two cases: (a) λ real and negative (for example, $\lambda = -2$); and (b) λ complex conjugate (for example $\lambda = -1 + j1$). What is the impact of the imaginary part of λ on the damped sinusoidal signal? What will happen to the original signal if λ is taken as real and positive (for example, $\lambda = 1$).

2.43. Use MATLAB to plot the train of rectangular pulses defined by

$$\sum_{k=0}^{\infty} p_\tau(t - kT_0)$$

Take $\tau = 0.1$ and $T_0 = 0.5$. Plot the signal in the time interval defined by $t \in [-0.3, 4.3]$.

2.44. Use MATLAB to plot $\text{sinc}(2t - 3)$ and $\text{sinc}[2k + 1]$. Choose appropriately continuous- and discrete-time intervals.

2.45. Use MATLAB to plot the signal defined in Figure 2.18.

PART ONE

FREQUENCY DOMAIN TECHNIQUES

CHAPTER 3

Fourier Series and Fourier Transform

The Fourier series and Fourier transform represent the first steps in the development of frequency domain techniques for analyzing continuous-time, time-invariant, linear dynamic systems. In the frequency domain, such dynamic systems are described by linear algebraic equations with complex coefficients, in contrast to the time domain, where they are described by constant coefficient linear differential equations. Systems of linear algebraic equations with complex coefficients can be solved easily, unless the number of equations is too great. The frequency domain representation simplifies the presentation of some fundamental linear system theory concepts. The Fourier series and Fourier transform are also important for frequency domain analysis of continuous-time signals.

In this chapter, we first present the basic results for the Fourier series in Section 3.1, and then study the Fourier transform in detail. For the purpose of this course, the Fourier series are first of all important for finding the response of linear dynamic systems driven by periodic inputs, and for the development of the Fourier transform. We present and prove all the properties of the Fourier transform in Section 3.2. Using the properties of the Fourier transform, in Section 3.3, we introduce the concept of the system transfer function and establish one of the fundamental results of linear system theory: *for a system at rest, the system output in the frequency domain is a product of the system transfer function and the Fourier transform of the system input*. The corresponding interpretation in the time domain leads to the famous convolution result, which states that *for a system at rest, the system output is the time domain convolution of the system input and the system impulse response, where the impulse response is the inverse Fourier transform of the system transfer function*.

The fundamental linear system theory results established using the Fourier transform preserve their forms in the Laplace transform, to be introduced in the next chapter. The Laplace transform is more important and more complete from the linear *dynamic* system theory point of view than the Fourier transform, since it can produce both components of the system response: zero-input and zero-state responses. However, the Fourier transform is a necessary step for developing and understanding the Laplace transform. It will be seen that the Fourier and Laplace transforms have many common properties. Though the Laplace transform is superior to the Fourier transform as far as the study of linear time-invariant *dynamic* systems is concerned, the Fourier transform is extremely powerful and useful for communication and signal processing systems, where only the zero-state component of the system response is taken into consideration.

3.1 FOURIER SERIES

The Fourier series, which are applicable to periodic signals, were discovered by French mathematician Joseph Fourier in 1807. By using the Fourier series, *a periodic signal satisfying certain conditions can be expanded into an infinite sum of sine and cosine functions.* Consider a real periodic signal of time, $x(t)$, with the period T, that is,

$$x(t) = x(t+T), \quad T < \infty \tag{3.1}$$

Define the *fundamental* angular frequency as $\omega_0 = 2\pi/T$. By the celebrated Fourier theorem, under certain mild conditions (to be stated subsequently), the following expansion for $x(t)$ holds:

$$x(t) = \frac{1}{2}a_0 + \sum_{n=1}^{\infty}[a_n \cos(n\omega_0 t) + b_n \sin(n\omega_0 t)] \tag{3.2}$$

The expansion formula (3.2) is known as the *real coefficient trigonometric form* of the Fourier series. The required real-number coefficients in (3.2) are obtained from

$$a_n = \frac{2}{T}\int_{-\frac{T}{2}}^{\frac{T}{2}} x(t)\cos(n\omega_0 t)\,dt = \frac{2}{T}\int_0^T x(t)\cos(n\omega_0 t)\,dt, \quad n = 0, 1, 2, \ldots$$

$$b_n = \frac{2}{T}\int_{-\frac{T}{2}}^{\frac{T}{2}} x(t)\sin(n\omega_0 t)\,dt = \frac{2}{T}\int_0^T x(t)\sin(n\omega_0 t)\,dt, \quad n = 1, 2, 3, \ldots$$

(3.3)

In the preceding integrals $n\omega_0$ plays the role of a parameter, hence the coefficients a_n and b_n are functions of $n\omega_0$; that is, $a_n = a_n(n\omega_0)$ and $b_n = b_n(n\omega_0)$. Due to periodicity, it is sometimes more convenient to use the integration limits from $t = 0$ to $t = T$. Note that the coefficient $a_0/2$ gives the average value of $x(t)$ in the interval of one period, that is,

$$\frac{a_0}{2} = \frac{1}{T}\int_{-\frac{T}{2}}^{\frac{T}{2}} x(t)\,dt \tag{3.3a}$$

If the period T is measured in seconds, denoted by (s), then the fundamental angular frequency ω_0 is measured in radians per second (rad/s). It is well known from elementary electrical circuits and physics courses that $\omega_0 = 2\pi f_0$, where f_0 is the frequency defined by $f_0 = 1/T$ and measured in Hertz (Hz = 1/s).

Formulas (3.3) can be justified by using the following known integrals of trigonometric functions:

$$\int_{-\frac{T}{2}}^{\frac{T}{2}} \cos(mt)\cos(nt)\,dt = \begin{cases} 0, & m \neq n \\ T/2, & m = n \geq 1 \\ T, & m = n = 0 \end{cases}$$

$$\int_{-\frac{T}{2}}^{\frac{T}{2}} \sin(mt)\sin(nt)\,dt = \begin{cases} 0, & m \neq n \\ T/2, & m = n \geq 1 \end{cases}$$

$$\int_{-\frac{T}{2}}^{\frac{T}{2}} \sin(mt)\cos(nt)\,dt = 0, \quad \text{for all } m \text{ and } n$$

Multiplying (3.2) by $\cos(n\omega_0 t)$, $n = 1, 2, \ldots$, and integrating from $-T/2$ to $T/2$, we obtain

$$\int_{-\frac{T}{2}}^{\frac{T}{2}} x(t)\cos(n\omega_0 t)\,dt = 0 + 0 + \cdots + 0 + a_n \frac{T}{2} + 0 + \cdots + 0$$

which gives the formula for a_n, $n = 1, 2, \ldots$. Integrating (3.2) directly from $-T/2$ to $T/2$ produces

$$\int_{-\frac{T}{2}}^{\frac{T}{2}} x(t)\,dt = \frac{1}{2}a_0 T + 0 + 0 + \cdots + 0$$

which gives formula (3.3a) for a_0. Similarly, multiplying (3.2) by $\sin(n\omega_0 t)$, $n = 1, 2, \ldots$, and integrating from $-T/2$ to $T/2$, we obtain the formula for b_n:

$$\int_{-\frac{T}{2}}^{\frac{T}{2}} x(t)\sin(n\omega_0 t)\,dt = 0 + 0 + \cdots + 0 + b_n \frac{T}{2} + 0 + \cdots + 0$$

In (3.2), $n = 1$ defines the fundamental harmonic, $n = 2$ gives the second harmonic, and so on; a sinusoidal signal of angular frequency $n\omega_0$ is called the nth harmonic of the sinusoidal signal of the fundamental angular frequency ω_0. Practical applications of (3.2) come from the fact that this infinite sum can be well approximated by a finite sum containing only a few first harmonics. In other words, for many real periodic signals, coefficients a_n and b_n decay very quickly to zero as n increases, hence periodic signals can be approximated by the truncated Fourier series

$$x_N(t) = \frac{1}{2}a_0 + \sum_{n=1}^{N}[a_n \cos(n\omega_0 t) + b_n \sin(n\omega_0 t)] \quad (3.4)$$

where N stands for the number of harmonics included in the approximation.

76 Chapter 3 Fourier Series and Fourier Transform

Calculations of the Fourier series coefficients can be simplified if we use the notions of *even* and *odd functions*. Recall that a function is even if $x(-t) = x(t)$ and odd if $x(-t) = -x(t)$. For example, since $\cos(-t) = \cos(t)$ and $\sin(-t) = -\sin(t)$, we conclude that $\cos(t)$ is an even function and $\sin(t)$ is an odd function. According to formula (3.2), periodic signals are expanded in terms of cosine and sine functions. Hence, the cosine terms represent the even part of $x(t)$ and the sine terms represent the odd part of $x(t)$. This indicates that *even periodic signals* will have no sine terms in their Fourier series expansions; that is, for them, the coefficients $b_n = 0$ for every n. Similarly, *odd periodic signals* will be represented only in terms of sine functions; that is, for odd periodic signals $a_n = 0$ for every n. The following example demonstrates this property of the coefficients of the Fourier series.

EXAMPLE 3.1

Consider an even periodic signal as presented in Figure 3.1. For this signal, the coefficients $b_n = 0$ for every n, which can be formally shown by performing the integration

$$b_n = \frac{2}{T} \int_0^T x(t) \sin(n\omega_0 t)\, dt = 0, \quad n = 1, 2, 3, \ldots$$

Since we know this fact, there is no need to evaluate the integral to find the coefficients b_n. The coefficients a_n are obtained from

$$a_n = \frac{2}{T} \int_0^T x(t) \cos(n\omega_0 t)\, dt$$

$$= \frac{E}{T} \int_0^{\frac{T}{4}} \cos(n\omega_0 t)\, dt - \frac{E}{T} \int_{\frac{T}{4}}^{\frac{3T}{4}} \cos(n\omega_0 t)\, dt + \frac{E}{T} \int_{\frac{3T}{4}}^{T} \cos(n\omega_0 t)\, dt$$

$$= \frac{E}{n\omega_0 T} \left[2\sin\left(n\omega_0 \frac{T}{4}\right) - 2\sin\left(n\omega_0 \frac{3T}{4}\right) + \sin(n\omega_0 T) \right]$$

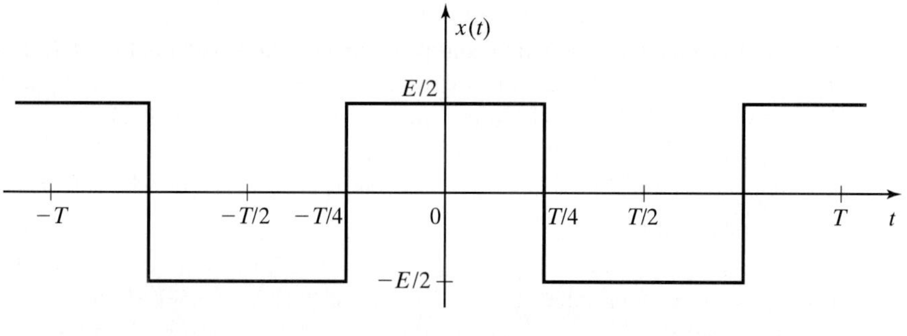

FIGURE 3.1: An even periodic signal

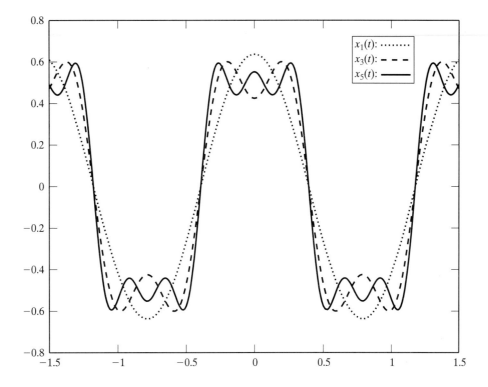

FIGURE 3.2: Signal approximation by the Fourier series

Note that $a_0/2$ represents the average value of the signal $x(t)$ over the time interval of one period. Hence, in this example, $a_0 = 0$. Replacing ω_0 by $2\pi/T$, we obtain

$$a_n = \frac{E}{n\pi}\left[\sin\left(n\frac{\pi}{2}\right) - \sin\left(n\frac{3\pi}{2}\right)\right] = \begin{cases} \frac{2E}{n\pi}, & n = 1, 5, 9, \ldots \\ -\frac{2E}{n\pi}, & n = 3, 7, 11, \ldots \\ 0, & \text{otherwise} \end{cases}$$

The original even periodic signal is represented by the Fourier series as follows:

$$x(t) = \frac{2E}{\pi}\left[\cos(\omega_0 t) - \frac{1}{3}\cos(3\omega_0 t) + \frac{1}{5}\cos(5\omega_0 t) - \frac{1}{7}\cos(7\omega_0 t) + \cdots\right]$$

Apparently, as more terms are included in this infinite series, a better approximation for the signal $x(t)$ is obtained. However, since the cosine function is bounded by ± 1, we see that $x(t)$ can be well approximated by a truncated series due to the fact that $a_n \to 0$ as n increases (theoretically as $n \to \infty$). In Figure 3.2 we present the approximations of the signal $x(t)$ for $E = 1$, $\omega_0 = 4$, $T = \pi/2$, and $N = 1, 3, 5$. In Figure 3.3, we compare the

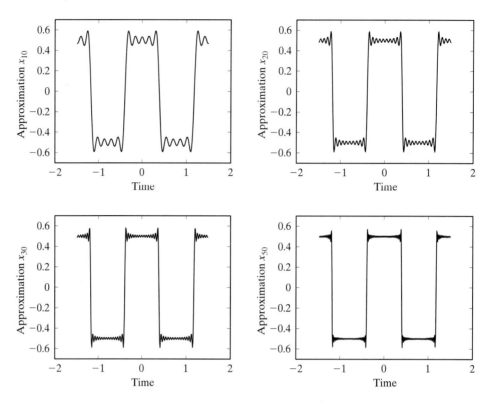

FIGURE 3.3: Higher order signal approximations by the Fourier series

higher order approximations for $N = 10, 20, 30, 50$. Figures 3.2 and 3.3 have been plotted using MATLAB.

The MATLAB program that calculates the corresponding approximations is given below.

```
t=-1.5:0.01:1.5;
wo=4; E=1; N=50
xN=0;
for n=1:N
an=(E/(n*pi))*(sin(n*pi/2)-sin(n*3*pi/2))
xN=xN+an.*cos(n*wo*t); % note pointwise multiplication
end
plot(t,xN)
```

Gibbs Phenomenon

It is expected that the signal approximation defined by (3.4) improves as N increases; theoretically, as $N \to \infty$, the signal $x(t)$ is perfectly approximated by the signal $x_N(t)$, $N \to \infty$. This observation is valid at all points where the signal $x(t)$ is continuous. However, according to the theory of Fourier series, at a point where a periodic signal has a jump discontinuity ($x(t_d^+) \neq x(t_d^-)$), $x_N(t_d)$ for $N \to \infty$ converges to the average value of the signal at the discontinuity point; that is, $x_N(t_d)$ for $N \to \infty$ converges to $(x(t_d^+) + x(t_d^-))/2$. Furthermore, in an arbitrary small neighborhood of the discontinuity point a small ripple always exists. This observation was first made by Willard Gibbs in 1899, hence this is known as the Gibbs phenomenon. No matter how many harmonics are included in the approximation $x_N(t)$, the ripple that exists in a very close neighborhood of t_d cannot be less than 9% (see Figure 3.3, approximation $x_{50}(t)$).

Another form of the Fourier series, known as the trigonometric Fourier series with *complex coefficients*, can be derived from (3.2) using the Euler formula

$$e^{\pm j\alpha} = \cos(\alpha) \pm j \sin(\alpha) \tag{3.5}$$

This formula can be easily justified with the help of standard expansions for exponential, sine, and cosine functions, as follows:

$$e^{j\alpha} = 1 + (j\alpha) + \frac{1}{2!}(j\alpha)^2 + \frac{1}{3!}(j\alpha)^3 + \frac{1}{4!}(j\alpha)^4 + \cdots$$

$$= \left(1 - \frac{\alpha^2}{2!} + \frac{\alpha^4}{4!} - \cdots\right) + j\left(\alpha - \frac{\alpha^3}{3!} + \frac{\alpha^5}{5!} - \cdots\right) = \cos(\alpha) + j \sin(\alpha)$$

Using Euler's formula (3.5), we can easily establish the following relationships:

$$\cos(\alpha) = \frac{1}{2}(e^{j\alpha} + e^{-j\alpha}), \quad \sin(\alpha) = \frac{1}{2j}(e^{j\alpha} - e^{-j\alpha}) \tag{3.6}$$

Using (3.6) with $\alpha = n\omega_0 t$ in (3.2), we obtain

$$x(t) = \frac{1}{2}a_0 + \sum_{n=1}^{\infty}\left[\frac{1}{2}a_n(e^{jn\omega_0 t} + e^{-jn\omega_0 t}) + \frac{1}{2j}b_n(e^{jn\omega_0 t} - e^{-jn\omega_0 t})\right]$$

$$= \frac{1}{2}a_0 + \sum_{n=1}^{\infty}\left[\frac{1}{2}e^{jn\omega_0 t}(a_n - jb_n) + \frac{1}{2}e^{-jn\omega_0 t}(a_n + jb_n)\right] \tag{3.7}$$

$$= X_0 + \sum_{n=1}^{\infty}(X_n e^{jn\omega_0 t} + X_n^* e^{-jn\omega_0 t})$$

where the asterisk denotes the complex conjugate, and

$$X_0 = \frac{1}{2}a_0, \quad X_n = \frac{1}{2}(a_n - jb_n) = |X_n|e^{j\angle X_n}, \quad n = 1, 2, 3, \ldots$$

$$|X_n| = \frac{1}{2}\sqrt{a_n^2 + b_n^2}, \quad \angle X_n = -\tan^{-1}\left(\frac{b_n}{a_n}\right) \tag{3.8}$$

Note that a_n and b_n coefficients are functions of $n\omega_0$. It follows that the complex coefficients are also functions of $n\omega_0$, that is, $X_n = X_n(jn\omega_0)$. It can be seen easily from (3.8) that the following relationships hold:

$$a_0 = 2X_0, \quad a_n = 2\operatorname{Re}\{X_n\}, \quad b_n = -2\operatorname{Im}\{X_n\}, \quad n = 1, 2, \ldots \tag{3.9}$$

The *trigonometric form of the Fourier series with complex coefficients* is obtained from (3.7) as follows:

$$\begin{aligned}
x(t) &= X_0 + \sum_{n=1}^{\infty} \left(|X_n| e^{j\angle X_n} e^{jn\omega_0 t} + |X_n| e^{-j\angle X_n} e^{-jn\omega_0 t} \right) \\
&= X_0 + \sum_{n=1}^{\infty} |X_n| \left(e^{j(n\omega_0 t + \angle X_n)} + e^{-j(n\omega_0 t + \angle X_n)} \right) \\
&= X_0 + 2 \sum_{n=1}^{\infty} |X_n| \cos(n\omega_0 t + \angle X_n)
\end{aligned} \tag{3.10}$$

The coefficients X_n, $n = 1, 2, \ldots$, in the preceding formula can be calculated from (3.3) and (3.8):

$$\begin{aligned}
X_n = X_n(jn\omega_0) &= \frac{1}{2}(a_n - jb_n) = \frac{1}{T} \int_{-\frac{T}{2}}^{\frac{T}{2}} x(t) e^{-jn\omega_0 t} \, dt \\
&= \frac{1}{T} \int_0^T x(t) e^{-jn\omega_0 t} \, dt, \quad n = 1, 2, \ldots, \quad X_0 = \frac{1}{2} a_0
\end{aligned} \tag{3.11}$$

It is possible to find the Fourier series coefficients for a given periodic function using formulas (3.3), but the infinite Fourier series obtained does not converge to the given periodic function. In general, not all periodic signals can be expanded into *convergent* Fourier series. The conditions that a periodic signal (function) must satisfy in order for the Fourier series to converge are known as Dirichlet's conditions.

Convergence Conditions for the Fourier Series (Dirichlet's Conditions)

A periodic function has a convergent Fourier series if the following conditions are satisfied:

1. The single-valued function $x(t)$ is bounded, and hence absolutely integrable over the finite period T; that is,

$$\int_{-\frac{T}{2}}^{\frac{T}{2}} |x(t)| \, dt < \infty \tag{3.12}$$

2. The function has a finite number of maxima and minima over the period T.
3. The function has a finite number of discontinuity points over the period T.

Note that a single-valued bounded function is absolutely integrable over a finite period T. Condition (3.12) is more general than signal boundedness over a finite period T, since (3.12) is satisfied even for bounded signals that contain unbounded delta impulse functions.

Under Dirichlet's conditions a Fourier series converges to $x(t)$ at all points where $x(t)$ is continuous, and to the average of the left- and right-hand limits of $x(t)$ at each point where $x(t)$ is discontinuous.

In practical applications, it is sufficient to check for condition (1), since it is almost impossible to find real physical signals that satisfy condition (1) yet do not satisfy conditions (2) and (3).

We have established so far two forms of the Fourier series, the real coefficient trigonometric form (3.2) and the complex coefficient trigonometric form (3.10). The third form, very useful in the passage from the Fourier series to the Fourier transform, can be derived from (3.7) and (3.11) by observing the following fact from (3.11):

$$X_n^* = X_n^*(jn\omega_0) = \frac{1}{T}\int_{-\frac{T}{2}}^{\frac{T}{2}} x(t) e^{jn\omega_0 t}\, dt$$

$$= \frac{1}{T}\int_{-\frac{T}{2}}^{\frac{T}{2}} x(t) e^{-j(-n)\omega_0 t}\, dt = X_{-n} = X_n(-jn\omega_0), \quad n = 1, 2, \ldots$$
(3.13)

Using (3.13) in (3.7), we obtain

$$x(t) = X_0 + \sum_{n=1}^{\infty} X_n(jn\omega_0) e^{jn\omega_0 t} + \sum_{n=1}^{\infty} X_n(-jn\omega_0) e^{j(-n)\omega_0 t}$$

$$= \sum_{n=-\infty}^{\infty} X_n(jn\omega_0) e^{jn\omega_0 t} = \sum_{n=-\infty}^{\infty} X_n e^{jn\omega_0 t}$$
(3.14)

with $X_n(jn\omega_0)$ as defined in (3.8) and (3.11). This form is called the *exponential form* of the Fourier series.

Line Spectra

Frequency plots of the magnitude and phase of the Fourier series coefficients $X_n(jn\omega_0) = |X_n(jn\omega_0)| e^{j\angle X_n(jn\omega_0)}$ define the *line spectra*. The line spectra can be plotted independently for the magnitude $|X_n(jn\omega_0)|$ (in this case the magnitude is also the amplitude, see (3.10)) and for the phase $\angle X_n(jn\omega_0)$. Hence, we distinguish between magnitude and phase line spectra.

From (3.13), we have $X_n^*(jn\omega_0) = X_n(-jn\omega_0)$ and

$$\left(|X_n(jn\omega_0)| e^{j\angle X_n(jn\omega_0)}\right)^* = |X_n(jn\omega_0)| e^{-j\angle X_n(jn\omega_0)}$$

$$= X_n(-jn\omega_0) = |X_n(-jn\omega_0)| e^{j\angle X_n(-jn\omega_0)}$$

82   Chapter 3   Fourier Series and Fourier Transform

It follows that $|X_n(jn\omega_0)|$ is an even function of $n\omega_0$, and $\angle X_n(jn\omega_0)$ is an odd function of $n\omega_0$, that is,

$$|X_n(jn\omega_0)| = |X_n(-jn\omega_0)|, \quad \angle X_n(jn\omega_0) = -\angle X_n(-jn\omega_0)$$

It is customary in engineering contexts to plot the amplitude (magnitude) and phase line spectra for all frequencies from $-\infty$ to ∞ at discrete frequency points defined at $\pm n\omega_0, n = 0, 1, 2, \ldots$. However, one can plot the line spectra only for positive frequencies, since negative frequencies do not exist in the physical world.

EXAMPLE 3.2

Consider the signal represented in Figure 3.4.

We will find the complex coefficients $X_n(jn\omega_0)$ using formula (3.8). Since this signal is odd it follows that $a_n = 0$, which implies that $X_0 = 0$. The coefficients b_n are obtained from (3.3) as

$$b_n = \frac{2}{T} \int_{-\frac{T}{2}}^{\frac{T}{2}} \frac{2E}{T} t \sin(n\omega_0 t)\, dt = \frac{4E}{T^2} \int_{-\frac{T}{2}}^{\frac{T}{2}} t \sin(n\omega_0 t)\, dt$$

$$= \frac{8E}{T^2} \int_0^{\frac{T}{2}} t \sin(n\omega_0 t)\, dt$$

This integral can be evaluated by using the standard integration tables (see, for example, [1], page 75) as

$$\int t \sin(n\omega_0 t)\, dt = \frac{1}{n^2 \omega_0^2} \sin(n\omega_0 t) - \frac{t}{n\omega_0} \cos(n\omega_0 t)$$

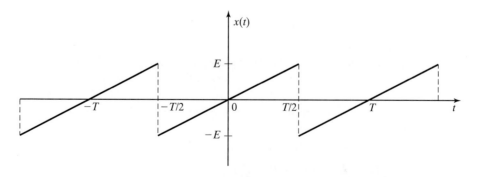

FIGURE 3.4: Sawtooth waveform

Substituting the corresponding integration limits, we have

$$b_n = \frac{8E}{T^2} \frac{T}{2n\omega_0}(-1)\cos(n\pi) = \frac{2E}{n\pi}(-1)^{n+1}$$

The coefficients $X_n(jn\omega_0)$ are given by formula (3.11) as

$$X_n(jn\omega_0) = \frac{1}{2}(a_n - jb_n) = \frac{-jb_n}{2} = \frac{E}{n\pi}j(-1)^n$$

Hence, in this example, the magnitude and phase line spectra are represented by

$$|X_n(jn\omega_0)| = \begin{cases} \frac{E}{n\pi}, & n = 1, 2, \ldots \\ 0, & n = 0 \end{cases}, \quad \angle X_n(jn\omega_0) = \begin{cases} (-1)^n \frac{\pi}{2}, & n = 1, 2, \ldots \\ 0, & n = 0 \end{cases}$$

(Note that $X_0 = a_0/2 = 0$.) The corresponding line spectra are presented in Figures 3.5 and 3.6.

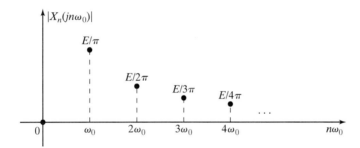

FIGURE 3.5: Magnitude line spectrum for Example 3.2

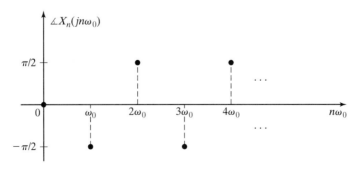

FIGURE 3.6: Phase line spectrum for Example 3.2

It is left as an exercise (see Problem 3.44) to establish that the *periodic signal power* can be evaluated in the frequency domain using its Fourier series coefficients as

$$\int_0^T |x(t)|^2 \, dt = \sum_{n=-\infty}^{\infty} |X_n(jn\omega_0)|^2 \qquad (3.15)$$

This relationship is known as the *Parseval formula for periodic signals*.

In summary, we have introduced three forms of the Fourier series, defined in (3.2), (3.10), and (3.14). The Fourier series trigonometric form with real coefficients defined in (3.2) is used for calculating the Fourier series of a given signal. The Fourier series trigonometric form with complex coefficients defined in (3.10) is used for signal spectral analysis (frequency domain analysis), since the corresponding complex coefficients obtained from (3.11) provide information about signal magnitude and phase in terms of frequency $n\omega_0$ (signal line spectra). Formula (3.10) will be also used in Section 3.4 for finding the zero-state system response due to periodic and sinusoidal inputs. The power of formula (3.10) will come from the fact that the superposition principle can be used to find the system response due to periodic inputs. The third form of the Fourier series, the exponential form defined in (3.14), is important for the development of the Fourier transform. The Fourier transform represents a generalization of the Fourier series that can be applied to both periodic and aperiodic signals. In the next subsection, starting with the Fourier series, we derive a more general signal transformation known as the Fourier transform.

The material presented so far about the Fourier series is sufficient for the purpose of this textbook. For more rigorous mathematical treatment of the Fourier series, the reader with standard engineering calculus background is referred, for example, to [2–3].

3.1.1 From Fourier Series to Fourier Transform

Since in linear system analysis we deal with general signals and the use of the Fourier series is limited to periodic signals, we need to derive a more general transform, which will be applicable to nonperiodic signals as well. This transform is known as the Fourier transform.

The Fourier transform can be derived from the Fourier series by using the following mathematical artifice: *any function is periodic with a period $T = \infty$*, that is, by stating

$$x(t) = x(t+T), \quad T = \infty \qquad (3.16)$$

By this means (assuming that it can be mathematically and practically justified), we pass from a discrete frequency representation $\omega_0, 2\omega_0, 3\omega_0, \ldots, n\omega_0, \ldots$, with $\omega_0 = 2\pi/T$, as used for the Fourier series, to a continuous frequency representation—since for $T \to \infty$ we have $\omega_0 \to d\omega$ (an infinitesimally small quantity) and $n\omega_0 \to \omega$ (evenly spaced discrete frequencies $n\omega_0$ become a continuum of frequencies ω). These limiting facts become obvious from the line spectra diagrams (see Figures 3.5 and 3.6, and observe that for $T \to \infty$ it follows that $\omega_0 \to d\omega$). Using these facts in (3.11) and (3.14), we obtain

$$x(t) = \lim_{T \to \infty} \left\{ \sum_{n=-\infty}^{n=\infty} \left(\frac{1}{T} \int_{-\frac{T}{2}}^{\frac{T}{2}} x(\tau) e^{-jn\omega_0 \tau} d\tau \right) e^{jn\omega_0 t} \right\}$$

that is,

$$x(t) = \lim_{n\omega_0 \to \omega} \left\{ \sum_{n=-\infty}^{n=\infty} \left(\int_{-\infty}^{\infty} x(\tau)e^{-jn\omega_0\tau} d\tau \right) \frac{d\omega}{2\pi} e^{jn\omega_0 t} \right\}$$

The infinite sum in the limit becomes the integral, hence

$$x(t) = \frac{1}{2\pi} \int_{-\infty}^{\infty} \left(\int_{-\infty}^{\infty} x(\tau)e^{-j\omega\tau} d\tau \right) e^{j\omega t} d\omega$$

The term in parentheses is a function of the angular frequency ω. It defines *the Fourier transform* of the signal $x(t)$, which we denote by $X(j\omega)$:

$$X(j\omega) = \mathcal{F}\{x(t)\} \triangleq \int_{-\infty}^{\infty} x(t)e^{-j\omega t} dt \qquad (3.17)$$

where the symbol $\triangleq$ means "equals by definition." The signal $x(t)$ is obtained from its Fourier transform by the inverse operation:

$$x(t) = \mathcal{F}^{-1}\{X(j\omega)\} \triangleq \frac{1}{2\pi} \int_{-\infty}^{\infty} X(j\omega)e^{j\omega t} d\omega \qquad (3.18)$$

We call expression (3.18) the *inverse Fourier transform* of $x(t)$. In other words, the signal $x(t)$ is obtained by finding the inverse Fourier transform of $X(j\omega)$. We say that the signals $x(t)$ and $X(j\omega)$ form a Fourier transform pair, and denote their relationship by $x(t) \leftrightarrow X(j\omega)$.

Some authors, especially in communication systems and signal processing, prefer to express the Fourier transform in terms of frequency f rather than angular frequency $\omega = 2\pi f$. In that case, the formulas corresponding to (3.17) and (3.18) would be

$$X(jf) = \mathcal{F}\{x(t)\} = \int_{-\infty}^{\infty} x(t)e^{-j2\pi ft} dt \qquad (3.19)$$

$$x(t) = \mathcal{F}^{-1}\{X(jf)\} = \int_{-\infty}^{\infty} X(jf)e^{j2\pi ft} df \qquad (3.20)$$

In this text we will use formulas (3.17) and (3.18) as the definitions of the Fourier transform and its inverse.

It is interesting to point out that the almost ridiculously simple mathematical trick (3.16), used more than two hundred years ago by the mathematician Fourier in an attempt to deal with infinity in a mathematical way, led to the development of the Fourier transform, which provided a foundation for the development of practical communication and signal processing systems, such as telephone, television, computer networks, and so on.

Existence Condition for the Fourier Transform

Since the Fourier transform is defined by an integral, the primary existence condition for the Fourier transform is in fact the existence condition of the corresponding integral; that is, the following integral must exist (must be convergent):

$$\left| \int_{-\infty}^{\infty} x(t) e^{-j\omega t} \, dt \right| \leq \int_{-\infty}^{\infty} |x(t) e^{-j\omega t}| \, dt \leq \int_{-\infty}^{\infty} |x(t)| \, |e^{-j\omega t}| \, dt < \infty$$

Using the fact that $|e^{-j\omega t}| = 1$, the preceding condition leads to

$$\int_{-\infty}^{\infty} |x(t)| \, dt < \infty \qquad (3.21)$$

Condition (3.21) is the primary existence condition for the Fourier transform. Some additional mathematical restrictions must also be imposed, similar to Dirichlet's conditions for the Fourier series. (The signal must have a finite number of minima and maxima and a finite number of discontinuities on any finite time interval.) For the purpose of this course, it is enough to test condition (3.21) to determine whether the corresponding Fourier transform exists, in terms of ordinary functions.

It should be pointed out that (3.21) *states a sufficient condition only,* which means that if (3.21) is satisfied, then the Fourier transform exists (in terms of ordinary functions), but not the other way around: the existence of the Fourier transform does not imply that (3.21) must hold. In subsequent sections, we will see that some standard signals do not satisfy (3.21) yet do have Fourier transforms given in terms of generalized functions.

3.2 FOURIER TRANSFORM AND ITS PROPERTIES

In this section we state and prove all important properties of the Fourier transform. The Fourier transform is given in an integral form, hence many of the well-known properties of integrals will hold in the case of the Fourier transform. Using these properties, we will derive Fourier transforms of common signals appearing in electrical engineering.

From the definition of the Fourier transform as given by (3.17), that is,

$$\mathcal{F}\{x(t)\} = X(j\omega) = \int_{-\infty}^{\infty} x(t) e^{-j\omega t} \, dt \qquad (3.22)$$

we can easily find Fourier transforms of some basic signals. Consider the following examples.

EXAMPLE 3.3

The Fourier transform of the *impulse delta signal* is directly obtained by using the mathematical definition of the impulse delta function given in (2.21):

$$\mathcal{F}\{\delta(t)\} = D(j\omega) = \int_{-\infty}^{\infty} \delta(t) e^{-j\omega t} \, dt = 1 \qquad (3.23)$$

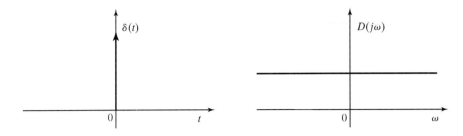

FIGURE 3.7: Impulse delta signal and its Fourier transform

This simple result indicates that it is easy to treat the impulse delta signal in the frequency domain, since its Fourier transform has an extremely simple form; that is, in the frequency domain, the corresponding Fourier transform is equal to 1 for all $\omega \in (-\infty, \infty)$. This result indicates the following fact: *signals that have an infinitesimally small width (band) in the time domain have an infinitely large width (band) in the frequency domain.* As a physical interpretation of this observation, note that the natural signal closest to $\delta(t)$, a lightning strike, creates noise at all radio frequency bands. The impulse delta signal and its Fourier transform are presented in Figure 3.7.

EXAMPLE 3.4

The Fourier transform of an *exponential signal* defined by

$$e(t) = \begin{cases} e^{-at}, a > 0, & t \geq 0 \\ 0, & t < 0 \end{cases} \quad (3.24)$$

is obtained by integrating (3.17) directly, that is,

$$\mathcal{F}\{e(t)\} = \int_0^\infty e^{-at} e^{-j\omega t}\, dt = \frac{1}{-a - j\omega} e^{-(a+j\omega)t} \Big|_{t=0}^{t=\infty} = \frac{1}{a + j\omega}$$

Using the same procedure, we establish also that

$$\mathcal{F}\{e^{-at} u_h(t)\} = \frac{1}{a + j\omega}, \quad a > 0 \quad (3.25)$$

EXAMPLE 3.5

The rectangular pulse, $p_\tau(t)$, defined in (2.9) and presented in Figure 2.4, also satisfies existence condition (3.21). Its Fourier transform can be obtained by direct integration as follows:

$$\mathcal{F}\{p_\tau(t)\} = \int_{-\infty}^{\infty} p_\tau(t) e^{-j\omega t}\, dt = \int_{-\frac{\tau}{2}}^{\frac{\tau}{2}} 1 \times e^{-j\omega t}\, dt$$

$$= -\frac{1}{j\omega}\left(e^{-j\omega\frac{\tau}{2}} - e^{j\omega\frac{\tau}{2}}\right) = \frac{2}{\omega}\sin\left(\omega\frac{\tau}{2}\right)$$

By using the definition of the sinc signal from Chapter 2, formula (2.16), we get

$$\mathcal{F}\{p_\tau(t)\} = \tau\, \mathrm{sinc}\left(\frac{\omega\tau}{2\pi}\right) = \tau\, \mathrm{sinc}(\tau f) \tag{3.26}$$

Using exactly the same derivations, it can also be shown that $\mathcal{F}\{p_\tau^h(t)\} = \tau\, \mathrm{sinc}(\frac{\omega\tau}{2\pi})$, with $p_\tau^h(t)$ as defined in (2.10), since the exact values of the rectangular pulse at the points $t = \pm\tau/2$ have no impact on the value of the corresponding integral. Hence, $p_\tau(t)$ and $p_\tau^h(t)$ have the same Fourier transform.

The results obtained in Examples 3.3–5 form time-frequency Fourier transform pairs of common signals. (The common Fourier transform pairs of standard signals used in this text are summarized in Table 3.4, at the end of this chapter.)

In general, it follows from the definition of integral (3.17) that Fourier transforms are complex functions whose real and imaginary parts are functions of ω. It is easy to show that Fourier transforms of even signals $(x(-t) = x(t))$ are real functions of ω and that Fourier transforms of odd signals are pure imaginary functions of ω (see Problem 3.19). In Example 3.5, the rectangular pulse is an even function, hence its Fourier transform is real.

It should be pointed out that when we try to find Fourier transforms of some basic signals (such as sine, cosine, and unit step), we will easily see that these signals do not satisfy existence condition (3.21), therefore their Fourier transforms do not exist. As a matter of fact, the Fourier transforms of these signals do not exist in an ordinary sense (the sense of ordinary functions that we are used to), but the Fourier transforms of these signals do exist in the generalized sense, the sense of generalized functions. It will be shown that the Fourier transforms of the sine, cosine, unit step, and certain other common signals can be obtained in terms of the frequency domain impulse delta signal.

Since many signals are not Fourier transformable (integrable) in the ordinary sense, we must start with the development of the properties of the Fourier transform. Using these properties, in combination with known common Fourier transform pairs, we will be able to cover quite a broad range of signals for which the Fourier transform can be found in either the ordinary or generalized sense.

In the following subsection we state the main properties of the Fourier transform and give their proofs.

3.2.1 Properties of the Fourier Transform

In this section, we present twelve main properties of the Fourier transform. Mastering these properties is indispensable for every undergraduate student in electrical/computer, mechanical, and biomedical engineering.

Property 1: Linearity

The Fourier transform is defined by the integral (3.17), and since integrals are linear operators, the linearity property holds for the Fourier transform as well; that is, for any set of constants $\alpha_i, i = 1, 2, \ldots, n$, and for any set of Fourier transform pairs $x_i(t) \leftrightarrow X_i(j\omega), i = 1, 2, \ldots, n$, we have

$$\mathcal{F}\{\alpha_1 x_1(t) \pm \alpha_2 x_2(t) \pm \cdots \pm \alpha_n x_n(t)\}$$

$$= \int_{-\infty}^{\infty} [\alpha_1 x_1(t) \pm \alpha_2 x_2(t) \pm \cdots \pm \alpha_n x_n(t)] e^{-j\omega t} \, dt$$

$$= \int_{-\infty}^{\infty} \alpha_1 x_1(t) e^{-j\omega t} \, dt \pm \int_{-\infty}^{\infty} \alpha_2 x_2(t) e^{-j\omega t} \, dt \pm \cdots \pm \int_{-\infty}^{\infty} \alpha_n x_n(t) e^{-j\omega t} \, dt \quad (3.27)$$

$$= \alpha_1 \mathcal{F}\{x_1(t)\} \pm \alpha_2 \mathcal{F}\{x_2(t)\} \pm \cdots \pm \alpha_n \mathcal{F}\{x_n(t)\}$$

$$= \alpha_1 X_1(j\omega) \pm \alpha_2 X_2(j\omega) \pm \cdots \pm \alpha_n X_n(j\omega)$$

This property will be demonstrated in Example 3.13 and several problems at the end of this chapter.

Property 2: Time Shifting

In signals and systems theory we often deal with signals that are shifted in time. Let the signal $x(t)$ have the Fourier transform $X(j\omega)$. Then, the Fourier transform of the shifted signal is given by

$$\mathcal{F}\{x(t - t_0)\} = e^{-j\omega t_0} X(j\omega) \quad (3.28)$$

Proof. In order to prove this property, we start with the definition of the Fourier transform of a shifted time signal,

$$\mathcal{F}\{x(t - t_0)\} = \int_{-\infty}^{\infty} x(t - t_0) e^{-j\omega t} \, dt$$

and introduce the change of variables as $t - t_0 = \sigma$, which leads to

$$\mathcal{F}\{x(t - t_0)\} = \int_{-\infty}^{\infty} x(\sigma) e^{-j\omega(\sigma + t_0)} \, d\sigma$$

$$= e^{-j\omega t_0} \int_{-\infty}^{\infty} x(\sigma) e^{-j\omega \sigma} \, d\sigma = e^{-j\omega t_0} X(j\omega) \qquad \square$$

Thus, in order to find the Fourier transform of a shifted signal we only need multiply the Fourier transform of the original signal by $e^{-j\omega t_0}$. This property can also be called the time delay ($t_0 > 0$) and time advance ($t_0 < 0$) property.

EXAMPLE 3.6

The Fourier transform of the rectangular pulse was derived in Example 3.5, formula (3.26). The Fourier transform of this pulse shifted, for example, by $t_0 = 19$, is

$$\mathcal{F}\{p_\tau(t-19)\} = e^{-j19\omega}\tau\,\mathrm{sinc}\left(\frac{\omega\tau}{2\pi}\right) = e^{-j19\omega}\frac{2}{\omega}\sin\left(\omega\frac{\tau}{2}\right)$$

Property 3: Time Scaling

This property is mathematically stated as follows:

$$\mathcal{F}\{x(at)\} = \frac{1}{|a|}X\left(\frac{j\omega}{a}\right), \quad a = \mathrm{const} \qquad (3.29)$$

Proof. We take the Fourier transform of the signal $x(at)$ and introduce the change of variables as $\sigma = at$. This leads to

$$\mathcal{F}\{x(at)\} = \int_{-\infty}^{\infty} x(at)e^{-j\omega t}\,dt = \frac{1}{a}\int_{-\infty}^{\infty} x(\sigma)e^{-j\frac{\omega}{a}\sigma}\,d\sigma = \frac{1}{a}X\left(\frac{j\omega}{a}\right), \quad a > 0$$

If $a < 0$, the change of variables as $\sigma = at$ implies

$$\int_{-\infty}^{\infty} x(at)e^{-j\omega t}\,dt = \frac{1}{a}\int_{+\infty}^{-\infty} x(\sigma)e^{-j\frac{\omega}{a}\sigma}\,d\sigma = -\frac{1}{a}\int_{-\infty}^{\infty} x(\sigma)e^{-j\frac{\omega}{a}\sigma}\,d\sigma$$

$$= -\frac{1}{a}X\left(\frac{j\omega}{a}\right) = \frac{1}{|a|}X\left(\frac{j\omega}{a}\right), \quad a < 0$$

Since $a = |a|, a > 0$, the last two formulas can be jointly written as (3.29), which establishes the time scaling property of the Fourier transform. □

EXAMPLE 3.7

The Fourier transform of a time scaled rectangular pulse is easily obtained, by using the fact that $\mathcal{F}\{p_\tau(t)\} = \frac{2}{\omega}\sin(\omega\frac{\tau}{2})$ combined with the time scaling property of the Fourier transform defined in (3.29), as follows:

$$\mathcal{F}\{p_\tau(3t)\} = \frac{1}{3}\frac{2}{\left(\frac{\omega}{3}\right)}\sin\left(\left(\frac{\omega}{3}\right)\frac{\tau}{2}\right) = \frac{2}{\omega}\sin\left(\omega\frac{\tau}{6}\right)$$

However, to solve the same problem by using the definition integral of the Fourier transform, we must first figure out the analytical expression for $p_\tau(3t)$ and then perform integration. Note that

$$p_\tau(3t) = \begin{cases} 1, & -\tau/2 \leq 3t \leq \tau/2 \\ 0, & \text{otherwise} \end{cases} = \begin{cases} 1, & -\tau/6 \leq t \leq \tau/6 \\ 0, & \text{otherwise} \end{cases}$$

Using the definition integral (3.17), we obtain the same result:

$$\mathcal{F}\{p_\tau(3t)\} = \int_{-\infty}^{\infty} p_\tau(3t)e^{-j\omega t}\,dt = \int_{-\tau/6}^{\tau/6} e^{-j\omega t}\,dt = -\frac{1}{j\omega}\left(e^{-j\omega\frac{\tau}{6}} - e^{j\omega\frac{\tau}{6}}\right)$$

$$= \frac{2}{\omega}\left[\frac{e^{j\omega\frac{\tau}{6}} - e^{-j\omega\frac{\tau}{6}}}{2j}\right] = \frac{2}{\omega}\sin\left(\omega\frac{\tau}{6}\right)$$

This method apparently involves more steps than the one obtained directly using the time scaling property of the Fourier transform.

We can combine both the time shifting and time scaling properties, and show similarly (see Problem 3.7) that

$$\mathcal{F}\{x(at - t_0)\} = e^{-j(\frac{\omega}{a})t_0}\frac{1}{|a|}X\left(\frac{j\omega}{a}\right), \quad a = \text{const} \tag{3.29a}$$

Property 4: Time Multiplication
Let the Fourier transform of the signal $x(t)$ be given by $X(j\omega)$. Then

$$\mathcal{F}\{t^n x(t)\} = (j)^n \frac{d^n}{d\omega^n} X(j\omega) \tag{3.30}$$

This property is sometimes called the time weighting property.

Proof. This useful property can easily be justified by starting with the definition of the Fourier transform and taking derivatives of both sides with respect to ω; that is,

$$X(j\omega) = \int_{-\infty}^{\infty} x(t)e^{-j\omega t}\,dt$$

$$\frac{dX(j\omega)}{d\omega} = \int_{-\infty}^{\infty} (-jt)x(t)e^{-j\omega t}\,dt$$

$$\vdots$$

$$\frac{d^n X(j\omega)}{d\omega^n} = \int_{-\infty}^{\infty} (-jt)^n x(t)e^{-j\omega t}\,dt$$

which indicates that $(-jt)^n x(t)$ and $d^n X(j\omega)/d\omega^n$ are the corresponding Fourier transform pair. Hence, we have

$$t^n x(t) \leftrightarrow \frac{1}{(-j)^n} \frac{d^n}{d\omega^n} X(j\omega) = j^n \frac{d^n}{d\omega^n} X(j\omega)$$

which proves the time multiplication property of the Fourier transform as defined by (3.30). □

Property 5: Frequency Shifting

Similarly to the time shifting property, we can formulate the frequency shifting property. It states that if $X(j\omega) = \mathcal{F}\{x(t)\}$, then

$$X(j(\omega - \omega_0)) = \mathcal{F}\{x(t)e^{j\omega_0 t}\} \qquad (3.31)$$

This property is sometimes called the exponential weighting property.

Proof. Let us start with the definition of the inverse Fourier transform of $X(j(\omega - \omega_0))$,

$$\mathcal{F}^{-1}\{X(j(\omega-\omega_0))\} = \frac{1}{2\pi} \int_{-\infty}^{\infty} X(j(\omega-\omega_0))e^{j\omega t} d\omega$$

and introduce the change of variables as $\omega - \omega_0 = \lambda$, which leads to

$$\frac{1}{2\pi} \int_{-\infty}^{\infty} X(j\lambda)e^{j(\lambda+\omega_0)t} d\lambda = e^{j\omega_0 t} \frac{1}{2\pi} \int_{-\infty}^{\infty} X(j\lambda)e^{j\lambda t} d\lambda = e^{j\omega_0 t} x(t)$$

A more elegant proof would start with the definition of the Fourier transform of a signal multiplied by a complex exponential, and interpret the result obtained according to the definition of the Fourier transform integral (3.17). The corresponding result is

$$\mathcal{F}\{x(t)e^{j\omega_0 t}\} = \int_{-\infty}^{\infty} x(t)e^{j\omega_0 t} e^{-j\omega t} dt = \int_{-\infty}^{\infty} x(t)e^{-j(\omega-\omega_0)t} dt \triangleq X(j(\omega-\omega_0))$$

□

It is important to note that the preceding Fourier transform properties, though simple, are very rich from the application point of view. For example, the simultaneous application of Properties 2 and 5, the time shifting and frequency shifting properties, led to the development of the modern *wavelets theory*. The wavelets theory is a rapidly growing research area in signal processing and communications.

Property 6: Modulation

This property is in fact a restatement of Property 5, but due to its importance to communication systems, it is considered independently. It states that

$$\mathcal{F}\{x(t)\cos(\omega_0 t)\} = \frac{1}{2}[X(j(\omega+\omega_0)) + X(j(\omega-\omega_0))]$$

$$\mathcal{F}\{x(t)\sin(\omega_0 t)\} = \frac{j}{2}[X(j(\omega+\omega_0)) - X(j(\omega-\omega_0))]$$

(3.32)

Proof. The proof of these relations follows from the Euler formulas for cosine and sine functions and the frequency shifting property:

$$\mathcal{F}\{x(t)\cos(\omega_0 t)\} = \mathcal{F}\left\{x(t)\left(\frac{e^{j\omega_0 t} + e^{-j\omega_0 t}}{2}\right)\right\}$$

$$= \frac{1}{2}\mathcal{F}\{x(t)e^{j\omega_0 t}\} + \frac{1}{2}\mathcal{F}\{x(t)e^{-j\omega_0 t}\}$$

$$= \frac{1}{2}X(j(\omega-\omega_0)) + \frac{1}{2}X(j(\omega+\omega_0))$$

Similarly,

$$\mathcal{F}\{x(t)\sin(\omega_0 t)\} = \mathcal{F}\left\{x(t)\left(\frac{e^{j\omega_0 t} - e^{-j\omega_0 t}}{2j}\right)\right\}$$

$$= \frac{1}{2j}\mathcal{F}\{x(t)e^{j\omega_0 t}\} - \frac{1}{2j}\mathcal{F}\{x(t)e^{-j\omega_0 t}\}$$

$$= -\frac{j}{2}X(j(\omega-\omega_0)) + \frac{j}{2}X(j(\omega+\omega_0))$$

The angular frequency ω_0 is called the *signal carrier angular frequency*. □

The application of the modulation property in communication systems for simultaneous transmission of multiple signals over a single physical communication channel will be discussed in detail in the section on signal modulation in Chapter 10. Here, we demonstrate signal modulation using an example of a telephone network that must simultaneously transmit many telephone signals (calls) over the same channel.

EXAMPLE 3.8

It is well known from basic physics and electrical engineering courses that speech frequencies fall approximately between 300–3400 Hz. In practice, the frequency band from 0–4000 Hz = 4 kHz is used for transmission of telephone signals. By modulating each telephone signal with the carrier frequencies, respectively given by 0 kHz, 4 kHz, 8 kHz, 12 kHz, ..., we are able to transmit many telephone signals over the same channel. Assuming that the channel can transmit signals without major distortion up to the frequencies of $N \times 4$ kHz, then it is possible to stock N telephone signals in the channel such that

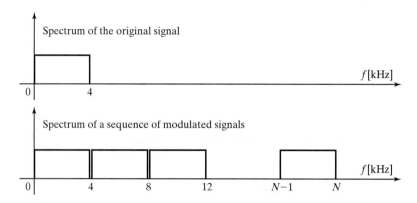

FIGURE 3.8: Transmission of N modulated telephone signals over the same channel

their modulated frequencies do not overlap, as demonstrated in Figure 3.8. For physical convenience, in this example we have given the presentation in terms of frequencies rather than angular frequencies (which we have likewise done in most of this chapter).

Property 7: Derivative

From the system theory point of view, the derivative property is one of the most important properties of the Fourier transform. It states that the *time derivative* of a signal $x(t)$ corresponds to the multiplication by $j\omega$ of the signal's frequency representation (Fourier transform) $X(j\omega)$, or in general,

$$\mathcal{F}\left\{\frac{d^n x(t)}{dt^n}\right\} = (j\omega)^n X(j\omega), \quad n = 1, 2, 3, \ldots \quad (3.33)$$

Proof. This useful property can easily be justified by starting with the definition of the inverse Fourier transform and taking derivatives of both sides with respect to time:

$$x(t) = \frac{1}{2\pi} \int_{-\infty}^{\infty} X(j\omega) e^{j\omega t} d\omega$$

$$\frac{dx(t)}{dt} = \frac{1}{2\pi} \int_{-\infty}^{\infty} (j\omega) X(j\omega) e^{j\omega t} d\omega$$

$$\vdots$$

$$\frac{d^n x(t)}{dt^n} = \frac{1}{2\pi} \int_{-\infty}^{\infty} (j\omega)^n X(j\omega) e^{j\omega t} d\omega$$

It can be observed from the last equation that $d^n x(t)/dt^n$ and $(j\omega)^n X(j\omega)$ are the Fourier transform pair, which proves the derivative property. □

Similarly to the time differentiation property of the Fourier transform given in (3.33), we can establish the corresponding frequency differentiation property as

$$\frac{d^n}{d\omega^n} X(j\omega) = \mathcal{F}\{(-jt)^n x(t)\} \tag{3.34}$$

which can be proved by taking derivatives of the definition integral (3.17) with respect to ω. (See also the proof of the time multiplication property.) The frequency differentiation property is far less important for linear dynamic systems than the time differentiation property. In fact, it is equivalent to the time multiplication property of the Fourier transform. That is why in this text, when we refer to the differentiation property of the Fourier transform, we mean the time differentiation property.

Property 8: Time Convolution
The time convolution of two signals is defined in Section 2.2 as

$$x_1(t) * x_2(t) = \int_{-\infty}^{\infty} x_1(t-\tau) x_2(\tau) \, d\tau, \quad -\infty \leq t \leq \infty$$

The Fourier transform of the convolution integral produces a very interesting result, namely

$$\mathcal{F}\{x_1(t) * x_2(t)\} = X_1(j\omega) X_2(j\omega) \tag{3.35}$$

Proof. This can be proved by starting with the definition integral of the Fourier transform,

$$\mathcal{F}\{x_1(t) * x_2(t)\} = \int_{-\infty}^{\infty} \left[\int_{-\infty}^{\infty} x_1(t-\tau) x_2(\tau) \, d\tau \right] e^{-j\omega t} \, dt$$

interchanging the order of integration,

$$\mathcal{F}\{x_1(t) * x_2(t)\} = \int_{-\infty}^{\infty} \left[\int_{-\infty}^{\infty} x_1(t-\tau) e^{-j\omega t} \, dt \right] x_2(\tau) \, d\tau$$

and using the time shifting property,

$$\mathcal{F}\{x_1(t) * x_2(t)\} = X_1(j\omega) \int_{-\infty}^{\infty} x_2(\tau) e^{-j\omega \tau} \, d\tau = X_1(j\omega) X_2(j\omega)$$

A simple and memorable fact, expressed in (3.35), is that *the convolution in the time domain corresponds to a product in the frequency domain*. □

Property 9: Frequency Convolution

A result analogous to the time domain convolution can be similarly established for the frequency domain convolution. It states that *the Fourier transform of a product of two signals in time is proportional to the convolution of their Fourier transforms in the frequency domain*:

$$\mathcal{F}\{x_1(t)x_2(t)\} = \frac{1}{2\pi} X_1(j\omega) * X_2(j\omega) = \frac{1}{2\pi} \int_{-\infty}^{\infty} X_1(j(\omega - \lambda)) X_2(j\lambda)\, d\lambda \quad (3.36)$$

Proof. The proof of this property is similar to the proof of the time convolution. It starts with

$$\mathcal{F}^{-1}\{X_1(j\omega) * X_2(j\omega)\} = \frac{1}{2\pi} \int_{-\infty}^{\infty} \left[\frac{1}{2\pi} \int_{-\infty}^{\infty} X_1(j(\omega - \lambda)) X_2(j\lambda)\, d\lambda \right] e^{j\omega t}\, d\omega$$

and requires the interchange of the integrals and the application of the frequency shifting property, that is,

$$\mathcal{F}^{-1}\{X_1(j\omega) * X_2(j\omega)\} = \frac{1}{2\pi} \int_{-\infty}^{\infty} X_2(j\lambda) \left[\frac{1}{2\pi} \int_{-\infty}^{\infty} X_1(j(\omega - \lambda)) e^{j\omega t}\, d\omega \right] d\lambda$$

$$= \frac{1}{2\pi} \int_{-\infty}^{\infty} X_2(j\lambda) x_1(t) e^{j\lambda t}\, d\lambda = x_1(t) x_2(t) \qquad \square$$

Parseval's Theorem From the frequency convolution property of the Fourier transform, we can establish the relationship between signal energy in the time and frequency domains. We have seen in Section 2.3, formula (2.52), that the total signal energy in the time domain is defined by

$$E_\infty = \int_{-\infty}^{\infty} |x(t)|^2\, dt \quad (3.37)$$

From (3.36), we have

$$\mathcal{F}\{x_1(t)x_2(t)\} = \int_{-\infty}^{\infty} x_1(t) x_2(t) e^{-j\omega t}\, dt = \frac{1}{2\pi} \int_{-\infty}^{\infty} X_1(j(\omega - \lambda)) X_2(j\lambda)\, d\lambda$$

Since this is valid for any ω, it must be valid for $\omega = 0$. Thus, we have

$$\int_{-\infty}^{\infty} x_1(t) x_2(t)\, dt = \frac{1}{2\pi} \int_{-\infty}^{\infty} X_1(-j\lambda) X_2(j\lambda)\, d\lambda \quad (3.38)$$

Let us assume that $x_2(t) = x_1(t) = x(t)$, with $x(t)$ being a real function such that $|x(t)|^2 = x^2(t)$. The expression that relates the signal energy in the time and frequency domains follows from

$$E_\infty = \int_{-\infty}^{\infty} x^2(t)\,dt = \frac{1}{2\pi} \int_{-\infty}^{\infty} X(-j\lambda)X(j\lambda)\,d\lambda$$

$$= \frac{1}{2\pi} \int_{-\infty}^{\infty} |X(j\lambda)|^2 d\lambda = \int_{-\infty}^{\infty} |X(jf)|^2 df \quad (3.39)$$

Note that $\lambda = 2\pi f$, a dummy variable of integration, plays the role of the angular frequency. The result established in (3.39) is also known as Rayleigh's energy theorem. The quantity $|X(jf)|^2$ is known as the *energy spectrum*.[†] In (3.39), we used the fact that $X(-j\lambda) = X^*(j\lambda)$, that is,

$$X(-j\lambda) = \int_{-\infty}^{\infty} x(t)e^{-j(-\lambda)t}\,dt = \int_{-\infty}^{\infty} x(t)e^{j\lambda t}\,dt = X^*(j\lambda)$$

and that $X(j\lambda)X^*(j\lambda) = |X(j\lambda)|^2$, where * stands for the complex conjugate. In some texts the relationship $X(-j\lambda) = X^*(j\lambda)$ is called the *conjugate symmetry property* of the Fourier transform (see, for example, [4], page 179). Formula (3.39) is known as *Parseval's theorem*. It has great importance in signal processing and communications.

Property 10: Time Reversal

In some applications, one must deal with the signal $x(-t)$. By the definition integral (3.17), its Fourier transform is given by

$$\mathcal{F}\{x(-t)\} = \int_{t=-\infty}^{t=\infty} x(-t)e^{-j\omega t}\,dt$$

Proof. Introducing the change of variables as $\sigma = -t$, this integral becomes

$$\mathcal{F}\{x(-t)\} = \int_{\sigma=+\infty}^{\sigma=-\infty} x(\sigma)e^{-j\omega(-\sigma)}d(-\sigma) = \int_{\sigma=-\infty}^{\sigma=\infty} x(\sigma)e^{-j(-\omega)\sigma}d\sigma = X(-j\omega)$$

Thus, we have established the Fourier transform pair

$$x(-t) \leftrightarrow X(-j\omega) \quad (3.40)$$

Note that (3.40) is a special case of (3.29) obtained for $a = -1$. □

[†] In the literature, $|X(j\omega)|^2$ is also called the *signal energy spectrum* (see Subsection 3.3.2).

Property 11: Duality

This very powerful property will help us to find the Fourier transforms of signals that are not absolutely integrable—those signals that do not satisfy existence condition (3.21). The duality property can be simply derived from the definition of the Fourier transform (3.17), that is,

$$X(j\omega) = \mathcal{F}\{x(t)\} \triangleq \int_{-\infty}^{\infty} x(t)e^{-j\omega t}\,dt$$

Since in this integral t is a dummy variable and ω is a parameter, they can be replaced by new variables. For the clarity of the derivations that follow, the integral is first represented as

$$X(jv) = \mathcal{F}\{x(t)\} \triangleq \int_{-\infty}^{\infty} x(t)e^{-jvt}\,dt$$

in which we have changed the parameter ω into the parameter v. In this integral, t is a dummy variable and v is a parameter, so that they can be replaced by new variables. If we introduce the change of variables $t = -\omega$ in the preceding integral, we will have

$$X(jv) = -\int_{\omega=+\infty}^{\omega=-\infty} x(-\omega)e^{-jv(-\omega)}\,d\omega = \frac{1}{2\pi}\int_{\omega=-\infty}^{\omega=\infty} 2\pi x(-\omega)e^{jv\omega}\,d\omega$$

Introducing another change of variables as $v = t$, we obtain

$$X(jt) = \frac{1}{2\pi}\int_{-\infty}^{\infty} 2\pi x(-\omega)e^{j\omega t}\,d\omega$$

Hence, the time domain signal $X(jt)$ and the frequency signal $2\pi x(-\omega)$ form the Fourier transform pair

$$X(jt) \leftrightarrow 2\pi x(-\omega) \tag{3.41}$$

Similarly, by setting $t = \omega$ and $v = -t$ in (3.17), we can obtain the Fourier transform pair

$$X(-jt) \leftrightarrow 2\pi x(\omega) \tag{3.42}$$

This also follows from (3.41) and the time reversal property. Relations (3.41) and (3.42) represent the duality property of the Fourier transform. This is also known in the literature as the *symmetry property* (see, for example, [5], page 14; [6], page 146). With this property we are able to find the Fourier transform of many signals that do not satisfy existence condition (3.21). Fourier transforms of signals that do not satisfy (3.21) are called *generalized Fourier transforms*.

The duality property is summarized in Table 3.1.

TABLE 3.1 Duality property of the Fourier transform.

Time domain function	Frequency domain function (Fourier transform)
$x(t)$	$X(j\omega)$
$X(jt)$	$2\pi x(-\omega)$
$X(-jt)$	$2\pi x(\omega)$

Note that for real time domain signals, $X(jt) = X(t)$ and $X(-jt) = X(-t)$.

EXAMPLE 3.9

In Example 3.3, we obtained $\mathcal{F}\{\delta(t)\} = 1$. By the duality property,

$$X(jt) = X(t) = 1 \leftrightarrow 2\pi\delta(-\omega) = 2\pi\delta(\omega) \tag{3.43}$$

From (3.43) we have the Fourier transform pair

$$\text{const} \leftrightarrow \text{const} \times 2\pi\delta(\omega) \tag{3.44}$$

This can be also justified by starting with the definition of the Fourier inverse

$$x(t) = \frac{1}{2\pi} \int_{-\infty}^{\infty} X(j\omega)e^{j\omega t} d\omega$$

and setting $X(j\omega) = 2\pi\delta(\omega)$, which leads to

$$x(t) = \frac{1}{2\pi} \int_{-\infty}^{\infty} 2\pi\delta(\omega)e^{j\omega t} d\omega = 1$$

indicating that $2\pi\delta(\omega)$ and 1 form the Fourier transform pair.

Comment: It is interesting to observe that the Fourier transform pair $1 \leftrightarrow 2\pi\delta(\omega)$ implies

$$2\pi\delta(\omega) = \int_{-\infty}^{\infty} 1 e^{-j\omega t} dt = \int_{-\infty}^{\infty} \cos(\omega t) dt - j \int_{-\infty}^{\infty} \sin(\omega t) dt$$

Since the integral involving the sine function is 0 (due to the fact that sine is an odd function), it follows that

$$\int_{-\infty}^{\infty} \cos(\omega \tau) d\tau = 2\pi\delta(\omega) \tag{3.45}$$

It is known from calculus that this integral does not exist in the sense of ordinary functions. However, in the sense of generalized functions, the integral exists and is given in terms of the frequency domain impulse delta signal. ∫

EXAMPLE 3.10

Using (3.43) and the modulation property, we are able to find generalized Fourier transforms of the cosine and sine functions,

$$1 \times \cos \omega_0 t \leftrightarrow \frac{1}{2}[2\pi \delta(\omega + \omega_0) + 2\pi \delta(\omega - \omega_0)] = \pi \delta(\omega + \omega_0) + \pi \delta(\omega - \omega_0)$$

(3.46)

and

$$1 \times \sin \omega_0 t \leftrightarrow \frac{j}{2}[2\pi \delta(\omega + \omega_0) - 2\pi \delta(\omega - \omega_0)] = j\pi \delta(\omega + \omega_0) - j\pi \delta(\omega - \omega_0)$$

(3.47)

We can verify these relations by going the other way around, that is,

$$\frac{1}{2\pi} \int_{-\infty}^{\infty} \pi[\delta(\omega + \omega_0) + \delta(\omega - \omega_0)]e^{j\omega t} d\omega = \frac{1}{2}(e^{j\omega_0 t} + e^{-j\omega_0 t}) = \cos(\omega_0 t)$$

Similarly, using the reverse order of arguments, we can verify the expression for the Fourier transform of the sine function as established in (3.47). ∫

Fourier Transform of Periodic Signals Periodic signals can be represented by the exponential form of the Fourier series defined in (3.14), that is

$$x(t) = \sum_{n=-\infty}^{\infty} X_n(jn\omega_0)e^{jn\omega_0 t}, \quad \omega_0 = \frac{2\pi}{T}$$

where $X_n(jn\omega_0)$ are constants (in general, complex). By the frequency shifting property and (3.44), we have

$$x(t) = \sum_{n=-\infty}^{\infty} X_n(jn\omega_0)e^{jn\omega_0 t} \leftrightarrow 2\pi \sum_{n=-\infty}^{\infty} X_n(jn\omega_0)\delta(\omega - n\omega_0)$$

(3.48)

Hence, in the frequency domain, a periodic signal is represented by an infinite sum of equidistantly distributed impulse delta signals.

EXAMPLE 3.11

The train of time domain impulse delta signals defined by

$$x(t) = \sum_{n=-\infty}^{n=\infty} \delta(t - nT_0), \quad T_0 = \frac{2\pi}{\omega_0}$$

(3.49)

is a periodic signal with the period T_0. This train is used in signal processing and discrete-time systems. Note that this signal satisfies condition (3.12), hence it can be expressed in terms of convergent Fourier series. The Fourier series coefficients of the train of impulse delta signals are given by

$$X_n(jn\omega_0) = \frac{1}{T_0} \int_{-T_0/2}^{T_0/2} \delta(t) e^{-jn\omega_0 t} \, dt = \frac{1}{T_0}$$

which implies that

$$x(t) = \sum_{n=-\infty}^{n=\infty} \delta(t - nT_0) = \frac{1}{T_0} \sum_{n=-\infty}^{n=\infty} e^{jn\omega_0 t}, \quad T_0 = \frac{2\pi}{\omega_0} \qquad (3.50)$$

The corresponding Fourier transform, according to formulas (3.48) and (3.50), is

$$\mathcal{F}\left\{\sum_{n=-\infty}^{n=\infty} \delta(t - nT_0)\right\} = \mathcal{F}\left\{\frac{1}{T_0} \sum_{n=-\infty}^{n=\infty} e^{jn\omega_0 t}\right\} = \omega_0 \sum_{n=-\infty}^{n=\infty} \delta(\omega - n\omega_0) \qquad (3.51)$$

Similarly, expanding the frequency domain train of impulse delta functions into the Fourier series, we have

$$\sum_{n=-\infty}^{n=\infty} \delta(\omega - n\omega_0) = \frac{1}{\omega_0} \sum_{n=-\infty}^{n=\infty} e^{-jnT_0\omega}, \quad T_0 = \frac{2\pi}{\omega_0} \qquad (3.52)$$

EXAMPLE 3.12 Fourier Transform of the Heaviside Unit Step Signal

The Fourier transform of the Heaviside unit step signal is extremely important; in addition to its use in system analysis, it helps to establish and prove the integral property of the Fourier transform. It is also used to derive the Hilbert transform (see Chapter 10), an important transform for communication systems. We consider first the Fourier transform of the sign function, defined in (2.3). Using the notion of the generalized derivative, it can be observed that

$$\frac{D \operatorname{sgn}(t)}{Dt} = [\operatorname{sgn}(0^+) - \operatorname{sgn}(0^-)]\delta(t) = 2\delta(t) \qquad (3.53)$$

By the time derivative property of the Fourier transform, we have

$$(j\omega)\mathcal{F}\{\operatorname{sgn}(t)\} = 2\mathcal{F}\{\delta(t)\} = 2 \implies \mathcal{F}\{\operatorname{sgn}(t)\} = \frac{2}{j\omega} \qquad (3.54)$$

102 Chapter 3 Fourier Series and Fourier Transform

Using formula (2.4), which relates the Heaviside unit step signal and the sign function, we obtain[†]

$$\mathcal{F}\{u_h(t)\} = \mathcal{F}\left\{\frac{1}{2} + \frac{1}{2}\mathrm{sgn}(t)\right\} = \pi\delta(\omega) + \frac{1}{j\omega} \quad (3.55)$$

Note that the time domain constant 0.5 produces the term $\pi\delta(\omega)$ in the frequency domain.

In Section 3.2.2 on the inverse Fourier transform, we will show that the Fourier inverse of $\pi\delta(\omega) + 1/j\omega$ indeed produces the value of $1/2 = u_h(0)$ in the time domain at $t = 0$. Thus, the Fourier inverse of $\pi\delta(\omega) + 1/j\omega$ gives neither $1 = u(0)$ nor an undefined value $u_1(0)$, as might be concluded if one uses the $u(t)$ and $u_1(t)$ definitions of the unit step signals as given in (2.1a–b).

Property 12: Integration

The integration property of the Fourier transform states: Let the signal $x(t)$ have the Fourier transform $X(j\omega)$, then the Fourier transform of its integral is given by

$$\mathcal{F}\left\{\int_{-\infty}^{t} x(\tau)\,d\tau\right\} = \frac{1}{j\omega}X(j\omega) + \pi X(j0)\delta(\omega) \quad (3.56)$$

Proof. The preceding integral can be represented by the convolution formula of the original signal and the Heaviside unit step signal as follows,

$$\int_{-\infty}^{t} x(\tau)\,d\tau = \int_{-\infty}^{\infty} x(\tau)u_h(t-\tau)\,d\tau = x(t) * u_h(t) \quad (3.57)$$

By using the time convolution property of the Fourier transform, we have

$$x(t) * u_h(t) \leftrightarrow X(j\omega)U_h(j\omega) \quad (3.58)$$

Hence

$$\mathcal{F}\left\{\int_{-\infty}^{t} x(\tau)\,d\tau\right\} = X(j\omega)\left[\frac{1}{j\omega} + \pi\delta(\omega)\right] = \frac{1}{j\omega}X(j\omega) + \pi X(j0)\delta(\omega) \quad (3.59)$$

□

EXAMPLE 3.13

In this nice example we find the Fourier transform of the triangular pulse by using knowledge about the Fourier transform of the rectangular pulse (which we obtained in Example 3.5), and the integration and time shifting properties of the Fourier transform.

[†]It should be pointed out that several standard undergraduate text on linear systems wrongly state that $u(t) = 0.5 + 0.5\,\mathrm{sgn}(t)$, with $u(t)$ as defined in (2.1a); see Problem 2.1.

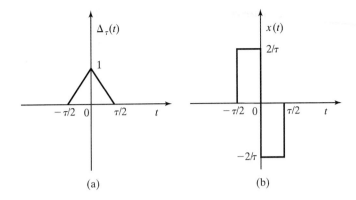

FIGURE 3.9: (a) Triangular pulse and (b) its derivative

We first define the signal $x(t)$, equal to the derivative of the triangular pulse, as

$$x(t) = \frac{d\Delta_\tau(t)}{dt} = \frac{1}{dt}d\left(1 - \frac{2|t|}{\tau}\right)p_\tau(t)$$

The triangular pulse and its derivative are presented in Figure 3.9. It is easy to see that this signal can be expressed in terms of rectangular pulses as

$$x(t) = \frac{2}{\tau}p_{\tau/2}\left(t + \frac{\tau}{4}\right) - \frac{2}{\tau}p_{\tau/2}\left(t - \frac{\tau}{4}\right)$$

Hence, its Fourier transform is given by

$$\mathcal{F}\{x(t)\} = \frac{2}{\tau}\left[e^{j\omega\frac{\tau}{4}}\mathcal{F}\{p_{\frac{\tau}{2}}(t)\} - e^{-j\omega\frac{\tau}{4}}\mathcal{F}\{p_{\frac{\tau}{2}}(t)\}\right]$$

$$= \frac{2}{\tau}[e^{j\omega\frac{\tau}{4}} - e^{-j\omega\frac{\tau}{4}}]\frac{\tau}{2}\frac{\sin(\omega\frac{\tau}{4})}{\omega\frac{\tau}{4}} = j2\sin\left(\omega\frac{\tau}{4}\right)\frac{\sin(\omega\frac{\tau}{4})}{\omega\frac{\tau}{4}} = X(j\omega)$$

The original triangular pulse is the integral of the signal $x(t)$. By the integral property of the Fourier transform, we have

$$\mathcal{F}\{\Delta_\tau(t)\} = \frac{1}{j\omega}X(j\omega) + \pi X(0)\delta(\omega)$$

$$= \frac{2}{\omega}\sin\left(\omega\frac{\tau}{4}\right)\frac{\sin(\omega\frac{\tau}{4})}{\omega\frac{\tau}{4}} + 0\pi\delta(\omega) = \frac{\tau}{2}\text{sinc}^2\left(\omega\frac{\tau}{4}\right) \quad (3.60)$$

With the integral property we have completed the list of the twelve important properties of the Fourier transform. (A summary of these properties is given in Table 3.3, at the end of this chapter.)

3.2.2 Inverse Fourier Transform

The definition of the inverse Fourier transform is given in formula (3.18). It can be seen from (3.18) that the inverse Fourier transform is given in terms of an integral involving the complex variable $j\omega$. Since complex variable integration is not an easy task, we will use Table 3.4 (common Fourier transform pairs) to find the inverse Fourier transform. Combined with Table 3.3 (properties of the Fourier transform), Table 3.4 will allow us to find the inverse Fourier transform of all important signals (functions) encountered in this book. Thus, the problem of finding the inverse Fourier transform will be replaced by a quite simple recognition problem involving tables of the common Fourier transform pairs and properties.

In the following example we demonstrate the procedure for finding the inverse Fourier transform by performing direct integration using formula (3.18) on the example of the Heaviside unit step signal.

EXAMPLE 3.14

Let us find the inverse Fourier transform by using the definition integral (3.18) of the frequency domain signal obtained in (3.55), which in fact represents the Fourier transform of the Heaviside unit step signal. It follows directly that

$$\mathcal{F}^{-1}\left\{\frac{1}{j\omega}+\pi\delta(\omega)\right\} = \frac{1}{2\pi}\int_{-\infty}^{\infty}\left[\frac{1}{j\omega}+\pi\delta(\omega)\right]e^{j\omega t}d\omega$$

$$= \frac{1}{2\pi}\pi + \frac{1}{2\pi}\int_{-\infty}^{\infty}\frac{e^{j\omega t}}{j\omega}d\omega$$

$$= \frac{1}{2} + \frac{1}{2\pi j}\int_{-\infty}^{\infty}\frac{\cos(\omega t)}{\omega}d\omega + \frac{1}{2\pi}\int_{-\infty}^{\infty}\frac{\sin(\omega t)}{\omega}d\omega$$

Since $\cos(\omega t)/\omega$ is an odd function (because the ratio of even and odd functions is an odd function), the corresponding integral is equal to 0. The integrand in the second integral is an even function, hence

$$\int_{-\infty}^{\infty}\frac{\sin(\omega t)}{\omega}d\omega = 2\int_{0}^{\infty}\frac{\sin(\omega t)}{\omega}d\omega$$

The last integral can be found in standard tables of integrals (see, for example, [1], page 96). It is given by

$$\int_{0}^{\infty}\frac{\sin(\omega t)}{\omega}d\omega = \begin{cases}\pi/2, & t>0\\ 0, & t=0 \\ -\pi/2, & t<0\end{cases} = \frac{\pi}{2}\,\text{sgn}(t) \qquad (3.61)$$

Using (3.61) in the original derivations, we obtain

$$\mathcal{F}^{-1}\left\{\frac{1}{j\omega}+\pi\delta(\omega)\right\}=\frac{1}{2}+0+\frac{1}{2\pi}2\frac{\pi}{2}\operatorname{sgn}(t)=\frac{1}{2}+\frac{1}{2}\operatorname{sgn}(t) \quad (3.62)$$

Since $\operatorname{sgn}(0) = 0$, the preceding Fourier inverse produces the value of 0.5 at $t = 0$. As a matter of fact, the corresponding Fourier inverse represents the Heaviside unit step signal defined in (2.2) (see also (2.4)).

Remark: The result obtained in Example 3.14 is consistent with the general result of the inverse Fourier transform theory for signals that have jump discontinuities. Namely, that result states that the inverse Fourier transform recovers in the time domain the average value of the signal at the point of jump discontinuity, that is, $(x(t_d^+) + x(t_d^-))/2$, where t_d indicates the point at which a jump discontinuity occurs. Since $u_h(0^-) = 0$ and $u_h(0^+) = 1$, the inverse Fourier transform produces at $t = 0$ the value $(u_h(0^+) + u_h(0^-))/2 = 1/2$.

Similarly, it can be shown that the Fourier inverse of $\tau\operatorname{sinc}(\frac{\omega\tau}{2\pi})$ implies in the time domain $p_\tau^h(t)$. Hence $p_\tau^h(t)$ and $\tau\operatorname{sinc}(\frac{\omega\tau}{2\pi})$ form the corresponding Fourier transform pair, that is,

$$p_\tau^h(t) \leftrightarrow \tau\operatorname{sinc}\left(\frac{\omega\tau}{2\pi}\right) \quad (3.63)$$

Both $p_\tau(t)$ and $p_\tau^h(t)$ have the same Fourier transform, as observed in Example 3.5, but

$$\mathcal{F}^{-1}\left\{\tau\operatorname{sinc}\left(\frac{\omega\tau}{2\pi}\right)\right\} \neq p_\tau(t)$$

It can be concluded that $u(t)$, $u_1(t)$, and $u_h(t)$ have the same Fourier transform, but only

$$u_h(t) \leftrightarrow \left(\frac{1}{j\omega}\right) + \pi\delta(\omega) \quad (3.64)$$

is the Fourier transform pair.

Note that the exponential signal defined in formula (3.24) of Example 3.4 also has a jump discontinuity at 0. Namely, $e(0^-) = 0$ and $e(0^+) = 1$. In that example, we found that $\mathcal{F}\{e(t)\} = 1/(1 + j\omega)$. The Fourier inverse of $1/(1 + j\omega)$ produces at $t = 0$ the value equal to 0.5. It follows that the signal $e^{-at}u_h(t)$ and $1/(1 + j\omega)$ are indeed the Fourier transform pair, even though both signals $e(t)$ and $e^{-at}u_h(t)$ have the same Fourier transform, equal to $1/(1 + j\omega)$. Hence, we have

$$e^{-at}u_h(t) \leftrightarrow \frac{1}{a + j\omega}, \quad a > 0 \quad (3.65)$$

In the next example we demonstrate how to use Tables 3.3 and 3.4 to find the inverse Fourier transform.

EXAMPLE 3.15

Consider the Fourier transform

$$X(j\omega) = \frac{5}{j(\omega - 3)}$$

From Table 3.4, we see that

$$\frac{2}{j\omega} \leftrightarrow \text{sgn}(t)$$

The frequency shifting property (Table 3.3, row 5) indicates that

$$\frac{2}{j(\omega - 3)} \leftrightarrow \text{sgn}(t)e^{j3t}$$

It remains to fix only the constant, which by the linearity property (Table 3.3, row 1) implies

$$\frac{5}{j(\omega - 3)} \leftrightarrow \frac{5}{2}\text{sgn}(t)e^{j3t} \qquad ◊$$

EXAMPLE 3.16

Let us find the Fourier transform of $p_2^h(\omega)$. In formula (3.63), we established the Fourier transform pair

$$p_\tau^h(t) \leftrightarrow \tau \text{sinc}\left(\frac{\tau}{2\pi}\omega\right)$$

Using the duality property, we have

$$X(jt) = \tau \text{sinc}\left(\frac{t\tau}{2\pi}\right) \leftrightarrow 2\pi p_\tau^h(-\omega) = 2\pi p_\tau^h(\omega)$$

which, for $\tau = 2$, implies

$$p_2^h(\omega) \leftrightarrow \frac{1}{\pi}\text{sinc}\left(\frac{t}{\pi}\right) \qquad ◊$$

In the next section, we will demonstrate the use of the Fourier transform linearity, derivative, and convolution properties in the analysis of linear time-invariant systems. We will develop the main linear system theory concept, the convolution concept, which will give the answer to the general problem of finding the linear time-invariant system response due to arbitrary input signals. We will also specialize the result obtained in Section 3.3 to periodic and sinusoidal input signals (Section 3.4).

3.3 FOURIER TRANSFORM IN SYSTEM ANALYSIS

In this section we study the use of the Fourier transform in linear dynamic system analysis. The main goal in analysis of any dynamic system is to find its response. We saw in Chapter 1 that the system response has two components: zero-input response due to system initial conditions, and zero-state response due to external forcing functions. In general, *the system response obtained by using the Fourier transform produces only the zero-state response,* since the *Fourier transform is not able to handle the system initial conditions.* Thus, in this section, and in this chapter, it is assumed that the *system initial conditions are* set to zero. We will see in the next section that only when the input signal is periodic or sinusoidal does the Fourier analysis produce the steady state system response.

Analysis of linear time-invariant systems in the frequency domain (complex domain) via the use of the Fourier transform is based on the concept of the system transfer function. The system transfer function will be introduced in the following subsection, in which we also derive the general expression for the linear system zero-state response due to any input. In addition, we introduce the notions of the magnitude and phase spectra of continuous-time linear dynamic systems and signals (in general, aperiodic signals).

3.3.1 System Transfer Function and System Response

The Fourier transform helps us to introduce a very important concept of linear system theory, the system transfer function. Taking the Fourier transform of both sides of a linear differential equation that describes the dynamical behavior of an nth-order system introduced in Chapter 1 in (1.9), that is,

$$\mathcal{F}\left\{\frac{d^n y(t)}{dt^n} + a_{n-1}\frac{d^{n-1} y(t)}{dt^{n-1}} + a_{n-2}\frac{d^{n-2} y(t)}{dt^{n-2}} + \cdots + a_1\frac{dy(t)}{dt} + a_0 y(t)\right\}$$
$$= \mathcal{F}\left\{b_m \frac{d^m x(t)}{dt^m} + b_{m-1}\frac{d^{m-1} x(t)}{dt^{m-1}} + \cdots + b_1\frac{dx(t)}{dt} + b_0 x(t)\right\}$$

and using the derivative property of the Fourier transform, we get

$$\{(j\omega)^n + a_{n-1}(j\omega)^{n-1} + \cdots + a_1(j\omega) + a_0\}Y(j\omega)$$
$$= \{b_m(j\omega)^m + b_{m-1}(j\omega)^{m-1} + \cdots + b_1(j\omega) + b_0\}X(j\omega)$$

This can be written in the form

$$Y(j\omega) = \frac{\{b_m(j\omega)^m + b_{m-1}(j\omega)^{m-1} + \cdots + b_1(j\omega) + b_0\}}{\{(j\omega)^n + a_{n-1}(j\omega)^{n-1} + \cdots + a_1(j\omega) + a_0\}} X(j\omega)$$

The quantity

$$H(j\omega) \triangleq \frac{b_m(j\omega)^m + b_{m-1}(j\omega)^{m-1} + \cdots + b_1(j\omega) + b_0}{(j\omega)^n + a_{n-1}(j\omega)^{n-1} + \cdots + a_1(j\omega) + a_0} \qquad (3.66)$$

defines the *system transfer function*.

It follows from the preceding derivations that the solution of the differential equation (1.9), which represents the system response (system output) due to the given input $x(t)$ and *zero system initial conditions*, can be obtained in the frequency domain by using a very simple formula,

$$Y_{zs}(j\omega) = H(j\omega)X(j\omega) \qquad (3.67)$$

Note that the solution obtained represents the *zero-state response* of the system described by differential equation (1.9). Bearing that in mind for the impulse delta signal $X(j\omega) = 1$, we see that the system response to the impulse delta signal is equal to $\mathcal{F}^{-1}\{H(j\omega)\}$. The system response to the impulse delta signal is called the *system impulse response*. It is mathematically defined by

$$h(t) \triangleq \mathcal{F}^{-1}\{H(j\omega)\} \tag{3.68}$$

By the convolution property of the Fourier transform, which states that a product in the frequency domain corresponds to the convolution in the time domain, we can conclude from (3.67) that

$$y_{zs}(t) = h(t) * x(t) = \int_{-\infty}^{\infty} h(\tau)x(t-\tau)\,d\tau \tag{3.69}$$

The formula (3.69) states that *the system output under zero system initial conditions is equal to the convolution of the system input and the system impulse response.*

In (3.67–69) we have established the most fundamental results of linear dynamic system theory. From these results, we see that in order to find the system response due to any input, we must first find the system response due to the impulse delta signal, and then convolve the obtained system impulse response with the given system input signal. Note that the *system impulse response is obtained (and defined) for the system at rest* (zero initial conditions). It should also be emphasized that for a given time-invariant system, the impulse response must be found only once. In other words, any linear time-invariant system is uniquely characterized by its impulse response (or by its transfer function in the frequency domain). In general, it is easy to find the system impulse response—see (3.68). A formula analogous to (3.68) for finding the system impulse response will be derived in the next chapter by using the Laplace transform—that formula represents the most efficient way for finding the impulse response of continuous-time linear time-invariant systems.

We have seen that the Fourier transforms of many common signals used in linear system theory are not defined in terms of ordinary functions. In the next chapter we will facilitate this problem by introducing the Laplace transform. The Laplace transform is basically the Fourier transform of the function $x(t)e^{-\sigma t}$, where the term $e^{-\sigma t}$ is known as the convergence factor. For example, for $t > 0$, $\sigma > 0$, the exponentially decaying factor multiplying the function $x(t)$ helps the convergence of the corresponding integral, hence many functions that are not Fourier integrable in the ordinary sense become Laplace integrable in the ordinary sense. Furthermore, from the system analysis point of view, it is extremely important that the Laplace transform is able to handle the system initial conditions and to produce the system response due to both initial conditions and external forcing functions. Since the Laplace transform has basically the same form as the Fourier transform, in the next chapter we will show that many properties of the Fourier transform extend to the Laplace transform. Especially, we will show that results analogous to (3.67–69) hold for the Laplace transform.

The Laplace transform will be our main tool for finding the complete (zero-input and zero-state) system response due to arbitrary system inputs and system initial conditions. In this chapter, in Sections 3.4.1 and 3.4.2, we will show how to use formula (3.67) to efficiently find the system steady state response due to periodic and sinusoidal inputs, respectively.

In the next example, we demonstrate the use of formulas (3.67–69) to find the zero-state response and to indicate its difference from the steady state system response.

EXAMPLE 3.17

Consider the following first-order system at rest:

$$\frac{dy(t)}{dt} + y(t) = x(t) = u_h(t)$$

driven by the step input. It is easy to see that the system transfer function and the corresponding system impulse response are given by

$$H(j\omega) = \frac{1}{j\omega + 1} \Rightarrow h(t) = e^{-t}u_h(t)$$

The corresponding step response in the frequency domain, obtained from (3.67), is

$$Y_{zs}(j\omega) = H(j\omega)X(j\omega) = \frac{1}{j\omega+1}\left(\frac{1}{j\omega} + \pi\delta(\omega)\right) = \frac{1}{(j\omega+1)j\omega} + \pi\delta(\omega)$$

$$= -\frac{1}{j\omega+1} + \frac{1}{j\omega} + \pi\delta(\omega)$$

Applying the inverse Fourier transformation, we have

$$y_{zs}(t) = -e^{-t}u_h(t) + u_h(t) = (1 - e^{-t})u_h(t)$$

Note that the same result can be obtained directly using the convolution integral (3.69), that is,

$$y_{zs}(t) = \int_{-\infty}^{\infty} e^{-\tau}u_h(\tau)u_h(t-\tau)\,d\tau = \int_0^t e^{-\tau}d\tau = 1 - e^{-t}, \quad t \geq 0$$

It is obvious that in this case $y_{ss}(t) = 1$, and that $-e^{-t}$, $t > 0$ represents the system transient response. It is interesting that despite the fact that all system initial conditions are zero, the system input has excited the term e^{-t} (often called the system natural mode). This term is in general excited by the system nonzero initial conditions.

The main purpose of this example is to demonstrate that the system response obtained via the Fourier transform is indeed the zero-state response, which in general differs from the steady state response. The methodology used in this example to find the system zero-state response will be further simplified in Chapter 4 with the use of the Laplace transform. ♣

Linear System Eigenfunction (Complex Exponential)

Often in linear theory and its applications we are faced with the term "system eigenfunction." This term is used for the complex exponential $e^{j\omega t}$, according to the terminology used within the theory of linear differential/integral operators. Though that theory is outside the scope of this text, we will explain the source of the name, and indicate the importance of the complex exponential in the study of linear dynamic systems.

Most undergraduate students are familiar with the eigenvalues and eigenvectors of a matrix, which are defined by

$$\mathbf{A}\mathbf{v}_i = \lambda_i \mathbf{v}_i, \quad \mathbf{v}_i \neq 0, \quad i = 1, 2, \ldots, n$$

where $\mathbf{A}$ is a square $n \times n$ matrix, $\mathbf{v}_i$ are n dimensional eigenvectors, and λ_i are scalars called the eigenvalues. Analogous to this static eigenvalue/eigenvector problem, a dynamic eigenvalue/eigenvector problem can be defined using linear differential/integral operators [2], as

$$\mathbf{L}\{x(t)\} = \lambda x(t)$$

where $\mathbf{L}$ stands for the linear differential/integral operator (in this case, our system), $x(t)$ is called the system eigenfunction, and λ is the system eigenvalue. In this formula, $\mathbf{L}\{x(t)\}$ should be read as "$\mathbf{L}$ acting on $x(t)$." With this brief introduction, we are ready to explain why $e^{j\omega t}$ is called the linear dynamic system eigenfunction.

Let us use the complex exponential $e^{j\omega t}$ as an input to the linear dynamic system (1.9). The corresponding zero-state response is $y_{zs}(t) = \mathbf{L}\{e^{j\omega t}\}$, with $\mathbf{L}$ representing the linear convolution operator defined in (3.69). It follows from (3.69) that

$$y_{zs}(t) = \mathbf{L}\{e^{j\omega t}\} = \int_{-\infty}^{\infty} h(\tau) e^{j\omega(t-\tau)} \, d\tau = e^{j\omega t} \int_{-\infty}^{\infty} h(\tau) e^{-j\omega \tau} \, d\tau = e^{j\omega t} H(j\omega) \quad (3.70)$$

Hence, for linear dynamic systems, we have the relationship

$$\mathbf{L}\{e^{j\omega t}\} = H(j\omega) e^{j\omega t} \quad (3.71)$$

indicating that $e^{j\omega t}$ is the system eigenfunction and $H(j\omega)$ plays the role of the system eigenvalue. Similarly, it can be shown that $e^{-j\omega t}$ is also the linear system eigenfunction with the eigenvalue equal to $H(-j\omega)$; see Problem 3.46.

It can be concluded from (3.70) that if the complex exponential $e^{j\omega t}$ is used as the system input, the system zero-state response will be equal to the same complex exponential multiplied by the complex number $H(j\omega)$ that in fact represents the system transfer function at the given angular frequency ω.

3.3.2 Frequency Spectra

The frequency spectra of a continuous-time system are represented by frequency plots of the magnitude and phase of its transfer function. Since the system transfer function for a given frequency is a complex number, its magnitude and phase (angle) depend on frequency; that is,

$$H(j\omega) = \text{Re}\{H(j\omega)\} + j\,\text{Im}\{H(j\omega)\} = |H(j\omega)| e^{j \angle H(j\omega)}$$

$$|H(j\omega)| = \sqrt{(\text{Re}\{H(j\omega)\})^2 + (\text{Im}\{H(j\omega)\})^2} \quad (3.72)$$

$$\angle H(j\omega) = \tan^{-1}\left(\frac{\text{Im}\{H(j\omega)\}}{\text{Re}\{H(j\omega)\}}\right)$$

Plotting $|H(j\omega)|$ and $\angle H(j\omega)$ in terms of frequencies ω produces the *system magnitude* and *phase* spectra. It can be shown that $|H(j\omega)|$ is an even function and $\angle H(j\omega)$ is an odd function of ω. These arguments are established as follows:

$$H(j\omega) = \int_{-\infty}^{\infty} h(t)e^{-j\omega t}\,dt \Rightarrow (H(j\omega))^* = \int_{-\infty}^{\infty} h(t)e^{j\omega t}\,dt = H(-j\omega)$$

Using (3.72), we have

$$H(-j\omega) = \text{Re}\{H(j\omega)\} - j\,\text{Im}\{H(j\omega)\}$$

$$\Rightarrow |H(-j\omega)| = \sqrt{(\text{Re}\{H(j\omega)\})^2 + (\text{Im}\{H(j\omega)\})^2} = |H(j\omega)|$$

$$\angle H(-j\omega) = -\tan^{-1}\left(\frac{\text{Im}\{H(j\omega)\}}{\text{Re}\{H(j\omega)\}}\right) = -\angle H(j\omega)$$

Many useful results can be obtained by studying the shapes of the magnitude and phase spectra. The spectra can be easily drawn using the MATLAB function `freqs` as demonstrated in the next example. The magnitude and phase spectra are very important for communications and signal processing.

EXAMPLE 3.18

Consider the system represented by the transfer function

$$H(j\omega) = \frac{1 + j\omega}{(j\omega)^3 + 3(j\omega)^2 + 2(j\omega) + 1}$$

The following MATLAB script calculates and plots the magnitude spectrum.

```
num=[1 1]
den=[1 3 2 1]
w=0:0.01:4              % defines the range of frequencies
H=freqs(num,den,w);     % calculates the values for H(jω)
magH=abs(H);
plot(w,magH)
semilogx(w,magH)
```

The corresponding MATLAB magnitude spectra in linear-linear and linear-logarithmic scales are presented in Figures 3.10 and 3.11, respectively.

It is customary in engineering to use the logarithmic scale, which has an equidistant distribution of ..., 0.1, 1, 10, 100, ... points along the ω-axis. Comparing Figures 3.10 and 3.11 we see clearly the impact of using the logarithmic scale: it stretches the graph

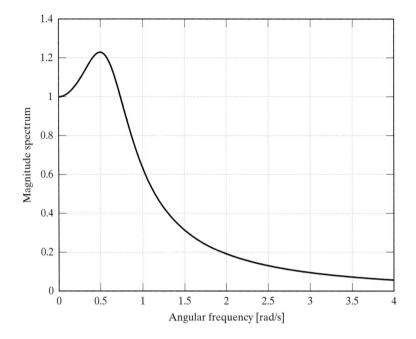

FIGURE 3.10: The magnitude spectrum in the linear-linear scale for Example 3.18

at lower frequencies and shrinks the graph at higher frequencies. The phase spectrum can be calculated and plotted using the same MATLAB script with a different frequency range, and using the MATLAB function `angle` to calculate the corresponding phase; that is,

```
w=0:0.1:10;
H=freqs(num,den,w);
phaseH=angle(H)*180/pi;
semilogx(w,phaseH)
```

The phase spectrum is presented in Figure 3.12.

There are other options for plotting the frequency spectra using the MATLAB function `freqs`, which can be obtained by typing `help freqs`. For example, `H=freqs(num,den)` plots directly the magnitude and phase spectra with the frequency range automatically determined by MATLAB.

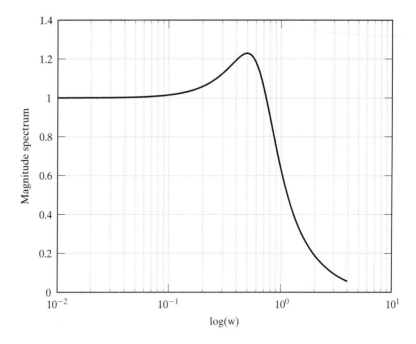

FIGURE 3.11: The magnitude spectrum in the linear-logarithmic scale for Example 3.18

In Example 3.18 we have drawn typical magnitude and phase spectra of a real linear dynamic system. Figure 3.11 indicates that the magnitude of the input signal at low frequencies is practically unchanged, since $H(j\omega) \approx 1$, and that at high frequencies the system drastically attenuates incoming signals, since at high frequencies $H(j\omega) \approx 0$. For this reason, real linear dynamic systems are called *low-pass filters*.

For a typical magnitude spectrum, we can define the frequency at which the maximum occurs by

$$\frac{d|H(j\omega)|}{d\omega} = 0 \Rightarrow \omega_r \tag{3.73}$$

Such a frequency is called the *resonant frequency*, and the corresponding maximum $|H(j\omega_r)|$ is called the *peak resonance*.

For real physical systems, the magnitude of the system transfer function almost always decays to zero as frequency increases to infinity ($n > m$). Due to the system attenuation of input signals at high frequencies—which follows from (3.66) for $n > m$ and the preceding fact about the shape of the system transfer function at low frequencies—we define the *system frequency bandwidth* by the frequency at which the system attenuation is

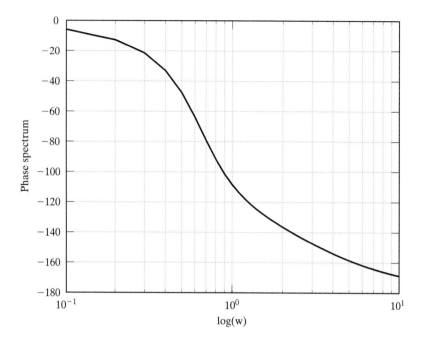

FIGURE 3.12: The phase spectrum in the linear-logarithmic scale for Example 3.18

no more than $1/\sqrt{2}$, that is,

$$|H(j\omega_{BW})| = \frac{1}{\sqrt{2}}|H(j0)| = 0.707|H(j0)| \Rightarrow \omega_{BW} \qquad (3.74)$$

In Example 3.18, we see from Figure 3.10 that $\omega_r \approx 0.6\,\text{rad/s}$, $|H(j\omega_r)| \approx 1.25$, and $\omega_{BW} \approx 0.9\,\text{rad/s}$.

From (3.67), that is,

$$Y_{zs}(j\omega) = H(j\omega)X(j\omega) = |H(j\omega)|e^{j\angle H(j\omega)}|X(j\omega)|e^{j\angle X(j\omega)}$$

we see that the magnitude and phase of the system output signal are given by

$$|Y_{zs}(j\omega)| = |H(j\omega)||X(j\omega)|, \quad \angle Y_{zs}(j\omega) = \angle H(j\omega) + \angle X(j\omega)$$

Similarly to system spectra, we can define the *signal spectra* from the signal Fourier transform $X(j\omega) = |X(j\omega)|\angle X(j\omega)$, with $|X(j\omega)|$ representing the signal magnitude spectrum and $\angle X(j\omega)$ representing the signal phase spectrum. We saw in (3.39) that the total signal energy satisfies

$$E_\infty = \int_{-\infty}^{\infty} |x(t)|^2 \, dt = \frac{1}{2\pi} \int_{-\infty}^{\infty} |X(j\omega)|^2 \, d\omega = \int_{-\infty}^{\infty} |X(jf)|^2 \, df$$

The quantities $|X(j\omega)|^2$ and $|X(jf)|^2$ are known as the *signal energy spectrum*. They are also called the signal energy *density* spectrum. Having in mind the definition of the signal power, we can also define the signal power spectrum. Since the signal energy and power spectra are more important for signal processing and communication systems than for linear dynamic systems, further treatment of them will be deferred to Chapters 9 and 10. In Chapter 12, we will present the Bode diagrams, which represent the plots of $|H(j\omega)|_{dB} \triangleq 20 \log_{10}(|H(j\omega)|)$ and $\angle H(j\omega)$ with respect to $\log_{10}(\omega)$. Bode diagrams are very useful for control theory and its applications. They can be plotted by using the MATLAB function bode.

The system frequency bandwidth provides information about the types of signals that a system can accommodate without producing a major loss of signal frequency content. Rapidly changing signals have most of their frequency content at high frequencies; thus, they can be passed through the system without major frequency content distortion only if the system has a large frequency bandwidth. On the other hand, the frequency content of slowly changing signals is concentrated at low frequencies, and they can be accommodated without loss of their frequency contents by linear dynamic systems with small frequency bandwidths. It can be observed from Example 3.3, Figure 3.7, that the infinitely fast delta impulse signal has an infinitely large frequency spectrum.

Another related observation can be drawn from the result established in (3.63), namely $p_\tau^h(t) \leftrightarrow \tau \operatorname{sinc}(\tau f)$. The waveforms of this time/frequency pair are presented in Figure 3.13.

It can be seen from Figure 3.13 that the main lobe of the frequency domain sinc signal is between $-2\pi/\tau$ and $2\pi/\tau$. The frequency bandwidth of the frequency domain sinc signal is proportional to $2\pi/\tau$. Apparently, if we make the signal time duration narrower by reducing τ, the corresponding frequency bandwidth will be increased; conversely, the wider signal the time duration, the shorter the signal frequency bandwidth. It can be concluded from this example that the *signal time duration and the signal frequency bandwidth are inversely proportional*. This observation holds in general.

Comment: Some real physical systems, like oscillatory large space flexible antennae, and some man-made systems like pass-band filters, may have magnitude frequency spectra much different than the low-pass frequency spectra of most real physical systems, whose typical shape is presented in Figures 3.10 and 3.11. Namely, the frequency contents of those systems may be concentrated around certain frequencies or set of frequencies, or in some frequency bands. In such cases, the frequency bandwidth has to be redefined with respect to the transfer function magnitude at the most appropriate frequencies that characterize such systems.

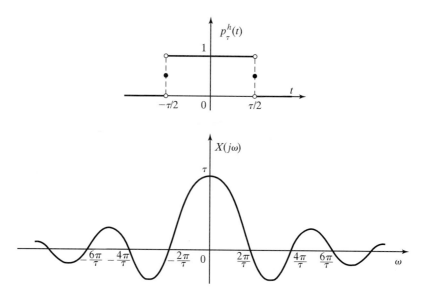

FIGURE 3.13: Time/frequency plots of a Fourier transform pair, demonstrating that the signal time duration and the frequency bandwidth are inversely proportional

3.4 FOURIER SERIES IN SYSTEM ANALYSIS

In this section we specialize the general results obtained in the previous section in formulas (3.67–69) to *periodic and sinusoidal inputs,* and show that the corresponding outputs are also periodic and sinusoidal. The main difference in the case of periodic and sinusoidal inputs, compared to general system inputs studied in the previous section, is that the system response obtained for periodic and sinusoidal inputs is the system *steady state* response. The main reason for this distinction is the fact that periodic signals are everlasting (from $-\infty$ to ∞) and are represented via the Fourier series in terms of everlasting sinusoidal signals. Such sinusoids pass through the linear system without exciting system natural modes (see Example 3.17), producing in the system output only another set of scaled and phase shifted sinusoids. In the general case, as indicated in the previous section, formula (3.69), the system response is the *zero-state* response.

3.4.1 System Response to Periodic Inputs

In this section we present a result that gives the answer to the linear time-invariant system response due to any periodic input. We have seen from formula (3.48) that a periodic signal, represented by the Fourier series, has the Fourier transform

$$x(t) = x(t+T) = \sum_{n=-\infty}^{\infty} X_n(jn\omega_0)e^{jn\omega_0 t} \leftrightarrow 2\pi \sum_{n=-\infty}^{\infty} X_n(jn\omega_0)\delta(\omega - n\omega_0) \quad (3.75)$$

where $X_n(jn\omega_0)$ are the Fourier series coefficients of the signal $x(t)$. The Fourier transform of the system output, according to formula (3.67), is

$$Y_{zs}(j\omega) = H(j\omega)X(j\omega) = 2\pi H(j\omega)\sum_{n=-\infty}^{\infty} X_n(jn\omega_0)\delta(\omega - n\omega_0)$$

$$= Y_{ss}(j\omega) = 2\pi \sum_{n=-\infty}^{\infty} X_n(jn\omega_0)H(jn\omega_0)\delta(\omega - n\omega_0) \quad (3.76)$$

From the preceding formula we see that the system output $y_{zs}(t)$ is also periodic, since its Fourier transform is represented as an infinite weighted sum of the frequency domain impulse delta signals. The corresponding weights are the coefficients of the Fourier series of $y_{zs}(t)$.

It should be observed that in the frequency domain, the input signal given by a frequency domain impulse delta signal has such a drastic impact on the system transfer function that the system transfer function loses its dynamic characteristics and becomes a constant complex number that must be evaluated at the given angular frequency, at the point where the frequency domain impulse delta signal is present, in this case $n\omega_0$ (this is the consequence of the impulse delta signal property defined in (2.30)). Such an input does not let the system exhibit its *natural modes* (in contrast see Example 3.17), so that the system response is fully dictated by the forcing function and has the same sinusoidal form as the forcing function (scaled and phase shifted)—it in fact represents the system steady state response, $y_{ss}(t)$. Note that the natural system modes are characterized by the terms $e^{p_i t}$, $i = 1, 2, \ldots, n$, where the p_i are the *system poles* found by solving the algebraic equation $p^n + a_{n-1}p^{n-1} + \cdots + a_1 p + a_0 = 0$, which is obtained by setting the transfer function denominator to zero. The natural system modes are always excited by the system's nonzero initial conditions. They can also be excited by the system inputs, as demonstrated in Example 3.17.

Denoting the coefficients of the Fourier series of $y_{ss}(t)$ by $Y_n(jn\omega_0)$, we have, from (3.76),

$$y_{ss}(t) = \sum_{n=-\infty}^{\infty} Y_n(jn\omega_0)e^{jn\omega_0 t} \leftrightarrow 2\pi \sum_{n=-\infty}^{\infty} X_n(jn\omega_0)H(jn\omega_0)\delta(\omega - n\omega_0) \quad (3.77)$$

It can be observed that the output signal $y_{ss}(t)$ has the same period $T = 2\pi/\omega_0$ as the input signal. From (3.77), we obtain the Fourier series coefficients of the output signal,

$$Y_n(jn\omega_0) = X_n(jn\omega_0)H(jn\omega_0)$$
$$= |X_n(jn\omega_0)|e^{j\angle X_n(jn\omega_0)}|H(jn\omega_0)|e^{j\angle H(jn\omega_0)}, \quad n = 0, \pm 1, \pm 2, \ldots \quad (3.78)$$

which leads to

$$|Y_n(jn\omega_0)| = |X_n(jn\omega_0)||H(jn\omega_0)|$$
$$\angle Y_n(jn\omega_0) = \angle X_n(jn\omega_0) + \angle H(jn\omega_0), \quad n = 0, \pm 1, \pm 2, \ldots \quad (3.79)$$

It can be concluded from (3.76–77) that *periodic system inputs produce periodic system outputs with the same period as the system inputs*. The corresponding Fourier series coefficients for the system output are obtained from (3.79). Formula (3.79) indicates how the system modifies the Fourier series coefficients of an incoming periodic input signal in order to produce the periodic output signal.

The trigonometric form of the Fourier series that corresponds to (3.77) is given (see (3.10)) by

$$y_{ss}(t) = Y_n(j0) + 2\sum_{n=1}^{\infty} |Y_n(jn\omega_0)| \cos(n\omega_0 t + \angle Y_n(jn\omega_0)) \qquad (3.80)$$

In general, the coefficients of the Fourier series decay to zero rapidly, so we can use a finite sum approximation of (3.80) and include only the first N harmonics, that is,

$$y_N(t) = Y_n(j0) + 2\sum_{n=1}^{N} |Y_n(jn\omega_0)| \cos(n\omega_0 t + \angle Y_n(jn\omega_0)) \qquad (3.81)$$

Such an approximation produces $y_{ss}(t) \approx y_N(t)$. Note that we can easily calculate the magnitude and the phase of the Fourier series coefficients (complex numbers). In addition, we can use the MATLAB functions `abs` and `angle` and write a simple program for calculating the required coefficients. The method for finding the linear system response due to a periodic input and the corresponding MATLAB procedures are demonstrated in the next example.

EXAMPLE 3.19

Consider the second-order dynamic system at rest given by

$$\frac{d^2 y(t)}{dt^2} + 2\frac{dy(t)}{dt} + 3y(t) = x(t)$$

with the periodic input signal $x(t)$ as presented in Figure 3.4 (sawtooth waveform). The system transfer function is obtained from (3.66) as

$$H(j\omega) = \frac{1}{(j\omega)^2 + 2(j\omega) + 3} = \frac{1}{3 - \omega^2 + j2\omega}$$

$$= \frac{1}{\sqrt{(3-\omega^2)^2 + 4\omega^2}} \angle -\tan^{-1}\left(\frac{2\omega}{3-\omega^2}\right)$$

The complex Fourier series coefficients of the periodic input signal are given (as in Example 3.2) by

$$X_n(j0) = 0, \quad X_n(jn\omega_0) = \frac{E}{n\pi} j(-1)^n = \frac{E}{n\pi} \angle (-1)^n \frac{\pi}{2}, \quad n = 1, 2, \ldots$$

The Fourier series coefficients of the periodic output signal are obtained from (3.79):

$$|Y_n(jn\omega_0)| = \frac{E}{n\pi} \frac{1}{\sqrt{(3 - n^2\omega_0^2)^2 + 4n^2\omega_0^2}}, \quad n = 1, 2, \ldots$$

$$\angle Y_n(jn\omega_0) = (-1)^n \frac{\pi}{2} - \tan^{-1}\left(\frac{2n\omega_0}{3 - n^2\omega_0^2}\right), \quad n = 1, 2, \ldots$$

For the given values of E and ω_0, we can calculate the actual values for $Y_n(jn\omega_0)$ coefficients, $n = 1, 2, \ldots$, and obtain the expression for the output signal Fourier series as defined by (3.80). Note that in practice we truncate the Fourier series (3.80) and use the corresponding approximation (3.81) that includes a sufficient number of harmonics. This is demonstrated by using the data $E = 5$, $\omega_0 = 1$, and the following MATLAB program:

```
t=0:0.01:4*pi;
E=5; w0=1; N=5;yN=0;
for n=1:N
Hm=1/sqrt((3-(n*w0)^2)^2+4*(n*w0)^2);
Hp=-atan(2*n*w0/(3-(n*w0)^2));
Xn=j*(-1)^n*E/(n*pi)
Xnm=abs(Xn);
Xnp=angle(Xn);
Ynm=Xnm*Hm
Ynp=Xnp+Hp
yN=yN+2*Ynm*cos(n*w0*t+Ynp)
end
```

The magnitudes of the Fourier series coefficients of the output periodic signal decay rapidly to zero as n increases, as demonstrated in Table 3.2.

The MATLAB plots of the system response for $N = 1, 2, 3, 4$ are presented in Figure 3.14. Note that the plot of the approximation $y_5(t)$ is hardly distinguishable from the plot of the approximation $y_4(t)$. We can use the MATLAB function max(y4-y5) to get the

TABLE 3.2 Magnitudes of the Fourier series coefficients for Example 3.19

n	$|Y_n(jn\omega_0)|$
1	0.5627
2	0.1930
3	0.0625
4	0.0261
5	0.0131
6	0.0076

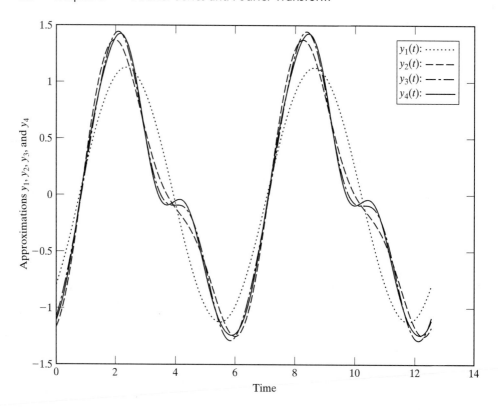

FIGURE 3.14: Approximations for the system response due to a periodic input for Example 3.19

maximal difference between these two approximations as a function of time, which yields 0.0263. Similarly, we find $\max(y_6(t) - y_7(t)) = 0.0151$.

EXAMPLE 3.20

Consider an electrical circuit forming a cascade connection of resistor, inductor, and capacitor (RLC), driven by a voltage source whose waveform is given in Figure 3.4. The circuit is presented in Figure 3.15. Assume that the circuit initial conditions are zero. Using Fourier series analysis, we will find the steady state value for the capacitor voltage.

From elementary circuit laws [7], the differential equation that describes the dynamics of the capacitor voltage can be derived as

$$\frac{d^2 v_C(t)}{dt^2} + \frac{R}{L}\frac{dv_C(t)}{dt} + \frac{1}{LC}v_C(t) = \frac{1}{LC}v_s(t)$$

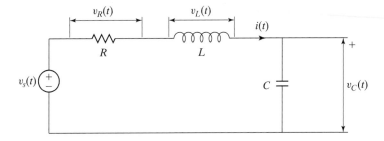

FIGURE 3.15: An RLC circuit driven by a voltage source

The circuit transfer function has the form

$$H(j\omega) = \frac{\frac{1}{LC}}{(j\omega)^2 + \frac{R}{L}(j\omega) + \frac{1}{LC}} = \frac{\frac{1}{LC}}{\frac{1}{LC} - \omega^2 + j\frac{R}{L}\omega}$$

$$= \frac{\frac{1}{LC}}{\sqrt{\left(\frac{1}{LC} - \omega^2\right)^2 + \left(\frac{R}{L}\right)^2 \omega^2}} \angle - \tan^{-1}\left(\frac{\frac{R}{L}\omega}{\frac{1}{LC} - \omega^2}\right)$$

The complex Fourier series coefficients of the periodic input signal are given as in Example 3.2 by

$$V_{sn}(j0) = 0, \quad V_{sn}(jn\omega_0) = \frac{E}{n\pi} j(-1)^n = \frac{E}{n\pi} \angle (-1)^n \frac{\pi}{2}, \quad n = 1, 2, \ldots$$

The Fourier series coefficients of the periodic output signal are obtained from (3.79) as

$$|V_{Cn}(jn\omega_0)| = \frac{E}{n\pi} \frac{\frac{1}{LC}}{\sqrt{\left(\frac{1}{LC} - n^2\omega_0^2\right)^2 + \left(\frac{R}{L}\right)^2 n^2 \omega_0^2}}, \quad n = 1, 2, \ldots$$

$$\angle V_{Cn}(jn\omega_0) = (-1)^n \frac{\pi}{2} - \tan^{-1}\left(\frac{\frac{R}{L} n\omega_0}{\frac{1}{LC} - n^2\omega_0^2}\right), \quad n = 1, 2, \ldots$$

The required steady state capacitor voltage is given by

$$v_{Css}(t) = V_{Cn}(j0) + 2\sum_{n=1}^{\infty} |V_{Cn}(jn\omega_0)| \cos(n\omega_0 t + \angle V_{Cn}(jn\omega_0))$$

3.4.2 System Response to Sinusoidal Inputs

In this section, we find the steady state response of a system at rest due to sinusoidal inputs, by using formula (3.67). The Fourier transforms of cosine and sine functions are given, respectively, in (3.46) and (3.47). Since cosine and sine functions are related by the very simple relationships $\cos(\omega_0 t) = \sin(\omega_0 t + \frac{\pi}{2})$ and $\sin(\omega_0 t) = \cos(\omega_0 t - \frac{\pi}{2})$, we will perform derivations for the cosine function only.

Using (3.67) with (3.46), that is,

$$\mathcal{F}\{\cos(\omega_0 t)\} = \pi\delta(\omega + \omega_0) + \pi\delta(\omega - \omega_0) = X(j\omega)$$

we have

$$Y_{ss}(j\omega) = H(j\omega)X(j\omega) = \pi H(j\omega)\delta(\omega + \omega_0) + \pi H(j\omega)\delta(\omega - \omega_0)$$

By the impulse delta signal property defined in (2.30), we obtain

$$Y_{ss}(j\omega) = \pi H(j\omega_0)\delta(\omega - \omega_0) + \pi H(-j\omega_0)\delta(\omega + \omega_0)$$
$$= \pi|H(j\omega_0)|e^{j\angle H(j\omega_0)}\delta(\omega - \omega_0) + \pi|H(j\omega_0)|e^{-j\angle H(j\omega_0)}\delta(\omega + \omega_0)$$

which implies

$$y_{ss}(t) = \mathcal{F}^{-1}\{Y_{ss}(j\omega)\}$$
$$= \pi|H(j\omega_0)|e^{j\angle H(j\omega_0)}\mathcal{F}^{-1}\{\delta(\omega - \omega_0)\} + \pi|H(j\omega_0)|e^{-j\angle H(j\omega_0)}\mathcal{F}^{-1}\{\delta(\omega + \omega_0)\}$$

Using (3.43)—that is, $1 \leftrightarrow 2\pi\delta(\omega)$ and the frequency shifting property—it follows that

$$y_{ss}(t) = \frac{1}{2}|H(j\omega_0)|e^{j\angle H(j\omega_0)}e^{j\omega_0 t} + \frac{1}{2}|H(j\omega_0)|e^{-j\angle H(j\omega_0)}e^{-j\omega_0 t} \quad (3.82)$$
$$= |H(j\omega_0)|\cos(\omega_0 t + \angle H(j\omega_0))$$

If we used as the system input $A\cos(\omega_0 t + \theta)$ instead of $\cos(\omega_0 t)$, the solution procedure would be exactly the same, leading to the system steady state response of the form

$$y_{ss}(t) = |H(j\omega_0)|A\cos(\omega_0 t + \theta + \angle H(j\omega_0)) \quad (3.83)$$

Another interesting way to derive formula (3.82) is to start with the expressions for the Fourier series coefficients of $x(t) = \cos(\omega_0 t)$, which (from (3.2)) are given by

$$a_0 = 0, \quad a_1 = 1, \quad a_n = 0, \quad \forall n \geq 2$$
$$b_n = 0, \quad \forall n \geq 1$$

This leads to the complex form Fourier series coefficients (defined in (3.11)) for the cosine function

$$X_1(j\omega_0) = \frac{1}{2}(a_1 - jb_1) = \frac{1}{2}, \quad X_n(jn\omega_0) = 0, \quad n \neq 1$$

Using (3.80), we obtain the system output response to the sinusoidal input signal $x(t) = \cos(\omega_0 t)$ as

$$y_{ss}(t) = 2|Y_1(j\omega_0)|\cos(\omega_0 t + \angle Y_1(j\omega_0))$$

with $Y_1(j\omega_0)$ defined by (3.79), that is,

$$Y_1(j\omega_0) = X_1(j\omega_0)H(j\omega_0) = \frac{1}{2}|H(j\omega_0)|\angle H(j\omega_0) = |Y_1(j\omega_0)|\angle Y_1(j\omega_0)$$

Hence, the final expression for the time-invariant linear system response due to the cosine input is given by

$$y_{ss}(t) = 2\frac{1}{2}|H(j\omega_0)|\cos(\omega_0 t + \angle H(j\omega_0)) = |H(j\omega_0)|\cos(\omega_0 t + \angle H(j\omega_0))$$

which is identical to (3.82).

Similarly, using the same technique, we can find the response of a linear time-invariant system due to the sine input, $x(t) = \sin(\omega_0 t)$, for which we have

$$a_n = 0, \quad \forall n$$
$$b_1 = 1, \quad b_n = 0, \quad \forall n \geq 2$$

so that

$$X_1(j\omega_0) = \frac{1}{2}(a_1 - jb_1) = -j\frac{1}{2}, \quad X_n(jn\omega_0) = 0, \quad n \neq 1$$

Hence

$$Y_1(j\omega_0) = -\frac{j}{2}H(j\omega_0) = \frac{1}{2}|H(j\omega_0)|\angle\left(H(j\omega_0) - \frac{\pi}{2}\right) = |Y_1(j\omega_0)|\angle Y_1(j\omega_0)$$

The system response due to the sine input is now given by

$$y_{ss}(t) = |H(j\omega_0)|\cos\left(\omega_0 t + \angle H(j\omega_0) - \frac{\pi}{2}\right) \tag{3.84}$$

EXAMPLE 3.21

Consider the second-order system at rest defined in Example 3.19, whose transfer function is given by

$$H(j\omega) = \frac{1}{(j\omega)^2 + 2(j\omega) + 3} = \frac{1}{\sqrt{(3-\omega^2)^2 + 4\omega^2}}\angle -\tan^{-1}\left(\frac{2\omega}{3-\omega^2}\right)$$

Let the input to this system be given by $x(t) = \cos(2t)$. The output signal is obtained from formula (3.82), and calculated for $\omega_0 = 2$ rad/s as

$$y_{ss}(t) = \frac{1}{\sqrt{(3-2^2)^2 + 4 \times 2^2}}\cos\left(2t - \angle\tan^{-1}\left(\frac{2 \times 2}{3-2^2}\right)\right)$$

$$= \frac{1}{\sqrt{17}}\cos(2t - \angle\tan^{-1}(-4)) = \frac{1}{\sqrt{17}}\cos(2t + 75.96°) \quad ♪$$

EXAMPLE 3.22

Consider a mechanical system as presented in Figure 3.16, whose mathematical model is given by

$$\frac{d^2 y(t)}{dt^2} + \frac{c}{m}\frac{dy(t)}{dt} + \frac{k}{m} y(t) = \frac{1}{m} x(t) = \frac{1}{m} A \cos\left(\omega_0 t + \frac{\pi}{3}\right)$$

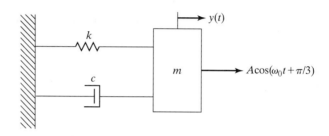

FIGURE 3.16: A mechanical system

Its transfer function is

$$H(j\omega) = \frac{\frac{1}{m}}{(j\omega)^2 + \frac{c}{m}(j\omega) + \frac{k}{m}} = \frac{\frac{1}{m}}{\frac{k}{m} - \omega^2 + j\omega \frac{c}{m}}$$

$$= \frac{\frac{1}{m}}{\sqrt{\left(\frac{k}{m} - \omega^2\right)^2 + \frac{c^2}{m^2}\omega^2}} \angle - \tan^{-1}\left(\frac{\frac{c}{m}\omega}{\frac{k}{m} - \omega^2}\right)$$

Using formula (3.83), we obtain the following expression for the system steady state response:

$$y_{ss}(t) = \frac{\frac{1}{m} A}{\sqrt{\left(\frac{k}{m} - \omega_0^2\right)^2 + \frac{c^2}{m^2}\omega_0^2}} \cos\left(\omega_0 t + \frac{\pi}{3} - \angle \tan^{-1}\left(\frac{\frac{c}{m}\omega_0}{\frac{k}{m} - \omega_0^2}\right)\right)$$

The preceding example demonstrates how simple it is to find the system steady state response due to sinusoidal inputs.

In this chapter we have presented in sufficient detail the material on the Fourier series and Fourier transform needed for an undergraduate course on linear dynamic systems and signals. The interested reader can find more about Fourier analysis and its use in linear system theory and applications in both classic [5, 8] and modern [9] texts.

3.5 FROM FOURIER TRANSFORM TO LAPLACE TRANSFORM

We have seen in the previous sections that many common signals have no Fourier transform, given in terms of ordinary functions. The reason for this is that the main existence condition (3.21) is not satisfied for those signals. In order to broaden the class of signals

Section 3.5 From Fourier Transform to Laplace Transform

for which the frequency representation in terms of ordinary functions is possible, we introduce the so-called Laplace transform. It is basically the Fourier transform of the function $x(t)e^{-\sigma t}$, that is,

$$\mathcal{L}\{x(t)\} = \int_{-\infty}^{\infty} x(t)e^{-\sigma t}e^{-j\omega t}\, dt = \int_{-\infty}^{\infty} x(t)e^{-st}\, dt \tag{3.85}$$

where $s = \sigma + j\omega$ is called the complex frequency, with σ chosen to facilitate the convergence of the integral. More precisely, the transform defined in (3.85) is known as the *double-sided Laplace transform*. The double-sided Laplace transform is important for the analysis of power systems, as well as for some applications in communication and signal processing systems.

For analysis of causal signals ($x(t) = 0$, $t < 0$) that pass through linear time-invariant causal (real physical) systems, the *one-sided Laplace transform*, defined in the following formula, plays a fundamental role:

$$\mathcal{L}\{x(t)\} \triangleq \int_{0}^{\infty} x(t)e^{-\sigma t}e^{-j\omega t}\, dt = \int_{0}^{\infty} x(t)e^{-st}\, dt = X(s) \tag{3.86}$$

Since the one-sided Laplace transform is extremely important for linear dynamic systems, it will be presented in detail in Chapter 4.

In view of (3.21), the existence condition of the one-sided Laplace transform is

$$\int_{0}^{\infty} |x(t)e^{-\sigma t}|\, dt < \infty \tag{3.87}$$

Note that the preceding integral converges in the half of the complex plane defined by $\sigma > \gamma$, where γ depends on $x(t)$ (see Figure 3.17).

In the case of the double-sided Laplace transform, the existence condition becomes

$$\int_{-\infty}^{\infty} |x(t)e^{-\sigma t}|\, dt < \infty \tag{3.88}$$

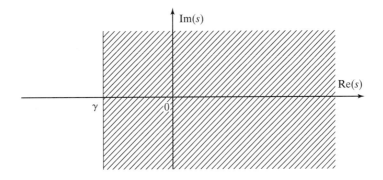

FIGURE 3.17: The region of convergence of the one-sided Laplace transform (shaded area)

This condition requires that the exponent σ satisfies $\gamma_1 < \sigma < \gamma_2$, where γ_1 and γ_2 depend on $x(t)$. For the double-sided Laplace transform (if it exists), the convergence region is a strip in the complex plane $s = \sigma + j\omega$ with $\gamma_1 < \sigma < \gamma_2$ and $\omega \in (-\infty, \infty)$.

Note that *under the assumption that the $j\omega$ axis ($\sigma = 0$) belongs to the convergence region of the Laplace transform, we can recover the Fourier transform from the Laplace transform by simply setting $s = j\omega$ in the corresponding expression for the Laplace transform* (see Figure 3.17). It follows from Figure 3.17 that for the functions for which $\gamma > 0$, the Fourier transform cannot be obtained from the Laplace transform by a simple substitution $s = j\omega$.

EXAMPLE 3.23

In this example, we demonstrate the procedure for finding the convergence regions of the one- and double-sided Laplace transforms. Consider the signal

$$x(t) = \begin{cases} e^{-2t}, & t > 0 \\ e^{3t}, & t < 0 \end{cases}$$

Finding the double-sided Laplace transform of this signal requires the following integration:

$$\mathcal{L}\{x(t)\} = \int_{-\infty}^{0} e^{3t} e^{-\sigma t} e^{-j\omega t} \, dt + \int_{0}^{\infty} e^{-2t} e^{-\sigma t} e^{-j\omega t} \, dt$$

The first integral converges for $3 - \sigma > 0 \Rightarrow \sigma < 3$ and the second integral converges for $-2 - \sigma < 0 \Rightarrow \sigma > -2$. Hence, both integrals exist (converge) in the strip $-2 < \sigma < 3$, which represents the domain of convergence of the double-sided Laplace transform. The corresponding double-sided Laplace transform is given by

$$\mathcal{L}\{x(t)\} = X(s) = \int_{-\infty}^{0} e^{3t} e^{-st} \, dt + \int_{0}^{\infty} e^{-2t} e^{-st} \, dt$$

$$= \frac{1}{3-s} + \frac{1}{2+s} = \frac{5}{(2+s)(3-s)}$$

Since the $j\omega$ axis ($\sigma = 0$) belongs to the convergence region of the double-sided Laplace transform ($-2 < 0 < 3$), we can get the Fourier transform from the double-sided Laplace transform by simply replacing s with $j\omega$ in the preceding formula, that is,

$$X(j\omega) = \frac{5}{(2+j\omega)(3-j\omega)}$$

Let us assume that we have only a causal signal ($x(t) = 0, t < 0$) defined by $x(t) = e^{-2t}, t > 0$. Then, from the preceding discussion, we see that the convergence region for the corresponding one-sided Laplace transform is given by $\sigma > -2$.

It should be emphasized that the exponential function $e^{-\sigma t}$ with $\sigma > 0$ is a rapidly decaying function that facilitates the convergence of the Laplace integral. For example, for the function $x(t) = t^n$ with n being an arbitrary large positive integer, the integral (3.87) is convergent and $t^n e^{-\sigma t} \to 0$ as $t \to \infty$. Thus, many common signals such as sine, cosine, and step have the Laplace transform given in terms of ordinary functions, since they satisfy condition (3.87).

In the next chapter, we will redefine the concepts of the system transfer function and the system impulse response in terms of the Laplace transform. The main results are basically the same as in the case of the Fourier transform, but the Laplace transform is more general and more powerful for the study of linear time-invariant dynamic systems—especially for the analysis of system transient behavior. Note that the Fourier transform can give the solution to the system response subject to forcing functions and zero initial conditions, but the Laplace transform produces the system response due to both the initial conditions and external forcing functions.

The Fourier transform remains an important tool is signal analysis, since the signal spectrum reveals the signal frequency contents. Many useful and important signal characteristics can be learned from the signal frequency spectrum. Communication systems have their foundation in the Fourier transform and its modulation property. Further studies of the Fourier transform and its applications are usually undertaken in signal processing and communication systems courses. We will return to signals in Chapters 9, where we will present the discrete-time Fourier series, discrete-time Fourier transform (DTFT), discrete Fourier transform (DFT), and fast Fourier transform (FFT).

3.6 FOURIER ANALYSIS MATLAB LABORATORY EXPERIMENT

Purpose: This experiment demonstrates approximations of periodic signals by truncated Fourier series as defined in formula (3.4). Using MATLAB, students will plot the actual approximate signals and observe, for large values of N, the Gibbs phenomenon at the jump discontinuity points. In addition, students will use MATLAB to plot the system frequency spectra, and to find the system response due to periodic inputs.

Part 1. Find the trigonometric form of the Fourier series for the periodic signal presented in Figure 3.18. Take $E = 1$, $T = 1$ and use MATLAB to calculate the coefficients of the Fourier series for $n = 0, 1, 2, \ldots, N$. Plot the approximations $x_N(t)$ as defined in (3.4) for $N = 5, 10, 20, 30, 40, 50$. Observe

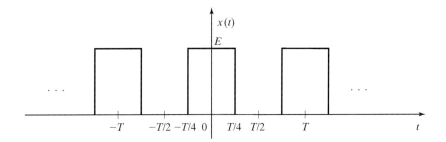

FIGURE 3.18: A square wave signal

the Gibbs phenomenon and for $N = 50$ estimate the relative magnitude of ripples at the jump discontinuity points.

Part 2. Use MATLAB to plot the amplitude and phase line spectra of the periodic signal from Part 1.

Part 3. Plot the magnitude and phase spectra of the system defined in Example 3.19 by using the MATLAB function `freqs(num,den)`, where the vectors num and den contain the coefficients of the transfer function numerator and denominator in descending order.

Part 4. For the system defined in Example 3.19, and $x_N(t)$ determined in Part 1 with $E = 5$, $\omega_0 = 1$ rad/s, calculate the Fourier series coefficients of the output signal for $n = 0, 1, 3, 5$. Print the values for the magnitudes of the Fourier series coefficients of the output signal. Plot the approximations $x_N(t)$ and observe the convergence of the output signal as N increases. Take $N = 1, 3, 5$ and $t \in [0, 4\pi]$. Comment on the frequency of the output signal and check its value from the plot obtained.

Submit all plots and comment on the results obtained.

3.7 SUMMARY

Study Guide for Chapter Three: Students should understand the fundamentals of the Fourier series, particularly their use for finding the linear system response due to sinusoidal and periodic inputs. Students must master all of the properties of the Fourier transform presented in Section 3.2 and summarized in Table 3.3. Familiarity with common Fourier transforms presented in Table 3.4 is useful. Standard problems: (1) find the Fourier series of a given periodic signal; (2) plot the line spectra; (3) use the Fourier series to find the system response due to periodic and sinusoidal inputs; (4) find the Fourier transform of a given time signal; (5) find the Fourier inverse of a given frequency domain signal; and (6) state and prove some of the properties of the Fourier transform.

TABLE 3.3 Properties of the Fourier transform

	Time	Frequency
1	$\alpha_1 x_1(t) \pm \alpha_2 x_2(t)$	$\alpha_1 X_1(j\omega) \pm \alpha_2 X_2(j\omega)$
2	$x(t - t_0)$	$e^{-j\omega t_0} X(j\omega)$
3	$x(at)$	$\frac{1}{|a|} X\left(\frac{j\omega}{a}\right)$
2 & 3	$x(at - t_0)$	$\frac{1}{|a|} e^{-j(\frac{\omega}{a})t_0} X\left(\frac{j\omega}{a}\right)$
4	$t^n x(t)$	$(j)^n \frac{d^n}{d\omega^n} X(j\omega)$
5	$x(t) e^{j\omega_0 t}$	$X(j(\omega - \omega_0))$
6	$x(t) \cos(\omega_0 t)$	$\frac{1}{2}[X(j(\omega + \omega_0)) + X(j(\omega - \omega_0))]$
	$x(t) \sin(\omega_0 t)$	$\frac{j}{2}[X(j(\omega + \omega_0)) - X(j(\omega - \omega_0))]$

TABLE 3.3 (Continued)

	Time	Frequency
7	$\dfrac{d^n}{dt^n} x(t)$	$(j\omega)^n X(j\omega)$
7a	$(-jt)^n x(t)$	$\dfrac{d^n X(j\omega)}{d\omega^n}$
8	$x_1(t) * x_2(t)$	$X_1(j\omega) X_2(j\omega)$
9	$x_1(t) x_2(t)$	$\dfrac{1}{2\pi} X_1(j\omega) * X_2(j\omega)$
10	$x(-t)$	$X(-j\omega)$
11	$X(jt)$	$2\pi x(-\omega)$
11a	$X(-jt)$	$2\pi x(\omega)$
12	$\displaystyle\int_{-\infty}^{t} x(\tau)\, d\tau$	$\dfrac{1}{j\omega} X(j\omega) + \pi X(j0) \delta(\omega)$
Parseval's Theorem	$\displaystyle\int_{-\infty}^{\infty} \|x(t)\|^2\, dt$	$\dfrac{1}{2\pi} \displaystyle\int_{-\infty}^{\infty} \|X(j\omega)\|^2\, d\omega$

TABLE 3.4 Common Fourier transform pairs

	Time	Frequency
1	$\delta(t)$	1
2	$e^{-at} u_h(t),\ a > 0$	$\dfrac{1}{a + j\omega}$
3	$p_\tau^h(t)$	$\tau\,\text{sinc}\left(\dfrac{\omega\tau}{2\pi}\right) = \tau \dfrac{\sin(\omega\tau/2)}{\omega\tau/2}$
4	$\Delta_\tau(t)$	$\dfrac{\tau}{2} \text{sinc}^2\left(\dfrac{\omega\tau}{4}\right)$
5	1	$2\pi \delta(\omega)$
6	const	const $\times\ 2\pi \delta(\omega)$
7	$\cos(\omega_0 t)$	$\pi[\delta(\omega + \omega_0) + \delta(\omega - \omega_0)]$
8	$\sin(\omega_0 t)$	$j\pi[\delta(\omega + \omega_0) - \delta(\omega - \omega_0)]$
9	$\text{sgn}(t)$	$\dfrac{2}{j\omega}$
10	$u_h(t)$	$\dfrac{1}{j\omega} + \pi \delta(\omega)$
11	$e^{-\alpha \|t\|},\ \alpha > 0$	$\dfrac{2\alpha}{\alpha^2 + \omega^2}$
12	$e^{j\omega_0 t}$	$2\pi \delta(\omega - \omega_0)$
13	$\displaystyle\sum_{n=-\infty}^{\infty} X_n e^{jn\omega_0 t},\ \omega_0 = \dfrac{2\pi}{T}$	$2\pi \displaystyle\sum_{n=-\infty}^{\infty} X_n \delta(\omega - n\omega_0)$
14	$\displaystyle\sum_{n=-\infty}^{\infty} \delta(t - nT_0)$	$\omega_0 \displaystyle\sum_{n=-\infty}^{\infty} \delta(\omega - n\omega_0),\ \omega_0 = \dfrac{2\pi}{T_0}$

Fourier Series:

$$x(t) = \frac{1}{2}a_0 + \sum_{n=1}^{\infty} [a_n \cos(n\omega_0 t) + b_n \sin(n\omega_0 t)]$$

$$x(t) = X_0 + 2\sum_{n=1}^{\infty} |X_n(jn\omega_0)| \cos(n\omega_0 t + \angle X_n(jn\omega_0))$$

$$x(t) = \sum_{n=-\infty}^{\infty} X_n(jn\omega_0) e^{jn\omega_0 t}$$

$$a_0 = \frac{2}{T} \int_{-\frac{T}{2}}^{\frac{T}{2}} x(t)\, dt, \quad a_n = \frac{2}{T} \int_{-\frac{T}{2}}^{\frac{T}{2}} x(t) \cos(n\omega_0 t)\, dt$$

$$b_n = \frac{2}{T} \int_{-\frac{T}{2}}^{\frac{T}{2}} x(t) \sin(n\omega_0 t)\, dt, \quad n = 1, 2, \ldots$$

$$X_n(jn\omega_0) = \frac{1}{T} \int_{-\frac{T}{2}}^{\frac{T}{2}} x(t) e^{-jn\omega_0 t}\, dt, \quad n = 1, 2, \ldots, \quad X_n^*(jn\omega_0) = X_n(-jn\omega_0)$$

$$X_0(j0) = X_0 = \frac{1}{2}a_0, \quad X_n(jn\omega_0) = \frac{1}{2}a_n - j\frac{1}{2}b_n$$

$$a_n = 2\,\mathrm{Re}\{X_n(jn\omega_0)\}, \quad b_n = -2\,\mathrm{Im}\{X_n(jn\omega_0)\}$$

Parseval's Theorem for Periodic Signals:

$$\int_0^T |x(t)|^2\, dt = \sum_{n=-\infty}^{\infty} |X_n(jn\omega_0)|^2$$

Fourier Series Formulas of Interest:

$$x(t) = \sum_{n=-\infty}^{n=\infty} \delta(t - nT_0) = \frac{1}{T_0} \sum_{n=-\infty}^{n=\infty} e^{jn\omega_0 t}, \quad T_0 = \frac{2\pi}{\omega_0}$$

$$\sum_{n=-\infty}^{n=\infty} \delta(\omega - n\omega_0) = \frac{1}{\omega_0} \sum_{n=-\infty}^{n=\infty} e^{jnT_0\omega}, \quad T_0 = \frac{2\pi}{\omega_0}$$

Fourier Transform and Inverse:

$$\mathcal{F}\{x(t)\} = X(j\omega) \triangleq \int_{-\infty}^{\infty} x(t) e^{-j\omega t}\, dt$$

$$x(t) = \mathcal{F}^{-1}\{X(j\omega)\} \triangleq \frac{1}{2\pi} \int_{-\infty}^{\infty} X(j\omega) e^{j\omega t}\, d\omega$$

System Transfer Function:

$$H(j\omega) = \frac{b_m(j\omega)^m + b_{m-1}(j\omega)^{m-1} + \cdots + b_1(j\omega) + b_0}{(j\omega)^n + a_{n-1}(j\omega)^{n-1} + \cdots + a_1(j\omega) + a_0} = |H(j\omega)| \angle H(j\omega)$$

System Impulse Response:

$$h(t) \triangleq \mathcal{F}^{-1}\{H(j\omega)\}$$

General System Response due to a Forcing Function and Zero Initial Conditions:

$$Y_{zs}(j\omega) = H(j\omega)X(j\omega) \Rightarrow y_{zs}(t) = h(t) * x(t)$$

System Response to Periodic Inputs:

$$x(t) = x(t+T) = \sum_{n=-\infty}^{\infty} X_n(jn\omega_0)e^{jn\omega_0 t} \Rightarrow y_{ss}(t) = \sum_{n=-\infty}^{\infty} Y_n(jn\omega_0)e^{jn\omega_0 t}$$

$$y_{ss}(t) = Y_n(j0) + 2\sum_{n=1}^{\infty} |Y_n(jn\omega_0)| \cos(n\omega_0 t + \angle Y_n(jn\omega_0))$$

$$y_N(t) = Y_n(j0) + 2\sum_{n=1}^{N} |Y_n(jn\omega_0)| \cos(n\omega_0 t + \angle Y_n(jn\omega_0))$$

$$|Y_n(jn\omega_0)| = |X_n(jn\omega_0)||H(jn\omega_0)|, \quad \angle Y_n(jn\omega_0) = \angle X_n(jn\omega_0) + \angle H(jn\omega_0)$$

System Response to Sinusoidal Inputs:

$$x(t) = \cos(\omega_0 t + \theta) \Rightarrow y_{ss}(t) = |H(j\omega_0)| \cos(\omega_0 t + \theta + \angle H(j\omega_0))$$

$$x(t) = \sin(\omega_0 t + \theta) \Rightarrow y_{ss}(t) = |H(j\omega_0)| \cos\left(\omega_0 t + \theta - \frac{\pi}{2} + \angle H(j\omega_0)\right)$$

Parseval's Theorem for Aperiodic Signals:

$$\int_{-\infty}^{\infty} x^2(t)\, dt = \int_{-\infty}^{\infty} |X(jf)|^2\, df = \frac{1}{2\pi} \int_{-\infty}^{\infty} |X(j\omega)|^2\, d\omega$$

3.8 REFERENCES

[1] M. Spiegel, *Mathematical Handbook of Formulas and Tables,* McGraw-Hill, New York, 1968.

[2] M. Greenberg, *Foundations of Applied Mathematics,* Prentice Hall, Englewood Cliffs, NJ, 1978.

[3] C. Wylie and L. Barrett, *Advanced Engineering Mathematics,* McGraw-Hill, New York, 1995.

[4] S. Soliman and M. Srinath, *Continuous and Discrete Signals and Systems,* Prentice Hall, Englewood Cliffs, NJ, 1990.

132 Chapter 3   Fourier Series and Fourier Transform

[5] A. Papoulis, *The Fourier Integral and Its Applications,* McGraw-Hill, New York, 1962.

[6] C. McGillem and G. Cooper, *Continuous and Discrete Signal and System Analysis,* Saunders College Publishing, Philadelphia, 1991.

[7] J. Nilsson and S. Reidel, *Electric Circuits,* Addison Wesley, Reading, MA, 1996.

[8] R. Bracewell, *The Fourier Transform and Its Applications,* McGraw-Hill, New York, 1999.

[9] R. Gray and J. Goodman, *Fourier Transforms: An Introduction for Engineers,* Kluwer Academic Publishers, Boston, 1995.

[10] Y. Sun et al., "Time dependent perturbation theory and tones in cascaded erbium-doped fiber amplifier systems," *Journal of Lightwave Technology,* 15:1083–87, 1997.

[11] D. Inman, *Engineering Vibration,* Prentice Hall, Englewood Cliffs, NJ, 1994.

3.9 PROBLEMS

3.1. Find the trigonometric form with real coefficients of the Fourier series of the signal presented in Figure 3.19.

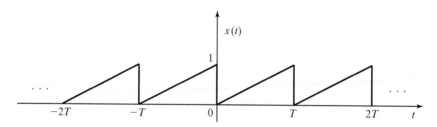

FIGURE 3.19: Periodic signal

Answer:
$$a_0 = 1, \quad a_n = 0, \quad b_n = -1/n\pi, \quad n = 1, 2, 3, \ldots$$

3.2. Find the trigonometric form with complex coefficients of the Fourier series for the signal presented in Figure 3.19, and sketch the corresponding magnitude and phase line spectra.

Answer:
$$X_0 = 0.5, \quad X_n = \frac{j}{n\pi}, \quad n = 1, 2, 3, \ldots$$

$$|X_0| = 0.5, \quad \angle X_0 = 0, \quad |X_n| = \frac{1}{n\pi}, \quad \angle X_n = \frac{\pi}{2}, \quad n = 1, 2, 3, \ldots$$

$$x(t) = 0.5 + 2 \sum_{n=1}^{\infty} \frac{1}{n\pi} \cos\left(n\omega_0 t + \frac{\pi}{2}\right)$$

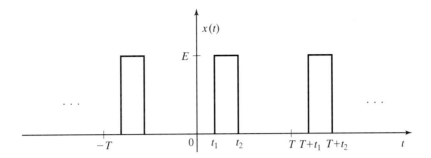

FIGURE 3.20: A train of rectangular signals

3.3. Find the exponential form of the Fourier series of a train of rectangular signals as presented in Figure 3.20. Evaluate the coefficients obtained for $t_1 = 0$ and $t_2 = T/2$.
Answer:
$$X_n(jn\omega_0) = \frac{jE}{2n\pi}((-1)^n - 1), \quad t_1 = 0, \quad t_2 = \frac{T}{2}$$

3.4. Sketch the magnitude and phase line spectra for the exponential Fourier series coefficients found in Problem 3.3.

3.5. Verify the trigonometric identity
$$a\cos(\alpha t) \mp b\sin(\alpha t) = \sqrt{a^2 + b^2} \cos\left(\alpha t \pm \tan^{-1}\left(\frac{b}{a}\right)\right)$$

Use this identity to derive directly the complex coefficient trigonometric Fourier series (3.10) from the real coefficient trigonometric Fourier series (3.2).

3.6. Find the Fourier series of the periodic signal presented in Figure 3.21, and plot the corresponding magnitude and phase line spectra.

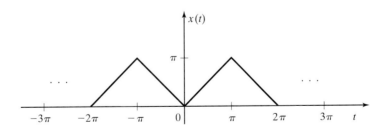

FIGURE 3.21: An even periodic signal

Answer:
$$a_0 = \pi, \quad a_n = (-2/\pi n^2)(1 - (-1)^n), \quad b_n = 0, \quad n = 1, 2, 3, \ldots$$

134 Chapter 3 Fourier Series and Fourier Transform

3.7. Find Fourier transforms of the following signals under the assumption that $\mathcal{F}\{x(t)\} = X(j\omega)$ is known.
(a) $x(at - t_0)$
(b) $tx(t - t_0)$
(c) $dx(at)/dt$

3.8. Find Fourier transforms of the following signals.
(a) $u_h(t)\cos(\omega_0 t)$
(b) $u_h(t)e^{-\alpha t}\sin(\omega_0 t)$ and $u_h(t)e^{-\alpha t}\cos(\omega_0 t)$
(c) $e^{j\omega_0 t}u_h(t)$
(d) $t^{n-1}e^{-\alpha t}u_h(t)$
(e) $e^{-t}[u_h(t) - u_h(t-1)]$ and $\cos(2\pi t)[u_h(t+1) - u_h(t-1)]$

Answers:

(b) $u_h(t)e^{-\alpha t}\cos(\omega_0 t) \leftrightarrow \dfrac{\alpha + j\omega}{(\alpha + j\omega)^2 + \omega_0^2}$, $\quad u_h(t)e^{-\alpha t}\sin(\omega_0 t) \leftrightarrow \dfrac{\omega_0}{(\alpha + j\omega)^2 + \omega_0^2}$

(c) $\dfrac{1}{j(\omega - \omega_0)} + \pi\delta(\omega - \omega_0)$

(d) $j^{n-1}\dfrac{d^{n-1}}{d\omega^{n-1}}\left(\dfrac{1}{\alpha + j\omega}\right) = \dfrac{(n-1)!}{(\alpha + j\omega)^n}$

(e) $e^{-t}[u_h(t) - u_h(t-1)] \leftrightarrow \dfrac{1}{1+j\omega}\left(1 - e^{-(1+j\omega)}\right)$

3.9. A time signal $x(t)$ has the Fourier transform $4/(4 + \omega^2)$. Find Fourier transforms of the following signals.
(a) $x_1(t) = x(2t - 3)$
(b) $x_2(t) = \displaystyle\int_{-\infty}^{t} x(\tau)\,d\tau$

3.10. Using the duality property, find the Fourier transform of the sinc signal, $\text{sinc}(t) = \sin(\pi t)/\pi t$.

3.11. Find Fourier transforms of the following signals.
(a) $x_1(t) = \displaystyle\sum_{n=-\infty}^{n=\infty} \dfrac{1}{n\pi} e^{jn\omega_0 t}$
(b) $x_2(t) = e^{-7t} u_h(t) \operatorname{sgn}(2t - 5)$
(c) $x_3(t) = \dfrac{1}{t}$
(d) $x_4(t) = \text{sinc}(2t - 5)$

Answer: (c) $-j\pi\operatorname{sgn}(\omega)$

3.12. (a) Show that the Fourier transform of $e^{-|t|}$ is $2/(1 + \omega^2)$, and find the Fourier transform of the following signals:

$$x_1(t) = e^{-2|t|}, \quad x_2(t) = e^{-2|t|}\sin(t), \quad \text{and} \quad x_3(t) = te^{-2|t|}$$

(b) Use the duality property to find the Fourier transform of the signal

$$x(t) = \frac{1}{(1+t^2)}$$

Answer: (b) $\pi e^{-|\omega|}$

3.13. A continuous-time signal $x(t)$ has the Fourier transform

$$X(j\omega) = \frac{2a}{\omega^2 + a^2}, \quad a > 0$$

Find the Fourier transform of the signal

$$x_1(t) = x(5t - 4) + tx(t - 1) + \frac{dx(3t)}{dt} + x(t - 1)\cos(4t)$$

Answer:

$$\frac{dx(at)}{dt} \leftrightarrow j\left(\frac{\omega}{|a|}\right) X\left(\frac{j\omega}{a}\right), \quad tx(t - t_0) \leftrightarrow j\frac{dX(j\omega)}{d\omega}e^{-j\omega t_0} + t_0 X(j\omega)e^{-j\omega t_0}$$

$$X_1(j\omega) = \frac{1}{5}e^{-j(\frac{\omega}{5})4} X\left(\frac{j\omega}{5}\right) + e^{-j\omega}\left(X(j\omega) + j\frac{dX(j\omega)}{d\omega}\right) + j\frac{\omega}{3}X\left(\frac{j\omega}{3}\right)$$

$$+ \frac{1}{2}\left[e^{-j(\omega+4)}X(j(\omega + 4)) + e^{-j(\omega-4)}X(j(\omega - 4))\right]$$

3.14. A signal $x(t)$ has the Fourier transform $X(j\omega) = \frac{1}{j\omega}$. Determine Fourier transforms of the following signals.
  (a) $x_1(t) = 5x(3t - 2)$
  (b) $x_2(t) = x(t - 1)\sin(2t)$
  (c) $x_3(t) = x^2(t - 5)$

3.15. Repeat Problem 3.13 with $X(\omega) = \frac{1}{j\omega} + \pi\delta(\omega)$.

3.16. Derive the expressions for the Fourier transforms of sine and cosine signals by using the fact that $e^{\pm j\omega_0 t} \leftrightarrow 2\pi\delta(\omega \pm \omega_0)$.

Hint: Use Euler's formula.

3.17. Show that the Fourier transform of $d^n\delta(t)/dt^n$ is given by $2\pi j^n(d^n\delta(\omega)/d\omega^n)$. Then use the duality property to find the Fourier transform of $x(t) = t^n$. Verify your result by using the time multiplication property.

3.18. Find Fourier transforms of the signals presented in Figure 3.22.
Hints: (a) See Example 3.13. (b) Use the definition of the Fourier transform.

3.19. Show that Fourier transforms of real even signals ($x(-t) = x(t)$) are real, and Fourier transforms of real odd signals ($x(-t) = -x(t)$) are pure imaginary.

Hints: The product of even functions is even, the product of odd functions is even, and the product of even and odd functions is odd. An integral of an odd function with integration limits from $-\infty$ to ∞ is zero.

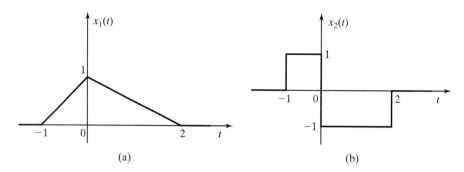

FIGURE 3.22: Fourier transformable signals

3.20. Find Fourier transforms of signals presented in Figure 3.23(a) and (b) by using the fact that $\mathcal{F}\{p_\tau(t)\} = \tau\operatorname{sinc}(\frac{\omega\tau}{2\pi})$.
Hint: See Example 3.13.

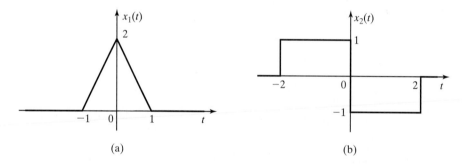

FIGURE 3.23: Time domain signals

3.21. Find Fourier transforms of the following signals.
(a) $\dfrac{5}{t} + e^{-2t}u_h(t)\cos(2t) + p_2(3t - 4)$
(b) $\dfrac{2}{t-1} + e^{-2t}u_h(t)\sin(3t) + p_4(3t - 2)$
(c) $\dfrac{2}{t+2} + p_4(2t+2) + e^{-2t}[u_h(t-1) - u_h(t-3)]$
Answer:
(c) $-2j\pi e^{2j\omega}\operatorname{sgn}(\omega) + 2e^{j\omega}\operatorname{sinc}\left(\dfrac{\omega}{\pi}\right) - \dfrac{1}{2+j\omega}\left(e^{-j3(2+j\omega)} - e^{-(2+j\omega)}\right)$

3.22. Find Fourier transforms of the signals presented in Figure 3.24.

3.23. Find the inverse Fourier transforms of the following signals.
(a) $\dfrac{\cos(\omega)}{j\omega+2}e^{-j3\omega}$ and $\dfrac{\cos(\omega+2)}{j\omega+2}e^{-j3\omega}$

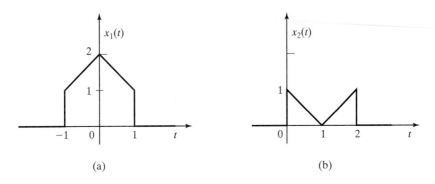

(a)   (b)

FIGURE 3.24: Time domain signals

(b) $p_2(\omega)\cos(\omega)$ and $p_2(2\omega)\sin(2\omega)$

(c) $\dfrac{j\omega}{5+j\omega} e^{-j3\omega}$ and $\dfrac{d}{dt}\left(e^{-5(t-3)}u_h(t-3)\right)$

(d) $\dfrac{1}{j(\omega-2)} \tau\,\text{sinc}\left(\dfrac{\omega\tau}{2}\right)$ and $\dfrac{1}{j\omega}\dfrac{1}{(1+j\omega)^2}$

Answers:

(a) $\dfrac{\cos(\omega+2)}{j\omega+2} e^{-j3\omega} \leftrightarrow \dfrac{e^{j2}}{2} e^{-2(t-2)} u_h(t-2) + \dfrac{e^{-j2}}{2} e^{-2(t-4)} u_h(t-4)$

(b) $p_2(\omega)\cos(\omega) \leftrightarrow \dfrac{1}{2\pi}\left[\text{sinc}\left(\dfrac{t+1}{\pi}\right) + \text{sinc}\left(\dfrac{t-1}{\pi}\right)\right]$

3.24. Find the inverse of the Fourier transform of the signal whose frequency representation is given in Figure 3.25.

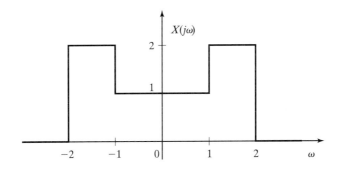

FIGURE 3.25: Frequency domain signal

Hint: $X(j\omega) = 2p_4(\omega) - p_2(\omega)$, with the inverse Fourier transform of $p_\tau(\omega)$ as defined in Example 3.16.

Chapter 3 · Fourier Series and Fourier Transform

3.25. Find the inverse Fourier transformation of the following signals.

(a) $\dfrac{1}{1+j\omega}\cos(2\omega)e^{-j5\omega} + p_6(\omega)\sin(\omega)$

(b) $\dfrac{1}{1+\omega^2}\cos(5\omega) + p_3(\omega)\cos(\omega)$

Answers:

(a) $\dfrac{0.5}{1+j\omega}(e^{-j3\omega} + e^{-j7\omega}) + \dfrac{3}{2j\pi}\left(\operatorname{sinc}\left(\dfrac{3(t+1)}{\pi}\right) - \operatorname{sinc}\left(\dfrac{3(t-1)}{\pi}\right)\right)$

(b) $\dfrac{1}{4}(e^{-|t+5|} + e^{-|t-5|}) + \dfrac{3}{4\pi}\left(\operatorname{sinc}\left(\dfrac{3(t+1)}{2\pi}\right) + \operatorname{sinc}\left(\dfrac{3(t-1)}{2\pi}\right)\right)$

3.26. Using direct integration, verify that

$$\frac{j}{2}\int_{-\infty}^{\infty}[\delta(\omega-\omega_0) - \delta(\omega+\omega_0)]e^{j\omega t}d\omega = \sin(\omega_0 t)$$

3.27. Erbium-doped optical fiber (photonic) amplifier, popularly known as EDFA, made a revolution in optical communication in the second half of the 1990s. It played a dominant role in increasing signal transmission speeds from several megabits to several gigabits per second. Assuming that there is only one signal to be amplified, and constant laser pump power (the laser pumps photons in the process of signal amplification), small sinusoidal input and output signal variations in the frequency domain are related by the following algebraic equation [10]:

$$Y_s(j\omega) = X_s(j\omega) - \frac{k\omega_c}{j\omega + \omega_c}X_s(j\omega)$$

where k is a known constant, ω_c is the amplifier's "corner frequency," and

$$X_s(j\omega) = \mathcal{F}(\sin(\omega_0 t)) = j\pi[\delta(\omega-\omega_0) - \delta(\omega+\omega_0)]$$

Using the inverse Fourier transform, find the time domain output signal variations $y_s(t) = \mathcal{F}^{-1}\{Y_s(j\omega)\}$.

Answer:

$$y_s(t) = x_s(t) + \frac{k\omega_c^2}{\omega_0^2+\omega_c^2}x_s(t) - \frac{k\omega_c\omega_0}{\omega_0^2+\omega_c^2}x_s\left(t+\frac{\pi}{2}\right), \quad x_s(t) = \sin(\omega_0 t)$$

3.28. Find the Fourier inverse of the Heaviside unit step signal defined in the frequency domain by

$$U_h(j\omega) = \begin{cases} 1, & \omega > 0 \\ 0.5, & \omega = 0 \\ 0, & \omega < 0 \end{cases}$$

Answer:

$$\frac{1}{2}\delta(t) + j\frac{1}{2\pi t}$$

3.29. Using Parseval's theorem, evaluate the energy of the sinc(t) signal; that is, evaluate the integral

$$\int_{-\infty}^{\infty} \text{sinc}^2(t) \, dt$$

3.30. It follows from the definition formula of the Fourier transform that, for $\omega = 0$,

$$X(0) = \int_{-\infty}^{\infty} x(t) \, dt$$

Using this formula and the corresponding Fourier transform pair, evaluate the integral

$$\int_{-\infty}^{\infty} \text{sinc}(t) \, dt$$

Answer: 1.

3.31. Can the spectrum of an original signal be recovered by modulating the modulated signal $x(t) \cos(\omega_0 t)$ with the cosine signal of the same angular frequency ω_0?

3.32. Find the transfer function of the system

$$\frac{d^2 y}{dt^2} + 7\frac{dy}{dt} + 12y = \frac{dx}{dt} + 2x$$

and calculate its impulse response.

3.33. Find the transfer function of the linear system

$$\frac{d^5 y}{dt^5} + 2\frac{d^4 y}{dt^4} + \frac{d^3 y}{dt^3} + 2\frac{d^2 y}{dt^2} + \frac{dy}{dt} + 3y = \frac{d^3 x}{dt^3} + 3\frac{d^2 x}{dt^2} + x$$

Answer: $H(s) = (s^3 + 3s^2 + s)/(s^5 + 2s^4 + s^3 + 2s^2 + s + 3)$

3.34. Find the response of the linear system at rest

$$\frac{dy(t)}{dt} + y(t) = x(t), \quad x(t) = \sin(t)$$

using the Fourier transform and formulas (3.67–69).

3.35. Repeat Problem 3.34 with $x(t) = e^{-2t} u_h(t)$.

3.36. Find the response of a linear system represented by

$$H(j\omega) = \frac{a}{j\omega + b}$$

due to a periodic input given by $x(t) = (c + jd)e^{j2\omega_0 t} + jq e^{j5\omega_0 t}$, where $a, b, c, d,$ and q are known positive constants.

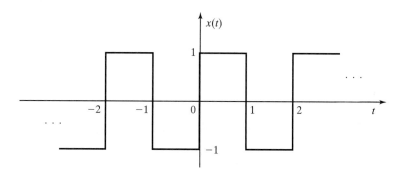

FIGURE 3.26: Periodic input signal

3.37. Find the response of the system whose transfer function is given in Example 3.19 to the periodic input defined in Figure 3.19. Use formula (3.78), and find analytical expressions for the magnitude and phase of the corresponding Fourier series coefficients.

3.38. Repeat Problem 3.37 for the periodic input given in Figure 3.20.

3.39. Find the response of the system defined in Example 3.19 to a sinusoidal input given by $x(t) = 5\sin(2t + \frac{\pi}{6})$.
Answer: $y(t) = (5/\sqrt{17})\cos(2t - \pi/3 - \tan^{-1}(-4)) = 1.2127\cos(2t + 15.96°)$

3.40. Show that
$$\mathcal{F}\{\cos(\omega_0 t + \theta)\} = e^{j\theta}\pi\delta(\omega - \omega_0) + e^{-j\theta}\pi\delta(\omega + \omega_0)$$
Using this fact, derive formula (3.80) for the linear system response due to $\cos(\omega_0 t + \theta)$.

3.41. Find the response of a system whose transfer function is
$$H(j\omega) = \frac{j\omega}{(1+j\omega)^2}$$
to the input signal given by $x(t) = 2\cos(t + \frac{\pi}{3})$.
Answer: $\cos(t + \pi/3)$

3.42. Find the response of a system whose transfer function is
$$H(j\omega) = \frac{1}{1+j\omega}$$
to the periodic input signal represented in Figure 3.26.

3.43. Find the response of a system whose transfer function is
$$H(j\omega) = \frac{1+j\omega}{j\omega(2+j\omega)}$$
to the periodic input signal represented in Figure 3.27.

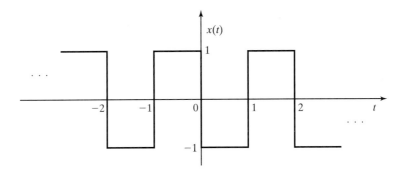

FIGURE 3.27: Periodic input signal

Answer:

$$y(t) = \sum_{1,3,\ldots}^{\infty} \left(\frac{2}{n^2\pi^2}\right) \frac{\sqrt{1+n^2\pi^2}}{\sqrt{4+n^2\pi^2}} \cos\left(n\pi t + \tan^{-1}(n\pi) - \tan^{-1}\left(\frac{n\pi}{2}\right)\right)$$

3.44. Establish the orthogonality condition for complex exponentials; that is, show that

$$\int_0^T e^{j(k-m)\omega_0 t}\, dt = \begin{cases} T, & k = m \\ 0, & k \neq m \end{cases}$$

Use this condition to derive the Parseval formula for periodic signals given in (3.15).

3.45. Verify the Parseval formula for periodic signals for the periodic signal presented in Figure 3.27.

Hint: Approximate the infinite sum by a finite sum, such that the summation error is less than 0.1.

3.46. Show that $e^{-j\omega t}$ is the eigenfunction of the time-invariant, continuous-time, linear dynamic system with the corresponding eigenvalue equal to $H(-j\omega)$.

3.47. Use the concept of the linear system eigenfunction established in (3.70) to derive formulas for linear system response due to sine and cosine input signals.

Hint: Use (3.70) and the result derived in Problem 3.46, and combine them with the Euler formula.

3.48. Reconsider the electrical circuit presented in Figure 1.10. Assume that $R_1 = R_2 = 10\,\mathrm{k}\Omega$, $L = 5\,\mathrm{mH}$, and $C = 2\,\mu\mathrm{F}$. Using Fourier series analysis and the circuit differential equation derived in (1.26), find the circuit steady state output voltage due to the periodic input presented in Figure 3.1.

3.49. Repeat Problem 3.48 for the periodic input signal whose waveform is given in Figure 3.18.

3.50. Find the output voltage due to a sinusoidal input $e_i(t) = \cos(3t + \pi/3)$ for the electrical circuit described in Problem 3.48.

3.51. Solve the problem defined in Example 3.20 using the square wave presented in Figure 3.18 as the circuit input.

3.52. Find analytically the frequency spectrum for the second-order system that models eye movement, as defined in (1.39).

3.53. Using formula (3.67), show that the system input and output average powers are related by

$$P_y(j\omega) = |H(j\omega)|^2 P_x(j\omega)$$

3.54. Find the steady state response of a linear harmonic oscillator defined by

$$\frac{d^2 y(t)}{dt^2} + \omega_n^2 y(t) = x(t)$$

to the square-wave input whose waveform is presented in Figure 3.18. Use MATLAB to plot the first five harmonics of the steady state response when $\omega_n^2 = 1$, $T = 1$, $E = 1$.

3.55. Using MATLAB, plot the frequency spectrum for oculomotor dynamics (eye movement dynamics), whose mathematical model is given in (1.39). Estimate graphically the frequency bandwidth.

3.56. Using MATLAB, plot the frequency spectrum of vehicle lateral error dynamics, whose mathematical model is given in (1.40).

3.57. Using MATLAB, plot the frequency spectrum of a commercial aircraft pitch angle, whose mathematical model is given in (1.43).

3.58. Using MATLAB, plot the first three terms of the steady state capacitor voltage derived in Example 3.20. Assume the numerical values $R = 1 \, \text{k}\Omega$, $L = 1 \, \text{mH}$, $C = 1 \, \mu\text{F}$, $E = \pi$, $\omega_0 = 100 \, \text{rad/s}$.

3.59. The dynamics of a car suspension system under road disturbances can be modeled by the second-order system

$$\frac{d^2 y(t)}{dt^2} + 2\zeta \omega_n \frac{dy(t)}{dt} + \omega_n^2 y(t) = 2\zeta \omega_n \omega_b A \cos(\omega_b t) + \omega_n^2 A \sin(\omega_b t)$$

Calculate the vibrations of this system based on the following data (taken from [11]): $A = 0.01$, $\zeta = 0.498$, $\omega_n = 19.93 \, \text{rad/s}$, $\omega_b = 5.818 \, \text{rad/s}$. Using MATLAB, plot the steady state response approximated by the first five harmonics; that is, simulate the car's vibrations including dominant harmonics. How quickly does the magnitude of the harmonics decay to zero?

3.60. Assume a model (based, for instance, on the viewing of a tennis match) in which the human eye is exposed to periodic stimuli that can be roughly represented by the waveform given in Figure 3.26, with the magnitude equal to 15° and the time period of 2 s. Using the eye-movement mathematical model given in (1.39), determine the steady state movement of the eyes. Using MATLAB, tabulate the magnitude of the first ten harmonics.

CHAPTER 4

Laplace Transform

The Laplace transform was introduced in Section 3.5 as the Fourier transform of a function multiplied by an exponentially decaying function. This modification aids the convergence of the corresponding integral, and produces very important features that make the Laplace transform one of the most powerful tools for the analysis of continuous-time linear time-invariant dynamic systems.

Both the Fourier and Laplace transforms are described by similar integrals, hence many properties valid for the Fourier transform extend to the Laplace transform. It is important to emphasize that the Laplace transform of a function is always obtained in terms of regular functions. Thus, it is much easier to work with the Laplace transform than with the Fourier transform. In addition, the Laplace transform is more general than the Fourier transform: it can be used to analyze both the zero-state and zero-input responses of linear time-invariant dynamic systems, since it is able to take into the account the system initial conditions and find the corresponding response. The Laplace transform preserves in the same manner the derivative and convolution properties of the Fourier transform, which facilitates the study of systems described by linear constant coefficient differential equations via the Laplace transform in the frequency domain, in terms of linear algebraic equations with complex coefficients.

The Laplace transform and its properties are presented in Section 4.1. The method for finding the inverse of the Laplace transform is given in Section 4.2. For completeness, we introduce the formal definition of the inverse Laplace transform, though that definition is not used to find the inverse Laplace transform. A simple method, based on the partial fraction expansion, is used to determine the inverse Laplace transform; this method is applicable to all complex domain functions encountered in this introductory course to linear dynamic systems. In Section 4.3, we show how to use the Laplace transform in the analysis of linear time-invariant dynamic systems. Similarly to the case of the Fourier transform, we define the system transfer function in the Laplace domain ($s = \sigma + j\omega$ domain), and the system impulse response as the inverse Laplace transform of the system transfer function. We state and derive (or—to be precise—rephrase, since this result has been already obtained in the Fourier $j\omega$ domain) that *for a system at rest, the system response due to any input is equal to the convolution of the system impulse response and the given input,* which constitutes one of the main results of linear system theory. Since linear dynamic systems are described by linear differential equations, the Laplace transform is also a very powerful tool for solving constant coefficient linear differential equations, especially when the forcing function is discontinuous and/or contains delta impulses. In Section 4.4 we present an introduction to block diagrams and use Simulink to find responses of some real-world continuous-time linear systems. In the same section we introduce the concept of linear feedback systems. Section 4.5 presents the passage from the Laplace transform to its discrete-time counterpart, the $\mathcal{Z}$-transform. A MATLAB laboratory experiment on the Laplace transform and its use in analysis of continuous-time linear time-invariant systems is presented in Section 4.6.

4.1 LAPLACE TRANSFORM AND ITS PROPERTIES

In this section, we give definitions of the Laplace transform and its inverse, establish the corresponding existence condition, and state and prove all important properties of the Laplace transform.

4.1.1 Definitions and Existence Condition

The Laplace transform of a continuous-time signal (function) $f(t)$ is defined by

$$\mathcal{L}\{f(t)\} = F(s) \triangleq \int_{0^-}^{\infty} f(t) e^{-st} \, dt \tag{4.1}$$

This is the *one-sided* or *unilateral Laplace transform* of $f(t)$, and depends only on the values of $f(t)$ for $t \geq 0$; that is, it is applicable to *causal* signals for which $f(t) = 0, t < 0$ and $f(t) \neq 0, t \geq 0$. In general, the *two-sided* or *bilateral* Laplace transform, with the lower limit in the integral equal to $-\infty$, can be defined (see (3.85)). For the purpose of this course, however, it is sufficient to define and use only the one-sided Laplace transform. Note that the definition integral of the Laplace transform given in (4.1) is a little different from the one introduced in (3.86) as

$$\mathcal{L}\{f(t)\} = F(s) \triangleq \int_{0}^{\infty} f(t) e^{-st} \, dt \tag{4.1a}$$

The difference comes from the fact that in (4.1) the integration is performed from 0^- instead of from 0. The reason for this change is the fact that the Laplace transform in linear system theory is used first of all to study the response of linear time-invariant systems, hence we must be able to find the system impulse response due to the impulse delta signal located at the time origin, $t = 0$, which requires integration from 0^- in order to completely include the impulse delta signal within the integration limits. We know from Chapter 1 that the system response has two components: zero-state and zero-input. The zero-input component is contributed by the system initial conditions (stored system energy) that are present in the system before the input is applied. Assuming that the initial time at which the system input is applied is $t = 0$, the system initial conditions must be defined at $t = 0^-$. In such a case, the Laplace transform defined by (4.1) will take into account and produce the system response due to the system initial conditions.

Note that in the case of functions (signals) that do not contain an impulse delta signal at the origin, the two definition formulas of the one-sided Laplace transform, (4.1) and (4.1a), are identical.

The Laplace transform $F(s)$ as defined either in (4.1) or (4.1a) depends on a complex variable $s = \sigma + j\omega$, called the complex frequency. The existence condition of the Laplace transform requires the convergence of the corresponding integral, that is,

$$\left| \int_{0^-}^{\infty} f(t) e^{-(\sigma + j\omega)t} \, dt \right| < \infty$$

This leads to

$$\left| \int_{0^-}^{\infty} f(t) e^{-(\sigma + j\omega)t} \, dt \right| \leq \int_{0^-}^{\infty} \left| f(t) e^{-(\sigma + j\omega)t} \right| dt \leq \int_{0^-}^{\infty} |f(t)| \left| e^{-(\sigma + j\omega)t} \right| dt$$

$$\leq \int_{0^-}^{\infty} |f(t)| e^{-\sigma t} \, dt < \infty \tag{4.2}$$

where σ is a nonnegative real number. The last inequality in (4.2) represents the existence condition for the Laplace transform.

The inverse Laplace transform can be obtained from the definition of the inverse Fourier transform (given by (3.18)) using the facts that $j\omega$ must be replaced by $s = \sigma + j\omega$ and that $j \, d\omega = ds$. This leads to the following definition of the inverse Laplace transform:

$$f(t) = \mathcal{L}^{-1}\{F(s)\} \triangleq \frac{1}{2\pi j} \int_{\gamma - j\infty}^{\gamma + j\infty} F(s) e^{st} \, ds \tag{4.3}$$

where γ is a real value chosen to the right of all singularities of the function $F(s)$ (most common singularities are the poles of $F(s)$, the values of s at which $F(s) = \infty$). The complete definition of the singularities of complex variable functions is outside of the scope of this course, but can be found in books on complex variables (for example, [1–3]).

The inverse Laplace transform, as defined by (4.3), represents a complex variable integral, which in general is not easy to calculate. In order to avoid integration of a complex variable function (using the method known as contour integration [1–4]), the procedure used in this text for finding the Laplace inverse combines the method of partial fraction expansion, properties of the Laplace transform to be derived in this section, and the tables of properties and common Laplace transform pairs, given at the end of the chapter. This procedure will be presented in detail in Section 4.2. The use of the partial fraction expansion method is sufficient for the purpose of this course; in general, however, to find the Laplace transform of any Laplace transformable function we must learn complex variable integration and obtain the result via the use of formula (4.3). Note that the inverse of the one-sided Laplace transform recovers the original function $f(t)$ for $t \geq 0$ and produces 0 for $t < 0$.

Using the definition of the Laplace transform, as given by (4.1), we are able to find directly the Laplace transform of some basic functions.

EXAMPLE 4.1

The unit step signal has the following Laplace transform:

$$\mathcal{L}\{u(t)\} = \int_0^\infty u(t)e^{-st}\,dt = \int_0^\infty e^{-st}\,dt = \frac{1}{s} \tag{4.4}$$

EXAMPLE 4.2

Laplace transforms of the impulse delta signal $\delta(t)$ and its shifted version $\delta(t - t_0)$ are easily obtained as

$$\mathcal{L}\{\delta(t)\} = \int_{0^-}^\infty \delta(t)e^{-st}\,dt = 1$$

$$\mathcal{L}\{\delta(t - t_0)\} = \int_{0^-}^\infty \delta(t - t_0)e^{-st}\,dt = e^{-st_0} \tag{4.5}$$

EXAMPLE 4.3

The exponential time decaying signal, $e^{-at}u(t)$, $a > 0$, easily produces

$$\mathcal{L}\{e(t)\} = \int_0^\infty e^{-at}e^{-st}\,dt = \frac{1}{s+a}, \quad a > 0 \tag{4.6}$$

4.1.2 Properties of the Laplace Transform

In the following we state and prove the main properties of the Laplace transform. Most of these properties are analogous to the corresponding properties of the Fourier transform, and the corresponding proofs are similar. For completeness (especially for students and instructors who will not cover the chapter on the Fourier transform), the proofs of all of the properties of the Laplace transform will be given in this section. In order to simplify the proofs, we will use the definition formula of the Laplace transform given in (4.1a) unless explicitly indicated otherwise.

Property 1: Linearity

Let $\mathcal{L}\{f_i(t)\} = F_i(s), i = 1, 2, \ldots, n$, be given Laplace transform pairs. Then for any constants $\alpha_i, i = 1, 2, \ldots, n$, the following holds:

$$\mathcal{L}\{\alpha_1 f_1(t) \pm \alpha_2 f_2(t) \pm \cdots \pm \alpha_n f_n(t)\}$$
$$= \alpha_1 \mathcal{L}\{f_1(t)\} \pm \alpha_2 \mathcal{L}\{f_2(t)\} \pm \cdots \pm \alpha_n \mathcal{L}\{f_n(t)\} \qquad (4.7)$$
$$= \alpha_1 F_1(s) \pm \alpha_2 F_2(s) \pm \cdots \pm \alpha_n F_n(s)$$

Proof. This property can be proved by using the known properties of integrals, as follows.

$$\mathcal{L}\{\alpha_1 f_1(t) \pm \alpha_2 f_2(t) \pm \cdots \pm \alpha_n f_n(t)\}$$
$$= \int_0^\infty [\alpha_1 f_1(t) \pm \alpha_2 f_2(t) \pm \cdots \pm \alpha_n f_n(t)] e^{-st} \, dt$$
$$= \alpha_1 \int_0^\infty f_1(t) e^{-st} \, dt \pm \alpha_2 \int_0^\infty f_2(t) e^{-st} \, dt \pm \cdots \pm \alpha_n \int_0^\infty f_n(t) e^{-st} \, dt$$
$$= \alpha_1 F_1(s) \pm \alpha_2 F_2(s) \pm \cdots \pm \alpha_n F_n(s) \qquad \square$$

Hence, the Laplace transform of a linear combination of continuous-time functions can be obtained by breaking them into elementary continuous-time functions whose Laplace transforms are given (are known or can be easily found). In other words, the Laplace transform of a linear combination of signals is a linear combination of the Laplace transforms of the signals.

Property 2: Time Shifting

This property is similar to the corresponding property of the Fourier transform. It states that

$$\mathcal{L}\{f(t)u(t)\} = F(s) \Rightarrow \mathcal{L}\{f(t-t_0)u(t-t_0)\} = e^{-t_0 s} F(s), \quad t_0 > 0 \qquad (4.8)$$

where t_0 is the positive time shifting parameter. It should be emphasized that shifting the signal left in time as defined by $f(t+t_0)u(t+t_0), t_0 > 0$, in general violates signal causality, so that the one-sided Laplace transform cannot be correctly applied. For that reason the stated time shifting property is also called the "right shift in time" property. Only if the signal remains causal under left time shifting will we be able to find the corresponding one-sided Laplace transform. For example, the rectangular pulse $p_2(t-3)$ can be shifted to the left by two time units and still remain causal.

Proof. Property (4.8) can be proved as follows. By the definition of the Laplace transform, we have

$$\mathcal{L}\{f(t - t_0)u(t - t_0)\} = \int_{t=0}^{t=\infty} f(t - t_0)u(t - t_0)e^{-st}\, dt$$

Using a change of variables $t - t_0 = \tau$, we obtain

$$\int_{\tau=-t_0}^{\tau=\infty} f(\tau)u(\tau)e^{-s(t_0+\tau)}\, d\tau = e^{-st_0}\int_{0}^{\infty} f(\tau)e^{-s\tau}\, d\tau = e^{-st_0}F(s) \qquad \square$$

We want to emphasize the importance of the proper use of time shifting because, in contrast to the Fourier transform where $t \in (-\infty, \infty)$, here we deal with the semi-infinite time axis $t \in [0, \infty)$. In Figure 4.1, we represent (a) the original function, (b) the function properly time-shifted according to (4.8), and (c) the function that analytically looks as though it has been time shifted, though it has only been multiplied by e^{t_0} and not shifted in time. It can be concluded from Figure 4.1 that the causal function (signal) $f(t)$ is properly shifted by finding $f(t - t_0)$ and multiplying it by $u(t - t_0)$, that is, by forming the signal $f(t - t_0)u(t - t_0)$.

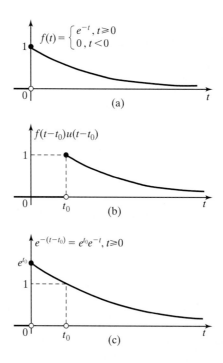

FIGURE 4.1: Proper shifting of a continuous-time function (signal)

EXAMPLE 4.4

Using the time shifting property, we find the Laplace transform of the signal $(t^2+1)u(t-1)$ in the following manner:

$$\begin{aligned}\mathcal{L}\{(t^2+1)u(t-1)\} &= \mathcal{L}\{(t^2+2t-2t+1)u(t-1)\} \\ &= \mathcal{L}\{[(t-1)^2+2t]u(t-1)\} \\ &= \mathcal{L}\{(t-1)^2 u(t-1)\} + \mathcal{L}\{2tu(t-1)\} \\ &= e^{-s}\mathcal{L}\{t^2 u(t)\} + \mathcal{L}\{2(t-1+1)u(t-1)\} \\ &= \frac{2e^{-s}}{s^3} + \mathcal{L}\{2(t-1)u(t-1)\} + \mathcal{L}\{2u(t-1)\} \\ &= \frac{2e^{-s}}{s^3} + \frac{2e^{-s}}{s^2} + \frac{2e^{-s}}{s}\end{aligned}$$

The preceding steps are self-explanatory. Note that we must bring each term into the form defined in (4.8). ∎

Property 3: Time Scaling

This property has already been established in the context of the Fourier transform. Here, it reads

$$\mathcal{L}\{f(t)\} = F(s) \Rightarrow \mathcal{L}\{f(at)\} = \frac{1}{a} F\left(\frac{s}{a}\right), \quad a > 0 \tag{4.9}$$

where a is a *positive* time scaling parameter. Note that if the parameter a is negative the original causal signal $f(t)$ is transformed into a noncausal signal $f(-|a|t)$ (through time reversal and time scaling), so that the single-sided Laplace transform is not applicable in this case.

Proof. We take the Laplace transform of the signal $f(at)$ and introduce the change of variables $\sigma = at$, $a > 0$. This leads to

$$\mathcal{L}\{f(at)\} = \int_0^\infty f(at)e^{-st}\, dt = \frac{1}{a}\int_0^\infty f(\sigma)e^{-\frac{s}{a}\sigma}\, d\sigma = \frac{1}{a} F\left(\frac{s}{a}\right), \quad a > 0 \quad \square$$

We can also combine both the time scaling and time shifting properties, which leads to

$$\mathcal{L}\{f(t)u(t)\} = F(s) \Rightarrow \mathcal{L}\{f(a(t-t_0))u(t-t_0)\} = \frac{1}{|a|} F\left(\frac{s}{a}\right) e^{-t_0 s}$$

$$a > 0, \quad t_0 > 0 \tag{4.10}$$

We leave to students to verify formula (4.10) in Problem 4.1.

Property 4: Time Multiplication

The time multiplication property is analogous to the corresponding property for the Fourier transform. In the case of the Laplace transform, it states that

$$\mathcal{L}\{f(t)\} = F(s) \Rightarrow \mathcal{L}\{t^n f(t)\} = (-1)^n \frac{d^n F(s)}{ds^n} \qquad (4.11)$$

where t represents time and s is the complex frequency.

Proof. The proof of this property is as follows:

$$F(s) = \int_0^\infty f(t) e^{-st} \, dt \Rightarrow \frac{dF(s)}{ds} = \int_0^\infty (-t) f(t) e^{-st} \, dt$$

or

$$(-1) \frac{dF(s)}{ds} = \int_0^\infty t f(t) e^{-st} \, dt = \mathcal{L}\{t f(t)\}$$

In general, taking the nth derivative, we have

$$\frac{d^n F(s)}{ds^n} = \int_0^\infty (-t)^n f(t) e^{-st} \, dt$$

which, after a multiplication by $(-1)^n$, produces (4.11). □

EXAMPLE 4.5

The Laplace transforms of the signals $t^n u(t), n = 1, 2, \ldots$, are obtained by applying the time multiplication property to the result of Example 4.1:

$$\mathcal{L}\{u(t)\} = \frac{1}{s} \Rightarrow tu(t) \leftrightarrow (-1) \frac{d}{ds}\left(\frac{1}{s}\right) = \frac{1}{s^2}$$

$$t^2 u(t) \leftrightarrow (-1)^2 \frac{d^2}{ds^2}\left(\frac{1}{s}\right) = \frac{2}{s^3} \qquad (4.12)$$

$$\vdots$$

$$t^n u(t) \leftrightarrow (-1)^n \frac{d^n}{ds^n}\left(\frac{1}{s}\right) = \frac{n!}{s^{n+1}}$$

Property 5: Frequency Shifting

The frequency shifting property has been established in the context of the Fourier transform. In the case of the Laplace transform, it is given by

$$\mathcal{L}\{f(t)\} = F(s) \Rightarrow \mathcal{L}\{f(t) e^{\lambda t}\} = F(s - \lambda) \qquad (4.13)$$

where λ represents the frequency shift.

Proof. A simple and a short proof of this property is as follows:

$$\mathcal{L}\{f(t)e^{\lambda t}\} = \int_0^\infty f(t)e^{\lambda t}e^{-st}\,dt = \int_0^\infty f(t)e^{-(s-\lambda)t}\,dt \triangleq F(s-\lambda) \qquad \square$$

The next property, important for communication and signal processing systems, is a direct consequence of the frequency shifting property.

Property 6: Modulation

This property in the Fourier domain ($j\omega$ domain) is very important for communication systems. The corresponding property in the Laplace domain has no great importance for linear system theory; however, it can be used to find the Laplace transform of some signals. The modulation property is directly derived from the frequency shifting property by using Euler's formula and $\pm j\omega_0$ frequency shifts of the function $F(s)$. That is,

$$f(t)e^{j\omega_0 t} = f(t)[\cos(\omega_0 t) + j\sin(\omega_0 t)] \leftrightarrow F(s - j\omega_0)$$

and

$$f(t)e^{-j\omega_0 t} = f(t)[\cos(\omega_0 t) - j\sin(\omega_0 t)] \leftrightarrow F(s + j\omega_0)$$

imply the modulation property, defined by

$$f(t)\cos(\omega_0 t) \leftrightarrow \frac{1}{2}[F(s+j\omega_0) + F(s-j\omega_0)]$$
$$f(t)\sin(\omega_0 t) \leftrightarrow \frac{j}{2}[F(s+j\omega_0) - F(s-j\omega_0)]$$
(4.14)

EXAMPLE 4.6

Laplace transforms of the cosine and sine functions can be found by using the modulation property as follows:

$$u(t) \leftrightarrow \frac{1}{s} \Rightarrow u(t)\cos(\omega_0 t) \leftrightarrow \frac{1}{2}\left[\frac{1}{s+j\omega_0} + \frac{1}{s-j\omega_0}\right] = \frac{s}{s^2 + \omega_0^2}$$

$$u(t) \leftrightarrow \frac{1}{s} \Rightarrow u(t)\sin(\omega_0 t) \leftrightarrow \frac{j}{2}\left[\frac{1}{s+j\omega_0} - \frac{1}{s-j\omega_0}\right] = \frac{\omega_0}{s^2 + \omega_0^2}$$
(4.15)

EXAMPLE 4.7

In this example, we find the Laplace transform of the signal $te^{-2t}\sin(\pi t)u(t-2)$. In the derivations, we will have to use the known trigonometric formula

$$\sin(\pi(t-2+2)) = \sin(\pi(t-2))\cos(2\pi) + \cos(\pi(t-2))\sin(2\pi)$$
$$= \sin(\pi(t-2))$$

The detailed derivations are given as follows:

$$\mathcal{L}\{te^{-2t}\sin(\pi t)u(t-2)\} = \mathcal{L}\{(t-2+2)e^{-2(t-2+2)}\sin(\pi(t-2+2))u(t-2)\}$$
$$= e^{-4}\mathcal{L}\{(t-2)e^{-2(t-2)}\sin(\pi(t-2))u(t-2)\}$$
$$+ 2e^{-4}\mathcal{L}\{e^{-2(t-2)}\sin(\pi(t-2))u(t-2)\}$$

Now we can apply the time shifting property, which leads to

$$\mathcal{L}\{te^{-2t}\sin(\pi t)u(t-2)\} = e^{-4}e^{-2s}\mathcal{L}\{te^{-2t}\sin(\pi t)u(t)\} + 2e^{-4}e^{-2s}\mathcal{L}\{e^{-2t}\sin(\pi t)u(t)\}$$
$$= e^{-4}e^{-2s}\frac{2\pi(s+2)}{[(s+2)^2+\pi^2]^2} + 2e^{-4}e^{-2s}\frac{\pi}{(s+2)^2+\pi^2}$$

Property 7: Time Derivatives

From the linear system theory point of view, the derivative is one of the most important properties of the Laplace transform. It helps in converting constant coefficient differential equations (which describe dynamics of continuous-time, linear time-invariant systems) into complex coefficient algebraic equations. The algebraic equations obtained are frequency (complex) domain representations of the considered linear dynamic systems. We can solve these algebraic equations rather easily and find the system output in the frequency domain. By using the procedure for finding the inverse of the Laplace transform we can recover the corresponding time domain system output. This procedure will be studied in detail in Section 4.3.

The time derivative property states that Laplace transforms of signal time derivatives are given by the expressions

$$\mathcal{L}\left\{\frac{df(t)}{dt}\right\} = sF(s) - f(0^-)$$

$$\mathcal{L}\left\{\frac{d^2 f(t)}{dt^2}\right\} = s^2 F(s) - sf(0^-) - f^{(1)}(0^-)$$

$$\vdots$$

$$\mathcal{L}\left\{\frac{d^n f(t)}{dt^n}\right\} = s^n F(s) - s^{n-1} f(0^-) - s^{n-2} f^{(1)}(0^-) - \cdots - f^{(n-1)}(0^-)$$

(4.16)

Proof. The expression for the first derivative can be proved by using the definition of the Laplace transform of the first derivative and performing integration by parts, as follows:

$$\mathcal{L}\left\{\frac{df(t)}{dt}\right\} = \int_{0^-}^{\infty} \frac{df(t)}{dt} e^{-st}\, dt = \int_{0^-}^{\infty} e^{-st}\, df(t)$$

$$= f(\infty)e^{-\infty} - f(0^-)e^0 + s\int_{0^-}^{\infty} f(t)e^{-st}\, dt = sF(s) - f(0^-)$$

Similarly, using the definition of the Laplace transform for the nth derivative and integrating n times by parts, we can verify the general formula for the Laplace transform of the nth derivative given in (4.16) (see Problem 4.2). □

The derivative property together with the convolution property, to be defined next, constitute the main linear system theory properties of the Laplace transform.

Property 8: Time Convolution

This property states exactly the same fact as the corresponding property of the Fourier transform. Namely, let

$$\mathcal{L}\{f_1(t)\} = F_1(s) \quad \text{and} \quad \mathcal{L}\{f_2(t)\} = F_2(s)$$

then

$$\mathcal{L}\{f_1(t) * f_2(t)\} = F_1(s) F_2(s) \tag{4.17}$$

Hence, the Laplace transform of the convolution of signals is equal to the product of the Laplace transforms of the signals.

Proof. The convolution of two causal signals is defined by

$$f_1(t) * f_2(t) = \int_0^\infty f_1(t - \tau) f_2(\tau) \, d\tau$$

The corresponding Laplace transform is

$$\mathcal{L}\{f_1(t) * f_2(t)\} = \int_0^\infty \left[\int_0^\infty f_1(t - \tau) f_2(\tau) \, d\tau \right] e^{-st} \, dt$$

$$= \int_0^\infty \left[\int_0^\infty f_1(t - \tau) u(t - \tau) f_2(\tau) \, d\tau \right] e^{-st} \, dt$$

We have inserted the shifted unit step function in the preceding formula in order to make clear the effect of an exchange of integrals. Interchanging the order of integrals, we have

$$\mathcal{L}\{f_1(t) * f_2(t)\} = \int_0^\infty f_2(\tau) \left[\int_0^\infty f_1(t - \tau) u(t - \tau) e^{-st} \, dt \right] d\tau$$

$$= \int_0^\infty f_2(\tau) \left[\int_\tau^\infty f_1(t - \tau) e^{-st} \, dt \right] d\tau$$

Introducing the change of variables $t - \tau = \sigma$ with $dt = d\sigma$, we obtain the stated result

$$\mathcal{L}\{f_1(t) * f_2(t)\} = \int_0^\infty f_2(\tau) \left[\int_0^\infty f_1(\sigma) e^{-s(\sigma+\tau)} d\sigma \right] d\tau$$

$$= F_1(s) \int_0^\infty f_2(\tau) e^{-s\tau} d\tau = F_1(s) F_2(s) \qquad \square$$

Property 9: Integral

The integral property of the Laplace transform states that

$$\mathcal{L}\{f(t)\} = F(s) \Rightarrow \mathcal{L}\left\{ \int_0^t f(\tau) d\tau \right\} = \frac{1}{s} F(s) \qquad (4.18)$$

Proof. The proof follows easily by the convolution property stated in (4.17), that is,

$$\int_0^t f(\tau) d\tau = \int_0^t f(\tau) u(t - \tau) d\tau = u(t) * f(t) \leftrightarrow \frac{1}{s} F(s) \qquad \square$$

Property 10: Initial Value Theorem

This theorem helps us to find the initial value of a time signal by finding a limit in the frequency domain. Let the function $f(t)$ be continuous for $t \geq 0^-$. For such a function, the initial value theorem states that

$$\mathcal{L}\{f(t)\} = F(s) \Rightarrow \lim_{t \to 0^-} \{f(t)\} = \lim_{s \to \infty} \{s F(s)\} \qquad (4.19)$$

Proof. The proof of this result is obtained by using the time derivative property and finding the corresponding limits

$$\mathcal{L}\left\{ \frac{df(t)}{dt} \right\} = \int_{0^-}^\infty \frac{df(t)}{dt} e^{-st} dt = s F(s) - f(0^-)$$

$$\Rightarrow \lim_{s \to \infty} \left\{ \int_{0^-}^\infty \frac{df(t)}{dt} e^{-st} dt \right\} = \lim_{s \to \infty} \{s F(s) - f(0^-)\}$$

$$0 = \lim_{s \to \infty} \{s F(s)\} - f(0^-) \Rightarrow f(0^-) = \lim_{t \to 0^-} \{f(t)\} = \lim_{s \to \infty} \{s F(s)\}$$

This proof is valid under the assumption that the function $f(t)$ does not have a jump discontinuity at zero, that is, $f(0^-) = f(0^+)$. In the case of functions that have jump discontinuities at the origin, say $f_\delta(t)$ has a jump discontinuity at $t = 0$, the rigorous proof of the initial value theorem ([5–6]) requires that we split the above integral as

$$\int_{0^-}^{\infty} \frac{df(t)}{dt} e^{-st} dt = \int_{0^-}^{0^+} \frac{df(t)}{dt} e^{-st} dt + \int_{0^+}^{\infty} \frac{df(t)}{dt} e^{-st} dt$$

$$= \int_{0^-}^{0^+} \frac{df(t)}{dt} e^{0} dt + \int_{0^+}^{\infty} \frac{df(t)}{dt} e^{-st} dt$$

$$= f(0^+) - f(0^-) = sF(s) - f(0^-)$$

Taking the limit when $s \to \infty$, we obtain

$$\mathcal{L}\{f_\delta(t)\} = F(s) \Rightarrow \lim_{t \to 0^+} \{f_\delta(t)\} = f_\delta(0^+) = \lim_{s \to \infty} \{sF(s)\} \quad (4.19a)$$

To be on the safe side, we can always use formula (4.19a) without checking whether or not the given function has a jump discontinuity at $t = 0$. That is why, in the table of properties given at the end of the chapter, we have stated the initial value theorem consistently with (4.19a). □

EXAMPLE 4.8

The preceding distinction is best demonstrated on the example of the unit step signal $u(t)$. Note that $u(0^+) = 1$ and $u(0^-) = 0$. Since the signal $u(t)$ has a jump discontinuity at zero, we must use formula (4.19a), which leads to

$$u(0^+) = \lim_{s \to \infty} \{sU(s)\} = \lim_{s \to \infty} \left\{s \frac{1}{s}\right\} = 1$$

and produces the correct answer. However, the application of (4.19) produces $u(0^-) = 1$, which is not correct.

Property 11: Final Value Theorem

This property is very useful since it allows the signal steady state value (final value) in time to be obtained from its Laplace transform without a need to first perform the inverse Laplace transform. The final value theorem states that

$$\lim_{t \to \infty} \{f(t)\} = \lim_{s \to 0} \{sF(s)\} \quad (4.20)$$

Proof. The proof is similar to the proof of the initial value theorem; that is,

$$\mathcal{L}\left\{\frac{df(t)}{dt}\right\} = \int_{0^-}^{\infty} \frac{df(t)}{dt} e^{-st} \, dt = sF(s) - f(0^-)$$

$$\Rightarrow \lim_{s \to 0} \left\{ \int_{0^-}^{\infty} \frac{df(t)}{dt} e^{-st} \, dt \right\} = \lim_{s \to 0} \{sF(s) - f(0^-)\}$$

$$f(\infty) - f(0^-) = \lim_{s \to 0} \{sF(s)\} - f(0^-)$$

$$\Rightarrow f(\infty) = \lim_{t \to \infty} \{f(t)\} = \lim_{s \to 0} \{sF(s)\} \qquad \square$$

The final value theorem is very often used in linear control system analysis and design, for example, in order to find the system response steady state errors (see Chapter 12 in general, and Section 12.3 in particular). Note that the final value theorem is applicable only to time functions for which the limit at infinity exists. For example, $\sin(t)$ has no limit at infinity, hence the final value theorem is not applicable to $\sin(t)$. Also note that from (4.15), the Laplace transform of $\sin(t)$ has a pair of complex conjugate poles on the imaginary axis. An easy test in the complex domain for whether a time function $f(t)$ has a limit at infinity is to examine the poles of its Laplace transform $F(s)$. This test says: if the function $sF(s)$ has no poles (the values of s at which $sF(s) = \infty$) on the imaginary axis or in the right half of the complex plane, then the final value theorem of the Laplace transform will be applicable. Hence, if a function $F(s)$ has all its poles in the left half plane and only a *simple (distinct) pole* at the origin, the final value theorem may be applied. (This will become more obvious once we introduce the concept of system stability.)

EXAMPLE 4.9

Given the Laplace transform

$$F(s) = \frac{2s^2 + s + 1}{s^3 + 3s^2 + 3s + 1}$$

by the initial value theorem, we have

$$\lim_{t \to 0^-} \{f(t)\} = \lim_{s \to \infty} \{sF(s)\} = \lim_{s \to \infty} \left\{ s \frac{2s^2 + s + 1}{s^3 + 3s^2 + 3s + 1} \right\} = 2$$

From the final value theorem, it follows that

$$\lim_{t \to \infty} \{f(t)\} = \lim_{s \to 0} \{sF(s)\} = \lim_{s \to 0} \left\{ s \frac{2s^2 + s + 1}{s^3 + 3s^2 + 3s + 1} \right\} = 0$$

Note that the function $F(s)$ has all its poles in the left half of the complex plane, $s_{1,2,3} = -1$, hence the final value theorem is applicable. Thus, without going through the procedure of finding the inverse Laplace transform, and by using the results of these two theorems, we can conclude that $f(0^-) = 2$ and $f(\infty) = 0$. Note that we have also assumed that the function $f(t)$ has no jump discontinuities at $t = 0$, which follows from the fact that the degree of the denominator of $F(s)$ is strictly greater than the degree of the numerator of $F(s)$.

4.2 INVERSE LAPLACE TRANSFORM

As pointed out previously, the use of definition formula (4.3) for finding the Laplace inverse is complicated. We will avoid that problem by employing the partial fraction expansion of $F(s)$. Combined with the properties of the Laplace transform (summarized in Table 4.1) and the table of common Laplace transform pairs (Table 4.2), this will allow us to find Laplace inverses of almost all functions that we are faced with in linear system theory and its applications. Hence, the partial fraction expansion method and Tables 4.1 and 4.2 are the main tools for finding the inverse Laplace transform.

The partial fraction expansion method may be applied to the *strictly proper rational functions* $F(s)$. Such functions are represented by the ratio of two polynomials:

$$F(s) = \frac{b_m s^m + b_{m-1} s^{m-1} + \cdots + b_1 s + b_0}{s^n + a_{n-1} s^{n-1} + \cdots + a_1 s + a_0} = \frac{N_m(s)}{D_n(s)} \quad (4.21)$$

with $m < n$. Most of the Laplace transforms coming from real signals and real physical linear systems are strictly proper rational functions. It can also happen that $m = n$, which defines the *proper* rational function. In such cases, one must first perform long division, as demonstrated in the next example. The long division method extracts from a proper rational function a strictly proper part, the part to be subjected to partial fraction expansion.

Note that the long division operation can be also performed in MATLAB by using its function deconv.

EXAMPLE 4.10

Let a Laplace transform be given by

$$F(s) = \frac{s^2 + s + 1}{s^2 + 2s + 1}$$

Performing long division, we obtain

$$F(s) = 1 - \frac{s}{s^2 + 2s + 1}$$

The first component on the right-hand side is a constant and its Laplace inverse is $\delta(t)$. The second term on the right-hand side is a proper rational function. Its Laplace inverse will be obtained by the procedure subsequently explained.

It can be concluded from Example 4.10 that when $F(s)$ has $n = m$, the corresponding time domain signal displays impulsive behavior, that is, the time domain signal contains an impulse delta signal.

Case 1: Distinct Real Poles

Let the function $F(s)$ have all real distinct poles, that is,

$$F(s) = \frac{N_m(s)}{D_n(s)} = \frac{N_m(s)}{(p-p_1)(p-p_2)\cdots(p-p_n)}, \quad p_1 \neq p_2 \neq \cdots \neq p_n$$

This function can be expanded as follows:

$$F(s) = \frac{k_1}{s-p_1} + \frac{k_2}{s-p_2} + \cdots + \frac{k_n}{s-p_n} \tag{4.22}$$

The coefficients in (4.22) are obtained by using a very simple formula given by

$$k_i = \lim_{s \to p_i} \{(s-p_i)F(s)\}, \quad i = 1, 2, \ldots, n \tag{4.23}$$

This formula can be justified easily by multiplying both sides of (4.22) by $(s-p_i)$ and taking the limit when $s \to p_i$.

It can be seen from Table 4.2 that every term on the right-hand side of (4.22) represents an exponential signal in the time domain. Hence, the inverse Laplace transform of $F(s)$ is given by

$$f(t) = \{F(s)\} = (k_1 e^{p_1 t} + k_2 e^{p_2 t} + \cdots + k_n e^{p_n t})u(t) \tag{4.24}$$

This procedure is demonstrated in the following example.

EXAMPLE 4.11

Consider the Laplace transform given by

$$F(s) = \frac{s+4}{(s+1)(s+2)(s+3)}$$

Its partial fraction expansion according to (4.22) is

$$F(s) = \frac{k_1}{s+1} + \frac{k_2}{s+2} + \frac{k_3}{s+3}$$

where the corresponding coefficients are evaluated using formula (4.23):

$$k_1 = \lim_{s \to -1} \left\{(s+1)\frac{(s+4)}{(s+1)(s+2)(s+3)}\right\} = \lim_{s \to -1} \left\{\frac{(s+4)}{(s+2)(s+3)}\right\} = \frac{3}{2}$$

$$k_2 = \lim_{s \to -2} \left\{(s+2)\frac{(s+4)}{(s+1)(s+2)(s+3)}\right\} = \lim_{s \to -2} \left\{\frac{(s+4)}{(s+1)(s+3)}\right\} = -2$$

$$k_3 = \lim_{s \to -3} \left\{(s+3)\frac{(s+4)}{(s+1)(s+2)(s+3)}\right\} = \lim_{s \to -3} \left\{\frac{(s+4)}{(s+1)(s+2)}\right\} = \frac{1}{2}$$

From (4.24), we see that the corresponding time signal is given by

$$f(t) = \left(\frac{3}{2}e^{-t} - 2e^{-2t} + \frac{1}{2}e^{-3t}\right)u(t)$$

Case 2: Multiple Real Poles

Consider the case when the function $F(s)$ has r multiple real poles and $n - r$ distinct real poles, that is,

$$F(s) = \frac{N_m(s)}{D_n(s)}$$

$$= \frac{N_m(s)}{(s - p_1)^r (s - p_{r+1})(s - p_{r+2}) \cdots (s - p_n)}, \quad p_1 \neq p_{r+1} \neq p_{r+2} \neq \cdots \neq p_n$$

This function can be expanded into partial fractions as

$$F(s) = \frac{k_{11}}{s - p_1} + \frac{k_{12}}{(s - p_1)^2} + \cdots + \frac{k_{1r}}{(s - p_1)^r} + \frac{k_{r+1}}{s - p_{r+1}}$$

$$+ \frac{k_{r+2}}{s - p_{r+2}} + \cdots + \frac{k_n}{s - p_n} \quad (4.25)$$

The coefficients $k_{r+1}, k_{r+2}, \ldots, k_n$ correspond to distinct real poles, thus they can be calculated by using formula (4.23). The coefficients corresponding to the multiple pole p_1 are calculated according to the formula

$$k_{1j} = \frac{1}{(r-j)!} \lim_{s \to p_1} \left\{ \frac{d^{r-j}}{ds^{r-j}} [(s - p_1)^r F(s)] \right\}, \quad j = r, r-1, \ldots, 2, 1 \quad (4.26)$$

This formula can be justified as follows. In order to find k_{1r}, we multiply both sides of (4.25) by $(s - p_1)^r$ and take the limit of both sides when $s \to p_1$, which leads to

$$\lim_{s \to p_1} \{(s - p_1)^r F(s)\} = \lim_{s \to p_1} \left\{ k_{11}(s - p_1)^{r-1} + k_{12}(s - p_1)^{r-2} + \cdots \right.$$

$$+ k_{1r-1}(s - p_1) + k_{1r} + (s - p_1)^r \frac{k_{r+1}}{(s - p_{r+1})} + \cdots$$

$$\left. + (s - p_1)^r \frac{k_n}{(s - p_n)} \right\} \quad (4.27)$$

It can be seen that all terms on the right-hand side tend to zero in the limit, except for the term k_{1r}, hence formula (4.26) is verified for $j = r$. If we first take the derivative with respect to s of both sides of (4.27), and then perform the limiting operation, we will obtain

$$\lim_{s \to p_1} \left\{ \frac{d}{ds}[(s-p_1)^r F(s)] \right\}$$

$$= \lim_{s \to p_1} \left\{ k_{11}(r-1)(s-p_1)^{r-2} + k_{12}(r-2)(s-p_1)^{r-3} + \cdots + k_{1r-1} + 0 \right.$$

$$+ r(s-p_1)^{r-1}\left(\frac{k_{r+1}}{(s-p_{r+1})} + \frac{k_{r+2}}{(s-p_{r+2})} + \cdots + \frac{k_n}{(s-p_n)} \right)$$

$$\left. - (s-p_1)^r \left(\frac{k_{r+1}}{(s-p_{r+1})^2} + \frac{k_{r+2}}{(s-p_{r+2})^2} + \cdots + \frac{k_n}{(s-p_n)^2} \right) \right\}$$

In the limit when $s \to p_1$, the right-hand side of this formula is equal to k_{1r-1}, which justifies (4.26) for $j = r - 1$. Similarly, if we first take the second derivative and then take the limit of both sides of (4.26), we get

$$\lim_{s \to p_1} \left\{ \frac{d^2}{ds^2}[(s-p_1)^r F(s)] \right\} = 0 + 0 + \cdots + 0 + 2!k_{1r-2} + 0 + \cdots + 0$$

Note that $2!$ appears in front of k_{1r-2} because we have taken the second derivative of $(s-p_1)^2 k_{1r-2}$. In general, by first taking the $(r-j)$th derivative of the right-hand side of (4.27) and then finding the corresponding limit, we obtain

$$\lim_{s \to p_1} \left\{ \frac{d^j}{ds^j}[(s-p_1)^r F(s)] \right\} = 0 + 0 + \cdots + 0 + (r-j)!k_{1r-j} + 0 + \cdots + 0$$

which proves formula (4.26).

Having obtained the expansion for $F(s)$ as defined in (4.25), we can now use Table 4.2 to find the Laplace inverse of $F(s)$. It can be seen from Table 4.2 that

$$\frac{1}{(s+\alpha)^{r+1}} \leftrightarrow \frac{1}{r!} t^r e^{-\alpha t} u(t)$$

which indicates that the inverse Laplace transform of $F(s)$ is given by

$$f(t) = \mathcal{L}^{-1}\{F(s)\}$$
$$= \left(k_{11} e^{p_1 t} + k_{12} t e^{p_1 t} + k_{13} \frac{1}{2!} t^2 e^{p_1 t} + \cdots + k_{1r} \frac{1}{(r-1)!} t^{r-1} e^{p_1 t} \right) u(t)$$
$$+ (k_{r+1} e^{p_{r+1} t} + k_{r+2} e^{p_{r+2} t} + \cdots + k_n e^{p_n t}) u(t) \quad (4.28)$$

EXAMPLE 4.12

Consider the function $F(s)$, with a simple pole at -2 and a triple pole at -1, given by

$$F(s) = \frac{1}{(s+1)^3(s+2)}$$

According to (4.25), this function is expanded as

$$F(s) = \frac{k_{11}}{(s+1)} + \frac{k_{12}}{(s+1)^2} + \frac{k_{13}}{(s+1)^3} + \frac{k_4}{(s+2)}$$

The coefficients corresponding to the triple pole are obtained via (4.26) as follows:

$$k_{13} = \lim_{s \to -1} \left\{ (s+1)^3 \frac{1}{(s+1)^3(s+2)} \right\} = 1$$

$$k_{12} = \lim_{s \to -1} \left\{ \frac{d}{ds}\left[(s+1)^3 \frac{1}{(s+1)^3(s+2)}\right] \right\} = \lim_{s \to -1} \left\{ \frac{d}{ds}\left[\frac{1}{(s+2)}\right] \right\}$$

$$= \lim_{s \to -1} \left\{ \frac{-1}{(s+2)^2} \right\} = -1$$

$$k_{11} = \frac{1}{2} \lim_{s \to -1} \left\{ \frac{d^2}{ds^2}\left[(s+1)^3 \frac{1}{(s+1)^3(s+2)}\right] \right\} = \frac{1}{2} \lim_{s \to -1} \left\{ \frac{d^2}{ds^2}\left[\frac{1}{(s+2)}\right] \right\}$$

$$= \lim_{s \to -1} \left\{ \frac{1}{(s+2)^3} \right\} = 1$$

The coefficient k_4, corresponding to a simple pole, is obtained from formula (4.23) as

$$k_4 = \lim_{s \to -2} \left\{ (s+2) \frac{1}{(s+1)^3(s+2)} \right\} = \frac{1}{(-2+1)^3} = -1$$

By using formula (4.28), it is easy to see that the inverse Laplace transform of $F(s)$ is given by

$$f(t) = \left(e^{-t} - te^{-t} + \frac{1}{2}t^2 e^{-t} - e^{-2t} \right) u(t)$$

Case 3: Complex Conjugate Distinct Poles

Let $p_1 = \alpha + j\beta$ and its complex conjugate $p_1^* = \alpha - j\beta$ represent a distinct complex conjugate pair. The corresponding partial fraction expansion has the form

$$F(s) = \frac{N_m(s)}{D_n(s)} = \frac{k_1}{s - p_1} + \frac{k_1^*}{s - p_1^*} + \frac{N_\mu(s)}{D_{n-2}(s)}, \quad \mu < m \quad (4.29)$$

Since p_1 and p_1^* are distinct poles, the coefficients k_1 and k_1^* can be evaluated by using formula (4.23). Thus, the required coefficients in (4.29) are obtained from

$$k_1 = \lim_{s \to p_1} \{(s - p_1) F(s)\} = a + jb \Rightarrow k_1^* = a - jb \quad (4.30)$$

Due to the linearity property of the Laplace transform, we can proceed to find the Laplace inverse of the complex conjugate terms only, which is

$$\mathcal{L}^{-1}\left\{ \frac{k_1}{s - p_1} + \frac{k_1^*}{s - p_1^*} \right\} = k_1 e^{p_1 t} + k_1^* e^{p_1^* t}$$

Using a simple and well-known formula from calculus with complex numbers,

$$c + c^* = 2\operatorname{Re}\{c\}$$

we obtain

$$k_1 e^{p_1 t} + k_1^* e^{p_1^* t} = 2\operatorname{Re}\{k_1 e^{p_1 t}\} = 2\operatorname{Re}\{|k_1| e^{j \angle k_1} e^{\alpha t + j\beta t}\} \\ = 2|k_1| e^{\alpha t} \operatorname{Re}\{e^{j(\beta t + \angle k_1)}\} = 2|k_1| e^{\alpha t} \cos(\beta t + \angle k_1) \quad (4.31)$$

Note that in formula (4.31),

$$|k_1| = \sqrt{a^2 + b^2}, \quad \angle k_1 = \tan^{-1}\left(\frac{b}{a}\right) \quad (4.32)$$

It can be concluded that a pair of complex conjugate poles produces in the time domain a damped (assuming $\alpha < 0$) sinusoidal function whose coefficients are calculated from (4.30) and (4.32).

It is also possible to find the inverse Laplace transform of a pair of complex conjugate poles by using the following simple algebra with complex numbers:

$$\mathcal{L}^{-1}\left(\frac{k_1}{s - p_1} + \frac{k_1^*}{s - p_1^*}\right) = \mathcal{L}^{-1}\left(\frac{a + jb}{s - \alpha - j\beta} + \frac{a - jb}{s - \alpha + j\beta}\right)$$

$$= \mathcal{L}^{-1}\left(\frac{(a + jb)(s - \alpha + j\beta) + (a - jb)(s - \alpha - j\beta)}{(s - \alpha - j\beta)(s - \alpha + j\beta)}\right)$$

$$= \mathcal{L}^{-1}\left(\frac{2a(s - \alpha) - 2b\beta}{(s - \alpha)^2 + \beta^2}\right) = 2a\mathcal{L}^{-1}\left(\frac{s - \alpha}{(s - \alpha)^2 + \beta^2}\right)$$

$$- 2b\mathcal{L}^{-1}\left(\frac{\beta}{(s - \alpha)^2 + \beta^2}\right)$$

$$= 2ae^{\alpha t} \sin(\beta t) - 2be^{\alpha t} \cos(\beta t)$$

In the last step, we have used our knowledge of the forms of the Laplace transforms of the sine and cosine functions, and the Laplace transform frequency shifting property.

EXAMPLE 4.13

Consider the problem of finding the inverse Laplace transform of a function that has one simple pole at the origin and a pair of complex conjugate poles, and whose partial fraction expansion is given by

$$\mathcal{L}^{-1}\left\{\frac{s^2 + 3}{s(s^2 + 2s + 3)}\right\} = \mathcal{L}^{-1}\left\{\frac{k_1}{s + 1 + j\sqrt{2}} + \frac{k_1^*}{s + 1 - j\sqrt{2}} + \frac{k_3}{s}\right\}$$

The corresponding coefficients are calculated using formula (4.23) as

$$k_3 = \frac{s^2 + 3}{s^2 + 2s + 3}\bigg|_{s=0} = 1$$

and

$$k_1 = \frac{s^2+3}{s(s+1-j\sqrt{2})}\bigg|_{s=-1-j\sqrt{2}}$$

$$= \frac{(1+j\sqrt{2})^2+3}{(-1-j\sqrt{2})(-2j\sqrt{2})} = \frac{1+j\sqrt{2}}{-2+j\sqrt{2}} \times \frac{-2-j\sqrt{2}}{-2-j\sqrt{2}}$$

$$= \frac{-j3\sqrt{2}}{6} = -j\frac{\sqrt{2}}{2} = a+jb = k_1 \Rightarrow |k_1| = \frac{\sqrt{2}}{2}, \quad \angle k_1 = -90°$$

According to (4.31–32), the inverse Laplace transform of the considered function is given by

$$\mathcal{L}^{-1}\left\{\frac{k_1}{s+1+j\sqrt{2}} + \frac{k_1^*}{s+1-j\sqrt{2}} + \frac{k_3}{s}\right\} = [2|k_1|e^{\alpha t}\cos(\beta t + \angle k_1) + k_3]u(t)$$

$$= \left[\sqrt{2}e^{-t}\cos(\sqrt{2}t + 90°) + 1\right]u(t)$$

$$p_1 = \alpha + j\beta = -1 - j\sqrt{2} \Rightarrow \alpha = -1, \beta = -\sqrt{2}$$

EXAMPLE 4.14

A shortcut is possible in the case when the function $F(s)$ is given by

$$F(s) = \frac{cs+d}{(s+a)^2+b^2}$$

This function can be written as

$$F(s) = \frac{c(s+b)-cb+d}{(s+a)^2+b^2} = \frac{c(s+b)}{(s+a)^2+b^2} + \frac{d-cb}{b}\frac{b}{(s+a)^2+b^2}$$

Using Table 4.2, we note that the first term corresponds in the time domain to the damped cosine signal and the second term corresponds to the damped sine signal; that is, we have

$$f(t) = e^{-at}\left[c\cos(bt) + \frac{(d-cb)}{b}\sin(bt)\right]u(t)$$

Note that, in general, the preceding form for $F(s)$ can be obtained using the completion of squares method, as demonstrated in the following example:

$$\frac{s+1}{s^2+4s+7} = \frac{s+1}{(s+2)^2+3} = \frac{s+2-1}{(s+2)^2+3}$$

$$\frac{s+2}{(s+2)^2+3} - \frac{1}{(s+2)^2+3} \leftrightarrow e^{-2t}\left[\cos(\sqrt{3}t) - \frac{1}{\sqrt{3}}\sin(\sqrt{3}t)\right]u(t)$$

Case 4: Multiple Complex Conjugate Poles

The general procedure for finding the inverse Laplace transforms for functions that have multiple complex conjugate poles basically combines the procedures presented in Cases 2 and 3. However, computations of the coefficients of the corresponding partial fraction expansions become very involved, and we suggest that in such cases MATLAB be used for finding partial fractions. In Sections 4.3.5 and 4.6, we will explain how to use MATLAB to find partial fractions of rational functions.

Case 5: Signals Containing Time Delay Elements

We have seen from the time delay property of the Laplace transform that

$$\mathcal{L}\{f(t)u(t)\} = F(s) \Rightarrow \mathcal{L}\{f(t-t_0)u(t-t_0)\} = e^{-t_0 s}F(s)$$

which indicates that time delayed signals have Laplace transforms that are not rational functions. In other words, they are not represented by the ratio of two polynomials. Note that the complex exponential term $e^{-t_0 s}$ indicates a time delay of t_0 time units. In such cases, we need only find the Laplace inverse of the function $F(s)$, and once we obtain its time domain representation $f(t)u(t)$, the signal $f(t)u(t)$ should be delayed by t_0. This procedure is demonstrated in the next example.

EXAMPLE 4.15

Consider the Laplace transform

$$F(s) = \frac{e^{-5s}}{(s+1)^3(s+2)} = F_1(s)e^{-5s}$$

Note that we have obtained the Laplace inverse of the function $F_1(s)$ in Example 4.12, that is,

$$F_1(s) \leftrightarrow \left(e^{-t} - te^{-t} + \frac{1}{2}t^2 e^{-t} - e^{-2t}\right)u(t)$$

By the time delay property of the Laplace transform, we obtain the Laplace transform pair

$$F(s) \leftrightarrow \left(e^{-(t-5)} - (t-5)e^{-(t-5)} + \frac{1}{2}(t-5)^2 e^{-(t-5)} - e^{-2(t-5)}\right)u(t-5)$$

In summary of this section, we can establish a set of steps for finding the inverse Laplace transform.

Step 1. Check whether $F(s)$ is a strictly proper ($n > m$) rational (ratio of two polynomials) function. If yes, go to Step 2; if no, go to Step 3.

Step 2. Perform the partial fraction expansion procedure, and use the table of common pairs (Table 4.2) to identify corresponding time domain signals.

Step 3. If $F(s)$ is a rational function but not strictly proper, perform long division and then go to Step 2 to work on the obtained strictly proper part. If $F(s)$ is not a rational function (contains complex exponentials of the form $e^{-t_0 s}$), go to Step 4.

Step 4. Drop the complex exponential(s) and work on the remainder, assuming that it represents a rational function. Repeat Steps 1–3, then introduce the corresponding time delay of t_0 units, according to the time delay property of the Laplace transform.

Let us emphasize again that these steps are sufficient for finding Laplace transform inverses used in this introductory undergraduate course on linear systems. If $F(s)$ does not belong to any of the preceding categories, we have no other choice than to directly evaluate the definition integral (4.3), which requires advanced knowledge of complex variables and the method known as contour integration (see [1–3]).

4.3 LAPLACE TRANSFORM IN LINEAR SYSTEM ANALYSIS

The main goal in analysis of any dynamic system is to find its response to a given input. We have seen in Chapter 1 that the system response in general has two components: the zero-state response due to external forcing signals, and the zero-input response due to system initial conditions. The Laplace transform will produce both the zero-input and zero-state components of the system response. Note that with the Fourier transform we found only the zero-state system response. In this section, we will also present procedures for obtaining the system impulse, step, and ramp responses.

The Laplace transform is very convenient for dealing with system input signals that have jump discontinuities (and delta impulses). Delta impulse inputs can come from the system differentiation of input signals that have jump discontinuities. Recall from Chapter 1, formula (1.9), that systems in general differentiate input signals. Recall also from Section 2.1.1 the definition of the generalized derivative, which indicates that at the point of a jump discontinuity, the generalized derivative generates the impulse delta signal. Furthermore, for the same reason, a signal that is continuous and differentiable for all $t > 0$, but has a jump discontinuity at $t = 0$ (for example, $e^{-t}u(t)$), will generate an impulse delta signal (after being differentiated) at $t = 0$. The Laplace transform can be used to solve differential equations that represent dynamics of linear time-invariant systems in a straightforward manner, despite jump discontinuities in the input. We notice few peculiar features of these signals while using the Laplace transform for finding the system response. On the other hand, we will see in Chapter 7 that the time domain technique based on the classical method for solving linear differential equations with constant coefficients will require a special treatment for this kind of input signal.

Analysis of linear time-invariant systems in the frequency domain (complex domain), via the use of the Laplace transform, is based on the concept of the system transfer function. This concept was introduced in Section 3.3.1 in the $j\omega$ domain. Here, we redefine the transfer function in the s domain, and derive the general expression for the system response due to arbitrary input signals and arbitrary initial conditions.

4.3.1 System Transfer Function and Impulse Response

Let us take the Laplace transform of both sides of a linear differential equation that describes the dynamical behavior of an nth-order linear system introduced in formula (1.9); that is,

$$\mathcal{L}\left\{\frac{d^n y(t)}{dt^n} + a_{n-1}\frac{d^{n-1}y(t)}{dt^{n-1}} + a_{n-2}\frac{d^{n-2}y(t)}{dt^{n-2}} + \cdots + a_1\frac{dy(t)}{dt} + a_0 y(t)\right\}$$

$$= \mathcal{L}\left\{b_m \frac{d^m f(t)}{dt^m} + b_{m-1}\frac{d^{m-1}f(t)}{dt^{m-1}} + \cdots + b_1 \frac{df(t)}{dt} + b_0 f(t)\right\} \quad (4.33)$$

Using the time derivative property of the Laplace transform, we have

$$(s^n + a_{n-1}s^{n-1} + \cdots + a_1 s + a_0)Y(s) - I(s) = (b_m s^m + b_{m-1}s^{m-1} + \cdots + b_1 s + b_0)F(s) \quad (4.34)$$

where $I(s)$ contains terms coming from the system initial conditions, obtained by using in (4.33) the following formulas for the time derivatives

$$\mathcal{L}\left\{\frac{dy(t)}{dt}\right\} = sY(s) - y(0^-)$$

$$\mathcal{L}\left\{\frac{d^2 y(t)}{dt^2}\right\} = s^2 Y(s) - sy(0^-) - y^{(1)}(0^-)$$

$$\vdots \quad (4.35)$$

$$\mathcal{L}\left\{\frac{d^n y(t)}{dt^n}\right\} = s^n Y(s) - s^{n-1} y(0^-) - s^{n-2} y^{(1)}(0^-) - \cdots - y^{(n-1)}(0^-)$$

Note that we assume that the input signal $f(t)$ represents a causal signal for which $f(t) = 0$, $t < 0$. Thus, *we must set* $f^{(i)}(0^-) = 0$, $i = 0, 1, 2, \ldots, m$. Hence, $I(s)$ is a function of only the coefficients a_i and the system initial conditions; that is, in general,

$$I(s) = \left(a_1 y(0^-) + a_2 y^{(1)}(0^-) + \cdots + a_{n-1} y^{(n-2)}(0^-) + y^{(n-1)}(0^-)\right)$$
$$+ s\left(a_2 y(0^-) + a_3 y^{(1)}(0^-) + \cdots + a_{n-1} y^{(n-3)}(0^-) + y^{(n-2)}(0^-)\right)$$
$$+ s^2\left(a_3 y(0^-) + a_4 y^{(1)}(0^-) + \cdots + a_{n-1} y^{(n-4)}(0^-) + y^{(n-3)}(0^-)\right)$$
$$+ \cdots + s^{n-2}\left(a_{n-1} y(0^-) + y^{(1)}(0^-)\right) + s^{n-1} y(0^-) \quad (4.36)$$

The input signal is applied to the system at $t = 0$, and we are interested in finding the complete (total) system response—the response due to both system initial conditions and input signals (forcing functions). From (4.34), we have

$$Y(s) = \frac{b_m s^m + b_{m-1}s^{m-1} + \cdots + b_1 s + b_0}{s^n + a_{n-1}s^{n-1} + \cdots + a_1 s + a_0} F(s)$$
$$+ \frac{I(s)}{s^n + a_{n-1}s^{n-1} + \cdots + a_1 s + a_0} \quad (4.37)$$

which produces the solution $Y(s)$ in the frequency domain of the original differential equation. Of course, to get the time domain solution, we must use the inverse Laplace transform, that is, $y(t) = \mathcal{L}^{-1}\{Y(s)\}$.

If the initial conditions are set to 0, then $I(s) = 0$. The quantity

$$H(s) = \frac{Y(s)}{F(s)}\Big|_{I(s)=0} = \frac{b_m s^m + b_{m-1} s^{m-1} + \cdots + b_1 s + b_0}{s^n + a_{n-1} s^{n-1} + \cdots + a_1 s + a_0} \quad (4.38)$$

defines the *system transfer function*. The transfer function can also be written in factored form as

$$H(s) = K \frac{(s - z_1)(s - z_2) \cdots (s - z_m)}{(s - p_1)(s - p_2) \cdots (s - p_n)}, \quad K = b_m \quad (4.38a)$$

where z_i, $i = 1, 2, \ldots, m$, are the *transfer function zeros* (note that $H(z_i) = 0$), and p_j, $j = 1, 2, \ldots, n$, are the *transfer function poles* at which $H(p_j) = \infty$. Very often, we call these the *system* zeros and poles; K is called the *static gain*. We have assumed that $z_i \neq p_j$ for all i, j so that we have n poles and m zeros. However, in the case when there are common factors in the transfer function numerator and denominator, they must be cancelled out before the system poles and zeros are identified.

EXAMPLE 4.16

We can get the factored form of the transfer function by using the MATLAB function tf2zp (which stands for "transfer function to zero pole") as demonstrated in the following example. Let

$$H(s) = \frac{2s^3 + 3s^2 + s + 2}{s^4 + 3s^3 + 5s^2 + s + 7}$$

The statement [z,p,k]=tf2zp(num,den) with num=[2 3 1 2] and den=[1 3 5 1 7] yields $k = 2$, $z_{1,2,3} = \{-1.5832, 0.0416 \pm j0.7937\}$ and $p_{1,2,3,4} = \{-1.7831 \pm 1.6236, 0.2831 \pm j1.0600\}$. In addition, the MATLAB function [num,den]=zp2tf(z,p,k) produces the opposite; that is, it gets the original transfer function form defined in (4.38) from its zero-pole form (4.38a).

EXAMPLE 4.17

Let the system transfer function be given by

$$H(s) = \frac{(s+1)(s+3)}{s(s+2)(s+3)(s+4)} = \frac{s+1}{s(s+2)(s+4)}$$

After the cancellation of common factors, the system zeros and poles are identified as $z_1 = -1$ and $p_{1,2,3} = \{0, -2, -4\}$. Hence, -3 is neither the system zero nor the system pole.

Having in mind the general structure of a linear time-invariant continuous-time linear system introduced in (1.9), we can also say that the system transfer function is obtained by applying the Laplace transform to a linear differential equation with constant coefficients that represents the system, assuming that system initial conditions are set to zero. Let us demonstrate this fact with the simple electrical circuit presented in Section 1.3, whose mathematical model is given in (1.26).

EXAMPLE 4.18

Denote the input voltage of the system by $f(t)$, that is $f(t) = e_i(t)$, and denote the output voltage by $y(t)$, that is $y(t) = e_0(t)$ (see Figure 1.10). The corresponding differential equation is given by

$$\frac{d^2y(t)}{dt^2} + \left(\frac{L + R_1 R_2 C}{R_2 LC}\right)\frac{dy(t)}{dt} + \left(\frac{R_1 + R_2}{R_2 LC}\right)y(t) = \frac{1}{LC}f(t)$$

To simplify notation, we introduce

$$a_1 = \frac{L + R_1 R_2 C}{R_2 LC}, \quad a_0 = \frac{R_1 + R_2}{R_2 LC}, \quad b_0 = \frac{1}{LC}$$

Assuming that the system initial conditions are zero, the Laplace transform produces

$$s^2 Y(s) + a_1 s Y(s) + a_0 Y(s) = b_0 F(s)$$

so that the system transfer function is given by

$$H(s) = \frac{Y(s)}{F(s)}\bigg|_{\text{I.C.}=0} = \frac{b_0}{s^2 + a_1 s + a_0}$$

The quantity

$$H(0) = K \frac{(-z_1)(-z_2)\cdots(-z_m)}{(-p_1)(-p_2)\cdots(-p_n)} = \frac{b_0}{a_0} \qquad (4.39)$$

is called the *system DC gain* (system gain at zero frequency). It is also called the *system steady state gain* since it shows how much the system multiplies a constant input signal at steady state. It follows from formula (4.38) and the final value theorem of the Laplace transform that, for $f(t) = au(t) \leftrightarrow F(s) = a/s$, we have

$$y_{ss} = \lim_{t \to \infty} y(t) = \lim_{s \to 0} sY(s) = \lim_{s \to 0}\left\{sH(s)\frac{a}{s}\right\} = H(0)a \qquad (4.40)$$

Note that the final value theorem is applicable under the assumption that the function $sY(s) = aH(s)$ has no poles on the imaginary axis or in the right half of the complex plane (that is, when all poles of $H(s)$ are strictly in the left half of the complex plane). In such a case, the system output can reach its steady state value, which can be found by using the extraordinary simple formula (4.40).

EXAMPLE 4.19

The steady state response to a constant input $f(t) = 5u(t)$ of a system whose transfer function is given by

$$H(s) = \frac{2s^3 + 3s^2 + s + 2}{s^4 + 3s^3 + 5s^2 + s + 1} \Rightarrow H(0) = 2$$

exists because all poles of $H(s)$ are in the left half of the complex plane (pole locations can be checked by MATLAB). The steady state system output value is obtained using formula (4.40) as

$$y_{ss} = 5H(0) = 10$$

Since for the impulse delta signal the Laplace transform is given by $F(s) = 1$, we conclude from (4.38) that under zero initial conditions, the system response to the impulse delta signal is equal to $\mathcal{L}^{-1}\{H(s)\}$. Hence, in the time domain, the *system impulse response* is defined by

$$h(t) \triangleq \mathcal{L}^{-1}\{H(s)\} \tag{4.41}$$

Note that whenever we must find the system impulse response, by definition, the system initial conditions must be set to zero.

EXAMPLE 4.20

The system impulse response for

$$y^{(2)}(t) + 3y^{(1)}(t) + 2y(t) = f^{(1)}(t) + 3f(t)$$

is obtained as follows:

$$h(t) = \mathcal{L}^{-1}\{H(s)\} = \mathcal{L}^{-1}\left\{\frac{s+3}{s^2+3s+2}\right\} = \mathcal{L}^{-1}\left\{\frac{s+3}{(s+1)(s+2)}\right\}$$

$$= \frac{2}{s+1} - \frac{1}{s+2} = (2e^{-t} - e^{-2t})u(t)$$

The system impulse response can be found and plotted easily by using MATLAB. It is represented in Figure 4.2 and obtained by the following simple MATLAB script.

```
num=[1 3]; den=[1 3 2]; impulse(num,den)
% This MATLAB function automatically determines the length of the time axis
% and plots the impulse response.
% Instead of the last statement, we could have used
% y=impulse(num,den); plot(y) or more specifically
% t=0:0.1:3; y=impulse(num,den,t); plot(t,y)
% which would have determined the length of the time axis.
```

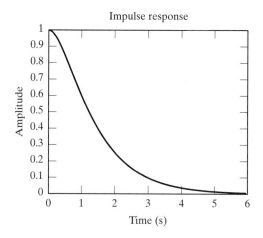

FIGURE 4.2: The system impulse response

Figure 4.2 represents a typical impulse response of a real physical linear system. The impulse delta signal at $t = 0$ brings the energy into the system (system excitation), which basically sets up system initial conditions to nonzero values at $t = 0^+$. As time passes, the system energy dissipates and the system response tends to zero. Note that in this particular example we had $h(0^-) = 0, h^{(1)}(0^-) = 0$. It is easy to conclude from the expression obtained for $h(t)$ that at $t = 0^+$ we have $h(0^+) = 1, h^{(1)}(0^+) = 0$. In Chapter 7, where we will present the method for finding the system impulse in the time domain, we will address this phenomenon of instantaneous system signal changes at the initial time in more detail.

Note that we can define the general problem of finding the complete system response due to both the impulse delta signal and the system initial conditions. Such a response is called the complete system response due to a particular input signal, $\delta(t)$, rather than the system impulse response, which is strictly defined by (4.41).

4.3.2 System Zero-State Response

The solution of the differential equation defined in (4.33), which represents the system response (system output) due to the given input $f(t)$ and zero system initial conditions ($I(s) = 0$), is obtained in the frequency domain from (4.37) in a very simple manner:

$$Y_{zs}(s) = H(s)F(s) \tag{4.42}$$

Applying the convolution property of the Laplace transform to (4.42), we obtain

$$y_{zs}(t) = h(t) * f(t) = \mathcal{L}^{-1}\{H(s)F(s)\} \tag{4.43}$$

Formula (4.43) states that the system output under zero system initial conditions is equal to the convolution of the system input and the system impulse response.

In (4.37–38) and (4.41–43) we have established the most fundamental results of linear time-invariant dynamic system theory. From these results, we can see that to find the system response to any input signal, one must first find the system response due to the impulse delta signal, and then convolve the obtained system impulse response with the given system input signal. Note that the *system impulse response is obtained (and defined) for the system at rest* (zero initial conditions). It should also be emphasized that for a given time-invariant system, the impulse response need be found only once. In other words, any linear time-invariant system is uniquely characterized by its impulse response (or by its transfer function in the frequency domain). In general, it is easy to find the system impulse response by using (4.41). As a matter of fact, this formula represents the most efficient means for finding the impulse response of continuous-time linear time-invariant systems. This issue will be further clarified in Chapter 7.

EXAMPLE 4.21

The zero-state response of the system defined by

$$y^{(3)}(t) + 3y^{(2)}(t) + 2y^{(1)}(t) = f^{(1)}(t) + 3f(t), \quad t \geq 0$$

due to the input signal $f(t) = e^{-5t}u(t)$, can be obtained by using formula (4.43) as follows. We first find the system transfer function and the Laplace transform of the input signal, that is,

$$H(s) = \frac{s+3}{s^3 + 3s^2 + 2s}, \quad F(s) = \frac{1}{s+5}$$

Then, from formula (4.43), we have

$$y_{zs}(t) = \mathcal{L}^{-1}\{H(s)F(s)\} = \mathcal{L}^{-1}\left\{\frac{(s+3)}{s(s+1)(s+2)(s+5)}\right\}$$

$$= \mathcal{L}^{-1}\left\{\frac{3/10}{s} - \frac{1/2}{s+1} + \frac{1/6}{s+2} + \frac{1/30}{s+5}\right\}$$

$$= \left(\frac{3}{10} - \frac{1}{2}e^{-t} + \frac{1}{6}e^{-2t} + \frac{1}{30}e^{-5t}\right)u(t)$$

This response can be also obtained by using the MATLAB function `lsim` (which stands for "linear system simulation") as follows.

```
num=[1 3]; den=[1 3 2 0];
t=0:0.1:10; f=exp(-5*t);
lsim(num,den,f,t)
```

The corresponding MATLAB plot is presented in Figure 4.3.

Let $F(s) = \mathcal{L}\{f(t)\}$ be a rational function; then the product of $H(s)$ and $F(s)$ is a rational function. In such cases, the system zero-state response can be obtained by using the MATLAB function `impulse(numhf,denhf)` with `numhf=numh*numf` and `denhf=denh*denf`, where `numh,numf` and `denh,denf` are, respectively, numerators and denominators of the transfer function and the Laplace transform of the input signal.

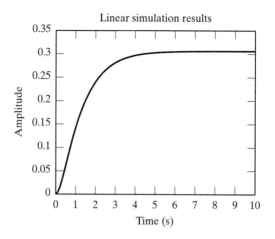

FIGURE 4.3: The zero-state response for Example 4.21

It should be pointed out that in Example 4.21 the initial conditions originally were given by $y_{zs}(0^-) = y_{zs}^{(1)}(0^-) = y_{zs}^{(2)}(0^-) = 0$. However, if we check the expression obtained for $y_{zs}(t)$, we will find that $y_{zs}(0^+) = y_{zs}^{(1)}(0^+) = 0$ and $y_{zs}^{(2)}(0^+) = 1$. *We know that only an impulse delta signal at zero is able to change the system initial conditions instantaneously.* In that example, the impulse delta function is generated by the system's differentiation of the input signal. Namely, we have

$$f^{(1)}(t) + 3f(t) = \frac{df(t)}{dt} + 3f(t) = \frac{d}{dt}\{e^{-5t}u(t)\} + 3e^{-5t}u(t)$$
$$= -5e^{-5t}u(t) + e^{-5t}\delta(t) + 3e^{-5t}u(t) = -2e^{-5t}u(t) + \delta(t)$$

Note that when no differentiation of the input signal takes place ($m = 0$ in formula (4.33)), the instantaneous change of the initial conditions could happen only in the case when the input signal is the impulse delta signal. Hence, *we can conclude that system input signal differentiation, in general, can produce an instantaneous change in the system initial conditions.*

It is also interesting to observe that

$$\mathcal{L}\{-2e^{-5t}u(t) + \delta(t)\} = \frac{-2}{s+5} + 1 = \frac{-2 + s + 5}{s+5} = \frac{s+3}{s+5}$$

which is identical to

$$\mathcal{L}\{f^{(1)}(t) + 3f(t)\} = sF(s) - f(0^-) + 3F(s) = (s+3)F(s) = \frac{s+3}{s+5}$$

The last expression can be viewed as a shortcut for finding the Laplace transform of the right-hand side of the original differential equation (4.33).

It can be concluded from the previous example and subsequent discussion that the Laplace transform has a mechanism that takes into the account the delta impulses generated by the system differentiation of the input signals. Note that the delta impulses are generated by taking the first- and higher-order derivatives of the input signals satisfying $f(0^-) \neq f(0^+)$. Also, in the case when $f(0^-) = f(0^+)$, but the input signal is not differentiable in the ordinary sense at $t = 0$ (for example, $\sin(t)u(t)$), the second- and higher-order generalized derivatives of this signal will produce the delta impulses (see Problem 4.21).

Another interesting phenomenon can be deduced from Example 4.21 by observing the expression obtained for the system *zero-state* response. It can be seen that the input signal $f(t) = e^{-5t}u(t)$ produces in the system output the component proportional to $e^{-5t}u(t)$ (which is expected), and components that correspond to system modes (poles) $p_1 = 0$, $p_2 = -1$, $p_2 = -2$. Hence, despite the fact that all initial conditions are zero, *the system input excites all system modes so that, in general, all of them appear in the system output*.

4.3.3 Unit Step and Ramp Responses

Finding the system response due to a unit step is a common problem in engineering. The unit step response can be related to the system impulse response by a very simple formula, assuming that in this case the *system initial conditions are also set to zero*. Namely, for $f(t) = u(t)$, formula (4.43) gives

$$h(t) * u(t) = \int_{0^-}^{t} u(t - \tau)h(\tau)\,d\tau = \int_{0^-}^{t} h(\tau)\,d\tau \triangleq y_{\text{step}}(t) \quad (4.44)$$

where $y_{\text{step}}(t)$ denotes the system step response under zero system initial conditions (system at rest). Note that we have taken the lower integration limit at $t = 0^-$ in order to completely include the delta impulse signal $\delta(\tau)$ within the integration limits since, in the case when $n = m$, the system impulse response contains the delta impulse signal at the origin. From formula (4.44) we have

$$h(t) = \frac{dy_{\text{step}}(t)}{dt} \quad (4.45)$$

This corresponds in the frequency domain to

$$Y_{\text{step}}(s) = \frac{1}{s}H(s) \quad (4.45a)$$

Note that by the definition of the system step response, the zero system initial conditions have been used. Very often, formula (4.45a) represents an easier way to find the system step response than the corresponding time domain formula (4.44).

A more general problem of finding the system response due to both a unit step input signal and system initial conditions can also be defined. Such a problem should be treated within the framework of the complete system response, using the methodology to be presented in Section 4.3.4.

Similarly, we can get the system unit *ramp response* subject to zero initial conditions. In this case, $f(t) = r(t)$, so that formula (4.43) produces

$$y_{\text{ramp}}(t) = h(t) * r(t) = \int_{0^-}^{t} (t - \tau)u(t - \tau)h(\tau)\,d\tau = \int_{0^-}^{t} (t - \tau)h(\tau)\,d\tau \quad (4.46)$$

Integrating by parts the last integral in (4.46) and using the result established in (4.44), we have

$$\int_{0^-}^{t}(t-\tau)h(\tau)\,d\tau = (t-\tau)y_{\text{step}}(\tau)|_{\tau=0^-}^{\tau=t} - \int_{0^-}^{t} y_{\text{step}}(\tau)(-d\tau) = \int_{0^-}^{t} y_{\text{step}}(\tau)\,d\tau \quad (4.47)$$

that is,

$$y_{\text{ramp}}(t) = h(t) * r(t) = \int_{0^-}^{t} y_{\text{step}}(\tau)\,d\tau \quad (4.48)$$

By taking the derivative of (4.48), we obtain

$$y_{\text{step}}(t) = \frac{dy_{\text{ramp}}(t)}{dt} \quad (4.49)$$

Combining (4.45) and (4.49), the relationship between the impulse response and the ramp response is established:

$$h(t) = \frac{dy_{\text{step}}(t)}{dt} = \frac{d^2 y_{\text{ramp}}(t)}{dt^2} \quad (4.50)$$

It follows also from the preceding discussion that

$$y_{\text{ramp}}(t) = \int_{0^-}^{t} y_{\text{step}}(\tau)\,d\tau = \int_{0^-}^{t}\int_{0^-}^{\tau} h(\sigma)\,d\sigma\,d\tau \quad (4.51)$$

Note that in the frequency domain (4.51) corresponds to

$$Y_{\text{ramp}}(s) = \frac{1}{s^2} H(s) \quad (4.51a)$$

EXAMPLE 4.22

The step response of the system at rest considered in Example 4.21 can be obtained by integrating the corresponding impulse response. The impulse response is obtained from

$$h(t) = \mathcal{L}^{-1}\{H(s)\} = \mathcal{L}^{-1}\left\{\frac{s+3}{s^3 + 3s^2 + 2s}\right\} = \mathcal{L}^{-1}\left\{\frac{s+3}{s(s+1)(s+2)}\right\}$$

$$= \mathcal{L}^{-1}\left\{\frac{1.5}{s} - \frac{2}{s+1} + \frac{0.5}{s+2}\right\} = (1.5 - 2e^{-t} + 0.5e^{-2t})u(t)$$

Hence, the step response is given by

$$y_{\text{step}}(t) = \int_0^t h(\tau)\,d\tau = \int_0^t (1.5 - 2e^{-\tau} + 0.5e^{-2\tau})\,d\tau$$

$$= (1.5t + 2e^{-t} - 2 - 0.25e^{-2t} + 0.25)u(t)$$

$$= (-1.75 + 1.5t + 2e^{-t} - 0.25e^{-2t})u(t)$$

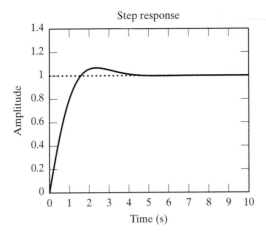

FIGURE 4.4: The system step response

The ramp response, obtained for zero initial conditions, is given by

$$y_{\text{ramp}}(t) = \int_0^t y_{\text{step}}(\tau)\,d\tau = \int_0^t (-1.75 + 1.5\tau + 2e^{-\tau} - 0.25e^{-2\tau})\,d\tau$$

$$= 1.75 - 1.75t + 0.75t^2 - 2e^{-t} + 0.25e^{-2t}, \quad t \geq 0$$

The system step response can be found by using MATLAB function step in the following simple code that produces the plot of the system step response, which is presented in Figure 4.4.

```
num=[1 2]; den=[1 2 2]; t=0:0.1:10; step(num,den,t)
% Alternative options exist for the step function as noted in Example 4.20
% for the impulse function. For more information use help step.
```

Note that MATLAB has no function for directly finding the system ramp response. However, we can use the MATLAB function lsim for that purpose as follows.

```
num=[1 2]; den=[1 2 2];
t=0:0.1:10;
r=lsim(num,den,t,t)
```

4.3.4 Complete System Response

From the system analysis point of view, it is extremely important that the Laplace transform is able to handle the system initial conditions and to produce the system response due to both initial conditions and external forcing functions. Let us define the *system characteristic polynomial* by

$$\Delta(s) = s^n + a_{n-1}s^{n-1} + \cdots + a_1 s + a_0 \qquad (4.52)$$

Formula (4.37) can be rewritten as

$$Y(s) = H(s)F(s) + \frac{I(s)}{\Delta(s)} = Y_{zs}(s) + Y_{zi}(s) \qquad (4.53)$$

Hence, the complete system response is obtained as the sum of the zero-state and zero-input responses. By taking the Laplace inverse of the last equation, we obtain the complete system response in the time domain:

$$y(t) = \mathcal{L}^{-1}\{H(s)F(s)\} + \mathcal{L}^{-1}\left\{\frac{I(s)}{\Delta(s)}\right\} = y_{zs}(t) + y_{zi}(t) \qquad (4.54)$$

In some texts, the zero-state response is also called the *forced system response* and the zero-input response is called the *natural system response*.

Note that there is no need to separately treat the problem of finding the system zero-input response by using the Laplace transform. It can be simply found, using the results presented in this section, as

$$y_{zi}(t) = \mathcal{L}^{-1}(Y_{zi}(s)) = \mathcal{L}^{-1}\left(\frac{I(s)}{\Delta(s)}\right) \qquad (4.55)$$

with $\Delta(s)$ determined by formula (4.52) and $I(s)$ obtained from the system initial conditions by formula (4.36).

The Laplace transform is one of the main techniques for finding the complete system response due to arbitrary system inputs and system initial conditions. The other techniques will be presented in Chapters 7 and 8.

EXAMPLE 4.23

The complete response of the system

$$y^{(2)}(t) + 6y^{(1)}(t) + 9y(t) = f(t), \quad f(t) = e^{-2t}u(t), \quad y(0^-) = -1, \quad y^{(1)}(0^-) = 2$$

can be obtained as follows. Applying the Laplace transform, we have

$$\left(s^2 Y(s) - sy(0^-) - y^{(1)}(0^-)\right) + 6(sY(s) - y(0^-)) + 9Y(s) = \frac{1}{s+2}$$

which implies

$$Y(s) = \frac{1}{(s+3)^2} \frac{1}{(s+2)} - \frac{s+4}{(s+3)^2} = H(s)F(s) + \frac{I(s)}{\Delta(s)}$$

Note that the system transfer function and the characteristic polynomial are given by

$$H(s) = \frac{1}{s^2 + 6s + 9}, \quad \Delta(s) = s^2 + 6s + 9$$

Using the Laplace inverse we obtain the zero-state response,

$$y_{zs}(t) = \mathcal{L}^{-1}\{Y_{zs}(s)\} = \mathcal{L}^{-1}\left\{\frac{1}{s+2} - \frac{1}{s+3} - \frac{1}{(s+3)^2}\right\}$$
$$= (e^{-2t} - e^{-3t} - te^{-3t})u(t)$$

and the zero-input response,

$$y_{zi}(t) = \mathcal{L}^{-1}\{Y_{zi}(s)\} = \mathcal{L}^{-1}\left\{-\frac{1}{s+3} - \frac{1}{(s+3)^2}\right\} = (-e^{-3t} - te^{-3t})u(t)$$

The complete system response is now given by

$$y(t) = y_{zs}(t) + y_{zi}(t) = (e^{-2t} - 2e^{-3t} - 2te^{-3t})u(t)$$

Note that in this case no differentiation of the input signal takes place, so that $y(0^-) = y(0^+)$ and $y^{(1)}(0^-) = y^{(1)}(0^+)$, which can be easily checked from the result obtained for $y(t)$ and from the expression for its derivative.

The MATLAB function `lsim` can be used to find and plot the complete system response for linear systems represented in state space form. (This form will be considered in Chapter 8, hence the use of `lsim` for finding the complete system response will be postponed until Chapter 8.)

In the case when the input signal has a nonstandard analytical form, for example,

$$f(t) = \begin{cases} t, & 0 \le t \le 1 \\ 0, & \text{otherwise} \end{cases} \quad (4.56)$$

the input signal should be represented in terms of elementary signals considered in Chapter 2 (in this case, $f(t) = r(t) - r(t-1) - u(t-1)$), and the linearity and time-invariance principles should be used. Note that the linearity principle is established for linear systems at rest (see Section 1.2.1), that is, for zero initial conditions. In this way we obtain the $y_{zs}(t)$ component of the system response. The system response due to initial conditions, $y_{zi}(t)$, must be found independently. Let $y_{\text{ramp}}(t)$ represent the system zero-state response due to $r(t)$, and let $y_{\text{step}}(t)$ represent the system zero-state response due to $u(t)$; then the system zero-state response due to $f(t)$ can be obtained, using linearity and time invariance, as

$$y_{zs}(t) = y_{\text{ramp}}(t) - y_{\text{ramp}}(t-1) - y_{\text{step}}(t-1) \quad (4.57)$$

$y_{\text{ramp}}(t)$ and $y_{\text{step}}(t)$ can be found using the results of Sections 4.3.1 and 4.3.3.

EXAMPLE 4.24

Find the complete response of the system defined by

$$y^{(2)}(t) + 3y^{(1)}(t) + 2y(t) = f(t), \quad y(0^-) = 1, y^{(1)}(0^-) = 1$$

due to the forcing function given in (4.56).

Since the complete system response is $y(t) = y_{zi}(t) + y_{zs}(t)$, we can find independently the zero-input and zero-state components.

The zero-input response satisfies

$$y_{zi}^{(2)}(t) + 3y_{zi}^{(1)}(t) + 2y_{zi}(t) = 0, \quad y_{zi}(0^-) = 1, y_{zi}^{(1)}(0^-) = 1$$

Applying the Laplace transform, we have

$$Y_{zi}(s) = \frac{s+4}{s^2 + 3s + 2} = \frac{I(s)}{\Delta(s)}$$

The time domain zero-input response is obtained by using the inverse Laplace transformation, that is,

$$y_{zi}(t) = \mathcal{L}^{-1}\{Y_{zi}(s)\} = \mathcal{L}^{-1}\left\{\frac{s+4}{s^2 + 3s + 2}\right\} = \mathcal{L}^{-1}\left\{\frac{3}{s+1} - \frac{2}{s+2}\right\}$$
$$= (3e^{-t} - 2e^{-2t})u(t)$$

In order to find the zero-state response, we first represent the input signal in terms of elementary step and ramp signals, and then use the linearity and time-invariance principles as discussed in the paragraph concerned with formulas (4.56) and (4.57). The desired zero-state response is a linear combination of step and ramp responses as given by (4.57). In that respect, we must find the impulse response and integrate it in order to obtain the step response. By integrating the step response we will get the ramp response. The impulse response is obtained as follows:

$$h(t) = \mathcal{L}^{-1}\left\{\frac{1}{(s+1)(s+2)}\right\} = \mathcal{L}^{-1}\left\{\frac{1}{s+1} - \frac{1}{s+2}\right\} = (e^{-t} - e^{-2t})u(t)$$

The step response is

$$y_{step}(t) = \int_0^t h(\tau)\,d\tau = \int_0^t (e^{-\tau} - e^{-2\tau})\,d\tau = \left(\frac{1}{2} - e^{-t} + \frac{1}{2}e^{-2t}\right)u(t)$$

The ramp response is obtained as

$$y_{ramp}(t) = \int_0^t y_{step}(\tau)\,d\tau = \int_0^t \left(\frac{1}{2} - e^{-\tau} + \frac{1}{2}e^{-2\tau}\right)d\tau$$
$$= \left(-\frac{3}{4} + \frac{1}{2}t + e^{-t} - \frac{1}{4}e^{-2t}\right)u(t)$$

The zero-state response is given by

$$y_{zs}(t) = y_{\text{ramp}}(t) - y_{\text{step}}(t-1) - y_{\text{ramp}}(t-1)$$

$$= \left(-\frac{3}{4} + \frac{1}{2}t + e^{-t} - \frac{1}{4}e^{-2t}\right)u(t) - \left(\frac{1}{2} - e^{-(t-1)} + \frac{1}{2}e^{-2(t-1)}\right)u(t-1)$$

$$- \left(-\frac{3}{4} + \frac{1}{2}(t-1) + e^{-(t-1)} - \frac{1}{4}e^{-2(t-1)}\right)u(t-1)$$

which simplifies to

$$y_{zs}(t) = \left(-\frac{3}{4} + \frac{1}{2}t + e^{-t} - \frac{1}{4}e^{-2t}\right)u(t) + \left(\frac{3}{4} - \frac{1}{2}t - \frac{1}{4}e^{-2(t-1)}\right)u(t-1)$$

The complete system response is obtained as the sum of the zero-input and zero-state responses:

$$y(t) = y_{zi}(t) + y_{zs}(t)$$

$$= \left(-\frac{3}{4} + \frac{1}{2}t + 4e^{-t} - \frac{9}{4}e^{-2t}\right)u(t) + \left(\frac{3}{4} - \frac{1}{2}t - \frac{1}{4}e^{-2(t-1)}\right)u(t-1)$$

Note that in this particular example, the system response is a continuous function at $t=1$ despite the fact that the system input signal has a jump at that point. This can be easily checked by observing that the coefficient that multiplies $u(t-1)$ is equal to zero at $t=1$, that is $3/4 - 1/2 - 1/4 = 0$. Also, the first derivative of $y(t)$ is continuous at $t=1$, since

$$\frac{d}{dt}\left\{\left(\frac{3}{4} - \frac{1}{2}t - \frac{1}{4}e^{-2(t-1)}\right)u(t-1)\right\}\bigg|_{t=1}$$

$$= \left(-\frac{1}{2} + \frac{1}{2}e^{-2(t-1)}\right)\bigg|_{t=1} u(t-1) + \left(\frac{3}{4} - \frac{1}{2}t - \frac{1}{4}e^{-2(t-1)}\right)\bigg|_{t=1} \delta(t-1)$$

$$= 0u(t-1) + 0\delta(t-1) = 0$$

However, the second derivative of $y(t)$ has a jump discontinuity at $t=1$ equal to -1, which is identical to a jump in the input signal at the same time instant. This observation can be easily confirmed by finding the second derivative of $y(t)$ and evaluating it at $t=1$ (see Problem 4.46). This observation is consistent with the general theory of linear constant coefficient differential equations with discontinuous forcing functions (see, for example, [7]).

The Laplace transform approach to finding the complete time-invariant linear system response will be compared to the approach using the classical method for solving constant coefficient linear differential equations in Chapter 7.

4.3.5 Case Studies

The procedures for finding linear continuous-time, time-invariant system responses will be demonstrated in this section on two real-world examples: heartbeat dynamics and commercial aircraft pitch angle dynamics. At the same time, while studying the dynamics of a heartbeat, we will demonstrate how to use the Laplace transform to solve a system of linear differential equations.

Heartbeat Dynamics

A system of three first-order differential equations that describe the dynamics of a heartbeat was presented in Section 1.3, in formula (1.38). This input-free system is driven by its initial conditions, so that we are only interested in the system zero-input response. First, we will use the Laplace transform to show that this system of three first-order differential equations is equivalent to one third-order differential equation. In that process, we will derive the third-order differential equation for the cardiac muscle fiber normalized length (variable $x_1(t)$ in (1.38)) during the systolic period. That variable will be the quantity of interest in this case study. Note that the systolic time period lasts approximately 0.4 s, and after that pumping interval the cardiac muscle is in the relaxed (diastolic) state for approximately 0.6 s. Similarly, we can derive the third-order differential equations for variables $x_2(t)$ and $x_3(t)$. Since we are interested only in $x_1(t)$, we assume that the system output is equal to $x_1(t)$, that is, $y(t) = x_1(t)$. In the following, we will obtain the required zero-input response by finding first $\Delta(s)$ and $I(s)$, and then using the formula $y_{zi}(t) = x_1(t) = \mathcal{L}^{-1}\{X_1(s)\} = \mathcal{L}^{-1}\{I(s)/\Delta(s)\}$.

Let us apply the Laplace transform to both sides of (1.38), that is,

$$\mathcal{L}\left\{\frac{dx_1(t)}{dt}\right\} = \mathcal{L}\left\{-\frac{2}{\epsilon}x_1(t) - \frac{1}{\epsilon}x_2(t) - \frac{1}{\epsilon}x_3(t)\right\}$$

$$\mathcal{L}\left\{\frac{dx_2(t)}{dt}\right\} = \mathcal{L}\{-2x_1(t) - 2x_2(t)\}$$

$$\mathcal{L}\left\{\frac{dx_3(t)}{dt}\right\} = \mathcal{L}\{-x_2(t)\}$$

Note that in this system ϵ is a small positive parameter. At the end of our derivations we will assume that $\epsilon = 0.1$, and for that value we will evaluate the values for the cardiac muscle fiber normalized length as a function of time. Using the notation $X_i(s) = \mathcal{L}\{x_i(t)\}, i = 1, 2, 3$, we obtain a set of three algebraic equations in the complex s-domain,

$$sX_1(s) - x_1(0^-) = -\frac{2}{\epsilon}X_1(s) - \frac{1}{\epsilon}X_2(s) - \frac{1}{\epsilon}X_3(s)$$

$$sX_2(s) - x_2(0^-) = -2X_1(s) - 2X_2(s)$$

$$sX_3(s) - x_3(0^-) = -X_2(s)$$

This system of algebraic equations can be solved with respect to $X_1(s)$ by evaluating $X_2(s)$ and $X_3(s)$ in the second and third equations and replacing them in the first equation; namely,

$$X_2(s) = \frac{1}{s+2}(x_2(0^-) - 2X_1(s)), \quad X_3(s) = \frac{1}{s}(x_3(0^-) - X_2(s))$$

which leads to

$$X_1(s) = \frac{s(s+2)x_1(0^-) - \frac{1}{\epsilon}(s-1)x_2(0^-) - \frac{1}{\epsilon}(s+2)x_3(0^-)}{s^3 + \left(2+\frac{2}{\epsilon}\right)s^2 + \frac{2}{\epsilon}s + \frac{2}{\epsilon}}$$

Assuming that $\epsilon = 0.1$ and using for the system normalized initial conditions $(x_1(0^-), x_2(0^-), x_3(0^-)) = (1, -1, 0)$, we obtain

$$X_1(s) = \frac{s^2 + 12s - 10}{s^3 + 22s^2 + 20s + 20} = \frac{I(s)}{\Delta(s)}$$

Note that this transfer function corresponds to the third-order differential equation

$$\frac{d^3 x_1(t)}{dt^3} + 22\frac{d^2 x_1(t)}{dt^2} + 20\frac{dx_1(t)}{dt} + 20x_1(t) = 0,$$

$$x_1(0^-) = 1, \quad x_1^{(1)}(0^-) = -10, \quad x_1^{(2)}(0^-) = 190$$

The values for the initial conditions are obtained using (4.36).

Using the MATLAB function `residue`, we perform the partial fraction expansion and get

$$X_1(s) = \frac{k_1}{s - p_1} + \frac{k_2}{s - p_2} + \frac{k_2^*}{s - p_2^*}$$

with

$k_1 = 0.4261, \quad k_2 = 0.2870 + j0.4361, \quad p_1 = -21.0969, \quad p_2 = -0.4515 + j0.8626$

Applying the Laplace inverse and using formula (4.31), we obtain the required time domain signal:

$$x_1(t) = \{0.4261 e^{-21.0969t} + 1.044 e^{-0.4515t} \cos(0.86t + 0.9887)\} u(t)$$

The plot of this signal is presented in Figure 4.5.

It can be seen from the figure that for this model of heartbeat dynamics, and for $\epsilon = 0.1$, the cardiac muscle fiber normalized length shrinks to roughly 0.2 (20% of its

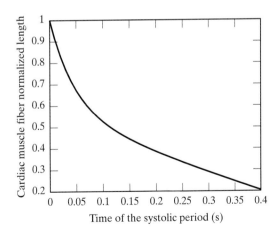

FIGURE 4.5: Cardiac muscle fiber normalized length as a function of time for $\epsilon = 0.1$

original length) at the end of the systolic period. At the beginning of the diastolic period the cardiac muscle goes quickly to its relaxed state and stays in it for approximately 0.6 s. Then, the heart pumping process is repeated periodically.

The MATLAB program that performs the preceding partial fraction and plots the corresponding time signal is given in the following script.

```
Td=0.4; t=0:0.01:Td;
num=[1 12 -10]; den=[1 22 20 20];
[k,p]=residue(num,den)
x1=k(1)*exp(p(1)*t)+2*abs(k(2))...
*exp(real(p(2))*t).*cos(imag(p(2))*t+angle(k(2)));
% ... stands for the concatenation MATLAB function
plot(t,x1)
xlabel('Time of the systolic period [s]')
ylabel('Cardiac muscle fiber normalized length')
print -deps fig4-5.eps
```

Commercial Aircraft Pitch Angle Dynamics

The mathematical model for a commercial aircraft's pitch angle was given in (1.43). For this third-order system, we assume that the initial conditions are given by $\theta(0^-) = 10°$, $\theta^{(1)}(0^-) = 0$, $\theta^{(2)}(0^-) = 0$, and that the elevator deflection angle (system input) is set to $f(t) = 3°$. We find the complete system response due to both the forcing function and the system initial conditions. From (4.54), we have

$$\theta(t) = \mathcal{L}^{-1}\left\{H(s)F(s) + \frac{I(s)}{\Delta(s)}\right\}$$

The system transfer function is given by

$$H(s) = \frac{1.151s + 0.1774}{s^3 + 0.739s^2 + 0.921s} = \frac{b_1 s + b_0}{s^3 + a_2 s^2 + a_1 s}$$

The Laplace transform of the forcing function is $F(s) = \mathcal{L}\{3u(t)\} = 3/s$. The polynomial $\Delta(s)$ is obviously given by $\Delta(s) = s^3 + 0.739s^2 + 0.921s$. It remains to find the polynomial $I(s)$, which can be done either directly from the corresponding differential equation (1.42), or using formula (4.36), for $n = 3$. Formula (4.36) implies

$$I(s) = a_1 \theta(0^-) + s a_2 \theta(0^-) + s^2 \theta(0^-)$$
$$= 0.921 \times 10 + s 0.739 \times 10 + s^2 10 = 10s^2 + 7.39s + 9.21$$

The complete response is given by

$$\theta(t) = \mathcal{L}^{-1}\left\{\frac{1.151s + 0.1774}{s(s^2 + 0.739s + 0.921)} \frac{3}{s} + \frac{10s^2 + 7.39s + 9.21}{s(s^2 + 0.739s + 0.921)}\right\}$$

Using the MATLAB function `residue` we obtain

$$\Theta_{zs}(s) = \frac{3.2855}{s} + \frac{0.5779}{s^2} + \frac{-1.6428 + j1.0115}{s + 0.3695 - j0.8857} + \frac{-1.6428 - j1.0115}{s + 0.3695 + j0.8857}$$

whose time domain equivalent is equal to

$$\theta_{zs}(t) = \{3.2855 + 0.5779t + 3.854e^{-0.3697t}\cos(0.8857t + 2.5897)\}u(t)$$

Due to cancellation of a common factor, the expression for the zero-input response simplifies to

$$\Theta_{zi}(s) = \frac{10}{s} \Rightarrow \theta_{zi}(t) = 10u(t)$$

It should be observed from the expression for $\theta_{zs}(t)$ that, for the given constant forcing function, the system zero-state response is unbounded in time (due to the presence of the ramp signal). Such a solution is unacceptable, and it reveals that this system is unstable. In Chapter 7, we will study system stability in detail. Here, we emphasize that this undesirable system property can be corrected by forming a *closed-loop* system. The concept of closed-loop (feedback) systems will be introduced in the next section, where we will also demonstrate for the aircraft example how the feedback loop can stabilize an unstable system.

4.4 BLOCK DIAGRAMS

In the previous section, we defined the linear time-invariant system transfer function. Using the Laplace transform linearity and convolution properties, we can easily extend the concept of the transfer function to configurations of several connected linear systems. In that way we will find the equivalent transfer functions for cascade and parallel connections of systems, introduce the feedback (closed-loop) configuration, and define the corresponding feedback system transfer function.

In Section 4.3.1, we defined the transfer function of a linear time-invariant continuous-time system. It can be seen from formula (4.38) that the corresponding system transfer function is the ratio of the Laplace transform of the system output to the Laplace transform of the system input, under the assumption that the system initial conditions are zero. The transfer function defined in (4.38) in fact represents the *open-loop* continuous-time *system transfer function*.

It follows from formula (4.38) that for a system at rest, the system input $F(s)$ produces in the system output the signal $Y_{zs}(s)$ given by

$$Y_{zs}(s) = H(s)F(s) \tag{4.58}$$

which is symbolically represented in Figure 4.6 using a block diagram.

Note that this block diagram can be used also in the case when the system initial conditions are nonzero. In such a case, an additive component coming from the system initial

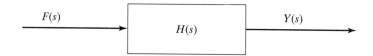

FIGURE 4.6: An open-loop transfer function

conditions should be added to the system output. For that reason, in all block diagrams presented in this section, we will denote the system output by $Y(s)$.

The open-loop transfer function $H(s)$ is derived from a differential equation that has been obtained using known physical laws (obtained through a mathematical modeling procedure). The accuracy of coefficients that appear in the open-loop system transfer function depends on the accuracy of the system coefficients (inductors, resistors, masses, friction coefficients, and so on). These coefficients are not always perfectly known. Furthermore, the coefficients change, due to aging or internal and/or external system disturbances. Due to changes in the system coefficients (system parameters), it can happen that the actual system output (in the open-loop system configuration) is significantly different from the output obtained analytically using formulas (4.37) and (4.38).

A way to cope with system parameter changes, and to reduce the impact of those changes on the system output, is to form a *closed-loop system* configuration, also known as a system feedback configuration. Assuming that it is feasible, we can feed the system output back and form a *closed loop* around the system, as presented in Figure 4.7(a). The directed path (as indicated by the arrows) from $F(s)$ to $Y(s)$ is called the *forward path*, and

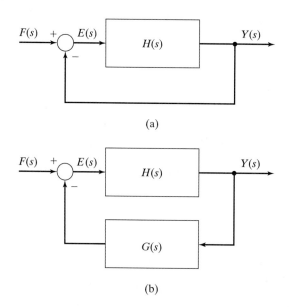

FIGURE 4.7: Closed-loop system configurations: (a) unity feedback; (b) nonunity feedback

the directed path from $Y(s)$ to $E(s)$ is called the *feedback path*. Such a feedback loop is called a unity feedback loop. In general, we can put a dynamic element $G(s)$ (another open-loop transfer function) in the feedback loop, as presented in Figure 4.7(b). For notational convenience, we will denote the transfer function in the feedback path by $G(s)$ and the transfer function in the forward path by $H(s)$. It should be pointed out that sometimes we put a static element equal to a constant in the feedback path; that is, $G(s) = \text{const.}$

In the feedback configuration presented in Figure 4.7(a), the output signal is fed back and compared to the input signal, and the difference between the input and output signals is used as a new input signal to the system. In practice, the feedback signal is taken with a negative sign since, in general, a positive feedback signal causes system instability. Using (4.58) and following signals in the block diagram in the direction of the arrows, we can find the closed-loop system transfer function from $F(s)$ to $Y(s)$, assuming zero initial conditions, as follows:

$$Y(s) = H(s)E(s), \quad E(s) = F(s) - Y(s) \Rightarrow Y(s) = H(s)(F(s) - Y(s))$$

$$\Rightarrow Y(s) = \frac{H(s)}{1 + H(s)} F(s) \triangleq M(s)F(s)$$

The *closed-loop system transfer function* for unity feedback, denoted by $M(s)$, is given by

$$M(s) = \frac{Y(s)}{F(s)}\bigg|_{\text{I.C.}=0} \triangleq \frac{H(s)}{1 + H(s)} \tag{4.59}$$

The closed-loop transfer function defined in (4.59) is called the closed-loop transfer function with unity feedback. In many applications, another transfer function is present in the feedback loop (see Figure 4.7(b)). The closed-loop transfer function with nonunity feedback is obtained similarly, as follows:

$$Y(s) = H(s)(F(s) - G(s)Y(s)) \Rightarrow Y(s) = \frac{H(s)}{1 + H(s)G(s)} F(s)$$

$$Y(s) \triangleq M(s)F(s), \quad M(s) = \frac{H(s)}{1 + H(s)G(s)} \tag{4.60}$$

Hence, we can introduce the following definition.

DEFINITION 4.1: The closed-loop system transfer function for nonunity feedback is defined by

$$M(s) \triangleq \frac{Y(s)}{F(s)}\bigg|_{\text{I.C.}=0} = \frac{H(s)}{1 + H(s)G(s)}$$

If we go around the loop of the nonunity feedback block diagram presented in Figure 4.7(b), we will encounter two transfer function $H(s)$ and $G(s)$. The product is called the loop transfer function. This can be formally stated in the form of a new definition.

DEFINITION 4.2: The *loop transfer function* for nonunity feedback (given the configuration of Figure 4.7(b)) is defined by the product $H(s)G(s)$.

We can form more complex configurations of open-loop transfer functions. In general, open-loop transfer functions can be connected in cascade or parallel, or they can form more complex feedback configurations containing several feedback loops.

The *cascade connection* of open-loop transfer functions is presented in Figure 4.8(a). It is easy to conclude that the equivalent open-loop transfer function is given by the product of elementary open-loop transfer functions,

$$H_{eq}^{cascade}(s) = H_1(s)H_2(s)\cdots H_n(s) \qquad (4.61)$$

Formula (4.61) can be called the *product rule for elementary open-loop transfer functions*.

The *parallel connection* of open-loop transfer functions is represented in Figure 4.8(b). Its equivalent open-loop transfer function is equal to the sum of elementary open-loop transfer functions, that is,

$$H_{eq}^{parallel}(s) = H_1(s) \pm H_2(s) \pm \cdots \pm H_n(s) \qquad (4.62)$$

The last formula can be called the *sum rule for elementary open-loop transfer functions*.

Using the basic transfer function formulas (rules) established in (4.58–62), we can simplify more complex feedback systems and represent them in the basic feedback form presented in Figure 4.7(b). The corresponding procedure is demonstrated in the next example.

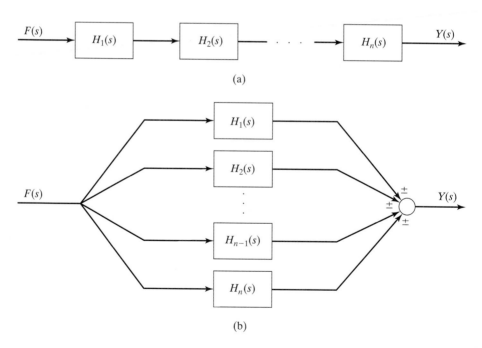

FIGURE 4.8: (a) Cascade and (b) parallel connections of transfer functions

EXAMPLE 4.25

Consider a feedback system as given in Figure 4.9. We find its closed-loop transfer function as follows. Using the product rule, (4.61), we have

$$E_2(s) = H_1(s)H_2(s)E_1(s)$$

$$Y(s) = H_3(s)E_3(s)$$

The product rule combined with the sum rule, (4.62), produces

$$E_1(s) = F(s) - G_2(s)Y(s)$$

$$E_3(s) = E_2(s) - G_1(s)Y(s)$$

Eliminating $E_i(s)$, $i = 1, 2, 3$, we obtain

$$\frac{Y(s)}{H_3(s)} = H_1(s)H_2(s)(F(s) - G_2(s)Y(s)) - G_1(s)Y(s)$$

This leads to the closed-loop transfer function of the form

$$M(s) = \frac{Y(s)}{F(s)} = \frac{H_1(s)H_2(s)H_3(s)}{1 + G_1(s)H_3(s) + G_2(s)H_1(s)H_2(s)H_3(s)}$$

Another way to find the closed-loop transfer function would be to identify the elements corresponding to those given in the block diagram in Figure 4.7(b). It can be easily seen that

$$G(s) = G_2(s), \quad H(s) = H_1(s)H_2(s)\frac{H_3(s)}{1 + G_1(s)H_3(s)}$$

Using the closed-loop transfer function formula (4.60), we can now obtain results for $M(s)$ identical to those given previously.

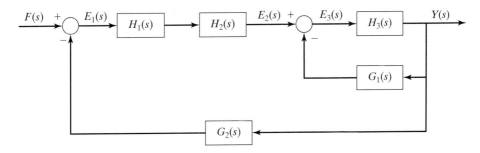

FIGURE 4.9: A feedback system

The algebra associated with transfer functions is simple and convenient for linear feedback systems composed of several loops. In the case of linear systems with many feedback loops, finding the closed-loop transfer function becomes a very tedious task. In such a case, we can use Mason's formula, obtained using elementary graph theory. (For more information on Mason's formula, the interested reader should consult an undergraduate text on control systems—example, [8]).

EXAMPLE 4.26

We consider the commercial aircraft pitch angle model from Section 4.3.5, and find first its unity feedback closed-loop transfer function. Using formula (4.59), we obtain easily the following expression for the closed-loop transfer function:

$$H(s) = \frac{1.151s + 0.1774}{s^3 + 0.739s^2 + 0.921s} \Rightarrow M(s) = \frac{H(s)}{1 + H(s)}$$

$$= \frac{1.151s + 0.1774}{s^3 + 0.739s^2 + 2.072s + 0.1774}$$

In the following, we use Simulink to simulate the unity feedback closed-loop zero-state response of the pitch angle model due to the same input used in Section 4.3.5, that is, due to $f(t) = 3u(t)$, and compare the result obtained to the corresponding result from Section 4.3.5.

Simulink is run on the MATLAB platform. It can be accessed from MATLAB by typing `simulink`. The corresponding Simulink block diagram is presented in Figure 4.10. (In Appendix D, we explain the procedures required to build this block diagram.) All elements in the block diagram are self-explanatory. Each block can be opened to set parameters. For example, after opening the step block we set the step initial time to 0 and the step height to 3. The signals flow in the direction of the arrows. The zero-state response can be monitored using an oscilloscope. Also, the obtained data can be passed to MATLAB via the block `to workspace`.

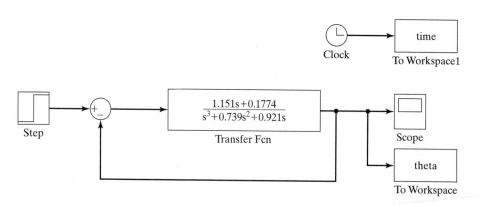

FIGURE 4.10: Simulink block diagram for the aircraft pitch angle feedback configuration

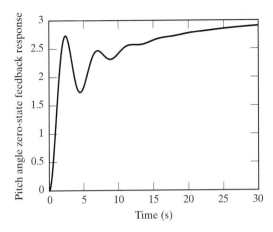

FIGURE 4.11: Aircraft pitch angle zero-state feedback response

In Figure 4.11, we plot the required zero-state using the MATLAB zero-state response statement plot(time,theta). It can be noticed from Figure 4.11 that in contrast to the open-loop zero-state response obtained in Section 4.3.5, the closed-loop zero-state response is not unbounded in time. Rather, as time passes, it settles down at the steady state value equal to $\theta_{ss} = 3$, which is an interesting result that indicates a very simple possibility for controlling the system output at steady state. Note that in this case the input signal $f(t) = 3u(t)$ produces the output signal at steady state equal to $\theta_{ss} = 3$, which is consistent with formula (4.40), since $\theta_{ss} = 3M(0) = 3$. Using the MATLAB function residue, we can find the partial fraction expansion, and obtain the analytical formula for the zero-state closed-loop system response as

$$\theta_{zs}(t) = \{3 - 1.319e^{-0.088t} + 2 \times 0.874e^{-0.841t} \cos(1.382t + 2.863)\}u(t)$$

4.5 FROM LAPLACE TRANSFORM TO $\mathcal{Z}$-TRANSFORM

The discrete-time counterpart to the Laplace transform is the $\mathcal{Z}$-transform. The $\mathcal{Z}$-transform is applicable to discrete-time signals. In this section we introduce the $\mathcal{Z}$-transform starting with the Laplace transform. Another approach that can be used to introduce the $\mathcal{Z}$-transform starts with the development of the discrete-time Fourier transform (the discrete-time counterpart to the Fourier transform presented in Chapter 3) via the sampling operation applied to a continuous-time signal. In that process, the $\mathcal{Z}$-transform is derived from the corresponding discrete-time (sampled) signal obtained using an ideal sampler. That approach to the $\mathcal{Z}$-transform is rather lengthy. In Chapter 9 we will present a detailed study of the sampling operation, and derive the sampling theorem and the discrete-time Fourier transform; the definition of the $\mathcal{Z}$-transform will come naturally as a byproduct of those derivations. In this section, we use a much shorter approach and derive the $\mathcal{Z}$-transform as the Laplace transform of a continuous-time signal sampled by a *zero-order hold*.

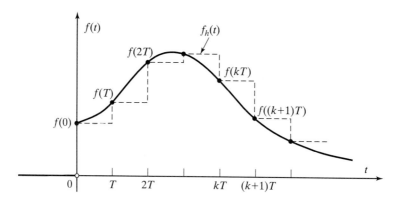

FIGURE 4.12: Signal "sampling" via the use of a zero-order hold

The zero-order hold "sampling" of a continuous-time signal $f(t)$ leads to another continuous-time signal that has a staircase form, as presented in Figure 4.12 (using dashed lines). The signal obtained, $f_h(t)$, can be considered a "discrete-time signal" with its values defined at $f(kT) \triangleq f[k]$.

The term "zero-order hold" comes from the fact that this dynamic element can make a continuous-time staircase signal from a discrete-time signal by holding the value of a discrete-time signal, $f(kT)$, during the given sampling interval kT. In practice, the zero-order hold element is used for that purpose—conversion of a discrete-time signal into a continuous-time signal. Here, we use the zero-order hold element for the purpose of deriving the $\mathcal{Z}$-transform from the Laplace transform. In the following, we will also derive the transfer function of the zero-order hold element.

Let T denote the sampling period. The obtained staircase signal is given by

$$f_h(t) = \sum_{k=0}^{\infty} f(kT)[u(t - kT) - u(t - (k+1)T)] \tag{4.63}$$

Applying the Laplace transform to this signal, we have

$$\mathcal{L}\{f_h(t)\} = \sum_{k=0}^{\infty} f(kT) \left[\frac{1}{s} e^{-skT} - \frac{1}{s} e^{-s(k+1)T} \right] = \frac{1 - e^{-sT}}{s} \sum_{k=0}^{\infty} f(kT) e^{-skT} \tag{4.64}$$

Introducing the notations

$$F_h(s) = \mathcal{L}\{f_h(t)\}, \quad F^*(s) = \sum_{k=0}^{\infty} f(kT) e^{-skT}, \quad H_h(s) = \frac{1 - e^{-sT}}{s} \tag{4.65}$$

we obtain

$$F_h(s) = H_h(s) F^*(s) \tag{4.66}$$

The quantity $F^*(s)$, defined in (4.65), is called the *starred Laplace transform*. $H_h(s)$ is the transfer function of the zero-order hold element.

The $\mathcal{Z}$-transform is defined by

$$F(z) \triangleq F^*(s)|_{s=\frac{1}{T}\ln(z)} = \sum_{k=0}^{\infty} f(kT)z^{-k}, \quad z = e^{sT} \qquad (4.67)$$

which, according to the notation used for discrete-time signals, can be written as

$$F(z) \triangleq \sum_{k=0}^{\infty} f[k]z^{-k} \qquad (4.68)$$

As a matter of fact, the definition formula (4.68) is obtained for causal signals, and it defines the *one-sided* or *unilateral* $\mathcal{Z}$-*transform*. We could have done the same derivations using a noncausal signal, $k \in (-\infty, \infty)$, in which case we would have obtained the *two-sided* or *bilateral* $\mathcal{Z}$-*transform* defined by

$$F(z) \triangleq \sum_{k=-\infty}^{\infty} f[k]z^{-k} \qquad (4.69)$$

4.6 MATLAB LABORATORY EXPERIMENT

Purpose: This experiment presents the frequency domain analysis of continuous-time linear systems using MATLAB. The impulse, step, sinusoidal, and exponential responses of continuous-time systems will be examined using the transfer function method based on the Laplace transform. In addition, MATLAB will be used to perform the partial fraction expansion and to find the inverse Laplace transform.

Part 1. Consider the linear system represented by the transfer function

$$H(s) = \frac{s+1}{s^2 + 5s + 6}$$

Using MATLAB, find and plot:

(a) The system impulse response.
(b) The system step response.
(c) The system zero-state response due to the input signal $f(t) = \sin(2t)u(t)$.
(d) The system zero-state response due to the input signal $f(t) = e^{-t}u(t)$.

Part 2. Consider the transfer function

$$H(s) = \frac{2s^5 + s^3 - 3s^2 + s + 4}{5s^8 + 2s^7 - s^6 - 3s^5 + 5s^4 + 2s^3 - 4s^2 + 2s - 1}$$

(a) Find the factored form of the transfer function by using the MATLAB function `[z,p,k]=tf2zp(num,den)`.
(b) The partial fraction expansion of rational functions can be performed using the MATLAB function `residue`. Find the Laplace inverse of the given transfer

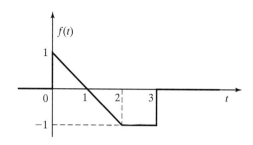

FIGURE 4.13: An input signal

function using the MATLAB function `residue`; that is, find analytically the system impulse response.

Part 3. Consider the system defined by

$$y^{(2)}(t) + 5y^{(1)}(t) + 4y(t) = f(t)$$

and the input signal represented in Figure 4.13. Use MATLAB to plot the zero-state response of this system. (*Hint:* See Example 4.24.)

Part 4. Find and plot the zero-input response of a flexible beam [9], whose transfer function is given by

$$H(s) = \frac{1.65s^4 - 0.331s^3 - 576s^2 + 90.6s + 19080}{s^6 + 0.996s^5 + 463s^4 + 97.8s^3 + 12131s^2 + 8.11s}$$

with the initial conditions $y^{(4)}(0^-) = 1$, $y^{(j)}(0^-) = 0$, $j = 0, 1, 2, 3, 5$. (*Hint:* Find $I(s)$ and $\Delta(s)$ as defined in formulas (4.36) and (4.52), and use the MATLAB function `impulse`.)

Submit four plots for Part 1, one plot for Part 3, and one plot for Part 4, and present analytical results obtained in Parts 2–4.

4.7 SUMMARY

Study Guide for Chapter Four: Students must master the properties of the Laplace transform and become familiar with the table of common pairs. The use of the Laplace transform in the analysis of linear time-invariant systems is the main thrust of this chapter. Knowledge of the techniques for finding impulse, step, zero-state, zero-input, and complete system responses is essential. Standard problems: (1) Find the Laplace transform of a given time signal; (2) find the inverse Laplace transform of a given frequency domain signal; (3) state and prove a particular property of the Laplace transform; (4) find the system transfer function and the system impulse response; (5) find the system step response; (6) find the zero-state response; (7) find the zero-input response; (8) find the complete system response.

Section 4.7 Summary

Laplace Transform and Its Inverse:

$$F(s) = \mathcal{L}\{f(t)\} \triangleq \int_{0^-}^{\infty} f(t)e^{-st}\,dt$$

$$f(t) = \mathcal{L}^{-1}\{F(s)\} \triangleq \frac{1}{2\pi j} \int_{\gamma-j\infty}^{\gamma+j\infty} F(s)e^{st}\,ds$$

Transfer Function and Its DC (Steady State) Gain:

$$H(s) = \frac{b_m s^m + b_{m-1}s^{m-1} + \cdots + b_1 s + b_0}{s^n + a_{n-1}s^{n-1} + a_{n-2}s^{n-2} + \cdots + a_1 s + a_0}, \quad H(0) = \frac{b_0}{a_0}$$

System Characteristic Polynomial:

$$\Delta(s) = s^n + a_{n-1}s^{n-1} + \cdots + a_1 s + a_0$$

System Steady State Response to a Constant Input:

$$y_{ss} = H(0)a, \quad f(t) = au(t)$$

System Impulse Response:

$$h(t) \triangleq \mathcal{L}^{-1}\{H(s)\}$$

System Step and Ramp Responses:

$$y_{\text{step}}(t) = \int_{0^-}^{t} h(\tau)\,d\tau, \quad h(t) = \frac{dy_{\text{step}}(t)}{dt} = \frac{d^2 y_{\text{ramp}}(t)}{dt^2}$$

$$y_{\text{step}}(t) = \frac{dy_{\text{ramp}}(t)}{dt}, \quad y_{\text{ramp}}(t) = \int_{0}^{t} y_{\text{step}}(\tau)\,d\tau$$

System Zero-State Response:

$$Y_{zs}(s) = H(s)F(s) \leftrightarrow y_{zs}(t) = h(t) * f(t)$$

System Zero-Input Response:

$$Y_{zi}(s) = \frac{I(s)}{\Delta(s)} \leftrightarrow y_{zi}(t) = \mathcal{L}^{-1}\left\{\frac{I(s)}{\Delta(s)}\right\}$$

System Complete Response:

$$y(t) = y_{zs}(t) + y_{zi}(t) = \mathcal{L}^{-1}\{Y_{zs}(s)\} + \mathcal{L}^{-1}\{Y_{zi}(s)\} = \mathcal{L}^{-1}\{H(s)F(s)\} + \mathcal{L}^{-1}\left\{\frac{I(s)}{\Delta(s)}\right\}$$

Partial Fraction Expansion Simple Pole Coefficients:

$$k_i = \lim_{s \to p_i} \{(s - p_i)F(s)\}, \quad i = 1, 2, \ldots, n$$

Partial Fraction Expansion Multiple Pole Coefficients:

$$k_{ij} = \frac{1}{(r-j)!} \lim_{s \to p_i} \left\{ \frac{d^{r-j}}{ds^{r-j}}[(s-p_i)^r F(s)] \right\}$$

$$j = r, r-1, \ldots, 2, 1; \quad i = 1, 2, \ldots, n$$

p_i is a multiple pole of multiplicity r

Formula for the Laplace Inverse of a Pair of Complex Conjugate Poles:

$$\mathcal{L}^{-1}\left\{ \frac{k_i}{s - p_i} + \frac{k_i^*}{s - p_i^*} \right\} = 2|k_i|e^{\alpha_i t}\cos(\beta_i t + \angle k_i), \quad p_i = \alpha_i + j\beta_i$$

TABLE 4.1 Properties of the Laplace Transform

$\mathcal{L}\{\alpha_1 f_1(t) \pm \alpha_2 f_2(t)\}$	$\alpha_1 F_1(s) \pm \alpha_2 F_2(s)$
$\mathcal{L}\{f(t - t_0)u(t - t_0)\}$	$e^{-st_0}F(s), \ t_0 > 0$
$\mathcal{L}\{f(at)\}$	$\frac{1}{a}F\left(\frac{s}{a}\right), \ a > 0$
$\mathcal{L}\{t^n f(t)\}$	$(-1)^n \frac{d^n}{ds^n} F(s)$
$e^{\lambda t} f(t)$	$F(s - \lambda)$
$f(t)\cos(\omega_0 t)$	$\frac{1}{2}[F(s + j\omega_0) + F(s - j\omega_0)]$
$f(t)\sin(\omega_0 t)$	$\frac{j}{2}[F(s + j\omega_0) - F(s - j\omega_0)]$
$\mathcal{L}\left\{\frac{d}{dt}f(t)\right\}$	$sF(s) - f(0^-)$
$\mathcal{L}\left\{\frac{d^2}{dt^2}f(t)\right\}$	$s^2 F(s) - sf(0^-) - f^{(1)}(0^-)$
$\mathcal{L}\left\{\frac{d^n}{dt^n}f(t)\right\}$	$s^n F(s) - s^{n-1}f(0^-) - s^{n-2}f^{(1)}(0^-) - \cdots - f^{(n-1)}(0^-)$
$\mathcal{L}\{f_1(t) * f_2(t)\}$	$F_1(s)F_2(s)$
$\mathcal{L}\left\{\int_0^t f(\tau)\,d\tau\right\}$	$\frac{1}{s}F(s)$
$\lim_{t \to 0^+}\{f(t)\}$	$\lim_{s \to \infty}\{sF(s)\}$
$\lim_{t \to \infty}\{f(t)\}$	$\lim_{s \to 0}\{sF(s)\}$

TABLE 4.2 Common Laplace Transform Pairs

$\delta(t)$	1
$u(t)$	$\dfrac{1}{s}$
$e^{-\alpha t}u(t)$	$\dfrac{1}{s+\alpha}$
$t^n u(t)$	$\dfrac{n!}{s^{n+1}}$
$t^n e^{-\alpha t} u(t)$	$\dfrac{n!}{(s+\alpha)^{n+1}}$
$u(t)\cos(\omega t)$	$\dfrac{s}{s^2+\omega^2}$
$u(t)\sin(\omega t)$	$\dfrac{\omega}{s^2+\omega^2}$
$e^{-\alpha t}u(t)\cos(\omega t)$	$\dfrac{s+\alpha}{(s+\alpha)^2+\omega^2}$
$e^{-\alpha t}u(t)\sin(\omega t)$	$\dfrac{\omega}{(s+\alpha)^2+\omega^2}$
$tu(t)\cos(\omega t)$	$\dfrac{s^2-\omega^2}{(s^2+\omega^2)^2}$
$tu(t)\sin(\omega t)$	$\dfrac{2\omega s}{(s^2+\omega^2)^2}$
$te^{-\alpha t}u(t)\cos(\omega t)$	$\dfrac{(s+\alpha)^2-\omega^2}{((s+\alpha)^2+\omega^2)^2}$
$te^{-\alpha t}u(t)\sin(\omega t)$	$\dfrac{2\omega(s+\alpha)}{((s+\alpha)^2+\omega^2)^2}$

4.8 REFERENCES

[1] R. Churchill, *Complex Variables and Applications,* McGraw-Hill, New York, 1990.

[2] M. Greenberg, *Foundations of Applied Mathematics,* Prentice Hall, Englewood Cliffs, NJ, 1978.

[3] C. Wylie and L. Barrett, *Advanced Engineering Mathematics,* McGraw-Hill, New York, 1995.

[4] M. Spiegel, *Laplace Transforms,* McGraw-Hill, New York, 1965.

[5] A. Zemanian, *Distribution Theory and Transform Analysis,* McGraw-Hill, New York, 1965.

[6] T. Kailath, *Linear Systems,* Prentice Hall, Englewood Cliffs, NJ, 1980.

[7] W. Boyce and R. DiPrima, *Elementary Differential Equations,* Wiley, New York, 1992.

[8] Z. Gajić and M. Lelić, *Modern Control Systems Engineering,* Prentice Hall, London, 1996.

[9] L. Qiu and E. Davison, "Performance limitations of non-minimum phase systems in the servomechanism problem," *Automatica,* 29:337–49, 1993.

[10] J. Schiff, *The Laplace Transform: Theory and Applications,* Springer-Verlag, New York, 1999.

[11] L. Pori, A. Bonini, and N. Oriente, "Impulse response measurement of balanced chains of EDFA's in a recirculating loop," *IEEE Photonics Technology Letters,* 11:1384–86, 1999.

[12] M. Spong and M. Vidyasagar, *Robot Dynamics and Control,* Wiley, New York, 1989.

[13] J. He, M. Maltenfort, Q. Wang, and T. Hamm, "Learning from biological systems: Modeling neural control," *IEEE Control Systems Magazine,* 21:55–69, 2001.

[14] K. Kadiman and D. Williamson, "Discrete minimax linear quadratic regulation of continuous-time systems," *Automatica,* 23:741–47, 1987.

[15] P. Kokotović, "Feedback design of large scale linear systems," in *Feedback Systems,* J. Cruz (ed.), McGraw-Hill, New York, 1972, 99–137.

4.9 PROBLEMS

4.1. Derive formula (4.10).

4.2. Derive formulas given in (4.16).

4.3. Let $f(t)u(t)$ and $F(s)$ form a Laplace transform pair. Find the Laplace transform of the signal $f(at - t_0)u(at - t_0)$, $a > 0$, $t_0 > 0$.

Answer:
$$f(at - t_0)u(t - t_0) \leftrightarrow \frac{1}{a}e^{-(\frac{s}{a})t_0}F\left(\frac{s}{a}\right)$$

4.4. Use the Euler formula for $\cos(\omega t)$ and the Laplace transform of the exponential signal to derive the Laplace transform for $\cos(\omega t)$. Repeat the same procedure for $\sin(\omega t)$.

4.5. Using the definition formula of the Laplace transform, derive the Laplace transform of a *semiperiodic signal* defined by

$$f(t) = \begin{cases} f(t - T), & t \geq 0 \\ 0, & t < 0 \end{cases}$$

Note that the Laplace transform of such a signal, in general, is not a rational function (ratio of two polynomials) [10]. Using the result obtained, find the Laplace transforms for (a) $u(t)\sin(\omega t)$ and (b) a semiperiodic train of rectangular pulses with $t_1 = 1$, $t_2 = 2$, and $E = 1$ (see Figure 3.20).

Answers:
$$F(s) = \frac{F_p(s)}{1 - e^{-sT}}, \quad F_p(s) = \int_0^T f(t)e^{-st}\,dt,$$

(b) $\dfrac{1}{s(1 + e^s)}$

4.6. Is the function $\tan(t)$ Laplace transformable?

4.7. Find the Laplace transforms of the following signals:

(a) $f_1(t) = 5\delta(t - 3) + 4u(t - 2) + 3e^{-5(t-3)}u(t - 4)$

(b) $f_2(t) = \dfrac{d}{dt}\{t^{10}e^{-t}u(t)\} + \displaystyle\int_{0^-}^{t} e^{-5\tau}\sin(5\tau)\,d\tau$

(c) $f_3(t) = (t+1)e^{-t}u(t-1) + e^{-t}\sin(2t)u(t)$

(d) $f_4(t) = \sin(t)[u(t) - u(t-1)]$

Answers:

(a) $F_1(s) = 5e^{-3s} + \dfrac{4}{s}e^{-2s} + 3e^{-5}\dfrac{1}{s+5}e^{-4s}$

(b) $F_2(s) = \dfrac{10!\,s}{(s+1)^{11}} + \dfrac{1}{s}\left(\dfrac{5}{(s+5)^2+5^2}\right)$

(c) $F_3(s) = \dfrac{e^{-(1+s)}}{(s+1)^2} + 2\dfrac{e^{-(1+s)}}{s+1}$

(d) $F_4(s) = \dfrac{1}{s^2+1} - \cos(1)e^{-s}\dfrac{1}{s^2+1} + \sin(1)e^{-s}\dfrac{s}{s^2+1}$

4.8. Use the time delay property to find the Laplace transform of the following signals.

(a) $f_1(t) = (t-3)^3 u(t-3)$

(b) $f_2(t) = (t-2)^2 e^{-5t} u(t-1)$

(c) $f_3(t) = te^{-t} u(2t-3)$

(d) $f_4(t) = 5t[u(t-1) + 2u(t-3)]$

Answers:

(b) $F_2(s) = e^{-5}e^{-s}\left[\dfrac{2}{(s+5)^3} - \dfrac{2}{(s+5)^2} + \dfrac{1}{s+5}\right]$

(c) $F_3(s) = \dfrac{1}{(s+1)^2}\left(2 - e^{-1.5(s+1)}\right) - \dfrac{1.5}{s+1}e^{-1.5(s+1)}$

4.9. Find the Laplace transform of the following time delayed signals.

(a) $f_1(t) = u(t-1)\sin(\pi t)$

(b) $f_2(t) = u(t)e^{-3t}\cos(\pi(t-2))$

(c) $f_3(t) = u(t-1)te^{-2t}\sin(\pi t)$

(d) $f_4(t) = te^{-2t}\sin(t-2)u(t-2)$

Answers:

(a) $F_1(s) = -e^{-s}\dfrac{\pi}{s^2+\pi^2}$

(b) $F_2(s) = \dfrac{s+3}{(s+3)^2+\pi^2}$

(c) $F_3(s) = -e^{-(s+2)}\dfrac{2\pi(s+2)}{[(s+2)^2+\pi^2]^2} + e^{-(s+2)}\dfrac{\pi}{(s+2)^2+\pi^2}$

4.10. Assuming that the Laplace transform of $f(t)$ is $F(s)$, find the Laplace transform of the signal

$$5e^{-2t}f(t) + f(3t-2)\cos(\omega t) + \int_{0^-}^{t} 2f(\tau)\,d\tau + \int_{0^-}^{\infty} f(t-\tau)\sin(5\tau)\,d\tau$$

Hint: See the answer to Problem 4.3.

4.11. Using the final and initial value theorems, find $f_i(0^+)$ and $f_i(\infty)$, $i = 1, 2, 3$, in the following cases.

(a) $F_1(s) = \dfrac{2s^2 + 2s + 3}{(s+1)(s+2)(s+3)}$

(b) $F_2(s) = \dfrac{s+4}{s(s+1)(s+2)}$

(c) $F_3(s) = \dfrac{s}{(s+1)(s^2+2s+2)}$

(d) $F_4(s) = \dfrac{s^2}{s^2 + 5s + 10}$

Answers:

(a) $f_1(0^+) = 2$, $\quad f_1(\infty) = 0$

(b) $f_2(0^+) = 0$, $\quad f_2(\infty) = 2$

(c) $f_3(0^+) = 0$, $\quad f_3(\infty) = 0$

(d) $f_4(0^+) = \infty$, $\quad f_4(\infty) = 0$

4.12. Can the final value theorem be applied to the unit step signal $u(t)$? Justify your answer.

4.13. Can the final value theorem be applied to the following Laplace transforms? If yes, find the signal final values. If no, explain why not.

(a) $F_1(s) = \dfrac{s^2}{(s+1)(s-2)}$

(b) $F_2(s) = \dfrac{1}{s^2(s+1)(s+2)}$

(c) $F_3(s) = \dfrac{s+2}{(s+1)(s^2+2)}$

(d) $F_4(s) = \dfrac{s+3}{(s+2)(s^2+2s+2)}$

(e) $F_5(s) = \dfrac{s+2}{(s+1)(s+3)(s+5)}$

4.14. Find the inverse Laplace transform of the following signals.

(a) $F_1(s) = \dfrac{5}{(s+1)(s+2)(s+3)(s+5)(s+10)(s+20)}$

(b) $F_2(s) = \dfrac{1}{(s+2)^5(s+4)}$

(c) $F_3(s) = \dfrac{s-1}{(s+1)^3(s+2)^4}$

4.15. Find the Laplace transform of the signal presented in Figure 2.18.

4.16. Find the inverse Laplace transform of the signals given in Problem 4.13, and check conclusions derived in Problem 4.13.

Answers:

(a) $f_1(t) = \delta(t) - \left(\dfrac{1}{3}e^{-t} - \dfrac{4}{3}e^{2t}\right)u(t)$

(b) $f_2(t) = \left(-\dfrac{3}{4} + \dfrac{1}{2}t + e^{-t} - \dfrac{1}{4}e^{-2t}\right)u(t)$

(c) $f_3(t) = \cos\left(-\sqrt{2}t + 109.47°\right)u(t) + \dfrac{1}{3}e^{-t}u(t)$

4.17. Find the Laplace transform of the signal presented in Figure 4.14.

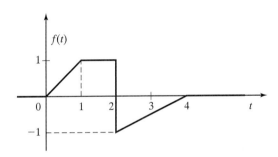

FIGURE 4.14

4.18. Find the inverse Laplace transform of the following complex domain signals.

(a) $F_1(s) = \dfrac{3s^2 + 2s + 1}{(s+1)(s+2)^2}$

(b) $F_2(s) = \dfrac{s+1}{(s^2+1)(s+2)}$

(c) $F_3(s) = \dfrac{2s+5}{(s^2+2s+2)(s+1)}$

(d) $F_4(s) = \dfrac{s+1}{(s^2+4)s}$

Answer: **(a)** $f_1(t) = (e^{-2t} - 9te^{-2t} + 2e^{-t})u(t)$

4.19. Apply the final and initial value theorems to the complex variable functions defined in Problem 4.18. Check that the results obtained are consistent with the results from Problem 4.18.

4.20. Find the inverse Laplace transform of the following functions containing complex exponentials.

(a) $F_1(s) = \dfrac{s + e^{-5s}}{s^2 + 2s + 1}$

(b) $F_2(s) = \dfrac{2 - e^{-5s}}{s^2(s + 3)}$

(c) $F_3(s) = \dfrac{s + 1}{s^2(s + 2)} e^{-3s}$

(d) $F_4(s) = \dfrac{(s + 1)e^{-s}}{(s^2 + 4)s}$

(e) $F_5(s) = \dfrac{s + 2}{s^2(s^2 + 2)} e^{-2s}$

Answers:

(a) $f_1(t) = (t - 5)e^{-(t-5)}u(t - 5)$

(b) $f_2(t) = \left(-\dfrac{2}{3} + \dfrac{2}{3}t + \dfrac{2}{9}e^{-3t}\right)u(t) - \left(-\dfrac{1}{3} + \dfrac{1}{3}(t - 5) + \dfrac{1}{9}e^{-3(t-5)}\right)u(t - 5)$

(c) $f_3(t) = \left(\dfrac{1}{4} + \dfrac{1}{2}(t - 3) - \dfrac{1}{4}e^{-2(t-3)}\right)u(t - 3)$

(d) $f_4(t) = \left[\dfrac{\sqrt{5}}{4}\cos(-2(t - 1) + 115.57°) + \dfrac{1}{4}\right]u(t - 1)$

4.21. Show that the following is true:

$$\dfrac{d^4}{dt^4}\{u(t)\sin(t)\} = \delta^{(2)}(t) - \delta(t) + u(t)\sin(t)$$

This formula indicates that the differentiation of a continuous signal can also produce delta impulses (note that the function is not differentiable in the regular sense at $t = 0$). Use this formula and the table of common pairs to find the Laplace transform of $d^4\{u(t)\sin(t)\}/dt^4$. Verify the result obtained using the fact that $u(t)\sin(t) \leftrightarrow 1/(s^2 + 1)$ and the derivative property of the Laplace transform.

4.22. Find the nth generalized derivative of the signal $e^{-\alpha t}u(t)$, and show that its Laplace transform is equal to $s^n/(s + \alpha)$.

4.23. In order to find the step response of a linear model of a chain of optical fiber amplifiers [11], one must find the following Laplace inverse,

$$\mathcal{L}^{-1}\left\{\dfrac{s}{\left(s^2 + \dfrac{1}{\tau_c}s + \Omega\right)^k}\right\}$$

where k stands for the number of amplifiers in the chain, τ_c is the chain time constant, and Ω is the relaxation angular frequency. Perform this inversion for $k = 1, 2, 3$.

4.24. Using the time domain convolution property of the Laplace transform, find Laplace inverses of the following frequency domain signals.

(a) $\dfrac{4}{(s^2+4)^2}$

(b) $\dfrac{4s}{(s^2+16)^2}$

Answer: (a) $0.25(\sin(2t) - 2t\cos(2t))$

4.25. Find the impulse response of the linear continuous-time systems represented by the following transfer functions.

(a) $H_1(s) = \dfrac{s+3}{s(s+1)(s+2)}$

(b) $H_2(s) = \dfrac{s}{(s^2+1)(s+2)}$

(c) $H_3(s) = \dfrac{1}{s^2(s+5)}$

(d) $H_4(s) = \dfrac{2}{(s+1)(s^2+s+1)}$

Answers:

(b) $h_2(t) = \left[\dfrac{\sqrt{5}}{5}\cos(-t + 26.57°) - \dfrac{2}{5}e^{-2t}\right]u(t)$

(c) $h_3(t) = \left(-\dfrac{1}{25} + \dfrac{1}{5}t + \dfrac{1}{25}e^{-5t}\right)u(t)$

(d) $h_4(t) = \left[2e^{-t} + \dfrac{4}{\sqrt{3}}e^{-0.5t}\cos\left(-\dfrac{\sqrt{3}}{2}t + 150°\right)\right]u(t)$

4.26. Find the impulse response of the systems represented by the following differential equations.

(a) $\dfrac{d^2y(t)}{dt^2} + \dfrac{dy(t)}{dt} + y(t) = f(t)$

(b) $\dfrac{d^2y(t)}{dt^2} + 3\dfrac{dy(t)}{dt} + 2y(t) = \dfrac{df(t)}{dt} + 4f(t)$

(c) $\dfrac{d^3y(t)}{dt^3} + 3\dfrac{d^2y(t)}{dt^2} + 3\dfrac{dy(t)}{dt} + y(t) = 2\dfrac{d^2f(t)}{dt^2} + \dfrac{df(t)}{dt} + 2f(t)$

4.27. Find the unit step responses of the linear systems whose transfer functions are given in Problem 4.25.

Answers:

(a) $y_{1\text{step}}(t) = \left(-\dfrac{7}{4} + \dfrac{3}{2}t + 2e^{-t} - \dfrac{1}{4}e^{-2t}\right)u(t)$

(b) $y_{2\text{step}}(t) = \left[\dfrac{\sqrt{5}}{5}(\sin(t-\theta)+\sin(\theta)) + \dfrac{1}{5}(e^{-2t}-1)\right]u(t), \quad \theta = 26.57°$

4.28. Find the unit ramp responses of the linear systems whose transfer functions are given in Problem 4.25.

4.29. Derive the relation between the system ramp and parabolic ($f(t) = t^2 u(t)$) responses, and find the parabolic responses of the systems defined in Problem 4.25.

4.30. Determine whether or not the steady state responses to constant inputs ($f(t) = au(t)$) exist for the systems whose transfer functions are defined in Problem 4.25. In the case when the steady state response exists, find its value for $a = 5$.

4.31. Find differential equations that describe the dynamics of the linear continuous-time systems whose transfer functions are given in Problem 4.25.

4.32. Find a third-order system whose impulse response is identical to the unit step response of a second-order system.

4.33. Find the complete response of the following systems.

(a) $y^{(2)}(t) + 2y^{(1)}(t) + y(t) = e^{-3t}u(t), \quad y(0^-) = 1, \quad y^{(1)}(0^-) = 0$

(b) $y^{(2)}(t) + 2y^{(1)}(t) + y(t) = \sin(3t)u(t), \quad y(0^-) = 1, \quad y^{(1)}(0^-) = 2$

Answers:

(a) $y(t) = (0.75e^{-t} + 1.5te^{-t} + 0.25e^{-3t})u(t)$

(b) $y(t) = [1.06e^{-t} + 3.3te^{-t} + 0.1\cos(-3t + 233.13°)]u(t)$

4.34. Find the complete response of the following system.

$$y^{(3)}(t) + 2y^{(1)}(t) + y(t) = u(t), \quad y(0^-) = 1, \quad y^{(1)}(0^-) = 1, \quad y^{(2)}(0^-) = 0$$

Hint: Use MATLAB and its function `residue` to perform long division (to find the Laplace inverse).

Answer:

$$y(t) = [1 - 0.1733e^{-4534t} + 2 \times 0.3128e^{0.2267t}\cos(0.2267t + 73.92°)]u(t)$$

4.35. Using MATLAB to perform the partial fraction expansion, find analytically the step response of the following system.

$$y^{(5)}(t) + 2y^{(4)}(t) + 3y^{(3)}(t) + 2y^{(2)}(t) + 3y^{(1)}(t) + y(t) = u(t)$$

4.36. For the continuous-time system given by

$$y^{(2)}(t) + 6y^{(1)}(t) + 9y(t) = 3f^{(1)}(t) - f(t), \quad y(0^-) = 1, \quad y^{(1)}(0^-) = 0$$

find the complete system response due to $f(t) = e^{-t}u(t)$. Determine the system transfer function and the system impulse response.

Answers:

$$y(t) = (2e^{-3t} + 8te^{-3t} - e^{-t})u(t), \quad H(s) = \dfrac{3s-1}{s^2 + 6s + 9}$$

4.37. Find the complete response of the system represented by

$$y^{(2)}(t) + 2y^{(1)}(t) + y(t) = u(t), \quad y(0^-) = 1, \quad y^{(1)}(0^-) = 2$$

and identify the zero-state and zero-input response components. Find the system transfer function and the system impulse response.

4.38. Find the complete response of the systems whose transfer functions are given by

(a) $H(s) = \dfrac{s+1}{(s+2)(s+4)(s+5)}$

(b) $H(s) = \dfrac{s}{(s+1)^2(s+3)}$

subject to the external input $f(t) = e^{-4t}u(t)$ and the initial conditions $y(0^-) = 1$, $y^{(1)}(0^-) = 0$, $y^{(2)}(0^-) = 2$.

4.39. Determine the complete response of the following system.

$$y^{(2)}(t) + 2y^{(1)}(t) + y(t) = e^{-t}u(t-1), \quad y(0^-) = 1, \quad y^{(1)}(0^-) = 1$$

4.40. Find the zero-state response of the system

$$y^{(2)}(t) + 4y^{(1)}(t) + 4y(t) = f(t)$$

due to the input signal presented in Figure 4.13.

Answer:

$$f(t) = u(t) - r(t) + r(t-2) + u(t-3)$$
$$y(t) = y_{\text{step}}(t) - y_{\text{ramp}}(t) + y_{\text{ramp}}(t-2) + y_{\text{step}}(t-3)$$

$$y_{\text{step}}(t) = \left(\frac{1}{4} - \frac{1}{4}e^{-2t} - \frac{1}{2}te^{-2t}\right)u(t), \quad y_{\text{ramp}}(t) = \frac{1}{4}(-1 + t + e^{-2t} + te^{-2t})u(t)$$

4.41. Repeat Problem 4.40 for the input signal given in Figure 4.14.

4.42. Find the complete response of the system considered in Problem 4.40 due to the input signal given in Figure 4.14 and the initial conditions given by $y(0^-) = 1$, $y^{(1)}(0^-) = 0$.

4.43. Find the complete response of the system

$$y^{(2)}(t) + 5y^{(1)}(t) + 4y(t) = f(t)$$

due to the input signal presented in Figure 4.13 and the initial conditions given by $y(0^-) = 2$, $y^{(1)}(0^-) = -1$.

4.44. Find the zero-input response of the systems represented by

(a) $H(s) = \dfrac{1}{s^2(s+2)(s+4)}$

(b) $H(s) = \dfrac{s+3}{(s+2)(s^2+2s+2)}$

(c) $\dfrac{d^3 y(t)}{dt^3} + 3\dfrac{d^2 y(t)}{dt^2} + 3\dfrac{dy(t)}{dt} + y(t) = \dfrac{df(t)}{dt} + 3f(t)$

subject to the initial conditions $y(0^-) = -1$, $y^{(1)}(0^-) = 2$, $y^{(2)}(0^-) = 1$.

4.45. Find the zero-state response of the second-order system represented by

$$y^{(2)}(t) + 3y^{(1)}(t) + 2y(t) = \sin(t)$$

Answer:

$$y(t) = \left(0.5e^{-t} - 0.2e^{-2t} + 0.1\sqrt{10}\cos(t + \tan^{-1}(1/3))\right)u(t)$$

4.46. Using knowledge of the generalized derivative, find the second derivative of the system output response signal obtained in Example 4.24, and show that it has a jump discontinuity equal to -1 at $t = 1$.

4.47. Repeat Example 4.24 assuming also that the system differentiates the input signal; that is, assume that the corresponding dynamic system is defined by

$$y^{(2)}(t) + 3y^{(1)}(t) + 2y(t) = f^{(1)}(t) + f(t), \quad y(0^-) = 1, \quad y^{(1)}(0^-) = 1$$

4.48. Find the response of the linear time-invariant system defined in Example 4.24 due to the saturation (often called "limiter") input signal defined by

$$\text{sat}(t) = \begin{cases} \alpha t, & 0 \leq t \leq t_0 \\ \beta, & t_0 \leq t \leq \infty, \\ 0, & t < 0 \end{cases} \quad \alpha, \beta = \text{constants}$$

Take $\alpha = 2$, $\beta = 1$ ($t_0 = 0.5$). Check continuity of the system response and its first and second derivatives.

4.49. Repeat Problem 4.48 for the system defined by

$$y^{(2)}(t) + 2y^{(1)}(t) + y(t) = 2f^{(1)}(t) + 3f(t)$$

Comment on the continuity of the system response and its derivatives, and compare with the results obtained in Problem 4.48. Conclude that the input signal differentiation (term $f^{(1)}(t)$) changes the nature of the system response.

4.50. Comment on the implementational difference between two transfer functions representing the same time-invariant linear dynamic system:

$$H(s) = \dfrac{s+1}{(s+2)(s+3)}, \quad H(s) = \dfrac{k_1}{s+2} + \dfrac{k_2}{s+3}$$

where the coefficients k_1 and k_2 are obtained using the partial fraction procedure. Which function do you prefer?

Hint: Signal differentiation can introduce instantaneous changes in system initial conditions, assuming that the input signal contains a jump discontinuity at the initial time. Signal differentiation amplifies noise in the system (see Problem 2.33).

4.51. Find the equivalent transfer function when the transfer functions

$$H_1(s) = \frac{s}{(s+1)(s+3)}, \quad H_2(s) = \frac{1}{s(s+2)}$$

are (a) connected in series; (b) form a parallel connection. Find the open-loop poles of the equivalent transfer functions.

4.52. Assuming unity feedback, find the closed-loop transfer function for the system

$$H(s) = \frac{s+1}{(s+2)(s+6)}$$

Find the closed-loop system zeros and poles.

4.53. Find the closed-loop system transfer function for the feedback system represented in Figure 4.15.

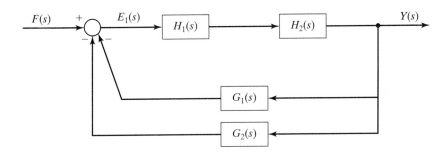

FIGURE 4.15

4.54. Find the closed-loop system transfer function for the feedback system represented in Figure 4.16.

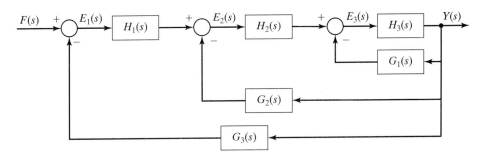

FIGURE 4.16

4.55. Find the closed-loop system transfer function for the feedback system represented in Figure 4.17.

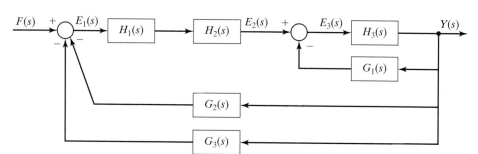

FIGURE 4.17

Answer:

$$\frac{Y(s)}{F(s)} = \frac{H_1(s)H_2(s)H_3(s)}{1 + G_1(s)H_3(s) + H_1(s)H_2(s)H_3(s)(G_2(s) + G_3(s))}$$

4.56. Find the closed-loop system transfer function for the feedback system represented in Figure 4.18.

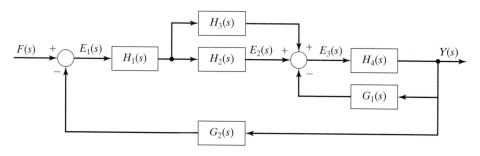

FIGURE 4.18

4.57. The open-loop transfer function of an aircraft is given by

$$H(s) = \frac{-6.81s^3 - 21.22s^2 + 809.16s + 582.25}{s^4 + 3.69s^3 + 2.07s^2 + 0.045s + 0.022}$$

Use MATLAB to find the closed-loop transfer function for unity feedback. Compare the open- and closed-loop aircraft poles. Do you recommend nonunity feedback for this aircraft?

4.58. A linearized model of a single-link manipulator with a flexible joint is given [12] by

$$J_l \ddot{\theta}_l(t) + B_l \dot{\theta}_l(t) + k(\theta_l(t) - \theta_m(t)) = 0$$
$$J_m \ddot{\theta}_m(t) + B_m \dot{\theta}_m(t) - k(\theta_l(t) - \theta_m(t)) = f(t)$$

where J_l, J_m are moments of inertia, B_l, B_m are damping factors, k is the spring constant, $f(t)$ is the input torque, and $\theta_m(t)$, $\theta_l(t)$ are angular positions. Using the Laplace transform, find the transfer functions $\Theta_m(s)/F(s)$ and $\Theta_l(s)/F(s)$. Assume that $J_l = J_m = 1$, $B_m = 0.1$, $B_l = 0.2$, and $k = 0.08$. Use MATLAB to find the zero-state response due to the saturation input signal defined in Problem 4.48. Simulate the system response using the Simulink block diagram.

4.59. In the process of a model development for a neuromuscular system [13], the transfer function of the motoneuron is empirically derived as

$$H(s) = \frac{Y(s)}{F(s)} = \frac{K\left(\left(\frac{s}{33}\right)^2 + \frac{s}{33} + 1\right)}{\left(\frac{s}{58}\right)^2 + 2\left(\frac{s}{58}\right) + 1}$$

where $Y(s)$ is motoneuron output, $F(s)$ is the synaptic current that plays the role of the input signal, and K is the static gain (equal to 1.5 pulses per second per nA of injected current). Using MATLAB, perform partial fraction expansion for the above transfer function. Find analytically the impulse and step responses. Plot the step response.

4.60. Find the complete response of the system defined in Problem 4.36 due to a semiperiodic train of rectangular pulses defined in Problem 4.5. Should we use the Laplace transform to find both the zero-state and zero-input (in this case also steady state) components? Note that it follows from the answer given to Problem 4.5 that the Laplace transform of this input signal is not a rational function. Suggest a method to solve the given problem and use it to find analytically the complete system response. Verify the results obtained using either MATLAB or Simulink.

4.61. Consider the electrical circuit presented in Figure 1.10 and its output voltage differential equation given in (1.26). Assume that $R_1 = R_2 = 100$ kΩ, $L = 10$ mH, and $C = 5$ μF with the initial output voltage conditions equal to $e_0(0) = 5$, $e_0^{(1)}(0) = 0$. Find the complete response of this electrical circuit due to the semiperiodic train of rectangular pulses defined in Problem 4.5 and the given initial conditions. Use MATLAB to plot the results obtained.

4.62. Repeat Problem 4.61 for the input signal given by $f(t) = |\sin(2t)|u(t)$.

4.63. Consider the vehicle lateral dynamics model defined in (1.40). Use MATLAB to find analytically its impulse and step responses. Plot the corresponding responses and observe that they are unbounded in time. Find the unity feedback closed-loop transfer function and find the closed-loop impulse and step responses. Plot the closed-loop impulse and step responses. Compare the open- and closed-loop impulse and step responses.

4.64. The swaying motion of a ship positioning system [14] can be described by the transfer function

$$H(s) = \frac{0.8424}{s(s + 0.0546)(s + 1.55)}$$

Find analytically the impulse and step responses. Use MATLAB to find analytically the zero-state response due to a sinusoidal input $f(t) = \sin(3t)u(t)$. Use Simulink to simulate the zero-state response due to the given sinusoidal input.

4.65. Consider the voltage regulator model of [15], defined by

$$H(s) = \frac{154280}{(s + 0.2)(s + 0.5)(s + 10)(s + 14.28)(s + 25)}$$

Find analytically the step response, and use Simulink to simulate the zero-state response due to a sinusoidal input of the form $f(t) = \cos(t)u(t)$.

CHAPTER 5

$\mathcal{Z}$-transform

The $\mathcal{Z}$-transform is the discrete-time counterpart of the Laplace transform. The Laplace transform was discovered more than two hundred years ago in 1779. The $\mathcal{Z}$-transform originated at the beginning of the 1950s and gained its maturity at the end of the 1950s in the works of several distinguished researchers (Ragazzini, Zadeh, Jury, Franklin, Tzypkin) who studied discrete-time signals, discrete-time control systems, and digital computers ([1–4]).

The $\mathcal{Z}$-transform operates on a discrete-time signal $f(kT)$, where T is the sampling period and kT stands for discrete time—in contrast to the Laplace transform, which acts on a continuous-time signal $f(t)$. The discrete-time signal $f(kT)$ arises in practice by sampling a continuous-time signal $f(t)$ every T seconds; another case is when the signal is inherently discrete and represented by its samples $f[k]$. For simplicity, while presenting discrete-time signals, we omit in this text the parameter T so that $f[k]$ *also stands for* $f(kT)$.

In Section 4.5 we showed rigorously how to derive the $\mathcal{Z}$-transform from the starred Laplace transform (see formula (4.68)). In Chapter 9, we will show rigorously how the $\mathcal{Z}$-transform can be derived in the process of sampling a continuous time signal, and establish the famous sampling theorem. Roughly speaking, the $\mathcal{Z}$-transform can be also introduced by using the following simple reasoning. The Laplace transform integral can be discretized with sampling period T, and approximated by the corresponding infinite sum, that is,

$$\int_0^\infty f(t)e^{-st}\,dt \approx T\sum_{k=0}^\infty f(kT)e^{-skT}$$

The quantity

$$\sum_{k=0}^\infty f(kT)e^{-skT} \triangleq \sum_{k=0}^\infty f[k]z^{-k} \triangleq F(z), \quad z = e^{sT}$$

defines the one-sided (unilateral) $\mathcal{Z}$-transform of the discrete-time signal (function) $f[k]$.

Notice that since $s = \sigma + j\omega$ is a complex variable, the $\mathcal{Z}$-transform can be considered as a function that maps the complex s-plane into another complex plane, using the mapping

$$z = e^{sT} = e^{(\sigma+j\omega)T} = e^{\sigma T}e^{j\omega T} = e^{\sigma T}(\cos(\omega T) + j\sin(\omega T)) \tag{5.1}$$

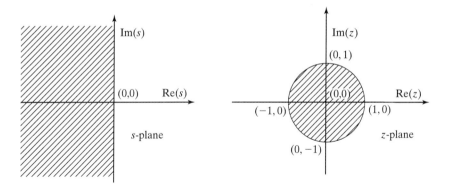

FIGURE 5.1: A mapping between the s- and z-planes

This mapping, for example, maps the left half-plane of the s-plane ($\sigma \leq 0$) into the interior of the unit circle in the z-complex plane, $|z| < 1$, with the imaginary axis of the s-plane being mapped into the unit circle, $|z| = 1$, in the z-plane (see Figure 5.1).

Many analogies can be drawn between the Laplace and $\mathcal{Z}$-transforms, and for linear time-invariant systems the results of these two transforms can be presented in parallel. However, there are some differences between the $\mathcal{Z}$-transform theory of discrete-time signals and the Laplace transform theory of continuous-time signals. The reader will become familiar with most of these differences by reading Sections 5.1 and 5.3–4.

This chapter is organized as follows: The main properties of the $\mathcal{Z}$-transform are introduced and derived in Section 5.1. The properties of the $\mathcal{Z}$-transform and the common $\mathcal{Z}$-transform pairs are given in Tables 5.1 and 5.2. In Section 5.2, we present methods for finding the inverse $\mathcal{Z}$-transform. Section 5.3 deals with applications of the $\mathcal{Z}$-transform for finding the response of discrete-time linear time-invariant systems. Block diagrams used for the representation of complex discrete-time linear systems are considered in Section 5.4. The frequency spectra of discrete-time signals and discrete-time systems (frequency response) obtained via the $\mathcal{Z}$-transform are considered in Section 5.5. In the same section, we study the response of discrete-time linear systems to sinusoidal inputs. In Section 5.6, a MATLAB laboratory experiment on discrete-time systems and signals is designed.

5.1 $\mathcal{Z}$-TRANSFORM AND ITS PROPERTIES

Consider a discrete time signal $f[k]$, where k is an integer that represents discrete-time instants. The *one-sided $\mathcal{Z}$-transform* of this signal is defined by (4.68), that is,

$$F(z) = \mathcal{Z}\{f[k]\} \triangleq \sum_{k=0}^{\infty} f[k] z^{-k} \tag{5.2}$$

This infinite sum has a very simple form, namely

$$F(z) = \mathcal{Z}\{f[k]\} = f[0] + f[1]z^{-1} + f[2]z^{-2} + \cdots \qquad (5.3)$$

We say that $f[k]$ and $F(z)$ form the $\mathcal{Z}$-transform pair, and we denote this fact by $f[k] \leftrightarrow F(z)$.

Similarly to the double-sided Laplace transform, the *double-sided $\mathcal{Z}$-transform* is defined in (4.69) as

$$F(z) = \mathcal{Z}\{f[k]\} \triangleq \sum_{k=-\infty}^{\infty} f[k]z^{-k}$$

where k is any integer. Since the double-sided $\mathcal{Z}$-transform is of marginal importance for discrete-time linear dynamic systems, it will not be studied in this chapter. Due to the fact that the double-sided $\mathcal{Z}$-transform is used in signal processing and digital communications (see, for example, [5]), its highlights will be considered in Chapter 9.

The $\mathcal{Z}$-transform is defined for those discrete-time signals (functions) for which the infinite series (5.2) is convergent. This will be the case if

$$\lim_{n \to \infty} \sum_{k=0}^{n} |f[k]|\rho^{-k} = C < \infty$$

where ρ is a real positive number, and C is a positive real constant, which may depend on ρ. This condition means that $f[k]$ has a one-sided $\mathcal{Z}$-transform if

$$|z| > \rho_{\min} \qquad (5.4)$$

where $\rho_{\min}$ denotes the minimal element in the set of real positive numbers such that the convergence condition is satisfied. The set of complex numbers z satisfying inequality (5.4) defines the *region of absolute convergence* of the $\mathcal{Z}$-transform $F(z)$, which means that $F(z)$ is not defined outside of that region (see Figure 5.2). We will denote the region of convergence by $\Re$, hence $\Re = \{z : |z| > \rho_{\min}\}$. In this text, we will not place emphasis

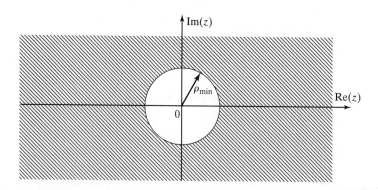

FIGURE 5.2: The region of convergence of the one-sided $\mathcal{Z}$-transform (shaded area)

on determining and studying the region of convergence of the $\mathcal{Z}$-transform. Rather, we will indicate it from time to time to remind students of the existence region of the $\mathcal{Z}$-transform under consideration. Detailed study of the properties of the region of convergence of the $\mathcal{Z}$-transform can be found in [6, Section 10.2].

It is trivial to find the $\mathcal{Z}$-transform of a discrete-time signal whose values are nonzero only at several discrete time instants. This can be done by using directly the definition sum of the $\mathcal{Z}$-transform as given in (5.2) and (5.3), as demonstrated in the following example.

EXAMPLE 5.1

Let $f[k]$ be defined by

$$f[k] = \begin{cases} 3, & k = 12 \\ 5, & k = 119 \\ 0, & \text{elsewhere} \end{cases}$$

then from (5.3), we have

$$F(z) = 3z^{-12} + 5z^{-119} = \frac{5 + 3z^{107}}{z^{119}}$$

Similarly, the problem of finding the $\mathcal{Z}$-transform of the discrete-time impulse delta signal is extremely simple, since from the definition of $\delta[k]$, (2.32), and the definition formula of the $\mathcal{Z}$-transform, we directly obtain

$$\mathcal{Z}\{\delta[k]\} = 1 \times z^0 = 1 \qquad (5.5)$$

♞

The $\mathcal{Z}$-transform of the discrete-time shifted impulse delta signal also can be easily found, as shown in the next example.

EXAMPLE 5.2

By the definition of the discrete-time shifted impulse delta signal, given in (2.34), we have

$$\mathcal{Z}\{\delta[k - k_0]\} = 1 \times z^{-k_0} = \frac{1}{z^{k_0}} \qquad (5.6)$$

♞

Finding the $\mathcal{Z}$-transform of discrete-time signals that have nonzero values over an infinite discrete-time interval requires summation of infinite series. Most of the expressions for the $\mathcal{Z}$-transforms of such signals are obtained by using the known expression for the sum of the *geometric series*

$$1 + q + q^2 + q^3 + \cdots = \frac{1}{1 - q} \quad \text{for } |q| < 1 \qquad (5.7)$$

Using (5.7) we are able to find $\mathcal{Z}$-transforms of the discrete-time unit step and exponential signals.

EXAMPLE 5.3

The $\mathcal{Z}$-transform of the discrete-time unit step signal is given by

$$\mathcal{Z}\{u[k]\} = 1 + z^{-1} + z^{-2} + z^{-3} + \cdots = \frac{1}{1 - \frac{1}{z}} = \frac{z}{z-1}, \quad \left|\frac{1}{z}\right| < 1 \quad (5.8)$$

The condition $|z| > 1$ is crucial for the convergence of the series, and thus for the existence of the corresponding $\mathcal{Z}$-transform; in other words, the $\mathcal{Z}$-transform of the unit step signal is defined only for $|z| > 1$ and not in the entire z-plane.

EXAMPLE 5.4

An exponential signal in the discrete-time domain is represented by $f[k] = a^k u[k]$. Its $\mathcal{Z}$-transform is obtained similarly to the $\mathcal{Z}$-transform of the unit step, that is,

$$\mathcal{Z}\{a^k u[k]\} = 1 + \frac{a}{z} + \frac{a^2}{z^2} + \frac{a^3}{z^3} + \cdots = \frac{1}{1 - \frac{a}{z}} = \frac{z}{z-a}, \quad |z| > |a| \quad (5.9)$$

Note that the $\mathcal{Z}$-transform of the discrete-time exponential signal is defined in the region $|z| > |a|$.

In the following, we define and prove the properties of the $\mathcal{Z}$-transform. Since the $\mathcal{Z}$-transform is, in general, an infinite sum of complex numbers, all of the properties of infinite sums will hold. Hence, most of the properties of the $\mathcal{Z}$-transform will be proved by using the known (and very simple) properties of infinite sums.

Property 1: Linearity

Let $\mathcal{Z}\{f_i[k]\} = F_i(z)$, $i = 1, 2, \ldots, n$, define $\mathcal{Z}$-transform pairs with regions of convergence respectively given by $\Re_1, \Re_2, \ldots, \Re_n$. Then,

$$\mathcal{Z}\{\alpha_1 f_1[k] \pm \alpha_2 f_2[k] \pm \cdots \pm \alpha_n f_n[k]\} = \alpha_1 F_1(z) \pm \alpha_2 F_2(z) \pm \cdots \pm \alpha_n F_n(z) \quad (5.10)$$

where $\alpha_1, \alpha_2, \ldots, \alpha_n$ are constants. The region of convergence of this linear combination of discrete-time signals is the intersection of the regions of convergence of the individual signals, that is, it is equal to $\Re_1 \cap \Re_2 \cap \cdots \cap \Re_n$.

Proof. The proof of this property follows directly from the definition of the $\mathcal{Z}$-transform:

$$\mathcal{Z}\{\alpha_1 f_1[k] \pm \alpha_2 f_2[k] \pm \cdots \pm \alpha_n f_n[k]\}$$
$$\triangleq \sum_{k=0}^{\infty} (\alpha_1 f_1[k] \pm \alpha_2 f_2[k] \pm \cdots \pm \alpha_n f_n[k]) z^{-k}$$
$$= \alpha_1 \sum_{k=0}^{\infty} f_1[k] z^{-k} \pm \alpha_2 \sum_{k=0}^{\infty} f_2[k] z^{-k} \pm \cdots \pm \alpha_n \sum_{k=0}^{\infty} f_n[k] z^{-k}$$
$$= \alpha_1 F_1(z) \pm \alpha_2 F_2(z) \pm \cdots \pm \alpha_n F_n(z) \qquad \square$$

Property 2: Right Shift in Time—"Integral" Property

The right shift in time of a discrete-time signal $f[k]u[k]$ shifted by $k_0 > 0$ discrete-time units, defined by $f[k - k_0]u[k - k_0]$, has the following $\mathcal{Z}$-transform

$$\mathcal{Z}\{f[k - k_0]u[k - k_0]\} = \frac{1}{z^{k_0}} F(z) \tag{5.11}$$

where $\mathcal{Z}\{f[k]u[k]\} = F(z)$. The region of convergence of the shifted signal is the same as the region of convergence of the original signal, except for possible addition or delation of infinity, [6].

Proof. The proof of this property is as follows:

$$\mathcal{Z}\{f[k - k_0]u[k - k_0]\} = \sum_{k=0}^{\infty} f[k - k_0]u[k - k_0]z^{-k}$$

$$= \sum_{k=k_0}^{\infty} f[k - k_0]u[k - k_0]z^{-k}$$

Introducing a change of variables $k - k_0 = i$, we obtain

$$\sum_{k=k_0}^{\infty} f[k - k_0]u[k - k_0]z^{-k} = \sum_{i=0}^{\infty} f[i]u[i]z^{-(i+k_0)}$$

$$= z^{-k_0} \sum_{i=0}^{\infty} f[i]z^{-i} = z^{-k_0} F(z)$$

which completes the proof. □

Another variant of the integral property of the $\mathcal{Z}$-transform is given by

$$\mathcal{Z}\{f[k - k_0]u[k]\} = z^{-k_0} F(z) + z^{-k_0} \left(\sum_{i=1}^{k_0} f[-i]z^i \right) \tag{5.11a}$$

It can be proved as follows. By the definition, we have

$$\mathcal{Z}\{f[k - k_0]u[k]\} = \sum_{k=0}^{\infty} f[k - k_0]u[k]z^{-k}$$

Introducing a change of variables $k - k_0 = i$, we obtain

$$\sum_{k=0}^{\infty} f[k - k_0]u[k]z^{-k} = \sum_{i=-k_0}^{\infty} f[i]u[i + k_0]z^{-(i+k_0)} = z^{-k_0} \sum_{i=-k_0}^{\infty} f[i]z^{-i}$$

$$= z^{-k_0} \left(\sum_{i=0}^{\infty} f[i]z^{-i} + \sum_{i=-k_0}^{i=-1} f[i]z^{-i} \right)$$

$$= z^{-k_0} F(z) + z^{-k_0} \sum_{i=-k_0}^{i=-1} f[i]z^{-i}$$

In the last sum we can replace the dummy variable of summation i by $-m$ and then rename m as i, which leads to formula (5.11a). In the expanded form, for $k_0 = 1, 2, \ldots,$ this property states the following:

$$\mathcal{Z}\{f[k-1]u[k]\} = z^{-1}F(z) + f[-1]$$

$$\mathcal{Z}\{f[k-2]u[k]\} = z^{-2}F(z) + z^{-1}f[-1] + f[-2]$$

$$\mathcal{Z}\{f[k-3]u[k]\} = z^{-3}F(z) + z^{-2}f[-1] + z^{-1}f[-2] + f[-3] \quad (5.11a)$$

$$\vdots$$

$$\mathcal{Z}\{f[k-k_0]u[k]\} = z^{-k_0}F(z) + z^{-k_0+1}f[-1] + \cdots + z^{-1}f[-k_0+1] + f[-k_0]$$

The property of the $\mathcal{Z}$-transform stated in (5.11a) will play a very important role in the analysis of linear time-invariant discrete systems, to be presented in Section 5.3.

Note that if $f[k]$ is a causal signal ($f[k] = 0, k < 0$), then all terms corresponding to the signal samples for negative discrete-time instants drop out, and we are left with

$$f[k-k_0]u[k] \leftrightarrow z^{-k_0}F(z), \quad f[k] = 0, \ k < 0 \quad (5.11b)$$

EXAMPLE 5.5

The discrete-time impulse delta signal can be represented in terms of unit step signals as

$$\delta[k] = u[k] - u[k-1]$$

By the linearity and right shift in time properties we have

$$\mathcal{Z}\{\delta[k]\} = \mathcal{Z}\{u[k]\} - \mathcal{Z}\{u[k-1]\} = \frac{z}{z-1} - z^{-1}\frac{z}{z-1} = 1$$

which verifies (5.5).

EXAMPLE 5.6

Using the right shift in time property of the $\mathcal{Z}$-transform, we have the following $\mathcal{Z}$-transform pair:

$$\mathcal{Z}\{2^k u[k]\} = \frac{z}{z-2} \Rightarrow \mathcal{Z}\{2^{k-k_0}u[k-k_0]\} = z^{-k_0}\frac{z}{z-2}$$

For $k_0 = 3$, we obtain

$$\mathcal{Z}\{2^{k-3}u[k-3]\} = z^{-3}\frac{z}{z-2} = \frac{1}{z^2(z-2)}$$

Property 3: Left Shift in Time—Derivative Property

Left shifts in time, which in fact stand for the discrete-time derivatives, satisfy the following:

$$\mathcal{Z}\{f[k+1]u[k]\} = zF(z) - zf[0]$$
$$\mathcal{Z}\{f[k+2]u[k]\} = z^2 F(z) - z^2 f[0] - zf[1]$$
$$\vdots \tag{5.12}$$
$$\mathcal{Z}\{f[k+k_0]u[k]\} = z^{k_0} F(z) - z^{k_0} f[0] - z^{k_0-1} f[1] - \cdots - zf[k_0-1]$$

Proof. We can establish the proof of (5.12) rather easily. For the first discrete-time derivative, we have

$$\mathcal{Z}\{f[k+1]u[k]\} = \sum_{k=0}^{\infty} f[k+1]u[k]z^{-k}$$

Introducing a change of variables $i = k+1$, we obtain

$$\mathcal{Z}\{f[k+1]u[k]\} = \sum_{i=1}^{\infty} f[i]u[i-1]z^{-(i-1)} = z \sum_{i=1}^{\infty} f[i]z^{-i}$$
$$= z \left[\sum_{i=1}^{\infty} f[i]z^{-i} + f[0] - f[0] \right]$$
$$= z \left[\sum_{i=0}^{\infty} f[i]z^{-i} - f[0] \right] = zF(z) - zf[0]$$

For the second discrete-time derivative, we have

$$\mathcal{Z}\{f[k+2]u[k]\} = \sum_{k=0}^{\infty} f[k+2]u[k]z^{-k}$$

which, after a change of variables $i = k+2$, becomes

$$\mathcal{Z}\{f[k+2]u[k]\} = \sum_{i=2}^{\infty} f[i]u[i-2]z^{-(i-2)} = z^2 \sum_{i=2}^{\infty} f[i]z^{-i}$$
$$= z^2 \left[\sum_{i=2}^{\infty} f[i]z^{-i} + z^{-1} f[1] + f[0] - z^{-1} f[1] - f[0] \right]$$
$$= z^2 \left[\sum_{i=0}^{\infty} f[i]z^{-i} - z^{-1} f[1] - f[0] \right] = z^2 F(z) - zf[1] - z^2 f[0]$$

It is left to the reader to complete the general proof for the k_0th derivative (see Problem 5.1). □

It can be concluded that the derivative property of the $\mathcal{Z}$-transform is analogous to the corresponding property of the Laplace transform. Namely, the nth derivative in discrete-time, which is represented by the left shift in time for n discrete-time instants, in the frequency domain corresponds to a multiplication by z^n; that is, *assuming that all initial conditions are zero*, we have

$$\mathcal{Z}\{f[k+n]u[k]\} = z^n F(z) \tag{5.13}$$

On the other hand, it is known from formula (4.16) that for the continuous-time derivatives the following holds:

$$\mathcal{L}\left\{\frac{d^n f(t)}{dt}\right\} = s^n F(s), \quad f^{(i)}(0^-) = 0, \quad i = 0, 1, 2, \ldots, n-1$$

Property 4: Time Multiplication

Assume that $f[k] \leftrightarrow F(z)$. The time multiplication property states that

$$\begin{aligned}\mathcal{Z}\{kf[k]\} &= -z\frac{d}{dz}F(z) \\ \mathcal{Z}\{k^2 f[k]\} &= z\frac{d}{dz}F(z) + z^2\frac{d^2}{dz^2}F(z)\end{aligned} \tag{5.14}$$

The regions of convergence for the newly formed signals $kf[k]$ and $k^2 f[k]$ are equal to the region of the convergence of the original signal $f[k]$.

Proof. The proof follows from the definition of the $\mathcal{Z}$-transform,

$$F(z) = \sum_{k=0}^{\infty} f[k]z^{-k}$$

by taking the derivative with respect to z:

$$\frac{dF(z)}{dz} = \sum_{k=0}^{\infty} f[k](-k)z^{-k-1} = -z^{-1}\sum_{k=0}^{\infty} kf[k]z^{-k} = -z^{-1}\mathcal{Z}\{kf[k]\}$$

Similarly, by taking the second derivative with respect to z, we prove the second expression in (5.14); (see Problem 5.2). $\square$

EXAMPLE 5.7

Using the time multiplication property and the $\mathcal{Z}$-transform of an exponential signal defined in (5.9), we are able to find the $\mathcal{Z}$-transform of the discrete-time unit ramp signal as follows:

$$\mathcal{Z}\{ka^k u[k]\} = -z\frac{d}{dz}\left(\frac{z}{z-a}\right) = \frac{az}{(z-a)^2} \tag{5.15}$$

Specializing this result for $a = 1$, we obtain the $\mathcal{Z}$-transform of the discrete-time unit ramp signal,

$$\mathcal{Z}\{ku[k]\} = \frac{z}{(z-1)^2} \tag{5.16}$$

Similarly, using the second expression in (5.14), we find that

$$\mathcal{Z}\{k^2 a^k u[k]\} = \frac{az(z+a)}{(z-a)^3} \tag{5.17}$$

For $a = 1$, we have

$$\mathcal{Z}\{k^2 u[k]\} = \frac{z(z+1)}{(z-1)^3} \tag{5.18}$$

which is the $\mathcal{Z}$-transform of the discrete-time parabolic signal.

Property 5: Frequency Scaling

The frequency scaling property of the $\mathcal{Z}$-transform is a consequence of the obvious fact that

$$\mathcal{Z}\{a^k f[k]\} = \sum_{k=0}^{\infty} a^k f[k] z^{-k} = \sum_{k=0}^{\infty} f[k] \left(\frac{z}{a}\right)^{-k} = F\left(\frac{z}{a}\right) \tag{5.19}$$

with the new region of convergence equal to $|a|\mathfrak{R}$, where $\mathfrak{R}$ is the region of convergence of the original signal. This property is an intermediate step in establishing the modulation property. In addition, it can be used for finding the $\mathcal{Z}$-transform of signals multiplied by an exponential function.

Property 6: Modulation

The modulation property is very useful for digital signal processing systems and communications theory and practice. It is directly derived from the frequency scaling property. Representing the sine and cosine functions by Euler's formulas,

$$\cos(\omega kT) = \frac{1}{2}(e^{j\omega kT} + e^{-j\omega kT}), \quad \sin(\omega kT) = \frac{1}{2j}(e^{j\omega kT} - e^{-j\omega kT}) \tag{5.20}$$

and using the frequency scaling property, we have

$$\mathcal{Z}\{f[k]\cos(\omega kT)\} = \frac{1}{2}[F(e^{j\omega T}z) + F(e^{-j\omega T}z)] \tag{5.21}$$

and

$$\mathcal{Z}\{f[k]\sin(\omega kT)\} = \frac{j}{2}[F(e^{j\omega T}z) - F(e^{-j\omega T}z)] \tag{5.22}$$

The relations (5.21) and (5.22) constitute the modulation property. We can also use this property to find $\mathcal{Z}$-transforms of the cosine and sine signals.

EXAMPLE 5.8

The $\mathcal{Z}$-transform of the discrete-time unit step signal is given by (5.8), that is, $U(z) = z/(z-1)$. From (5.21), we have

$$\mathcal{Z}\{u[k]\cos(\omega kT)\} = \frac{1}{2}[U(e^{j\omega T}z) + U(e^{-j\omega T}z)]$$

$$= \frac{1}{2}\left[\frac{e^{j\omega T}z}{e^{j\omega T}z - 1} + \frac{e^{-j\omega T}z}{e^{-j\omega T}z - 1}\right] \quad (5.23)$$

$$= \frac{z^2 - (\cos\omega T)z}{z^2 - (2\cos\omega T)z + 1} = \mathcal{Z}\{\cos(\omega kT)\}, \, k > 0$$

Similarly, (5.8) and (5.22) produce

$$\mathcal{Z}\{u[k]\sin(\omega kT)\} = \frac{j}{2}[U(e^{j\omega T}z) - U(e^{-j\omega T}z)]$$

$$= \frac{j}{2}\left[\frac{e^{j\omega T}z}{e^{j\omega T}z - 1} - \frac{e^{-j\omega T}z}{e^{-j\omega T}z - 1}\right] \quad (5.24)$$

$$= \frac{(\sin\omega T)z}{z^2 - (2\cos\omega T)z + 1} = \mathcal{Z}\{\sin(\omega kT)\}, \, k > 0$$

EXAMPLE 5.9

Let us find the $\mathcal{Z}$-transform of the discrete-time signal represented by

$$f[k] = 2^{-k}\cos\left(k\frac{\pi}{3}\right)u[k-1]$$

The detailed derivations are given in the following, where we first use the right shift in time property, as defined in (5.11), and combine it with a known trigonometric formula.

$$\mathcal{Z}\left\{2^{-k}\cos\left(k\frac{\pi}{3}\right)u[k-1]\right\} = \frac{1}{2}\mathcal{Z}\left\{\left(\frac{1}{2}\right)^{k-1}\cos\left((k-1+1)\frac{\pi}{3}\right)u[k-1]\right\}$$

$$= \frac{1}{2}\mathcal{Z}\left\{\left(\frac{1}{2}\right)^{k-1}\cos\left((k-1)\frac{\pi}{3}\right)\cos\left(\frac{\pi}{3}\right)u[k-1]\right.$$

$$\left. - \sin\left((k-1)\frac{\pi}{3}\right)\sin\left(\frac{\pi}{3}\right)u[k-1]\right\}$$

$$= \frac{1}{2}\mathcal{Z}\left\{\left(\frac{1}{2}\right)^{k-1}\frac{1}{2}\cos\left((k-1)\frac{\pi}{3}\right)u[k-1]\right.$$

$$\left. - \frac{\sqrt{3}}{2}\sin\left((k-1)\frac{\pi}{3}\right)u[k-1]\right\}$$

In the next step, we find $\mathcal{Z}$-transforms of the obtained time-shifted sine and cosine signals multiplied by corresponding discrete-time exponential signals (time shifting and frequency scaling):

$$\mathcal{Z}\{f[k]\} = \frac{1}{4}z^{-1}\mathcal{Z}\left\{\left(\frac{1}{2}\right)^k \cos\left(k\frac{\pi}{3}\right)u[k]\right\} - \frac{\sqrt{3}}{4}z^{-1}\mathcal{Z}\left\{\left(\frac{1}{2}\right)^k \sin\left(k\frac{\pi}{3}\right)u[k]\right\}$$

$$= \frac{1}{4}z^{-1}\frac{z^2 - \frac{1}{2}z\cos\left(\frac{\pi}{3}\right)}{z^2 - 2\frac{1}{2}z\cos\left(\frac{\pi}{3}\right) + \frac{1}{4}} - \frac{\sqrt{3}}{4}z^{-1}\frac{\frac{1}{2}z\sin\left(\frac{\pi}{3}\right)}{z^2 - 2\frac{1}{2}z\cos\left(\frac{\pi}{3}\right) + \frac{1}{4}}$$

$$= \frac{\frac{z}{4} - \frac{1}{16}}{z^2 - \frac{1}{2}z + \frac{1}{4}} - \frac{\frac{3}{16}}{z^2 - \frac{1}{2}z + \frac{1}{4}} = \frac{\frac{z}{4} - \frac{1}{4}}{z^2 - \frac{1}{2}z + \frac{1}{4}} = \frac{z - 1}{4z^2 - 2z + 1} \qquad \blacklozenge$$

Property 7: Convolution

This is the most important property of the one-sided $\mathcal{Z}$-transform from the linear system theory point of view. It states that

$$\mathcal{Z}\{f_1[k] * f_2[k]\} = F_1(z)F_2(z) \qquad (5.25)$$

The region of convergence of the convolution signal is equal to $\mathfrak{R}_1 \cap \mathfrak{R}_2$, where $\mathfrak{R}_1$ and $\mathfrak{R}_2$ are the regions of convergence of the original signals. Note that the discrete-time convolution is defined in (2.39). Since the one-sided $\mathcal{Z}$-transform is defined only for nonnegative values of k, we will use the following definition for the discrete-time convolution:

$$f_1[k] * f_2[k] = \sum_{m=0}^{\infty} f_1[m]f_2[k-m] \qquad (5.26)$$

This definition will play a very important role in the analysis of discrete-time, time-invariant, linear systems.

Proof. Using the definitions of the $\mathcal{Z}$-transform and the discrete-time signal convolution, we have

$$\mathcal{Z}\{f_1[k] * f_2[k]\} = \sum_{k=0}^{\infty}\sum_{m=0}^{\infty} f_1[m]f_2[k-m]z^{-k}$$

$$= \sum_{m=0}^{\infty} f_1[m]\sum_{k=0}^{\infty} f_2[k-m]z^{-k} \qquad (5.27)$$

$$= \sum_{m=0}^{\infty} f_1[m]z^{-m}F_2(z) = F_1(z)F_2(z)$$

In the last step of this proof we have interchanged the order of summation and used the $\mathcal{Z}$-transform property established in (5.11b). $\square$

Property 8: Initial Value Theorem

From the definition of the $\mathcal{Z}$-transform, the expression for the signal initial value follows directly, since

$$F(z) = \mathcal{Z}\{f[k]\} = f[0] + f[1]z^{-1} + f[2]z^{-2} + \cdots$$

implies that

$$f[0] = \lim_{z \to \infty} \{F(z)\} \tag{5.28}$$

Using simple algebra, we can also recover the other signal values. For example, $f[1]$ is obtained as

$$\lim_{z \to \infty} \{z[F(z) - f[0]]\} = f[1] \tag{5.29}$$

It is left as an exercise for the student to derive the expressions for $f[2]$, $f[3]$, $f[n-1]$ (see Problem 5.3).

Property 9: Final Value Theorem

Assuming that we know the $\mathcal{Z}$-transform of a signal, we are able to find the signal's steady state value (theoretically $f[\infty]$) without actually going back to the time domain (via the inverse $\mathcal{Z}$-transform) and then finding the corresponding limit. This can be achieved as follows:

$$\lim_{k \to \infty} f[k] = f[\infty] = \lim_{z \to 1} \left\{ \frac{z-1}{z} F(z) \right\} \tag{5.30}$$

Proof. To justify (5.30), we start with

$$\mathcal{Z}\{f[k] - f[k-1]\} = F(z) - \frac{1}{z}F(z) = \frac{z-1}{z}F(z)$$

and take the limit when $z \to 1$, which produces

$$\lim_{z \to 1} \left\{ \sum_{k=0}^{\infty} (f[k] - f[k-1])z^{-k} \right\} = \sum_{k=0}^{\infty} (f[k] - f[k-1])$$

$$= f[0] + (f[1] - f[0]) + (f[2] - f[1]) + \cdots$$

$$= f[\infty] = \lim_{z \to 1} \frac{z-1}{z} F(z)$$

Since all terms in this infinite sum except for the last cancel out, we obtain formula (5.30). □

Note that the final value theorem is applicable only to signals for which the limit at infinity exists. For example, it can not be applied to the sine and cosine signals since they have no limits at infinity. However, that information is not obvious from $F(z)$. Given $F(z)$, *the applicability condition (test) of the final value theorem requires that the complex variable function $(z-1)F(z)$ has no poles outside or on the unit circle* (recall that the poles are the values at which $(p_j - 1)F(p_j) = \infty$). This condition (test) will become clearer after we learn about the stability of discrete-time linear systems in Chapter 7.

EXAMPLE 5.10

Let us apply the initial and final value theorems to the unit step signal. We know that $u[0] = u[\infty] = 1$. By the initial value theorem, we have

$$u[0] = \lim_{z \to \infty} U(z) = \lim_{z \to \infty} \left\{ \frac{z}{z-1} \right\} = 1$$

The final value theorem implies that

$$u[\infty] = \lim_{z \to 1} \left\{ \frac{z-1}{z} U(z) \right\} = \lim_{z \to 1} \left\{ \frac{z-1}{z} \frac{z}{z-1} \right\} = 1$$

Thus, the conclusions of the initial and final value theorems are correct in this case.

Note that we cannot apply the final value theorem to the sine and cosine functions because the corresponding limits at infinity do not exist. This can also be observed by checking the poles of their $\mathcal{Z}$-transforms. It can be seen that in both cases the poles are given by

$$z^2 + 2z\cos(\omega T) + 1 = 0 \quad \Rightarrow \quad p_{1,2} = \cos(\omega T) \pm j\sin(\omega T)$$

that is, the poles are located on the unit circle, hence the final value theorem is not applicable. ∫

EXAMPLE 5.11

Assume that the $\mathcal{Z}$-transform of a certain signal is given by

$$F(z) = \frac{z - 0.5}{z(z + 0.5)(z - 1)}$$

The initial and final value theorems produce

$$f[0] = \lim_{z \to \infty} \left\{ \frac{z - 0.5}{z(z + 0.5)(z - 1)} \right\} = 0$$

$$f[\infty] = \lim_{z \to 1} \left\{ \frac{z-1}{z} \frac{z - 0.5}{z(z + 0.5)(z - 1)} \right\} = \frac{1 - 0.5}{1 + 0.5} = \frac{1}{3}$$

Note that the final value theorem is applicable in this case, and that it produces the correct answer, since $(z - 1)F(z) = (z - 0.5)/(z(z + 0.5))$ has both poles inside the unit circle (at 0 and -0.5). ∫

Tables 5.1 and 5.2, given at the end of this chapter, present a summary of the properties, and common pairs of the $\mathcal{Z}$-transform, respectively.

5.2 INVERSE OF THE $\mathcal{Z}$-TRANSFORM

It is not straightforward to derive the formula for the inverse $\mathcal{Z}$-transform. The inverse $\mathcal{Z}$-transform is derived using complex variable contour integration. Since integration involving complex variables is unfamiliar, in general, to sophomore and junior engineering students, we give only the definition formula for the inverse $\mathcal{Z}$-transform. Students who had a course in complex variables could refer, for example, to [7–8] for the contour integration procedure for deriving the inverse $\mathcal{Z}$-transform.

As in the case of the Laplace transform, the inverse $\mathcal{Z}$-transforms of all discrete-time functions used in this text can be obtained by combining the tables of properties and common pairs (Tables 5.1 and 5.2) with the partial fraction expansion method. Thus, knowledge of the actual formula for the inverse $\mathcal{Z}$-transform is not required for the purpose of finding the inverse $\mathcal{Z}$-transform. For completeness, however, we give in the next paragraph the formula for the inverse $\mathcal{Z}$-transform.

The definition formula for the inverse $\mathcal{Z}$-transform is given by the following complex variable contour integral,

$$f[k] \triangleq \frac{1}{2\pi j} \oint_\Gamma F(z) z^{k-1} \, dz \tag{5.31}$$

where Γ is a circle of radius greater than $\rho_{\min}$ (defined in (5.4)) that encircles all singularities in the region of convergence. (The most common singularities are the poles of $F(z)z^{k-1}$, at which $F(z)z^{k-1} = \infty$.) This integral can be evaluated using Cauchy's residue theorem, [7–8]. The result of this theorem is that a contour integral of a function of z that is analytic inside the contour Γ, except at a finite number of isolated singularities p_i (poles), is given by

$$f[k] = \frac{1}{2\pi j} \oint_\Gamma F(z) z^{k-1} \, dz = \sum_i \text{residues of } \left\{ F(p_i) p_i^{k-1} \right\} \tag{5.32}$$

The general method for finding the inverse transformation using the definition integral (5.31) is computationally involved. For rational functions (represented by ratios of two polynomials), two additional methods can be employed to obtain the inverse $\mathcal{Z}$-transform. These methods are known as the long division expansion and partial fraction expansion methods.

Long Division Expansion

The idea is to calculate $f[k]$ by expanding the function $F(z)$ into a series of powers of z^{-1}, that is, as

$$F(z) = f[0] + f[1]z^{-1} + f[2]z^{-2} + \cdots + f[k]z^{-k} + \cdots \tag{5.33}$$

The coefficients of this series are the values of $f[k]$ at $k = 0, 1, 2, \ldots$. Though this method is very simple, it is not very efficient. Furthermore, for signals that have infinite time duration, the method is not capable of producing all of the signal values.

Partial Fraction Expansion

Since both $\mathcal{Z}$-transforms and Laplace transforms are most often given in terms of ratios of two polynomials with complex variables, the problem of finding the inverse $\mathcal{Z}$-transform is exactly the same as the problem of finding the inverse Laplace transform. The main technique for finding the inverse Laplace transform, based on partial fraction expansion and presented in Section 4.2, holds in the case of the $\mathcal{Z}$-transform as well.

Examining the table of common pairs (Table 5.2), we notice that all of the $\mathcal{Z}$-transforms presented, except for $\delta[k]$, contain in the numerators a product of the variable z and some other quantity. This indicates that it will be wise to perform the partial fraction expansion of the function

$$\frac{1}{z}F(z) = \frac{1}{z}\frac{N(z)}{D(z)} \tag{5.34}$$

Thus, given the $\mathcal{Z}$-transform $F(z)$, we first form

$$F_1(z) = \frac{F(z)}{z} \tag{5.35}$$

The function $F_1(z)$ is then expanded in partial fractions. For example, for the case of distinct poles, we have

$$F_1(z) = \frac{c_1}{z - p_1} + \frac{c_2}{z - p_2} + \cdots + \frac{c_n}{z - p_n}$$

Now, each element of the expanded version of $F(z)$,

$$F(z) = \frac{c_1 z}{z - p_1} + \frac{c_2 z}{z - p_2} + \cdots + \frac{c_n z}{z - p_n}$$

has its inverse given by a discrete-time exponential signal (see Table 5.2), that is,

$$f[k] = \{c_1(p_1)^k + c_2(p_2)^k + \cdots + c_n(p_n)^k\}u[k]$$

EXAMPLE 5.12

Let us find the inverse $\mathcal{Z}$-transform of the function

$$F(z) = \frac{z(z - 1)}{(z + 0.5)(z - 0.5)(z + 1)}$$

According to (5.35), we must perform the following partial fraction expansion:

$$\frac{F(z)}{z} = F_1(z) = \frac{(z - 1)}{(z + 0.5)(z - 0.5)(z + 1)} = \frac{c_1}{z + 0.5} + \frac{c_2}{z - 0.5} + \frac{c_3}{z + 1}$$

The coefficients are evaluated using the formula that corresponds to formula (4.23), that is,

$$c_i = \lim_{z \to p_i} \left\{ (z - p_i) \frac{F(z)}{z} \right\} \tag{5.36}$$

leading to

$$c_1 = \frac{(z-1)}{(z-0.5)(z+1)}\bigg|_{z=-0.5} = 3$$

$$c_2 = \frac{(z-1)}{(z+0.5)(z+1)}\bigg|_{z=0.5} = -\frac{1}{3}$$

$$c_3 = \frac{(z-1)}{(z+0.5)(z-0.5)}\bigg|_{z=-1} = -\frac{8}{3}$$

The original function $F(z)$ is now expanded as

$$F(z) = \frac{3z}{z+0.5} - \frac{(1/3)z}{z-0.5} - \frac{(8/3)z}{z+1}$$

Since each term in the last expansion corresponds to an exponential signal in time, we have

$$f[k] = \left\{3(0.5)^k - \frac{1}{3}(-0.5)^k - \frac{8}{3}(-1)^k\right\}u[k]$$

EXAMPLE 5.13

Consider the case with multiple poles given by

$$F(z) = \frac{z(z+1)}{(z-0.5)^2(z-1)}$$

The required partial fraction expansion has the form

$$F_1(z) = \frac{F(z)}{z} = \frac{z+1}{(z-0.5)^2(z-1)} = \frac{c_{11}}{(z-0.5)} + \frac{c_{12}}{(z-0.5)^2} + \frac{c_2}{(z-1)}$$

where the coefficients c_{11} and c_{12}, corresponding to the multiple pole, are obtained using the formula that corresponds to formula (4.26). In the case of the $\mathcal{Z}$-transform, the corresponding formula for a pole of multiplicity r located at p_1 is given by

$$c_{1j} = \frac{1}{(r-j)!}\lim_{z \to p_1}\left\{\frac{d^{r-j}}{dz^{r-j}}(z-p_1)^r\frac{F(z)}{z}\right\} \tag{5.37}$$

Note that in this case, the formula is applied to the function $F_1(z)$. In our problem we have $r = 2$, hence

$$c_{12} = \frac{z+1}{z-1}\bigg|_{z=0.5} = -3, \quad c_{11} = \frac{d}{dz}\left(\frac{z+1}{z-1}\right)\bigg|_{z=0.5} = \frac{-2}{(z-1)^2}\bigg|_{z=0.5} = -8$$

The coefficient corresponding to the simple pole is obtained using formula (5.36),

$$c_2 = \frac{z+1}{(z-0.5)^2}\bigg|_{z=1} = 8$$

The original function $F(z)$ has the expansion

$$F(z) = \frac{-8z}{(z-0.5)} - \frac{3z}{(z-0.5)^2} + \frac{8z}{z-1}$$

Using Table 5.2, we obtain

$$f[k] = \{-8(0.5)^k - 6k(0.5)^k + 8\}u[k]$$ ∎

EXAMPLE 5.14

The inverse $\mathcal{Z}$-transform of the function

$$F(z) = \frac{2z^2}{z^2 - \sqrt{3}z + 1}$$

can be found through simple algebra and the recognition of the fact that the original function corresponds in the time domain to a sine type function:

$$F(z) = \frac{2z^2}{z^2 - \sqrt{3}z + 1} = \frac{2z}{\sin\left(\frac{\pi}{6}\right)} \times \frac{z\sin\left(\frac{\pi}{6}\right)}{z^2 - 2z\cos\left(\frac{\pi}{6}\right) + 1}$$

$$= 4z\left\{\mathcal{Z}\left\{\sin\left(k\frac{\pi}{6}\right)u[k]\right\} - \sin(0)u[0]\right\} \leftrightarrow 4\sin\left((k+1)\frac{\pi}{6}\right)u[k+1]$$ ∎

The complex conjugate distinct poles of $F(z)$ functions can be handled similarly to the complex conjugate distinct poles of $F(s)$ functions, as demonstrated in Section 4.2. The corresponding discrete-time result can be derived as follows.

Let $p_1 = |p_1|e^{j\angle p_1}$ be a complex conjugate pole, such that

$$\frac{F(z)}{z} = \frac{c_1}{z - p_1} + \frac{c_1^*}{z - p_1^*}$$

with the complex coefficient $c_1 = |c_1|e^{j\angle c_1}$ obtained using formula (5.36). The expression for the $\mathcal{Z}$-transform of an exponential signal (row three of Table 5.2) implies that

$$F(z) = \frac{c_1 z}{z - p_1} + \frac{c_1^* z}{z - p_1^*} \leftrightarrow \left(c_1 p_1^k + c_1^*(p_1^*)^k\right)u[k]$$

Using the facts that

$$p_1^k = |p_1|^k e^{jk\angle p_1}$$

and

$$c_1 p_1^k + c_1^*(p_1^k)^* = 2\operatorname{Re}\{c_1 p_1^k\} = 2\operatorname{Re}\{|c_1|e^{j\angle c_1}|p_1|^k e^{jk\angle p_1}\}$$
$$= 2|c_1||p_1|^k \operatorname{Re}\{e^{j(k\angle p_1 + \angle c_1)}\} = 2|c_1||p_1|^k \cos(k\angle p_1 + \angle c_1)$$

we obtain
$$\frac{c_1 z}{z - p_1} + \frac{c_1^* z}{z - p_1^*} \leftrightarrow 2|c_1||p_1|^k \cos(k \angle p_1 + \angle c_1) \tag{5.38}$$

EXAMPLE 5.15

Consider the function
$$F(z) = \frac{z(z+1)}{\left(z^2 + \frac{1}{4}\right)\left(z - \frac{1}{2}\right)}$$

which can be expanded as
$$\frac{F(z)}{z} = \frac{c_1}{z - j\frac{1}{2}} + \frac{c_1^*}{z + j\frac{1}{2}} + \frac{c_2}{z - \frac{1}{2}}$$

with
$$c_1 = \lim_{z \to j\frac{1}{2}} \left\{ \left(z - j\frac{1}{2}\right) \frac{F(z)}{z} \right\} = \frac{z+1}{\left(z + j\frac{1}{2}\right)\left(z - \frac{1}{2}\right)} \Big|_{z=j\frac{1}{2}} = -\frac{1 + j\frac{1}{2}}{\frac{1}{2} + j\frac{1}{2}} = -3 + j$$

and
$$c_2 = \lim_{z \to \frac{1}{2}} \left\{ \left(z - \frac{1}{2}\right) \frac{F(z)}{z} \right\} = \frac{z(z+1)}{\left(z^2 + \frac{1}{4}\right)} \Big|_{z=\frac{1}{2}} = \frac{3}{2}$$

Since
$$|c_1| = \sqrt{10}, \quad \angle c_1 = \tan^{-1}\left(-\frac{1}{3}\right) = 161.57°,$$
$$p_1 = j\frac{1}{2} \Rightarrow |p_1| = \frac{1}{2}, \quad \angle p_1 = \frac{\pi}{2} = 90°$$

we have from formula (5.38) the following result:
$$f[k] = \left(2\sqrt{10}\left(\frac{1}{2}\right)^k \cos\left(k\frac{\pi}{2} + 161.57°\right) + \frac{3}{2}\left(\frac{1}{2}\right)^k\right) u[k]$$

EXAMPLE 5.16

For the function defined in Example 5.14, we have
$$\frac{F(z)}{z} = \frac{2z}{z^2 - \sqrt{3}z + 1} = \frac{c_1}{z - \frac{\sqrt{3}}{2} - j\frac{1}{2}} + \frac{c_1^*}{z - \frac{\sqrt{3}}{2} + j\frac{1}{2}}$$

with
$$c_1 = 1 - j\sqrt{3} \Rightarrow |c_1| = 2, \quad \angle c_1 = -\frac{\pi}{3}$$

Formula (5.38) produces

$$F(z) = \frac{2z^2}{z^2 - \sqrt{3}z + 1} \leftrightarrow 2 \times 2 \times 1^k \times \cos\left(k\frac{\pi}{6} - \frac{\pi}{3}\right)u[k]$$

$$= 4\sin\left(k\frac{\pi}{6} - \frac{\pi}{3} + \frac{\pi}{2}\right)u[k] = 4\sin\left((k+1)\frac{\pi}{6}\right)u[k]$$

Since $u[k] = u[k+1]$ for $k \geq 0$, we conclude that the result obtained is identical to the result found in Example 5.14.

Note that the MATLAB function `residue` can be used for performing the partial fraction expansion procedure, and `deconv` can be used for polynomial division.

To end this section, we would like to indicate that the $\mathcal{Z}$-transform can be used for summing infinite series, due to the fact that

$$F(z)|_{z=1} = F(1) = \sum_{k=0}^{\infty} f[k]1^{-k} = \sum_{k=0}^{\infty} f[k]$$

This result is valid under the assumption that the corresponding sum is convergent, which requires that $z = 1$ belongs to the domain of convergence of the $\mathcal{Z}$-transform. This feature of the $\mathcal{Z}$-transform is demonstrated in the next example.

EXAMPLE 5.17   Summation of Infinite Series

Consider the infinite series

$$S_1 = \sum_{k=0}^{\infty} ka^k \quad \text{and} \quad S_2 = \sum_{k=0}^{\infty} k^2 a^k$$

The $\mathcal{Z}$-transform of the signals $ka^k u[k]$ and $k^2 a^k u[k]$ are respectively given (see Example 5.7) by

$$F_1(z) = \frac{az}{(z-a)^2} \quad \text{and} \quad F_2(z) = \frac{az(z+a)}{(z-a)^3}$$

with the regions of convergence equal, in both cases, to $|z| > |a|$. It can be observed that for $|a| < 1$, the point $z = 1$ belongs to the corresponding regions of convergence, hence

$$S_1 = \sum_{k=0}^{\infty} ka^k = F_1(1) = \frac{a}{(1-a)^2}, \quad |a| < 1$$

$$S_2 = \sum_{k=0}^{\infty} k^2 a^k = F_2(1) = \frac{a(1+a)}{(1-a)^3}, \quad |a| < 1$$

For example, for $a = 0.5$, these sums are respectively equal to

$$S_1 = \sum_{k=0}^{\infty} k(0.5)^k = \frac{0.5}{(1-0.5)^2} = 2, \quad S_2 = \sum_{k=0}^{\infty} k^2 (0.5)^k = \frac{0.5(1+0.5)}{(1-0.5)^3} = 6$$

5.3 $\mathcal{Z}$-TRANSFORM IN LINEAR SYSTEM ANALYSIS

In this section, we present the use of the $\mathcal{Z}$-transform in the analysis of discrete-time linear time-invariant systems. This section parallels the presentation of Section 4.3 (done in the context of Laplace transform analysis of continuous-time linear time-invariant systems), and indicates differences and similarities in the procedures for finding responses of continuous- and discrete-time linear systems.

The main goal in the analysis of any dynamic system is to find its response. We indicated in formula (1.8) of Chapter 1 that the complete discrete-time linear system response has two components: zero-state response due to external forcing signals (system inputs) and zero-input response due to system initial conditions; that is,

$$y[k] = y_{zs}[k] + y_{zi}[k] \leftrightarrow Y(z) = Y_{zs}(z) + Y_{zi}(z) \tag{5.39}$$

The $\mathcal{Z}$-transform will produce both the zero-state and zero-input components of the system response. In this section, we will define the discrete-time transfer function, and show how to find system impulse, step, zero-input, zero-state, and complete responses in the discrete-time domain.

5.3.1 Two Formulations of Discrete-Time Linear Systems

Linear time-invariant discrete systems can be described by linear difference equations with constant coefficients using two alternative formulations. We call these the *derivative (time forwarded)* and *integral (time delayed)* formulations of linear discrete-time systems.

The original formulation introduced in (1.10), as

$$y[k+n] + a_{n-1}y[k+n-1] + \cdots + a_1 y[k+1] + a_0 y[k]$$
$$= b_m f[k+m] + b_{m-1} f[k+m-1] + \cdots + b_1 f[k+1] + b_0 f[k], \quad k \geq k_0 + n \tag{5.40}$$

represents in fact the *derivative formulation* of a discrete-time time-invariant linear system. A set of n initial conditions is associated with (5.40):

$$y[k_0], y[k_0 + 1], \ldots, y[k_0 + n - 1] \quad \text{must be known} \tag{5.40a}$$

Without loss of generality, we can assume that the initial discrete-time instant is taken as $k_0 = 0$. The derivative formulation is consistent with the general mathematical theory of linear difference equations with constant coefficients. It also comes into the picture in the discretization of linear differential equations with constant coefficients, and it is used in numerical methods for solving differential equations. In addition, the state space technique to be studied in Chapter 8, a powerful tool for the analysis of high-dimensional time-invariant linear discrete-time systems, is developed exclusively for the derivative formulation.

The integral formulation of linear discrete-time systems is given by

$$y[k] + a_{n-1} y[k-1] + \cdots + a_1 y[k-(n-1)] + a_0 y[k-n]$$
$$= b_m f[k-(n-m)] + b_{m-1} f[k-(n-m+1)] + \cdots$$
$$+ b_1 f[k-(n-1)] + b_0 f[k-n], \quad k \geq 0 \tag{5.41}$$

The initial conditions associated with the integral formulation are defined by

$$y[-n], y[-(n-1)], \ldots, y[-2], y[-1] \quad \text{are known} \tag{5.41a}$$

The integral formulation is more engineering oriented. It is convenient for studying linear discrete-time systems in the frequency domain, using the concept of the discrete-time transfer function. The integral formulation is also used in digital signal processing systems (linear digital filters).

By appropriately propagating system initial conditions forward (backward) in time through the given difference equation, we can go from the integral formulation to the derivative formulation and vice versa. *The two formulations produce the same results only if the initial conditions are correctly propagated.* If the initial conditions are incorrectly propagated through the given difference equation, or not propagated at all, *the results obtained using these two formulations will be, in general, different, since they will be obtained for different initial conditions and different time intervals ($k \geq 0$ for the integral formulation and $k \geq n$ for the derivative formulation).* The new set of initial conditions should be consistently determined, as demonstrated in the next example.

EXAMPLE 5.18

Consider a discrete-time linear system that has a derivative formulation given by

$$y[k+2] - 3y[k+1] + 2y[k] = f[k+1], \quad y[0] = 0, \quad y[1] = -4, \quad f[k] = 3^k u[k]$$

We can go backward in time by two units and represent this difference equation in its integral formulation,

$$y[k] - 3y[k-1] + 2y[k-2] = f[k-1], \quad y[-2] = ?, \quad y[-1] = ?, \quad f[k] = 3^k u[k]$$

The initial conditions $y[-2]$ and $y[-1]$ must be determined from the original difference equation (derivative formulation) and the corresponding initial conditions $y[0]$ and $y[1]$. For $k = -1$, the derivative formulation implies that

$$y[1] - 3y[0] + 2y[-1] = f[0] = 1 \Rightarrow y[-1] = 0.5(1 + 4 + 0) = 5/2$$

For $k = -2$, we have

$$y[0] - 3y[-1] + 2y[-2] = f[-1] = 0 \Rightarrow y[-2] = 0.5(-y[0] + 3y[-1]) = 15/4$$

We will show in Example 5.27 that the system responses of these integral and derivative formulations give exactly the same result. It is interesting to point out that we get the same system response even if we use a mixed derivative-integral system formulation, which can be obtained by shifting backward in time by one unit the original derivative formulation, that is, by finding the response of the system

$$y[k+1] - 3y[k] + 2y[k-1] = f[k], \quad y[-1] = 5/2, \quad y[0] = 0, \quad f[k] = 3^k u[k]$$

We will also show in Example 5.27 that the output response of this system is identical to the output response of the systems represented in the derivative and integral formulations.

230 Chapter 5   Z-transform

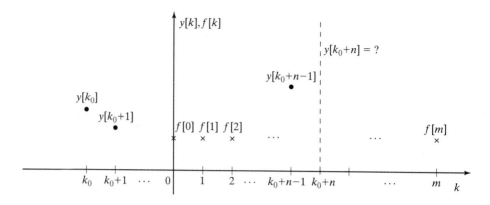

FIGURE 5.3: Time distribution of the system input and output values for the derivative formulation

For a complete understanding of the way in which linear time-invariant discrete-time systems process input signals with (or without) system initial conditions, it would be beneficial that *both formulations* be covered. However, for the sake of time, instructors may choose to cover only one of the presented formulations. In such a case, we suggest that the integral formulation be covered, since it is more engineering oriented than the derivative formulation, and it parallels completely the presentation in Chapter 4 about finding the response of continuous-time linear systems. If the integral formulation alone is covered, the rest of this subsection may be skipped.

Relation Between the Integral and Derivative Formulations[†]

It can be seen from (5.40) and (5.40a) that the system initial conditions are spread along the initial discrete-time interval of length equal to n, during which they completely determine the system time evolution. Let us assume that the system input signal is causal, $f[k] = 0, k < 0$, and that the input is applied to the system starting with $k = 0$. Our goal is to *find the system response $y[k]$ for all discrete time values $k \geq k_0 + n$*. According to the left shift in time property of the $\mathcal{Z}$-transform, the mth left shift in time (mth derivative) is completely determined in the frequency domain only if the signal values are known for $k = 0, 1, 2, \ldots, m - 1$ (see formula (5.12)). Due to the system input signal differentiation, the application of the $\mathcal{Z}$-transform to the right-hand side of difference equation (5.40) produces polynomial terms with respect to z whose coefficients are functions of $f[0], f[1], \ldots, f[m - 1]$. It can be observed that these input signal values will interfere with the system initial conditions unless k_0 is chosen as $k_0 \leq -n$, so that all system initial conditions are defined for negative times (see Figure 5.3). For $k_0 = -n$, the system initial conditions are specified at $y[-n], y[-n + 1], \ldots, y[-2], y[-1]$. Note that the values of the input signal $f[0], f[1], \ldots, f[m - 1]$ play the same role as the values of the

---

[†]May be skipped if the integral formulation alone is covered only.

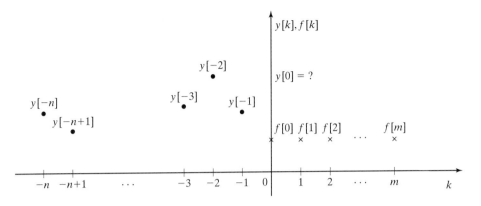

FIGURE 5.4: Time distribution of the system input and output values for the integral formulation

system initial conditions. Namely, they will introduce another component into the system response, which we call the *input-in transient response*.

The input-in transient response can be eliminated, assuming that the original difference equation is shifted backward in time by n discrete time units and represented using the integral formulation as given in (5.41), with the corresponding initial conditions defined for negative discrete time instants only. Note that in this formulation we also assume that all past input signal values are zero, $f[k] = 0$, $k < 0$ (causal input signal; see Figure 5.4). Our problem here is to *find $y[k]$ for all $k \geq 0$*. The right shifts in time in (5.41) can be evaluated using the variant of the right shift in time property of the $\mathcal{Z}$-transform, given in formulas (5.11a) and (5.11b). *Formula (5.11a) must be used for the system output right shifts in time,* and *formula (5.11b) must be used for the system input right shifts in time,* since the input signal has all zero values for negative time instants.

The integral formulation is used in many undergraduate linear systems texts, and it is built into MATLAB as the method for finding the response of linear time-invariant discrete-time systems. This formulation better fits the physical reality of determining the system response using the concept of the discrete-time system transfer function. Namely, for a given system whose initial conditions are set during some initial time interval, we must find the response due to the external input signal applied to the system at some time, say $k = 0$. The system response is due to both the *past* system initial conditions and the *present* and future input signal values. It is important to emphasize that *in the integral formulation, the present and future input signal values do not interfere with the past system initial conditions* (see Figures 5.4 and 5.5).

Concluding Remarks: The integral and derivative formulations produce the same results only if they are obtained from each other by correctly propagating the initial conditions through the given difference equation. The integral formulation produces the system response for $k \geq 0$ and the derivative formulation for $k \geq n$. The system response in the derivative formulation from $k = 0$ to $k = n - 1$ is determined by the system initial conditions.

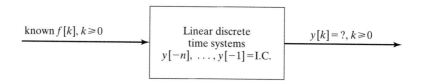

FIGURE 5.5: Finding the response of a discrete-time linear system in its integral formulation (I.C.: initial conditions)

5.3.2 System Response Using the Integral Formulation

In this section, we show how to find responses (impulse, step, zero-state, zero-input, and complete) of linear discrete-time, time-invariant systems, using the $\mathcal{Z}$-transform via the integral (time-delayed) formulation. We first define the system transfer function and show that the discrete-time system impulse response can be simply obtained by finding the inverse $\mathcal{Z}$-transform of the system transfer function.

System Transfer Function and Impulse Response

Let us apply the $\mathcal{Z}$-transform to both sides of the linear difference equation in the integral formulation defined in (5.41), that is,

$$\mathcal{Z}\{y[k] + a_{n-1}y[k-1] + \cdots + a_1 y[k-(n-1)] + a_0 y[k-n]\}$$
$$= \mathcal{Z}\{b_m f[k-(n-m)] + b_{m-1} f[k-(n-m+1)] + \cdots$$
$$+ b_1 f[k-(n-1)] + b_0 f[k-n]\}, \quad k \geq 0 \quad (5.42)$$

As stated in (5.41a), the associated initial conditions for the integral formulation are specified as $y[-n], y[-(n-1)], \ldots, y[-2], y[-1]$. It is assumed that the forcing signal $f[k]$ is defined for all nonnegative values of discrete-time instants k, and that corresponding values are known. In addition, we assume that the input signal $f[k]$ is causal, which means that $f[k] = 0, k < 0$. We are interested to find the system response $y[k]$ for all time instants $k \geq 0$. Hence, (5.42) is valid for all $k \geq 0$ (both sides of (5.42) can be multiplied by $u[k]$), so that we must use the right shift in time property of the $\mathcal{Z}$-transform defined in formulas (5.11a) and (5.11b).

Applying (5.11a) to the left-hand side and (5.11b) to the right-hand side of (5.42), we obtain

$$\frac{1}{z^n}(z^n + a_{n-1}z^{n-1} + \cdots + a_1 z + a_0)Y(z) + I_y^i(z)$$
$$= \frac{1}{z^n}(b_m z^m + b_{m-1} z^{m-1} + \cdots + b_1 z + b_0)F(z)$$

which simplifies into

$$(z^n + a_{n-1}z^{n-1} + \cdots + a_1 z + a_0)Y(z) + z^n I_y^i(z)$$
$$= (b_m z^m + b_{m-1} z^{m-1} + \cdots + b_1 z + b_0)F(z) \quad (5.43)$$

where $I_y^i(z)$ contains terms coming from the system initial conditions, obtained by using in (5.42) the following formulas:

$$\mathcal{Z}\{y[k-1]u[k]\} = \frac{1}{z}Y(z) + y[-1]$$

$$\mathcal{Z}\{y[k-2]u[k]\} = \frac{1}{z^2}Y(z) + \frac{1}{z}y[-1] + y[-2]$$

$$\vdots \quad (5.44)$$

$$\mathcal{Z}\{y[k-n]u[k]\} = \frac{1}{z^n}Y(z) + \frac{1}{z^{n-1}}y[-1] + \frac{1}{z^{n-2}}y[-2] + \cdots$$

$$+ \frac{1}{z^2}y[-(n-2)] + \frac{1}{z}y[-(n-1)] + y[-n]$$

(The superscript "i" in formula (5.43), in its $I_y^i(z)$ term, stands for the integral formulation.) It can be shown after simple algebra that the general form for the $I_y^i(z)$ term is given by

$$I_y^i(z) = (a_{n-1}z^{n-1} + a_{n-2}z^{n-2} + \cdots + a_1z + a_0)z^{-(n-1)}y[-1]$$

$$+ (a_{n-2}z^{n-2} + a_{n-3}z^{n-3} + \cdots + a_1z + a_0)z^{-(n-2)}y[-2]$$

$$+ (a_{n-3}z^{n-3} + a_{n-4}z^{n-4} + \cdots + a_1z + a_0)z^{-(n-3)}y[-3]$$

$$+ \cdots + (a_1z + a_0)z^{-1}y[-(n-1)] + a_0y[-n] \quad (5.45)$$

Note that we need not use the general formula (5.45) to find $I_y^i(z)$. A direct application of the $\mathcal{Z}$-transform properties (5.11a) to the left-hand side of (5.42) simply produces all terms in $I_y^i(z)$.

The input signal is applied to the system at $k = 0$, and we are interested in finding the complete (total) system response—the response due to both the system initial conditions and the forcing function for $k = 0, 1, 2, \ldots$. From (5.43), we have

$$Y(z) = Y_{zs}^i(z) + Y_{zi}^i(z)$$

$$= \frac{b_mz^m + b_{m-1}z^{m-1} + \cdots + b_1z + b_0}{z^n + a_{n-1}z^{n-1} + \cdots + a_1z + a_0}F(z) \quad (5.46)$$

$$- \frac{z^n I_y^i(z)}{z^n + a_{n-1}z^{n-1} + \cdots + a_1z + a_0}$$

where $Y_{zs}^i(z)$ and $Y_{zi}^i(z)$ denote, respectively, the zero-state and zero-input components of the system response in the frequency domain for the integral formulation. If the system initial conditions are set to zero, we will have $I_y^i(z) = 0$. Then, the quantity

$$H(z) = \frac{Y(z)}{F(z)}\bigg|_{I_y^i(z)=0} = \frac{b_mz^m + b_{m-1}z^{m-1} + \cdots + b_1z + b_0}{z^n + a_{n-1}z^{n-1} + \cdots + a_1z + a_0} \quad (5.47)$$

defines the *discrete-time system transfer function*. In words, we can say that the *discrete-time system transfer function represents the ratio of the $\mathcal{Z}$-transform of the system output* over the $\mathcal{Z}$-transform of the system input, under the assumption that the system initial conditions are equal to zero. This transfer function can be written in an alternative form,

$$H(z) = z^{-(n-m)} \frac{b_m + b_{m-1}z^{-1} + \cdots + b_1 z^{-(m-1)} + b_0 z^{-m}}{1 + a_{n-1}z^{-1} + \cdots + a_1 z^{-(n-1)} + a_0 z^{-n}} \qquad (5.47a)$$

This form is particularly common in digital signal processing (linear digital filters). The term z^{-1} is often called the time delay element of one discrete time unit.

The transfer function (5.47) can be also written in factored form as

$$H(z) = K \frac{(z - z_1)(z - z_2) \cdots (z - z_m)}{(z - p_1)(z - p_2) \cdots (z - p_n)} \qquad (5.48)$$

where z_i, $i = 1, 2, \ldots, m$, are *zeros* at which $H(z_i) = 0$; and p_j, $j = 1, 2, \ldots, n$, are the transfer function *poles* at which $H(p_j) = 0$. The transfer function zeros and poles are also called the *system* zeros and poles. The constant K is known as the transfer function *static gain*. We have tacitly assumed that $z_i \neq p_j$ for all i, j, so that we have n poles and m zeros. However, in the case when there are common factors in the transfer function numerator and denominator, they must be cancelled out before the system poles and zeros are identified (see the corresponding continuous-time Example 4.17).

The quantity $H(1)$, equal to

$$H(1) = \frac{b_m + b_{m-1} + \cdots + b_1 + b_0}{1 + a_{n-1} + a_{n-2} + \cdots + a_1 + a_0} = \frac{\sum_{i=0}^{m} b_i}{1 + \sum_{j=0}^{n-1} a_j} \qquad (5.49)$$

is called the *system steady state gain*. It shows how much the system multiplies a constant input signal at steady state. This result can be seen by applying the final value theorem to $Y(z)$ from (5.47), with $F(z) = az/(z-1) \leftrightarrow f[k] = au[k]$, which leads to

$$\begin{aligned} y_{ss} &= \lim_{k \to \infty} y[k] = \lim_{z \to 1} \left\{ \frac{z-1}{z} Y(z) \right\} = \lim_{z \to 1} \left\{ \frac{z-1}{z} H(z) \frac{az}{z-1} \right\} \\ &= \lim_{z \to 1} \{H(z)a\} = H(1)a \end{aligned} \qquad (5.50)$$

Note that the final value theorem is applicable under the assumption that the function $(z-1)Y(z)$ has all its poles strictly inside the unit circle. In such a case, the system output can reach its steady state constant value, which can be found by using formula (5.50).

EXAMPLE 5.19

Consider a system represented by the transfer function

$$H(z) = \frac{z+2}{z^2 + z + \frac{1}{4}} \Rightarrow H(1) = \frac{1+2}{1+1+\frac{1}{4}} = \frac{4}{3}$$

Assume that a constant input signal equal to $f[k] = -3u[k]$ is applied to the system. Since both poles of $H(z)$ are inside the unit circle, the system output can reach the steady state which, according to formula (5.50), is given by

$$y_{ss} = H(1)a = \frac{4}{3}(-3) = -4$$

Considering that, for the discrete-time impulse delta signal, the $\mathcal{Z}$-transform is given by $F(z) = 1$, we conclude from (5.47) that the linear discrete-time system response to the discrete-time impulse delta signal is equal to $\mathcal{Z}^{-1}\{H(z)\}$. Hence, in the discrete-time domain, the *system impulse response is obtained from*

$$h[k] = \mathcal{Z}^{-1}\{H(z)\} \qquad (5.51)$$

EXAMPLE 5.20

In this example, we find the impulse response of the discrete-time system

$$y[k] - \frac{1}{6}y[k-1] - \frac{1}{6}y[k-2] = f[k-1] + f[k-2]$$

using the $\mathcal{Z}$-transform. The discrete-time system transfer function is given by

$$H(z) = \frac{z+1}{z^2 - \frac{1}{6}z - \frac{1}{6}} = \frac{z+1}{(z-\frac{1}{2})(z+\frac{1}{3})}$$

Its partial fraction expansion is

$$\frac{1}{z}H(z) = -\frac{6}{z} + \frac{\frac{18}{5}}{z-\frac{1}{2}} + \frac{\frac{12}{5}}{z+\frac{1}{3}}$$

Taking the inverse $\mathcal{Z}$-transform, the discrete-time impulse response is obtained in the analytical form

$$h[k] = \mathcal{Z}^{-1}\{H(z)\} = -6\delta[k] + \frac{18}{5}\left(\frac{1}{2}\right)^k u[k] + \frac{12}{5}\left(-\frac{1}{3}\right)^k u[k]$$

The system impulse response for this example can also be obtained using the MATLAB function dimpulse, as follows:

```
num=[1 1]; den=[1 -1/6 -1/6];
h=dimpulse(num,den); d=length(h);
k=0:1:d-1;
plot(k,h,'*'); grid;
xlabel('Discrete time k');
ylabel('System impulse response')
```

236   Chapter 5   Z-transform

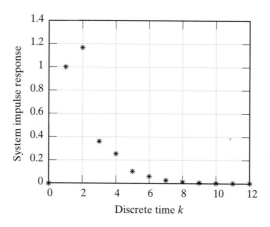

FIGURE 5.6: System impulse response

Note that since MATLAB *does not allow a zero index,* `dimpulse` produces the impulse response that starts at $k = 1$, and has a continuous, piecewise constant staircase form. In order to get the correct plot, we have shifted the result for $h[k]$ by one discrete-time instant to the left. This has been achieved with `k=0:1:d-1` and `plot(k,h,'*')`. The corresponding discrete-time impulse response is presented in Figure 5.6.

Zero-State Response

The zero-state response for the *integral formulation* is obtained for

$$y[-n] = y[-(n-1)] = \cdots = y[-2] = y[-1] = 0 \Leftrightarrow I_y^i(z) = 0 \qquad (5.52)$$

We can conclude from formulas (5.46) and (5.47) that the zero-state system response, the system response solely contributed by the given input signal $f[k]$, is obtained in the frequency domain by using the simple formula

$$Y_{zs}^i(z) = H(z)F(z) \qquad (5.53)$$

Applying the convolution property of the Z-transform to (5.53), we obtain

$$y_{zs}^i[k] = h[k] * f[k] \qquad (5.54)$$

Formula (5.54) states that the output response of a discrete-time linear system, whose initial conditions are set to zero, is equal to the convolution of the discrete-time system input and the discrete-time system impulse response.

In formulas (5.53–54), we have established the most fundamental results of the theory of linear discrete-time systems. From these results, we see that to find the system response to any input signal, one must first find the system response due to the discrete-time impulse delta signal, and then convolve the obtained discrete-time system impulse response with the given discrete-time system input signal. Note that the discrete-time system impulse response is obtained (and defined) for a system whose initial conditions are set to zero (discrete-time system at rest). It should also be emphasized that for the given discrete-time linear time-invariant system, the impulse response need only be found once. In other words, any discrete-time linear time-invariant system is uniquely characterized by its impulse response (or by its transfer function in the frequency domain). In general, it is easy to find the discrete-time system impulse response using (5.51). As a matter of fact, formula (5.51) represents the most efficient means for finding the impulse response of discrete-time linear time-invariant systems. This issue will be further clarified in Chapter 7.

EXAMPLE 5.21

Consider the discrete-time linear system

$$y[k] + 4y[k-1] + 4y[k-2] = f[k-1] + 4f[k-2]$$

whose input function is $f[k] = (-1)^k u[k]$. The corresponding zero-state response is obtained as

$$Y_{zs}^i(z) = H(z)F(z) = \frac{z+4}{(z+2)^2} \frac{z}{z+1}$$

$$\frac{1}{z}Y_{zs}^i(z) = \frac{-3}{z+2} + \frac{-2}{(z+2)^2} + \frac{3}{z+1}$$

$$y_{zs}^i[k] = \mathcal{Z}^{-1}\{Y_{zs}^i(z)\} = \{-3(-2)^k + k(-2)^k + 3(-1)^k\}u[k]$$

The MATLAB function dlsim can be used for finding the zero-state response. To find the values of the zero-state response during the time interval of six discrete time instants, we will use the following MATLAB script:

```
num=[1 4]; den=[1 4 4];
k=0:1:5;
f=(-1).^k;
yzs=dlsim(num,den,f)
```

Note that the dot after "(−1)" indicates in MATLAB a pointwise operation; that is, (−1) is exponentiated by each component of the vector k, producing the corresponding six components of the vector f.

Zero-Input Response

When no system input is present, the system response is driven by the system initial conditions only. Such a response is called the zero-input response.

For the integral formulation of a linear discrete-time system, with initial conditions defined at $y[-n], y[-(n-1)], \ldots, y[-2], y[-1]$, the zero-input response has been derived in (5.46) as

$$y_{zi}^i[k] = \mathcal{Z}^{-1}\{Y_{zi}^i(z)\}$$
$$= \mathcal{Z}^{-1}\left\{\frac{-z^n I_y^i(z)}{z^n + a_{n-1}z^{n-1} + \cdots + a_1 z + a_0}\right\} = \mathcal{Z}^{-1}\left\{\frac{-z^n I_y^i(z)}{\Delta(z)}\right\} \quad (5.55)$$

where $I_y^i(z)$ is determined by (5.45). Note that $I_y^i(z)$ is obtained by applying the $\mathcal{Z}$-transform property (5.11a) to the corresponding input-free difference equation. Hence, we need not use (or memorize) formula (5.45). We can simply apply (5.11a) to the difference equation and collect terms that depend on the initial conditions. Such terms form $I_y^i(z)$.

EXAMPLE 5.22

Consider the discrete-time input-free system in the integral formulation

$$y[k] + y[k-1] + 0.25y[k-2] = 0, \quad y[-2] = 44, \quad y[-1] = -12$$

By taking the $\mathcal{Z}$-transform of the difference equation in the integral formulation (using (5.11a)), we obtain

$$Y_{zi}^i(z) + \frac{1}{z}Y_{zi}^i(z) + y[-1] + 0.25\left(\frac{1}{z^2}Y_{zi}^i(z) + \frac{1}{z}y[-1] + y[-2]\right) = 0$$

which implies

$$Y_{zi}^i(z) = \frac{-z^2\left(y[-1] + \frac{0.25}{z}y[-1] + 0.25y[-2]\right)}{z^2 + z + 0.25} = \frac{z(z+3)}{(z+0.5)^2}$$

The inverse $\mathcal{Z}$-transform leads to the zero-input response

$$y_{zi}^i[k] = \mathcal{Z}^{-1}\{Y_{zi}^i(z)\} = (1 - 5k)\left(-\frac{1}{2}\right)^k u[k]$$

Note that MATLAB has no functions for finding either the complete system response or the zero-input response of discrete-time systems represented by their transfer functions (difference equations). However, its functions `dlsim` and `dinitial` can find the complete and zero-input responses of discrete-time systems represented in state space form (to be studied in Chapter 8). Hence, plotting and calculating the corresponding system responses via MATLAB will be postponed until Chapter 8.

Complete Response
From the system analysis point of view, it is extremely important that the $\mathcal{Z}$-transform is able to handle the system initial conditions and to produce the system response due to both initial conditions and external forcing functions.

Section 5.3 Z-transform in Linear System Analysis

We have shown in formula (5.46), derived for the integral system formulation, that in the frequency domain the complete system response is given by

$$Y(z) = Y_{zs}^i(z) + Y_{zi}^i(z) = H(z)F(z) - \frac{z^n I_y^i(z)}{\Delta(z)} \quad (5.56)$$

By taking the inverse $\mathcal{Z}$-transform of the last equation, we obtain the complete system response in time,

$$y[k] = \mathcal{Z}^{-1}\{H(z)F(z)\} + \mathcal{Z}^{-1}\left\{\frac{-z^n I_y^i(z)}{\Delta(z)}\right\} = y_{zs}^i[k] + y_{zi}^i[k] \quad (5.57)$$

which has two components: $y_{zs}^i[k]$, representing the zero-state response (response due to the forcing function), and $y_{zi}^i[k]$, representing the response due to system initial conditions.

EXAMPLE 5.23

Consider the discrete-time linear system

$$y[k] - y[k-1] + \frac{1}{4}y[k-2] = f[k-2],$$

$$f[k] = (-2)^k u[k], \quad y[-2] = -20, \quad y[-1] = -4$$

Applying the $\mathcal{Z}$-transform properties (5.11a) and (5.11b), respectively, to the left- and right-hand sides of the preceding difference equation, we obtain

$$Y(z) - \frac{1}{z}Y(z) - y[-1] + \frac{1}{4}\left(\frac{1}{z^2}Y(z) + \frac{1}{z}y[-1] + y[-2]\right) = \frac{1}{z^2}F(z)$$

This algebraic equation produces the following solution for $Y(z)$:

$$Y(z) = \frac{1}{(z-0.25)^2}F(z) + \frac{z(zy[-1] - 0.25zy[-2] - 0.25y[-1])}{(z-0.25)^2}$$

Substituting the expression for $F(z)$ and the values for $y[-1]$ and $y[-2]$, we have

$$Y(z) = \frac{1}{(z-0.5)^2} \frac{z}{(z+2)} + \frac{z(z+1)}{(z-0.5)^2}$$

which can be written as

$$\frac{1}{z}Y(z) = \frac{21}{25}\frac{1}{z-0.5} + \frac{19}{10}\frac{1}{(z-0.5)^2} + \frac{76}{9}\frac{1}{z+2}$$

Hence, the complete system response is given by

$$y[k] = \mathcal{Z}^{-1}\{Y(z)\} = \mathcal{Z}^{-1}\{Y_{zs}^i(z)\} + \mathcal{Z}^{-1}\{Y_{zi}^i(z)\}$$

$$= \mathcal{Z}^{-1}\left\{\frac{21}{25}\frac{z}{(z-0.5)} + \frac{19}{10}\frac{z}{(z-0.5)^2}\right\} + \mathcal{Z}^{-1}\left\{\frac{76}{9}\frac{z}{z+2}\right\}$$

$$= y_{zs}^i[k] + y_{zi}^i[k] = \left(\frac{21}{25}(0.5)^k + \frac{19}{5}k(0.5)^k\right)u[k] + \frac{76}{9}(-2)^k u[k], \quad k \geq 0 \quad §$$

Step Response

Very often we are faced with the problem of finding a system response due to a constant input. Without loss of generality, having in mind the linearity principle and the form of the unit step function, we can treat this problem as the problem of finding the system unit step response.

Using the integral system formulation, we have introduced the concept of the system transfer function in (5.47), and defined the system impulse response as the inverse $\mathcal{Z}$-transform of the system transfer function, (5.51). Furthermore, the same system formulation is used in the preceding treatment of the zero-state response to derive the result that states that, for a system at rest, the system response is equal to the convolution of the system input and the system impulse response (see (5.53–54)).

We know that the step response is defined for systems at rest. Using (5.54), that is, convolving the discrete-time system impulse response and the discrete-time unit step function, we obtain

$$y_{\text{step}}[k] = \sum_{m=0}^{k} u[k-m]h[m] = \sum_{m=0}^{k} h[m] \tag{5.58}$$

and

$$y_{\text{step}}[k] = \sum_{m=0}^{k} h[m], \quad y_{\text{step}}[k-1] = \sum_{m=0}^{k-1} h[m] \Rightarrow h[k] = y_{\text{step}}[k] - y_{\text{step}}[k-1] \tag{5.59}$$

Formulas (5.58) and (5.59) give the relationships between the discrete-time system impulse and step responses.

The step response can also be directly obtained from formula (5.53) with $F(z) = z/(z-1)$, that is, as

$$Y_{\text{step}}(z) = H(z)\frac{z}{z-1} \Rightarrow y_{\text{step}}[k] = \mathcal{Z}^{-1}\left\{\frac{z}{z-1}H(z)\right\} \tag{5.60}$$

EXAMPLE 5.24

Let us find the step response for the system from Example 5.20, using formula (5.60). It follows that

$$Y_{\text{step}}(z) = \frac{z+1}{\left(z-\frac{1}{2}\right)\left(z+\frac{1}{3}\right)}\frac{z}{(z-1)} = \frac{-\frac{18}{5}z}{z-\frac{1}{2}} + \frac{\frac{3}{5}z}{z+\frac{1}{3}} + \frac{3z}{z-1}$$

which implies that

$$y_{\text{step}}[k] = \left(3 - \frac{18}{5}\left(\frac{1}{2}\right)^k + \frac{3}{5}\left(-\frac{1}{3}\right)^k\right)u[k]$$

Section 5.3 $\mathcal{Z}$-transform in Linear System Analysis

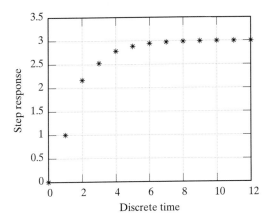

FIGURE 5.7: Step response for Example 5.24

On the other hand, formula (5.59) and the result for the corresponding impulse response from Example 5.20 produce

$$y_{\text{step}}[k] = \sum_{m=0}^{k} h[m] = \sum_{m=0}^{k} \left\{ -6\delta[m] + \left[\frac{18}{5} \left(\frac{1}{2} \right)^m + \frac{12}{5} \left(-\frac{1}{3} \right)^m \right] u[m] \right\}$$

It is left as an exercise to show that the last two formulas are identical (see Problem 5.37).

The discrete-time step response can be calculated and plotted using the following MATLAB script. The corresponding plot is given in Figure 5.7.

Note that $H(1) = 3$, hence the unit step response at steady state takes the value equal to $y_{ss} = H(1) \times 1 = 3$ (see formula (5.50)).

```
num=[1 1]; den=[1 -1/6 -1/6];
ystep=dstep(num,den);
d=length(ystep);
k=0:1:d-1;
plot(k,ystep,'*');
xlabel('Discrete time'); ylabel('Step response');
grid; axis([0 12 0 3.5])
```

As pointed out in the case of the discrete-time impulse response, in order to get the plot of the discrete-time step response, we could have used here only the statement ystep(num,den). However, such a plot represents a continuous staircase function (signal).

It should be emphasized that MATLAB uses only the integral discrete-time system representation. Hence all data provided to MATLAB about a linear discrete-time system must be consistent with the system integral formulation.

5.3.3 System Response Using the Derivative Formulation†

In this section we show how to use the $\mathcal{Z}$-transform to find responses (zero-state, zero-input, and complete) of linear discrete-time, time-invariant systems represented in the derivative (time-forwarded) formulation.

Applying the $\mathcal{Z}$-transform property (5.12) to both sides of the linear difference equation in the derivative form (5.40), that is,

$$\mathcal{Z}\{y[k+n] + a_{n-1}y[k+n-1] + \cdots + a_1 y[k+1] + a_0 y[k]\}$$
$$= \mathcal{Z}\{b_m f[k+m] + b_{m-1} f[k+m-1] + \cdots + b_1 f[k+1] + b_0 f[k]\}, \quad k \geq n$$
(5.61)

with the associated initial conditions $y[0], y[1], \ldots, y[n-1]$, we obtain the following algebraic equation in the complex domain:

$$(z^n + a_{n-1}z^{n-1} + \cdots + a_1 z + a_0)Y(z) - I_y^d(z)$$
$$= (b_m z^m + b_{m-1} z^{m-1} + \cdots + b_1 z + b_0)F(z) - I_f(z) \quad (5.62)$$

$I_y^d(z)$ contains terms that come from the system initial conditions, obtained by using in (5.61) the corresponding left shift in time property of the $\mathcal{Z}$-transform, which is given in formula (5.12). The superscript "d" stands for the derivative formulation. Similarly, $I_f(z)$ contains terms coming from $f[0], f[1], \ldots, f[m-1]$. These terms come from the application of the left shift in time property of the $\mathcal{Z}$-transform to the right-hand side of (5.61). For $m = 0$, *this component in the system transient response is equal to zero:* $I_f(z) = 0$ (the system does not differentiate the input signal). Note that we assume that the forcing function $f[k]$ is defined for all nonnegative values of discrete-time instants k, and that the corresponding values are known. We also assume that the input signal $f[k]$ is causal, which means that $f[k] = 0, k < 0$.

Using the time derivative property of the $\mathcal{Z}$-transform, we have from (5.12) the following formulas:

$$\mathcal{Z}\{y[k+1]u[k]\} = zY(z) - zy[0]$$
$$\mathcal{Z}\{y[k+2]u[k]\} = z^2 Y(z) - z^2 y[0] - zy[1]$$
$$\vdots \qquad (5.63)$$
$$\mathcal{Z}\{y[k+n]u[k]\} = z^n Y(z) - z^n y[0] - z^{n-1} y[1] - \cdots - zy[n-1]$$

It is easy to show, using (5.12) in (5.61), that

$$I_y^d(z) = (z^n + a_{n-1}z^{n-1} + \cdots + a_2 z^2 + a_1 z)y[0]$$
$$+ (z^{n-1} + a_{n-1}z^{n-2} + \cdots + a_3 z^2 + a_2 z)y[1]$$
$$+ (z^{n-2} + a_{n-1}z^{n-3} + \cdots + a_3 z)y[2]$$
$$+ \cdots + (z^2 + a_{n-1}z)y[n-2] + zy[n-1], \quad n = 1, 2, 3, \ldots \quad (5.64)$$

---

†This section may be skipped without loss of continuity.

Similarly, (5.12) in (5.61) also implies that

$$I_f(z) = (b_m z^m + b_{m-1} z^{m-1} + \cdots + b_2 z^2 + b_1 z) f[0]$$
$$+ (b_m z^{m-1} + b_{m-1} z^{m-2} + \cdots + b_3 z^2 + b_2 z) f[1]$$
$$+ (b_m z^{m-2} + b_{m-1} z^{m-3} + \cdots + b_3 z) f[2]$$
$$+ \cdots + (b_m z^2 + b_{m-1} z) f[m-2] + b_m z f[m-1], \quad m = 1, 2, 3, \ldots$$
(5.65)

The input signal is applied to the system at $k = 0$, and we are interested in finding the complete (total) system response—the response due to both the system initial conditions and the forcing function for $k = n, n+1, n+2, \ldots$. Note that the system initial conditions determine the system output response at $k = 0, 1, 2, \ldots, n-1$. From (5.62), we have

$$Y(z) = \frac{b_m z^m + b_{m-1} z^{m-1} + \cdots + b_1 z + b_0}{z^n + a_{n-1} z^{n-1} + \cdots + a_1 z + a_0} F(z)$$
$$- \frac{I_f(z)}{z^n + a_{n-1} z^{n-1} + \cdots + a_1 z + a_0} + \frac{I_y^d(z)}{z^n + a_{n-1} z^{n-1} + \cdots + a_1 z + a_0}$$
(5.66)

Using the definition of the discrete-time linear time-invariant system transfer function introduced in (5.47), and defining the discrete-time system characteristic polynomial by

$$\Delta(z) = z^n + a_{n-1} z^{n-1} + a_{n-2} z^{n-2} + \cdots + a_1 z + a_0 \qquad (5.67)$$

the expression for the system response in the frequency domain given in (5.66) can be recorded in a compact form as

$$Y(z) = Y_{zs}^d(z) + Y_{zi}^d(z) = \left(H(z) F(z) - \frac{I_f(z)}{\Delta(z)} \right) + \frac{I_y^d(z)}{\Delta(z)} \qquad (5.68)$$

The first two components in the system output are contributed by the system input. They form the signal $Y_{zs}^d(s)$. The second component generates the transient response due to system input (its first m values, that is, $f[0], f[1], \ldots, f[m-1]$). We call this component the input-in transient response. It follows from (5.65) that *the input-in transient response is present ($I_f(z) \neq 0$) only when $m \geq 1$; that is, it is present only in the case in which the system differentiates input signals.* The third component in the system output is contributed by the system initial conditions, $Y_{zi}^d(z)$.

It can be observed from (5.68) that, for $m \geq 1$, we will encounter difficulties if we attempt to introduce the concept of the discrete-time system transfer function using the derivative formulation. Namely, in this case, we will need to assume that both the system initial conditions and the first m values of the system input are equal to zero,

$$H(z) = \frac{Y(z)}{F(z)} \bigg|_{I_y^d(z)=0, I_f(z)=0}$$

which is a rather artificial definition that also depends on the input signal values. Hence, we will use the discrete-time transfer definition as given in (5.47), with the system impulse response represented by the inverse $\mathcal{Z}$-transform of the system transfer function, (5.51).

The result presented in (5.68) is convenient for finding the zero-state and zero-input system responses, as well as for obtaining the transient input-in response component in the frequency domain. Using the inverse $\mathcal{Z}$-transform, the corresponding quantities can be obtained in the time domain. A direct time domain solution of linear difference equations, with constant coefficients (5.40) and initial conditions (5.40a), will be studied in detail in Chapter 7.

Comment: It seems from the derivative formulation that discrete-time linear systems process input signals differently than continuous-time linear systems. This would imply that the general problem of finding the response of discrete-time systems is different than the corresponding continuous-time problem. However, the actual signal processing in both continuous- and discrete-time domains is basically the same. Note that in several examples in Chapter 4, we indicated that the original linear continuous-time system initial conditions given at $t = 0^-$ were instantaneously changed when the system differentiated the incoming input signal, so that another component in the system transient response was generated due to the new values of the system initial conditions at $t = 0^+$. The direct application of the Laplace transform to a linear constant coefficient differential equation tacitly incorporates the mechanism of generating the input-in transient component, so that we hardly notice the presence of this component in the continuous-time system response. Due to the fact that the system initial conditions are defined in the discrete-time domain over an initial discrete-time interval of length n, the observed input-in impact on the system initial conditions becomes more obvious, and it is explicitly given by the system response component equal to $-I_f(s)/\Delta(s)$.

Zero-State Response

It is known from the derivative formulation, formula (5.68), that the system response due to the input signal and zero initial conditions defined at

$$y[0] = y[1] = \cdots = y[n-2] = y[n-1] = 0 \Leftrightarrow I_y^d(z) = 0 \tag{5.69}$$

is given by

$$Y_{zs}^d(z) = H(z)F(z) - \frac{I_f(z)}{\Delta(z)} \tag{5.70}$$

Using the inverse $\mathcal{Z}$-transform, we have

$$y_{zs}^d[k] = \mathcal{Z}^{-1}\{Y_{zs}^d(z)\} = \mathcal{Z}^{-1}\left\{H(z)F(z) - \frac{I_f(z)}{\Delta(z)}\right\} \tag{5.71}$$

It is obvious from (5.53) and (5.70) that $Y_{zs}^i(z) \neq Y_{zs}^d(z)$. The reason for this discrepancy is the fact that the initial conditions are set at zero in different time intervals; compare (5.52) and (5.69). Note that, as mentioned, we can use any of the formulations (integral or derivative) to find the system response and we should get the same answer. In that process, we also must appropriately propagate the system initial conditions, as demonstrated in Example 5.18. If we start with the integral formulation and propagate its zero initial conditions defined in the interval from $k = -n$ to $k = -1$ forward in time, we will get (in general) nonzero initial conditions in the time interval from $k = 0$ to $k = n - 1$. Hence, the requirement to keep the initial conditions equal to zero in the time interval from $k = 0$ to $k = n - 1$ yields $Y_{zs}^i(z) \neq Y_{zs}^d(z)$. Note that if we intend to obtain the same response using the derivative formulation, which is identical to the zero-state response obtained via the integral formulation, $Y_{zs}^i(z)$, we must first propagate the initial condition $y[-n] = y[-(n - 1)] = \cdots = y[-2] = y[-1] = 0$ through the given difference equation, which will give rise to nonzero values for $y[1], y[2], \ldots, y[n - 1]$ and a nonzero value for $I_y^d(z)$. Then, the corresponding complete response of the derivative formulation will produce numerical values that are identical to those obtained in $y_{zs}^i[k] \leftrightarrow Y_{zs}^i(z)$.

Zero-Input Response

For discrete-time systems represented in the derivative formulation, with initial conditions defined at $y[0], y[1], \ldots, y[n - 1]$, the zero-input response has been derived in (5.68) as

$$y_{zi}^d[k] = \mathcal{Z}^{-1}\{Y_{zi}^d(z)\}$$
$$= \mathcal{Z}^{-1}\left\{\frac{I_y^d(z)}{z^n + a_{n-1}z^{n-1} + \cdots + a_1 z + a_0}\right\} = \mathcal{Z}^{-1}\left\{\frac{I_y^d(z)}{\Delta(z)}\right\} \quad (5.72)$$

with $I_y^d(z)$ defined in (5.64). Note that we need not memorize formula (5.64). Instead, we apply the left shift in time property of the $\mathcal{Z}$-transform defined in (5.12) to the corresponding difference equation, and collect terms that come from the initial conditions. Those terms form $I_y^d(z)$.

EXAMPLE 5.25

Consider the discrete-time input-free system

$$y[k + 2] + y[k + 1] + 0.25y[k] = 0, \quad y[0] = 1, \quad y[1] = 2$$

By taking the $\mathcal{Z}$-transform, we obtain

$$\left(z^2 Y_{zi}^d(z) - z^2 y[0] - zy[1]\right) + \left(z Y_{zi}^d(z) - zy[0]\right) + 0.25 Y_{zi}^d(z) = 0$$

which implies

$$Y_{zi}^d(z) = \frac{z^2 y[0] + zy[1] + zy[0]}{z^2 + z + 0.25} = \frac{z^2 + 3z}{z^2 + z + 0.25} = \frac{z}{z + 0.5} + \frac{\frac{5}{2}z}{(z + 0.5)^2}$$

The inverse $\mathcal{Z}$-transform leads to the zero-input response

$$y_{zi}^d[k] = \mathcal{Z}^{-1}\{Y_{zi}^d(z)\} = (1 - 5k)\left(-\frac{1}{2}\right)^k u[k]$$

246 Chapter 5 $\mathcal{Z}$-transform

Note that the initial conditions $y[0] = 1$ and $y[1] = 2$ have been obtained by propagating forward through the difference equation the initial conditions defined in Example 5.22 for the corresponding integral formulation. This has been done as follows. For $k = 0$ and $k = 1$, we have

$$y[0] + y[-1] + 0.25y[-2] = 0 \Rightarrow y[0] = -(-12) - 0.25(44) = 1$$

and

$$y[1] + y[0] + 0.25y[-1] = 0 \Rightarrow y[-2] = -1 - 0.25(-12) = 2$$

Hence, we have $y_{zi}^d[k] = y_{zi}^i[k]$, with $y_{zi}^i[k]$ derived in Example 5.22.

Complete Response

We can obtain the complete system response for the derivative formulation by using (5.68), whose time domain solution is given by

$$y[k] = \mathcal{Z}^{-1}\{Y(z)\} = \mathcal{Z}^{-1}\{Y_{zs}^d(z) + Y_{zi}^d(z)\}$$
$$= \mathcal{Z}^{-1}\left\{\left(H(z)F(z) - \frac{I_f(z)}{\Delta(z)}\right)\right\} + \mathcal{Z}^{-1}\left\{\frac{I_y^d(z)}{\Delta(z)}\right\} = y_{zs}^d[k] + y_{zi}^d[k] \quad (5.73)$$

In the following examples we find the complete system response using the derivative formulation, and show how to derive the initial conditions for the corresponding integral formulation considered in Example 5.23.

EXAMPLE 5.26

Let us find the complete response of the discrete-time system given in the derivative formulation

$$y[k+2] - y[k+1] + \frac{1}{4}y[k] = f[k], \quad f[k] = (-2)^k u[k], \quad y[0] = 1, \quad y[1] = 2$$

By taking the $\mathcal{Z}$-transform of both sides of this equation, we obtain

$$z^2 Y(s) - z^2 y[0] - zy[1] - zY(z) + zy[0] + \frac{1}{4}Y(z) = \frac{z}{z+2}$$

which implies

$$Y(z) = \frac{1}{(z^2 - z + 0.25)(z+2)} + \frac{z^2 + z}{z^2 - z + 0.25}$$

Expanding into partial fractions the function

$$\frac{1}{z}Y(z) = \frac{1}{(z-0.5)^2(z+2)} + \frac{z+1}{(z-0.5)^2}$$

we have

$$\frac{1}{z}Y(z) = \frac{21}{25}\frac{1}{z-0.5} + \frac{19}{10}\frac{1}{(z-0.5)^2} + \frac{76}{9}\frac{1}{z+2}$$

Hence, the complete system response is obtained as

$$y[k] = \mathcal{Z}^{-1}\{Y(z)\} = \mathcal{Z}^{-1}\{Y_{zs}^d(z)\} + \mathcal{Z}^{-1}\{Y_{zi}^d(z)\}$$

$$= \mathcal{Z}^{-1}\left\{\frac{21}{25}\frac{z}{(z-0.5)} + \frac{19}{10}\frac{z}{(z-0.5)^2}\right\} + \mathcal{Z}^{-1}\left\{\frac{76}{9}\frac{z}{z+2}\right\}$$

$$= y_{zs}^d[k] + y_{zi}^d[k] = \left(\frac{21}{25}(0.5)^k + \frac{19}{5}k(0.5)^k\right)u[k] + \frac{76}{9}(-2)^k u[k], \quad k \geq 2$$

We could have solved the same problem using the integral formulation obtained by shifting the derivative formulation backwards by two discrete time instants, that is,

$$y[k] - y[k-1] + \frac{1}{4}y[k-2] = f[k-2]$$

and determining the initial conditions $y[-1]$ and $y[-2]$. The new initial conditions must be consistently determined using the known initial conditions and propagating them backwards through the original difference equations. In this case, for $k = -1$, the derivative formulation implies

$$y[1] - y[0] + \frac{1}{4}y[-1] = f[-1] = 0 \Rightarrow y[-1] = 4(y[0] - y[1]) = -4$$

Similarly, for $k = -2$ we obtain the second initial condition as

$$y[0] - y[-1] + \frac{1}{4}y[-2] = f[-2] = 0 \Rightarrow y[-2] = 4(y[-1] - y[0]) = -20$$

The preceding integral formulation with the initial conditions $y[-2] = -20$ and $y[-1] = -4$ was considered in Example 5.23. It is clear that the complete system response is identical for both formulations, since they are obtained from each other by propagating the initial conditions. ∫

EXAMPLE 5.27

Consider the case in which the discrete-time system differentiates the input signal, that is,

$$y[k+2] - 3y[k+1] + 2y[k] = f[k+1], \quad f[k] = (3)^k u[k], \quad y[0] = 0, \quad y[1] = -4$$

By taking the $\mathcal{Z}$-transform of both sides of this equation, we have

$$z^2 Y(z) - z^2 y[0] - zy[1] - 3(zY(z) - zy[0]) + 2Y(z) = zF(z) - zf[0]$$

Since $F(z) = z/(z-3)$ and $f[0] = 1$, the equation simplifies into

$$(z^2 - 3z + 2)Y(z) = z(-4) + z\frac{z}{z-3} - z$$

which implies

$$Y(z) = \frac{-5z}{z^2 - 3z + 2} + \frac{z}{(z^2 - 3z + 2)} \frac{z}{(z - 3)}$$

Expanding into partial fractions the function

$$\frac{1}{z}Y(z) = \frac{-5}{z^2 - 3z + 2} + \frac{1}{(z^2 - 3z + 2)} \frac{z}{(z - 3)}$$

we get

$$\frac{1}{z}Y(z) = \frac{5}{z - 1} - \frac{5}{z - 2} + \frac{0.5}{z - 1} - \frac{2}{z - 2} + \frac{1.5}{z - 3}$$

Hence, the complete system response is obtained as

$$y[k] = \mathcal{Z}^{-1}\{Y(z)\} = \left(\frac{11}{2} - 7(2)^k + \frac{3}{2}(3)^k\right) u[k]$$

We showed in Example 5.18 that for this system the integral formulation is given by

$$y[k] - 3y[k-1] + 2y[k-2] = f[k-1], \quad y[-2] = \frac{15}{4}, \quad y[-1] = \frac{5}{2}$$

By applying $\mathcal{Z}$-transform properties (5.11a) and (5.11b), we obtain

$$Y(z) - \frac{3}{z}Y(z) - 3y[-1] + \frac{2}{z^2}Y(z) + \frac{2}{z}y[-1] + 2y[-2] = \frac{1}{z}F(z)$$

or

$$Y(z) = \frac{z}{z^2 - 3z + 2} F(z) + \frac{z^2 (3y[-1] - 2y[-2] - 2y[-1]/z)}{z^2 - 3z + 2}$$

Substituting $F(z)$ and the values for $y[-1]$ and $y[-2]$, we obtain exactly the same expression for $Y(z)$ as in the case of the derivative formulation.

It is interesting to point out that the same solution is obtained for the mixed derivative-integral formulation defined by

$$y[k+1] - 3y[k] + 2y[k-1] = f[k], \quad y[-1] = \frac{5}{2}, \quad y[0] = 0$$

Using $\mathcal{Z}$-transform properties (5.11a) and (5.12), we have

$$zY(z) - zy[0] - 3Y(z) + \frac{2}{z}Y(z) + y[-1] = F(z)$$

or

$$Y(z) = \frac{z}{z^2 - 3z + 2} F(z) + \frac{z(-2y[-1])}{z^2 - 3z + 2} = \frac{z}{(z^2 - 3z + 2)} \frac{z}{(z - 3)} - \frac{5z}{z^2 - 3z + 2}$$

which is exactly the same expression for $Y(z)$ as that obtained using either the derivative or integral formulations.

For further studies of the use of the $\mathcal{Z}$-transform in the analysis of linear discrete-time dynamic systems described by linear difference equations, interested students are referred to the very comprehensive books [9–10].

In concluding this section, we point out that we have demonstrated the use of the $\mathcal{Z}$-transform in system analysis on lower-dimensional systems (order two or three), since for higher-dimensional systems the algebra gets cumbersome. We will come back to this point in Chapters 7 and 8, where we will present, respectively, the time domain technique for solving difference equations and the discrete-time state space technique. The state space technique can deal with systems of very high dimensions (several hundreds or even several thousands).

In the next section we will demonstrate the use of the $\mathcal{Z}$-transform for *solving systems of linear difference equations* with the example of an ATM computer communication network switch introduced in Section 1.3. The application of the $\mathcal{Z}$-transform to a system of linear difference equations produces in the complex domain a system of linear algebraic equations, which can be solved rather easily. The solution obtained in the z-domain is then converted to the time domain via the inverse $\mathcal{Z}$-transform.

5.3.4 Case Study: An ATM Computer Network Switch

An ATM computer communication network switch was introduced in Section 1.3, formula (1.41). Its mathematical model, under the simplifying idealistic assumption that there is no time delay between the source and the switch ($d = 0$), is easily obtained from (1.41) as

$$q[k+1] = q[k] + y[k+1] - f[k]$$
$$y[k+1] = y[k] - \alpha_0(q[k] - q^0) - \alpha_1(q[k-1] - q^0) \quad (5.74)$$

where $q[k]$ is the data packet queue length (stored in the buffer), q^0 is the desired queue length steady state value, $y[k]$ is the data packet arrival rate into the switch, $f[k]$ is the service rate of the buffer, and α_j, $j = 0, 1$, are gains to be chosen by a network engineer to assure a smooth passage of data packets through the network (and to satisfy additional requirements such as network stability and elimination of network congestion). It is usually assumed that the service rate $f[k] = \mu$ is constant.

The mixed integral-derivative formulation given in (5.74) can be replaced by the integral formulation as follows:

$$q[k] = q[k-1] + y[k] - f[k-1]$$
$$y[k] = y[k-1] - \alpha_0(q[k-1] - q^0) - \alpha_1(q[k-2] - q^0) \quad (5.74a)$$

If we introduce the variable $e[k-i] = q[k-i] - q^0$, $i = 0, 1, 2$, which represents the queue length deviation from its desired steady state value, we have

$$e[k] = e[k-1] + y[k] - f[k-1]$$
$$y[k] = y[k-1] - \alpha_0 e[k-1] - \alpha_1 e[k-2] \quad (5.75)$$

First, we find the ATM switch transfer functions. Note that there are two transfer functions in this system, since we have one input, $f[k]$, and two outputs $e[k]$ and $y[k]$.

Assuming that all initial conditions are equal to zero and applying the $\mathcal{Z}$-transform to (5.75), we obtain a system of two algebraic equations,

$$E(z) = \frac{1}{z}E(z) + Y(z) - \frac{1}{z}F(z)$$

$$Y(z) = \frac{1}{z}Y(z) - \frac{\alpha_0}{z}E(z) - \frac{\alpha_1}{z^2}E(z)$$

The first equation can be solved with respect to $E(z)$,

$$E(z) = \frac{z}{z-1}Y(z) - \frac{1}{z-1}F(z) \tag{5.76}$$

Substituting this expression into the second equation, we have

$$z(z-1)Y(z) = -\frac{\alpha_0 z + \alpha_1}{z} \frac{z}{z-1} Y(z) + \frac{\alpha_0 z + \alpha_1}{z} \frac{1}{z-1} F(z)$$

From this relation we obtain the expression for the transfer function between the system input $F(z)$ and the system output $Y(z)$,

$$Y(z) = \frac{\alpha_0 z + \alpha_1}{z^3 + (\alpha_0 - 2)z^2 + (\alpha_1 + 1)z} F(z) = H_Y(z) F(z)$$

The second transfer function between $F(z)$ and $E(z)$ can be easily obtained by eliminating $Y(z)$ from (5.76) using the last expression, which leads to

$$E(z) = \frac{z}{z-1} H_Y(z) F(z) - \frac{1}{z-1} F(z)$$

$$= \left(z \frac{\alpha_0 z + \alpha_1}{z^3 + (\alpha_0 - 2)z^2 + (\alpha_1 + 1)z} - 1 \right) \frac{1}{z-1} F(z)$$

$$= \frac{-z^2 + 2z - 1}{[z^2 + (\alpha_0 - 2)z + (\alpha_1 + 1)](z - 1)} F(z)$$

$$= \frac{-(z-1)}{z^2 + (\alpha_0 - 2)z + (\alpha_1 + 1)} = H_E(z) F(z)$$

In summary, the transfer functions obtained are given by

$$H_Y(z) = \frac{\alpha_0 z + \alpha_1}{z^3 + (\alpha_0 - 2)z^2 + (\alpha_1 + 1)z}$$

$$H_E(z) = \frac{-(z-1)}{z^2 + (\alpha_0 - 2)z + (\alpha_1 + 1)} \tag{5.77}$$

In the following, we will find the ATM switch response when the buffer service rate is equal to $\mu = 10$ packets per time unit, that is, when $f[k] = 10u[k]$. Note that this is also an idealistic assumption; if the buffer has less than 10 packets, that is $q[k] < 10$, then the service rate is equal to $f[k] = q[k] < 10$ (see Problem 5.49). Before we proceed, we must choose the values for the constants α_0 and α_1 such that the ATM switch can store and forward packets without buffer overflow (without losing packets). As pointed out in Section 1.3, network engineers must provide the values for the coefficients α_j and β_i to assure smooth packet flow through the network. In the general case, that is not an easy task and rather sophisticated techniques must be used. In our idealistic case, obtained by neglecting time delays in the network ($d = 0$), we need choose only two parameters to achieve that goal. By guessing several value for α_j we have found that $\alpha_0 = 2.5$ and $\alpha_1 = -0.5$ produce satisfactory results. The transfer functions are given by

$$H_Y(z) = \frac{2.5z - 0.5}{z^3 + 0.5z^2 + 0.5z}, \quad H_E(z) = \frac{-(z-1)}{z^2 + 0.5z + 0.5}$$

The corresponding responses are found via MATLAB using the following code. The results obtained are plotted in Figures 5.8 and 5.9.

```
% ATM Switch - Idealized (d=0)
numy=[2.5 -0.5]; deny=[1 0.5 0.5 0];
nume=[-1 1]; dene=[1 0.5 0.5];
k=0:1:30;
ystep=dstep(10*numy,deny,k);
figure (1); plot(k, ystep,'*')
xlabel('Discrete time');
ylabel('Packet arrival rate');
grid; print -deps figure5-8.eps
estep=dstep(10*nume,dene,k);
figure (2); plot(k, estep,'*')
xlabel('Discrete time');
ylabel('Deviations from the desired queue length');
grid; print -deps figure5-9.eps
```

It can be observed from Figure 5.9 that for the proposed set of coefficients, the deviation of the buffer queue length from the desired value goes to zero rather quickly, irrespective of the desired value for q^0, which is a nice feature.

Note that we have solved this problem under the assumption that all network initial conditions are equal to zero. In Problem 5.48, students will be asked to solve the same problem under nonzero initial conditions.

The above example can be also simulated using Simulink. (In Appendix D, we provide more details about building Simulink block diagrams and running simulations.) Let us indicate here that the blocks with the transfer function $1/z$ provide a time delay of one unit. These blocks can be opened by double-clicking on the left mouse button, and initial conditions can be set to nonzero values. Note that in order to run simulations for discrete-time systems, the step size in the simulation menu called "parameters" should be

252 Chapter 5 $\mathcal{Z}$-transform

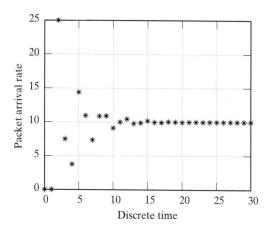

FIGURE 5.8: ATM switch arrival rate

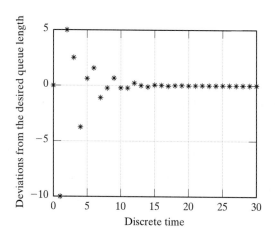

FIGURE 5.9: Deviation of ATM switch buffer queue length from desired value

set to "fixed-step". Of course, the simulation results obtained using Simulink are identical to those obtained using MATLAB (presented in Figures 5.8 and 5.9). If one reads the data directly from the oscilloscope, the discrete-time data will be presented using the piecewise staircase representation discussed in Example 5.20.

The Simulink block diagram presented in Figure 5.10 simulates the integral formulation of the ATM discrete-time model given in (5.74a), which corresponds to a new set of parameters, namely, $\alpha_0 = 0.6$ and $\alpha_1 = -0.5$. This choice of parameters also gives satisfactory results regarding the steady state values for the queue length and the arrival rate. The block diagram outputs produce information about the queue length and the arrival rate at any discrete-time instant. Students will be asked in Problem 5.50 to build this block diagram and provide the corresponding simulation results. Note that

Section 5.3 Z-transform in Linear System Analysis 253

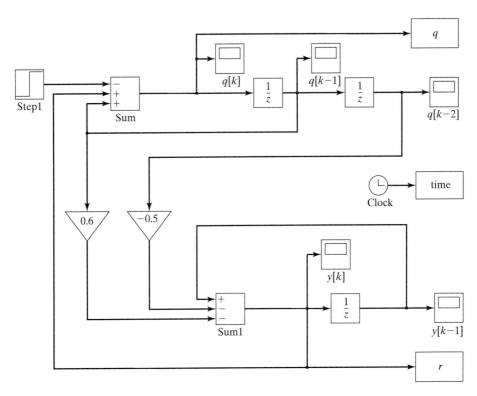

FIGURE 5.10: Simulink block diagram for a simplified ATM switch model

Figure 5.10 is generated using the MATLAB command (typed in the MATLAB window) print -s -deps figure5-10.eps, while keeping the corresponding Simulink block diagram open.

In this ATM model, we have used drastic simplifying assumptions. First of all, we have assumed that the round-trip communication time delay is equal to zero so that we end up with a model of order three. If we use real data, we will find that the dimension of the ATM switch model is much higher. Assume that the data packet propagation time is 5 μs/km (which is typical for an optical fiber transmission link), and that the source is 1000 km from the switch and connected by a link whose speed is 100 Mb/s. This corresponds to a round-trip time delay of $\tau_d = 10$ ms. Let the update time interval be equal to 200 data packet times (note that an ATM packet has 53 Bytes), which implies $T = (200 \times 53 \times 8)/(100 \times 10^6) = 0.848$ ms. Hence, $\tau_d/T \approx 12 = d$ discrete-time units. The discrete-time linear system that describes the ATM switch corresponding to the preceding data is of order at least twelve (see formula (1.41)). It will be shown in Chapter 8 that the system of difference equations (1.41) with $d = 12$ determines the linear time-invariant discrete-time system of order 14.

5.4 BLOCK DIAGRAMS

In formula (5.47) we defined the discrete-time linear time-invariant system transfer function. Using the $\mathcal{Z}$-transform linearity and convolution properties, we can easily extend the concept of the transfer function to configurations of several connected linear systems. To that end, we will find the equivalent transfer functions for cascade and parallel connections of systems, introduce the feedback (closed-loop) configuration, and define the corresponding feedback system transfer function. The transfer function defined in (5.47) in fact represents the *open-loop* discrete-time system transfer function.

It follows from formula (5.53) that for a system at rest, the system input $F(z)$ produces in the system output the signal $Y_{zs}(z)$, given by

$$Y_{zs}(z) = H(z)F(z) \qquad (5.78)$$

which is symbolically represented in Figure 5.11 using a block diagram.

Note that this block diagram can also be used in the case when the system initial conditions are nonzero. In such a case, an additive component coming from the system initial conditions should be added to the system output. For that reason, in all block diagrams presented in this section, we will denote the system output by $Y(z)$.

The open-loop system transfer function $H(z)$ is derived from a difference equation that is obtained using known physical laws (from a mathematical modeling procedure). The accuracy of coefficients that appear in the open-loop system transfer function depends on the accuracy of the system coefficients. These coefficients are not always perfectly known. Furthermore, the coefficients change, due to aging or internal and/or external system disturbances. Due to changes in the system coefficients (system parameters), it can happen that the actual system output (in the open-loop system configuration) is significantly different from the output obtained analytically.

It was indicated in Section 4.4 on the continuous-time block diagrams that a way to cope with system parameter changes, and a way to reduce the impact of those changes on the system output, is to form the *closed-loop* system configuration, also known as the system feedback configuration. Assuming that it is feasible, we can feed the system output back and form a *closed-loop* around the system, as presented in Figure 5.12(a). The directed path (as indicated by the arrows) from $F(z)$ to $Y(z)$ is called the *forward path*, and the directed path from $Y(z)$ to $E(z)$ is called the *feedback path*. Such a feedback loop is called a unity feedback loop. In general, we can put a dynamic element $G(z)$ (another open-loop transfer function) in the feedback loop, as presented in Figure 5.12(b). For notational convenience, we will denote the transfer function in the feedback path by $G(z)$ and the transfer function in the forward path by $H(z)$. It should be pointed out that sometimes we put a static element in the feedback path equal to a constant, that is $G(z) = $ const.

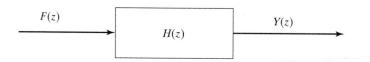

FIGURE 5.11: An open-loop system transfer function

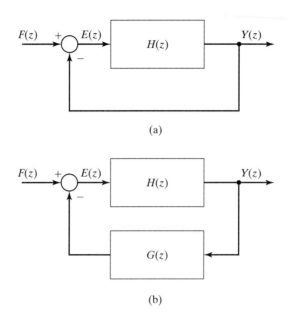

FIGURE 5.12: Closed-loop system configurations: (a) unity feedback; (b) nonunity feedback

In the feedback configuration presented in Figure 5.12(a), the output signal is fed back and compared to the input signal, and the difference between the input and output signals is used as a new input signal to the system. In practice, the feedback signal is taken with the negative sign since, in general, positive feedback signals cause system instability. Using (5.78) and following signals in the block diagram in the direction of the arrows, we can find the closed-loop system transfer function from $F(z)$ to $Y(z)$, assuming zero initial conditions, as follows

$$Y(z) = H(z)E(z), \quad E(z) = F(z) - Y(z)$$
$$\Rightarrow Y(z) = H(z)(F(z) - Y(z))$$
$$\Rightarrow Y(z) = \frac{H(z)}{1 + H(z)} F(z) \triangleq M(z) F(z)$$

The *closed-loop system transfer function* for unity feedback, denoted by $M(z)$, is given by

$$M(z) = \frac{Y(z)}{F(z)}\bigg|_{\text{I.C.}=0} \triangleq \frac{H(z)}{1 + H(z)} \qquad (5.79)$$

The function defined in (5.79) is called the closed-loop transfer function with unity feedback. In many applications, another transfer function is present in the feedback loop (see Figure 5.12(b)). The closed-loop transfer function with nonunity feedback is obtained similarly, as follows:

$$Y(z) = H(z)(F(z) - G(z)Y(z)) \Rightarrow Y(z) = \frac{H(z)}{1 + H(z)G(z)} F(z)$$

$$Y(z) \triangleq M(z)F(z), \quad M(z) = \frac{H(z)}{1 + H(z)G(z)} \tag{5.80}$$

We can form more complex configurations of open-loop transfer functions. In general, open-loop transfer functions can be connected in cascade or parallel, or they can form more complex feedback configurations containing several feedback loops.

The *cascade connection* of open-loop transfer functions is presented in Figure 5.13(a). It is easy to conclude that the equivalent open-loop transfer function is given by the product of elementary open-loop transfer functions,

$$H_{eq}^{cascade}(z) = H_1(z)H_2(z) \cdots H_n(z) \tag{5.81}$$

Formula (5.81) can be called the product rule for elementary open-loop transfer functions.

The *parallel connection* of the open-loop transfer functions is represented in Figure 5.13(b). Its equivalent open-loop transfer function is equal to the sum of the elementary open-loop transfer functions, that is,

$$H_{eq}^{parallel}(z) = H_1(z) \pm H_2(z) \pm \cdots \pm H_n(z) \tag{5.82}$$

The last formula can be called the sum rule for elementary open-loop transfer functions.

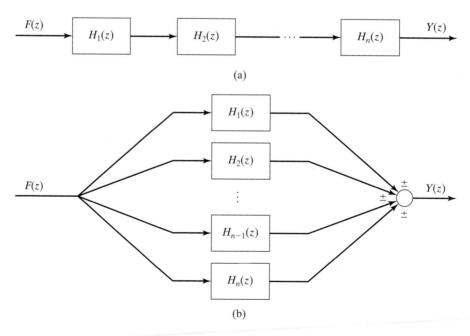

FIGURE 5.13: (a) Cascade and (b) parallel connections of transfer functions

Using the basic transfer function formulas (rules) established in (5.78–82), we can simplify more complex feedback systems and represent them in the basic feedback form presented in Figure 5.12(b). The corresponding procedure is demonstrated in the next example.

EXAMPLE 5.28

Consider a feedback system as given in Figure 5.14.

We find its closed-loop transfer function as follows. Using the product rule, we have

$$E_2(z) = H_1(z)H_2(z)E_1(z)$$

$$Y(z) = H_3(z)E_3(z)$$

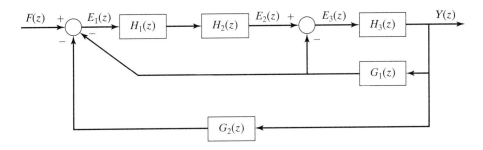

FIGURE 5.14: A feedback system

The product rule combined with the sum rule produces

$$E_1(z) = F(z) - G_1(z)Y(z) - G_2(z)Y(z)$$
$$E_3(z) = E_2(z) - G_1(z)Y(z)$$

Eliminating $E_i(z)$, $i = 1, 2, 3$, we obtain

$$Y(z) = H_1(z)H_2(z)H_3(z)F(z) - G_1(z)H_3(z)Y(z)$$
$$- (G_1(z) + G_2(z))H_1(z)H_2(z)H_3(z)Y(z)$$

This leads to the closed-loop transfer function of the form

$$M(z) = \frac{Y(z)}{F(z)} = \frac{H_1(z)H_2(z)H_3(z)}{1 + G_1(z)H_3(z) + (G_1(z) + G_2(z))H_1(z)H_2(z)H_3(z)}$$

The algebra associated with transfer functions is simple and convenient for linear feedback systems composed of several loops. In the case of linear systems with many feedback loops, finding the closed-loop transfer function becomes a very tedious task. In such a case, we can use Mason's formula, obtained using elementary knowledge of graph theory. For more information about that formula, the interested reader should consult undergraduate texts on control systems (see, for example, [11]).

5.5 DISCRETE-TIME FREQUENCY SPECTRA

The frequency spectra of a discrete-time system are the frequency plots of the magnitude and phase of its transfer function with respect to $z|_{\sigma=0} = e^{(\sigma+j\omega)T}|_{\sigma=0} = e^{j\omega T}$, that is,

$$H(z)|_{z=e^{j\omega T}} = H(e^{j\omega T})$$
$$= \text{Re}\{H(e^{j\omega T})\} + j\,\text{Im}\{H(e^{j\omega T})\} = |H(e^{j\omega T})|\angle H(e^{j\omega T}) \quad (5.83)$$

where

$$|H(e^{j\omega T})| = \sqrt{(\text{Re}\{H(e^{j\omega T})\})^2 + (\text{Im}\{H(e^{j\omega T})\})^2} \quad (5.84)$$

and

$$\angle H(e^{j\omega T}) = \tan\left\{\frac{\text{Im}\{H(e^{j\omega T})\}}{\text{Re}\{H(e^{j\omega T})\}}\right\} \quad (5.85)$$

The plot $|H(e^{j\omega T})|$ with respect to frequency defines the *discrete-time system magnitude spectrum*, and the plot $\angle H(e^{j\omega T})$ with respect to frequency defines the *discrete-time system phase spectrum*.

One of the major differences between continuous- and discrete-time linear systems comes from the properties of their spectra. In the continuous-time domain, the magnitude and phase spectra are plotted for a continuum of frequencies from $\omega = 0$ to $\omega = \infty$ (theoretically, from $\omega = -\infty$ to $\omega = \infty$). However, in the discrete-time domain, we need to plot both the magnitude and phase spectra from $\omega = 0$ to $\omega = 2\pi$ only, due to the fact that both are 2π periodic functions of frequency ω. Note that from (5.1) we have

$$z|_{\sigma=0} = e^{(\sigma+j\omega)T}|_{\sigma=0} = e^{j\omega T} = e^{j(\omega T+2\pi)} \quad (5.86)$$

since $e^{j2\pi} = 1$. Hence, the arguments of the functions $|H(e^{j\omega T})|$ and $\angle H(e^{j\omega T})$ are 2π periodic, which implies that the functions themselves are 2π periodic. This *periodicity property* of the discrete-time linear system magnitude and phase spectra can be analytically recorded as

$$|H(e^{j\omega T})| = |H(e^{j(\omega T+2\pi)})| \quad (5.87)$$

and

$$\angle H(e^{j\omega T}) = \angle H(e^{j(\omega T+2\pi)}) \quad (5.88)$$

Similarly, for a given discrete-time *signal* $f[k]$, we can define the discrete-time *signal* spectra by first finding $F(z) = \mathcal{Z}\{f[k]\}$ and evaluating its magnitude and phase, that is,

$$F(z)|_{\sigma=0} = F(e^{j\omega T}) = |F(e^{j\omega T})|\angle F(e^{j\omega T}) \quad (5.89)$$

with respect to frequency in the range $\omega \in [0, 2\pi]$. The corresponding frequency plot $|F(e^{j\omega T})|$ represents the *discrete-time signal magnitude spectrum*, and the frequency plot $\angle F(e^{j\omega T})$ is the *discrete-time signal phase spectrum*.

We can plot the magnitude and phase frequency spectra in the discrete-time domain by using the MATLAB function `freqz`, as demonstrated in the next example.

EXAMPLE 5.29

Consider a discrete-time system represented by the transfer function

$$H(z) = \frac{z(z+0.5)}{(z-0.3)(z+0.3)(z+1)}$$

The following MATLAB script finds both the magnitude and phase spectra of this system.

```
num=[1 0.5 0]; den=[1 1 -0.09 -0.09];
w=0:0.01:4*pi;
freqz(num,den,w)
```

The corresponding normalized spectra, generated by MATLAB, are presented in Figure 5.15. Note that normalization is obtained by dividing the horizontal axis by π so that the span from 0 to 4 in fact represents the span from 0 to 4π. It is obvious from this figure that both the system magnitude and phase spectra are 2π periodic functions.

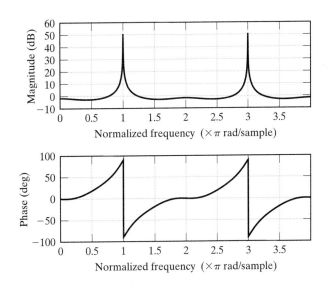

FIGURE 5.15: Magnitude and phase spectra of a discrete-time system

The magnitude and phase signal spectra and their study are fundamentally important for digital signal processing and digital communications. In linear control systems, discrete-time Bode diagrams are used. Discrete-time Bode diagrams are represented by the plots of $20\log_{10}(|H(e^{j\omega T})|)$ and $\angle H(e^{j\omega T})$ with respect to $\log_{10}(\omega)$ [12]; they can be plotted using the MATLAB function dbode as dbode(num,den,T), where num and den contain coefficients of the system transfer function and T represents the value of the sampling period.

5.5.1 System Response to Sinusoidal Inputs

In this section we will derive a general formula for the discrete-time linear system response to sinusoidal inputs. Consider a system at rest, and assume that its input is a hypothetical signal defined by $f[k] = z^k$. Then, by the convolution formula (5.54), the response of a linear discrete-time system characterized by the impulse response $h[k]$ to this signal is given by

$$y_{zs}[k] = h[k] * z^k = \sum_{m=0}^{\infty} h[m] z^{k-m} = z^k \sum_{m=0}^{\infty} h[m] z^{-m} = z^k H(z) \quad (5.90)$$

If we take $z = e^{j\omega T}$, it follows from (5.90) that the system response due to the input signal $f_1[k] = (e^{j\omega T})^k = e^{jk\omega T}$ is given by $e^{jk\omega T} H(e^{j\omega T})$. Similarly, the system response due to $f_2[k] = (e^{-j\omega T})^k = e^{-jk\omega T}$ is given by $e^{-jk\omega T} H(e^{-j\omega T})$. A cosine signal can be represented using Euler's formula as

$$\cos(k\omega_0 T) = \frac{1}{2}(e^{jk\omega_0 T} + e^{-jk\omega_0 T}) = \frac{1}{2}(f_1[k] + f_2[k])$$

Applying the linearity principle, we see that the system response due to the cosine signal $\cos(k\omega_0 T)$ is

$$\frac{1}{2}(e^{jk\omega_0 T} H(e^{j\omega_0 T}) + e^{-jk\omega_0 T} H(e^{-j\omega_0 T})) = \frac{1}{2} \times 2 \times \text{Re}\{e^{jk\omega_0 T} H(e^{j\omega_0 T})\}$$

$$= \text{Re}\{e^{jk\omega_0 T} |H(e^{j\omega_0 T})| e^{j\angle H(e^{j\omega_0 T})}\}$$

$$= |H(e^{j\omega_0 T})| \text{Re}\{e^{j(k\omega_0 T + \angle H(e^{j\omega_0 T}))}\}$$

$$= |H(e^{j\omega_0 T})| \cos(k\omega_0 T + \angle H(e^{j\omega_0 T}))$$

$$(5.91)$$

Formula (5.91) indicates the way in which the system processes the input cosine signal. Namely, the system multiplies the incoming cosine signal by $|H(e^{j\omega_0 T})|$ and introduces a phase shift of $\angle H(e^{j\omega_0 T})$, while keeping the cosine frequency unchanged.

In the case of the phase-shifted input sinusoidal signal $f[k] = \cos(k\omega_0 T + \varphi)$, the derivations remain practically the same, which leads to

$$y_{zs}[k] = |H(e^{j\omega_0 T})| \cos(k\omega_0 T + \varphi + \angle H(e^{j\omega_0 T})) \quad (5.92)$$

It can be observed from (5.92) that a sinusoidal input does not excite system natural modes, so that the zero-state response obtained in (5.92) represents at the same time the system *steady state response*, that is, $y_{ss}[k] = y_{zs}[k]$. By analogy with the continuous-time system natural modes considered in Section 3.4.1, the discrete-time system natural modes have the form p_i^k, $i = 1, 2, \ldots, n$, where the p_i are the system poles found by solving the algebraic equation $\Delta(p) = 0$, where $\Delta(p)$ is the system characteristic polynomial defined by (5.67).

EXAMPLE 5.30

Consider the input signal $f[k] = \cos(10kT + \pi/4)$, which passes through a linear discrete-time system defined by

$$H(z) = \frac{z+1}{(z-0.5)(z+0.5)} = \frac{z+1}{z^2-1}$$

Assume that the sampling time is $T = 0.1$, hence $\omega T = 1$ rad. The magnitude of the transfer function is given by

$$|H(e^{j\omega T})| = \left|\frac{e^{j\omega T}+1}{e^{j2\omega T}-1}\right| = \frac{\sqrt{(1+\cos(\omega T))^2 + (\sin(\omega T))^2}}{\sqrt{(-1+\cos(2\omega T))^2 + (\sin(2\omega T))^2}}$$

which leads to

$$|H(e^{j1})| = \frac{\sqrt{(1+\cos(1))^2 + (\sin(1))^2}}{\sqrt{(-1+\cos(2))^2 + (\sin(2))^2}} = \frac{1.7552}{1.6829} = 1.043$$

The phase of the transfer function is

$$\angle H(e^{j\omega T}) = \tan^{-1}\left(\frac{\sin(\omega T)}{1+\cos(\omega T)}\right) - \tan^{-1}\left(\frac{\sin(2\omega T)}{-1+\cos(2\omega T)}\right)$$

producing

$$\angle H(e^{j1}) = \tan^{-1}\left(\frac{\sin(1)}{1+\cos(1)}\right) - \tan^{-1}\left(\frac{\sin(2)}{-1+\cos(2)}\right) = 1.071 \text{ rad}$$

Hence, the system response due to the original input signal is given by

$$y_{zs}[k] = 1.043 \times \cos\left(k \times 1 + \frac{\pi}{4} + 1.071\right)$$

Note that the argument of the cosine function is given in radians.

5.6 $\mathcal{Z}$-TRANSFORM MATLAB LABORATORY EXPERIMENT

Purpose: This experiment presents the frequency domain analysis of discrete-time systems using MATLAB. The impulse, step, sinusoidal, and exponential responses of discrete-time systems will be examined using the system transfer function method based on the $\mathcal{Z}$-transform. In addition, MATLAB will be used to perform the partial fraction expansion and to find the inverse $\mathcal{Z}$-transform. Note that MATLAB *uses only the integral representation of linear discrete-time systems*.

Part 1. Consider the linear discrete-time system represented by the system transfer function

$$H(z) = \frac{z - 0.5}{z(z + \frac{1}{2})(z - \frac{1}{3})(z + \frac{1}{4})}$$

Using MATLAB, find and plot:

(a) The impulse response of the system.
(b) The step response of the system.
(c) The zero-state response due to the input signal $f[k] = \sin[2k]u[k]$.
(d) The zero-state response due to the input signal $f[k] = (-2)^k u[k]$.

Part 2. Consider the system transfer function

$$H(z) = \frac{z^4 - 3z^3 + 5z^2 + 2z}{2z^7 + 5z^5 - 3 + z^3 - 2z^2 + 3z - 1}$$

(a) Find the factored form of the transfer function using the MATLAB function `[z,p,k]=tf2zp(num,den)`.
(b) The partial fraction expansion of rational functions can be performed using the MATLAB function `residue`. Find the inverse $\mathcal{Z}$-transform of the given transfer function using `residue`, that is, find analytically the system impulse response. Plot the system impulse response.
(c) Repeat appropriately the steps outlined in (b) such that the system step response is obtained.

Part 3. Find and plot the zero-input response of the system whose transfer function is given in Part 2, and whose initial conditions are given by

$$y[-4] = 1, \quad y[j] = 0, \quad j = -7, -6, -5, -3, -2, -1$$

(*Hint:* Find $I_y^i(z)$ and $\Delta(z)$ as defined in formulas (5.45) and (5.67), respectively. Then, use the MATLAB function `dimpulse` with the coefficients of $I_y^i(z)$ and $\Delta(z)$ representing, respectively, the corresponding numerator and denominator coefficients.)

Submit four plots for Part 1, one plot for Part 2, and one plot for Part 3, and present analytical results obtained in Parts 2–3.

Comment: Instructors can design additional MATLAB laboratory experiments or additional parts of laboratory experiments by selecting from the problem section (Section 5.9) any set of problems that require the use of MATLAB or Simulink. (This comment is in general applicable to any chapter in this text.)

5.7 SUMMARY

Study Guide for Chapter Five: Students must master the properties of the $\mathcal{Z}$-transform and become familiar with the table of common pairs. The use of the $\mathcal{Z}$-transform in the analysis of linear time-invariant systems is the main thrust of this chapter. Knowledge of the techniques for finding impulse, step, zero-state, zero-input, and complete system responses is essential. Standard problems: (1) Find the $\mathcal{Z}$-transform of a given discrete-time signal; (2) find the inverse $\mathcal{Z}$-transform of a given frequency domain signal; (3) state and prove the properties of the $\mathcal{Z}$-transform; (4) find the system transfer function and the system impulse response; (5) find the system step response; (6) find the zero-input response; (7) find the zero-state response; (8) find the complete system response; (9) find the discrete-time system response to sinusoidal inputs.

The $\mathcal{Z}$-transform and Its Inverse:

$$F(z) = \mathcal{Z}\{f[k]\} \triangleq \sum_{k=0}^{\infty} f[k] z^{-k}$$

$$f[k] \triangleq \frac{1}{2\pi j} \oint_\Gamma F(z) z^{k-1} \, dz$$

Partial Fraction Expansion Simple Pole Coefficients:

$$c_i = \lim_{z \to p_i} \left\{ (z - p_i) \frac{F(z)}{z} \right\}, \quad i = 1, 2, \ldots, n$$

Partial Fraction Expansion Multiple Pole Coefficients:

$$c_{ij} = \frac{1}{(r-j)!} \lim_{s \to p_i} \left\{ \frac{d^{r-j}}{dz^{r-j}} \left[(z - p_i)^r \frac{F(z)}{z} \right] \right\}$$

$$j = r, r-1, \ldots, 2, 1; \quad i = 1, 2, \ldots, n$$

p_i is a multiple pole of multiplicity r

Formula for the $\mathcal{Z}^{-1}$ Inverse of a Pair of Complex Conjugate Poles:

$$\frac{c_i z}{z - p_i} + \frac{c_i^* z}{z - p_i^*} \leftrightarrow 2 |c_i| |p_i|^k \cos(k \angle p_i + \angle c_i)$$

Transfer Function and Its DC (Steady State) Gain:

$$H(z) = \frac{b_m z^m + b_{m-1} z^{m-1} + \cdots + b_1 z + b_0}{z^n + a_{n-1} z^{n-1} + a_{n-2} z^{n-2} + \cdots + a_1 z + a_0}, \quad H(1) = \frac{\sum_{i=0}^{m} b_i}{1 + \sum_{j=0}^{n-1} a_j}$$

System Characteristic Polynomial:

$$\Delta(z) = z^n + a_{n-1}z^{n-1} + \cdots + a_1 z + a_0$$

System Steady State Response to a Constant Input:

$$y_{ss} = H(1)a, \quad f[k] = au[k]$$

System Impulse Response:

$$h[k] \triangleq \mathcal{Z}^{-1}\{H(z)\}$$

System Step Response:

$$y_{\text{step}}[k] = \sum_{m=0}^{k} h[m], \quad h[k] = y_{\text{step}}[k] - y_{\text{step}}[k-1]$$

System Zero-Input Response—Integral Formulation:

$$Y_{zi}^i(z) = \frac{-z^n I_y^i(z)}{\Delta(z)} \leftrightarrow y_{zi}^i[k] = \mathcal{Z}^{-1}\left\{\frac{-z^n I_y^i(z)}{\Delta(z)}\right\}$$

$$y[-n], y[-(n-1)], \ldots, y[-2], y[-1] \quad \text{are given}$$

System Zero-State Response—Integral Formulation:

$$Y_{zs}^i(z) = H(z)F(z) \leftrightarrow y_{zs}^i[k] = h[k] * f[k]$$

Complete System Response—Integral Formulation:

$$y[k] = y_{zs}^i[k] + y_{zi}^i[k], \quad k \geq 0$$

$$\mathcal{Z}^{-1}\{Y_{zs}^i(z)\} + \mathcal{Z}^{-1}\{Y_{zi}^i(z)\} = \mathcal{Z}^{-1}\{H(z)F(z)\} + \mathcal{Z}^{-1}\left\{\frac{-z^n I_y^i(z)}{\Delta(z)}\right\}$$

System Zero-Input Response—Derivative Formulation:

$$Y_{zi}^d(z) = \frac{I_y^d(z)}{\Delta(z)} \leftrightarrow y_{zi}^d[k] = \mathcal{Z}^{-1}\left\{\frac{I_y^d(z)}{\Delta(z)}\right\}$$

$$y[0], y[1], \ldots, y[n-1] \quad \text{are given}$$

System Zero-State Response—Derivative Formulation:

$$Y_{zs}^d(z) = H(z)F(z) - \frac{I_f(z)}{\Delta(z)} \leftrightarrow y_{zs}^d[k] = h[k] * f[k] - \mathcal{Z}^{-1}\left\{\frac{I_f(z)}{\Delta(z)}\right\}$$

Complete System Response—Derivative Formulation:

$$y[k] = y_{zs}^d[k] + y_{zi}^d[k] = \mathcal{Z}^{-1}\{Y_{zs}^d(z)\} + \mathcal{Z}^{-1}\{Y_{zi}^d(z)\}$$

$$= \mathcal{Z}^{-1}\left\{H(z)F(z) - \frac{I_f(z)}{\Delta(z)}\right\} + \mathcal{Z}^{-1}\left\{\frac{I_y^d(z)}{\Delta(z)}\right\}, \quad k \geq n$$

System Zero-State Response to a Sinusoidal Input, $f[k] = \cos(k\omega_0 T + \varphi)$:

$$y_{zs}[k] = y_{ss}[k] = |H(e^{j\omega_0 T})| \cos(k\omega_0 T + \varphi + \angle H(e^{j\omega_0 T}))$$

Summation Formulas:

$$S_1 = \sum_{k=0}^{\infty} k a^k = F_1(1) = \frac{a}{(1-a)^2}, \quad |a| < 1$$

$$S_2 = \sum_{k=0}^{\infty} k^2 a^k = F_2(1) = \frac{a(1+a)}{(1-a)^3}, \quad |a| < 1$$

TABLE 5.1 Properties of the $\mathcal{Z}$-transform

$\mathcal{Z}\{a_1 f_1[k] \pm a_2 f_2[k]\}$	$a_1 F_1(z) \pm a_2 F_2(z)$
$\mathcal{Z}\{f[k-k_0]u[k-k_0]\}$	$\dfrac{1}{z^{k_0}} F(z)$
$\mathcal{Z}\{f[k-1]u[k]\}$	$\dfrac{1}{z} F(z) + f[-1]$
$\mathcal{Z}\{f[k-2]u[k]\}$	$\dfrac{1}{z^2} F(z) + \dfrac{1}{z} f[-1] + f[-2]$
$\mathcal{Z}\{f[k-k_0]u[k]\}$	$\dfrac{1}{z^{k_0}} F(z) + \dfrac{1}{z^{k_0-1}} f[-1] + \cdots + \dfrac{1}{z} f[-k_0+1] + f[-k_0]$
$\mathcal{Z}\{f[k+1]u[k]\}$	$zF(z) - zf[0]$
$\mathcal{Z}\{f[k+2]u[k]\}$	$z^2 F(z) - z^2 f[0] - zf[1]$
$\mathcal{Z}\{f[k+k_0]u[k]\}$	$z^{k_0} F(z) - z^{k_0} f[0] - z^{k_0-1} f[1] - \cdots - zf[k_0-1]$
$\mathcal{Z}\{kf[k]\}$	$-z \dfrac{d}{dz} F(z)$
$\mathcal{Z}\{k^2 f[k]\}$	$z \dfrac{d}{dz} F(z) + z^2 \dfrac{d^2}{dz^2} F(z)$
$\mathcal{Z}\{a^k f[k]\}$	$F\left(\dfrac{z}{a}\right)$
$\mathcal{Z}\{f[k]\cos(\omega kT)\}$	$\dfrac{1}{2}[F(ze^{j\omega T}) + F(ze^{-j\omega T})]$
$\mathcal{Z}\{f[k]\sin(\omega kT)\}$	$\dfrac{j}{2}[F(ze^{j\omega T}) - F(ze^{-j\omega T})]$
$\mathcal{Z}\{f_1[k] * f_2[k]\}$	$F_1(z) F_2(z)$
$\lim_{k \to 0} f[k]$	$\lim_{z \to \infty} \{F(z)\}$
$\lim_{k \to \infty} f[k]$	$\lim_{z \to 1} \left\{\dfrac{z-1}{z} F(z)\right\}$

TABLE 5.2 Common $\mathcal{Z}$-transform pairs

$\delta[k]$	1
$u[k]$	$\dfrac{z}{z-1}$
$a^k u[k]$	$\dfrac{z}{z-a}$
$k u[k]$	$\dfrac{z}{(z-1)^2}$
$k^2 u[k]$	$\dfrac{z(z+1)}{(z-1)^3}$
$k a^k u[k]$	$\dfrac{az}{(z-a)^2}$
$k^2 a^k u[k]$	$\dfrac{az(z+1)}{(z-a)^3}$
$\dfrac{1}{m!} k(k-1)(k-2)\cdots(k-m+1) u[k]$	$\dfrac{z}{(z-1)^m}$
$\dfrac{1}{m!} k(k+1)(k+2)\cdots(k+m) a^k u[k]$	$\dfrac{z^{m+1}}{(z-a)^{m+1}}$
$u[k] \cos(\omega k T)$	$\dfrac{z^2 - z\cos(\omega T)}{z^2 - 2z\cos(\omega T) + 1}$
$u[k] \sin(\omega k T)$	$\dfrac{z \sin(\omega T)}{z^2 - 2z\cos(\omega T) + 1}$
$a^k u[k] \cos(\omega k T)$	$\dfrac{z^2 - az\cos(\omega T)}{z^2 - 2az\cos(\omega T) + a^2}$
$a^k u[k] \sin(\omega k T)$	$\dfrac{az \sin(\omega T)}{z^2 - 2az\cos(\omega T) + a^2}$

5.8 REFERENCES

[1] J. Ragazzini and G. Franklin, *Sampled-Data Control Systems,* McGraw-Hill, New York, 1958.

[2] E. Jury, *Sampled-Data Control Systems,* Wiley, New York, 1958.

[3] E. Jury, *The Theory and Application of the Z-Transform Method,* Wiley, New York, 1964.

[4] Y. Tzypkin, *Theory of Pulse Systems* (in Russian), State Press for Physics and Mathematics Literature, Moscow, 1958.

[5] S. Orfanidis, *Introduction to Signal Processing,* Prentice Hall, Upper Saddle River, NJ, 1996.

[6] A. Oppenheim and A. Willsky, with H. Nawab, *Systems & Signals,* Prentice Hall, Upper Saddle River, NJ, 1997.

[7] R. Churchill, *Complex Variables and Applications,* McGraw-Hill, New York, 1990.

[8] S. Karni and W. Byatt, *Mathematical Methods in Continuous and Discrete Systems,* Holt, Rinehart, and Winston, New York, 1982.

[9] R. Vich, *Z Transform Theory and Applications,* Reidel Publishing Company, Dordrecht, The Netherlands, 1987.

[10] A. Jerri, *Linear Difference Equations with Discrete Transform Methods,* Kluwer Academic Publishers, Dordrecht, The Netherlands, 1996.

[11] Z. Gajić and M. Lelić, *Modern Control Systems Engineering,* Prentice Hall, London, 1996.

[12] B. Kuo, *Digital Control Systems,* Saunders College Publishing, Orlando, FL, 1992.

5.9 PROBLEMS

5.1. Derive the formula for the $\mathcal{Z}$-transform of the k_0th left shift in time (k_0th discrete-time derivative) as given in (5.12).

5.2. Derive the formula for the $\mathcal{Z}$-transform of $k^2 f[k]$.

Hint: See (5.14) and the corresponding proof of the time multiplication property.

5.3. Find an expression for $f[2]$ in terms of $F(z)$ and values of $f[k]$. Derive the similar result for $f[3], \ldots, f[n-1]$.

Hint: Follow the proof of the initial value theorem, Property 8.

5.4. Use the fact that $a^k u[k] \leftrightarrow z/(z-a)$ and Euler's formula to derive $\mathcal{Z}$-transforms of the sine and cosine functions.

5.5. Let $f[k] = 2^k u[k]$. Find the $\mathcal{Z}$-transform of the following signal.

$$kf[k] + 3^k f[k] + f[k-2]u[k-2] + f[k+1]u[k]$$

Answer:
$$\frac{2z}{(z-2)^2} + \frac{z}{z-6} + \frac{1}{z(z-2)} + \frac{2z}{z-2}$$

5.6. Given $f[k] = ku[k]$, find the $\mathcal{Z}$-transform of the following signal.

$$(k-1)f[k] + 5^k f[k] + f[k-1]u[k-1] + 5f[k+3]u[k]$$

Answer:
$$\frac{2z}{(z-1)^3} + \frac{5z}{(z-5)^2} + \frac{1}{(z-1)^2} + 5\frac{4z^2 - 2z}{(z-1)^2}$$

5.7. Find $\mathcal{Z}$-transforms of the signals represented by

(a) $f_1[k] = \begin{cases} 11, & k=7 \\ -16, & k=12 \\ 5, & k=19 \\ 0, & \text{otherwise} \end{cases}$

(b) $f_2[k] = \begin{cases} 7, & k=8 \\ -9, & k=12 \\ 0, & \text{otherwise} \end{cases}$

Answers:

(a) $F_1(z) = \dfrac{11}{z^7} - \dfrac{16}{z^{12}} + \dfrac{5}{z^{19}}$

(b) $F_2(z) = \dfrac{7}{z^8} - \dfrac{9}{z^{12}}$

5.8. Find $\mathcal{Z}$-transforms of the following signals.

(a) $2\delta[k-1] + 3^{k+1} k u[k] + \sin\left(k\dfrac{\pi}{2}\right) u[k-2]$

(b) $5^k \{u[k] - u[k-11]\}$

(c) $k u[k-1] - \dfrac{1}{k}(u[k-2] - u[k-3]) + e^{-k-2} \sin\left(k\dfrac{\pi}{2}\right) u[k] + 5\delta[k-2]$

(d) $(k-1)u[k] - k u[k-3] + \sin\left(k\dfrac{\pi}{2}\right) u[k-2]$

Answers:

(a) $\dfrac{2}{z} + \dfrac{9z}{(z-3)^2} - z^{-1} \dfrac{1}{z^2+1}$

(b) $\dfrac{z}{z-5} - 5^{11} z^{-11} \dfrac{z}{z-5}$

(c) $\dfrac{-z+2}{(z-1)^2} - \dfrac{1}{z} - \dfrac{0.5}{z^2} + \dfrac{z}{e^3 z^2 + e} + \dfrac{5}{z^2}$

5.9. Show that $\mathcal{Z}$-transforms of $k^3 u[k]$ and $k^4 u[k]$ are respectively given by

$$\dfrac{z(z^2 + 4z + 1)}{(z-1)^4} \quad \text{and} \quad \dfrac{z(z^3 + 11z^2 + 11z + 1)}{(z-1)^5}$$

5.10. Find $\mathcal{Z}$-transforms of the following discrete-time signals.

(a) $f_1[k] = k^3 a^k u[k]$

(b) $f_2[k] = k^4 a^k u[k]$

Hint: See Problem 5.9.

Answers:

(a) $F_1(z) = \dfrac{az(z^2 + 4az + a^2)}{(z-a)^4}$

(b) $F_2(z) = \dfrac{az(z^3 + 11az^2 + 11a^2 z + a^3)}{(z-a)^5}$

5.11. Using the partial fraction expansion method, find inverse $\mathcal{Z}$-transforms of the following functions.

(a) $F_1(z) = \dfrac{z(z+1)}{(z+2)(z-0.5)}$

(b) $F_2(z) = \dfrac{6z}{(z-1)^2(z+0.5)}$

(c) $F_3(z) = \dfrac{2}{z(z+1)(z-1)}$

(d) $F_4(z) = \dfrac{5z^3}{(z+0.5)^2(z-0.5)}$

(e) $F_5(z) = \dfrac{5z}{z^2 - z + 1}$

Answers:

(a) $f_1[k] = 0.4(-2)^k u[k] + 0.6(0.5)^k u[k]$

(b) $f_2[k] = \left(-\dfrac{8}{3} + 4k + \dfrac{8}{3}(-0.5)^k\right) u[k]$

(c) $f_3[k] = -2\delta[k-1] - (-1)^k u[k] + u[k]$

(d) $f_4[k] = \dfrac{5}{4}\left\{3\left(-\dfrac{1}{2}\right)^k - 2k\left(-\dfrac{1}{2}\right)^k + \left(\dfrac{1}{2}\right)^k\right\} u[k]$

(e) $f_5[k] = \dfrac{10}{\sqrt{3}} \sin\left(k\dfrac{\pi}{3}\right) u[k]$

5.12. Find inverse $\mathcal{Z}$-transforms of the following functions that have complex conjugate poles.

(a) $F_1(z) = \dfrac{z}{(z^2 + z + 1)\left(z + \tfrac{1}{3}\right)}$

(b) $F_2(z) = \dfrac{z^2}{\left(z^2 + \tfrac{1}{4}\right)\left(z^2 + \tfrac{1}{9}\right)}$

(c) $F_3(z) = \dfrac{1}{z^2 + 2z + 2}$

(d) $F_4(z) = \dfrac{z\left(z + \tfrac{1}{2}\right)}{\left(z^2 - z + \tfrac{1}{2}\right)(z - 1)}$

5.13. Find inverse $\mathcal{Z}$-transforms of the following functions

(a) $\dfrac{2z - 1}{z^2 + 1}$

(b) $\dfrac{z(22 - 5z)}{(z+1)(z-2)^2}$

(c) $\dfrac{1}{z^2 + 4}$

(d) $\dfrac{z}{16z^2 + 1}$

(e) $\dfrac{2z}{z^2+4}$

(f) $\dfrac{2}{z}+\dfrac{5}{z^3}-\dfrac{3}{z^7}$

Answers:

(a) $2\cos\left((k-1)\dfrac{\pi}{2}\right)u[k-1]-\cos\left((k-2)\dfrac{\pi}{2}\right)u[k-2]$

(b) $(-3(2)^k+2k(2)^k+3(-1)^k)u[k]$

(c) $2^{k-2}\cos\left((k-2)\dfrac{\pi}{2}\right)u[k-2]$

5.14. Find initial values of the signals whose $\mathcal{Z}$-transforms are given in Problem 5.11.

Answers:

(a) $f_1[0]=1$

(b) $f_2[0]=0$

(c) $f_3[0]=0$

(d) $f_4[0]=5$

(e) $f_5[0]=0$

5.15. Check the applicability of the final value theorem to the $\mathcal{Z}$-transforms given in Problem 5.11. Find the corresponding signal steady state values in cases where the final value theorem is applicable.

Answers: (d) $f_4[\infty]=0$. Not applicable in other cases.

5.16. Use the results from Problem 5.10 in order to sum the following infinite series.

(a) $\displaystyle\sum_{k=0}^{\infty} k^3\left(\dfrac{1}{3}\right)^k$

(b) $\displaystyle\sum_{k=0}^{\infty} k^4\left(-\dfrac{1}{2}\right)^k$

5.17. Show that

$$\mathcal{Z}\left\{\sum_{k=0}^{n-1} f[k]\right\}=\dfrac{1}{z-1}F(z)$$

5.18. Show that the discrete-time *signal correlation* has the following $\mathcal{Z}$-transform

$$\mathcal{Z}\{f[k]\}\triangleq \mathcal{Z}\left\{\sum_{m=0}^{\infty} f_1[m]f_2[m+k]\right\}=F_1(z)F_2\left(\dfrac{1}{z}\right)$$

Hint: See the proof for the convolution property.

5.19. Derive the formula for the $\mathcal{Z}$-transform of a periodic discrete-time signal defined by $f[k] = f[k+N]$, where N is the discrete-time period.
Answer:
$$F(z) = \frac{z^N}{z^N - 1} F_1(z), \quad F_1(z) = \mathcal{Z}\{f[k], \quad k = 0, 1, \ldots, N-1\}$$

5.20. Using the formula derived in Problem 5.19, find the $\mathcal{Z}$-transform of the periodic signal defined by
$$f[k] = f[k+N] = \begin{cases} 1, & k = 0, 1, 2, 6, 7, 8, \ldots \\ -1, & k = 3, 4, 5, 9, 10, 11, \ldots \end{cases}, \quad N = 6$$

5.21. Consider an input-free discrete-time system represented in its derivative formulation by
$$y[k+2] + \frac{1}{2}y[k+1] + y[k] = 0, \quad y[0] = 1, \quad y[1] = 2$$
Write the corresponding integral formulation with the initial conditions obtained by propagating the original initial conditions through the difference equation.

5.22. Consider a discrete-time system driven by the input $f[k] = (-0.5)^k u[k]$, represented in its derivative formulation as
$$y[k+2] + \frac{1}{2}y[k+1] + y[k] = f[k], \quad y[0] = 1, \quad y[1] = 2$$
Write the corresponding integral formulation with the initial conditions obtained by propagating the original initial conditions through the difference equation.

5.23. Starting with the integral formulation of a discrete-time linear system given by
$$y[k] + \frac{1}{3}y[k-1] + \frac{1}{2}y[k-2] = f[k-1] + \frac{1}{2}f[k-2], \quad y[-2] = -2, \quad y[-1] = 0$$
obtain its derivative formulation by propagating the original initial conditions. Assume that $f[k] = u[k]$.

5.24. Find the impulse and step responses for the discrete-time linear system defined in Problem 5.23.

5.25. Find the impulse response of the following discrete-time linear systems.
(a) $y[k+2] + 4y[k+1] + 4y[k] = f[k+1] + 4f[k]$
(b) $y[k+2] + \frac{1}{6}y[k+1] - \frac{1}{6}y[k] = f[k+1] + f[k]$

Hint: Represent this system in the integral formulation, and then find the system transfer function and take the inverse $\mathcal{Z}$-transform.

Answers:

(a) $h[k] = \delta[k] - (-2)^k u[k] + \dfrac{1}{2}k(-2)^k u[k]$

(b) $h[k] = -6\delta[k] + 1.2\left(-\dfrac{1}{2}\right)^k u[k] + 4.8\left(\dfrac{1}{3}\right)^k u[k]$

5.26. Find the impulse responses of the systems whose transfer functions are given by

(a) $H_1(z) = \dfrac{z + 0.5}{(z+1)(z-0.5)(z+0.1)}$

(b) $H_2(z) = \dfrac{z}{z^2 + 0.25z + 0.25}$

5.27. Find the step responses of the discrete-time linear systems defined in Problem 5.25.

Answer:

(a) $h[k] = \mathcal{Z}^{-1}\{H(z)\} = \delta[k] - (-2)^k u[k] + \dfrac{1}{2}k(-2)^k u[k]$

$y_{\text{step}}[k] = \sum_{m=0}^{k} h[m] = \sum_{m=0}^{k}\left\{\delta[m] - (-2)^m + \dfrac{1}{2}m(-2)^m\right\}$

The closed formula for $y_{\text{step}}[k]$ can be obtained by using $y_{\text{step}}[k] = \mathcal{Z}^{-1}\{H(z)z/(z-1)\}$

5.28. Find the step responses of the discrete-time linear systems whose transfer functions are defined in Problem 5.26.

5.29. Determine whether steady state responses to constant inputs ($f[k] = au[k]$) exist for the systems whose transfer functions are defined in Problem 5.26. When the steady state response exists, find its value for $a = 5$.

5.30. Find the transfer function of the following discrete-time system.

$$y[k] - 3y[k-2] + 4y[k-3] - y[k-4] + 2y[k-5]$$
$$= f[k-2] - 2f[k-3] + 5f[k-4] + f[k-5]$$

Answer:

$$H(z) = \dfrac{z^3 - 2z^2 + 5z + 1}{z^5 - 3z^3 + 4z^2 - z + 2}$$

5.31. For the discrete-time system given by

$$y[k+2] + 4y[k] + 3y[k] = f[k] = (-1)^k u[k], \quad y[0] = y[1] = 1$$

find the complete system response.

5.32. Show that the integral formulation leads to the same complete system response in Problem 5.31. To that end, first form the integral formulation and appropriately evaluate new initial conditions at $y[-1]$ and $y[-2]$, and then find $Y(z)$.

5.33. Find the complete response of the system represented by

$$y[k+2] + 5y[k+1] + 4y[k] = u[k+1], \quad y[0] = 1, \quad y[1] = 2$$

Find the system transfer function and the system impulse response.

Answers:

$$y[k] = \frac{1}{10}u[k] + \frac{11}{6}(-1)^k u[k] - \frac{14}{15}(-4)^k u[k]$$

$$H(z) = \frac{z}{z^2 + 5z + 4}, \quad h[k] = \frac{1}{3}(-1)^k u[k] - \frac{1}{3}(-4)^k u[k]$$

5.34. Find the complete response of the system represented by

$$y[k] + 5y[k-1] + 4y[k-2] = u[k-1], \quad y[-2] = 1, \quad y[-1] = 2$$

Compare the result obtained with the corresponding result from Problem 5.33.

5.35. Determine the complete response of the following system.

$$y[k+2] + 4y[k+1] + 4y[k] = (-1)^k u[k], \quad y[0] = 2, \quad y[1] = -2$$

Identify the system zero-state and zero-input responses.

5.36. Find the zero-input responses of the discrete-time systems from Problems 5.33–35.

5.37. Use the summation formula given in (1.35) to show that the two formulas for the unit step responses derived in Example 5.24 are identical.

5.38. Using the $\mathcal{Z}$-transform, solve the following system of difference equations.

$$y_1[k] = y_1[k-1] + \frac{1}{2}y_2[k-1]$$

$$y_2[k] = -\frac{1}{3}y_1[k-1] + f[k-1]$$

$$f[k] = (-1)^k u[k], \quad y_1[-1] = 2, \quad y_2[-1] = -1$$

5.39. Using the $\mathcal{Z}$-transform, find the transfer functions for the mixed integral-derivative difference equations representing the dynamics of an ATM computer network switch, and show that they are identical to those derived for the corresponding integral formulation in Section 5.3.4.

5.40. Find the closed-loop system transfer function for the feedback system represented in Figure 5.16.

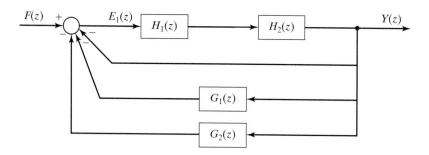

FIGURE 5.16

5.41. Find the closed-loop system transfer function for the feedback system represented in Figure 5.17.

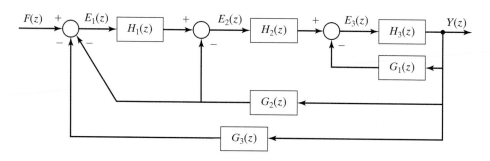

FIGURE 5.17

5.42. Find the closed-loop system transfer function for the feedback system represented in Figure 5.18.

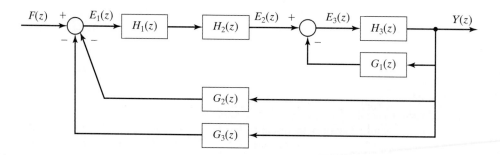

FIGURE 5.18

Answer:
$$\frac{Y(z)}{F(z)} = \frac{H_1(z)H_2(z)H_3(z)}{1 + G_1(z)H_3(z) + H_1(z)H_2(z)H_3(z)(G_2(z) + G_3(z))}$$

5.43. Find the closed-loop system transfer function for the feedback system represented in Figure 5.19.

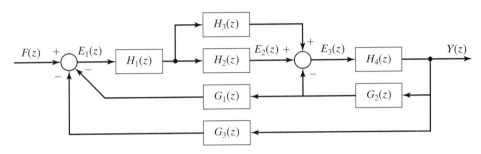

FIGURE 5.19

5.44. Find the response of a discrete-time system, whose transfer function is given by
$$H(z) = \frac{z}{z^2 + 0.5z + 0.25}$$
to a sinusoidal input of the form $f[k] = \cos(2kT + \pi/3)$ with $T = 0.1$.

5.45. Assuming that $\omega T = 2$ rad, repeat Problem 5.44 for a sinusoidal input function of the form $f[k] = \sin(10k\omega T + \pi/6)$.

5.46. Use MATLAB to find and plot the impulse and step responses for the system whose transfer function is defined in Problem 5.30.

5.47. Use MATLAB to plot the spectra of the discrete-time systems defined in Problem 5.26.

5.48. Consider the ATM switch from Section 5.3.4, and its representation using the integral formulation given in (5.75). Find the complete response due to the input signal $f[k] = 12u[k]$ and the initial conditions $e[-2] = 5$, $e[-1] = 5$, $y[-1] = 10$. Use MATLAB to plot the responses for the average arrival rate, and the deviation of buffer queue length from the desired value.

5.49. Use MATLAB and Simulink to study the dynamics of the ATM switch considered in Section 5.3.4, assuming that in addition the buffer service rate satisfies
$$f[k] = \begin{cases} q[k], & q[k] < 10 = \mu \\ \mu, & q[k] \geq 10 = \mu \end{cases}$$
Assume that the initial conditions are defined as in Problem 5.48.

5.50. Build the Simulink block diagram for the ATM switch as presented in Figure 5.10. Simulate the values for the queue length and the arrival rate, assuming that the initial conditions are given by $q[-2] = 0$, $q[-1] = 0$, $y[-1] = 10$. Comment on the results obtained.

5.51. Consider the mathematical model for national income given in formula (1.32). Find analytically the impulse and step responses. Choose several sets of values for the parameters α and β, and use MATLAB to plot the corresponding step responses over a period of ten years, assuming that the discrete-time period is three months (quarterly calculations), which implies $k = 0, 1, \ldots, 35$. Can you suggest any choices for the parameters that will assure a steady increase in the national income?

5.52. Repeat Problem 5.51, assuming that the input to the national income model is given by $f[k] = u[k] + \gamma r[k]$. For simplicity, assume zero initial conditions. Experiment with several numerical values for γ (for example, $\gamma = 0.01, 0.05, 0.1, 0.2$), and use MATLAB to plot the corresponding responses. Compare the results obtained to the results from Problem 5.51 using the same values for α and β parameters.

5.53. Use MATLAB/Simulink to simulate the amortization process model defined in (1.33). Plot the response for the car purchase problem defined in Example 1.7 for a period of 48 months.

5.54. Use MATLAB/Simulink to simulate the amortization process model defined in (1.33). Plot the response for the house purchase problem defined in Problem 1.15, for a period of twenty years. Find the model response due to the inputs $f_1[k] = pu[k] + 0.01 pr[k]$ and $f_2[k] = pu[k] + 0.02 pr[k]$. How much will you save over the period of twenty years if you can increase your payments to either 1% ($f_1[k]$) or 2% ($f_2[k]$) every month?

PART TWO
TIME DOMAIN TECHNIQUES

CHAPTER 6

Convolution

Convolution is one of the primary concepts of linear system theory. It answers the problem of finding the system zero-state response due to any input—the most important problem for linear systems. The main convolution theorem states that *the response of a system at rest (zero initial conditions) due to any input is the convolution of that input and the system impulse response*. We derived this result in the frequency domain in Chapters 3–5. Hence, the main convolution theorem is applicable to $j\omega$, s, and z domains; that is, it is applicable to both continuous- and discrete-time linear systems.

In this chapter, we study the *convolution concept in the time domain* for both continuous- and discrete-time linear time-invariant systems. We state and prove the properties of signal convolution, present techniques for finding the convolution of signals, and specialize the signal convolution to two particular classes of signals: the system impulse response and system input signals. We emphasize the graphical convolution procedure for the continuous-time domain and the sliding tape method for the discrete-time domain. These two methods give complete insights into all steps required for finding the continuous- and discrete-time convolutions. Note that looking at the continuous-time convolution only as a simple integral (2.38), and the discrete-time convolution only as a simple sum (2.39), very often gives a superficial understanding of these extremely important linear system theory concepts.

In Section 6.1, we present theoretical results for continuous-time signal convolution. The use of convolution in continuous-time linear systems, and the time domain derivations of the main theorem of linear system theory via the use of the convolution concept appear in Section 6.2. In Sections 6.3 and 6.4, we introduce discrete-time counterparts to the results of Sections 6.1 and 6.2, study discrete-time signal convolution, and consider its use for discrete-time linear systems. In Section 6.5, a method for evaluating continuous-time convolution via MATLAB is presented. MATLAB laboratory experiments on convolution are designed in Section 6.6. Using and modifying given MATLAB programs to perform these experiments, students will be able to master every step of the continuous- and discrete-time convolution procedures.

6.1 CONVOLUTION OF CONTINUOUS-TIME SIGNALS

The continuous-time convolution of two signals $f_1(t)$ and $f_2(t)$ is defined by

$$f(t) = f_1(t) * f_2(t) = \int_{-\infty}^{\infty} f_1(\tau) f_2(t - \tau) \, d\tau, \quad -\infty < t < \infty \quad (6.1)$$

In this integral, τ plays the role of a dummy variable of integration, and t is a parameter.

Before we state the properties of the convolution integral, we first introduce the notion of the signal duration. The duration of a signal $f_i(t)$ is defined by the time instants t_i and T_i for which the signal is equal to zero for every t outside the interval $[t_i, T_i]$; that is, $f_i(t) = 0$, $t \notin [t_i, T_i]$. Signals that have finite duration are often called *time-limited signals*. For example, rectangular and triangular pulses are time-limited signals, but $u(t)$, $\sin(t)$, $\cos(t)$ have infinite durations.

The properties of the convolution integral (6.1) follow.

1. *Commutativity*

$$f_1(t) * f_2(t) = f_2(t) * f_1(t) \quad (6.2)$$

2. *Distributivity*

$$f_1(t) * \{f_2(t) + f_3(t)\} = f_1(t) * f_2(t) + f_1(t) * f_3(t) \quad (6.3)$$

3. *Associativity*

$$f_1(t) * \{f_2(t) * f_3(t)\} = \{f_1(t) * f_2(t)\} * f_3(t) \quad (6.4)$$

4. *Duration*

Let the signals $f_1(t)$ and $f_2(t)$ have the durations, respectively, defined by the time intervals $[t_1, T_1]$ and $[t_2, T_2]$; then,

$$f(t) = f_1(t) * f_2(t) = \begin{cases} 0, & t \leq t_1 + t_2 \\ \int_{t_1+t_2}^{T_1+T_2} f_1(\tau) f_2(t-\tau) \, d\tau, & t_1 + t_2 \leq t \leq T_1 + T_2 \\ 0, & t \geq T_1 + T_2 \end{cases} \quad (6.5)$$

5. *Time Shifting*

Let $f(t) = f_1(t) * f_2(t)$. Then, convolutions of shifted signals are given by

$$f_1(t - \sigma_1) * f_2(t) = f(t - \sigma_1)$$
$$f_1(t) * f_2(t - \sigma_2) = f(t - \sigma_2) \quad (6.6)$$
$$f_1(t - \sigma_1) * f_2(t - \sigma_2) = f(t - \sigma_1 - \sigma_2)$$

6. *Continuity*

This property simply states that the convolution is a continuous function of the parameter t. The continuity property is useful for plotting convolution graphs and checking obtained convolution results.

Section 6.1 Convolution of Continuous-Time Signals

Now we give some of the proofs of the preceding convolution properties.

Proof of Property 1. We introduce the change of variables $\sigma = t - \tau$ into (6.1). This leads to

$$f_1(t) * f_2(t) = \int_{\infty}^{-\infty} f_1(t-\sigma) f_2(\sigma)(-d\sigma)$$

$$= \int_{-\infty}^{\infty} f_2(\sigma) f_1(t-\sigma) d\sigma = f_2(t) * f_1(t) \quad \square$$

Which of the two forms of the convolution integral should we choose? Definitely, the one that requires fewer computations. For example, while convolving $e^{-t}u(t)$ and $\sin(t)$, we may use either of the integrals,

$$\int_{-\infty}^{\infty} e^{-(t-\tau)} u(t-\tau) \sin(\tau) d\tau = \int_{-\infty}^{\infty} e^{-\tau} u(\tau) \sin(t-\tau) d\tau$$

but the first integral requires less computational effort than the second. Thus, the first integral is the better choice.

Proof of Property 2. This proof follows from the well-known integral addition property; that is,

$$f_1(t) * \{f_2(t) + f_3(t)\} = \int_{-\infty}^{\infty} f_1(\tau) [f_2(t-\tau) + f_3(t-\tau)] d\tau$$

$$= \int_{-\infty}^{\infty} f_1(\tau) f_2(t-\tau) d\tau + \int_{-\infty}^{\infty} f_1(\tau) f_3(t-\tau) d\tau$$

$$= f_1(t) * f_2(t) + f_1(t) * f_3(t) \quad \square$$

Proof of Property 3. We can obtain this proof by showing the equivalence of the corresponding double integrals, by introducing the appropriate change of variables. Since the double convolution is not used in the basic course on linear system theory, the proof of this property is omitted. The double convolution is often used in communication and signal processing systems [1]. $\square$

Proof of Property 4. We can easily verify this property by examining the integration limits for the case when both signals are time limited. It can be observed that the signals $f_1(\tau)$ and $f_2(t-\tau)$, in the case of time-limited signals, overlap only in the interval $t \in [t_1 + t_2, T_1 + T_2]$, hence the convolution is equal to zero outside of this

282 Chapter 6 Convolution

time interval. Note that this property can be extended to signals that have infinite duration, but in such cases the existence of the corresponding convolution integral is much more crucial than the integration limits. □

Proof of Property 5. This proof follows directly from the definition of the convolution integral. This property is used to simplify the graphical convolution procedure, which will be demonstrated in Section 6.1.1. □

The proof of convolution continuity is mathematically involved and hence omitted.
In the following, we solve analytically several examples, to better our understanding of the convolution integral.

EXAMPLE 6.1

Consider the convolution of the impulse delta (singular) signal and any other regular signal $f(t)$,

$$f(t) * \delta(t) = \int_{-\infty}^{\infty} f(\tau)\delta(t-\tau)\,d\tau, \quad -\infty < t < \infty$$

Based on the sifting property of the delta impulse signal, we conclude that

$$f(t) * \delta(t) = f(t) \qquad (6.7)$$

♣

EXAMPLE 6.2

We have already seen in the context of the integral property of the Fourier transform that the convolution of the unit step signal with a regular function (signal) produces function's integral in the specified limits, that is, that

$$f(t) * u(t) = \int_{-\infty}^{\infty} f(\tau)u(t-\tau)\,d\tau = \int_{-\infty}^{t} f(\tau)\,d\tau, \quad -\infty \le t \le \infty$$

Note that $u(t-\tau) = 0$ for $\tau > t$. ♣

EXAMPLE 6.3

Consider the convolution of $e^{-t}u(t)$ and $\sin(t)$, which can be obtained by using either of the following integrals:

$$e^{-t}u(t) * \sin(t) = \int_{-\infty}^{\infty} e^{-(t-\tau)}u(t-\tau)\sin(\tau)\,d\tau = \int_{-\infty}^{\infty} e^{-\tau}u(\tau)\sin(t-\tau)\,d\tau$$

We will evaluate both integrals to show the difference in the computations required. In that respect, we will need the following formulas.

$$\int e^{at} \sin(bt)\,dt = \frac{e^{at}}{a^2+b^2}(a\sin(bt) - b\cos(bt))$$

$$\int e^{at} \cos(bt)\,dt = \frac{e^{at}}{a^2+b^2}(a\cos(bt) + b\sin(bt))$$

(6.8)

It follows that the first convolution integral easily produces

$$e^{-t}u(t) * \sin(t) = e^{-t}\int_{-\infty}^{t} e^{\tau}\sin(\tau)\,d\tau$$

$$= e^{-t}\left[\frac{e^t}{2}(\sin(t) - \cos(t)) - 0\right] = \frac{1}{2}(\sin(t) - \cos(t))$$

The evaluation of the second integral requires first an expansion of the $\sin(t-\tau)$ term, that is,

$$\int_0^\infty e^{-\tau}\sin(t-\tau)\,d\tau = \int_0^\infty e^{-\tau}[\sin(t)\cos(\tau) - \cos(t)\sin(\tau)]\,d\tau$$

which, with the help of the second formula in (6.8), gives

$$\sin(t)\int_0^\infty e^{-\tau}\cos(\tau)\,d\tau - \cos(t)\int_0^\infty e^{-\tau}\sin(\tau)\,d\tau$$

$$= \sin(t)\left[0 - \frac{1}{2}(-1\cos(0) + \sin(0))\right] - \cos(t)\left[0 - \frac{1}{2}(-\sin(0) - \cos(0))\right]$$

$$= \frac{1}{2}\sin(t) - \frac{1}{2}\cos(t) = e^{-t}u(t) * \sin(t)$$

Thus, both convolution integrals produce the same result, but the first is obviously less computationally involved.

6.1.1 Graphical Convolution

The graphical presentation of the convolution integral helps in the understanding of every step in the convolution procedure. In this section, we solve several signal convolution problems using the graphical convolution method.

According to the definition integral (6.1), the convolution procedure must be performed using the following steps:

Step 1. Apply the convolution duration property to identify intervals in which the convolution is equal to zero.

Step 2. Flip about the vertical axis one of the signals (the one with the simpler form or shape, since commutativity holds); that is, represent one of the signals in the time scale $-\tau$.

Step 3. Vary the parameter t from $-\infty$ to ∞; that is, slide the flipped signal from the left to the right, look for the intervals where it overlaps with the other signal, and evaluate the integral of the product of the two signals in the corresponding intervals.

In these steps one can also incorporate (if applicable) the convolution time shifting property, such that all signals start at the origin. In such a case, after the final convolution result is obtained, the convolution time shifting formula (6.6) should be applied appropriately. In addition, the obtained convolution result may be checked by the convolution continuity property, which requires that at the boundaries of adjacent intervals the convolution remains a continuous function of the parameter t.

In the following, we present examples of several graphical convolution problems, starting with the simplest.

EXAMPLE 6.4

Consider the convolution of two rectangular pulses as represented in Figure 6.1.

Since the durations of the signals $f_1(t)$ and $f_2(t)$ are respectively given by $[t_1, T_1] = [0, 3]$ and $[t_2, T_2] = [0, 1]$, we conclude from (6.5) that the convolution of these two signals is zero in the following intervals (Step 1):

$$f_1(t) * f_2(t) = 0, \quad t \leq t_1 + t_2 = 0 + 0 = 0$$
$$f_1(t) * f_2(t) = 0, \quad t \geq T_1 + T_2 = 1 + 3 = 4$$

Thus, we need only evaluate the convolution integral in the interval $0 \leq t \leq 4$.

In Step 2, we flip the signal that has the simpler shape about the vertical axis. Since in this case both signals are rectangular pulses, it is irrelevant which one is flipped. Let us flip $f_2(t)$. Note that the convolution is performed in the time scale τ. In Figure 6.2, we present the signals $f_1(\tau)$ and $f_2(-\tau)$. This figure corresponds to the convolution for $t = 0$.

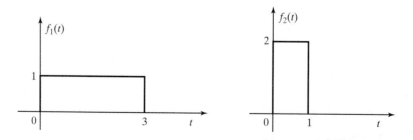

FIGURE 6.1: Two rectangular signals

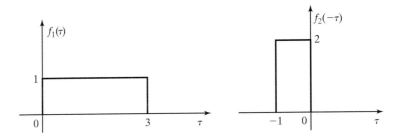

FIGURE 6.2: Signals $f_1(\tau)$ and $f_2(-\tau)$

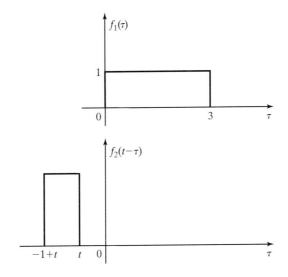

FIGURE 6.3: Signals $f_1(\tau)$ and $f_2(t-\tau)$, $t < 0$

In Step 3, we shift the signal $f_2(-\tau)$ to the left and to the right, that is, we form the signal $f_2(t - \tau)$ for $t \in (-\infty, 0]$ and $t \in [0, \infty)$. A shift of the signal $f_2(t - \tau)$ to the left ($t < 0$) produces no overlapping between the signals $f_1(\tau)$ and $f_2(t - \tau)$, thus the convolution integral is equal to zero for $t < 0$ (see Figure 6.3). Note that the same conclusion has been already made in Step 1.

Let us start shifting the signal $f_2(t-\tau)$ to the right ($t > 0$). Consider first the interval $0 \leq t \leq 1$ (see Figure 6.4).

It can be seen from Figure 6.4 that in the interval from zero to t the signals overlap, hence their product is different from zero in this interval, which implies that the convolution integral is given by

$$f_1(t) * f_2(t) = \int_0^t 1 \times 2 \, d\tau = 2t, \quad 0 \leq t \leq 1$$

286   Chapter 6   Convolution

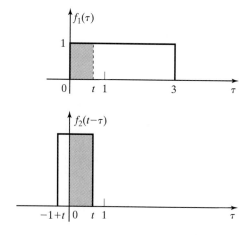

FIGURE 6.4: Signals $f_1(\tau)$ and $f_2(t-\tau)$, $0 \leq t \leq 1$

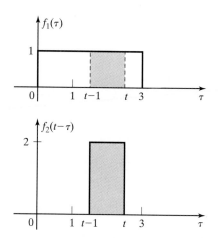

FIGURE 6.5: Signals $f_1(\tau)$ and $f_2(t-\tau)$, $1 \leq t \leq 3$

By shifting the signal $f_2(t-\tau)$ further to the right, we get the same "kind of overlap" for $1 \leq t \leq 3$ (see Figure 6.5).

From this figure, we see that the actual convolution integration limits are from $t-1$ to t, that is,

$$f_1(t) * f_2(t) = \int_{t-1}^{t} 1 \times 2 \, d\tau = 2, \quad 1 \leq t \leq 3$$

By shifting $f_2(t-\tau)$ further to the right, for $3 \leq t \leq 4$, we get the situation presented in Figure 6.6.

Section 6.1 Convolution of Continuous-Time Signals

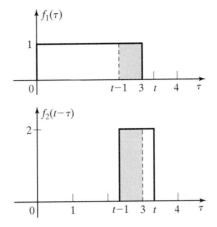

FIGURE 6.6: Signals $f_1(\tau)$ and $f_2(t-\tau)$, $3 \le t \le 4$

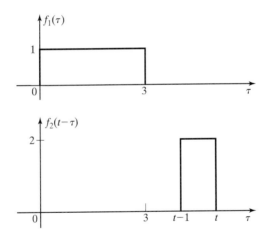

FIGURE 6.7: Signals $f_1(\tau)$ and $f_2(t-\tau)$, $t > 4$

In this interval, the convolution integral is given by

$$f_1(t) * f_2(t) = \int_{t-1}^{3} 1 \times 2\, d\tau = 8 - 2t, \quad 3 \le t \le 4$$

For $t > 4$, the convolution is equal to zero, as determined in Step 1. This can be justified by the fact that the signals $f_1(\tau)$ and $f_2(t-\tau)$ do not overlap for $t > 4$, that is, their product is equal to zero for $t > 4$, which implies that the corresponding integral is equal to zero in the same interval (see Figure 6.7).

In summary, the convolution of the considered signals is given by

$$f_1(t) * f_2(t) = \begin{cases} 0, & t \leq 0 \\ 2t, & 0 \leq t \leq 1 \\ 2, & 1 \leq t \leq 3 \\ 8 - 2t, & 3 \leq t \leq 4 \\ 0, & t \geq 4 \end{cases}$$

Note that from the convolution continuity property, the convolution signal obtained must be a continuous function of t. This can be easily checked as follows. For $t = 0$ the expression $2t$ produces zero. At $t = 1$ we see that $2t = 2 \times 1 = 2$, also for $t = 3$ we have $8 - 2t = 8 - 2 \times 3 = 2$, and finally for $t = 4$ we get $8 - 2t = 8 - 2 \times 4 = 0$. Thus the function obtained, $f_1(t) * f_2(t)$, is a continuous function of the parameter t.

EXAMPLE 6.5

Let us convolve the signals represented in Figure 6.8.

Since both signals have duration intervals from 0 to 2, we conclude from (6.5) that the convolution integral is zero for $t \leq 0$ and $t \geq 4$. In the next step we flip the rectangular signal about the vertical axis, since it apparently has a simpler shape (see Figure 6.9(a)). In Step 3, we slide the rectangular signal to the right for $t \in [0, 2]$ (Figure 6.9(b)), and for $t \in [2, 4]$ (Figure 6.9(c)). The convolution integral in these two intervals, evaluated according to information given in Figures 6.9(b) and 6.9(c), is respectively given by

$$f_1(t) * f_2(t) = \int_0^t 2(-\tau + 2) \, d\tau = 4t - t^2, \quad 0 \leq t \leq 2$$

$$f_1(t) * f_2(t) = \int_{t-2}^2 2(-\tau + 2) \, d\tau = 16 - 8t + t^2, \quad 2 \leq t \leq 4$$

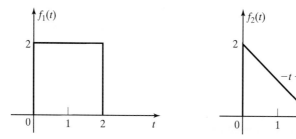

FIGURE 6.8: Two signals: rectangular and triangular pulses

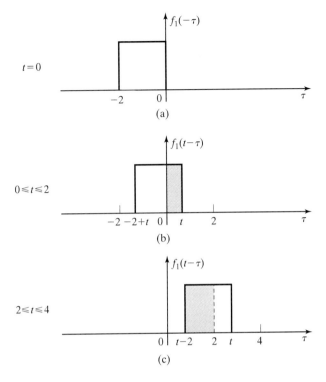

FIGURE 6.9: Convolution procedure for signals in Example 6.5

In summary, we have obtained

$$f_1(t) * f_2(t) = \begin{cases} 0, & t \leq 0 \\ 4t - t^2, & 0 \leq t \leq 2 \\ 16 - 8t + t^2, & 2 \leq t \leq 4 \\ 0, & t \geq 4 \end{cases}$$

It can be easily checked that the obtained convolution result represents a continuous function of the parameter t.

EXAMPLE 6.6

We now consider the slightly more difficult problem of convolving two signals with triangular shapes, as presented in Figure 6.10. Note that the signals are represented in the time scale τ. The problem is more difficult in the sense that we must flip one of these two triangularly shaped signals about the vertical axis and find its analytical expression. The remaining part of the problem is the standard convolution technique.

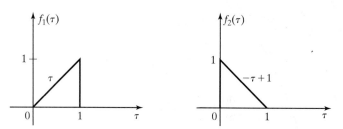

FIGURE 6.10: Two triangularly shaped signals

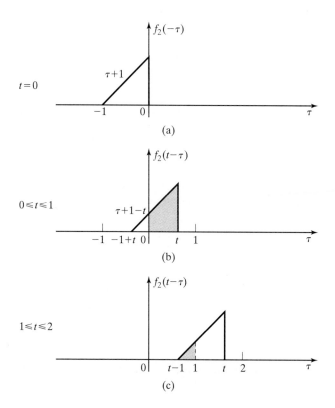

FIGURE 6.11: Convolution procedure for signals in Example 6.6

Let us flip the signal $f_2(\tau)$ about the vertical axis. The flipped signal for $t=0$ is presented in Figure 6.11(a). Note its new analytical expression, now given by $f_2(-\tau) = \tau + 1$.

From the convolution duration property, we conclude that the convolution is equal to zero for $t \leq 0$ and $t \geq 2$. Thus, we need work only in the interval $0 \leq t \leq 2$.

Consider the interval $0 \leq t \leq 1$. In this interval, the signal $f_2(t - \tau)$ is given by $f_2(t - \tau) = \tau - t + 1$ (see Figure 11(b)). Since the signal $f_1(\tau)$ overlaps with the signal

$f_2(t - \tau)$ in the interval from zero to t, the convolution is given by

$$f_1(t) * f_2(t) = \int_0^t \tau(\tau - t + 1)\, d\tau = -\frac{1}{6}t^3 + \frac{1}{2}t^2$$

For $1 \leq t \leq 2$, the signal $f_2(t - \tau)$, presented in Figure 6.11(c), overlaps with the signal $f_1(\tau)$ in the interval from $t - 1$ to 1. Here, the convolution is given by

$$f_1(t) * f_2(t) = \int_{t-1}^1 \tau(\tau - t + 1)\, d\tau = \frac{1}{3} - \frac{1}{2}(t-1) + \frac{1}{6}(t-1)^3$$

EXAMPLE 6.7

Consider the convolution problem that involves shifted signals, as presented in Figure 6.12. According to the convolution time shifting property, we can shift the signal $f_1(t)$ to the origin and find the convolution of the shifted signal $f_1(t + 1)$ and the signal $f_2(t)$. Let $f(t + 1)$ represent the convolution of $f_1(t + 1)$ and $f_2(t)$. In order to find the required original convolution result, the convolution obtained through the regular convolution procedure with $f_1(t + 1)$ and $f_2(t)$ must be shifted backward by one unit.

However, we can convolve these two signal without applying the convolution time shifting property, as demonstrated in the following. Since the durations of these signals are respectively given by $[t_1, T_1] = [1, 3]$ and $[t_2, T_2] = [0, 2]$, we conclude that the corresponding convolution is equal to zero for $t \leq t_1 + t_2 = 1$ and $t \geq T_1 + T_2 = 5$.

Let us flip the rectangular signal $f_1(t)$ about the vertical axis (Figure 6.13(a)). In the interval $1 \leq t \leq 3$, the convolution is given (see Figure 6.13(b)) by

$$f(t) = f_1(t) * f_2(t) = \int_0^{t-1} \frac{1}{2}\tau\, d\tau = \frac{(t-1)^2}{4}$$

In the interval $3 \leq t \leq 5$, we have (from Figure 6.13(c))

$$f(t) = f_1(t) * f_2(t) = \int_{t-3}^2 \frac{1}{2}\tau\, d\tau = -\frac{1}{4}(t-1)(t-5)$$

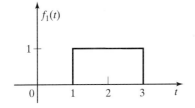
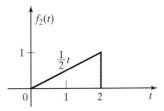

FIGURE 6.12: Convolution with a shifted signal

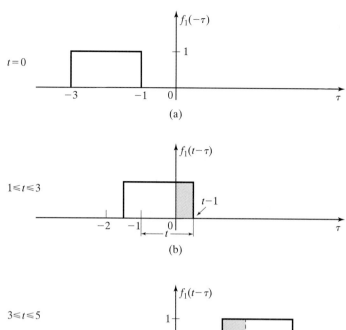

FIGURE 6.13: Convolution process for signals in Example 6.7

It is left as an exercise to students to solve the same convolution problem by applying the convolution time shifting property (see Problem 6.1).

In the next section, we apply the convolution formula to continuous-time linear time-invariant systems, and show that the system response to any input is given in terms of the convolution integral. To that end, we will use the concepts of system transfer function and system impulse response introduced in Chapters 3 and 4.

6.2 CONVOLUTION FOR LINEAR CONTINUOUS-TIME SYSTEMS

Consider the problem of finding the response of a system at rest (zero initial conditions) due to any input, say $f(t)$. In the chapters on frequency domain techniques, we saw that every linear system is characterized by the system transfer function and/or the system impulse response. The problem that we are faced with is symbolically presented in Figure 6.14.

Recall that $H(s) = \mathcal{L}\{h(t)\}$. We assume that the system initial conditions are zero (system at rest) and that $f(t)$ is causal ($f(t) = 0, t < 0$). Hence, we are interested in finding the *system zero-state response* for $t \geq 0$ due to the input signal $f(t)$ applied to the system at $t = 0$.

Section 6.2 Convolution for Linear Continuous-Time Systems

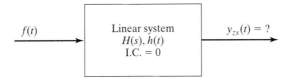

FIGURE 6.14: System response due to an arbitrary input

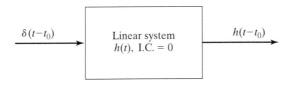

FIGURE 6.15: System response due to the impulse delta function

Since the system impulse response is known, we do know the answer to the system impulse response problem, which is symbolically presented in Figure 6.15. If we introduce **L** to denote the system action (system operation)[†] on the known input, we will have

$$h(t) = \mathbf{L}\{\delta(t)\} \tag{6.9}$$

or, similarly, by assuming time invariance,

$$h(t - t_0) = \mathbf{L}\{\delta(t - t_0)\} \tag{6.10}$$

Note that the linear system action **L** is performed in the time scale t.

We can present any input signal $f(t)$ in terms of the delta impulse signal (see Example 6.1) as

$$f(t) = \int_{-\infty}^{\infty} f(\tau)\delta(t - \tau)\,d\tau = f(t) * \delta(t) \tag{6.11}$$

This follows from the sifting property of the impulse delta function and from the definition of the convolution integral.

Applying the linear system action **L** to the input $f(t)$, we get

$$y_{zs}(t) = \mathbf{L}\{f(t)\} = \mathbf{L}\left\{\int_{-\infty}^{\infty} f(\tau)\delta(t - \tau)\,d\tau\right\}$$
$$= \int_{-\infty}^{\infty} f(\tau)\mathbf{L}\{\delta(t - \tau)\}\,d\tau = \int_{-\infty}^{\infty} f(\tau)h(t - \tau)\,d\tau = f(t) * h(t) \tag{6.12}$$

[†]To get a solution of an nth order differential equation, we basically must integrate the differential equation n times. It follows that since linear dynamic systems integrate inputs signals, **L** represents an n-tuple integral.

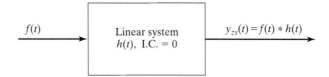

FIGURE 6.16: Zero-state system response is the convolution of the system input and the system impulse response

This formula establishes in the time domain the most fundamental result of linear system theory, which can be stated in the form of the following theorem.

THEOREM 6.1. The response of a continuous-time linear system at rest (zero-state response) due to any input is the convolution of that input and the system impulse response. □

The result of this theorem is symbolically presented in Figure 6.16.

Derivations of the discrete-time result corresponding to the one stated in Theorem 6.1 will be even more evident, due to the nice and simple structure of the discrete-time impulse delta signal (Section 6.4 and Figures 6.24–6.26).

Note that formula (6.12) can be represented as a sum of three integrals,

$$y_{zs}(t) = \int_{-\infty}^{0} f(\tau)h(t-\tau)\,d\tau + \int_{0}^{t} f(\tau)h(t-\tau)\,d\tau + \int_{t}^{\infty} f(\tau)h(t-\tau)\,d\tau \quad (6.13)$$

Since $f(t) = 0$ for $t < 0$ (causal input signal), the first integral is equal to zero. In the third integral, the integration is performed in the region $\tau > t$ where $h(t-\tau) = 0$ (causal linear system), hence, the third integral is also equal to zero. Thus, we are left with

$$y_{zs}(t) = \int_{0}^{t} f(\tau)h(t-\tau)\,d\tau \quad (6.14)$$

which produces the zero-state system response at time t due to an input signal $f(t)$. By introducing the change of variable $\sigma = t - \tau$, it can be easily shown that

$$y_{zs}(t) = \int_{0}^{t} f(\tau)h(t-\tau)\,d\tau = \int_{0}^{t} f(t-\sigma)h(\sigma)\,d\sigma \quad (6.15)$$

In the next two examples we use the convolution formula (6.15) to find the system zero-state response.

EXAMPLE 6.8

Consider a linear dynamic system represented by the differential equation

$$\frac{d^2y(t)}{dt^2} + 4\frac{dy(t)}{dt} + 3y(t) = \frac{df(t)}{dt} + 2f(t) \quad f(t) = e^{-2t}u(t), \quad y(0^-) = \frac{dy(0^-)}{dt} = 0$$

Its transfer function is given by

$$H(s) = \frac{(s+2)}{(s+1)(s+3)} = \frac{1/2}{s+1} + \frac{1/2}{s+3}$$

Applying the inverse Laplace transform to $H(s)$, we get the system impulse response as

$$h(t) = \frac{1}{2}(e^{-t} + e^{-3t})u(t)$$

The system zero-state response due to $f(t) = e^{-2t}u(t)$ is given by

$$y_{zs}(t) = \int_0^t h(\tau)f(t-\tau)\,d\tau = \frac{1}{2}\int_0^t (e^{-\tau} + e^{-3\tau})e^{-2(t-\tau)}\,d\tau$$

$$= \frac{1}{2}e^{-2t}\int_0^t (e^{\tau} + e^{-\tau})\,d\tau = \frac{1}{2}(e^{-t} - e^{-3t}), \quad t \geq 0 \qquad \blacklozenge$$

EXAMPLE 6.9

Assume that the system from Example 6.8 is driven by another external forcing function, for example $f(t) = \sin(t)u(t)$. In order to find the system zero-state response due to a new input, we need only evaluate the corresponding convolution integral (6.15), since the system impulse response is already known. In this case, we have

$$y_{zs}(t) = \int_0^t h(t-\tau)f(\tau)\,d\tau = \int_0^t \frac{1}{2}\left(e^{-(t-\tau)} + e^{-3(t-\tau)}\right)u(t-\tau)\sin(\tau)u(\tau)\,d\tau$$

$$= \frac{1}{2}e^{-t}\int_0^t e^{\tau}\sin(\tau)\,d\tau + \frac{1}{2}e^{-3t}\int_0^t e^{3\tau}\sin(\tau)\,d\tau$$

Using the integral formula given in (6.8), we obtain

$$y_{zs}(t) = \frac{1}{4}\left(e^{-t} + \frac{1}{5}e^{-3t}\right) + \frac{2}{5}\sin(t) - \frac{3}{10}\cos(t), \quad t \geq 0 \qquad \blacklozenge$$

6.3 CONVOLUTION OF DISCRETE-TIME SIGNALS

The discrete-time convolution of two signals $f_1[k]$ and $f_2[k]$ is defined in Section 2.2 as the infinite sum

$$f[k] = f_1[k] * f_2[k] = \sum_{m=-\infty}^{m=\infty} f_1[m]f_2[k-m], \quad -\infty < k < \infty \quad (6.16)$$

where k is an integer parameter and m is a dummy variable of summation.

In this section, we state the properties of the discrete-time convolution and present an efficient method for evaluating the convolution sum—the sliding tape method. In the next section, we show how to use the discrete-time convolution formula for finding the zero-state response of discrete-time linear time-invariant systems.

Most of the properties presented for continuous-time convolution are valid in the case of discrete-time convolution. However, some of them must be slightly modified. The properties of the discrete-time convolution are as follows.

1. *Commutativity*

$$f_1[k] * f_2[k] = f_2[k] * f_1[k] \quad (6.17)$$

2. *Distributivity*

$$f_1[k] * \{f_2[k] + f_3[k]\} = f_1[k] * f_2[k] + f_1[k] * f_3[k] \quad (6.18)$$

3. *Associativity*

$$f_1[k] * \{f_2[k] * f_3[k]\} = \{f_1[k] * f_2[k]\} * f_3[k] \quad (6.19)$$

4. *Duration*

The duration of a discrete-time signal $f[k]$ is defined by the discrete time instants k_0 and k_f for which the discrete-time signal $f[k] = 0$ for every k outside the interval $[k_0, k_f]$. We use M to denote the discrete-time signal duration. It follows that $M = k_f - k_0$.

Let the signals $f_1[k]$ and $f_2[k]$ have durations respectively given by M_1 and M_2, then the duration of their convolution, $f[k] = f_1[k] * f_2[k]$, is given by $M_1 + M_2$.

The discrete-time convolution duration property can be also expressed in terms of the number of signal samples. Let the number of samples in the signal (signal size) be denoted by L, then $L = M + 1$. Consider two signals $f_1[k]$ and $f_2[k]$, with the number of samples respectively given by L_1 and L_2. The number of samples in their convolution signal is equal to $L_1 + L_2 - 1$, which corresponds to the duration of $L_1 + L_2 - 2 = M_1 + M_2$.

5. *Time Shifting*

Let $f[k] = f_1[k] * f_2[k]$. Then, convolutions of the shifted functions are given by

$$\begin{aligned} f_1[k-k_1] * f_2[k] &= f[k-k_1] \\ f_1[k] * f_2[k-k_2] &= f[k-k_2] \\ f_1[k-k_1] * f_2[k-k_2] &= f[k-k_1-k_2] \end{aligned} \quad (6.20)$$

The proofs of these properties are based on the same concepts as the proofs of the corresponding properties of continuous-time convolution. For example, in order to establish the commutativity property, we must introduce the change of variables $k - m = n$ in (6.16), which leads to

$$f[k] = f_1[k] * f_2[k] = \sum_{m=-\infty}^{m=\infty} f_1[m]f_2[k-m]$$

$$= \sum_{n=-\infty}^{n=\infty} f_1[k-n]f_2[n] = f_2[k] * f_1[k]$$

The discrete-time convolution can be calculated analytically in some cases, as demonstrated in the examples given below. The general method for evaluating the convolution sum efficiently, known as the sliding tape method, is presented in the next section.

EXAMPLE 6.10

Consider the convolution of the discrete-time impulse delta function with a general function $f[k]$, that is,

$$f[k] * \delta[k] = \sum_{m=-\infty}^{m=\infty} f[m]\delta[k-m], \quad -\infty < k < \infty$$

Since the discrete-time impulse delta function is equal to zero everywhere except at $k = m$, where it is equal to 1, we conclude that

$$f[k] * \delta[k] = f[k] \qquad (6.21)$$

EXAMPLE 6.11

Consider the convolution of two causal exponential functions defined by $f_1[k] = a^k u[k]$ and $f_2[k] = b^k u[k]$. Using the convolution definition formula, we have

$$a^k u[k] * b^k u[k] = \sum_{m=-\infty}^{m=\infty} a^m u[m] b^{k-m} u[k-m] = \sum_{m=0}^{m=k} a^m b^{k-m} = b^k \sum_{m=0}^{m=k} \left(\frac{a}{b}\right)^m$$

Using the known summation formula (see Appendix B),

$$\sum_{m=0}^{k} \alpha^m = \frac{1 - \alpha^{k+1}}{1 - \alpha} \qquad (6.22)$$

we obtain

$$a^k u[k] * b^k u[k] = b^k \frac{1 - (a/b)^{k+1}}{1 - (a/b)} = \frac{b^{k+1} - a^{k+1}}{b - a}$$

EXAMPLE 6.12

The convolution of the unit step signal and any causal signal ($f[k] = 0, k < 0$) produces

$$f[k] * u[k] = \sum_{m=0}^{\infty} f[m]u[k-m] = \sum_{m=0}^{k} f[m] = f[0] + f[1] + \cdots + f[k]$$ ∫

EXAMPLE 6.13

Convolution of two causal signals $f_1[k]$ and $f_2[k]$ produces

$$\begin{aligned} f_1[k] * f_2[k] &= \sum_{m=-\infty}^{\infty} f_1[m]f_2[k-m] = \sum_{m=0}^{k} f_1[m]f_2[k-m] \\ &= f_1[0]f_2[k] + f_1[1]f_2[k-1] + \cdots \\ &\quad + f_1[k-1]f_2[1] + f_1[k]f_2[0], \quad k = 0, 1, 2, \ldots \end{aligned}$$

which represents an easy-to-remember formula. ∫

In the next section we present a method used for actual evaluation of the convolution sum. Known as the sliding tape method, it can be easily programmed for computer applications.

6.3.1 Sliding Tape Method

The sliding tape method represents an efficient procedure for evaluating the convolution sum. As with the continuous-time convolution, the discrete-time convolution requires the "flip and slide" steps. For simplicity, we will explain the method using two causal signals. However, the method is applicable to any two discrete-time signals. Note that by using the discrete-time convolution shifting property (6.20), this method can be also applied to noncausal signals. The sliding tape method is presented in the following three steps.

Step 1. The signal values are recorded on two tapes, one tape for the values of the signal $f_1[m]$ and another tape for the values of the signal $f_2[m]$. In Figure 6.17, the sliding tape method is demonstrated on an example of two causal signals represented by

$$f_1[0], \ f_1[1], \ f_1[2], \ \ldots, \ f_1[M_1-1], \ f_1[M_1]$$

$$f_2[0], \ f_2[1], \ f_2[2], \ \ldots, \ f_2[M_2-1], \ f_2[M_2]$$

Note that the durations of these signals, which contain $L_1 = M_1 + 1$ and $L_2 = M_2 + 1$ samples (values), are M_1 and M_2.

Step 2. One of the tapes, say the second tape, is flipped about its value at $f_2[0]$ to form the signal $f_2[-m]$ (see Figure 6.17(b)). It should be pointed out that the signal $f_2[m]$ is flipped such that the signal value $f_2[0]$ remains in the same position.

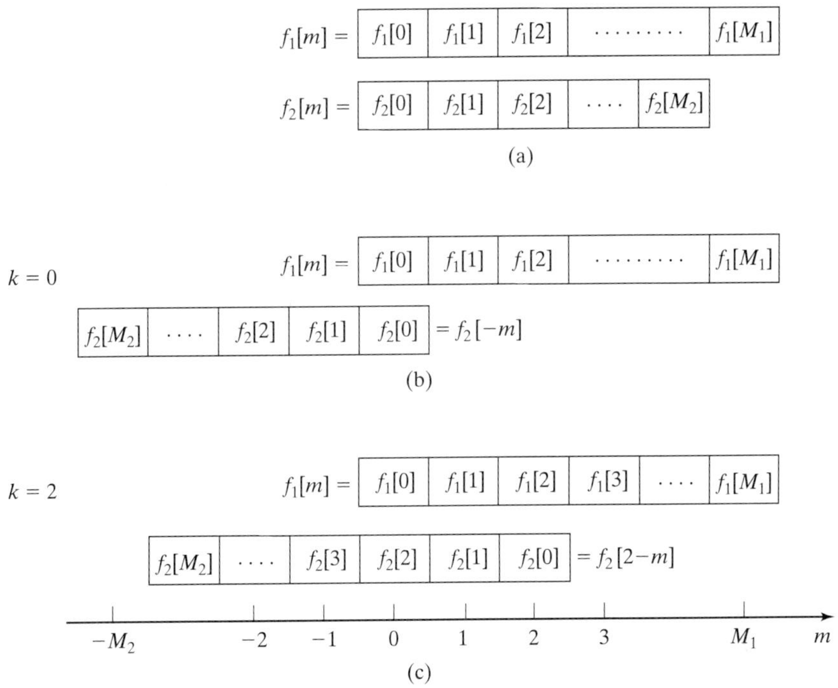

Section 6.3 Convolution of Discrete-Time Signals

$f_1[m] = \boxed{f_1[0] \mid f_1[1] \mid f_1[2] \mid \cdots\cdots \mid f_1[M_1]}$

$f_2[m] = \boxed{f_2[0] \mid f_2[1] \mid f_2[2] \mid \cdots \mid f_2[M_2]}$

(a)

$k = 0$

$f_1[m] = \boxed{f_1[0] \mid f_1[1] \mid f_1[2] \mid \cdots\cdots \mid f_1[M_1]}$

$\boxed{f_2[M_2] \mid \cdots \mid f_2[2] \mid f_2[1] \mid f_2[0]} = f_2[-m]$

(b)

$k = 2$

$f_1[m] = \boxed{f_1[0] \mid f_1[1] \mid f_1[2] \mid f_1[3] \mid \cdots \mid f_1[M_1]}$

$\boxed{f_2[M_2] \mid \cdots \mid f_2[3] \mid f_2[2] \mid f_2[1] \mid f_2[0]} = f_2[2-m]$

$-M_2 \quad -2 \quad -1 \quad 0 \quad 1 \quad 2 \quad 3 \quad M_1 \quad m$

(c)

FIGURE 6.17: Graphical representation of the sliding tape method

Step 3. The second tape is shifted to the left and right; that is, a traveling signal $f_2[k-m]$ is formed. The parameter k is an integer that theoretically takes all values from $-\infty$ to ∞. Practically, we must shift the second signal only for those values of k for which the convolution sum is different from zero. In that respect, the duration property of the discrete-time convolution plays an important role. After we shift the second tape for the given value of k, we evaluate the products of the corresponding overlapping signal values on the tapes. The sum of all products gives the convolution value for the chosen value of the parameter k (see Figure 6.17(c)). This procedure is repeated for all values of k for which the convolution sum may be different from zero.

Let $f[k] = f_1[k] * f_2[k]$. From Figure 6.17(b), we see that for $k = 0$ only $f_1[0]$ and $f_2[0]$ overlap, hence $f[0] = f_1[0]f_2[0]$. From Figure 6.17(c), drawn for $k = 2$, we obtain three pairs of the overlapped signal values, hence the convolution of these two signals for $k = 2$ is given by $f[2] = f_1[0]f_2[2] + f_1[1]f_2[1] + f_1[2]f_2[0]$. Similarly, we evaluate the discrete-time convolution for other values of k. Note that for $k \leq -1$ the signals $f_1[m]$ and $f_2[k-m]$ do not overlap, hence the convolution is equal to zero for $k \leq -1$. Also, no overlapping between the values of $f_1[m]$ and $f_2[k-m]$ exists for $k \geq M_1 + M_2 + 1$, and the corresponding discrete-time convolution is equal to zero in this interval.

In Examples 6.14 and 6.15, we will demonstrate the sliding tape method on signals represented by numerical values. We assume that the signals have finite durations.

EXAMPLE 6.14

Let two signals be defined as follows:

$$f_1[k] = \begin{cases} 1, & k=0 \\ 2, & k=1 \end{cases}, \qquad f_2[k] = \begin{cases} -1, & k=0 \\ 3, & k=1 \end{cases}$$

The durations of these signals are $M_1 = M_2 = 1$. By the convolution duration property, the convolution sum may be different from zero in the time interval of length $M = M_1 + M_2 = 1 + 1 = 2$. Tapes for $f_1[m]$ and $f_2[-m]$ are shown in Figure 6.18.

The convolution of these two signals, $f[k] = f_1[k] * f_2[k]$, for $k = 0$, is easily obtained from Figure 6.18 as $f[0] = f_1[0] * f_2[0] = 1 \times (-1) = -1$. If we slide the second tape to the left, which corresponds to $k \leq -1$, we see that the convolution is equal to zero. Sliding the second tape to the right for $k = 1$, we obtain $f_1[1] = 1 \times 3 + 2 \times (-1) = 1$ (see Figure 6.19).

For $k = 2$, according to Figure 6.20, the convolution is given by $f[2] = 2 \times 3 = 6$. For $k \geq 3$, the signals $f_1[m]$ and $f_2[k-m]$ do not overlap, hence, the convolution is equal to zero in this interval.

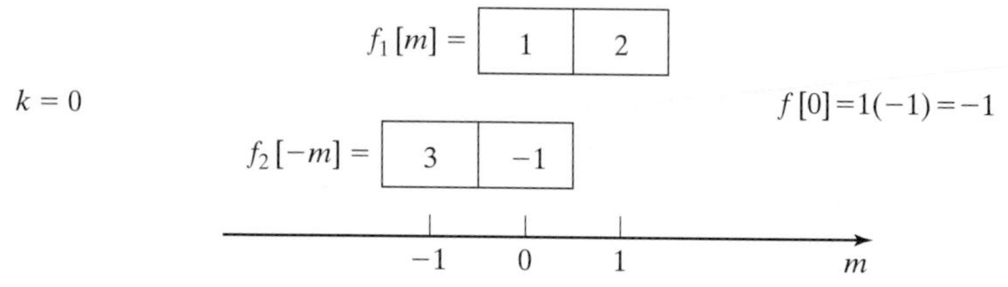

FIGURE 6.18: The sliding tape method for Example 6.14, $k = 0$

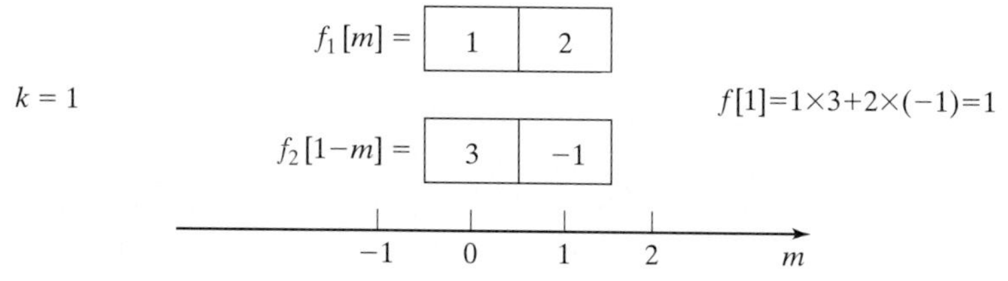

FIGURE 6.19: The sliding tape method for Example 6.14, $k = 1$

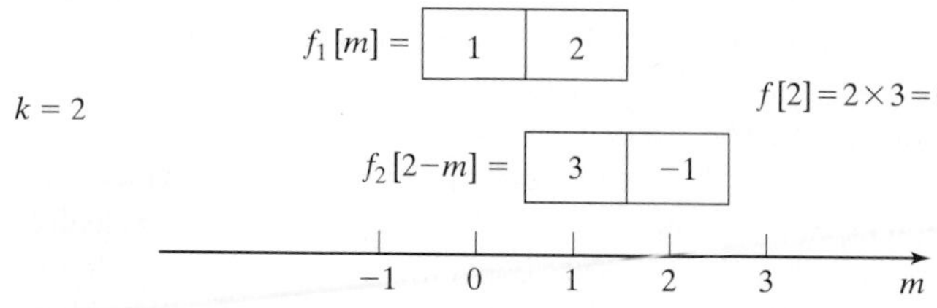

FIGURE 6.20: The sliding tape method for Example 6.14, $k = 2$

In summary, we have obtained

$$f[k] = f_1[k] * f_2[k] = \begin{cases} 0, & k \leq -1 \\ -1, & k = 0 \\ 1, & k = 1 \\ 6, & k = 2 \\ 0, & k \geq 3 \end{cases}$$

EXAMPLE 6.15

Let us find the convolution of the following two signals by using the sliding tape method.

$$f_1[k] = \begin{cases} -2, & k = -1 \\ 1, & k = 0 \\ 3, & k = 1 \\ 0, & \text{otherwise} \end{cases}, \quad f_2[k] = \begin{cases} 2, & k = 0 \\ 3, & k = 1 \\ -1, & k = 2 \\ 1, & k = 3 \\ 0, & \text{otherwise} \end{cases}$$

Note that the first signal is noncausal. However, the sliding tape procedure to be applied is exactly the same as in the case of causal signals. The durations of these signals are respectively given by $M_1 = 2$ and $M_2 = 3$, hence the duration of their convolution is equal to $M_c = M_1 + M_2 = 5$, which means that at most six ($L_c = M_c + 1$) discrete-time instants the convolution sum may be different from zero.

In Figure 6.21, we present the signals $f_1[m]$ and $f_2[-m]$. It can be seen from this figure that $f[0] = 1 \times 2 + (-2) \times 3 = -4$. It can be also concluded that the convolution is equal to zero for $k \leq -2$ and $k \geq 5$, which is consistent with the discrete-time convolution duration property.

In Figure 6.22, we present the sliding tapes for $k = -1$ and $k = 1$. It can be seen that $f[-1] = (-2) \times 2 = -4$ and $f[1] = (-2) \times (-1) + 1 \times 3 + 3 \times 2 = 11$.

Figure 6.23 presents the situation for $k = 2, 3, 4$. It can be seen from this figure that $f[2] = (-2) \times 1 + 1 \times (-1) + 3 \times 3 = 6$, $f[3] = 1 \times 1 + 3 \times (-1) = -2$, and $f[4] = 3 \times 1 = 3$.

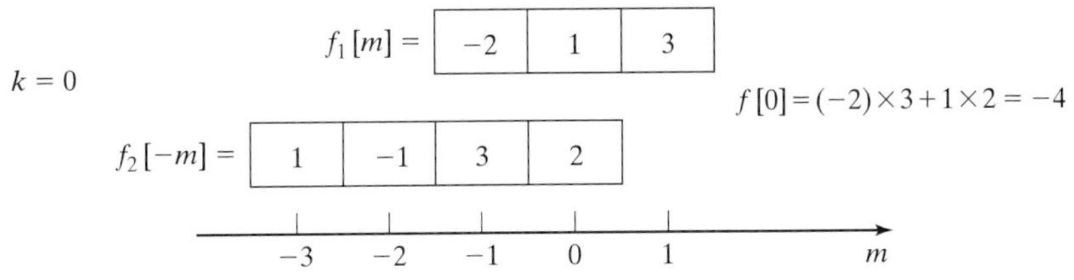

FIGURE 6.21: The sliding tapes for Example 6.15, $k = 0$

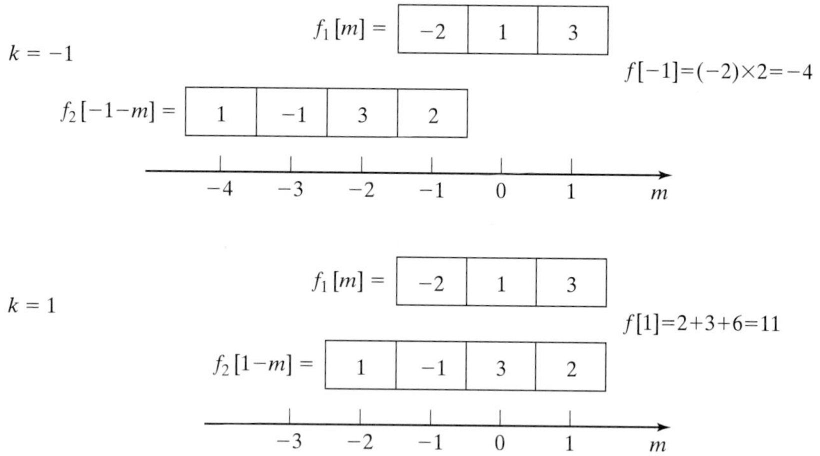

302 Chapter 6 Convolution

$k = -1$

$f_1[m] = \boxed{-2 \mid 1 \mid 3}$

$f_2[-1-m] = \boxed{1 \mid -1 \mid 3 \mid 2}$

$f[-1] = (-2) \times 2 = -4$

$k = 1$

$f_1[m] = \boxed{-2 \mid 1 \mid 3}$

$f_2[1-m] = \boxed{1 \mid -1 \mid 3 \mid 2}$

$f[1] = 2 + 3 + 6 = 11$

FIGURE 6.22: The sliding tapes for Example 6.15, $k = -1, 1$

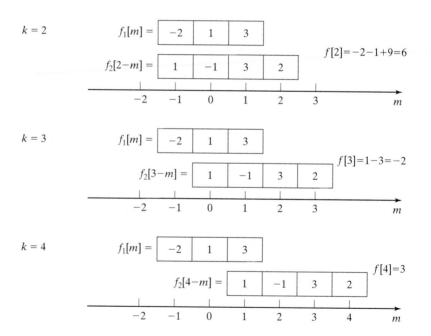

FIGURE 6.23: The sliding tape method for Example 6.15, $k = 2, 3, 4$

In summary, we have the following values for the convolution of the considered signals:

$$f[k] = f_1[k] * f_2[k] = \begin{cases} 0, & k \leq -2 \\ -4, & k = -1 \\ -4, & k = 0 \\ 11, & k = 1 \\ 6, & k = 2 \\ -2, & k = 3 \\ 3, & k = 4 \\ 0, & k \geq 5 \end{cases}$$

6.4 CONVOLUTION FOR LINEAR DISCRETE-TIME SYSTEMS

In this section we show how to use the discrete-time convolution in order to find the zero-state response of discrete-time linear time-invariant systems. We saw in Chapter 5 that every discrete-time linear time-invariant system is uniquely characterized by either its transfer function or its impulse response. We also saw that the system transfer function is the $\mathcal{Z}$-transform of the system impulse response, and that the system impulse response is obtained as the inverse $\mathcal{Z}$-transform of the system transfer function. Let us assume that the system initial conditions are set to zero and that the system impulse response is known. This can be symbolically represented as in Figure 6.24 (where I.C. stands for initial conditions). By the time-invariance principle, $\delta[k-m]$ produces the system output $h[k-m]$. By the linearity principle, a weighted impulse delta function $f[m]\delta[k-m]$, where $f[m]$ is a constant, produces the system output signal $f[m]h[k-m]$ (see Figure 6.25). Note that we assume that the system in Figures 6.24 and 6.25 is represented in its integral formulation, for which we have defined the system transfer function and the system impulse response, respectively, in formulas (5.47) and (5.51).

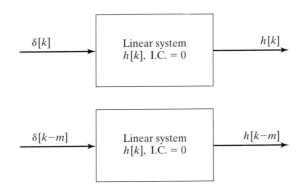

FIGURE 6.24: Discrete-time system impulse response

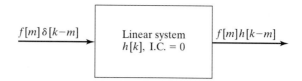

FIGURE 6.25: Weighted system impulse response

Our goal is to find the system zero-state response due to any input function $f[k]$. In that respect, we have assumed that the system initial conditions are equal to zero, that is,

$$y[-n] = y[-(n-1)] = \cdots = y[-2] = y[-1] = 0 \qquad (6.23)$$

Let the system input signal be causal ($f[k] = 0$, $k < 0$), and defined by its values at discrete-time instants

$$f[0], \ f[1], \ f[2], \ \ldots, \ f[k]$$

For any value of k, we can represent $f[k]$ by a weighted sum of impulse delta signals,

$$f[k] = \sum_{m=0}^{m=k} f[m]\delta[k-m] = f[k] * \delta[k] \qquad (6.24)$$

which is a special case of formula (6.21). By the linearity and time-invariance principles, we know that a sum of weighted and shifted impulse delta signals produces in the system output a sum of weighted and shifted system impulse response signals. Hence, using $f[k]$ as the system input, we obtain the system output in the form

$$y_{zs}^i[k] = \sum_{m=0}^{m=k} f[m]h[k-m] = f[k] * h[k] \qquad (6.25)$$

which is symbolically presented in Figure 6.26. Note that the superscript "i" indicates the integral formulation of a discrete-time linear system (see Section 5.3.1) that is consistent with the initial conditions defined in (6.23). The preceding derivations establish the fundamental result of theory of linear discrete-time systems, which is restated in the following theorem.

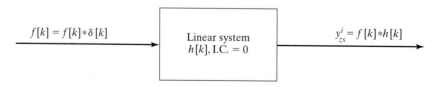

FIGURE 6.26: Discrete-time system response as the convolution of a system input and the system impulse response

THEOREM 6.2. The response of a linear discrete-time system at rest (zero initial condition response) due to any input is the convolution of that input and the system impulse response. □

In addition to its importance for linear discrete-time dynamic systems, the discrete-time convolution is also very important for digital signal processing. For applications of discrete-time convolution in signal processing, the reader is referred to any undergraduate book on signal processing, for example [1].

6.5 NUMERICAL CONVOLUTION USING MATLAB[†]

In this section we discuss the use of a computer in evaluating continuous-time convolution through an approximate numerical integration technique. In addition, we demonstrate the use of the MATLAB function conv for finding the discrete-time convolution.

Consider the convolution integral (6.14), which has a particular importance for linear dynamic systems; that is,

$$y_{zs}(t) = \int_0^t f(\tau)h(t-\tau)\,d\tau \qquad (6.26)$$

Sometimes evaluation of the convolution integral can be quite tedious. In such cases, it may be useful to incorporate the use of a digital computer to evaluate the convolution integral numerically. This can be accomplished via the use of a programming language like C or FORTRAN or a mathematical software package such as MATLAB. In this discussion, the focus will be on numerical integration using MATLAB. To evaluate the convolution integral with MATLAB, the integral must be approximated by a summation. When the input signal and the impulse response are continuous-time signals, this implies that the signals must be discretized by sampling before the convolution summation can be evaluated. The result of the summation can closely approximate the exact expression of the convolution integral if the sampling period T of the signals is kept small. Let $t = kT$ and $\tau = iT$, then the convolution summation (approximation of integral (6.26)) can be expressed as

$$y_{zs}(kT) \approx T\sum_{i=0}^{kT} f(iT)h((k-i)T) \triangleq y_{zs}[k] \approx T\sum_{i=0}^{k} f[i]h[k-i] \qquad (6.27)$$

In this formula, k indicates the sample number in the discrete-time signal $y[\cdot]$, and i denotes a dummy variable of summation.

Before the convolution can be performed, the discrete-time signals must be represented such that MATLAB can easily perform operations on the signals required in the numerical convolution. MATLAB will represent signals by vectors. The contents of each vector will be the actual value of the signal at a particular discrete-time instant. Because *all vectors in* MATLAB *begin with an index of one,* the first available sample of each signal

---

[†]This section is based on the notes of Mark Wehle, who was a teaching assistant for the course in Linear Systems and Signals at Rutgers University Department of Electrical Engineering, 1995.

is the first element in its respective MATLAB vector. Thus, $f[1]$ indicates the first available sample in the input signal. Similarly, $h[1]$ represents the first available sample in the system impulse response. Note that the vector index indicates the sample number of the signal. For example, suppose that the very first time sample in the input signal occurs at time 5 seconds and that a sampling period of $T = 0.1$ is used. This means that the value of the input signal at time 5 seconds is stored in $f[1]$ and the value of the input signal at 5.1 seconds is stored in $f[2]$. It is important to note that when the convolution is performed, the value stored in $f[1]$ must occur at the same time as that stored in $h[1]$, and the value stored in $f[2]$ must also occur at the same time as that stored in $h[2]$, and so on.

To begin the convolution, as with any graphical convolution process, one of the two signals being convolved must be chosen as the traveling signal. In this case, h will be chosen as the MATLAB vector representation of the traveling signal. Note that because the convolution is commutative, f could have been chosen as the traveling signal with the same system output result. Choosing h as a traveling signal means that before the convolution can be performed, h must first be time-reversed and then shifted in time. This denotes a "flip and slide" operation.

Let the number of samples (elements) in the vector f be L_f and the number of elements in h be L_h. Due to the fact that signals in MATLAB are represented by vectors that begin with an index of one, formula (6.27) must be modified appropriately to

$$y_{zs}[k] \approx T \sum_{i=1}^{k} f[i]h[k-(i-1)], \quad 1 \leq k \leq L_f + L_h - 1 \tag{6.28}$$

The amount of sliding is controlled by the value of k, which can be thought of as the index of element in the output vector y_{zs} that is currently being computed by the convolution sum. As k changes, there will be values for which h and f "overlap" in time. It is for these values of k that the convolution sum may be nonzero. Due to the "flip and slide" operation, and for the purpose of writing a MATLAB program to implement the convolution summation, we must define more precisely the range of values that the index i takes. It is not difficult to conclude from the convolution process in which f is the stationary signal that contains L_f samples, and h is the traveling signal containing L_h samples, that the required summation limits are

$$y_{zs}[k] \approx T \times \sum_{i=\max(1,k-(L_h-1))}^{\min(k,L_f)} f[i]h[k-(i-1)], \quad 1 \leq k \leq L_f + L_h - 1 \tag{6.29}$$

Assuming that the sampling period T and the vectors representing the signals f and h are specified, formula (6.29) can be easily programed in MATLAB as follows:

```
Lf=length(f)
Lh=length(h)
for k=1:Lf+Lh-1
y(k)=0;
for i=max(1,k-(Lh-1)):min(k,Lf)
```

```
y(k)=y(k)+f(i)*h(k-i+1);
end
yzsappr(k)=T*y(k);
end
```

In the presented program, it is important to note that L_h is the length of the traveling signal (in this case h). The first for loop represents the values of k for which the convolution may be nonzero. The second for loop denotes the limits of the i variable.

Note that MATLAB has the built-in function conv that performs discrete-time convolution given two vectors f and h that represent discrete-time (or discretized) signals. We can use this function as well to solve the convolution problem considered in this section. In such a case, the preceding MATLAB convolution program can be replaced by only one statement: yzsappr=T*conv(f,h). However, writing and using the presented MATLAB program gives a better insight into the mechanism of the convolution process.

It should be emphasized that in both cases (the use of the given program and the use of the MATLAB function conv) we must interpret the convolution results obtained and properly place them in the convolution duration interval. In that respect, the obtained value yzsappr[1] is in fact $y_{\text{zsappr}}(0)$, yzsappr[2] is $y_{\text{zsappr}}(1T)$, yzsappr[3] is $y_{\text{zsappr}}(2T)$, and so on, which is the consequence of MATLAB's representation of signals (vectors whose first index is one).

The remaining issue is how to perform sampling of a continuous-time signal. This can be easily done in MATLAB as shown in the next example, where the procedure for the approximate evaluation of the convolution integral is demonstrated for the problem solved analytically in Example 6.9.

EXAMPLE 6.16

It is known from Example 6.9 that the system, represented by its impulse response $h(t) = 0.5(e^{-t} + e^{-3t})u(t)$, driven by the input signal equal to $f(t) = \sin(t)u(t)$, has the zero-state response

$$y_{zs}(t) = \frac{1}{4}\left(e^{-t} + \frac{1}{5}e^{-3t}\right) + \frac{2}{5}\sin(t) - \frac{3}{10}\cos(t), \quad t \geq 0$$

Consider the time interval from 0 to 10 seconds, and take the sampling period $T = 0.1$ s. The approximate zero-state response can be obtained using MATLAB as follows.

```
% choose the sampling period T
T=0.1
t=0:T:10
f=sin(t); % calculates f[k], k=1,2,..., 101
h=0.5*(exp(-t)+exp(-3t)); % calculates h[k]
% the above statements perform the sampling operation
```

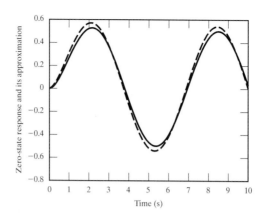

FIGURE 6.27: Exact (solid line) and approximate for $T = 0.1$ s (dashed line) zero-state responses

```
% calculations of the approximate response
yzsappr=T*conv(f,h);
% calculations of the exact response
yzs=0.25*(exp(-t)+0.2*exp(-3*t))+0.4*sin(t)-0.3*cos(t)
plot(t,yzs,t,yzsappr(1:length(t)),'- -')
xlabel('Time (s)')
ylabel('Zero-state response and its approximation')
```

In Figure 6.27 we show the exact and approximate (for $T = 0.1$ s) zero-state responses.

It can be seen that the approximation obtained is pretty good. To improve the approximation we must take more samples, that is, we must decrease the sampling period T. Try $T = 0.01$ and $T = 0.001$.

Note that the continuous-time convolution MATLAB program presented in this section for approximating the zero-state system response can also be used in an approximate procedure for finding the convolution of continuous-time signals. This will be demonstrated in the MATLAB laboratory experiments.

The MATLAB function conv can be used for calculation of the discrete-time convolution, which is now demonstrated for the problems solved in Examples 6.14 and 6.15.

EXAMPLE 6.17

Consider the discrete-time convolution problem from Example 6.14. A very simple MATLAB code,

```
f1=[1 2]; f2=[-1 3]; f=conv(f1, f2)
```

produces the result $f = [-1\ 1\ 6]$. Bearing in mind the way MATLAB treats vector indices (all vectors begin with an index equal to one) and using the convolution duration property, we interpret these values as the convolution results for $k = 0, 1, 2$. All other values of the convolution are equal to zero.

Similarly, we can solve the problem from Example 6.15 with the MATLAB statements

```
f1=[-2 1 3]; f2=[2 3 -1 1]; f=conv(f1,f2)
```

which produce $f = [-4\ -4\ 11\ 6\ -2\ 3]$. We must use the convolution duration property in order to relate these values to the discrete-time axis k. According to Example 6.15, the values obtained are respectively the convolution values at the discrete-time instants $k = -1, 0, 1, 2, 3, 4$.

6.6 MATLAB LABORATORY EXPERIMENTS ON CONVOLUTION

Purpose: In this section we design two experiments dealing with continuous- and discrete-time convolutions and their applications to linear continuous- and discrete-time dynamic systems. The purpose of the first experiment is to present the convolution operator, and to demonstrate some of its properties in both continuous- and discrete-time domains. By writing and modifying the corresponding MATLAB programs, students will master every step of the convolution process. In the second experiment, the convolution method will be used to determine the zero-state responses of both continuous- and discrete-time linear dynamic systems by using the famous formula that states that *the response of a linear system at rest due to an arbitrary input is the convolution of that input with the system impulse response.*

6.6.1 Convolution of Signals

In this experiment, students must verify the commutativity and associativity properties of continuous- and discrete-time convolutions.

Part 1. Continuous-Time Signals

The convolution of two signals is defined by

$$f_1(t) * f_2(t) = \int_{-\infty}^{\infty} f_1(\tau) f_2(t-\tau)\, d\tau, \quad -\infty < t < \infty \quad (6.30)$$

Consider the continuous-time signals

$$f_1(t) = p_2(t), \quad f_2(t) = \Delta_2(t), \quad f_3(t) = [u(t) - u(t-5)]$$

Write a MATLAB program to perform and verify the continuous-time convolution commutativity and associativity properties:

$$f_1(t) * f_2(t) = f_2(t) * f_1(t)$$
$$f_1(t) * (f_2(t) + f_3(t)) = f_1(t) * f_2(t) + f_1(t) * f_3(t)$$

While writing the program you must discretize the convolution integral given in (6.30); that is, the integral in (6.30) must be approximated by a finite sum of the form

$$f_1(kT) * f_2(kT) \approx T \sum_{i=k_0}^{k} f_1(iT) f_2[(k-i)T] \tag{6.31}$$

Take $T = 0.1$, plot the results obtained, and comment on the durations of the signals obtained through the convolution procedure. Observe that the commutativity and associativity properties hold for the approximate continuous-time convolution, and conclude that when $T \to 0$ these properties also hold for the convolution integral (6.30).

Part 2. Discrete-Time Signals
In the discrete-time domain, the convolution of two signals is defined by

$$f_1[k] * f_2[k] = \sum_{m=-\infty}^{m=\infty} f_1[m] f_2[k-m], \quad -\infty < k < \infty \tag{6.32}$$

Consider the discrete-time signals

$$f_1[k] = \text{sinc}[k]\{u[k+5] - u[k-5]\}$$
$$f_2[k] = 1 - p_2[k], \quad 0 \leq k \leq 5$$
$$f_3[k] = u[k] - u[k-5]$$

Write a MATLAB program to perform the operations defined in (6.17–18). Plot the results obtained and comment on the convolution signal durations. Verify the results by using the MATLAB function conv.

Hint: Use the MATLAB program given in Section 6.5 and write its modifications. The program can be used for calculating both the continuous-time and discrete-time convolutions. Note that you must plot the convolution results obtained in the proper time intervals. Use plot(t,fapr) with t=t1+t2:T:T1+T2, where $[t_i, T_i], i = 1, 2$, define the duration intervals of the signals being convolved.

6.6.2 Convolution for Linear Dynamic Systems

In this experiment, students are required to use the convolution operator to find the system zero-state response for both continuous- and discrete-time linear dynamic systems. MATLAB programs developed for the convolution of two signals in the previous experiment can be used in this experiment, subject to minor modifications.

Part 1. Continuous-Time Systems

The output of a linear continuous-time system at rest due to any input $f(t)$ is given by the convolution formula

$$y_{zs}(t) = \int_0^t f(\tau) h(t-\tau)\, d\tau \qquad (6.33)$$

Consider the continuous-time system at rest represented by the differential equation

$$\ddot{y}(t) + 4\dot{y}(t) + 3y(t) = \dot{f}(t) + 5f(t)$$

Take $t = 5$ s and discretize integral (6.33) as given in (6.31) with $T = 0.1$. Use MATLAB and the convolution procedure to find and plot

(a) the impulse response of the system;
(b) the step response of the system; and
(c) the system zero-state response due to the input $\sin(2t)$.

For (b) and (c) use the MATLAB program from Section 6.5.
Verify the results obtained by using the MATLAB functions `step` and `lsim`.

Part 2. Discrete-Time Systems

Consider the discrete-time system represented by the difference equation

$$y[k+2] - 0.5y[k+1] + 0.06y[k] = f[k+1] - 4f[k]$$

Use the convolution technique and MATLAB to find and plot

(a) the impulse response of the system;
(b) the step response of the system; and
(c) the zero-state response to the input $\sin[2k]$.

Plot the zero-state responses during the first ten discrete-time instants. Verify the results using the MATLAB functions `dstep` and `dlsim`. Submit all plots and comment on the results obtained.

6.7 SUMMARY

Study Guide for Chapter Six: The knowledge that *the response of a linear system at rest (zero-state response) due to an arbitrary input is the convolution of that input and the system impulse response* is essential. Understanding the graphical convolution and the sliding tape method is important. Knowledge of the properties of the continuous- and discrete-time convolutions is useful. Standard problems: (1) find continuous-time signal convolution; (2) find discrete-time signal convolution; (3) find the zero-state response of continuous-time linear systems; (4) find the zero-state response of discrete-time linear systems.

Continuous-Time Convolution:

$$f(t) = f_1(t) * f_2(t) = \int_{-\infty}^{\infty} f_1(\tau) f_2(t - \tau) \, d\tau = f_2(t) * f_1(t), \quad -\infty < t < \infty$$

Continuous-Time Convolution Duration Property:

Let $f_1(t)$ and $f_2(t)$ be two time-limited signals with the duration intervals of $[t_1, T_1]$ and $[t_2, T_2]$, respectively. Then, the duration of their convolution, $f(t) = f_1(t) * f_2(t)$, is given by $[t, T] = [t_1 + t_2, T_1 + T_2]$.

Zero-State Response of a Linear Continuous-Time System:

$$y_{zs}(t) = \int_0^t f(\tau) h(t - \tau) \, d\tau = \int_0^t f(t - \sigma) h(\sigma) \, d\sigma$$

Discrete-Time Convolution:

$$f[k] = f_1[k] * f_2[k] = \sum_{m=-\infty}^{m=\infty} f_1[m] f_2[k - m] = f_2[k] * f_1[k], \quad -\infty < k < \infty$$

Discrete-Time Convolution Duration Property:

Let $f_1[k]$ and $f_2[k]$ be two discrete-time time-limited signals with the durations (number of samples) M_1 ($L_1 = M_1 + 1$) and M_2 ($L_2 = M_2 + 1$), respectively. Then, the duration (number of samples) of their convolution, $f[k] = f_1[k] * f_2[k]$, is given by $M_c = M_1 + M_2$ ($L_c = L_1 + L_2 - 1$).

Zero-State Response of a Linear Discrete-Time System:

$$y_{zs}^i[k] = \sum_{m=0}^{m=k} f[m] g[k - m] = \sum_{m=0}^{m=k} f[k - m] g[m]$$

6.8 REFERENCE

[1] S. Orfanidis, *Introduction to Signal Processing*, Prentice Hall, Upper Saddle River, NJ, 1996.

6.9 PROBLEMS

6.1. Solve the convolution problem from Example 6.7 by using the graphical convolution method combined with the convolution time shifting property.

6.2. Use the graphical method to convolve $p_2(t)$ with $p_2(t)$.
Answer:

$$p_2(t) * p_2(t) = \begin{cases} 0, & t \leq -2 \\ t + 2, & -2 \leq t \leq 0 \\ 2 - t, & 0 \leq t \leq 2 \\ 0, & t \geq 2 \end{cases}$$

6.3. Use the graphical method to convolve the signals presented in Figure 6.28.

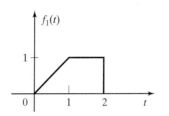
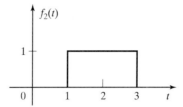

FIGURE 6.28

Answer:

$$f_1(t) * f_2(t) = \begin{cases} 0, & t \leq 2 \\ \frac{1}{2}(t-2)^2, & 2 \leq t \leq 3 \\ t - \frac{5}{2}, & 3 \leq t \leq 4 \\ \frac{3}{2} - \frac{1}{2}(t-4)^2, & 4 \leq t \leq 5 \\ 6 - t, & 5 \leq t \leq 6 \\ 0, & t \geq 6 \end{cases}$$

6.4. Use the graphical method to convolve the triangularly shaped signals presented in Figure 6.29.

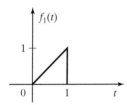
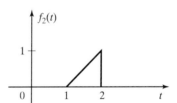

FIGURE 6.29

6.5. Convolve graphically the signals given in Figure 6.30.

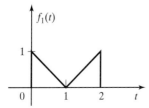
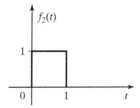

FIGURE 6.30

Answer:

$$f_1(t) * f_2(t) = \begin{cases} 0, & t \leq 0 \\ -\frac{1}{2}t^2 + t, & 0 \leq t \leq 1 \\ t^2 - 3t + \frac{5}{2}, & 1 \leq t \leq 2 \\ \frac{1}{2}(-t^2 + 4t - 3), & 2 \leq t \leq 3 \\ 0, & t \geq 3 \end{cases}$$

6.6. Convolve graphically the signals given in Figure 6.31.

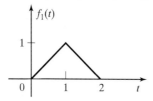 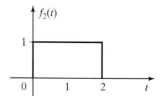

FIGURE 6.31

6.7. Using the convolution method, find the zero-state response of the following systems due to the input signal given by $f(t) = e^{-3t}u(t)$.

(a) $\dfrac{d^2 y(t)}{dt^2} + 2\dfrac{dy(t)}{dt} + y(t) = f(t)$

(b) $\dfrac{d^2 y(t)}{dt^2} + 7\dfrac{dy(t)}{dt} + 10y(t) = 2\dfrac{df(t)}{dt} + 3f(t)$

(c) $\dfrac{d^3 y(t)}{dt^3} + 3\dfrac{d^2 y(t)}{dt^2} + 3\dfrac{dy(t)}{dt} + y(t) = \dfrac{d^2 f(t)}{dt^2} + 2\dfrac{df(t)}{dt} + 3f(t)$

6.8. Repeat Problem 6.7 for the input signal $f(t) = p_2(t-1) = u(t) - u(t-2)$.

6.9. Use the convolution method to find the zero-state response of the system

$$\frac{d^2 y(t)}{dt^2} + 5\frac{dy(t)}{dt} + 4y(t) = f(t)$$

due to the input signal given in Figure 4.13.

6.10. Repeat Problem 6.9 for the system

$$\frac{d^2 y(t)}{dt^2} + 5\frac{dy(t)}{dt} + 4y(t) = \frac{df(t)}{dt} + 2f(t)$$

6.11. Repeat Problem 6.9 for the input signal given in Figure 4.14.

6.12. Convolve $f_1[k] = 2^k \{u[k-1] - u[k-3]\}$ and $f_2[k] = 3^k \delta[k]$.

Answer:

$$f_1[k] * f_2[k] = \begin{cases} 0, & k \leq 0 \\ 2, & k = 1 \\ 4, & k = 2 \\ 0, & k \geq 3 \end{cases}$$

6.13. Convolve $f_1[k] = u[k]$ and $f_2[k] = p_2[k]$.

Answer:
$$u[k] * p_2[k] = \begin{cases} 0, & k \leq -2 \\ 1, & k = -1 \\ 2, & k = 0 \\ 3, & k \geq 1 \end{cases}$$

6.14. Convolve the discrete-time signals

$$f_1[k] = \begin{cases} 2, & k = 1 \\ 1, & k = 2 \\ 0, & \text{otherwise} \end{cases}, \quad f_2[k] = \begin{cases} 1, & k = 0 \\ 1, & k = 2 \\ 0, & \text{otherwise} \end{cases}$$

Answer:
$$f_1[k] * f_2[k] = \begin{cases} 0, & k \leq 0 \\ 2, & k = 1 \\ 1, & k = 2 \\ 2, & k = 3 \\ 1, & k = 4 \\ 0, & k \geq 5 \end{cases}$$

6.15. Use the sliding tape method to convolve the discrete-time signals

$$f_1[k] = \begin{cases} 1, & k = 0 \\ -1, & k = 1 \\ 1, & k = 2 \\ 0, & \text{otherwise} \end{cases}, \quad f_2[k] = \begin{cases} 3, & k = 0 \\ -2, & k = 2 \\ 0, & \text{otherwise} \end{cases}$$

6.16. Repeat Problem 6.15 for

$$f_1[k] = \begin{cases} -2, & k = -1 \\ 2, & k = 0 \\ 1, & k = 1 \\ -1, & k = 2 \\ 4, & k = 3 \\ 0, & \text{otherwise} \end{cases}, \quad f_2[k] = \begin{cases} 1, & k = 0 \\ 2, & k = 1 \\ 3, & k = 2 \\ 2, & k = 3 \\ 0, & \text{otherwise} \end{cases}$$

Answer:
$$f_1[k] * f_2[k] = \begin{cases} 0, & k \leq -2 \\ -2, & k = -1 \\ -2, & k = 0 \\ -1, & k = 1 \\ 3, & k = 2 \\ 9, & k = 3 \\ 7, & k = 4 \\ 10, & k = 5 \\ 8, & k = 6 \\ 0, & k \geq 7 \end{cases}$$

6.17. Using the sliding tape method, convolve the discrete-time signals given in Problem 6.16 by first shifting the first signal one step ahead (making the newly obtained system causal) and then convolving the newly obtained signal with the second signal. Second, apply appropriately the discrete-time convolution shifting property (6.20) to obtain the convolution of the original signals.

6.18. Find the discrete-time convolution of the signals

$$f_1[k] = \begin{cases} 2, & k=1 \\ 3, & k=2 \\ 0, & \text{otherwise} \end{cases}, \quad f_2[k] = \begin{cases} 1, & k=0 \\ -2, & k=1 \\ 5, & k=5 \\ 0, & \text{otherwise} \end{cases}$$

6.19. Using the $\mathcal{Z}$-transform, find the impulse response of the following discrete-time linear systems.

(a) $y[k+2] + \dfrac{5}{6}y[k+1] + \dfrac{1}{6}y[k] = f[k+1] + 3f[k]$

(b) $y[k+2] + y[k+1] + \dfrac{1}{4}y[k] = f[k]$

(c) $y[k+2] + y[k] = f[k+1]$

Use the convolution method to find the system zero-state responses due to an input $f[k] = (-2)^k u[k]$.

6.20. Compare the impulse responses of the following two linear discrete-time systems.

(a) $y[k+2] + \dfrac{2}{3}y[k+1] + \dfrac{1}{9}y[k] = f[k]$

(b) $y[k+2] + \dfrac{2}{3}y[k+1] + \dfrac{1}{9}y[k] = f[k+1] + f[k]$

Comment on the impact of the term $f[k+1]$ on the system impulse response. Using the convolution method, find the zero-state responses of these systems due to

$$f[k] = \begin{cases} 2, & k=0 \\ -1, & k=1 \\ 3, & k=2 \\ 1, & k=3 \\ 0, & \text{otherwise} \end{cases}$$

6.21. Find analytically the impulse response for the eye movement dynamics whose mathematical model is given in formula (1.39). Find analytically the zero-state response due to $f(t) = (1 - e^{-t})u(t)$, using the convolution formula.

6.22. Consider the national income model defined in (1.32) and find analytically its impulse response. Using the convolution formula, write the expression for the zero-state response. Assuming that $f[k] = u[k] + \gamma r[k]$, where γ is a given constant, find the closed formula for the corresponding zero-state response for the period of ten years.

6.23. Using MATLAB, find the impulse response for the commercial aircraft problem considered in Section 4.3.5. Find and plot the zero-state response due to $f(t) = e^{-2.5t}u(t)$ using the convolution formula.

6.24. Using MATLAB, find the impulse response for the flexible beam problem considered in Section 4.6. Find and plot the zero-state response due to $f(t) = e^{-t}u(t)$ using the convolution formula.

6.25. Using MATLAB, find the impulse response for the aircraft problem considered in Problem 4.57. Find and plot the zero-state response due to the external force $f(t) = e^{-t}\sin(2t)u(t)$ using the convolution formula.

6.26. Using MATLAB, find the impulse response for the neuromuscular system problem considered in Problem 4.59. Find and plot the zero-state response due to $f(t) = \sin(t)u(t)$ using the convolution formula.

6.27. Using MATLAB, find the impulse response for the electrical circuit problem presented in Figure 1.10, with the circuit parameters defined in Problem 4.61. Find and plot the zero-state response due to $f(t) = \sin(t)u(t)$ using the convolution formula.

6.28. Using MATLAB, find the impulse response for the ship positioning system considered in Problem 4.64. Find and plot the zero-state response due to $f(t) = e^{-t}\cos(t)u(t)$ using the convolution formula.

6.29. Write a MATLAB program for finding a particular solution of a general linear difference equation of order n using the convolution method. Note that the particular solution obtained represents the zero-state response of the corresponding linear discrete-time dynamic system.

6.30. Find the impulse response for the ATM computer network switch transfer function $H_Y(z)$ defined in (5.77), assuming that $\alpha_0 = 0.6$ and $\alpha_1 = -0.5$. Using the convolution formula, find the switch zero-state response due to the input defined by

$$f[k] = \begin{cases} k, & k < 10 \\ 10, & k \geq 10 \end{cases}$$

Use MATLAB to plot the response obtained.

6.31. Repeat Problem 6.30 for the input signal defined by

$$f[k] = \begin{cases} 2^k, & k < 4 \\ 16, & k \geq 4 \end{cases}$$

CHAPTER 7

System Response in the Time Domain

In this chapter, we study methods for the analysis of continuous- and discrete-time, time-invariant, linear systems. The analysis of continuous-time systems uses elementary results from theory of linear ordinary *differential equations*. This theory is known to all undergraduate engineering students, either from basic calculus courses or specialized courses in differential equations. The linear discrete-time systems are analyzed using the corresponding theory of linear *difference equations,* which is first introduced to engineering students either in digital signal processing or linear systems and signals courses. In both the continuous- and discrete-time domains, we limit our attention to linear *time-invariant* systems that are represented by linear differential/difference equations with *constant coefficients*. The general forms of these equations were introduced in Chapter 1, in formulas (1.9–10). The methods for solving linear differential and difference equations with constant coefficients are very similar. They are presented, respectively, in Sections 7.1 and 7.2.

We have seen from Chapters 4–6 that one of the most elementary and important problems in the process of finding the system response to any input signal is the problem of finding the system impulse response. In Sections 7.3 and 7.4, we show how to find the system impulse response in both continuous- and discrete-time domains. Sections 7.5 and 7.6 present procedures for finding complete system responses for, respectively, continuous- and discrete-time linear systems.

In Sections 7.7 and 7.8, we introduce formal definitions of system stability in continuous- and discrete-time domains, respectively, and relate them to the locations of the system eigenvalues/poles. The stability concept is fundamentally important for linear system theory and its applications, and unstable systems are rarely used in practice. We present the concepts of both internal system stability and bounded-input bounded-output system stability. At the end of this chapter, in Sections 7.9 and 7.10, several MATLAB experiments are designed in continuous- and discrete-time domains, respectively.

7.1 SOLVING LINEAR DIFFERENTIAL EQUATIONS

In this section, we present the basic technique for solving linear differential equations with constant coefficients. That technique will be used in this chapter to study the response of continuous-time, time-invariant, linear systems to external forcing functions and system initial conditions. Here, *we present only the most elementary results about linear* differential equations needed for the purpose of this course. Detailed study of ordinary linear differential equations can be found in specialized undergraduate textbooks, for example, [1].

The dynamics of time-invariant, linear, continuous-time systems are described by nth order linear differential equations with constant coefficients, of the form

$$\frac{d^n y(t)}{dt^n} + a_{n-1}\frac{d^{n-1} y(t)}{dt^{n-1}} + a_{n-2}\frac{d^{n-2} y(t)}{dt^{n-2}} + \cdots + a_1\frac{dy(t)}{dt} + a_0 y(t)$$
$$= b_m \frac{d^m f(t)}{dt^m} + b_{m-1}\frac{d^{m-1} f(t)}{dt^{m-1}} + \cdots + b_1\frac{df(t)}{dt} + b_0 f(t), \quad t \geq 0 \quad (7.1)$$

where $f(t)$ and $y(t)$ represent, respectively, the system input and output signals, and $a_i, i = 0, 1, 2, \ldots, n-1$, and $b_j, j = 0, 1, 2, \ldots, m$, are constant coefficients. For (7.1) to correspond to a real physical system it is required that $n \geq m$. The initial conditions that carry information about $y(t)$ and its $n - 1$ derivatives at the initial time $t = 0^-$, that is, $y(0^-)$, $y^{(1)}(0^-)$, $y^{(2)}(0^-)$, ..., $y^{(n-1)}(0^-)$, are assumed to be known. The reasons for specifying the system initial conditions at $t = 0^-$ instead of at $t = 0$ are twofold. First, we must be able to find the system impulse response due to the impulse delta signal located at the time origin, $t = 0$, which requires integration from 0^- in order to completely include the impulse delta signal within the integration limits. Second, the system zero-input response is contributed by the system initial conditions that are present in the system before the input is applied to the system. Assuming that the initial time at which the system input is applied is $t = 0$, the system initial conditions must be defined before $t = 0$, that is at $t = 0^-$. Our task is to find the system output $y(t)$ for every $t \geq 0$ due to both the system initial conditions and the system input signal (forcing function).

It is known from the elementary theory of differential equations (and the elementary theory of electrical circuits and basic engineering dynamics courses) that the solution of differential equation (7.1) has two components: the $y_p(t)$ or *particular solution* and the $y_h(t)$ or *homogeneous solution;* that is,

$$y(t) = y_p(t) + y_h(t) \quad (7.2)$$

The homogeneous solution satisfies the corresponding homogeneous differential equation, obtained by setting $f(t) = 0$ in (7.1), that is,

$$\frac{d^n y_h(t)}{dt^n} + a_{n-1}\frac{d^{n-1} y_h(t)}{dt^{n-1}} + a_{n-2}\frac{d^{n-2} y_h(t)}{dt^{n-2}} + \cdots + a_1\frac{dy_h(t)}{dt} + a_0 y_h(t) = 0, \quad t \geq 0$$
$$(7.3)$$

The particular solution is any solution that satisfies (7.1), regardless of the system initial conditions. In what follows, we will discuss in details the methods for finding the homogeneous and particular solutions.

Homogeneous Solution

The *characteristic equation* of (7.1) and (7.3) is defined by

$$\Delta(\lambda) = \lambda^n + a_{n-1}\lambda^{n-1} + a_{n-2}\lambda^{n-2} + \cdots + a_1\lambda + a_0 = 0 \quad (7.4)$$

The solutions of the characteristic equation (roots of the characteristic polynomial $\Delta(\lambda)$) are called the system *characteristic values*. The characteristic values are also known as *system eigenvalues*. In most cases, the system eigenvalues are identical to the *system poles* (see the formal definition of system poles in Section 4.3.1). This distinction between the system poles and the system eigenvalues will be further explained in Section 7.5.

It is well known from the elementary theory of differential equations that, depending on the nature of the solutions of the characteristic equation (distinct or multiple roots), we can get in a systematic way the expression for the homogeneous solution.

Let us assume that the solutions of the characteristic equation are *distinct*, that is,

$$\Delta(\lambda) = (\lambda - \lambda_1)(\lambda - \lambda_2) \cdots (\lambda - \lambda_n) = 0 \qquad (7.5)$$

with $\lambda_1 \neq \lambda_2 \neq \cdots \neq \lambda_n$; then the homogeneous differential equation (7.3) has a solution given by

$$y_h(t) = C_1 e^{\lambda_1 t} + C_2 e^{\lambda_2 t} + \cdots + C_n e^{\lambda_n t}, \quad t \geq 0 \qquad (7.6)$$

The constant coefficients $C_i, i = 1, 2, \ldots, n$, are determined from the system initial conditions using formula (7.12), which requires knowledge of the particular solution. Note that the form of the homogeneous solution given in (7.6) guarantees the existence of the unique constants $C_i, i = 1, 2, \ldots, n$ [1].

EXAMPLE 7.1

Consider a continuous-time linear system whose mathematical model is described by the homogeneous linear differential equation

$$\frac{d^2 y_h(t)}{dt^2} + 4\frac{dy_h(t)}{dt} + 3y_h(t) = 0, \quad t \geq 0$$

The characteristic equation of this differential equation is given by

$$\Delta(\lambda) = \lambda^2 + 4\lambda + 3 = (\lambda + 1)(\lambda + 3) = 0$$

which implies the characteristic values $\lambda_1 = -1, \lambda_2 = -3$. The corresponding homogeneous solution is given by

$$y_h(t) = C_1 e^{-t} + C_2 e^{-3t}, \quad t \geq 0$$

Note that a *pair of complex characteristic values*, say $\lambda_{1,2} = \alpha \pm j\beta$, represents two distinct characteristic values. Hence, we can use formula (7.6) to find the corresponding homogeneous solution. It seems that in such a case we must play algebra with complex numbers, that is,

$$y_h(t) = \hat{C}_1 e^{(\alpha + j\beta)t} + \hat{C}_2 e^{(\alpha - j\beta)t} = e^{\alpha t}(\hat{C}_1 e^{j\beta t} + \hat{C}_2 e^{-j\beta t})$$

However, this solution can be further simplified by using Euler's formula and determining the constants $\hat{C}_1$ and $\hat{C}_2$ from the initial conditions. Since $y_h(t)$ is a real function of time, its final expression cannot contain complex numbers. In general, a pair of complex conjugate characteristic values implies oscillatory system response, so that we must obtain sine and cosine functions in the final expression for $y_h(t)$. For that reason, a more convenient form for the homogeneous solution in the case of complex conjugate characteristic values is

$$y_h(t) = C_1 e^{\alpha t} \cos(\beta t) + C_2 e^{\alpha t} \sin(\beta t) \tag{7.7}$$

EXAMPLE 7.2

The homogeneous linear differential equation

$$\frac{d^2 y_h(t)}{dt^2} + 2 \frac{dy_h(t)}{dt} + 2 y_h(t) = 0, \quad t \geq 0$$

has the characteristic equation

$$\Delta(\lambda) = \lambda^2 + 2\lambda + 2 = (\lambda + 1 + j1)(\lambda + 1 - j1) = 0$$

which implies the complex conjugate characteristic values $\lambda_1 = -1 - j1, \lambda_2 = -1 + j1$. The corresponding homogeneous solution is given by

$$y_h(t) = C_1 e^{-t} \cos(t) + C_2 e^{-t} \sin(t), \quad t \geq 0$$

In the case when we have a *multiple root* (say the characteristic value λ_1 has multiplicity r, and the remaining characteristic values are distinct), the homogeneous solution has the form

$$y_h(t) = (C_1 + C_2 t + C_3 t^2 + \cdots + C_r t^{r-1}) e^{\lambda_1 t} + C_{r+1} e^{\lambda_{r+1} t} + \cdots + C_n e^{\lambda_n t}, \quad t \geq 0 \tag{7.8}$$

In general, every characteristic value λ_j of multiplicity r_j generates a term of the form

$$(C_j + C_{j+1} t + \cdots + C_{j+r_j-2} t^{r_j-2} + C_{j+r_j-1} t^{r_j-1}) e^{\lambda_j t} \tag{7.9}$$

in the homogeneous solution of the corresponding differential equation. It can be shown (see, for example, [1]) that the form for $y_h(t)$ given in (7.8) facilitates the existence of unique coefficients $C_i, i = 1, 2, \ldots, n$.

In the case of *multiple complex characteristic values*—for example the characteristic value $\lambda_j = \alpha_j + j\beta_j$ with multiplicity r—the corresponding part of the homogeneous solution has the following form (similar to the multiple real characteristic values form):

$$y_h^j(t) = e^{\alpha_j t} \left[C_1^j \cos(\beta_j t) + C_2^j \sin(\beta_j t) \right] + t e^{\alpha_j t} \left[C_3^j \cos(\beta_j t) + C_4^j \sin(\beta_j t) \right]$$
$$+ t^2 e^{\alpha_j t} \left[C_5^j \cos(\beta_j t) + C_6^j \sin(\beta_j t) \right] + \cdots$$
$$+ t^{r-1} e^{\alpha_j t} \left[C_{2r-1}^j \cos(\beta_j t) + C_{2r}^j e^{\alpha_j t} \sin(\beta_j t) \right] \tag{7.10}$$

EXAMPLE 7.3

The homogeneous linear differential equation

$$\frac{d^3 y_h(t)}{dt^3} + 5\frac{d^2 y_h(t)}{dt^2} + 8\frac{dy_h(t)}{dt} + 4y_h(t) = 0, \quad t \geq 0$$

has the characteristic equation

$$\Delta(\lambda) = \lambda^3 + 5\lambda^2 + 8\lambda + 4 = (\lambda + 1)(\lambda + 2)^2 = 0$$

and the characteristic values $\lambda_1 = -1, \lambda_2 = \lambda_3 = -2$. The corresponding homogeneous solution is given by

$$y_h(t) = C_1 e^{-t} + C_2 e^{-2t} + C_3 t e^{-2t}, \quad t \geq 0$$

EXAMPLE 7.4

Consider the following homogeneous linear differential equation (with a triple characteristic value):

$$\frac{d^3 y_h(t)}{dt^3} + 3\frac{d^2 y_h(t)}{dt^2} + 3\frac{dy_h(t)}{dt} + y_h(t) = 0, \quad t \geq 0$$

The characteristic equation is given by

$$\Delta(\lambda) = \lambda^3 + 3\lambda^2 + 3\lambda + 1 = (\lambda + 1)^3 = 0$$

The triple characteristic values are $\lambda_1 = \lambda_2 = \lambda_3 = -1$, so that the corresponding homogeneous solution, obtained using (7.8), is

$$y_h(t) = C_1 e^{-t} + C_2 t e^{-t} + C_3 t^2 e^{-t}, \quad t \geq 0$$

Particular Solutions

In Chapters 4 and 6, we dealt with *a special form of a particular solution given by the convolution integral*

$$y_p(t) = f(t) * h(t) = \int_{0^-}^{t} f(t - \tau) h(\tau) \, d\tau = y_{zs}(t) \quad (7.11)$$

where $h(t)$, representing the system impulse response, is the solution of (7.1) for $f(t) = \delta(t)$. The lower integration limit in the convolution integral is taken at 0^- in order to accommodate the impulse delta function at $t = 0$. Note that when $n = m$, the system impulse response contains the impulse delta function at the origin, as demonstrated in the next example. It should be emphasized that the particular solution defined by the convolution integral (7.11) represents the system's *zero-state response*, $y_{zs}(t)$ (see (4.43)).

The particular solution, $y_p(t)$, is any solution that satisfies (7.1), regardless of the system initial conditions [1]. However, for any $y_p(t)$, it follows from formula (7.2) that at the initial time the following relations must be satisfied [1]:

$$y(0) = y_p(0) + y_h(0)$$
$$y^{(1)}(0) = y_p^{(1)}(0) + y_h^{(1)}(0)$$
$$\vdots$$
$$y^{(n-1)}(0) = y_p^{(n-1)}(0) + y_h^{(n-1)}(0)$$

(7.12)

This formula requires a bit more explanation, since the system initial conditions are specified at $t = 0^-$. Let us solve a simple first-order example in order to get a better understanding of a *potential problem*.

EXAMPLE 7.5

Consider the linear differential equation

$$\frac{dy(t)}{dt} + 2y(t) = \frac{df(t)}{dt} + 3f(t), \quad f(t) = e^{-5t}u(t), \quad y(0^-) = 1, \quad t \geq 0$$

We first solve this differential equation using the Laplace transform method from Chapter 4, which leads to

$$sY(s) - y(0^-) + 2Y(s) = sF(s) + 3F(s)$$

$$\Rightarrow Y(s) = \frac{y(0^-)}{s+2} + \frac{s+3}{s+2}F(s) = \frac{1}{s+2} + \frac{(s+3)}{(s+2)(s+5)}\cdot 4$$

$$\Rightarrow y(t) = e^{-2t} + \frac{4}{3}e^{-2t} + \frac{8}{3}e^{-5t} = \frac{7}{3}e^{-2t} + \frac{8}{3}e^{-5t}, \quad t \geq 0$$

Note that at $t = 0^+$ we have $y(0^+) = 7/3 + 8/3 = 5$. Hence, an *instantaneous jump* occurs at the initial time from $y(0^-) = 1$ to $y(0^+) = 5$. This jump in the initial condition is contributed by the *system differentiation of the input signal*, which in general generates the impulse delta function as an additional input to the system. The impulse delta function is able to change system initial conditions instantaneously. Note that $d(e^{-5t}u(t))/dt = -5e^{-5t}u(t) + e^{-5t}\delta(t)$.

Let us now find the homogeneous and particular solutions of the given differential equation. The homogeneous differential equation is given by

$$\frac{dy_h(t)}{dt} + 2y_h(t) = 0, \quad t \geq 0$$

Its solution is easily obtained as $y_h(t) = C_1 e^{-2t}$, $t \geq 0$. In order to find a particular solution, we first guess its form as $y_p(t) = \alpha e^{-5t}$. Substituting this particular solution into the original differential equation, we find that $\alpha = 8/3$, hence $y_p(t) = 8e^{-5t}/3$. It follows that

$$y(t) = y_h(t) + y_p(t) = C_1 e^{-2t} + \frac{8}{3}e^{-5t}$$

If we use the information that $y(0^-) = 1$, we will get either $1 = C_1$ (if we assume that the particular solution is not present at $t = 0^-$) or $1 = C_1 + 8/3$, which implies $C_1 = -5/3$. Obviously, both values for C_1 are different from $7/3$ and hence they produce wrong expressions for $y(t)$. The way to get the correct answer for C_1 is to use information about $y(0^+)$, which (in this particular example) is equal to $y(0^+) = 5$ (see the solution obtained using the Laplace transform). That information will produce $5 = C_1 + 8/3 \Rightarrow C_1 = 7/3$, leading to the correct answer for $y(t)$. However, information about the value of $y(0^+)$ is not given, in general.

It seems that using the classical method for solving a linear differential equation that corresponds to a real physical linear system is problematic, since we need to know the system initial conditions at $t = 0^+$. In general, it is difficult to determine the system initial conditions at $t = 0^+$ based on information about the system differential equation and the system initial conditions at $t = 0^-$. However, this problem can be completely avoided if *we use the convolution formula to obtain the particular solution*. In that case, the particular solution represents the zero-state solution (it does not depend on the system initial conditions) and the homogeneous solution represents the zero-input solution (it depends only on the system initial conditions); see formula (4.54).

In this example, the system impulse response is easily found as

$$h(t) = \mathcal{L}^{-1}\left\{\frac{s+3}{s+2}\right\} = \mathcal{L}^{-1}\left\{1 + \frac{1}{s+2}\right\} = \delta(t) + e^{-2t}$$

so that the particular solution obtained from the convolution integral is given by

$$y_p(t) = \int_{0^-}^{t} h(\tau)f(t-\tau)\,d\tau = 4\int_{0^-}^{t}(\delta(\tau) + e^{-2\tau})e^{-5(t-\tau)}\,d\tau$$

$$= \frac{4}{3}e^{-2t} + \frac{8}{3}e^{-5t} = y_{zs}(t)$$

Now we can find the constant C_1 in the expression for the homogeneous solution as

$$y_h(0^-) = C_1 = y(0^-) = 1 \Rightarrow C_1 = 1 \Rightarrow y_h(t) = e^{-2t} = y_{zi}(t)$$

so that

$$y(t) = y_h(t) + y_p(t) = y_{zi}(t) + y_{zs}(t)$$

$$= e^{-2t} + \left(\frac{4}{3}e^{-2t} + \frac{8}{3}e^{-5t}\right) = \frac{7}{3}e^{-2t} + \frac{8}{3}e^{-5t}$$

which is identical to the result obtained using the Laplace transform method.

We can conclude from the preceding example that *when the system differentiates the input signal,* the classical method for solving linear differential equations with constant coefficients can be used to find the response of time-invariant linear systems only if the particular solution is obtained by using the convolution formula, or the system initial conditions are known at $t = 0^+$. (The latter is rarely the case, and system initial conditions at $t = 0^+$ are difficult to determine from information about the initial conditions at $t = 0^-$.) This general conclusion comes from formula (4.53), where the split of the system response into its zero-input and zero-state components is clearly indicated, with the zero-state component being represented by the convolution of $h(t)$ and $f(t)$. Hence, in finding the system response, *we must use the particular solution obtained through the convolution formula* (7.11). In that case, $y_h(t) = y_{zi}(t)$, so that all constants in $y_h(t)$ are obtained from

$$y_p(t) = y_{zs}(t) \Rightarrow \begin{cases} y_h(0^-) = y(0^-) \\ y_h^{(1)}(0^-) = y^{(1)}(0^-) \\ \vdots \\ y_h^{(n-1)}(0^-) = y^{(n-1)}(0^-) \end{cases} \quad (7.13)$$

In this course, we find the particular solutions of linear differential equations that represent continuous-time linear systems through the convolution procedure. To that end, we will present in Section 7.4 a time domain method for finding the continuous-time system impulse response. Note that the frequency domain technique for finding the continuous-time system impulse response using the inverse Laplace transform was presented in Chapter 4.

Note that a procedure for obtaining a particular solution of a linear differential equation with constant coefficients, known as the *method of undetermined coefficients,* is studied in detail in elementary differential equations courses [1]. That method is based on guessing a particular solution, usually as a linear combination of polynomial, exponential, and sinusoidal functions (a linear combination of the forcing function and its derivatives). However, the procedure is not systematic and sometimes has difficulty finding a particular solution. Another systematic method for finding particular solutions of linear differential equations with constant coefficients is the *method of variation of parameters* [1]. That method is computationally much more involved than the method of undetermined coefficients.

For completeness, in the next two examples we indicate difficulties that may face while attempting to find a particular solution using the method of undetermined coefficients.

EXAMPLE 7.6

In this example, we demonstrate difficulties one faces in the process of guessing a particular solution of a second-order differential equation. Consider first a differential equation given by

$$\frac{d^2 y(t)}{dt^2} + 2\frac{dy(t)}{dt} = t^2, \quad t > 0$$

A natural choice would be to take as a candidate for a particular solution the expression $y_p(t) = \alpha + \beta t + \gamma t^2$ (a linear combination of the forcing function and its derivatives), and to try to determine the unknown constant coefficients α, β, γ. By taking the first and second derivatives of $y_p(t)$ and plugging them into the differential equation, it can be checked easily that no α, β, γ exist such that the guessed solution $\alpha + \beta t + \gamma t^2$ is a particular solution of this differential equation. We must try a more complex form that also includes an integral of the forcing function, that is, $y_p(t) = \alpha + \beta t + \gamma t^2 + \theta t^3$. This will produce the required solution as

$$y_p(t) = \alpha + \frac{1}{4}t - \frac{1}{4}t^2 + \frac{1}{6}t^3, \quad t > 0$$

with α being an arbitrary constant. For simplicity, we may set $\alpha = 0$, so that

$$y_p(t) = \frac{1}{4}t - \frac{1}{4}t^2 + \frac{1}{6}t^3, \quad t > 0$$

The particular solution obtained through convolution formula (7.11) is given by

$$y_p^c(t) = -\frac{1}{8} + \frac{1}{4}t - \frac{1}{4}t^2 + \frac{1}{6}t^3 + \frac{1}{8}e^{-2t}, \quad t > 0$$

(In this formula the superscript c indicates that the particular solution is obtained through the convolution method.) Note that the system impulse response is $h(t) = 0.5(1 - e^{-2t})u(t)$. It should be emphasized that the particular solution obtained through the convolution formula, by integrating a product of the system input and the system impulse response, conveys information about both the actual system input and the internal system dynamics represented in the system impulse response by means of a weighted sum of $e^{p_i t}$ terms. (Note the term $(1/8)(e^{-2t})$ in the particular solution obtained using the convolution formula, and its absence from the particular solution obtained using the method of undetermined coefficients.)

The preceding example was not particularly difficult. However, if we change the original differential equation slightly, into

$$\frac{d^2 y(t)}{dt^2} + 2\frac{dy(t)}{dt} + y(t) = 5t + t^2 e^{-t}, \quad t > 0$$

then the problem of obtaining a particular solution by guessing its form becomes much more difficult. Students may try several choices. However, we can obtain the particular solution of this differential equation using the convolution formula in a pretty straightforward manner (see Problem 7.2).

EXAMPLE 7.7

Consider a continuous-time linear system represented by the differential equation defined in Example 7.6, that is,

$$\frac{d^2 y(t)}{dt^2} + 2\frac{dy(t)}{dt} = t^2 = f(t), \quad t > 0$$

with the initial conditions given by $y(0^-) = 1$, $y^{(1)}(0^-) = 2$. The characteristic values of this differential equation are $\lambda_1 = 0$, $\lambda_2 = -2$, so that its homogeneous solution is $y_h(t) =$

$C_1 + C_2 e^{-2t}$. Using the particular solution obtained in Example 7.6 via the method of undetermined coefficients, the solution of this differential equation has the form

$$y(t) = y_h(t) + y_p(t) = C_1 + C_2 e^{-2t} + \frac{1}{4}t - \frac{1}{4}t^2 + \frac{1}{6}t^3$$

From the given initial conditions, we obtain a set of linear algebraic equations for determining the constants C_1 and C_2,

$$y(0^-) = 1 = C_1 + C_2$$
$$y^{(1)}(0^-) = 2 = C_2 - \frac{1}{4}$$

which leads to $C_2 = -7/8$ and $C_1 = 1 - C_2 = 15/8$. The required solution, obtained via the method of undetermined coefficients, is given by

$$y(t) = \frac{15}{8} - \frac{7}{8}e^{-2t} + \frac{1}{4}t - \frac{1}{4}t^2 + \frac{1}{6}t^3$$

If we use the particular solution obtained through convolution, then the constants C_1 and C_2 must be determined from the following set of linear algebraic equations:

$$y(0^-) = 1 = y_h(0^-) = C_1 + C_2$$
$$y^{(1)}(0^-) = 2 = y_h^{(1)}(0^-) = -2C_2$$

which produces $C_2 = -1$ and $C_1 = 2$. The solution obtained this way is identical to the solution previously obtained using the method of undetermined coefficients, since

$$y(t) = y_h^c(t) + y_p^c(t) = 2 - e^{-2t} - \frac{1}{8} + \frac{1}{4}t - \frac{1}{4}t^2 + \frac{1}{6}t^3 + \frac{1}{8}e^{-2t}$$
$$= \frac{15}{8} + \frac{1}{4}t - \frac{1}{4}t^2 + \frac{1}{6}t^3 - \frac{7}{8}e^{-2t}$$

In concluding this section, we repeat that in the case when the linear time-invariant system does not differentiate the input signal, we can find the system response using any method for determining a particular solution of the corresponding linear constant coefficient differential equation. However, *in the case when the system differentiates the input signal, the particular solution must be found through the convolution procedure.*

7.2 SOLVING LINEAR DIFFERENCE EQUATIONS

In this section, we present a technique for solving linear difference equations with constant coefficients. That technique will be used in this chapter for finding the response of discrete-time, time-invariant, linear systems. *We present only elementary results about solving linear difference equations that are needed for the purpose of this course.* Detailed study of linear difference equations can be found in several books, for example, [2–5].

328 Chapter 7 System Response in the Time Domain

An nth order discrete-time, time-invariant, linear dynamic system is represented by a constant coefficient linear difference equation,

$$y[k+n] + a_{n-1}y[k+n-1] + \cdots + a_1 y[k+1] + a_0 y[k]$$
$$= b_m f[k+m] + b_{m-1} f[k+m-1] + \cdots + b_1 f[k+1] + b_0 f[k] \quad (7.14)$$

where $f[k]$ and $y[k]$ represent, respectively, the system input and output signals, and $a_i, i = 0, 1, 2, \ldots, n-1$, and $b_j, j = 0, 1, 2, \ldots, m$, are constant coefficients. The initial conditions for (7.14), defined by the given values for $y[0], y[1], y[2], \ldots, y[n-1]$, are assumed to be known. Due to system causality (for real physical systems, the present output does not depend on the future inputs), $n \geq m$ must hold. Our goal is to find the system response $y[k]$ for any nonnegative k due to both the system forcing function $f[k]$ and the system initial conditions. Note that (7.14) represents the derivative formulation of a linear time-invariant discrete system, which is exclusively used in the mathematical literature on difference equations.

Remark 7.1: Strictly speaking, the order of difference equation (7.14) is equal to n, assuming that the coefficient $a_0 \neq 0$. In general, it can be shown that for $a_0 = 0, a_1 = 0, \ldots, a_i = 0$, (7.14) degenerates into a difference equation of order $n - i - 1$. In such a case, (7.14) represents a real physical system if $m \leq n - i - 1$. For simplicity, we assume for all difference equations of this chapter that the coefficient a_0 is nonzero.

It should be pointed out that (7.14) is a recursive formula, hence we can get $y[k]$ for any k by performing iterations for $k = 0, 1, 2, \ldots$. However, that procedure is not very efficient, since to find, say, $y[1000]$, we must first calculate all previous values of $y[k]$, $k = 0, 1, 2, \ldots, 999$. For example, for $k = 0$, we have, from (7.14),

$$y[n] = -a_{n-1}y[n-1] - a_{n-2}y[n-2] - \cdots - a_1 y[1] - a_0 y[0]$$
$$+ b_m f[m] + b_{m-1} f[m-1] + \cdots + b_1 f[1] + b_0 f[0]$$

Since all terms on the right-hand side are assumed to be known, $y[n]$ is easily determined by a simple summation. Having obtained the value for $y[n]$, we can set $k = 1$ in (7.14) and similarly find $y[n+1]$, and so on; by repeating this iterative procedure we can obtain $y[k]$ for any k. On the other hand, the time-domain method for solving difference equations, to be presented in this section, produces closed formulas for the solution of (7.14) dependent on discrete-time k. By plugging a specific value for k into such formulas, we can easily get the system response at any given time instant without evaluating all previous values of $y[k]$.

As in the continuous-time domain, the solution of the linear difference equation (7.14) has two components, particular and homogeneous:

$$y[k] = y_p[k] + y_h[k] \quad (7.15)$$

In the following, we present procedures for determining homogeneous and particular solutions.

Homogeneous Solution

The homogeneous solution satisfies the homogeneous difference equation, obtained by setting all terms corresponding to the forcing function to zero in formula (7.14), that is,

$$y_h[k+n] + a_{n-1} y_h[k+n-1] + \cdots + a_1 y_h[k+1] + a_0 y_h[k] = 0 \quad (7.16)$$

The homogeneous solution has a form that depends on the nature of the solutions of the characteristic equation of (7.14) and (7.16). To that end, we first define the characteristic equation of a linear discrete-time system.

The *discrete-time characteristic equation* is defined by

$$\Delta(\rho) = \rho^n + a_{n-1}\rho^{n-1} + a_{n-2}\rho^{n-2} + \cdots + a_1 \rho + a_0 = 0 \quad (7.17)$$

The solutions of the characteristic equation (roots of the characteristic polynomial $\Delta(\rho)$) are called the *characteristic values*. They are also known as the *system eigenvalues*.

In the case when the characteristic values are distinct, $\rho_1 \neq \rho_2 \neq \cdots \neq \rho_n$, the homogeneous solution of (7.14), obtained by solving difference equation (7.16), is given by

$$y_h[k] = C_1 (\rho_1)^k + C_2 (\rho_2)^k + \cdots + C_n (\rho_n)^k \quad (7.18)$$

The constant coefficients C_i, $i = 1, 2, \ldots, n$, must be determined from the system initial conditions. (The corresponding discrete-time formulas that must be satisfied by the discrete-time linear system initial conditions are derived from (7.15) and given in (7.23) and (7.25).)

EXAMPLE 7.8

Consider the homogeneous difference equation

$$y_h[k+2] - \frac{1}{6} y_h[k+1] - \frac{1}{6} y_h[k] = 0$$

Its characteristic equation,

$$\Delta(\rho) = \rho^2 - \frac{1}{6}\rho - \frac{1}{6} = 0$$

produces the characteristic values $\rho_1 = 1/2$ and $\rho_2 = -1/3$. Hence, the homogeneous solution is given by

$$y_h[k] = C_1 \left(\frac{1}{2}\right)^k + C_2 \left(-\frac{1}{3}\right)^k \qquad \text{\textlbrackdbl}$$

In the case when we have a multiple characteristic value (say the characteristic value ρ_1 has multiplicity r, that is, $\rho_1 = \rho_2 = \cdots = \rho_r$, and the remaining characteristic values are distinct), the homogeneous solution has the form

$$y_h[k] = (C_1 + C_2 k + C_3 k^2 + \cdots + C_r k^{r-1})(\rho_1)^k + C_{r+1}(\rho_{r+1})^k + \cdots + C_n (\rho_n)^k \quad (7.19)$$

In general, every characteristic value ρ_j of multiplicity r_j generates in the discrete-time domain a term of the form

$$(C_j + C_{j+1}k + \cdots + C_{j+r_j-2}k^{r_j-2} + C_{j+r_j-1}k^{r_j-1})(\rho_j)^k \tag{7.20}$$

in the homogeneous solution of the corresponding difference equation.

Note that the forms of the homogeneous solutions given in (7.18) and (7.19) facilitate the realization that the constants $C_1, C_2, \ldots, C_n$ can be uniquely determined from a system of linear algebraic equations that are generated using the known initial conditions [2–4].

EXAMPLE 7.9

The homogeneous difference equation

$$y_h[k+3] + \frac{3}{2}y_h[k+2] + \frac{3}{4}y_h[k+1] + \frac{1}{8}y_h[k] = 0$$

has the characteristic equation

$$\Delta(\rho) = \rho^3 + \frac{3}{2}\rho^2 + \frac{3}{4}\rho + \frac{1}{4} = \left(\rho + \frac{1}{2}\right)^3 = 0$$

which implies $\rho_1 = \rho_2 = \rho_3 = -1/2$. According to formula (7.19), the homogeneous solution has the form

$$y_h[k] = C_1\left(-\frac{1}{2}\right)^k + C_2 k\left(-\frac{1}{2}\right)^k + C_3 k^2\left(-\frac{1}{2}\right)^k$$

EXAMPLE 7.10

Let the characteristic equation of a difference equation be given by

$$\Delta(\rho) = \left(\rho - \frac{1}{2}\right)\left(\rho + \frac{1}{4}\right)^3 (\rho + 1)^2 = 0$$

Then, the corresponding homogeneous solution has the form

$$y_h[k] = C_1\left(\frac{1}{2}\right)^k + C_2\left(-\frac{1}{4}\right)^k + C_3 k\left(-\frac{1}{4}\right)^k + C_4 k^2\left(-\frac{1}{4}\right)^k + C_5(-1)^k + C_6 k(-1)^k$$

A *pair of complex characteristic values*, say $\rho_{1,2} = \alpha_1 \pm j\beta_1 = \mu_1 e^{\pm j\varphi_1}$, represents two distinct characteristic values. In general, this pair produces an oscillatory system response, so that we obtain sine and cosine functions in the final expression for $y_h[k]$. The corresponding homogeneous solution should be sought in the form

$$y_h[k] = C_1\mu_1^k \cos[\varphi_1 k] + C_2\mu_1^k \sin[\varphi_1 k] \tag{7.21}$$

In the case of *multiple complex characteristic values* (for example, the characteristic value $\rho_j = \alpha_j \pm j\beta_j = \mu_j e^{\pm j\varphi_j}$ has multiplicity r), the corresponding part of the homogeneous solution has the form

$$y_h^j[k] = \mu_j^k \{C_1^j \cos[\varphi_j k] + C_2^j \sin[\varphi_j k]\} + k\mu_j^k \{C_3^j \cos[\varphi_j k] + C_4^j \sin[\varphi_j k]\}$$
$$+ k^2 \mu_j^k \{C_5^j \cos[\varphi_j k] + C_6^j \sin[\varphi_j k]\} + \cdots$$
$$+ k^{r-1} \mu_j^k \{C_{2r-1}^j \cos[\varphi_j k] + C_{2r}^j \sin[\varphi_j k]\} \qquad (7.22)$$

It can be shown that all constants in (7.22) can be uniquely determined from the initial conditions [3–5].

Particular Solution

The particular solution, $y_p[k]$, is any solution that satisfies difference equation (7.14), regardless of the system initial conditions. However, at the initial time instants we must have (see (7.15))

$$y[0] = y_p[0] + y_h[0]$$
$$y[1] = y_p[1] + y_h[1]$$
$$\vdots \qquad (7.23)$$
$$y[n-1] = y_p[n-1] + y_h[n-1]$$

A common classroom procedure for obtaining a particular solution of a difference equation guesses the form of a particular solution (usually as a linear combination of polynomial, exponential, and sinusoidal functions)—the *method of undetermined coefficients*, [2, 3]. It is not easy in general to guess the form of a particular solution of a difference equation [6]. Another systematic procedure for determining particular solutions of difference equations is known as the *variation of parameters method*. That method is computationally very involved, and mostly used for finding particular solutions of linear difference equations with time-varying coefficients [2–4].

The method of undetermined coefficients is demonstrated in the following three examples.

EXAMPLE 7.11

Consider a discrete-time linear system represented by the difference equation

$$y[k+2] + 2y[k+1] + y[k] = (-2)^k, \quad y[0] = 0, \quad y[1] = 1$$

The characteristic values are $\rho_1 = \rho_2 = -1$, so that the homogeneous solution is given by

$$y_h[k] = C_1(-1)^k + C_2 k(-1)^k$$

A particular solution can be obtained through the method of undetermined coefficients, by guessing its form as a linear combination of the forcing function and its derivatives, that is,

$$y_p[k] = \alpha_1(-2)^k + \alpha_2(-2)^{k+1} - (\alpha_1 - 2\alpha_2)(-2)^k = \alpha(-2)^k$$

A substitution of this solution into the original difference equation produces

$$\alpha(-2)^{k+2} + 2\alpha(-2)^{k+1} + \alpha(-2)^k = (-2)^k$$

From this equation we have $\alpha = 1$, so that the sought particular solution is given by $y_p[k] = (-2)^k$. Using formula (7.23), we can determine the constants C_1 and C_2 as follows:

$$0 = y[0] = y_h[0] + y_p[0] = C_1 + 0 + 1$$
$$1 = y[1] = y_h[1] + y_p[1] = -C_1 - C_2 - 2$$

which implies $C_1 = -1$ and $C_2 = -2$. Hence, the solution of the difference equation is given by

$$y[k] = y_h[k] + y_p[k] = -(-1)^k - 2k(-1)^k + (-2)^k, \quad k \geq 0$$

In the next example, we demonstrate a peculiar feature that appears when the forcing function is equal to a term in the homogeneous solution, so that a particular solution must be sought taking into the account the effect of multiple characteristic values.

EXAMPLE 7.12

Consider the difference equation

$$y[k+2] + \frac{1}{2}y[k+1] - \frac{1}{2}y[k] = (-1)^k, \quad y[0] = 0, \quad y[1] = 1$$

The characteristic equation is

$$\Delta(\rho) = \rho^2 + \frac{1}{2}\rho - \frac{1}{2} = \left(\rho - \frac{1}{2}\right)(\rho + 1) = 0 \Rightarrow \rho_1 = \frac{1}{2}, \quad \rho_2 = -1$$

The homogeneous solution is given by

$$y_h[k] = C_1\left(\frac{1}{2}\right)^k + C_2(-1)^k$$

A natural guess for a particular solution would be $y_p[k] = \alpha(-1)^k$. However, it can be easily shown that there is no constant coefficient α such that this choice will satisfy the original difference equation. Since the term $(-1)^k$ has already appeared in the homogeneous solution, we must look for a particular solution in the form $y_p[k] = \alpha k(-1)^k$, as we have in the case of multiple roots. Substituting this particular solution into the original equation, we obtain

$$\alpha(k+2)(-1)^{k+2} + \frac{1}{2}\alpha(k+1)(-1)^{k+1} - \frac{1}{2}\alpha k(-1)^k = (-1)^k$$

From this equation we have $\alpha = 2/3$, so that the particular solution sought is given by $y_p[k] = 2k(-1)^k/3$. The constants C_1 and C_2 are obtained using formula (7.23) as

$$0 = y[0] = y_h[0] + y_p[0] = C_1 + C_2 + 0$$
$$1 = y[1] = y_h[1] + y_p[1] = \frac{1}{2}C_1 - C_2 - \frac{2}{3}$$

which implies $C_1 = 10/9$ and $C_2 = -10/9$. The solution of the difference equation is given by

$$y[k] = y_h[k] + y_p[k] = \frac{10}{9}\left(\frac{1}{2}\right)^k - \frac{10}{9}(-1)^k + \frac{2}{3}k(-1)^k$$

EXAMPLE 7.13

Consider the following linear difference equation

$$y[k+2] + y[k+1] + \frac{1}{4}y[k] = k + \left(-\frac{1}{3}\right)^k, \quad y[0] = 1, \quad y[1] = 2$$

Its characteristic values are $\rho_1 = \rho_2 = -1/2$, so that the homogeneous solution is

$$y_h[k] = C_1\left(-\frac{1}{2}\right)^k + C_2 k\left(-\frac{1}{2}\right)^k$$

We look for a particular solution that has the form

$$y_p[k] = \alpha + \beta k + \gamma\left(-\frac{1}{3}\right)^k$$

Substituting this solution into the original difference equation, we obtain

$$\alpha + \beta(k+2) + \gamma\left(-\frac{1}{3}\right)^{k+2} + \alpha + \beta(k+1) + \gamma\left(-\frac{1}{3}\right)^{k+1}$$

$$+ \frac{1}{4}\left(\alpha + \beta k + \gamma\left(-\frac{1}{3}\right)^k\right) = k + \left(-\frac{1}{3}\right)^k$$

By matching the coefficients with respect to $(-1/3)^k$, we have

$$\frac{1}{9}\gamma - \frac{\gamma}{3} + \frac{\gamma}{4} = 1 \Rightarrow \gamma = 36$$

Similarly, the coefficients multiplying k produce

$$2\beta + \frac{1}{4}\beta = 1 \Rightarrow \beta = \frac{4}{9}$$

and the constant coefficients satisfy

$$\alpha + 2\beta + \alpha + \beta + \frac{1}{4}\alpha = 0 \Rightarrow \alpha = -\frac{16}{27}$$

Hence, the particular solution is given by

$$y_p[k] = -\frac{16}{27} + \frac{4}{9}k + 36\left(-\frac{1}{3}\right)^k$$

Since no input is applied before $k = -m$, we conclude that $h[k] = 0$ for $k < -m$. Furthermore, since $n \geq m$, we can see from (7.27) that only $h[i]$, $i = 0, 1, 2, \ldots$, can be affected by the sequence of discrete-time impulse delta functions defined on the right-hand side of (7.26) at $k = -m, -m+1, -m+2, \ldots, -2, -1, 0$. More precisely, it follows from (7.26) and (7.27) that only $h[i]$, $i = n - m, n - m + 1, \ldots$, are affected by the delta impulse functions. For $k = -m + 1$, we obtain from (7.26) that

$$h[-m + n + 1] + a_{n-1}h[-m + n] + 0 + \cdots + 0 = b_{m-1}$$
$$\Rightarrow h[n - m + 1] = b_{m-1} - a_{n-1}b_m \qquad (7.28)$$

By repeating the same procedure for $k = -m + 2, \ldots, -1, 0$, we can find $m + 1$ new (in general, nonzero) initial conditions $h[n - m], h[n - m + 1], \ldots, h[n]$. In addition to these $m + 1$ new initial conditions, we can use the previous initial conditions that are equal to zero, that is, $h[n - m - i] = 0$, $i = 1, 2, \ldots$.

Having found the new initial conditions set by the impulse delta signals for $k \leq 0$, we now have an input-free (homogeneous) system whose solution for $k \geq 1$ determines the system impulse response, that is,

$$h[k + n] + a_{n-1}h[k + n - 1] + \cdots + a_1h[k + 1] + a_0h[k] = 0, \quad k \geq 1 \qquad (7.29)$$

with initial conditions given by

$$h[n - m] = b_m, \quad h[n - m + 1] = b_{m-1} - a_{n-1}b_m, \ldots, \quad h[n] = b_0 - (*),$$
$$h[n - m - i] = 0, \quad i = 1, 2, \ldots \qquad (7.30)$$

There is no need to identify the term denoted in (7.30) by $(*)$ since we need not use these formulas to find the new initial conditions. Instead, *we need only remember that $h[n - m] = b_m$, and then iteratively evaluate the remaining m new initial conditions.* In addition to these $m + 1$ new (in general, nonzero) initial conditions, we must use the previous initial conditions that are equal to zero.

We can conclude that the impulse response of a linear discrete-time system is obtained by solving the homogeneous difference equation (7.29) subject to the initial conditions defined in (7.30). The homogeneous solution of (7.29) can be found using the methodology presented in Section 7.2.

The procedure outlined for finding the new system initial conditions and the entire procedure for finding the system impulse response are demonstrated in the following examples.

EXAMPLE 7.14

Consider the following linear discrete-time system and find its impulse response:

$$y[k + 2] + y[k + 1] + \frac{1}{4}y[k] = f[k + 1] + f[k]$$

Note that $n = 2$ and $m = 1$. For $f[k] = \delta[k]$, we obtain from (7.30) that $h[2 - 1] = h[1] = b_1 = 1$. Now, for $k = 0$, we have

$$h[2] + h[1] + 0 = 0 + 1 \Rightarrow h[2] = 1 - h[1] = 0$$

Hence, the system impulse response is obtained from

$$h[k+2] + h[k+1] + \frac{1}{4}h[k] = 0, \quad h[1] = 1, \quad h[2] = 0, \quad k \geq 1$$

Since the roots of the characteristic polynomial are $\rho_{1,2} = -0.5$, we have

$$h[k] = C_1 \left(-\frac{1}{2}\right)^k + C_2 k \left(-\frac{1}{2}\right)^k$$

Using the newly obtained initial conditions, we have

$$h[1] = 1 = -\frac{1}{2}C_1 - \frac{1}{2}C_2$$

$$h[2] = 0 = \frac{1}{4}C_1 + \frac{1}{2}C_2 \Rightarrow C_1 = -2C_2$$

This leads to $1 = C_2 - 0.5C_2 \Rightarrow C_2 = 2 \Rightarrow C_1 = -4$. Hence, the impulse response is given by

$$h[k] = -4\left(-\frac{1}{2}\right)^k + 2k\left(-\frac{1}{2}\right)^k = 2\left(-\frac{1}{2}\right)^k (k-2), \quad k \geq 1 \qquad §$$

EXAMPLE 7.15

Consider the fifth-order linear discrete-time system

$$y[k+5] + 2y[k+4] + 3y[k+3] + 2y[k+2] + y[k+1] + y[k]$$
$$= 3f[k+2] + 2f[k+1] + f[k]$$

and determine the new initial conditions for the impulse response.

Since $n = 5$ and $m = 2$, we have from formula (7.30) the result $h[5-2] = h[3] = b_2 = 1$, which implies $h[2] = h[1] = 0$. For $k = -1$, the original system equation for $f[k] = \delta[k]$ produces

$$h[4] + 2h[3] + 0 + 0 + 0 = 0 + 2\delta[0] + 0 \Rightarrow h[4] = 2 - 6 = -4$$

Plugging in $k = 0$ and using $f[k] = \delta[k]$, we obtain

$$h[5] + 2h[4] + 3h[3] + 0 + 0 = 0 + 0 + \delta[0] \Rightarrow h[5] = 1 - (-8) - 9 = 0$$

Hence, the initial conditions set by the input impulse delta functions are given by

$$h[5] = 0, \quad h[4] = -4, \quad h[3] = 1, \quad h[2] = 0, \quad h[1] = 0 \qquad §$$

7.3.2 Impulse Response by Linearity and Time Invariance

The second method for finding the discrete-time linear system impulse response uses the linearity and time-invariance principles (introduced in Sections 1.2.1 and 1.2.2). The linearity principle requires that the system initial conditions are zero, which is the case in the impulse response problem.

Under the linearity and time-invariance principles, the solution of (7.26) can be sought in the form

$$h[k] = \sum_{i=0}^{m} b_i h_0[k+i] u[k+i] \qquad (7.31)$$

where $h_0[k]$ is defined by

$$h_0[k+n] + a_{n-1} h_0[k+n-1] + \cdots + a_1 h_0[k+1] + a_0 h_0[k] = \delta[k] \qquad (7.32)$$

with all initial conditions of $h_0[k]$ being equal to zero. This procedure is symbolically represented in Figure 7.1.

We are faced now with a simpler problem than the original problem defined by (7.26). We call (7.32) the *elementary* discrete-time linear system impulse response problem. The system defined in (7.32) has an input equal to 1 for $k = 0$ and no input for all $k \neq 0$. At $k = 0$, from (7.32) we have

$$h_0[n] + a_{n-1} h_0[n-1] + \cdots + a_1 h_0[1] + a_0 h_0[0] = \delta[0] \qquad (7.33)$$

and for $k = -1$, we obtain

$$h_0[n-1] + a_{n-1} h_0[n-2] + \cdots + a_1 h_0[0] + a_0 h_0[-1] = 0 \qquad (7.34)$$

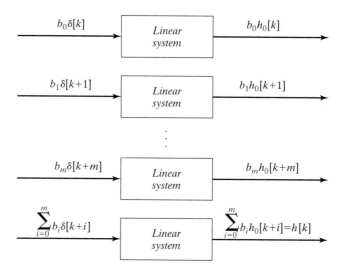

FIGURE 7.1: Procedure for determining the discrete-time linear system impulse response

Since no input is present in the system for $k \leq -1$, it follows from (7.34) that none of the elementary impulse response values can be nonzero up to the discrete-time instant $k = n - 1$, that is, $h_0[n - i] = 0, i = 1, 2, \ldots$. Using this observation in (7.33), we conclude that $h_0[n] = 1$.

For $k \geq 1$, no input is present in (7.32), so that we have the following linear homogeneous difference equation, whose solution produces the *elementary impulse response:*

$$h_0[k+n] + a_{n-1}h_0[k+n-1] + \cdots + a_1h_0[k+1] + a_0h_0[k] = 0, \quad k \geq 1$$
$$h_0[1] = h_0[2] = \cdots = h_0[n-1] = 0, \quad h_0[n] = 1 \quad (7.35)$$

Note also that $h_0[0] = 0$, but that this information is redundant for solving (7.35). It can be observed from (7.32) and (7.35) that it takes n discrete-time instants for the impulse delta input signal to excite a linear time-invariant discrete-time system of order n. We can conclude that the higher the order of the system, the longer the time needed for an impulse input signal to pass through the system.

The solution of (7.35) can be obtained in a standard manner, using the methodology outlined in Section 7.2 for solving homogeneous linear difference equations with constant coefficients.

This procedure for finding the discrete-time impulse response is demonstrated in the next example, using the linear system from Example 7.14.

EXAMPLE 7.16

The linear discrete-time impulse response problem of Example 7.14 is defined by

$$h[k+2] + h[k+1] + \frac{1}{4}h[k] = \delta[k+1] + \delta[k]$$

The elementary impulse response problem for this system is obtained from

$$h_0[k+2] + h_0[k+1] + \frac{1}{4}h_0[k] = 0, \quad k \geq 1$$
$$h_0[1] = 0, \quad h_0[2] = 1$$

The solution of this difference equation is given by

$$h_0[k] = C_1 \left(-\frac{1}{2}\right)^k + C_2 k \left(-\frac{1}{2}\right)^k$$

Using the known initial conditions, we have

$$h_0[1] = 0 = -\frac{1}{2}C_1 - \frac{1}{2}C_2 \Rightarrow C_1 = -C_2$$
$$h_0[2] = 1 = \frac{1}{4}C_1 + \frac{1}{2}C_2 \Rightarrow C_1 = 4 - 2C_2$$

This leads to $C_1 = -4$, $C_2 = 4$. The elementary impulse response is given by

$$h_0[k] = -4\left(-\frac{1}{2}\right)^k + 4k\left(-\frac{1}{2}\right)^k, \quad k \geq 1$$

Note that due to the time-invariance principle we have

$$h_0[k+1] = -4\left(-\frac{1}{2}\right)^{k+1} + 4(k+1)\left(-\frac{1}{2}\right)^{k+1}, \quad k \geq 0$$

Using formula (7.31), the impulse response of the original linear discrete-time system becomes

$$h[k] = h_0[k] + h_0[k+1] = -4\left(-\frac{1}{2}\right)^k + 4k\left(-\frac{1}{2}\right)^k$$

$$-4\left(-\frac{1}{2}\right)^{k+1} + 4k\left(-\frac{1}{2}\right)^{k+1} = \left(-\frac{1}{2}\right)^k (2k-4), \quad k \geq 1$$

which is identical to the result obtained in Example 7.14.

The preceding example demonstrates that we must sum discrete-time signals that do not start at the same discrete-time instants (see formula (7.31)). We see from (7.35) that the elementary system impulse response is equal to zero from the zeroth up to the $n-1$ discrete-time instants. By shifting the elementary impulse response back in time—that is, by forming $h_0[k+i]u[k+i]$ signals—we can note from (7.31) and (7.35) that the value of $h[0]$ is zero unless $m = n$. For $k = 0$ and $n = m$, (7.30) implies that $h[0] = b_m$. This observation and the fact that $h_0[k]$ is obtained from (7.35) for $k \geq 1$ help us to rewrite formula (7.31) in a more explicit form, namely

$$h[k] = \begin{cases} \sum_{i=0}^{m} b_i h_0[k+i], & k \geq 1 \\ 0, & k = 0, n > m \\ b_m, & k = 0, n = m \end{cases} \quad (7.31a)$$

EXAMPLE 7.17

Consider the problem of finding the impulse response for

$$h[k+3] + 1.5h[k+2] + 0.75h[k+1] + 0.125h[k] = \delta[k+2] + 2\delta[k+1] + 3\delta[k]$$

The elementary impulse response problem for this system satisfies

$$h_0[k+3] + 1.5h_0[k+2] + 0.75h_0[k+1] + 0.125h_0[k] = 0, \quad k \geq 1$$

$$h_0[1] = 0, \quad h_0[2] = 0, \quad h_0[3] = 1$$

Section 7.3 Discrete-Time System Impulse Response

The characteristic polynomial is $\Delta(\rho) = (\rho + 0.5)^3$, which implies that the characteristic values are given by $\rho_1 = \rho_2 = \rho_3 = -0.5$. The solution for the elementary impulse response has the form

$$h_0[k] = C_1\left(-\frac{1}{2}\right)^k + C_2 k\left(-\frac{1}{2}\right)^k + C_3 k^2\left(-\frac{1}{2}\right)^k$$

Using the values for known initial conditions at $k = 1, 2, 3$, we obtain the following system of algebraic equations:

$$h_0[1] = 0 = C_1 + C_2 + C_3$$
$$h_0[2] = 0 = \frac{1}{4}C_1 + \frac{1}{2}C_2 + C_3$$
$$h_0[3] = 1 = -\frac{1}{8}C_1 - \frac{3}{8}C_2 - \frac{9}{8}C_3$$

whose solutions are $C_1 = -8$, $C_2 = 12$, $C_3 = -4$. The *elementary* impulse response is

$$h_0[k] = -8\left(-\frac{1}{2}\right)^k + 12k\left(-\frac{1}{2}\right)^k - 4k^2\left(-\frac{1}{2}\right)^k, \quad k \geq 1$$

From formula (7.31a), the system impulse response is given by

$$h[k] = \begin{cases} 3h_0[k] + 2h_0[k+1] + h_0[k], & k \geq 1 \\ 0, & k = 0 \end{cases}$$

which further simplifies into

$$h[k] = 3\left(-\frac{1}{2}\right)^k (-8 + 12k - 4k^2) + 2\left(-\frac{1}{2}\right)^{k+1} \{-8 + 12(k+1) - 4(k+1)^2\}$$
$$+ \left(-\frac{1}{2}\right)^{k+2} \{-8 + 12(k+2) - 4(k+2)^2\}, \quad k \geq 1$$
$$= \left(-\frac{1}{2}\right)^k (-24 + 31k - 9k^2), \quad k \geq 1$$

Note that the impulse response of this system can be obtained and plotted using the MATLAB function `dimpulse`:

```
num=[1 2 3]; den=[1 1.5 0.75 0.125];
h=dimpulse(num,den); kf=length(h);
k=0:1:kf-1; plot(k,h,'*'); grid;
xlabel('Discrete time k'); ylabel('Impulse response')
```

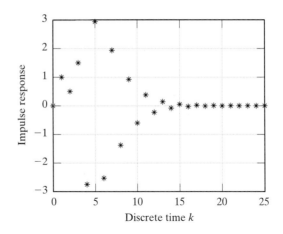

FIGURE 7.2: Discrete-time impulse response for Example 7.17

The corresponding impulse response plot is presented in Figure 7.2. It can be seen from this figure that for $k = 10$ the impulse response is roughly equal to -0.6. By plugging $k = 10$ into the result obtained for the system impulse response, we can calculate this value exactly, that is, $h[10] = -0.05996$.

EXAMPLE 7.18

Consider again the problem of finding the system impulse response given in Example 7.17, but this time use the direct method. Given

$$h[k+3] + 1.5h[k+2] + 0.75h[k+1] + 0.125h[k] = \delta[k+2] + 2\delta[k+1] + 3\delta[k]$$

it follows from the direct method that $n = 3, m = 2, h[n-m] = h[1] = b_m = 1$, and $h[0] = 0, h[-1] = 0$. The original system implies that, for $k = -1$,

$$h[2] + 1.5h[1] + 0 + 0 = 0 + 2\delta[0] + 0 = 2 \Rightarrow h[2] = 2 - 1.5 = 0.5$$

Also, for $k = 0$, it follows that

$$h[3] + 1.5h[2] + 0.75h[1] + 0 = 0 + 0 + 3\delta[0] = 2 \Rightarrow h[3] = 3 - 0.75 - 0.75 = 1.5$$

The impulse response has the following form (note the triple characteristic value at -0.5):

$$h[k] = C_1\left(-\frac{1}{2}\right)^k + C_2 k\left(-\frac{1}{2}\right)^k + C_3 k^2\left(-\frac{1}{2}\right)^k, \quad k \geq 1$$

Using the obtained values for $h[1], h[2], h[3]$, the unknown constants C_1, C_2, C_3 can be found from

$$h[1] = 1 = -\frac{1}{2}C_1 - \frac{1}{2}C_2 - \frac{1}{2}C_3$$

$$h[2] = \frac{1}{2} = \frac{1}{4}C_1 + \frac{1}{2}C_2 + C_3$$

$$h[3] = \frac{3}{2} = -\frac{1}{8}C_1 - \frac{3}{8}C_2 - \frac{9}{8}C_3$$

This system of linear algebraic equations has the unique solution given by $C_1 = -24$, $C_2 = 31$, $C_3 = -9$. Hence, the result obtained is identical to the result obtained in Example 7.17.

Comment: Since the impulse response is defined in the frequency domain by

$$h[k] = \mathcal{Z}^{-1}\{H(z)\} = \mathcal{Z}^{-1}\left\{\frac{b_m z^m + b_{m-1} z^{m-1} + \cdots + b_1 z + b_0}{z^n + a_{n-1} z^{n-1} + \cdots + a_1 z + a_0}\right\}$$

the procedures for finding the discrete-time impulse response presented in this section can be used also for finding the inverse $\mathcal{Z}$-transform of rational complex functions (represented by a ratio of two polynomials).

7.4 CONTINUOUS-TIME SYSTEM IMPULSE RESPONSE

The continuous-time system impulse response is obtained by using the impulse delta function as the system input signal and assuming that the system initial conditions are zero. It is important to observe that the impulse delta function is present only at the initial time ($t = 0$), and that during that time it sets the system initial conditions to nonzero values.

From (7.1), the linear continuous-time system impulse response is defined by

$$\frac{d^n h(t)}{dt^n} + a_{n-1}\frac{d^{n-1}h(t)}{dt^{n-1}} + \cdots + a_1 \frac{dh(t)}{dt} + a_0 h(t)$$
$$= b_m \frac{d^m \delta(t)}{dt^m} + b_{m-1}\frac{d^{m-1}\delta(t)}{dt^{m-1}} + \cdots + b_1 \frac{d\delta(t)}{dt} + b_0 \delta(t) \qquad (7.36)$$

with

$$h(0^-) = 0, \ h^{(1)}(0^-) = 0, \ldots, h^{(n-1)}(0^-) = 0 \qquad (7.37)$$

For $t \geq 0^+$, we have from (7.36) the homogeneous differential equation (an input-free system)

$$\frac{d^n h(t)}{dt^n} + a_{n-1}\frac{d^{n-1}h(t)}{dt^{n-1}} + \cdots + a_1 \frac{dh(t)}{dt} + a_0 h(t) = 0 \qquad (7.38)$$

whose initial conditions $h^{(i)}(0^+)$, $i = 0, 1, 2, \ldots, n-1$, must be determined.

The initial conditions $h^{(i)}(0^+)$, $i = 0, 1, 2, \ldots, n-1$, can be found as in the discrete-time domain, using either the direct method or system linearity and time invariance. In the continuous-time domain, in general, the direct method is fairly cumbersome, so we will limit our presentation to the method based on the linearity and time-invariance principles.

To that end, we define first the *elementary* continuous-time system impulse response by

$$\frac{d^n h_0(t)}{dt^n} + a_{n-1}\frac{d^{n-1} h_0(t)}{dt^{n-1}} + \cdots + a_1\frac{dh_0(t)}{dt} + a_0 h_0(t) = \delta(t), \quad t \geq 0$$

$$h_0(0^-) = h_0^{(1)}(0^-) = h_0^{(2)}(0^-) = \cdots = h_0^{(n-1)}(0^-) = 0 \quad (7.39)$$

Since the impulse delta function at $t=0$ sets some of the system initial conditions to nonzero values and then disappears, the system in (7.39) becomes the homogeneous system for $t>0$, that is,

$$\frac{d^n h_0(t)}{dt^n} + a_{n-1}\frac{d^{n-1} h_0(t)}{dt^{n-1}} + \cdots + a_1\frac{dh_0(t)}{dt} + a_0 h_0(t) = 0, \quad t > 0 \quad (7.40)$$

The initial conditions for this system, $h^{(i)}(0^+)$, $i=0,1,2,\ldots,n-1$, can be obtained using the following reasoning. The right-hand side of (7.39) at $t=0$ is equal to the impulse delta function, therefore the left-hand side of (7.39) must also be equal to the impulse delta function. Recalling the notion of generalized derivatives from Chapter 2, the preceding observation means that none of the terms $h_0(t), h_0^{(1)}(t), \ldots, h_0^{(n-2)}(t)$ can have a jump discontinuity at $t=0$, since the derivative of a function with a jump discontinuity produces the impulse delta function. Hence, we must have for all of them $h_0^{(i)}(0^-) = h_0^{(i)}(0^+) = 0$, $i=0,1,2,\ldots,n-2$. However, in order to get the impulse delta function on the left-hand side of (7.39), the term $h_0^{(n-1)}(t)$ must contain a jump discontinuity at $t=0$. In addition, that jump must be equal to 1, since after being differentiated in the $h_0^{(n)}(t)$ term, it must produce the impulse delta function multiplied by 1 on the left-hand side of (7.39) for the equality between the left- and right-hand sides of (7.39) to hold. We conclude that the initial conditions for the elementary system impulse response at $t=0^+$ are given by

$$h_0(0^+) = h_0^{(1)}(0^+) = \cdots = h_0^{(n-2)}(0^+) = 0, \quad h_0^{(n-1)}(0^+) = 1 \quad (7.41)$$

Using linearity and time invariance, the original impulse response defined in (7.36) can be obtained as

$$h(t) = \sum_{i=0}^{m} b_i \frac{d^i h_0(t)}{dt^i}, \quad t \geq 0, \quad n > m \quad (7.42)$$

Note that we have assumed that $n > m$; the case in which $n = m$ will be discussed briefly at the end of this exposition. The result given in (7.42) is graphically presented in Figure 7.3.

The elementary continuous-time impulse response can now be determined in a straightforward manner using the methodology from Section 7.1 for solving the homogeneous differential equation (7.40) subject to its initial conditions given in (7.41).

We need to justify a general claim that we have used in (7.42), that the linear time-invariant system at rest under the input signal equal to the ith derivative of the system input produces in its output the ith derivative of the original system output. This can be

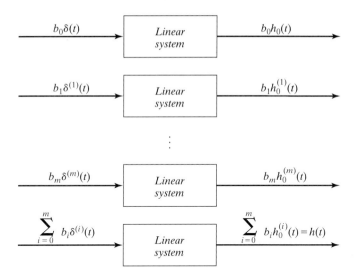

FIGURE 7.3: Procedure for determining the continuous-time linear system impulse response

justified as follows. Let $h_0(t)$ be the solution of (7.39). If we take the ith derivative of the corresponding differential equation, we will obtain

$$\frac{d^n}{dt^n}\left\{\frac{d^i}{dt^i}h_0(t)\right\} + a_{n-1}\frac{d^{n-1}}{dt^{n-1}}\left\{\frac{d^i}{dt^i}h_0(t)\right\} + \cdots + a_1\frac{d}{dt}\left\{\frac{d^i}{dt^i}h_0(t)\right\} + a_0\left\{\frac{d^i}{dt^i}h_0(t)\right\}$$
$$= \frac{d^i}{dt^i}\delta(t), \quad t \geq 0$$

Assuming that all initial conditions are zero, which is the case for the impulse response, it follows that $d^i h_0(t)/dt^i$ is the solution (the system output) of the preceding differential equation due to the forcing function given by $d^i \delta(t)/dt^i$.

The preceding derivations are also valid when $f(t)$ is any differentiable signal (in the generalized sense), so that we have the following general result, valid for the continuous-time, time-invariant linear system (7.1)

$$\begin{aligned} f(t) &\Rightarrow y_{zs}(t) \\ \frac{df(t)}{dt} &\Rightarrow \frac{dy_{zs}(t)}{dt} \\ \frac{d^2 f(t)}{dt^2} &\Rightarrow \frac{d^2 y_{zs}(t)}{dt^2} \\ &\vdots \\ \frac{d^i f(t)}{dt^i} &\Rightarrow \frac{d^i y_{zs}(t)}{dt^i} \end{aligned} \quad (7.43)$$

Note that in the case when $n = m$, formula (7.42) must be slightly modified. Namely, integrating (7.36) n times, we conclude that in such a case at $t = 0$ we have $y(t) = b_n \delta(t)$. Formula (7.36), modified for $n = m$, becomes

$$h(t) = \begin{cases} \sum_{i=0}^{m} b_i \frac{d^i}{dt^i} h_0(t), & t \geq 0, \quad n > m \\ b_n \delta(t) + \sum_{i=0}^{m} b_i \frac{d^i}{dt^i} h_0(t), & t \geq 0, \quad n = m \end{cases} \tag{7.44}$$

EXAMPLE 7.19

Consider the system impulse response problem solved in Example 4.20 using the Laplace transform. The system is defined by

$$\frac{d^2 y(t)}{dt^2} + 3\frac{dy(t)}{dt} + 2y(t) = \frac{df(t)}{dt} + 3f(t)$$

so that the system impulse response $h(t)$ satisfies

$$\frac{d^2 h(t)}{dt^2} + 3\frac{dh(t)}{dt} + 2h(t) = \frac{d\delta(t)}{dt} + 3\delta(t),$$

$$h(0^-) = 0, \quad \frac{dh(0^-)}{dt} = 0$$

The solution to this differential equation can be obtained, using formula (7.44), as

$$h(t) = \frac{d}{dt} h_0(t) + 3h_0(t)$$

where the elementary system impulse response $h_0(t)$ is obtained from

$$\frac{d^2 h_0(t)}{dt^2} + 3\frac{dh_0(t)}{dt} + 2h_0(t) = 0,$$

$$h_0(0^+) = 0, \quad \frac{dh_0(0^+)}{dt} = 1$$

The solution for $h_0(t)$ is given by

$$h_0(t) = C_1 e^{-t} + C_2 e^{-2t}$$

Using the initial conditions at $t = 0^+$ set by the impulse delta function, we obtain $C_1 = 1$, $C_2 = -1$. Hence, the elementary system impulse response is given by

$$h_0(t) = e^{-t} - e^{-2t}$$

Section 7.4 Continuous-Time System Impulse Response 347

so that the system impulse response sought is

$$h(t) = \frac{d}{dt}h_0(t) + 3h_0(t) = -e^{-t} + 2e^{-2t} + 3e^{-t} - 3e^{-2t} = 2e^{-t} - e^{-2t}$$

which is identical to the result obtained in Example 4.20.

EXAMPLE 7.20

We consider now the case when $n = m = 2$, and demonstrate the use of formula (7.44). The impulse response of the system

$$\frac{d^2 y(t)}{dt^2} + 2\frac{dy(t)}{dt} + y(t) = \frac{d^2 f(t)}{dt^2} + \frac{df(t)}{dt} + f(t)$$

satisfies the differential equation

$$\frac{d^2 h(t)}{dt^2} + 2\frac{dh(t)}{dt} + h(t) = \frac{d^2 \delta(t)}{dt^2} + \frac{d\delta(t)}{dt} + \delta(t),$$

$$h(0^-) = 0, \quad \frac{dh(0^-)}{dt} = 0$$

It follows from (7.44) that

$$h(t) = \delta(t) + \frac{d^2 h_0(t)}{dt^2} + \frac{dh_0(t)}{dt} + h_0(t)$$

where the elementary system impulse response $h_0(t)$, obtained using (7.40) and (7.41), is

$$\frac{d^2 h_0(t)}{dt^2} + 2\frac{dh_0(t)}{dt} + h_0(t) = 0,$$

$$h_0(0^+) = 0, \quad \frac{dh_0(0^+)}{dt} = 1$$

The solution for $h_0(t)$ is given by

$$h_0(t) = C_1 e^{-t} + C_2 t e^{-2t}$$

Using the initial conditions at $t = 0^+$, we obtain $C_1 = 0$, $C_2 = 1$. Hence, the elementary system impulse response is given by

$$h_0(t) = t e^{-t}$$

so that the system impulse response is

$$h(t) = \delta(t) + h_0(t) + \frac{d}{dt}h_0(t) + \frac{d^2}{dt^2}h_0(t)$$

$$= \delta(t) + te^{-t} + (e^{-t} - te^{-t}) + (-e^{-t} - e^{-t} + te^{-t}) = \delta(t) - e^{-t} + te^{-t}$$

7.5 COMPLETE CONTINUOUS-TIME SYSTEM RESPONSE

In Chapter 1, we initiated discussion of the relationship between the zero-state and zero-input components of the system response and the particular and homogeneous solutions of the corresponding differential equation. Now, we know that the particular solution, obtained through the convolution procedure, is equal to the zero-state system response. In such a case, the homogeneous solution that satisfies the system initial conditions is the zero-input system response.

Based on the discussion presented at the end of Section 7.1, we conclude that the classical method for solving linear differential equations with constant coefficients can be used to find the complete response of linear time-invariant systems, even in the case when the system differentiates the input signal. Furthermore, the same method can be used to find the zero-state and zero-input components of the complete system response. To do so, we must perform the following steps.

Step 1. Find the impulse response $h(t)$, using formula (7.44) with the elementary system impulse response determined from (7.40) and (7.41).

Step 2. Find the particular solution using the convolution formula (7.11), that is, $y_p(t) = h(t) * f(t)$. That particular solution is the system zero-state response, $y_p(t) = y_{zs}(t)$.

Step 3. Find the general form of a homogeneous solution and determine its constants using the initial conditions defined in (7.13). The homogeneous solution obtained represents the system zero-input response, that is, $y_h(t) = y_{zi}(t)$.

EXAMPLE 7.21

The preceding steps are demonstrated on the second-order system considered in Example 7.20, that is,

$$\frac{d^2 y(t)}{dt^2} + 2\frac{dy(t)}{dt} + y(t) = \frac{d^2 f(t)}{dt^2} + \frac{df(t)}{dt} + f(t),$$

$$f(t) = e^{-3t} u(t), \quad y(0^-) = 1, \quad y^{(1)}(0^-) = 2$$

The impulse response of this system (Step 1) has already been found in Example 7.20 as $h(t) = \delta(t) - e^{-t} + te^{-t}$, so that the particular solution (Step 2) obtained using the convolution formula is given by

$$y_p(t) = \int_{0^-}^{t} h(\tau) f(t-\tau) d\tau = \int_{0^-}^{t} (\delta(\tau) - e^{-\tau} + \tau e^{-\tau}) e^{-3(t-\tau)} u(t-\tau) d\tau$$

$$= e^{-3t} \int_{0^-}^{t} (e^{3\tau}\delta(\tau) - e^{2\tau} + \tau e^{2\tau}) d\tau = e^{-3t}\left(1 - \frac{1}{2}e^{2t} + \frac{1}{2} + \frac{1}{2}te^{2t} - \frac{1}{4}e^{2t} + \frac{1}{4}\right)$$

$$= \frac{7}{4}e^{-3t} + \frac{1}{2}te^{-t} - \frac{3}{4}e^{-t} = y_{zs}(t)$$

In Step 3, we find the homogeneous solution that satisfies the system initial conditions. The homogeneous solution is given by

$$y_h(t) = C_1 e^{-t} + C_2 t e^{-t}$$

Using the system initial conditions according to (7.13), we have

$$y_h(0^-) = y(0^-) = 1 = C_1$$

and

$$y_h^{(1)}(0^-) = y^{(1)}(0^-) = 2 = -C_1 + C_2 \Rightarrow C_2 = 3$$

hence,

$$y_h(t) = y_{zi}(t) = e^{-t} + 3te^{-t}$$

The complete system response is now given by

$$y(t) = y_{zs}(t) + y_{zi}(t) = \frac{1}{4} e^{-t} + \frac{7}{2} t e^{-t} + \frac{7}{4} e^{-3t}$$

Very often in linear system theory and practice, we use the notions of the system *transient response* and the system *steady state response*. Sometimes these are identified, respectively, as the system zero-input and zero-state responses, which is only partially true as explained in the following. The system transient response is determined by $y(t)$—the complete system response during the short time interval immediately following the initial time, that is, $y_{tr}(t) = y(t), t \in [0, t_1]$, for some small t_1 (see Figure 1.4). How small t_1 should be depends on the system. The transient response is usually characterized by fast signal changes, mostly contributed by $y_{zi}(t)$. The system steady state response is defined *theoretically* as

$$y_{ss} = \lim_{t \to \infty} \{y(t)\} \tag{7.45}$$

Practically, this means that the system response after some long interval of time is considered the system steady state response, $y_{ss}(t) = y(t), t \in [t_1, \infty]$. Usually, the zero-input response decays rapidly to zero, hence the steady state response is predominantly determined by the system zero-state response (the forced response).

7.6 COMPLETE DISCRETE-TIME SYSTEM RESPONSE

It can be concluded from Section 7.2 that the complete system response of a discrete-time time-invariant system—represented by a difference equation with constant coefficients, such as (7.14), and initial conditions specified at $y[0], y[1], \ldots, y[n-1]$—can be obtained simply using the general methodology presented in Section 7.2 for solving that kind of difference equation. However, the complete system response obtained in that way does not distinguish between the system response zero-state and zero-input components.

Independently obtaining the zero-state and zero-input components of the complete system response using the time domain technique for solving linear difference equations with constant coefficients is a little more complicated. At the end of Section 7.2, we indicated that the particular solution can be obtained through the convolution procedure using formula (7.24). That formula requires knowledge of the discrete-time system impulse response, which can be obtained using the systematic procedure presented in Section 7.3. This particular solution, obtained through the convolution procedure, represents the *zero-state system response* assuming that the homogeneous solution is obtained from the initial conditions specified as $y[-n], y[-n+1], \ldots, y[-2], y[-1]$ (see formula (7.25)). In order to determine those initial conditions we must propagate backwards in time the initial conditions $y[0], y[1], \ldots, y[n-1]$, using the given difference equation. The procedure for independently determining the zero-input and zero-state components of the complete system response can be performed using the following steps.

Step 1. Determine the discrete-time system impulse response, using any of the techniques presented in Section 7.3.

Step 2. Find the particular solution using the convolution formula (7.24). Such a solution represents the system zero-state response, $y_{zs}^i[k]$ (assuming that the homogeneous solution is obtained using the methodology presented in Step 4).

Step 3. Propagate backwards through the difference equations the initial conditions $y[0], y[1], \ldots, y[n-1]$, and determine the new set of initial conditions $y[-n], y[-n+1], \ldots, y[-2], y[-1]$.

Step 4. Find the homogeneous solution of the given difference equation and determine unknown constants using the initial conditions $y[-n], y[-n+1], \ldots, y[-2], y[-1]$, as specified in (7.25). The homogeneous solution obtained in this way represents the system zero-input response, $y_h[k] = y_{zi}^i[k]$.

These steps are demonstrated in the next example.

EXAMPLE 7.22

Consider the following system, studied in Example 7.11:

$$y[k+2] + 2y[k+1] + y[k] = (-2)^k u[k], \quad y[0] = 0, \quad y[1] = 1$$

This difference equation is easily solved directly, using the methodology for solving linear difference equations with constant coefficients presented in Section 7.2. The solution obtained in Example 7.11 as

$$y[k] = \{-(-1)^k - 2k(-1)^k + (-2)^k\} u[k]$$

in fact represents the complete system response due to the given forcing function and the given system initial conditions. However, from the result obtained for the complete system response, we are unable to determine the zero-state and zero-input components of the system response. This can be done by following the four steps of the algorithm outlined in this section.

Step 1. The system impulse response is obtained from the input-free system

$$h[k+2] + 2h[k+1] + h[k] = 0$$

Section 7.6 Complete Discrete-Time System Response

whose initial conditions are given by ($n = 2, m = 0$)

$$h[2 - 0] = h[2] = b_0 = 1, \quad h[1] = 0$$

The impulse response has the form

$$h[k] = C_1(-1)^k + C_2 k(-1)^k, \quad k \geq 1$$

Using the preceding initial conditions, we have

$$0 = -C_1 - C_2$$
$$1 = C_1 + 2C_2$$

which implies $C_1 = -1, C_2 = 1$. Hence,

$$h[k] = -(-1)^k + k(-1)^k = (k - 1)(-1)^k, \quad k \geq 1$$

Note that $h[0] = 0$.

Step 2. The particular solution obtained from convolution formula (7.24) is given by

$$y_p[k] = h[k] * f[k] = y_{zs}^i[k] = \sum_{m=0}^{k} h[m]f[k - m]$$

$$= \sum_{m=1}^{k} (m - 1)(-1)^m (-2)^{k-m}$$

$$= (-2)^k \sum_{m=1}^{k} m \left(\frac{1}{2}\right)^m - (-2)^k \sum_{m=1}^{k} m \left(\frac{1}{2}\right)^m$$

Note that because $h[0] = 0$ the preceding sums start with $m = 1$. The superscript i indicates that the zero-state response corresponds to the zero-state response of the integral formulation of discrete-time linear time-invariant systems (see Section 5.3.1). The last expression can be rewritten in the form

$$y_{zs}^i[k] = (-2)^k \sum_{m=0}^{k} m \left(\frac{1}{2}\right)^m - (-2)^k \left\{ \sum_{m=1}^{k} m \left(\frac{1}{2}\right)^m - 1 \right\}$$

Using the known formulas for finite summation,

$$\sum_{m=0}^{k} \rho^m = \frac{\rho^{k+1} - 1}{\rho - 1}, \quad \sum_{m=0}^{k} m\rho^m = \frac{\rho}{(1 - \rho)^2}[1 + (\rho k - k - 1)\rho^k], \quad \rho \neq 1$$

(7.46)

we obtain the closed formula for the zero-state response as

$$y_{zs}^i[k] = \{-(-1)^k - k(-1)^k + (-2)^k\}u[k]$$

Step 3. In order to find the zero-input response we must first propagate backwards the system initial conditions. For $k = -1$, we have

$$y[1] + 2y[0] + y[-1] = 0 \Rightarrow y[-1] = -y[1] - 2y[0] = -1$$

and for $k = -2$, we obtain

$$y[0] + 2y[-1] + y[-2] = 0 \Rightarrow y[-2] = -y[0] - 2y[-1] = 2$$

Step 4. The homogeneous solution subject to the initial conditions found in Step 3 represents the system zero-input response, that is,

$$y_{zi}^i[k+2] + 2y_{zi}^i[k+1] + y_{zi}^i[k] = 0, \quad y_{zi}^i[-2] = 2, \quad y_{zi}^i[-1] = -1$$

Its solution is given by

$$y_{zi}^i[k] = C_1(-1)^k + C_2 k(-1)^k, \quad k \geq 0$$

which, with the use of the corresponding initial conditions, implies the expression for the system zero-input response as $y_{zi}^i[k] = -k(-1)^k u[k]$.

It is obvious that the sum $y_{zs}^i[k] + y_{zi}^i[k]$ produces the same value for the complete system response as that obtained in Example 7.11 using the methodology for solving linear difference equations with constant coefficients. ∫

7.7 STABILITY OF CONTINUOUS-TIME LINEAR SYSTEMS

In this section we introduce the stability concept for continuous-time linear systems. To that end, we study *internal system stability* of input-free systems, and *bounded-input bounded-output (BIBO) stability*. Internal system stability can be interpreted as the stability of the system zero-input response, and BIBO system stability can be considered the stability of the zero-state response. We will relate internal system stability to system eigenvalues, and BIBO stability to system impulse response. In addition, we will show that internal system stability implies BIBO system stability, but the reverse implication is not true in general. Given the concerns of this text, the corresponding stability concepts will be presented for time-invariant linear systems.

7.7.1 Internal Stability of Continuous-Time Linear Systems

The concept of internal system stability is related to the motion (system response) of an input-free system. The motion of an input-free system is contributed by its initial conditions and dictated by the nature of the system's characteristic values. The formal definition of the stability of continuous-time, time-invariant linear systems can be stated as follows.

DEFINITION 7.1: A continuous-time input-free system is *stable* if its zero-input response is bounded in time, that is, if

$$|y_{zi}(t)| < M = \text{const} < \infty, \quad \forall t \tag{7.47}$$

Section 7.7 Stability of Continuous-Time Linear Systems

The system is *asymptotically stable* if, in addition to (7.47), the zero-input response tends to zero as time increases, which can be theoretically written as

$$\lim_{t \to \infty} \{y_{zi}(t)\} \to 0 \qquad (7.48)$$

Stable systems as defined by (7.47) are also called *marginally stable* systems. If a system is neither asymptotically nor marginally stable, it is unstable. It follows from (7.47) and (7.48) that for unstable systems the zero-state response escapes to infinity as time increases, that is,

$$\lim_{t \to \infty} \{y_{zi}(t)\} \to \infty \qquad (7.49)$$

Our primary concern is to establish criteria under which linear continuous-time dynamic systems are stable, asymptotically stable, and unstable.

Internal system stability is strictly related to the nature of system characteristic values (eigenvalues). The system characteristic values are the solutions of the system characteristic equation defined in (7.4). Consider first the case when the characteristic values are *distinct*. In such a case, the solution of the corresponding differential equation is given by (7.6), which means that for a zero-input system the zero-input response is represented by

$$y_{zi}(t) = C_1 e^{\lambda_1 t} + C_2 e^{\lambda_2 t} + \cdots + C_n e^{\lambda_n t}, \quad \lambda_1 \neq \lambda_2 \neq \cdots \neq \lambda_n, \quad t \geq 0 \qquad (7.50)$$

Since the eigenvalues are in general complex conjugate, it is easy to observe that a term of the form $e^{(\alpha \pm j\beta)t} = e^{\alpha t} e^{\pm j\beta t}$, as a function of time $t \geq 0$, will decay to zero as time increases for $\alpha < 0$, will tend to infinity as time increases for $\alpha > 0$, and will remain bounded by 1 for all times when $\alpha = 0$ ($|e^{\pm j\beta}| = |\cos(\beta) \pm j \sin(\beta)| = 1$). Note that $\alpha = 0$ indicates an imaginary axis eigenvalue. Hence, it can be easily concluded that asymptotic stability requires $\text{Re}\{\lambda_i\} < 0, \forall i$. For marginal stability, it is required that $\text{Re}\{\lambda_i\} \leq 0, \forall i$. In the case when only one $\text{Re}\{\lambda_i\} > 0, \exists i$, the system is unstable. These simple observations can be stated in the form of the following continuous-time stability theorem.

THEOREM 7.1. A continuous-time linear time-invariant system with distinct eigenvalues is asymptotically stable if $\text{Re}\{\lambda_i\} < 0, \forall i$; it is stable (marginally stable) for $\text{Re}\{\lambda_i\} \leq 0, \forall i$; and it is unstable if there exists an eigenvalue such that $\text{Re}\{\lambda_i\} > 0$. □

Very often, we interpret this theorem by saying that the system is asymptotically stable if all its eigenvalues are located in the open left half of the complex plane (not including the imaginary axis); marginally stable if all eigenvalues are *distinct* and located in the closed left half of the complex plane (including the imaginary axis); and unstable if only one of its eigenvalues is in the open right half of the complex plane (not including the imaginary axis)—see Figure 7.4.

In the case of a *multiple eigenvalue*, the zero-input system response has a term of the form given in either (7.9) or (7.10) depending on whether the multiple eigenvalue is real or complex conjugate. Note that

$$t^i e^{\lambda t} \to \begin{cases} 0, & \text{Re}\{\lambda\} < 0, \\ \infty, & \text{Re}\{\lambda\} \geq 0, \end{cases} \quad t \to \infty, \quad i = 1, 2, \ldots, r-1$$

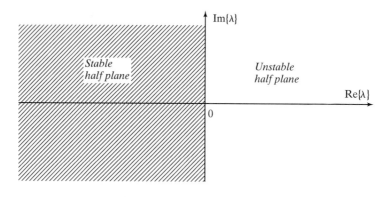

FIGURE 7.4: Stable and unstable half planes

where r indicates the multiplicity of the given eigenvalue. It follows that in the case of multiple eigenvalues, system asymptotic stability requires $\text{Re}\{\lambda_i\} < 0$, $\forall i$, and that if there exists a multiple eigenvalue with $\text{Re}\{\lambda_i\} \geq 0$, the system is unstable. This can be formulated in the following theorem.

THEOREM 7.2. A continuous-time linear time-invariant system with multiple eigenvalues is asymptotically stable if $\text{Re}\{\lambda_i\} < 0$, $\forall i$; marginally stable if all multiple eigenvalues satisfy $\text{Re}\{\lambda_i\} < 0$ and for all distinct eigenvalues $\text{Re}\{\lambda_i\} \leq 0$; and unstable if there exists either a distinct eigenvalue with $\text{Re}\{\lambda_i\} > 0$ or a multiple eigenvalue such that $\text{Re}\{\lambda_i\} \geq 0$. □

Remark 7.2: It should be pointed out that the statement of Theorem 7.2 is superficial. It claims that all multiple eigenvalues on the imaginary axis are unstable. This is true in most cases. However, there are systems with multiple eigenvalues on the imaginary axis that are marginally stable (for example, flexible space structures, antennas, oscillators). A deeper study of this phenomenon requires first a system presentation in the state space form (see Chapter 8), and then advanced knowledge of linear algebra, including the Jordan canonical form. The interested reader is referred to [7], where the Jordan form and its relation to the stability of linear dynamic systems with multiple eigenvalues on the imaginary axis is fully explained.

EXAMPLE 7.23

The linear dynamic system represented in Example 7.1 has the eigenvalues $\lambda_1 = -1$, $\lambda_2 = -3$. Both eigenvalues are in the open left half of the complex plane, and the system is asymptotically stable. The system in Example 7.2 is asymptotically stable, since it has a pair of complex conjugate eigenvalues $\lambda_{1,2} = -1 \pm j1$ located in the open left half of the complex plane. In Example 7.4, we have a triple eigenvalue, $\lambda_1 = \lambda_2 = \lambda_3 = -1$, in the open left half of the complex plane, and this system is also asymptotically stable. A linear dynamic system with the characteristic equation

$$(\lambda + 1)^2(\lambda^2 + 1) = 0 \Rightarrow \lambda_{1,2} = -1, \ \lambda_{3,4} = \pm j$$

Section 7.7 Stability of Continuous-Time Linear Systems

is marginally stable due to the fact that it has simple (distinct) eigenvalues on the imaginary axis and a double eigenvalue in the open left half of the complex plane.

Two examples of unstable systems are represented below by the following characteristic equations and corresponding eigenvalues:

$$(\lambda+1)(\lambda+3)(\lambda^2-4)=0 \Rightarrow \lambda_1=-1,\ \lambda_2=-2,\ \lambda_3=-3,\ \lambda_4=2$$

$$(\lambda^2+1)^2(\lambda+5)=0 \Rightarrow \lambda_{1,2}=-j,\ \lambda_{3,4}=j,\ \lambda_5=-5$$

The first system is unstable since $\lambda_4=2$ lies in the open right half plane. Instability of the second system follows from the multiple eigenvalue on the imaginary axis.

7.7.2 Routh–Hurwitz Stability Criterion†

Algebraic tests for examining internal system stability without actually finding the eigenvalues of linear systems were obtained independently by Routh and Hurwitz. Due to the fact that both methods produce similar results, the method to be presented is known in the literature as the Routh–Hurwitz method. It comes down to a simple algebraic game involving only coefficients of the characteristic polynomial. In this section we show only the mechanics of the test. Its proof, with complete explanation of all steps involved, can be found in [7].

Before we proceed to the Routh–Hurwitz test, we introduce the definition of the Hurwitz polynomial and establish simply a nonstability (instability) result, which is sometimes useful for quickly determining whether a given system is unstable.

DEFINITION 7.2: A polynomial whose roots are in the closed left half of the complex plane is called the Hurwitz polynomial.

Note that in our terminology Hurwitz polynomials are asymptotically stable.

Consider now the following polynomial, which corresponds to the system characteristic equation:

$$s^n + a_{n-1}s^{n-1} + \cdots + a_1 s + a_0$$

Since this polynomial can have either real or complex conjugate roots, it can be factored as

$$s^n + a_{n-1}s^{n-1} + \cdots + a_1 s + a_0 = \prod_{k=1}^{n_1}(s+\alpha_k) \prod_{i=n_1+1}^{(n-n_1)/2}(s+\alpha_i+j\beta_i)(s+\alpha_i-j\beta_i)$$

$$= \prod_{k=1}^{n_1}(s+\alpha_k) \prod_{i=n_1+1}^{(n-n_1)/2}\left(s^2+2\alpha_i s+\alpha_i^2+\beta_i^2\right)$$

For a Hurwitz polynomial, it is valid that $\alpha_k>0,\ \forall k$ and $\alpha_i>0,\ \forall i$, hence, if we multiply factors in the last expression, we see that all coefficients $a_i,\ i=0,1,\ldots,n-1$, must

---

†This section is adopted in part from *Modern Control Systems Engineering* by Lelić and Gajić, Prentice Hall International, London, 1996 (see [7]).

be positive. Thus, if only one of the coefficients a_i is zero or negative, the given polynomial cannot be Hurwitz. Since the existence of the Hurwitz polynomial implies system asymptotic stability, we have the following instability theorem.

THEOREM 7.3. INSTABILITY THEOREM If there exists a coefficient of the characteristic polynomial whose value is zero or whose sign differs from all the remaining coefficients, then the given system is not asymptotically stable. □

Note that Theorem 7.3 provides a *sufficient condition* for system instability and at the same time gives the *necessary condition* for system asymptotic stability. That is, not all polynomials that have all coefficients strictly positive are asymptotically stable, but *all asymptotically stable polynomials have all coefficients strictly positive*.

EXAMPLE 7.24

The following are obvious examples of systems that are not asymptotically stable according to the statement of Theorem 7.3:

$$s^3 + 2s + 3$$
$$s^4 + 2s^3 - s^2 + 2s + 1$$
$$s^6 + s^4 + s^3 + s^2 + s - 1$$

EXAMPLE 7.25

The following polynomial is unstable despite the fact that all its coefficients are strictly positive:
$$s^8 + 2s^7 + 3s^6 + 4s^5 + 5s^4 + 4s^3 + 3s^2 + 2s + 1$$

(see Problem 7.37).

Routh Table

For a given characteristic polynomial, we evaluate the following coefficients and form the Routh table, as given in Table 7.1.

$$A_1 = \frac{a_{n-1}a_{n-2} - a_n a_{n-3}}{a_{n-1}}, \quad A_2 = \frac{a_{n-1}a_{n-4} - a_n a_{n-5}}{a_{n-1}}, \quad \ldots$$

$$B_1 = \frac{A_1 a_{n-3} - a_{n-1} A_2}{A_1}, \quad B_2 = \frac{A_1 a_{n-5} - a_{n-1} A_3}{A_1}, \quad \ldots$$

$$C_1 = \frac{B_1 A_2 - A_1 B_2}{B_1}, \quad C_2 = \frac{B_1 A_3 - A_1 B_3}{B_1}, \quad \ldots$$

$$\ldots \qquad \ldots \qquad \ldots$$

The rationale for the way the coefficients of the Routh table are calculated becomes clear from the corresponding proof of the Routh–Hurwitz stability criterion, as presented in [7]. Using the table, the Routh stability criterion can be stated as follows.

Section 7.7   Stability of Continuous-Time Linear Systems

TABLE 7.1   The Routh Table

s^n	a_n	a_{n-2}	a_{n-4}	...
s^{n-1}	a_{n-1}	a_{n-3}	a_{n-5}	...
s^{n-2}	A_1	A_2	A_3	...
s^{n-3}	B_1	B_2	B_3	...
s^{n-4}	C_1	C_2	C_3	...
$\vdots$	$\vdots$	$\vdots$	$\vdots$	0
s^1	...	...	0	
s^0	...	0		

ROUTH'S CRITERION: For asymptotically stable polynomials (Hurwitz polynomials) all coefficients in the first column of the Routh table must be positive.   □

The following two examples demonstrate the procedure for forming the Routh table and drawing conclusions about a system's asymptotic stability.

EXAMPLE 7.26

A stable system with double poles at -1 and -2 has the characteristic polynomial

$$\Delta(s) = (s+1)^2(s+2)^2 = s^4 + 6s^3 + 13s^2 + 12s + 4$$

The Routh table for this polynomial has the form

s^4	1	13	4
s^3	6	12	0
s^2	11	4	0
s^1	$\dfrac{108}{11}$	0	
s^0	4	0	

Since all coefficients in the first column of the Routh table are positive, we conclude that this polynomial is asymptotically stable, which agrees with previous information about the location of the system poles.

EXAMPLE 7.27

The following polynomial is an example of a system that is not asymptotically stable:

$$\Delta(s) = s^6 + s^5 + 3s^4 + 2s^3 + s^2 + 2s + 1$$

Its Routh table is obtained as

s^6	1	3	1	1
s^5	1	2	2	0
s^4	1	-1	1	0
s^3	3	1	0	
s^2	$-\dfrac{4}{3}$	1	0	
s^1	$\dfrac{13}{4}$	0		
s^0	1	0		

Since one coefficient in the first column of the Routh table is negative, this polynomial is not asymptotically stable. For this example, we can use the MATLAB function roots to find the actual location of the poles of the system. This is achieved by the MATLAB statements

```
c=[1 1 3 2 1 2 1];
p=roots(c)
```

The obtained result shows the following poles:

$$p_{1,2} = -0.2609 \pm j1.6067, \quad p_{3,4} = -0.6742 \pm j0.1154,$$
$$p_{5,6} = 0.4352 \pm j0.7857$$

which indicates four asymptotically stable and two unstable poles.

It is interesting to point out that it is possible to determine exactly the number of poles located in the closed right half of the complex plane (unstable poles) by checking the number of sign changes in the first column of the Routh table. As a matter of fact, this result constitutes the original Routh theorem [8].

THEOREM 7.4. *The number of sign changes in the first column of the Routh table determines the number of unstable poles.* ☐

In the preceding example, the sign in the first column is changed twice: first from 3 to $-4/3$, and then from $-4/3$ to $13/4$. Thus, according to Theorem 7.4, we have two unstable poles, which has been confirmed by MATLAB.

An important advantage of the Routh–Hurwitz method over the direct procedure for finding and examining eigenvalues lies in the fact that, for linear systems containing parameters, the Routh–Hurwitz method gives the range of values for the given parameters such that the system under consideration is asymptotically stable. Of course, in that case the eigenvalue method is not applicable: unless we know exactly all coefficients, the eigenvalue method is useless, except in trivial cases of low-order dimensional systems.[†]

[†]Polynomial algebraic equations can be analytically solved only for $n \leq 4$.

Section 7.7   Stability of Continuous-Time Linear Systems   359

The following example illustrates this feature.

EXAMPLE 7.28

Consider a polynomial whose coefficients contain a parameter; for example,
$$\Delta(s) = s^4 + 2s^3 + Ks^2 + s + 3$$
The Routh table is given by

s^4	1	K	3
s^3	2	1	0
s^2	$\dfrac{2K-1}{2}$	3	0
s^1	$\dfrac{2K-13}{2K-1}$	0	
s^0	3	0	

Since all elements in the first column must be positive for asymptotic stability, it follows that
$$2K - 1 > 0 \quad \text{and} \quad 2K - 13 > 0 \Rightarrow K > 6.5$$
preserves the polynomial's asymptotic stability.

While forming the Routh table, singular cases may appear. That is, zero elements may appear in the first column of the Routh table, and/or everywhere else in the table. A zero element in the first column does not allow completion of the procedure for forming the Routh table as outlined so far in this section.

Singular Cases
While forming the Routh table, two types of singular cases may appear.

1. The first column of the Routh table contains a zero element.
2. An all-zero row appears in the table.

The second singular case may happen if the polynomial has either: (a) pure imaginary axis poles, (b) pairs of real poles symmetrically distributed with respect to the origin, or (c) quadruples of complex poles symmetrical with respect to the origin. These forms of the second singular case are presented in Figure 7.5.

Note that in either singular case, the polynomials under consideration cannot be Hurwitz (asymptotically stable). However, it is interesting to study these cases to determine whether or not the systems are *stable* (see Example 7.30), and to find the exact number of unstable poles.

In order to remedy the first singular case, Routh suggested the ϵ-method. In that method, a zero in the first column is replaced by a small positive parameter $\epsilon > 0$, and the standard procedure for forming the Routh table is continued. After the table is completed, the signs of all elements in the first row (which now depend on ϵ) are determined by taking limits as $\epsilon \to 0^+$. The next example demonstrates the ϵ-method.

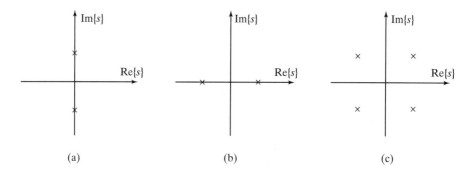

FIGURE 7.5: Possible locations of system poles for the singular case of an all-zero row

EXAMPLE 7.29

Applying the ϵ-method to the polynomial

$$\Delta(s) = s^5 + 2s^4 + 3s^3 + 4s^2 + 5s + 6$$

we get the Routh table

s^5	1	3	5	0
s^4	2	4	6	0
s^3	1	2	0	
s^2	ϵ	6	0	
s^1	$\dfrac{2\epsilon - 6}{\epsilon}$	0		
s^0	6	0		

In the limit when $\epsilon \to 0^+$, we have

$$\lim_{\epsilon \to 0^+} \frac{2\epsilon - 6}{\epsilon} \to -\infty$$

so that there are two sign changes in the first column of the Routh table and the given polynomial has two unstable poles.

Routh proposed to resolve the second singular case, with an all-zero row, by using the following scheme:

1. Form an auxiliary polynomial $d(s)$, as an even polynomial whose coefficients are obtained from the row above the all-zero row.
2. Take the derivative of $d(s)$ and use the coefficients of the polynomial $\frac{d}{ds}d(s)$ instead of the all-zero row.
3. Proceed with the standard way of forming the Routh table.

Section 7.7   Stability of Continuous-Time Linear Systems   361

The next two examples demonstrate the procedure for handling the second singular case.

EXAMPLE 7.30

For the polynomial
$$\Delta(s) = s^5 + s^4 + 2s^3 + 2s^2 + s + 1$$
we get all zeros in the third row, that is,

s^5	1	2	1	0
s^4	1	2	1	0
s^3	0	0	0	

An auxiliary *even* polynomial is formed by using the coefficients above the all-zero row, that is,
$$d(s) = s^4 + 2s^2 + 1$$
whose derivative is given by
$$\frac{d}{ds} d(s) = 4s^3 + 4s$$
so that the coefficients 4, 4 are used in the table instead of the all-zero row, and the procedure is continued in the standard way:

s^5	1	2	1	0
s^4	1	2	1	0
s^3	4	4	0	
s^2	1	1	0	
s^1	0	0		

Since an all-zero row is again encountered, a new auxiliary even polynomial is formed,
$$d_1(s) = s^2 + 1 \Rightarrow \frac{d}{ds} d_1(s) = 2s$$
and the Routh table is completed as

s^5	1	2	1	0
s^4	1	2	1	0
s^3	4	4	0	
s^2	1	1	0	
s^1	2	0		
s^0	1	0		

It can be seen that there are no sign changes in the first column of the Routh table, so there are no poles in the right half of the complex plane, but the polynomial is not asymptotically stable because this is a *singular case and the original Routh criterion for asymptotic stability is not applicable.* In order to find the locations of the poles, we can solve the two auxiliary equations

$$d(s) = s^4 + 2s^2 + 1 = 0, \quad d_1(s) = s^2 + 1 = 0$$

whose solutions are $s_{1,2} = j$, $s_{3,4} = -j$; that is, this polynomial has a pair of double poles at $\pm j$.

EXAMPLE 7.31

Consider the polynomial

$$\Delta(s) = s^7 + 2s^6 + 3s^5 + 4s^4 + 4s^3 + 3s^2 + 2s + 1$$

whose Routh table is given by

s^7	1	3	4	2	0
s^6	2	4	3	1	0
s^5	1	$\frac{5}{2}$	$\frac{3}{2}$	0	
s^4	-1	0	1	0	
s^3	$\frac{5}{2}$	$\frac{5}{2}$	0		
s^2	1	1	0		
s^1	0(2)	0			
s^0	1	0			

An auxiliary second-order polynomial has been formed as $d(s) = s^2 + 1$. Note that the polynomial under consideration has two unstable poles, due to the two sign changes in the first column of the Routh table, and a pair of poles on the imaginary axis at $\pm j$.

It has been shown in the literature that even for complex cases, such as the simultaneous appearance of a row with a zero element in the first column and an all-zero row, the ϵ-method, subject to minor modifications, is efficient [9]. Note that singular cases can be eliminated by forming an extended Routh table according to the procedure presented in [10]. However, the extended Routh table is more complicated than the standard Routh table.

Finally, we point out that the Routh–Hurwitz test can be used to obtain a measure of relative system stability, that is, to determine how far system poles are to the left of the imaginary axis. This can be achieved by working with the shifted characteristic polynomial obtained by replacing the complex frequency s by $s - \gamma$, where $\gamma > 0$ indicates the distance from the imaginary axis.

7.7.3 Continuous-Time Linear System BIBO Stability

The zero-state response of a linear continuous-time system is given by the convolution formula (7.11), that is,

$$y_{zs}(t) = \int_{0^-}^{t} f(t-\tau)h(\tau)\,d\tau = \int_{0^-}^{t} f(\tau)h(t-\tau)\,d\tau \tag{7.51}$$

Bounded-input bounded-output (BIBO) stability means that a bounded input signal $|f(t)| < M_f = \text{const} < \infty, \forall t$, produces a bounded signal in the system output, that is $|y_{zs}(t)| < M_y = \text{const} < \infty, \forall t$. It follows from (7.51) that

$$\begin{aligned} |y_{zs}(t)| &= \left| \int_{0^-}^{t} f(t-\tau)h(\tau)\,d\tau \right| \leq \int_{0^-}^{t} |f(t-\tau)h(\tau)|\,d\tau \\ &\leq \int_{0^-}^{t} |f(t-\tau)|\,|h(\tau)|\,d\tau \leq M_f \int_{0^-}^{t} |h(\tau)|\,d\tau \end{aligned} \tag{7.52}$$

Since this condition must hold for any t, we conclude that

$$\int_{0^-}^{\infty} |h(\tau)|\,d\tau \leq \text{const} < \infty \;\Rightarrow\; |y_{zs}(t)| \leq M_y = \text{const} < \infty, \quad \forall t \tag{7.53}$$

Formula (7.53) produces the following BIBO stability theorem.

THEOREM 7.5. A continuous-time linear system is bounded-input bounded-output stable if and only if its impulse response is absolutely integrable. □

We know from Chapter 4, from formula (4.38), that the system poles (at which the system transfer function is equal to infinity, $H(p_i) = \infty$) are defined by

$$\begin{aligned} H(s) &= \frac{b_m s^m + b_{m-1} s^{m-1} + \cdots + b_1 s + b_0}{s^n + a_{n-1} s^{n-1} + \cdots + a_1 s + a_0} \\ &= \frac{b_m (s-z_1)(s-z_2)\cdots(s-z_m)}{(s-p_1)(s-p_2)\cdots(s-p_n)} \end{aligned} \tag{7.54}$$

It follows from (7.4), (7.5), and (7.54) that the *system poles are equal to the system eigenvalues, except for those eigenvalues that disappear from the system transfer function due to cancellations of common factors in the system transfer function numerator and denominator*. Since the impulse response is the inverse Laplace transform of the system transfer function, we conclude that the system impulse response is a linear combination of the exponential terms $e^{p_i t}$. The system impulse response will be absolutely integrable if all these exponential terms decay to zero as time increases, that is, if all the system poles are located in the open left half of the complex plane (or, in other words, if all the system eigenvalues that appear in the system impulse response are located in the open left half of the complex plane).

In summary, we have the following theorem.

THEOREM 7.6. A continuous-time linear system is bounded-input bounded-output stable if and only if all its poles are located in the open left half plane. □

Note that an asymptotically stable system has all eigenvalues in the open left half plane, which implies that all system poles are also in the open left half of the complex plane. Hence, *an asymptotically stable system is at the same time BIBO stable*. However, it is possible for all system poles to be located in the open left half plane but not all system eigenvalues to be in the open left half plane (those eigenvalues cancelled in the system transfer function are either on the imaginary axis or in the open right half plane). Hence, *BIBO stability does not imply in general asymptotic system stability*. Only in the case in which there is no cancellation of common factors in the system transfer function (when the number of the system poles is identical to the number of the system eigenvalues) is system asymptotic stability equivalent to system BIBO stability.

These observations are demonstrated in the following example.

EXAMPLE 7.32

Let the system transfer function be given by

$$H(s) = \frac{2(s+2)(s-4)}{(s+1)(s+3)(s-4)} = \frac{2(s+2)}{(s+1)(s+3)}$$

The eigenvalues of this system are $\lambda_1 = -1, \lambda_2 = -3, \lambda_3 = 4$, and the system poles are given by $p_1 = -1, p_2 = -3$. It follows that this system is BIBO stable, since all its poles are in the open left half plane, but the system is not asymptotically stable since it has an eigenvalue in the open right half plane.

The system transfer function given by

$$H(s) = \frac{10}{s(s+1)(s+2)}$$

indicates that $\lambda_1 = p_1 = 0, \lambda_2 = p_2 = -1, \lambda_3 = p_3 = -2$. The corresponding system is marginally stable, but it is not BIBO stable since the pole $p_1 = 0$ is on the imaginary axis, so that the corresponding impulse response is not absolutely integrable. Note that $h(t)$ contains a unit step function contributed by the pole $p_1 = 0$, which implies that the corresponding integral in (7.53) diverges, that is, the integral is equal to ∞. ∫

7.8 STABILITY OF DISCRETE-TIME LINEAR SYSTEMS

As in the continuous-time domain, in the discrete-time domain we distinguish between internal system stability (stability of the system zero-input response) and BIBO stability (stability of the system zero-state response). And, as in the continuous-time domain, discrete-time internal system stability depends on the location of the system eigenvalues, and system BIBO stability is determined by nature of the system impulse response, that is, BIBO stability is determined by the location of the system poles in the complex plane.

7.8.1 Internal Stability of Discrete-Time Linear Systems

The definition of discrete-time internal system stability is analogous to Definition 7.1.

DEFINITION 7.3: A linear discrete-time input-free system is *stable* if its zero-input response is bounded in time, that is,

$$|y_{zi}[k]| < M = \text{const} < \infty, \quad \forall k \tag{7.55}$$

The system is *asymptotically stable* if, in addition to (7.55), its zero-input response tends to zero as time increases, that is,

$$\lim_{k \to \infty} \{y_{zi}[k]\} \to 0 \tag{7.56}$$

Internal system stability is related to the system eigenvalues. Discrete-time linear system eigenvalues are the solutions of the corresponding system characteristic equation defined in (7.17). Consider first the case when the characteristic values are *distinct*. The homogeneous solution of the corresponding difference equation is given by (7.18), which means that for a zero-input system the zero-input response is represented by

$$y_{zi}[k] = C_1(\rho_1)^k + C_2(\rho_2)^k + \cdots + C_n(\rho_n)^k, \quad \rho_1 \neq \rho_2 \neq \cdots \neq \rho_n, \quad k \geq 0 \tag{7.57}$$

Let us observe that ρ^k will decay to zero as k increases, if $|\rho| < 1$. The term ρ^k will tend to infinity as time increases if $|\rho| > 1$. For $|\rho| = 1$, which represents the unit circle in the complex plane, ρ^k remains on the unit circle since $(1e^{j\varphi})^k = 1^k e^{jk\varphi} = e^{jk\varphi}$. It can be concluded that asymptotic stability requires that $|\rho_i| < 1$, $\forall i$. For marginal stability, it is needed that $|\rho_i| \leq 1$, $\forall i$. In the case when *only one* $|\rho_i| > 1$, the system is unstable. These simple observations can be stated in the form of the following discrete-time stability theorem valid for the case of distinct system eigenvalues.

THEOREM 7.7. A discrete-time linear time-invariant system with distinct eigenvalues is asymptotically stable if $|\rho_i| < 1$, $\forall i$; stable (marginally stable) for $|\rho_i| \leq 1$, $\forall i$; and unstable if there exists an eigenvalue such that $|\rho_i| > 1$. □

Theorem 7.7 can be interpreted as follows: a discrete-time linear system with distinct eigenvalues is asymptotically stable if all its eigenvalues are located in the unit circle of the complex plane; marginally stable if all eigenvalues are either inside or on the unit circle; and unstable if only one of its eigenvalues is outside the unit circle (see Figure 7.6).

In the case of a *multiple eigenvalue,* the zero-input system response has a term of the form given by either (7.20) or (7.22) depending on whether the multiple eigenvalue is real or complex conjugate. Note that

$$k^i(\rho)^k \to \begin{cases} 0, & |\rho| < 1 \\ \infty, & |\rho| > 1 \end{cases}, \quad k \to \infty, \quad i = 1, 2, \ldots, r-1$$

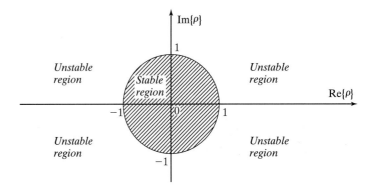

FIGURE 7.6: Unit circle eigenvalue stability criterion for discrete-time linear systems

where r indicates the multiplicity of the given eigenvalue. It follows that, in the case of multiple eigenvalues, discrete-time system asymptotic stability requires $|\rho_i| < 1, \forall i$; and that if there exists a multiple eigenvalue with $|\rho_i| \geq 1$, the system is unstable. This can be formulated in the following theorem.

THEOREM 7.8. A discrete-time linear time-invariant system with multiple eigenvalues is asymptotically stable if $|\rho_i| < 1, \forall i$; marginally stable if all multiple eigenvalues satisfy $|\rho_i| < 1$, and for all distinct eigenvalues $|\rho_i| \leq 1$; and unstable if there exists either a distinct eigenvalue with $\text{Re}\{\lambda_i\} > 0$ or a multiple eigenvalue such that $|\rho_i| \geq 1$. □

Remark 7.3: As in the case of Theorem 7.2, it should be pointed out that the statement of Theorem 7.8 is superficial. It claims that all multiple eigenvalues of a discrete-time linear system that lie on the unit circle are unstable. This is true in most cases. However, it is possible for multiple eigenvalues on the unit circle to be marginally stable. A deeper study of this phenomenon requires discrete-time linear system representation in the state space form (see Chapter 8) and the use of the corresponding Jordan canonical form. This observation indicates that the system representation in the state space form is more general than the system representation using difference equations.

EXAMPLE 7.33

The linear discrete-time system in Example 7.8 has the eigenvalues $\rho_1 = 1/2$, $\rho_2 = -1/3$. Since both eigenvalues are inside the unit circle, the system is asymptotically stable. The same is true for the discrete-time linear system from Example 7.9, which has a triple eigenvalue within the unit circle, $\rho_1 = \rho_2 = \rho_3 = -0.5$. The system in Example 7.10 has four eigenvalues inside the unit circle, $\rho_1 = 0.5$, $\rho_2 = \rho_3 = \rho_4 = -0.25$, and a double eigenvalue on the unit circle, $\rho_5 = \rho_6 = -1$. According to Theorem 7.8, this system is unstable, due to a double eigenvalue at -1.

7.8.2 Algebraic Stability Tests for Discrete Systems[†]

In this section, we study the stability of time-invariant linear discrete-time systems and present two algebraic methods: Jury's test and the bilinear transformation method.

Jury's Stability Test

An algebraic method corresponding to the Routh–Hurwitz test, for examining stability of time-invariant linear discrete-time systems, was developed in a series of papers by Jury in the early sixties [11]. (A simplified form of that method is presented in [12].)

Consider a polynomial represented in the z-domain by the general expression

$$a_n z^n + a_{n-1} z^{n-1} + \cdots + a_1 z + a_0$$

The algebraic test to be presented is based on Table 7.2, the simplified Jury table [13, 14], which can be easily obtained by playing simple algebra with the polynomial's coefficients.

The newly defined coefficients are obtained by a simple pattern:

$$b_{n-1} = a_n - k_0 a_0, \quad b_{n-2} = a_{n-1} - k_0 a_1, \quad \ldots, \quad b_0 = a_1 - k_0 a_{n-1}$$

$$k_0 = \frac{a_0}{a_n}$$

$$c_{n-2} = b_{n-1} - k_1 b_0, \quad c_{n-3} = b_{n-2} - k_1 b_1, \quad \ldots, \quad c_0 = b_1 - k_1 b_{n-2}$$

$$k_1 = \frac{b_0}{b_{n-1}}$$

and so on.

We can state the following theorem, corresponding to the result from the previous section [12–14].

THEOREM 7.9. Assume that $a_n > 0$. Then the polynomial under consideration is asymptotically stable if and only if all coefficients in the first column of Table 7.2 are positive. In addition, the number of negative coefficients in the first column indicates the number of poles outside the unit circle. □

[†]This section is adopted in part from *Modern Control Systems Engineering* by Gajić and Lelić [7].

TABLE 7.2 Simplified Jury Table

a_n	a_{n-1}	a_{n-2}	...	a_2	a_1	a_0
b_{n-1}	b_{n-2}	...	...	b_1	b_0	
c_{n-2}	c_{n-3}	...	...	c_0		
...	...	...	...			
...	...	...				
u_1	u_0					
w_0						

The proof of this theorem is omitted. It follows a similar pattern to that presented in the proof of the Routh–Hurwitz criterion; interested readers can find the complete proof in [13, 14].

In the next example, we demonstrate the procedure for forming Table 7.2.

EXAMPLE 7.34

A polynomial under consideration is given by

$$z^3 + 0.5z^2 + 0.3z + 0.1$$

The simplified Jury table for this example has the form

1	0.5	0.3	0.1
0.99	0.47	0.25	
0.93	0.35		
0.80			

with coefficients $k_0 = 0.1$, $k_1 = 0.25$, $k_2 = 0.24$. From this table and Theorem 7.9, we conclude that the considered polynomial is stable. Using MATLAB, we find the eigenvalues as $p_1 = -0.3893$, $p_{2,3} = -0.0554 \pm j0.53038$; that is, all of them are inside the unit circle.

For clarification of singular cases in regard to the stability of time-invariant linear discrete-time systems, the reader is referred to [11, 12].

Note that a new stability table for discrete-time linear systems, which is very similar to the Routh table, has recently been proposed [15].

Bilinear Transformation

It is important to point out that the Routh–Hurwitz method can be used for studying the stability of discrete-time linear systems as well. That is, the very well-known bilinear transformation defined by

$$z = \frac{s+1}{s-1}$$

maps the unit circle in the z-domain into the left complex plane in the s-domain.

For the given discrete-time characteristic equation $\Delta(z) = 0$, by using the bilinear transformation, we get another characteristic equation in the s-domain, $\Delta_d(s) = 0$, so that the Routh–Hurwitz criterion can be directly applied to $\Delta_d(s)$. The stability conclusion reached for the polynomial $\Delta_d(s)$ is valid also for the polynomial $\Delta(z)$. The following example demonstrates this procedure.

EXAMPLE 7.35

Consider the discrete-time characteristic equation

$$\Delta(z) = z^3 + z^2 + z + 2 = 0$$

Using the bilinear transformation, this is mapped into

$$\left(\frac{s+1}{s-1}\right)^3 + \left(\frac{s+1}{s-1}\right)^2 + \left(\frac{s+1}{s-1}\right) + 2 = 0$$

which implies

$$\Delta_d(s) = 5s^3 - 3s^2 + 7s - 1 = 0$$

Using knowledge from the previous section, we immediately conclude that this polynomial is unstable (it has coefficients of opposite signs). The same instability conclusion is valid for the polynomial $\Delta(z)$. If we form the Routh table, we obtain

s^3	5	7	0	
s^2	-3	-1	0	
s^1	$\frac{16}{3}$	0		
s^0	-1	0		

The first column of this table indicates the existence of three s-domain unstable roots (three sign changes), which means also that in the z-domain there are three roots outside of the unit circle. This can be confirmed by MATLAB, which produces the following roots: $p_1 = -1.3532$, $p_{2,3} = 0.1766 \pm j1.2028$. Also, by applying the Jury test, we obtain the following table.

1	1	1	2
-3	-1	-1	
$-\frac{8}{3}$	$-\frac{2}{3}$		
$-\frac{15}{6}$			

Using Theorem 7.9, we conclude that all three roots are outside the unit circle. ∎

7.8.3 Discrete-Time Linear System BIBO Stability

BIBO stability requires that a bounded input signal $|f[k]| < M_f = \text{const} < \infty$, $\forall k$, produce a bounded signal in the system output, that is, $|y_{zs}[k]| < M_y = \text{const} < \infty$, $\forall k$. The zero-state response of a linear discrete-time system is given by the convolution formula (see (5.54)) and (7.24–25), that is,

$$y_{zs}^i[k] = \sum_{m=0}^{k} f[k-m]h[m] = \mathcal{Z}^{-1}\{F(z)H(z)\} \tag{7.58}$$

where $h[k]$ is the discrete-time system impulse response and $H(z)$ is the discrete-time system transfer function. We know from (5.51) that $h[k] = \mathcal{Z}^{-1}\{H(z)\}$.

It follows from (7.58) that

$$|y_{zs}[k]| = \left|\sum_{m=0}^{k} f[k-m]h[m]\right| \le \sum_{m=0}^{k} |f[k-m]h[m]|$$
$$\le \sum_{m=0}^{k} |f[k-m]||h[m]| \le M_f \sum_{m=0}^{k} |h[m]| \tag{7.59}$$

Since this condition must hold for any k, we conclude that

$$\sum_{m=0}^{\infty} |h(m)| \le \text{const} < \infty \;\Rightarrow\; |y_{zs}[k]| \le M_y = \text{const} < \infty, \;\; \forall k \tag{7.60}$$

Formula (7.60) produces the following BIBO stability theorem for discrete-time linear systems.

THEOREM 7.10. A discrete-time linear system is bounded-input bounded-output stable if and only if its impulse response is absolutely summable. □

We know from Chapter 5, from formulas (5.47–48), that the system poles (at which the discrete-time system transfer function is equal to infinity, $H(p_i) = \infty$) are defined by

$$H(z) = \frac{b_m z^m + b_{m-1} z^{m-1} + \cdots + b_1 z + b_0}{z^n + a_{n-1} z^{n-1} + \cdots + a_1 z + a_0}$$
$$= \frac{b_m(z-z_1)(z-z_2)\cdots(z-z_m)}{(z-p_1)(z-p_2)\cdots(z-p_n)} \tag{7.61}$$

It follows from (7.17) and (7.61) that the discrete-time system poles are equal to the system eigenvalues, except for those eigenvalues that disappear from the system transfer function due to cancellations of common factors in the system transfer function numerator and denominator. Since the discrete-time impulse response is the inverse $\mathcal{Z}$-transform of the discrete-time system transfer function, we conclude that the discrete-time system impulse response is a linear combination of the discrete-time exponential terms p_i^k. The discrete-time system impulse response will be absolutely summable if all these exponential terms decay to zero as time increases, that is, if all the system poles are located in the unit circle of the complex plane.

In summary, we have the following theorem.

THEOREM 7.11. A discrete-time linear system is bounded-input bounded-output stable if and only if all its poles are in the unit circle of the complex plane. □

Note that an asymptotically stable discrete-time system has all eigenvalues in the unit circle, which implies that all system poles are also in the unit circle of the complex plane. Hence, *an asymptotically stable discrete-time system is at the same time BIBO stable*. However, it is possible for all system poles to be located in the unit circle, but not all system eigenvalues to be in the unit circle (those eigenvalues cancelled in the system transfer function are either on the unit circle or outside the unit circle). Hence, *BIBO stability does not imply in general asymptotic system stability*. Only in the case when there is no cancellation of common factors in the discrete-time system transfer function (when the number of the discrete-time system poles is identical to the number of the system eigenvalues) is discrete-time system asymptotic stability equivalent to discrete-time system BIBO stability.

EXAMPLE 7.36

Let a discrete-time system transfer function be given by

$$H(z) = \frac{(z-2)(z+1.5)}{(z+0.3)(z-0.4)(z+1.5)} = \frac{z-2}{(z+0.3)(z-0.4)}$$

It can be observed that the eigenvalues of this discrete-time system are given by $\rho_1 = -0.3$, $\rho_2 = 0.4$, $\rho_3 = -1.5$. The discrete-time system poles are $p_1 = -0.3$, $p_2 = 0.4$. It follows that this system is BIBO stable, since all its poles are in the unit circle, but the system is not asymptotically stable since it has an eigenvalue outside the unit circle, $\rho_3 = -1.5$.

The discrete-time system transfer function given by

$$H(z) = \frac{z+3}{z(z+1)(z-1)}$$

indicates that $\rho_1 = p_1 = 0$, $\rho_2 = p_2 = -1$, $\rho_3 = p_3 = 1$. The corresponding system is marginally stable since the eigenvalues are distinct and $|p_i| \leq 1$, $i = 1, 2, 3$. However, the system is not BIBO stable, since the system poles $p_2 = -1$, $p_3 = 1$ are on the unit circle so that the corresponding impulse response is not absolutely summable. Note that $h[k]$ contains a term $(-1)^k u[k]$ contributed by the pole $p_2 = -1$ and a unit step function contributed by the pole $p_3 = 1$, which implies that the corresponding sum in (7.60) diverges, that is, that sum is equal to ∞. ∫

7.9 MATLAB EXPERIMENT ON CONTINUOUS-TIME SYSTEMS

Purpose: In this experiment, we analyze time responses of a real physical system using MATLAB. We study impulse, step, and sinusoidal responses of the yaw rate dynamics under the influence of the rudder for a commercial aircraft, and draw some useful conclusions about the aircraft's dynamic behavior. In addition, by examining the location of the aircraft eigenvalues and poles we make conclusions about its internal and BIBO stability. By performing this experiment, students will realize how simple and easy it is to analyze higher-order linear continuous-time dynamic systems using MATLAB.

A mathematical model that describes the lateral dynamics of a commercial aircraft is given by a fourth-order differential equation [16]. Using the numerical data from [16], the corresponding differential equation is given by

$$y^{(4)}(t) + 0.6363 y^{(3)}(t) + 0.9396 y^{(2)}(t) + 0.5123 y^{(1)}(t) + 0.0037 y(t)$$
$$= -0.475 f^{(3)}(t) - 0.248 f^{(2)}(t) - 0.1189 f^{(1)}(t) - 0.0564 f(t)$$
(7.62)

where $y(t)$ is the yaw rate and $f(t)$ stands for the changes in the rudder.

Part 1. Using the MATLAB function `impulse`, find the impulse response, that is, observe the yaw rate changes due to an impulse delta signal disturbance acting on the rudder, $f(t) = \delta(t)$. Plot the impulse time response in the range $t \in [0, 200]$ seconds. Find the impulse response analytically by using the methodology from Section 7.4. Plot the obtained analytical result using MATLAB, and compare it with the obtained simulation result (using the MATLAB function `impulse`).

Part 2. Find the step response using the MATLAB function `step`. Plot the step response during the first 200 seconds. Comment on the physical meaning of the obtained results. Do you expect that the aircraft moves to the right when the rudder moves to the left, and vice versa? Check the obtained steady state value for the yaw rate by using formulas (4.39–40).

Part 3. Assume that the rudder is under wind disturbances that can be approximated by a sinusoidal function, $f(t) = \sin(t)$ for the first 200 seconds. Find and plot the aircraft yaw rate dynamics during that time interval. Estimate the maximal yaw rate change due to a sinusoidal disturbance whose maximal magnitude is equal to 1. (*Hint:* Use the MATLAB function `lsim(num,den,f,t)` with `t=0:0.1:200` and `f=sin(t)`.)

Part 4. Examine the aircraft's internal stability by finding its eigenvalues. Use the MATLAB function `eig`. Find the aircraft's transfer function zeros and poles and check whether common factors of the transfer function numerator and denominator can be cancelled. Comment on the aircraft's BIBO stability.

Comment: *Examining system stability should be the first task* in any analysis of a linear system. If the system is found to be unstable, the system should not be subjected to any input; instead, stabilization techniques should first be applied (for example, by introducing a feedback loop as presented in Section 4.4), and then the system can be tested for various input signals. In this experiment, we examine the system stability at the end to be consistent with the order of presentation in chapter.

7.10 MATLAB EXPERIMENT ON DISCRETE-TIME SYSTEMS

Purpose: In this experiment, we analyze time responses of a higher-order linear discrete-time system using MATLAB. We study system impulse, step, and sinusoidal responses. In addition, system internal and BIBO stability are examined. By performing this experiment, the students will realize how simple and easy is to analyze higher-order linear discrete-time dynamic systems using MATLAB.

Consider the linear discrete-time system

$$y[k+5] + 2.2833y[k+4] + 1.2500y[k+3] + 0.0625y[k+2]$$
$$- 0.0833y[k+1] - 0.0125y[k]$$
$$= f[k+2] + 1.6f[k+1] + 0.15f[k]$$

Part 1. Examine the system's internal stability by finding its eigenvalues. Use the MATLAB function `eig`. Find the system's transfer function zeros and poles and check whether common factors of the transfer function numerator and denominator can be cancelled. Comment on the system's BIBO stability.

Part 2. Using the MATLAB function `dimpulse` find the system impulse response. Plot the system impulse response for k=0:1:14 and k=0:1:40. Comment on the differences of the two plots, and explain the reason for the time behavior of the system step response. Find the impulse response analytically using the methodology from Section 7.3. Plot the obtained analytical result using MATLAB and compare it with the obtained simulation result (using the MATLAB function `dimpulse`). Form the reduced-order system by cancelling the zero at −1.5 and the pole at −1.4999, and find the impulse response of the reduced order system for k=0:1:40.

Part 3. Find the step response using the MATLAB function `dstep`. Plot the step response k=0:1:14 and k=0:1:40. Comment on the differences of the two plots, and explain the reason for the time behavior of the system step response. Find the step response of the reduced order system for k=0:1:40. Check the obtained steady state value for the step response of the reduced-order system using formulas (5.49–50).

Part 4. Find and plot the system sinusoidal response. (*Hint:* Use the MATLAB function `dlsim(num,den,f)` with f=sin(k) for k=0:1:14 and k=0:1:40.) Find and plot the sinusoidal reduced order system response for k=0:1:40.

7.11 SUMMARY

Study Guide for Chapter Seven: The students should know how to find homogeneous and particular solutions of linear differential/difference equations with constant coefficients, and how to interpret that knowledge from the linear dynamic system theory point of view. To that end, the problems of finding system impulse response, complete system response, zero-input response, and zero-state response (the particular solution obtained through the convolution process) are fundamentally important. The stability of continuous- and discrete-time linear systems can be easily examined by checking the location of system poles (eigenvalues). Standard questions: (1) find the system impulse response; (2) find the complete system response; (3) find the zero-state system response; (4) find the zero-input system response; (5) examine the stability of a continuous-time system; (6) examine the stability of a discrete-time system.

Solving Differential Equations

General nth-Order Linear Differential Equation with Constant Coefficients:

$$\frac{d^n y(t)}{dt^n} + a_{n-1}\frac{d^{n-1} y(t)}{dt^{n-1}} + a_{n-2}\frac{d^{n-2} y(t)}{dt^{n-2}} + \cdots + a_1 \frac{dy(t)}{dt} + a_0 y(t)$$

$$= b_m \frac{d^m f(t)}{dt^m} + b_{m-1}\frac{d^{m-1} f(t)}{dt^{m-1}} + \cdots + b_1 \frac{df(t)}{dt} + b_0 f(t)$$

General nth-Order Linear Homogeneous Differential Equation:

$$\frac{d^n y_h(t)}{dt^n} + a_{n-1}\frac{d^{n-1} y_h(t)}{dt^{n-1}} + \cdots + a_1 \frac{dy_h(t)}{dt} + a_0 y_h(t) = 0$$

Characteristic Equation in Continuous Time:

$$\Delta(\lambda) = \lambda^n + a_{n-1}\lambda^{n-1} + \cdots + a_1\lambda + a_0 = (\lambda - \lambda_1)(\lambda - \lambda_2)\cdots(\lambda - \lambda_n) = 0$$

Solution of the Homogeneous Equation when $\lambda_1 \neq \lambda_2 \neq \cdots \neq \lambda_n$ and all λ_i are Real:

$$y_h(t) = C_1 e^{\lambda_1 t} + C_2 e^{\lambda_2 t} + \cdots + C_n e^{\lambda_n t}$$

Homogeneous Solution when $\lambda_1 \neq \lambda_2 \neq \cdots \neq \lambda_n$, $\lambda_1 = \alpha_1 + j\beta_1 = \lambda_2^*$, and $\lambda_3, \ldots, \lambda_n$ are Real:

$$y_h(t) = C_1 e^{\alpha_1 t}\cos(\beta_1 t) + C_2 e^{\alpha_1 t}\sin(\beta_1 t) + C_3 e^{\lambda_3 t} + \cdots + C_n e^{\lambda_n t}$$

Homogeneous Solution when $\lambda_1 = \lambda_2 = \cdots = \lambda_r \neq \lambda_{r+1} \neq \cdots \neq \lambda_n$ and all λ_i are Real:

$$y_h(t) = (C_1 + C_2 t + C_3 t^2 + \cdots + C_r t^{r-1})e^{\lambda_1 t} + C_{r+1} e^{\lambda_{r+1} t} + \cdots + C_n e^{\lambda_n t}$$

Homogeneous Solution when $\lambda_1 = \lambda_2 = \cdots = \lambda_r = \alpha_1 + j\beta_1 = \lambda_{r+1}^* = \lambda_{r+2}^* = \cdots = \lambda_{2r}^* \neq \lambda_{2r+1} \neq \cdots \neq \lambda_n$ and $\lambda_{2r+1}, \ldots, \lambda_n$ are Real:

$$y_h(t) = e^{\alpha_1 t}[C_1 \cos(\beta_1 t) + C_2 \sin(\beta_1 t)] + te^{\alpha_1 t}[C_3 \cos(\beta_1 t) + C_4 \sin(\beta_1 t)]$$
$$+ t^2 e^{\alpha_1 t}[C_5 \cos(\beta_1 t) + C_6 \sin(\beta_1 t)] + \cdots$$
$$+ t^{r-1} e^{\alpha_1 t}\left[C_{2r-1} \cos(\beta_1 t) + C_{2r} \sin(\beta_1 t)\right]$$
$$+ C_{2r+1} e^{\lambda_{2r+1} t} + \cdots + C_n e^{\lambda_n t}$$

General Solution of nth-Order Linear Differential Equations:
$$y(t) = y_p(t) + y_h(t)$$
General Initial Conditions:
$$y(0) = y_p(0) + y_h(0)$$
$$y^{(1)}(0) = y_p^{(1)}(0) + y_h^{(1)}(0)$$
$$\vdots$$
$$y^{(n-1)}(0) = y_p^{(n-1)}(0) + y_h^{(n-1)}(0)$$

Solving Difference Equations

General nth-Order Linear Difference Equation with Constant Coefficients:
$$y[k+n] + a_{n-1} y[k+n-1] + \cdots + a_1 y[k+1] + a_0 y[k]$$
$$= b_m f[k+m] + b_{m-1} f[k+m-1] + \cdots + b_1 f[k+1] + b_0 f[k]$$

General nth-Order Linear Homogeneous Difference Equation:
$$y_h[k+n] + a_{n-1} y_h[k+n-1] + \cdots + a_1 y_h[k+1] + a_0 y_h[k] = 0$$

Characteristic Equation in Discrete Time:
$$\Delta(\rho) = \rho^n + a_{n-1}\rho^{n-1} + \cdots + a_1\rho + a_0 = (\rho - \rho_1)(\rho - \rho_2)\cdots(\rho - \rho_n) = 0$$

Solution of the Homogeneous Equation when $\rho_1 \neq \rho_2 \neq \cdots \neq \rho_n$ and all ρ_i are Real:
$$y_h[k] = C_1(\rho_1)^k + C_2(\rho_2)^k + \cdots + C_n(\rho_n)^k$$

Homogeneous Solution when $\rho_1 \neq \rho_2 \neq \cdots \neq \rho_n$, $\rho_1 = \mu_1 e^{j\varphi_1} = \rho_2^*$, and $\rho_3, \ldots, \rho_n$ are Real:
$$y_h[k] = C_1 \mu_1^k \cos[\varphi_1 k] + C_2 \mu_1^k \sin[\varphi_1 k] + C_3 \rho_3^k + \cdots + C_n \rho_n^k$$

Homogeneous Solution when $\rho_1 = \rho_2 = \cdots = \rho_r \neq \rho_{r+1} \neq \cdots \neq \rho_n$ and all ρ_i are Real:
$$y_h[k] = (C_1 + C_2 k + C_3 k^2 + \cdots + C_r k^{r-1})(\rho_1)^k + C_{r+1}(\rho_{r+1})^k + \cdots + C_n(\rho_n)^k$$

Homogeneous Solution when $\rho_1 = \rho_2 = \cdots = \rho_r = \mu_1 e^{j\varphi_1} = \rho_{r+1}^* = \rho_{r+2}^* = \cdots = \rho_{2r}^* \neq \rho_{2r+1}^* \neq \cdots \neq \rho_n^*$ and $\rho_{2r+1}, \ldots, \rho_n$ are Real:
$$y_h[k] = \mu_1^k \{C_1 \cos[\varphi_1 k] + C_2 \sin[\varphi_1 k]\} + k\mu_1^k \{C_3 \cos[\varphi_1 k] + C_4 \sin[\varphi_1 k]\}$$
$$+ k^2 \mu_1^k \{C_5 \cos[\varphi_1 k] + C_6 \sin[\varphi_1 k]\} + \cdots$$
$$+ k^{r-1} \mu_1^k \{C_{2r-1} \cos[\varphi_1 k] + C_{2r} \sin[\varphi_1 k]\}$$
$$+ C_{2r+1} \rho_{2r+1}^k + \cdots + C_n \rho_n^k$$

General Solution of nth-Order Linear Difference Equations:

$$y[k] = y_p[k] + y_h[k]$$

General Initial Conditions:

$$y[0] = y_p[0] + y_h[0]$$
$$y[1] = y_p[1] + y_h[1]$$
$$\vdots$$
$$y[n-1] = y_p[n-1] + y_h[n-1]$$

Response of Continuous-Time Linear Dynamic Systems

Definition of the Continuous-Time Linear System Impulse Response:

$$\frac{d^n h(t)}{dt^n} + a_{n-1}\frac{d^{n-1} h(t)}{dt^{n-1}} + \cdots + a_1 \frac{dh(t)}{dt} + a_0 h(t)$$
$$= b_m \frac{d^m \delta(t)}{dt^m} + b_{m-1}\frac{d^{m-1} \delta(t)}{dt^{m-1}} + \cdots + b_1 \frac{d\delta(t)}{dt} + b_0 \delta(t),$$
$$h(0^-) = 0, \ h^{(1)}(0^-) = 0, \ \ldots, \ h^{(n-1)}(0^-) = 0$$

Solution for the Continuous-Time Linear System Impulse Response:

$$h(t) = \begin{cases} \sum_{i=0}^{m} b_i \frac{d^i}{dt^i} h_0(t), & t \geq 0, \quad n > m \\ b_n \delta(t) + \sum_{i=0}^{m} b_i \frac{d^i}{dt^i} h_0(t), & t \geq 0, \quad n = m \end{cases}$$

$$\frac{d^n h_0(t)}{dt^n} + a_{n-1}\frac{d^{n-1} h_0(t)}{dt^{n-1}} + \cdots + a_1 \frac{dh_0(t)}{dt} + a_0 h_0(t) = 0, \quad t > 0,$$
$$h_0(0^+) = h_0^{(1)}(0^+) = \cdots = h_0^{(n-2)}(0^+) = 0, \quad h_0^{(n-1)}(0^+) = 1$$

Zero-State System Response, Equal to the Particular Solution Obtained through the Convolution Procedure:

$$y_{zs}(t) = y_p(t) = f(t) * h(t) = \int_{0^-}^{t} f(t-\tau) h(\tau)\, d\tau$$

Initial Conditions for the Zero-Input Response (when $y_{zs}(t) = y_p(t)$):

$$y_p(t) = y_{zs}(t) \Rightarrow \begin{cases} y_h(0^-) = y_{zi}(0^-) = y(0^-) \\ y_h^{(1)}(0^-) = y_{zi}^{(1)}(0^-) = y^{(1)}(0^-) \\ \vdots \\ y_h^{(n-1)}(0^-) = y_{zi}^{(n-1)}(0^-) = y^{(n-1)}(0^-) \end{cases}$$

(The zero-input system response is equal to the system's homogeneous solution under the preceding initial conditions.)

Response of Discrete-Time Linear Dynamic Systems

Definition of the Discrete-Time Linear System Impulse Response:

$$h[k+n] + a_{n-1}h[k+n-1] + \cdots + a_1 h[k+1] + a_0 h[k]$$
$$= b_m \delta[k+m] + b_{m-1}\delta[k+m-1] + \cdots + b_1 \delta[k+1] + b_0 \delta[k],$$
$$h[i] = 0, \quad \forall i$$

Solution for the Discrete-Time Linear System Impulse Response:

$$h[k] = \begin{cases} \sum_{i=0}^{m} b_i h_0[k+i], & k \geq 1 \\ 0, & k=0, \ n > m \\ b_m, & k=0, \ n = m \end{cases}$$

$$h_0[k+n] + a_{n-1}h_0[k+n-1] + \cdots + a_1 h_0[k+1] + a_0 h_0[k] = 0, \quad k \geq 1,$$
$$h_0[1] = h_0[2] = \cdots = h_0[n-1] = 0, \quad h_0[n] = 1$$

Direct Method for Finding Discrete-Time Linear System Impulse Response:

$$h[k+n] + a_{n-1}h[k+n-1] + \cdots + a_1 h[k+1] + a_0 h[k] = 0, \quad k \geq 1,$$
$$h[n-m] = b_m$$

(with other initial values for $h[.]$ recursively obtained from the definition equation and $h[n-m-i] = 0, i = 1, 2, \ldots$)

Zero-State Response for the Integral System Formulation, equal to the Particular Solution Obtained through the Convolution Procedure:

$$y_{zs}^i[k] = y_p[k] = f[k] * h[k] = \sum_{m=0}^{k} f[k-m]h[m]$$

Useful Finite Summation Formulas:

$$\sum_{m=0}^{k} \rho^m = \frac{\rho^{k+1} - 1}{\rho - 1}, \quad \sum_{m=0}^{k} m\rho^m = \frac{\rho}{(1-\rho)^2}[1 + (\rho k - k - 1)\rho^k], \quad \rho \neq 1$$

Zero-Input Initial Conditions for the Integral System Formulation (in this case, $y_p[k] = y_{zs}^i[k]$):

$$y_{zs}^i[k] = y_p[k] = f[k] * h[k] \Rightarrow \begin{cases} y_h[-n] = y_{zi}^i[-n] = y[-n] \\ y_h[-n+1] = y_{zi}^i[-n+1] = y[-n+1] \\ \cdots \\ y_h[-1] = y_{zi}^i[-1] = y[-1] \end{cases}$$

Chapter 7 System Response in the Time Domain

(The zero-input system response for the integral system formulation is equal to the system's homogeneous solution under the preceding initial conditions.)

Stability of Continuous-Time Linear Dynamic Systems

Internal Stability in Continuous Time:

$$\text{Re}\{\lambda_i\} < 0, \quad \forall i \Rightarrow \text{asymptotic stability} \left(\lim_{t \to \infty}\{y_{zi}(t)\} \to 0\right)$$

$$\text{Re}\{\lambda_i\} \leq 0, \quad \forall i \quad \text{and those on the imaginary axis are distinct eigenvalues}$$
$$\Rightarrow \text{marginal stability } (|y_{zi}(t)| < \text{const} < \infty, \forall t)$$

$$\exists \lambda_i \text{ such that } \text{Re}\{\lambda_i\} > 0, \text{ or such that } \text{Re}\{\lambda_i\} = 0 \text{ and } \lambda_i \text{ is multiple}$$
$$\Rightarrow \text{instability} \left(\lim_{t \to \infty}\{y_{zi}(t)\} \to \infty\right)$$

Bounded-Input Bounded-Output (BIBO) Stability in Continuous Time:

$$\int_{0^-}^{\infty} |h(\tau)|\, d\tau < \text{const} < \infty \Leftrightarrow \text{Re}\{p_i\} < 0, \quad \forall i$$

$$\Rightarrow \text{BIBO stability } (|f(t)| < M_f = \text{const} \Rightarrow |y_{zs}(t)| < M_y = \text{const})$$

Stability of Discrete-Time Linear Dynamic Systems

Internal Stability in Discrete Time:

$$|\rho_i| < 1, \quad \forall i \Rightarrow \text{asymptotic stability} \left(\lim_{k \to \infty}\{y_{zi}[k]\} \to 0\right)$$

$$|\rho_i| \leq 1, \quad \forall i \text{ and those on the unit circle are distinct eigenvalues}$$
$$\Rightarrow \text{marginal stability } (|y_{zi}[k]| < \text{const} < \infty, \forall k)$$

$$\exists \rho_i \text{ such that } |\rho_i| > 1, \text{ or such that } |\rho_i| = 1 \text{ and } \rho_i \text{ is multiple}$$
$$\Rightarrow \text{instability}\left(\lim_{k \to \infty}\{y_{zi}[k]\} \to \infty\right)$$

Bounded-Input Bounded-Output (BIBO) Stability in Discrete Time:

$$\sum_{m=0}^{\infty} h[m] < \text{const} < \infty \Leftrightarrow |p_i| < 1, \quad \forall i$$

$$\Rightarrow \text{BIBO stability } (|f[k]| < M_f = \text{const} \Rightarrow |y_{zs}[k]| < M_y = \text{const})$$

7.12 REFERENCES

[1] W. Boyce and R. DiPrima, *Elementary Differential Equations,* Wiley, New York, 1992.

[2] M. Spiegel, *Finite Differences and Difference Equations,* McGraw-Hill, New York, 1971.

[3] A. Jerri, *Linear Difference Equations with Discrete Transform Methods,* Kluwer Academic Publishers, Dordrecht, The Netherlands, 1996.

[4] S. Elaydi, *An Introduction to Difference Equations,* Springer-Verlag, New York, 1996.

[5] R. Mickens, *Difference Equations: Theory and Applications,* Van Nostrand Reinhold, New York, 1990.

[6] R. Gupta, "On particular solutions of linear difference equations with constant coefficients," *SIAM Review,* 40:680–84, 1998.

[7] Z. Gajić and M. Lelić, *Modern Control Systems Engineering,* Prentice Hall International, London, 1996.

[8] E. Routh, *A Treatise on the Stability of a Given State Motion,* Macmillan, London, 1887.

[9] M. Fahmy and J. O'Reilly, "A note on the Routh-Hurwitz test," *IEEE Transactions on Automatic Control,* 27:483–85, 1982.

[10] M. Benidir and B. Picinbobo, "Extended table for eliminating singularities in Routh's array," *IEEE Transactions on Automatic Control,* 35:218–22, 1990.

[11] E. Jury, *The Theory and Application of the Z-Transform Method,* Wiley, New York, 1964.

[12] R. Raible, "A simplification of Jury's tabular form," *IEEE Transactions on Automatic Control,* 19:248–50, 1974.

[13] Y. Horn and C. Chen, "A proof of discrete stability test via the Lyapunov theorem," *IEEE Transactions on Automatic Control,* 26:733–34, 1981.

[14] T. Chen, *Linear Systems Theory and Design,* Holt, Rinehart and Winston, New York, 1984.

[15] X. Hu, "A new stability table for discrete-time systems," *Systems & Control Letters,* 22: 385–92, 1994.

[16] G. Franklin, J. Powell, and A. Emami-Naemi, *Feedback Control of Dynamic Systems,* Addison Wesley, Reading, MA, 1994.

[17] M. Spong and M. Vidyasagar, *Robot Dynamics and Control,* Wiley, New York, 1989.

[18] M. Mahmoud, "Order reduction and control of discrete systems," *Proceedings of IEE Part D,* 129:129–35, 1982.

7.13 PROBLEMS

7.1. Find the constants C_1 and C_2 in the expression for the homogeneous solution in Example 7.2, assuming that $y_h(0) = a$ and $y_h^{(1)}(0) = b$.

7.2. Using the convolution formula, find the particular solution of the dynamic system defined in Example 7.6.

7.3. Find homogeneous and particular solutions of the following differential equations.

(a) $\dfrac{d^3y(t)}{dt^3} + 3\dfrac{d^2y(t)}{dt^2} + 2\dfrac{dy(t)}{dt} = e^{-3t}$

(b) $\dfrac{d^3y(t)}{dt^3} + 2\dfrac{d^2y(t)}{dt^2} + \dfrac{dy(t)}{dt} + 2y(t) = \sin(2t)$

(c) $\dfrac{d^3y(t)}{dt^3} + 5\dfrac{d^2y(t)}{dt^2} = t^2$

(d) $\dfrac{d^3y(t)}{dt^3} + 2\dfrac{d^2y(t)}{dt^2} + 2\dfrac{dy(t)}{dt} + y(t) = 2$

(e) $\dfrac{d^3y(t)}{dt^3} + 6\dfrac{d^2y(t)}{dt^2} + 12\dfrac{dy(t)}{dt} + 8y(t) = t+1$

Hint: The characteristic values in part (b) are given by $-2, \pm j1$ and in part (d) by $-1, -0.5 \pm j0.5\sqrt{3}$.

7.4. Find impulse responses of linear continuous-time systems represented by the following differential equations.

(a) $\dfrac{d^3y(t)}{dt^3} + 3\dfrac{d^2y(t)}{dt^2} + 2\dfrac{dy(t)}{dt} = \dfrac{df(t)}{dt} + 3f(t)$

(b) $\dfrac{d^3y(t)}{dt^3} + 2\dfrac{d^2y(t)}{dt^2} + \dfrac{dy(t)}{dt} + 2y(t) = 2\dfrac{df(t)}{dt}$

(c) $\dfrac{d^3y(t)}{dt^3} + 5\dfrac{d^2y(t)}{dt^2} = f(t)$

(d) $\dfrac{d^3y(t)}{dt^3} + 2\dfrac{d^2y(t)}{dt^2} + 2\dfrac{dy(t)}{dt} + y(t) = 2f(t)$

(e) $\dfrac{d^3y(t)}{dt^3} + 6\dfrac{d^2y(t)}{dt^2} + 12\dfrac{dy(t)}{dt} + 8y(t) = \dfrac{df(t)}{dt} - 2f(t)$

Hint: The characteristic values in part (b) are $-2, \pm j1$ and in part (d) are $-1, -0.5 \pm j0.5\sqrt{3}$.

Answers:

(b) $h(t) = 2\dfrac{dh_0(t)}{dt}$, $\dfrac{d^3h_0(t)}{dt^3} + 2\dfrac{d^2h_0(t)}{dt^2} + \dfrac{dh_0(t)}{dt} + 2h_0(t)$,

$h_0(0^+) = h_0^{(1)}(0^+) = 0$, $h_0^{(2)}(0^+) = 1$, $h_0(t) = C_1\cos(t) + C_2\sin(t) + C_3 e^{-2t}$

(c) $h(t) = \left(-\dfrac{1}{25} + \dfrac{1}{5}t + \dfrac{1}{25}e^{-5t}\right)u(t)$

7.5. Using the convolution formula, find particular solutions for the differential equations defined in Problem 7.4.

7.6. Find impulse responses of the continuous-time linear systems represented by the following.

(a) $\dfrac{d^2 y(t)}{dt^2} + \dfrac{dy(t)}{dt} + y(t) = f(t)$

(b) $\dfrac{d^2 y(t)}{dt^2} + 3\dfrac{dy(t)}{dt} + 2y(t) = \dfrac{df(t)}{dt} + 4f(t)$

(c) $\dfrac{d^3 y(t)}{dt^3} + 3\dfrac{d^2 y(t)}{dt^2} + 2\dfrac{dy(t)}{dt} = \dfrac{df(t)}{dt} + 3f(t)$

(d) $\dfrac{d^3 y(t)}{dt^3} + 3\dfrac{d^2 y(t)}{dt^2} + 3\dfrac{dy(t)}{dt} + y(t) = 2\dfrac{d^2 f(t)}{dt^2} + \dfrac{df(t)}{dt} + 2f(t)$

7.7. Find the unit step responses of the linear systems defined in Problem 7.6. Note that the initial conditions for the step response are assumed to be zero.

Hint: $y_{\text{step}}(t) = h(t) * u(t)$.

Answer:

(b) $y_{\text{step}}(t) = \left(-\dfrac{7}{4} + \dfrac{3}{2}t + 2e^{-t} - \dfrac{1}{4}e^{-2t}\right) u(t)$

7.8. Find the unit ramp responses of the linear systems defined in Problem 7.6. Note that the initial conditions for the ramp response are assumed to be zero.

Hint: $y_{\text{ramp}}(t) = h(t) * tu(t)$.

7.9. Find the parabolic responses of the linear systems defined in Problem 7.6. Note that the initial conditions for the parabolic response are assumed to be zero.

7.10. Find the complete response of the following systems

(a) $y^{(2)}(t) + 2y^{(1)}(t) + y(t) = e^{-3t}u(t), \quad y(0^-) = 1, \quad y^{(1)}(0^-) = 0$

(b) $y^{(2)}(t) + 2y^{(1)}(t) + y(t) = \sin(3t)u(t), \quad y(0^-) = 1, \quad y^{(1)}(0^-) = 2$

Answers:

(a) $y(t) = (0.75e^{-t} + 1.5te^{-t} + 0.25e^{-3t})u(t)$

(b) $y(t) = y_{zs}(t) + y_{zi}(t), \quad y_{zi}(t) = C_1 e^{-t} + C_2 t e^{-t}, \quad y_{zi}(0^-) = 1, \quad y_{zi}^{(1)}(0^-) = 0,$

$y_{zs}(t) = \displaystyle\int_{0^-}^{t} h(t-\tau) \sin(3\tau)\, d\tau, \quad h(t) = te^{-t}u(t)$

7.11. Find the complete response of the system

$y^{(3)}(t) + 2y^{(1)}(t) + y(t) = u(t), \quad y(0^-) = 1, \quad y^{(1)}(0^-) = 1, \quad y^{(2)}(0^-) = 0$

Hint: Use MATLAB and its function `roots` to find the system characteristic values.

7.12. Use MATLAB to find the characteristic values of the system

$y^{(5)}(t) + 2y^{(4)}(t) + 3y^{(3)}(t) + 2y^{(2)}(t) + 3y^{(1)}(t) + y(t) = u(t)$

Find analytically the system step response.

382 Chapter 7 System Response in the Time Domain

7.13. For the continuous-time system

$$y^{(2)}(t) + 6y^{(1)}(t) + 9y(t) = 3f^{(1)}(t) - f(t), \quad y(0^-) = 1, \quad y^{(1)}(0^-) = 0$$

find the complete system response due to $f(t) = e^{-t}u(t)$.

Answer:

$$y_{zi}(t) = C_1 e^{-3t} + C_2 t e^{-3t}, \quad y_{zi}(0^-) = 1, \quad y_{zi}^{(1)}(0^-) = 0,$$

$$h(t) = 3h_0^{(1)}(t) - h_0(t), \quad h_0(t) = t e^{-3t} u(t), \quad y_{zs}(t) = h(t) * e^{-t} u(t)$$

$$\Rightarrow y(t) = y_{zi}(t) + y_{zs}(t) = (2e^{-3t} + 8te^{-3t} - e^{-t})u(t)$$

7.14. Find the complete response of the system represented by

$$y^{(2)}(t) + 2y^{(1)}(t) + y(t) = u(t), \quad y(0^-) = 1, \quad y^{(1)}(0^-) = 2$$

and identify the zero-state and zero-input response components.

7.15. Find the complete response of the following linear systems subject to the external input $f(t) = e^{-4t}u(t)$ and the initial conditions $y(0^-) = 1, y^{(1)}(0^-) = 0, y^{(2)}(0^-) = 2$.

(a) $\dfrac{d^3 y(t)}{dt^3} + 11 \dfrac{d^2 y(t)}{dt^2} + 38 \dfrac{dy(t)}{dt} + 40 y(t) = \dfrac{df(t)}{dt} + f(t)$

(b) $\dfrac{d^3 y(t)}{dt^3} + 5 \dfrac{d^2 y(t)}{dt^2} + 7 \dfrac{dy(t)}{dt} + 3y(t) = \dfrac{df(t)}{dt}$

Hint: The characteristic values in part (a) are $-2, -4, -5$ and in part (b) are $-1, -1, -3$.

7.16. Find the zero-input response of the following systems, subject to the initial conditions $y(0^-) = -1, y^{(1)}(0^-) = 2, y^{(2)}(0^-) = 1$, and where required, $y^{(3)}(0^-) = 0$.

(a) $\dfrac{d^4 y(t)}{dt^4} + 6 \dfrac{d^3 y(t)}{dt^3} + 8 \dfrac{d^2 y(t)}{dt^2} = f(t)$

(b) $\dfrac{d^3 y(t)}{dt^3} + 4 \dfrac{d^2 y(t)}{dt^2} + 6 \dfrac{dy(t)}{dt} + 4y(t) = \dfrac{df(t)}{dt} + 3f(t)$

(c) $\dfrac{d^3 y(t)}{dt^3} + 3 \dfrac{d^2 y(t)}{dt^2} + 3 \dfrac{dy(t)}{dt} + y(t) = \dfrac{df(t)}{dt} + 3f(t)$

Hint: The characteristic values in part (b) are $-2, -1 \pm j1$.

7.17. Consider a linear continuous-time system

$$\dfrac{d^3 y(t)}{dt^3} + \dfrac{d^2 y(t)}{dt^2} - \dfrac{dy(t)}{dt} - y(t) = \dfrac{df(t)}{dt} - f(t)$$

(a) Find its impulse response.

(b) Find its step response.

(c) Find the response of this system subject to the input signal $f(t) = e^{-t}u(t)$ and the initial conditions $y(0) = 0, \ dy(0)/dt = 1, \ d^2 y(0)/dt^2 = 2$.

7.18. Find homogeneous and particular solutions of the following difference equations.

  (a) $y[k+2] + 2y[k+1] + 2y[k] = k(-2)^k$
  (b) $y[k+3] + 2y[k+2] + y[k+1] + 2y[k] = 2(-1)^k$
  (c) $y[k+3] + 3y[k+2] + 3y[k+1] + y[k] = k+5$
  (d) $y[k+2] + 0.75y[k+1] + 0.125y[k] = 2\sin[3k]$
  (e) $y[k+3] + 6y[k+2] + 12y[k+1] + 8y[k] = 2k + 3k^2$

  Hint: The characteristic values in part (b) are $-2, \pm j1$ and in part (e) are $-2, -2, -2$.

7.19. Find the impulse response of the following discrete-time linear systems.

  (a) $y_1[k+2] + 1.5y_1[k+1] - y_1[k] = f[k+2] + f[k+1]$
  (b) $y_2[k+3] - 1.5y_2[k+2] + y_2[k] = 6f[k+1]$
  (c) $y_3[k+3] + \frac{1}{2}y_3[k+2] - \frac{1}{4}y_3[k+1] - \frac{1}{8}y_3[k] = 5f[k+3]$
  (d) $y_4[k+2] - y_4[k+1] + y_4[k] = 5f[k+1]$

  Answers:

  (a) $h_1[k] = 0.4(-2)^k u[k] + 0.6(0.5)^k u[k]$
  (b) $h_2[k] = \left(-\frac{8}{3} + 4k + \frac{8}{3}(-0.5)^k\right)u[k]$
  (c) $h_3[k] = \frac{5}{4}\left\{3\left(-\frac{1}{2}\right)^k - 2k\left(-\frac{1}{2}\right)^k + \left(\frac{1}{2}\right)^k\right\}u[k]$

7.20. Using the convolution formula, find particular solutions for the difference equations defined in Problem 7.19.

7.21. Find the impulse response of the following discrete-time linear systems.

  (a) $y[k+2] + 4y[k+1] + 4y[k] = f[k+1] + 4f[k]$
  (b) $y[k+2] + \frac{1}{6}y[k+1] - \frac{1}{6}y[k] = f[k+1] + f[k]$

  Answers:

  (a) $h[k] = -(-2)^k + \frac{1}{2}k(-2)^k, \quad k \geq 1$
  (b) $h[k] = 1.2\left(-\frac{1}{2}\right)^k u[k] + 4.8\left(\frac{1}{3}\right)^k, \quad k \geq 1$

7.22. Find initial conditions set by the impulse delta input signal, $f[k] = \delta[k]$, for the discrete-time linear system

$$y[k+5] - 3y[k+3] + 4y[k+2] - y[k+1] + 2y[k]$$
$$= f[k+3] - 2f[k+2] + 5f[k+1] + f[k]$$

Answer:
$$h[2] = 1, \quad h[3] = -2, \quad h[4] = 3h[2] + 5\delta[0] = 8,$$
$$h[5] = 3h[3] - 4h[2] + \delta[0] = -9, \quad h[1] = 0$$

7.23. Find the step responses of the discrete-time linear systems defined in Problem 7.21.

7.24. For the discrete-time system given by

$$y[k+2] + 4y[k+1] + 3y[k] = f[k] = (-1)^k u[k], \quad y[0] = y[1] = 1$$

find the complete system response.

7.25. Find the impulse response of the system represented by

$$y[k+2] + 5y[k+1] + 4y[k] = f[k+1], \quad y[0] = 1, \quad y[1] = 2$$

Find the complete system response for $f[k] = u[k]$.

Answers:

$$h[k] = \frac{1}{3}(-1)^k u[k] - \frac{1}{3}(-4)^k u[k]$$

$$y[k] = \left(\frac{1}{10} + \frac{11}{6}(-1)^k - \frac{14}{15}(-4)^k \right) u[k]$$

7.26. Determine the complete response of the system

$$y[k+2] + 4y[k+1] + 4y[k] = (-1)^k u[k], \quad y[0] = 2, \quad y[1] = -2$$

Identify the system zero-state and zero-input responses.

7.27. Consider a linear discrete-time system represented by

$$y[k+2] + \frac{1}{2}y[k+1] - \frac{1}{2}y[k] = f[k+1] + f[k]$$

(a) Find its discrete-time impulse response.

(b) Using the expression obtained for the impulse response, write the expression for the step response.

(c) Find the response of this system subject to the input signal $f[k] = (-1)^k u[k]$ and the initial conditions $y[0] = 1, y[1] = -1$.

7.28. Examine the internal and BIBO stability of the linear continuous-time systems defined in Problem 7.4.

7.29. Examine the internal and BIBO stability of the linear continuous-time systems defined in Problem 7.6.

7.30. Use MATLAB to examine the stability of the continuous-time system defined in Problem 7.12.

7.31. Use MATLAB to examine the stability of the continuous-time system defined in Problem 7.17.

7.32. Examine the internal and BIBO stability of the linear discrete-time systems defined in Problem 7.19.

7.33. Examine the internal and BIBO stability of the linear discrete-time systems defined in Problem 7.21.

7.34. Use MATLAB to examine the stability of the discrete-time system defined in Problem 7.22.

7.35. Using the Routh–Hurwitz method, examine the stability of the following systems, represented by their characteristic polynomials.

(a) $\Delta_1(s) = s^2 + 2s + 1$

(b) $\Delta_2(s) = s^3 + 2s^2 + 2s + 1$

(c) $\Delta_3(s) = s^4 + 2s^3 + 3s^2 + 2s + 1$

(d) $\Delta_4(s) = s^5 + 2s^4 + 3s^3 + 3s^2 + 2s + 1$

(e) $\Delta_5(s) = s^6 + 2s^5 + 3s^4 + 4s^3 + 3s^2 + 2s + 1$

Answers:

(c) $\Delta_3(s)$ is asymptotically stable

(d) $\Delta_4(s)$ is marginally stable

(e) $\Delta_5(s)$ has a pair of double poles on the imaginary axis

7.36. Using instability Theorem 7.3, determine whether the following polynomials are unstable.

(a) $\Delta_1(s) = s^4 + 2s^3 - 2s^2 + s + 2$

(b) $\Delta_2(s) = s^5 + 3s^3 + 2s^2 + s + 1$

(c) $\Delta_3(s) = s^6 + s^5 + s^4 + 3s^3 + 2s^2 + s$

(d) $\Delta_2(s) = s^7 - s^5 + s^3 - s + 1$

7.37. Using the Routh–Hurwitz stability criterion, show that the polynomial given in Example 7.25 is unstable.

7.38. Examine the stability of the following polynomials, and determine the number of unstable poles.

(a) $\Delta_1(s) = s^6 + 2s^5 + 3s^4 + s^3 + 3s^2 + 2s + 1$

(b) $\Delta_2(s) = s^7 + s^6 + 2s^5 + 3s^4 + 6s^3 + 2s^2 + s + 1$

(c) $\Delta_3(s) = s^7 + 2s^6 + 2s^5 + s^4 + s^3 + 3s^2 + 2s + 3$

(d) $\Delta_4(s) = s^8 - 2s^7 + s^6 - 3s^5 + 3s^4 - s^3 + 2s^2 - s + 1$

Answers:

(a) $\Delta_1(s)$ has two unstable poles

(b) $\Delta_2(s)$ has four unstable poles

(c) $\Delta_3(s)$ has two unstable poles

(d) $\Delta_4(s)$ has four unstable poles

7.39. Examine whether the following systems can be made asymptotically stable by properly choosing the values of unknown parameters.

(a) $\Delta_1(s) = s^5 + 3s^4 + s^3 + Ks^2 + 2s + 1$

(b) $\Delta_2(s) = s^6 + 2s^5 + 3s^4 + 6s^3 + 2s^2 + s + K$

(c) $\Delta_3(s) = s^4 + s^3 + Ks^2 + 2s + 1$

Answer: (c) $\Delta_3(s)$ is asymptotically stable for $K > 2.5$.

7.40. For the following polynomials, find the number of poles located to the right of the vertical line parallel to the imaginary axis passing through the point $(-1, j0)$; in other words, examine the relative stability of the polynomials.

(a) $\Delta_1(s) = s^3 + 2s^2 + 3s + 4$

(b) $\Delta_2(s) = s^4 + s^3 + 2s^2 + 3s + 6$

Hint: Form the shifted polynomials by using a change of variables $s = s_1 - 1$, and apply the Routh–Hurwitz test to the shifted polynomials.

7.41. Using the Routh–Hurwitz test, examine the stability of the single-link robotic manipulator [17], whose characteristic equation is given by

$$\Delta(s) = s^4 + 4.49s^2 + 0.3464$$

7.42. Determine the stability of the voltage regulator whose transfer function is given in Problem 4.65.

7.43. Examine the stability of the vehicle lateral dynamic model whose differential equation is given in formula (1.40).

7.44. Consider the national income model defined in (1.32). Determine the values for parameters α and β such that the corresponding model is unstable, marginally stable, and asymptotically stable. Do you prefer stability or instability for this system?

7.45. Consider the characteristic polynomial of a discrete-time model of a steam power system [18],

$$\Delta(z) = z^5 - 2.317z^4 + 1.833z^3 - 0.570z^2 + 0.067z - 0.002$$

Examine the stability of this system using Jury's test.

7.46. Using the simplified Jury table and Theorem 7.9, examine the stability of the following polynomial [12].

$$\Delta(z) = z^4 + 2.8z^3 - 4.35z^2 + 1.75z - 0.2$$

Answer: Unstable system with one pole outside of the unit circle.

7.47. Examine the stability of the ATM computer switch whose transfer functions are given in (5.77). Assume that $\alpha_0 = 0.6$ and $\alpha_1 = -0.5$. Note that you must examine the stability of both characteristic polynomials. How will you make a decision about the switch's stability based on information about the stability of its transfer functions?

7.48. Using MATLAB, examine the stability of the flexible beam considered in Section 4.6. Find the beam's differential equation; and find and plot the beam's impulse and step responses.

7.49. Using MATLAB, examine the stability of the aircraft considered in Problem 4.57. Find the corresponding differential equation. Find its impulse and step responses. Find and plot the zero-state response due to the external force $f(t) = e^{-t}\cos(t)u(t)$.

7.50. Find the differential equation that corresponds to the transfer function of the ship positioning system considered in Problem 4.64. Using MATLAB, find the impulse response. Find and plot the complete response due to the forcing function $f(t) = e^{-t}\cos(t)u(t)$ and the initial conditions given by $y(0) = 1$, $y^{(1)}(0) = -1$, $y^{(2)}(0) = 0$.

7.51. Find the difference equation that corresponds to the ATM computer network switch transfer function $H_Y(z)$ defined in (5.77), assuming that $\alpha_0 = 0.6$ and $\alpha_1 = -0.5$. Using the methodology presented in this chapter, find the corresponding impulse and step responses. Find the switch zero-state response due to the input defined by

$$f[k] = \begin{cases} (k+5)u[k-5], & k < 20 \\ 20, & k \geq 20 \end{cases}$$

Using MATLAB, plot the responses obtained.

7.52. Repeat Problem 7.51, assuming that the initial conditions are given by $y[0] = 10$, $y[1] = 15$, $y[2] = 25$.

CHAPTER 8

State Space Approach[†]

The state space technique represents the modern approach to linear system theory and its applications. The state space approach is particularly convenient for representation of higher-order linear systems and linear systems with several inputs and outputs (multi-input multi-output systems). It is extremely efficient for numerical calculations, since many efficient and reliable numerical algorithms developed in mathematics, especially within the area of numerical linear algebra, can be used directly for solving problems defined in state space form. In addition, the state space form is the basis for introducing new system concepts (like system observability and controllability) and new system analysis and design techniques. It should also be pointed out that the state space representation of continuous- and discrete-time linear dynamic systems is more general than the system representation using scalar nth-order differential/difference equations (see Remarks 7.2 and 7.3 in Chapter 7).

The state space model of a continuous-time linear system is represented by a system of n linear differential equations. In matrix form, it is given by

$$\frac{d}{dt}\mathbf{x}(t) = \dot{\mathbf{x}}(t) = \mathbf{A}\mathbf{x}(t) + \mathbf{B}\mathbf{f}(t), \quad \mathbf{x}(0) = \mathbf{x}_0 \quad (8.1)$$

$$\mathbf{y}(t) = \mathbf{C}\mathbf{x}(t) + \mathbf{D}\mathbf{f}(t) \quad (8.2)$$

where the vectors $\mathbf{x}(t) \in \Re^n$, $\mathbf{f}(t) \in \Re^r$, and $\mathbf{y}(t) \in \Re^p$ are, respectively, the state vector, input vector, and output vector, that is,

$$\mathbf{x}(t) = \begin{bmatrix} x_1(t) \\ x_2(t) \\ \vdots \\ x_n(t) \end{bmatrix}, \quad \mathbf{f}(t) = \begin{bmatrix} f_1(t) \\ f_2(t) \\ \vdots \\ f_r(t) \end{bmatrix}, \quad \mathbf{y}(t) = \begin{bmatrix} y_1(t) \\ y_2(t) \\ \vdots \\ y_p(t) \end{bmatrix}$$

(Note that vectors and matrices are denoted by boldface letters.)

We call n the dimension of the system, r the dimension of the system input, and p the dimension of the system output. The matrix $\mathbf{A}^{n \times n}$ describes the *internal* behavior of the system, while matrices $\mathbf{B}^{n \times r}$, $\mathbf{C}^{p \times n}$, and $\mathbf{D}^{p \times r}$ represent connections between the

---

[†] Some parts of this chapter are adopted in revised form from *Modern Control Systems Engineering*, Gajić and Lelić [1].

external world and the system. If there are no direct paths between inputs and outputs, which is often the case, the matrix $\mathbf{D}^{p \times r}$ is zero. These matrices represent arrays of real scalar numbers, as

$$\mathbf{A}^{n \times n} = \begin{bmatrix} a_{11} & a_{12} & \cdots & a_{1n} \\ a_{21} & a_{22} & \cdots & \vdots \\ \vdots & \cdots & \cdots & \vdots \\ a_{n1} & \cdots & \cdots & a_{nn} \end{bmatrix}, \quad \mathbf{B}^{n \times r} = \begin{bmatrix} b_{11} & b_{12} & \cdots & b_{1r} \\ b_{21} & b_{22} & \cdots & \vdots \\ \vdots & \cdots & \cdots & \vdots \\ b_{n1} & \cdots & \cdots & b_{nr} \end{bmatrix}$$

$$\mathbf{C}^{p \times n} = \begin{bmatrix} c_{11} & c_{12} & \cdots & c_{1n} \\ c_{21} & c_{22} & \cdots & \vdots \\ \vdots & \cdots & \cdots & \vdots \\ c_{p1} & \cdots & \cdots & c_{pn} \end{bmatrix}, \quad \mathbf{D}^{p \times r} = \begin{bmatrix} d_{11} & d_{12} & \cdots & d_{1r} \\ d_{21} & d_{22} & \cdots & \vdots \\ \vdots & \cdots & \cdots & \vdots \\ d_{p1} & \cdots & \cdots & d_{pr} \end{bmatrix}$$

It is assumed in this book that all matrices in (8.1) and (8.2) are time invariant, which implies that all scalars a_{ij}, b_{ij}, c_{ij}, and d_{ij} are constant. Studying linear systems with time-varying coefficient matrices requires knowledge of advanced topics in mathematics.

In Section (8.1), we will show how to relate differential equations describing linear dynamical systems to the state space form, and how to determine, in general, scalars $a_{ij}, b_{ij}, c_{ij}, d_{ij}$.

In this chapter, we will encounter mathematical models of electrical networks, electrical machines, aircraft, antennas, industrial reactors, and robots, described by state space forms. Real-world linear dynamic systems can be represented by a compact form known as state space form, defined by (8.1–2), with corresponding dimensions n, r, p and corresponding entries $a_{ij}, b_{ij}, c_{ij}, d_{ij}$.

The state space model for linear discrete-time systems has the same form as (8.1) and (8.2), with the vector differential equation replaced by a vector difference equation, that is,

$$\mathbf{x}[k+1] = \mathbf{A}_d \mathbf{x}[k] + \mathbf{B}_d \mathbf{f}[k], \quad \mathbf{x}[0] = \mathbf{x}_0 \tag{8.3}$$

$$\mathbf{y}[k] = \mathbf{C}_d \mathbf{x}[k] + \mathbf{D}_d \mathbf{f}[k] \tag{8.4}$$

All vectors and matrices defined in (8.3) and (8.4) have the same dimensions as the corresponding vectors and matrices given in (8.1) and (8.2). In this chapter, we present and derive in detail the main state space concepts for continuous-time linear systems and then give the corresponding interpretations in the discrete-time domain.

In the next two examples, we demonstrate how to obtain state space forms for two standard electrical and mechanical linear systems.

EXAMPLE 8.1

Consider a simple RLC electrical circuit as given in Figure 1.10 of Section 1.3, whose mathematical model is represented by the second-order differential equation (1.26), that is,

$$\frac{d^2 e_0(t)}{dt^2} + \left(\frac{L + R_1 R_2 C}{R_2 LC}\right) \frac{de_0(t)}{dt} + \left(\frac{R_1 + R_2}{R_2 LC}\right) e_0(t) = \frac{1}{LC} e_i(t) \tag{8.5}$$

In this mathematical model, $e_i(t)$ represents the system input and $e_0(t)$ the system output.

Introducing the following change of variables,

$$x_1(t) = e_o \Rightarrow \frac{dx_1(t)}{dt} = \frac{de_o(t)}{dt} = x_2(t)$$

$$x_2(t) = \frac{de_o(t)}{dt} \tag{8.6}$$

$$f(t) = e_i(t)$$

$$y(t) = e_o(t) \Rightarrow y(t) = x_1(t)$$

and combining it with (8.5), we obtain

$$\frac{dx_2(t)}{dt} + \left(\frac{L + R_1 R_2 C}{R_2 LC}\right) x_2(t) + \left(\frac{R_1 + R_2}{R_2 LC}\right) x_1(t) = \frac{1}{LC} f(t) \tag{8.7}$$

The first equation in (8.6) and equation (8.7) can be put into matrix form as

$$\begin{bmatrix} \dot{x}_1(t) \\ \dot{x}_2(t) \end{bmatrix} = \begin{bmatrix} 0 & 1 \\ -\dfrac{R_1 + R_2}{R_2 LC} & -\dfrac{L + R_1 R_2 C}{R_2 LC} \end{bmatrix} \begin{bmatrix} x_1(t) \\ x_2(t) \end{bmatrix} + \begin{bmatrix} 0 \\ \dfrac{1}{LC} \end{bmatrix} f(t) \tag{8.8}$$

The last equation from (8.6), in matrix form, is written as

$$y(t) = [1 \ \ 0] \begin{bmatrix} x_1(t) \\ x_2(t) \end{bmatrix} \tag{8.9}$$

Equations (8.8) and (8.9) represent the state space form for the system whose mathematical model is defined by (8.5). The corresponding state space matrices for this system are given by

$$\mathbf{A} = \begin{bmatrix} 0 & 1 \\ -\dfrac{R_1 + R_2}{R_2 LC} & -\dfrac{L + R_1 R_2 C}{R_2 LC} \end{bmatrix}, \quad \mathbf{B} = \begin{bmatrix} 0 \\ \dfrac{1}{LC} \end{bmatrix}, \quad \mathbf{C} = [1 \ \ 0], \quad \mathbf{D} = 0 \tag{8.10}$$

The state space form of a system is not unique. Using another change of variables, we can obtain, for the same system, another state space form (as demonstrated in Problem 8.1). This issue will be further clarified in Section 8.4.

EXAMPLE 8.2

Consider a translational mechanical system as represented in Figure 1.11 of Section 1.3. This system has two inputs, F_1 and F_2, and two outputs, $y_1(t)$ and $y_2(t)$. Its mathematical model is derived in (1.29–30), and given by

$$m_1 \frac{d^2 y_1(t)}{dt^2} + B_1 \frac{dy_1(t)}{dt} + k_1 y_1(t) - B_1 \frac{dy_2(t)}{dt} - k_1 y_2(t) = F_1 \tag{8.11}$$

and
$$-B_1\frac{dy_1(t)}{dt} - k_1 y_1(t) + m_2\frac{d^2 y_2(t)}{dt^2} + (B_1 + B_2)\frac{dy_2(t)}{dt} + (k_1 + k_2)y_2(t) = F_2 \quad (8.12)$$

From equations (8.11) and (8.12), the state space form can be obtained easily by choosing the state space variables

$$x_1(t) = y_1(t), \quad x_2(t) = \frac{dy_1(t)}{dt}, \quad x_3(t) = y_2(t), \quad x_4(t) = \frac{dy_2(t)}{dt} \quad (8.13)$$

$$f_1(t) = F_1, \quad f_2(t) = F_2$$

The state space form of this two-input two-output system is given by

$$\begin{bmatrix} \dot{x}_1(t) \\ \dot{x}_2(t) \\ \dot{x}_3(t) \\ \dot{x}_4(t) \end{bmatrix} = \begin{bmatrix} 0 & 1 & 0 & 0 \\ -\frac{k_1}{m_1} & -\frac{B_1}{m_1} & \frac{k_1}{m_1} & \frac{B_1}{m_1} \\ 0 & 0 & 0 & 1 \\ \frac{k_1}{m_2} & \frac{B_1}{m_2} & -\frac{k_1 + k_2}{m_2} & -\frac{B_1 + B_2}{m_2} \end{bmatrix} \begin{bmatrix} x_1(t) \\ x_2(t) \\ x_3(t) \\ x_4(t) \end{bmatrix} + \begin{bmatrix} 0 & 0 \\ \frac{1}{m_1} & 0 \\ 0 & 0 \\ 0 & \frac{1}{m_2} \end{bmatrix} \begin{bmatrix} f_1(t) \\ f_2(t) \end{bmatrix}$$

(8.14)

and

$$\begin{bmatrix} y_1(t) \\ y_2(t) \end{bmatrix} = \begin{bmatrix} 1 & 0 & 0 & 0 \\ 0 & 0 & 1 & 0 \end{bmatrix} \begin{bmatrix} x_1(t) \\ x_2(t) \\ x_3(t) \\ x_4(t) \end{bmatrix} \quad (8.15)$$

The rest of this chapter is organized as follows. In Section 8.1, a systematic method for obtaining the state space form from differential equations (transfer functions) is developed. The time response of linear systems given in the state space form is considered in Section 8.2. The corresponding results for discrete-time systems, and the procedure for discretization of continuous-time systems leading to discrete-time models, are given in Section 8.3. The characteristic equation, eigenvalues, and eigenvectors, and their use in linear system theory are presented in Section 8.4. In Section 8.5 we present the Cayley–Hamilton theorem and derive methods for finding both continuous- and discrete-time state transition matrices. Section 8.6 considers the linearization problem of nonlinear time-invariant systems. The obtained linearized models of nonlinear systems can be studied by using techniques presented in this book. In Section 8.7, three MATLAB laboratory experiments are designed. A working knowledge of undergraduate linear algebra and the basic theory of differential equations is helpful for complete understanding of this chapter. Some useful results from linear algebra are reviewed in Appendix A.

8.1 STATE SPACE MODELS

In this section, we study state space models of continuous-time linear systems. The corresponding results for discrete-time systems, obtained via analogy with the continuous-time models, are given in Section 8.3.

Chapter 8 State Space Approach

Consider a general nth-order model of a dynamic system, represented by an nth-order differential equation

$$\frac{d^n y(t)}{dt^n} + a_{n-1}\frac{d^{n-1} y(t)}{dt^{n-1}} + \cdots + a_1 \frac{dy(t)}{dt} + a_0 y(t)$$

$$= b_n \frac{d^n f(t)}{dt^n} + b_{n-1}\frac{d^{n-1} f(t)}{dt^{n-1}} + \cdots + b_1 \frac{df(t)}{dt} + b_0 f(t) \qquad (8.16)$$

We assume that all initial conditions for this differential equation, that is, $y(0^-), dy(0^-)/dt, \ldots, d^{n-1}y(0^-)/dt^{n-1}$, are equal to zero.

In order to derive a systematic procedure that transforms a differential equation of order n to a state space form representing a system of n first-order differential equations, we first start with a simplified version of (8.16); namely, we study the case in which no derivatives with respect to the input are present:

$$\frac{d^n y(t)}{dt^n} + a_{n-1}\frac{d^{n-1} y(t)}{dt^{n-1}} + \cdots + a_1 \frac{dy(t)}{dt} + a_0 y(t) = f(t) \qquad (8.17)$$

We introduce the (easy to remember) change of variables

$$x_1(t) = y(t), \quad x_2(t) = \frac{dy(t)}{dt}, \quad x_3(t) = \frac{d^2 y(t)}{dt^2}, \quad \ldots, \quad x_n(t) = \frac{d^{n-1} y(t)}{dt^{n-1}} \qquad (8.18)$$

which, after taking derivatives, leads to

$$\frac{dx_1(t)}{dt} = \dot{x}_1(t) = \frac{dy(t)}{dt} = x_2(t)$$

$$\frac{dx_2(t)}{dt} = \dot{x}_2(t) = \frac{d^2 y(t)}{dt^2} = x_3(t)$$

$$\frac{dx_3(t)}{dt} = \dot{x}_3(t) = \frac{d^3 y(t)}{dt^3} = x_4(t)$$

$$\vdots \qquad (8.19)$$

$$\frac{dx_n(t)}{dt} = \dot{x}_n(t) = \frac{d^n y(t)}{dt^n}$$

$$= -a_0 y(t) - a_1 \frac{dy(t)}{dt} - a_2 \frac{d^2 y(t)}{dt^2} - \cdots - a_{n-1}\frac{d^{n-1} y(t)}{dt^{n-1}} + f(t)$$

$$= -a_0 x_1(t) - a_1 x_2(t) - a_2 x_3(t) - \cdots - a_{n-1} x_n(t) + f(t)$$

The state space form of (8.19) is given by

$$\begin{bmatrix} \dot{x}_1(t) \\ \dot{x}_2(t) \\ \vdots \\ \vdots \\ \dot{x}_{n-1}(t) \\ \dot{x}_n(t) \end{bmatrix} = \begin{bmatrix} 0 & 1 & 0 & \cdots & \cdots & 0 \\ 0 & 0 & 1 & 0 & \cdots & 0 \\ \vdots & \vdots & \ddots & \ddots & \ddots & \vdots \\ \vdots & \vdots & & \ddots & \ddots & 0 \\ 0 & 0 & \cdots & \cdots & 0 & 1 \\ -a_0 & -a_1 & -a_2 & \cdots & \cdots & -a_{n-1} \end{bmatrix} \begin{bmatrix} x_1(t) \\ x_2(t) \\ \vdots \\ \vdots \\ x_{n-1}(t) \\ x_n(t) \end{bmatrix} + \begin{bmatrix} 0 \\ 0 \\ \vdots \\ \vdots \\ 0 \\ 1 \end{bmatrix} f(t) \quad (8.20)$$

with the corresponding output equation obtained from (8.18) as

$$y(t) = \begin{bmatrix} 1 & 0 & \cdots & 0 \end{bmatrix} \begin{bmatrix} x_1(t) \\ x_2(t) \\ \vdots \\ x_{n-1}(t) \\ x_n(t) \end{bmatrix} \quad (8.21)$$

The state space form in (8.20) and (8.21) is known in the literature as the *phase variable canonical form*.

In order to extend this technique to the general case defined by (8.16), which includes derivatives with respect to the input, we form an auxiliary differential equation of (8.16) having the form of (8.17), as

$$\frac{d^n \xi(t)}{dt^n} + a_{n-1} \frac{d^{n-1} \xi(t)}{dt^{n-1}} + \cdots + a_1 \frac{d\xi(t)}{dt} + a_0 \xi(t) = f(t) \quad (8.22)$$

for which the change of variables (8.18) is applicable,

$$x_1(t) = \xi(t)$$
$$x_2(t) = \frac{d\xi(t)}{dt}$$
$$x_3(t) = \frac{d^2 \xi(t)}{dt^2} \quad (8.23)$$
$$\vdots$$
$$x_n(t) = \frac{d^{n-1} \xi(t)}{dt^{n-1}}$$

and then apply the superposition principle to (8.16) and (8.22). Since $\xi(t)$ is the response of (8.22), by the superposition principle the response of (8.16) is given by

$$y(t) = b_0 \xi(t) + b_1 \frac{d\xi(t)}{dt} + b_2 \frac{d^2 \xi(t)}{dt^2} + \cdots + b_n \frac{d^n \xi(t)}{dt^n} \quad (8.24)$$

Equations (8.23) produce the state space equations in the form already given by (8.20). The output equation can be obtained by eliminating $d^n \xi(t)/dt^n$ from (8.24), using (8.22):

$$\frac{d^n \xi(t)}{dt^n} = f(t) - a_{n-1} x_n(t) - \cdots - a_1 x_2(t) - a_0 x_1(t)$$

This leads to the output equation

$$y(t) = [(b_0 - a_0 b_n) \quad (b_1 - a_1 b_n) \quad \cdots \quad (b_{n-1} - a_{n-1} b_n)] \begin{bmatrix} x_1(t) \\ x_2(t) \\ \vdots \\ x_n(t) \end{bmatrix} + b_n f(t) \quad (8.25)$$

It is interesting to point out that for $b_n = 0$, which is almost always the case, the output equation also has an easily remembered form given by

$$y(t) = [b_0 \quad b_1 \quad \cdots \quad b_{n-1}] \begin{bmatrix} x_1(t) \\ x_2(t) \\ \vdots \\ x_n(t) \end{bmatrix} \quad (8.26)$$

Thus, in summary, for a given dynamic system modeled by differential equation (8.16), one can write immediately its state space form, given by (8.20) and (8.26), just by identifying coefficients a_i and b_i, $i = 0, 1, 2, \ldots, n-1$, and using them to form the corresponding entries in matrices **A** and **C**.

There are several other important canonical forms used in linear systems; for more information, the interested reader may consult [1, 2].

EXAMPLE 8.3

Consider a dynamic system represented by the differential equation

$$y^{(6)}(t) + 6y^{(5)}(t) - 2y^{(4)}(t) + y^{(2)}(t) - 5y^{(1)}(t) + 3y(t) = 7f^{(3)}(t) + f^{(1)}(t) + 4f(t)$$

where $y^{(i)}(t)$ stands for the ith derivative, that is, $y^{(i)}(t) = d^i y(t)/dt^i$. According to (8.20) and (8.26), the state space model of this system is described by the matrices

$$\mathbf{A} = \begin{bmatrix} 0 & 1 & 0 & 0 & 0 & 0 \\ 0 & 0 & 1 & 0 & 0 & 0 \\ 0 & 0 & 0 & 1 & 0 & 0 \\ 0 & 0 & 0 & 0 & 1 & 0 \\ 0 & 0 & 0 & 0 & 0 & 1 \\ -3 & 5 & -1 & 0 & 2 & -6 \end{bmatrix}, \quad \mathbf{B} = \begin{bmatrix} 0 \\ 0 \\ 0 \\ 0 \\ 0 \\ 1 \end{bmatrix},$$

$$\mathbf{C} = [4 \quad 1 \quad 0 \quad 7 \quad 0 \quad 0], \quad \mathbf{D} = 0$$

Note that between the system differential equation (8.16) and the system transfer function, defined by

$$H(s) = \frac{b_n s^n + b_{n-1} s^{n-1} + \cdots + b_1 s + b_0}{s^n + a_{n-1} s^{n-1} + \cdots + a_1 s + a_0} \tag{8.27}$$

there is a unique correspondence. Hence, the method just described can be used for obtaining the state space form from the system transfer function.

EXAMPLE 8.4

The transfer function of a flexible beam [3] is given by

$$H(s) = \frac{1.65s^4 - 0.331s^3 - 576s^2 + 90.6s + 19080}{s^6 + 0.996s^5 + 463s^4 + 97.8s^3 + 12131s^2 + 8.11s}$$

Using formulas (8.20) and (8.26), the state space phase variable canonical form is given by

$$\dot{\mathbf{x}}(t) = \begin{bmatrix} 0 & 1 & 0 & 0 & 0 & 0 \\ 0 & 0 & 1 & 0 & 0 & 0 \\ 0 & 0 & 0 & 1 & 0 & 0 \\ 0 & 0 & 0 & 0 & 1 & 0 \\ 0 & 0 & 0 & 0 & 0 & 1 \\ 0 & -8.11 & -12131 & -97.8 & -463 & -0.996 \end{bmatrix} \mathbf{x}(t) + \begin{bmatrix} 0 \\ 0 \\ 0 \\ 0 \\ 0 \\ 1 \end{bmatrix} f(t)$$

and

$$y(t) = [19080 \quad 90.6 \quad -576 \quad -0.331 \quad 1.65 \quad 0]\mathbf{x}(t)$$

MATLAB has a built-in function called tf2ss (transfer function to state space), which produces the state space form from the coefficients of the system transfer function. For this example, the following MATLAB script finds the state space form:

```
num=[1.65 -0.331 -576 90.6 19080];
den=[1 0.996 463 97.8 12131 8.11 0];
[A,B,C,D]=tf2ss(num,den)
```

Also, the MATLAB function ss2tf (state space to transfer function) produces the system transfer function given the system state space form.

Note that the state space form obtained using the MATLAB function tf2ss differs from the state space form derived in this section and represented by formulas (8.20) and (8.26), in the sense that the state space variables are swapped: $x_1(t)$ for $x_n(t)$, $x_2(t)$ for $x_{n-1}(t)$, and so on. This corresponds to the following choice of state variables, which corresponds to (8.18):

$$x_n(t) = y(t), \quad x_{n-1}(t) = \frac{dy(t)}{dt}, \quad x_{n-2}(t) = \frac{d^2 y(t)}{dt^2}, \quad \ldots, \quad x_1(t) = \frac{d^{n-1} y(t)}{dt^{n-1}}$$

which is another confirmation that the state space form is not unique.

8.2 TIME RESPONSE FROM THE STATE SPACE EQUATION

The solution of the state space equations (8.1) and (8.2) can be obtained either in the time domain, by solving the corresponding matrix differential equation directly, or in the frequency domain, by exploiting the power of the Laplace transform. Both methods will be presented in this section.

8.2.1 Time Domain Solution

For the purpose of solving the state space equation (8.1), let us first suppose that the system is in the scalar form

$$\dot{x}(t) = ax(t) + bf(t) \tag{8.28}$$

with a known initial condition $x(0) = x_0$. It is very well known from the elementary theory of differential equations that the solution of (8.28) is

$$x(t) = e^{at}x_0 + \int_0^t e^{a(t-\tau)} bf(\tau)\, d\tau \tag{8.29}$$

The exponential term e^{at} can be expressed using the Taylor series expansion about $t_0 = 0$ as

$$e^{at} = 1 + at + \frac{1}{2!}a^2 t^2 + \frac{1}{3!}a^3 t^3 + \cdots = \sum_{i=0}^{\infty} \frac{1}{i!}(at)^i \tag{8.30}$$

Analogously, in the following we prove that the solution of a general nth-order matrix state space differential equation (8.1) is given by

$$\mathbf{x}(t) = e^{\mathbf{A}t}\mathbf{x}(0) + \int_0^t e^{\mathbf{A}(t-\tau)} \mathbf{B} \mathbf{f}(\tau)\, d\tau \tag{8.31}$$

For simplicity, we first consider the homogeneous system without an input, that is,

$$\dot{\mathbf{x}}(t) = \mathbf{A}\mathbf{x}(t), \quad \mathbf{x}(0) = \mathbf{x}_0 \tag{8.32}$$

By analogy with the scalar case, we expect the solution of this differential equation to be

$$\mathbf{x}(t) = e^{\mathbf{A}t}\mathbf{x}(0) \tag{8.33}$$

We shall prove that this is indeed a solution if (8.33) satisfies differential equation (8.32), where *the matrix exponential is defined using the Taylor series expansion* as

$$e^{\mathbf{A}t} = \mathbf{I} + \mathbf{A}t + \frac{1}{2!}\mathbf{A}^2 t^2 + \frac{1}{3!}\mathbf{A}^3 t^3 + \cdots = \sum_{i=0}^{\infty} \frac{1}{i!}\mathbf{A}^i t^i \tag{8.34}$$

Section 8.2 Time Response from the State Space Equation

The proof is simple and is obtained by taking the derivative of the right-hand side of (8.34), that is,

$$\begin{aligned}
\frac{de^{\mathbf{A}t}}{dt} &= \frac{d}{dt}\left(\mathbf{I} + \mathbf{A}t + \frac{1}{2!}\mathbf{A}^2 t^2 + \cdots\right) \\
&= \mathbf{A} + \frac{2}{2!}\mathbf{A}^2 t + \frac{3}{3!}\mathbf{A}^3 t^2 + \cdots = \mathbf{A}\left(\mathbf{I} + \frac{1}{1!}\mathbf{A}t + \frac{1}{2!}\mathbf{A}^2 t^2 + \cdots\right) \\
&= \left(\mathbf{I} + \frac{1}{1!}\mathbf{A}t + \frac{1}{2!}\mathbf{A}^2 t^2 + \cdots\right)\mathbf{A} = \mathbf{A}e^{\mathbf{A}t} = e^{\mathbf{A}t}\mathbf{A}
\end{aligned}$$

Now, substitution of (8.33) into differential equation (8.32) yields

$$\dot{\mathbf{x}}(t) = \frac{d}{dt}\mathbf{x}(t) = \frac{d}{dt}e^{\mathbf{A}t}\mathbf{x}(0) = \mathbf{A}e^{\mathbf{A}t}\mathbf{x}(0) = \mathbf{A}\mathbf{x}(t)$$

so that the matrix differential equation (8.32) is satisfied, and hence $\mathbf{x}(t) = e^{\mathbf{A}t}\mathbf{x}(0)$ is its solution. Note that for $t = 0$ we have $\mathbf{x}(0) = e^{\mathbf{A}0}\mathbf{x}(0) = \mathbf{I}\mathbf{x}(0) = \mathbf{x}(0)$, so the initial condition is also satisfied.

The matrix $e^{\mathbf{A}t}$ is known as the *state transition matrix* because it relates the system state at time t to that at time zero, and is denoted by

$$\Phi(t) = e^{\mathbf{A}t} \tag{8.35}$$

The state transition matrix as a time function depends only on the matrix $\mathbf{A}$. Therefore $\Phi(t)$ completely describes the internal behavior of the system, when the external influence (system input $\mathbf{f}(t)$) is absent. The system transition matrix plays a fundamental role in the theory of linear dynamic systems. In the following, we state and verify the main properties of this matrix, which is represented in the symbolic form by $e^{\mathbf{A}t}$ and (so far) defined only by (8.34).

Properties of the State Transition Matrix

It can be easily verified, by taking the derivative of

$$\mathbf{x}(t) = \Phi(t)\mathbf{x}(0)$$

that the state transition matrix satisfies the linear homogeneous state space equation (8.1) with the initial condition equal to an identity matrix, that is,

$$\frac{d\Phi(t)}{dt} = \mathbf{A}\Phi(t), \quad \Phi(0) = \mathbf{I} \tag{8.36}$$

The main properties of the matrix $\Phi(t)$, which follow from (8.34) and (8.36), are as follows:

(a) $\Phi(0) = \mathbf{I}$
(b) $\Phi^{-1}(t) = \Phi(-t) \Rightarrow \Phi(t)$ is nonsingular for every t

(c) $\Phi(t_2 - t_0) = \Phi(t_2 - t_1)\Phi(t_1 - t_0)$
(d) $\Phi(t)^i = \Phi(it)$, for $i \in N$

The proofs are straightforward. Property (a) is obtained when $t = 0$ is substituted into (8.34).

Property (b) holds, since

$$(e^{\mathbf{A}t})^{-1}e^{\mathbf{A}t} = \mathbf{I}$$

which, after multiplication from the right-hand by $e^{-\mathbf{A}t}$, implies

$$(e^{\mathbf{A}t})^{-1}e^0 = e^{-\mathbf{A}t} \Rightarrow \Phi^{-1}(t) = \Phi(-t)$$

Property (c) follows from

$$\Phi(t_2 - t_0) = e^{\mathbf{A}(t_2-t_0)} = e^{\mathbf{A}(t_2-t_1+t_1-t_0)}$$
$$= e^{\mathbf{A}(t_2-t_0)}e^{\mathbf{A}(t_1-t_0)} = \Phi(t_2 - t_1)\Phi(t_1 - t_0)$$

Property (d) is proved by using the fact that

$$\Phi(t)^i = (e^{\mathbf{A}t})^i = e^{\mathbf{A}(it)} = \Phi(it)$$

In addition to properties (a), (b), (c), and (d), we have already established one additional property, namely the derivative property, as

(e) $\dfrac{d}{dt}\Phi(t) = \mathbf{A}\Phi(t) \Leftrightarrow \dfrac{d}{dt}e^{\mathbf{A}t} = e^{\mathbf{A}t}\mathbf{A} = \mathbf{A}e^{\mathbf{A}t}$

The state transition matrix $\Phi(t)$ can be found using several methods. Two of them are given in this section in formulas (8.34) and (8.47). The third, very popular in linear algebra, is based on the Cayley–Hamilton theorem and is given in Section 8.5.

In the case when the input $\mathbf{f}(t)$ is present in the system (forced response), that is,

$$\dot{\mathbf{x}}(t) = \mathbf{A}\mathbf{x}(t) + \mathbf{B}\mathbf{f}(t), \quad \mathbf{x}(0) = \mathbf{x}_0$$

we look for the solution of the state space equation in the form

$$\mathbf{x}(t) = e^{\mathbf{A}t}\mathbf{g}(t) \tag{8.37}$$

Then

$$\dot{\mathbf{x}}(t) = \mathbf{A}e^{\mathbf{A}t}\mathbf{g}(t) + e^{\mathbf{A}t}\dot{\mathbf{g}}(t) = \mathbf{A}\mathbf{x}(t) + e^{\mathbf{A}t}\dot{\mathbf{g}}(t) \tag{8.38}$$

It follows from (8.1) and (8.38) that

$$e^{\mathbf{A}t}\dot{\mathbf{g}}(t) = \mathbf{B}\mathbf{f}(t) \tag{8.39}$$

From (8.39), we have

$$\dot{\mathbf{g}}(t) = (e^{\mathbf{A}t})^{-1}\mathbf{B}\mathbf{f}(t) = e^{-\mathbf{A}t}\mathbf{B}\mathbf{f}(t) \tag{8.40}$$

Integrating this equation, bearing in mind that $\mathbf{x}(0) = e^{\mathbf{A}\cdot 0}\mathbf{g}(0) = \mathbf{g}(0)$, we get

$$\mathbf{g}(t) - \mathbf{g}(0) = \int_0^t e^{-\mathbf{A}\tau}\mathbf{B}\mathbf{f}(\tau)\,d\tau \tag{8.41}$$

Substitution of the expression in (8.41) into (8.37) gives the required solution,

$$\mathbf{x}(t) = e^{\mathbf{A}t}\mathbf{x}(0) + \int_0^t e^{\mathbf{A}(t-\tau)}\mathbf{B}\mathbf{f}(\tau)\,d\tau \tag{8.42}$$

or

$$\mathbf{x}(t) = \Phi(t)\mathbf{x}(0) + \int_0^t \Phi(t-\tau)\mathbf{B}\mathbf{f}(\tau)\,d\tau \tag{8.43}$$

When the initial state of the system is known at time t_0, rather than at time $t=0$, the solution of the state space equation is similarly obtained as

$$\mathbf{x}(t) = \Phi(t - t_0)\mathbf{x}(t_0) + \int_{t_0}^t \Phi(t-\tau)\mathbf{B}\mathbf{f}(\tau)\,d\tau$$

$$= e^{\mathbf{A}(t-t_0)}\mathbf{x}(t_0) + \int_{t_0}^t e^{\mathbf{A}(t-\tau)}\mathbf{B}\mathbf{f}(\tau)\,d\tau \tag{8.44}$$

This can be easily verified by repeating steps for (8.37–42) with $\mathbf{x}(t_0) = \mathbf{x}_0$ and $\mathbf{x}(t_0) = e^{\mathbf{A}t_0}\mathbf{g}(t_0)$.

The solution derived in formulas (8.42–44) represents the *system state response*. The *system output response* is obtained from (8.2) and (8.44) as

$$\mathbf{y}(t) = \mathbf{C}e^{\mathbf{A}(t-t_0)}\mathbf{x}(t_0) + \int_{t_0}^t e^{\mathbf{A}(t-\tau)}\mathbf{B}\mathbf{f}(\tau)\,d\tau + \mathbf{D}\mathbf{f}(t) \tag{8.45}$$

EXAMPLE 8.5

For the system given by

$$\dot{\mathbf{x}}(t) = \begin{bmatrix} -1 & 0 & 0 \\ 0 & -2 & 0 \\ 0 & 0 & -3 \end{bmatrix}\mathbf{x}(t) + \begin{bmatrix} 1 \\ 1 \\ 1 \end{bmatrix}f(t)$$

$$y(t) = \begin{bmatrix} 6 & -6 & 1 \end{bmatrix}\mathbf{x}(t)$$

400 Chapter 8 State Space Approach

we will find the state transition matrix $\Phi(t)$ and evaluate $\Phi(1)$; find the system state response to a unit step; and check the obtained solution using the MATLAB function step.

At this point, we are able to find the state transition matrix (matrix exponential) using formula (8.34) only, which deals with an infinite series, and hence is not very convenient for calculations. We are able only in rare cases to obtain an analytical form for $\Phi(t)$ using formula (8.34) (an example is given in Problem 8.28). Better ways to find $\Phi(t)$ are the method based on the Cayley–Hamilton theorem (see Section 8.5) and the formula based on the Laplace transform (see formula (8.47)). However, in this problem, since the system matrix is diagonal (uncoupled state space form), we can avoid using either of these methods to find the state transition matrix. Namely, *for diagonal matrices only,* it can be easily shown (see Problem 8.2) that

$$\Phi(t) = e^{\begin{bmatrix} -1 & 0 & 0 \\ 0 & -2 & 0 \\ 0 & 0 & -3 \end{bmatrix} t} = \begin{bmatrix} e^{-t} & 0 & 0 \\ 0 & e^{-2t} & 0 \\ 0 & 0 & e^{-3t} \end{bmatrix}$$

For $t = 1$, the state transition matrix is given by

$$\Phi(1) = \begin{bmatrix} e^{-1} & 0 & 0 \\ 0 & e^{-2} & 0 \\ 0 & 0 & e^{-3} \end{bmatrix} = \begin{bmatrix} 0.3679 & 0 & 0 \\ 0 & 0.1353 & 0 \\ 0 & 0 & 0.0498 \end{bmatrix}$$

It should be pointed out that, in general, the matrix exponential (8.34), representing the state transition matrix, can be evaluated at any given time instant t using the MATLAB function expm(A*t).

The state response to a unit step is found by (8.42), which leads to

$$\mathbf{x}(t) = \int_0^t \Phi(t-\tau)\mathbf{B}\mathbf{f}(\tau)\,d\tau = \int_0^t \begin{bmatrix} e^{-(t-\tau)} & 0 & 0 \\ 0 & e^{-2(t-\tau)} & 0 \\ 0 & 0 & e^{-3(t-\tau)} \end{bmatrix} \begin{bmatrix} 1 \\ 1 \\ 1 \end{bmatrix} \cdot 1\,d\tau$$

$$= \int_0^t \begin{bmatrix} e^{-(t-\tau)} \\ e^{-2(t-\tau)} \\ e^{-3(t-\tau)} \end{bmatrix} d\tau = \begin{bmatrix} 1 - e^{-t} \\ 0.5(1 - e^{-2t}) \\ 0.333(1 - e^{-3t}) \end{bmatrix}$$

Note that for the step response, by definition, the initial conditions must be set to zero (see Section 4.3.3). It can be seen from the last expression that the steady state values for the system state space variables are 1, 0.5, 0.333. The step responses of system states, obtained using MATLAB statements [y,x,t]=step(A,B,C,D) and plot(t,x), with

$$\mathbf{A} = \begin{bmatrix} -1 & 0 & 0 \\ 0 & -2 & 0 \\ 0 & 0 & -3 \end{bmatrix}, \quad \mathbf{B} = \begin{bmatrix} 1 \\ 1 \\ 1 \end{bmatrix}, \quad \mathbf{C} = [6 \quad -6 \quad 1], \quad \mathbf{D} = 0$$

are shown in Figure 8.1.

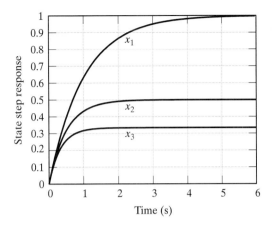

FIGURE 8.1: The state step responses for Example 8.5

The plots of the analytical results obtained for $x_1(t), x_2(t), x_3(t)$ overlap with the corresponding plots in Figure 8.1 obtained using MATLAB to find the step response.

8.2.2 Solution Using the Laplace Transform

The time trajectory of the state vector $\mathbf{x}(t)$ can also be found using the Laplace transform method. (The main properties of the Laplace transform and common Laplace transform pairs are given in Tables 4.1 and 4.2 of Chapter 4.)

The Laplace transform applied to the state equation (8.1) gives

$$s\mathbf{X}(s) - \mathbf{x}(0^-) = \mathbf{A}\mathbf{X}(s) + \mathbf{B}\mathbf{F}(s)$$

or

$$(s\mathbf{I} - \mathbf{A})\mathbf{X}(s) = \mathbf{x}(0^-) + \mathbf{B}\mathbf{F}(s)$$

which implies

$$\mathbf{X}(s) = (s\mathbf{I} - \mathbf{A})^{-1}\mathbf{x}(0^-) + (s\mathbf{I} - \mathbf{A})^{-1}\mathbf{B}\mathbf{F}(s) \qquad (8.46)$$

where $\mathbf{I}$ is identity matrix of order n. Comparing equations (8.43) and (8.46), it is easy to conclude that the term $(s\mathbf{I} - \mathbf{A})^{-1}$ is the Laplace transform of the state transition matrix, that is,

$$\Phi(s) = (s\mathbf{I} - \mathbf{A})^{-1} = \frac{1}{\det(s\mathbf{I} - \mathbf{A})} \operatorname{adj}(s\mathbf{I} - \mathbf{A}) = \mathcal{L}\{\Phi(t)\} \qquad (8.47)$$

or

$$\Phi(t) = \mathcal{L}^{-1}\{\Phi(s)\} = \mathcal{L}^{-1}\{(s\mathbf{I} - \mathbf{A})^{-1}\} \qquad (8.47a)$$

Let us assume that $\mathbf{x}(0) = \mathbf{x}(0^-)$. The time form of the state vector is obtained by applying the inverse Laplace transform to

$$\mathbf{X}(s) = \Phi(s)\mathbf{x}(0) + \Phi(s)\mathbf{B}\mathbf{F}(s) \qquad (8.48)$$

Note that the second term on the right-hand side corresponds in the time domain to the convolution integral, so that we have

$$\mathbf{x}(t) = \mathbf{x}_{zi}(t) + \mathbf{x}_{zs}(t) = e^{\mathbf{A}t}\mathbf{x}(0) + \int_0^t e^{\mathbf{A}(t-\tau)}\mathbf{B}\mathbf{f}(\tau)\,d\tau \tag{8.49}$$

where $\mathbf{x}_{zi}(t)$ and $\mathbf{x}_{zs}(t)$ are the *zero-input* and *zero-state* components of the system state response.

Once the state vector $\mathbf{x}(t)$ is determined, the system output vector $\mathbf{y}(t)$ is simply obtained by substitution of $\mathbf{x}(t)$ into equation (8.2), that is,

$$\mathbf{y}(t) = \mathbf{y}_{zi}(t) + \mathbf{y}_{zs}(t) = \mathbf{C}e^{\mathbf{A}t}\mathbf{x}(0) + \mathbf{C}\int_0^t e^{\mathbf{A}(t-\tau)}\mathbf{B}\mathbf{f}(\tau)\,d\tau + \mathbf{D}\mathbf{f}(t) \tag{8.50}$$

where $\mathbf{y}_{zi}(t)$ and $\mathbf{y}_{zs}(t)$ represent the *zero-input* and *zero-state* components of the system output response. By taking the Laplace transform of Formula (8.50), we get the corresponding formula in the frequency domain,

$$\mathbf{Y}(s) = \mathbf{Y}_{zi}(s) + \mathbf{Y}_{zs}(s) = \mathbf{C}\Phi(s)\mathbf{x}(0) + [\mathbf{C}\Phi(s)\mathbf{B} + \mathbf{D}]\mathbf{F}(s) \tag{8.51}$$

EXAMPLE 8.6

Consider the linear system whose state space matrices are

$$\mathbf{A} = \begin{bmatrix} 0 & 1 \\ -6 & -5 \end{bmatrix}, \quad \mathbf{B} = \begin{bmatrix} 0 \\ 1 \end{bmatrix}, \quad \mathbf{C} = [1 \quad 0], \quad \mathbf{D} = 0$$

The state transition matrix of this system is obtained as follows:

$$\Phi(s) = (s\mathbf{I} - \mathbf{A})^{-1} = \left\{\begin{bmatrix} s & 0 \\ 0 & s \end{bmatrix} - \begin{bmatrix} 0 & 1 \\ -6 & -5 \end{bmatrix}\right\}^{-1} = \begin{bmatrix} s & -1 \\ 6 & s+5 \end{bmatrix}^{-1}$$

$$= \frac{1}{s(s+5)+6}\begin{bmatrix} s+5 & -6 \\ 1 & s \end{bmatrix}^T = \begin{bmatrix} \dfrac{s+5}{(s+2)(s+3)} & \dfrac{1}{(s+2)(s+3)} \\ \dfrac{-6}{(s+2)(s+3)} & \dfrac{s}{(s+2)(s+3)} \end{bmatrix}$$

which implies

$$\Phi(t) = e^{\mathbf{A}t} = \mathcal{L}^{-1}\{\Phi(s)\} = \mathcal{L}^{-1}\left\{\begin{bmatrix} \dfrac{3}{s+2} - \dfrac{2}{s+3} & \dfrac{1}{s+2} - \dfrac{1}{s+3} \\ \dfrac{-6}{s+2} + \dfrac{6}{s+3} & \dfrac{-2}{s+2} + \dfrac{3}{s+3} \end{bmatrix}\right\}$$

$$= \begin{bmatrix} 3e^{-2t} - 2e^{-3t} & e^{-2t} - e^{-3t} \\ -6e^{-2t} + 6e^{-3t} & -2e^{-2t} + 3e^{-3t} \end{bmatrix}$$

Section 8.2 Time Response from the State Space Equation

Let the system input function and initial conditions be given by
$$f(t) = e^{-4t}u(t), \quad \mathbf{x}(0) = [1 \quad 0]^T$$

Then the state response of this system is obtained using formula (8.49):

$$\mathbf{x}(t) = e^{\mathbf{A}t}\begin{bmatrix}1\\0\end{bmatrix} + e^{\mathbf{A}t}\int_0^t e^{-\mathbf{A}\tau}\begin{bmatrix}0\\1\end{bmatrix}e^{-4\tau}\,d\tau$$

$$= \begin{bmatrix}3e^{-2t} - 2e^{-3t}\\-6e^{-2t} + 6e^{-3t}\end{bmatrix} + e^{\mathbf{A}t}\int_0^t \begin{bmatrix}e^{2\tau} - e^{3\tau}\\-2e^{2\tau} + 3e^{3\tau}\end{bmatrix}e^{-4\tau}\,d\tau$$

$$= \begin{bmatrix}3e^{-2t} - 2e^{-3t}\\-6e^{-2t} + 6e^{-3t}\end{bmatrix} + \begin{bmatrix}\tfrac{1}{2}e^{-2t} - e^{-3t} + \tfrac{1}{2}e^{-4t}\\-e^{-2t} + 3e^{-3t} - 2e^{-4t}\end{bmatrix}$$

$$= \begin{bmatrix}\tfrac{7}{2}e^{-2t} - 3e^{-3t} + \tfrac{1}{2}e^{-4t}\\-7e^{-2t} + 9e^{-3t} - 2e^{-4t}\end{bmatrix} = \begin{bmatrix}x_1(t)\\x_2(t)\end{bmatrix}, \quad t \geq 0$$

The system output response is obtained as

$$y(t) = \mathbf{Cx}(t) = [1 \quad 0]\mathbf{x}(t) = x_1(t) = \tfrac{7}{2}e^{-2t} - 3e^{-3t} + \tfrac{1}{2}e^{-4t}, \quad t \geq 0$$

The system state and output responses can be also obtained in the frequency domain by using formulas (8.48) and (8.51), and applying the Laplace inverse to $X(s)$ and $Y(s)$, which seems to be easier (at least for this example, and lower-order systems) than the direct time domain method used previously. This procedure is demonstrated in the following:

$$\mathbf{X}(s) = \Phi(s)\mathbf{x}(0) + \Phi(s)\mathbf{B}F(s) = \Phi(s)\begin{bmatrix}1\\0\end{bmatrix} + \Phi(s)\begin{bmatrix}0\\1\end{bmatrix}\frac{1}{(s+4)}$$

$$= \begin{bmatrix}\dfrac{s+5}{(s+2)(s+3)}\\[6pt] \dfrac{-6}{(s+2)(s+3)}\end{bmatrix} + \begin{bmatrix}\dfrac{1}{(s+2)(s+3)(s+4)}\\[6pt] \dfrac{s}{(s+2)(s+3)(s+4)}\end{bmatrix} = \begin{bmatrix}X_1(s)\\X_2(s)\end{bmatrix}$$

$$Y(s) = \mathbf{CX}(s) = [1 \quad 0]\mathbf{X}(s) = X_1(s)$$

The time domain solution is given by

$$\mathbf{x}(t) = \mathcal{L}^{-1}(\mathbf{X}(s)) = \mathcal{L}^{-1}\left\{\begin{bmatrix}\dfrac{3}{s+2} - \dfrac{2}{s+3}\\[6pt] \dfrac{-6}{s+2} + \dfrac{6}{s+3}\end{bmatrix} + \begin{bmatrix}\dfrac{0.5}{s+2} - \dfrac{1}{s+3} + \dfrac{0.5}{s+4}\\[6pt] -\dfrac{1}{s+2} + \dfrac{3}{s+3} - \dfrac{2}{s+4}\end{bmatrix}\right\}$$

$$= \begin{bmatrix}\tfrac{7}{2}e^{-2t} - 3e^{-3t} + \tfrac{1}{2}e^{-4t}\\-7e^{-2t} + 9e^{-3t} - 2e^{-4t}\end{bmatrix} = \begin{bmatrix}x_1(t)\\x_2(t)\end{bmatrix}$$

$$y(t) = [1 \quad 0]\mathbf{x}(t) = x_1(t)$$

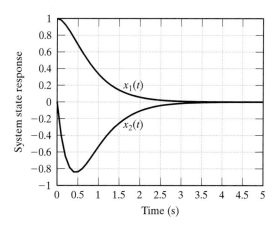

FIGURE 8.2: System state response

We can use MATLAB and its function lsim to find the system state and output responses subject to both the system initial conditions and the forcing function. For this example, this can be done using the following MATLAB script.

```
A=[0 1;-6 -5]; B=[0;1]; C=[1 0]; D=0; x0=[1;0];
t=0:0.1:5;
f=exp(-4*t);
[y,x]=lsim(A,B,C,D,f,t,x0)
plot(t,x); grid
xlabel('Time (s)')
ylabel('System state response')
```

This program produces the values for the system output response and the state space variables in the time interval from $t_0 = 0$ s to $t_f = 5$ s. The plot of the system state response is presented in Figure 8.2. The labels on the curves are produced by the following MATLAB statements:

```
text(1.6,0.16,'x1(t)')
text(1.6,-0.25,'x2(t)')
```

8.2.3 State Space Model and Transfer Function

The matrix that establishes a relationship between the output vector $\mathbf{Y}(s)$ and the input vector $\mathbf{F}(s)$, for the zero initial conditions, $\mathbf{x}(0) = 0$, is called the *system matrix transfer function*. From (8.51), it is given by

$$\mathbf{H}(s) = \mathbf{C}\Phi(s)\mathbf{B} + \mathbf{D} = \mathbf{C}(s\mathbf{I} - \mathbf{A})^{-1}\mathbf{B} + \mathbf{D} \tag{8.52}$$

Note that (8.52) represents the open-loop system matrix transfer function.

EXAMPLE 8.7

We find the transfer function for the system given in Example 8.5. Formula (8.52) leads to

$$H(s) = \begin{bmatrix} 6 & -6 & 1 \end{bmatrix} \begin{bmatrix} s+1 & 0 & 0 \\ 0 & s+2 & 0 \\ 0 & 0 & s+3 \end{bmatrix}^{-1} \begin{bmatrix} 1 \\ 1 \\ 1 \end{bmatrix}$$

$$= \begin{bmatrix} 6 & -6 & 1 \end{bmatrix} \begin{bmatrix} \frac{1}{s+1} & 0 & 0 \\ 0 & \frac{1}{s+2} & 0 \\ 0 & 0 & \frac{1}{s+3} \end{bmatrix} \begin{bmatrix} 1 \\ 1 \\ 1 \end{bmatrix}$$

$$= \frac{6}{s+1} - \frac{6}{s+2} + \frac{1}{s+3} = \frac{(s+5)(s+4)}{(s+1)(s+2)(s+3)}$$

Note that the MATLAB statement [num,den]=ss2tf(A,B,C,D) can be used to solve this problem. ♪

8.2.4 Impulse and Step Responses

Recall from Chapter 4 that the impulse and step responses of single-input single-output systems are defined for zero initial conditions. We keep the same assumption for multi-input multi-output systems, hence the impulse and step responses can be obtained from the formulas

$$\mathbf{x}(t) = \int_{0^-}^{t} e^{\mathbf{A}(t-\tau)} \mathbf{B}\mathbf{f}(\tau)\, d\tau \tag{8.53}$$

$$\mathbf{y}(t) = \mathbf{C}\mathbf{x}(t) + \mathbf{D}\mathbf{f}(t)$$

Since the input forcing function is a vector of dimension $r \times 1$, we can define the impulse and step responses for every input of the system. We introduce the input function all of whose components are zero except for the jth component,

$$\mathbf{f}^j(t) = \begin{bmatrix} 0 & \cdots & 0 & f_j(t) & 0 & \cdots & 0 \end{bmatrix}^T \tag{8.54}$$

Note that

$$\mathbf{B}\mathbf{f}^j(t) = f_j(t)\mathbf{b}_j, \quad \mathbf{B} = \begin{bmatrix} \mathbf{b}_1 & \mathbf{b}_2 & \cdots & \mathbf{b}_{j-1} & \mathbf{b}_j & \mathbf{b}_{j+1} & \cdots & \mathbf{b}_{r-1} & \mathbf{b}_r \end{bmatrix} \tag{8.55}$$

where $\mathbf{b}_j$ is the jth column of the matrix $\mathbf{B}$. The state and output responses due to the jth component of the input signal are given by

$$\mathbf{x}^j(t) = \int_{0^-}^{t} e^{\mathbf{A}(t-\tau)} \mathbf{b}_j f_j(\tau)\, d\tau \qquad (8.56)$$

$$\mathbf{y}^j(t) = \mathbf{C}\mathbf{x}^j(t) + \mathbf{D}\mathbf{f}^j(t) = \mathbf{C}\mathbf{x}^j(t) + \mathbf{d}_j f_j(t)$$

where $\mathbf{d}_j$ is the jth column of the matrix $\mathbf{D}$.

If $f_j(t) = \delta(t)$, then the preceding formulas will produce the *system output impulse response*, $\mathbf{h}^j(t)$, due to the impulse delta function on the jth system input and all other inputs equal to zero, as follows:

$$\mathbf{x}^j(t) = \int_{0^-}^{t} e^{\mathbf{A}(t-\tau)} \mathbf{b}_j \delta(\tau)\, d\tau = e^{\mathbf{A}t} \mathbf{b}_j \qquad (8.57)$$

$$\mathbf{y}^j(t) = \mathbf{C} e^{\mathbf{A}t} \mathbf{b}_j + \mathbf{d}_j \delta(t) = \mathbf{h}^j(t)$$

Similarly, with $f_j(t) = u(t)$, we can define the *system output step response* due to the unit step function on the jth system input and with all other inputs set to zero,

$$\mathbf{x}_{\text{step}}^j(t) = \int_{0^-}^{t} e^{\mathbf{A}(t-\tau)} \mathbf{b}_j u(\tau)\, d\tau = \int_{0^-}^{t} e^{\mathbf{A}(t-\tau)}\, d\tau\, \mathbf{b}_j = \left(\int_{0^-}^{t} e^{\mathbf{A}\sigma}\, d\sigma \right) \mathbf{b}_j$$

$$\mathbf{y}_{\text{step}}^j(t) = \mathbf{C}\mathbf{x}_{\text{step}}^j(t) + \mathbf{d}_j u(t) = \mathbf{C}\left(\int_{0^-}^{t} e^{\mathbf{A}\sigma}\, d\sigma \right) \mathbf{b}_j + \mathbf{d}_j u(t) \qquad (8.58)$$

It is easy to show from (8.57) and (8.58) that

$$\mathbf{y}_{\text{step}}^j(t) = \int_{0^-}^{t} \mathbf{h}^j(\tau)\, d\tau \qquad (8.59)$$

Also, it can be shown (see Problem 8.31) that

$$\mathbf{h}^j(t) = \frac{d}{dt} \mathbf{y}_{\text{step}}^j(t) \qquad (8.60)$$

Note that the vector differentiation and integration in formulas (8.59) and (8.60) must be performed component-wise; that is, the derivative (integral) operator must be applied to every component of the corresponding vector. The last two formulas obtained for multi-input multi-output systems are the vector versions of formulas (4.44–45), derived in Chapter 4 for single-input single-output systems.

The impulse and step responses of multi-input multi-output systems can be found using the MATLAB statements

```
t=ti:dt:tf;
[y,x]=impulse(A,B,C,D,in,t)
[y,x]=step(A,B,C,D,in,t)
```

where in is the number of the system input, and t determines the response time interval, with ti, dt, and tf representing, respectively, the initial time, sampling time, and final time.

8.3 DISCRETE-TIME MODELS

Discrete-time systems are either inherently discrete (e.g., models of bank accounts, national economy growth models, population growth models, digital words) or they are obtained as a result of sampling (discretization) of continuous-time systems. In such systems, inputs, state space variables, and outputs have the discrete form, and the system models can be represented in the form of transition tables.

The mathematical model of a discrete-time system can be written in terms of a recursive formula using linear matrix difference equations, as

$$\mathbf{x}((k+1)T) = \mathbf{A}_d \mathbf{x}(kT) + \mathbf{B}_d \mathbf{f}(kT)$$
$$\mathbf{y}(kT) = \mathbf{C}_d \mathbf{x}(kT) + \mathbf{D}_d \mathbf{f}(kT)$$
(8.61)

Here T represents the constant sampling interval, which may be omitted for brevity; that is, we use the notation

$$\mathbf{x}[k+1] = \mathbf{A}_d \mathbf{x}[k] + \mathbf{B}_d \mathbf{f}[k]$$
$$\mathbf{y}[k] = \mathbf{C}_d \mathbf{x}[k] + \mathbf{D}_d \mathbf{f}[k]$$
(8.61a)

As with continuous-time linear systems, discrete-time state space equations can be derived from difference equations (Section 8.3.1). In Section 8.3.2, we show how to discretize continuous-time linear systems in order to obtain discrete-time linear systems. In the remainder of Section 8.3, we parallel most of the results obtained in previous sections for continuous-time linear systems.

8.3.1 Difference Equations and State Space Form

An nth-order difference equation is defined by

$$y[k+n] + a_{n-1} y[k+n-1] + \cdots + a_1 y[k+1] + a_0 y[k]$$
$$= b_n f[k+n] + b_{n-1} f[k+n-1] + \cdots + b_1 f[k+1] + b_0 f[k] \quad (8.62)$$

This equation expresses all values in terms of discrete-time k. The corresponding state space equation can be derived using the same technique as in the continuous-time case.

For phase variable canonical form in discrete time, we have

$$\begin{bmatrix} x_1[k+1] \\ x_2[k+1] \\ \vdots \\ \vdots \\ x_n[k+1] \end{bmatrix} = \begin{bmatrix} 0 & 1 & 0 & \cdots & 0 \\ 0 & 0 & 1 & \cdots & 0 \\ \vdots & \vdots & \vdots & \ddots & \vdots \\ 0 & 0 & 0 & \cdots & 1 \\ -a_0 & -a_1 & -a_2 & \cdots & -a_{n-1} \end{bmatrix} \begin{bmatrix} x_1[k] \\ x_2[k] \\ \vdots \\ \vdots \\ x_n[k] \end{bmatrix} + \begin{bmatrix} 0 \\ 0 \\ \vdots \\ 0 \\ 1 \end{bmatrix} f[k]$$

$$y[k] = [(b_0 - a_0 b_n) \quad (b_1 - a_1 b_n) \quad \cdots \quad (b_{n-1} - a_{n-1} b_n)] \begin{bmatrix} x_1[k] \\ x_2[k] \\ \vdots \\ x_n[k] \end{bmatrix} + b_n f[k]$$

(8.63)

Note that the transformation equations, analogous to the continuous-time case (8.22–24), are given in the discrete-time domain by

$$\xi[k+n] + a_{n-1}\xi[k+n-1] + \cdots + a_1 \xi[k+1] + a_0 \xi[k] = f[k] \qquad (8.64)$$

$$\begin{aligned} x_1[k] &= \xi[k] \\ x_2[k] &= \xi[k+1] \\ x_3[k] &= \xi[k+2] \\ &\vdots \\ x_n[k] &= \xi[k+n-1] \end{aligned} \qquad (8.65)$$

$$y[k] = b_0 \xi[k] + b_1 \xi[k+1] + b_2 \xi[k+2] + \cdots + b_n \xi[k+n] \qquad (8.66)$$

Eliminating $\xi[k+n]$ from (8.66), using (8.64) and (8.65), the output equation given in (8.63) is obtained.

8.3.2 Discretization of Continuous-Time Systems

Real physical dynamic systems are continuous in nature. In this section, we show how to obtain discrete-time state space models from continuous-time system models. In the dynamic linear system literature, several methods for discretization of continuous-time linear systems can be found. Two methods, known as the integral approximation method and Euler's method, are considered in this text.

Integral Approximation Method

In this section, we show how to perform discretization of a continuous-time state space model of the form in (8.1–2) and obtain a discrete-time state space model having the form of (8.3–4), together with the corresponding expressions for matrices $\mathbf{A}_d$, $\mathbf{B}_d$, $\mathbf{C}_d$, and

$\mathbf{D}_d$. The integral approximation method for discretization of a continuous-time linear system is based on the assumption that the system input is constant during the given sampling period. The method approximates the input signal by its staircase form, that is,

$$\mathbf{f}(t) = \mathbf{f}(kT), \quad kT \leq t < (k+1)T, \quad k = 0, 1, 2, \ldots \tag{8.67}$$

The solution for the state space variables in continuous time is given by (8.42). In the following, we study the impact of approximation (8.67) on (8.42). Consider formula (8.42) with $t = T$,

$$\mathbf{x}(T) = e^{\mathbf{A}T}\mathbf{x}(0) + \int_0^T e^{\mathbf{A}(T-\tau)}\mathbf{B}\mathbf{f}(0)\, d\tau$$

$$= e^{\mathbf{A}T}\mathbf{x}(0) + e^{\mathbf{A}T}\int_0^T e^{-\mathbf{A}\tau}\, d\tau\, \mathbf{B}\mathbf{f}(0) = \Phi(T)x(0) + \int_0^T \Phi(T-\tau)\, d\tau\, \mathbf{B}\mathbf{f}(0) \tag{8.68}$$

which can be written in the form

$$\mathbf{x}(T) = \mathbf{A}_d\mathbf{x}(0) + \mathbf{B}_d\mathbf{f}(0) \tag{8.69}$$

Comparing (8.68) and (8.69), we can find expressions for $\mathbf{A}_d$ and $\mathbf{B}_d$. They are given by

$$\mathbf{A}_d = e^{\mathbf{A}T} = \Phi(T)$$

$$\mathbf{B}_d = e^{\mathbf{A}T}\int_0^T e^{-\mathbf{A}\tau}\, d\tau\, \mathbf{B} = \int_0^T e^{\mathbf{A}(T-\tau)}\, d\tau \cdot \mathbf{B} = \int_0^T e^{\mathbf{A}\sigma}\, d\sigma \cdot \mathbf{B} \tag{8.70}$$

Note that $\mathbf{A}_d$ and $\mathbf{B}_d$ are obtained for the time interval from 0 to T. However, it can easily be shown that, due to system time invariance, the same expressions for $\mathbf{A}_d$ and $\mathbf{B}_d$ are obtained for any time interval. Namely, steps (8.68–70) can be repeated for succeeding time intervals $2T, 3T, \ldots, (k+1)T$ with initial conditions taken, respectively, as $\mathbf{x}(T), \mathbf{x}(2T), \ldots, \mathbf{x}(kT)$. Therefore, for the time instant $t = (k+1)T$ and for $t_0 = kT$, we have from (8.44) that

$$\mathbf{x}((k+1)T) = \Phi((k+1)T - kT)\mathbf{x}(kT) + \int_{kT}^{(k+1)T} \Phi((k+1)T - \tau)\, d\tau\, \mathbf{B}\mathbf{f}(kT) \tag{8.71}$$

$$= \mathbf{A}_d\mathbf{x}(kT) + \mathbf{B}_d\mathbf{f}(kT)$$

From equation (8.71), we see that the matrices $\mathbf{A}_d$ and $\mathbf{B}_d$ are given by

$$\mathbf{A}_d = \Phi((k+1)T - kT) = \Phi(T) = e^{\mathbf{A}T}$$

$$\mathbf{B}_d = \int_{kT}^{(k+1)T} \Phi((k+1)T - \tau)\, d\tau\, \mathbf{B} = \int_0^T \Phi(\sigma)\, d\sigma\, \mathbf{B} = \int_0^T e^{\mathbf{A}\sigma}\, d\sigma\, \mathbf{B} \tag{8.72}$$

The last equality is obtained using a change of variables $\sigma = (k+1)T - \tau$. Since (8.70) and (8.72) are identical, we conclude that for a time-invariant continuous-time linear system, the discretization procedure yields a time-invariant discrete-time linear system whose matrices $\mathbf{A}_d$ and $\mathbf{B}_d$ depend only on $\mathbf{A}$, $\mathbf{B}$, and the sampling interval T.

In a similar manner, the formula for output in (8.2) at $t = kT$ implies

$$\mathbf{y}(kT) = \mathbf{C}\mathbf{x}(kT) + \mathbf{D}f(kT) \tag{8.73}$$

Comparing this equation with the general output equation of linear discrete-time systems (8.61a), we conclude that

$$\mathbf{C}_d = \mathbf{C}, \quad \mathbf{D}_d = \mathbf{D} \tag{8.74}$$

In the case of discrete-time linear systems obtained by sampling continuous-time linear systems, the matrix $\mathbf{A}_d$, given by (8.70), can be determined from the infinite series

$$\mathbf{A}_d = e^{\mathbf{A}T} = \mathbf{I} + \mathbf{A}T + \frac{1}{2!}\mathbf{A}^2 T^2 + \cdots = \sum_{i=0}^{\infty} \frac{1}{i!}\mathbf{A}^i T^i \tag{8.75}$$

The matrix $\mathbf{A}_d$ can be also obtained using either the Laplace transform method or the method based on the Cayley–Hamilton theorem, and setting $t = T$ in $\Phi(t) = e^{\mathbf{A}t}$. Note that we can use MATLAB function expm(A*T) to evaluate $e^{\mathbf{A}T}$.

To find $\mathbf{B}_d$, the second expression in (8.70) is integrated (see Appendix A on matrix integrals) to give

$$\mathbf{B}_d = e^{\mathbf{A}T}(-e^{-\mathbf{A}T}\mathbf{A}^{-1} + \mathbf{A}^{-1})\mathbf{B} = (\mathbf{A}_d - \mathbf{I})\mathbf{A}^{-1}\mathbf{B} \tag{8.76}$$

which is valid under the assumption that $\mathbf{A}$ is *invertible*.

EXAMPLE 8.8

We find the discrete-time state space model of the continuous-time system

$$\dot{\mathbf{x}}(t) = \begin{bmatrix} 0 & 1 \\ -2 & -3 \end{bmatrix}\mathbf{x}(t) + \begin{bmatrix} 0 \\ 1 \end{bmatrix}f(t)$$

$$y(t) = \begin{bmatrix} 1 & 0 \end{bmatrix}\mathbf{x}(t)$$

The sampling period T is equal to 0.1.

According to (8.70) and (8.74), we have, from formula (8.47a),

$$\mathbf{A}_d = \Phi(T) = \begin{bmatrix} 2e^{-T} - e^{-2T} & e^{-T} - e^{-2T} \\ 2e^{-2T} - 2e^{-T} & 2e^{-2T} - e^{-T} \end{bmatrix} = \begin{bmatrix} 0.9909 & 0.0861 \\ -0.1722 & 0.7326 \end{bmatrix}$$

$$\mathbf{B}_d = (\mathbf{A}_d - \mathbf{I})\mathbf{A}^{-1}\mathbf{B} = \begin{bmatrix} \frac{1}{2}(1 + e^{-2T}) - e^{-T} \\ e^{-T} - e^{-2T} \end{bmatrix} = \begin{bmatrix} 0.0045 \\ 0.0861 \end{bmatrix}$$

$$\mathbf{C}_d = \mathbf{C} = \begin{bmatrix} 1 & 0 \end{bmatrix}, \quad \mathbf{D}_d = \mathbf{D} = 0$$

The same result is obtained using the MATLAB function for discretization of a continuous state space model as [Ad,Bd]=c2d(A,B,T).

Euler's Method

Less accurate but simpler than the integral approximation method is Euler's method. Euler's method is based on the following simple approximation of the first derivative at the time instant $t = kT$:

$$\dot{\mathbf{x}}(t) = \frac{d\mathbf{x}(t)}{dt} \approx \frac{1}{T}(\mathbf{x}((k+1)T) - \mathbf{x}(kT)) \qquad (8.77)$$

Applying this approximative formula to the state space system equation (8.1), we have

$$\frac{1}{T}(\mathbf{x}((k+1)T) - \mathbf{x}(kT)) \approx \mathbf{A}\mathbf{x}(kT) + \mathbf{B}\mathbf{f}(kT)$$

which implies

$$\mathbf{x}((k+1)T) \approx (\mathbf{I} + T\mathbf{A})\mathbf{x}(kT) + T\mathbf{B}\mathbf{f}(kT)$$

or

$$\mathbf{x}[k+1] \approx (\mathbf{I} + T\mathbf{A})\mathbf{x}[k] + T\mathbf{B}\mathbf{f}[k]$$

From the last equation, we conclude that for the Euler approximation the system state and input matrices are given by

$$\mathbf{A}_d = \mathbf{I} + T \cdot \mathbf{A}, \quad \mathbf{B}_d = T \cdot \mathbf{B} \qquad (8.78)$$

The Euler approximation is less accurate than the integral approximation, and for Euler's approximation the sampling interval T must be chosen sufficiently small to get satisfactory results.

8.3.3 Solution of the Discrete-Time State Space Equation

The objective of this section is to find the solution of the state space difference equation (8.61) for the given initial state $\mathbf{x}[0]$ and the input signal $\mathbf{f}[k]$.

From the state space equation

$$\mathbf{x}[k+1] = \mathbf{A}_d \mathbf{x}[k] + \mathbf{B}_d \mathbf{f}[k]$$

for $k = 0, 1, 2 \ldots$, it follows that

$$\begin{aligned}
\mathbf{x}[1] &= \mathbf{A}_d \mathbf{x}[0] + \mathbf{B}_d \mathbf{f}[0] \\
\mathbf{x}[2] &= \mathbf{A}_d \mathbf{x}[1] + \mathbf{B}_d \mathbf{f}[1] = \mathbf{A}_d^2 \mathbf{x}[0] + \mathbf{A}_d \mathbf{B}_d \mathbf{f}[0] + \mathbf{B}_d \mathbf{f}[1] \\
\mathbf{x}[3] &= \mathbf{A}_d \mathbf{x}[2] + \mathbf{B}_d \mathbf{f}[2] = \mathbf{A}_d^3 \mathbf{x}[0] + \mathbf{A}_d^2 \mathbf{B}_d \mathbf{f}[0] + \mathbf{A}_d \mathbf{B}_d \mathbf{f}[1] + \mathbf{B}_d \mathbf{f}[2] \\
&\vdots
\end{aligned} \qquad (8.79)$$

$$\mathbf{x}[k] = \mathbf{A}_d \mathbf{x}[k-1] + \mathbf{B}_d \mathbf{f}[k-1] = \mathbf{A}_d^k \mathbf{x}[0] + \sum_{i=0}^{k-1} \mathbf{A}_d^{k-i-1} \mathbf{B}_d \mathbf{f}[i]$$

Using the notion of the *discrete-time state transition matrix,* defined by

$$\Phi_d[k] = \mathbf{A}_d^k \tag{8.80}$$

we get

$$\mathbf{x}[k] = \Phi_d[k]\mathbf{x}[0] + \sum_{i=0}^{k-1} \Phi_d[k - i - 1]\mathbf{B}_d\mathbf{f}[i] \tag{8.81}$$

Note that the discrete-time state transition matrix relates the state of an input-free system at initial time ($k = 0$) to the state of the system at any other time $k > 0$, that is,

$$\mathbf{x}[k] = \Phi_d[k]\mathbf{x}[0] = \mathbf{A}_d^k\mathbf{x}[0] \tag{8.82}$$

It is easy to verify that the discrete-time state transition matrix has the following properties:

(a) $\Phi_d[0] = \mathbf{A}_d^0 = \mathbf{I} \Leftarrow \mathbf{x}[0] = \Phi_d[0]\mathbf{x}[0]$
(b) $\Phi_d[k_2 - k_0] = \Phi_d[k_2 - k_1]\Phi_d[k_1 - k_0] = \mathbf{A}_d^{k_2 - k_1}\mathbf{A}_d^{k_1 - k_0} = \mathbf{A}_d^{k_2 - k_0}$
(c) $\Phi_d^i[k] = \Phi_d[ik] \Leftarrow (\mathbf{A}_d^k)^i = \mathbf{A}_d^{ik}$
(d) $\Phi_d[k + 1] = \mathbf{A}_d\Phi_d[k], \ \Phi_d[0] = \mathbf{I}$

Property (d) follows from

$$\mathbf{x}[k+1] = \mathbf{A}_d\mathbf{x}[k] \Rightarrow \Phi_d[k+1]\mathbf{x}[0] = \mathbf{A}_d\Phi_d[k]\mathbf{x}[0]$$

It is important to point out that the discrete-time state transition matrix may be singular, which follows from the fact that $\mathbf{A}_d^k$ is nonsingular if and only if the matrix $\mathbf{A}_d$ is nonsingular. In the case of inherent discrete-time systems, the matrix $\mathbf{A}_d$ may be singular in general. However, if $\mathbf{A}_d$ is obtained through the discretization of a continuous-time linear system, as in (8.70), then

$$(\mathbf{A}_d)^{-1} = (e^{\mathbf{A}T})^{-1} = e^{-\mathbf{A}T}$$

so that the discrete-time state transition matrix is nonsingular in this case.

The output of the system at sampling instant k is obtained by substituting $\mathbf{x}[k]$ from (8.81) into the output equation, producing

$$\mathbf{y}[k] = \mathbf{C}_d\Phi_d[k]\mathbf{x}[0] + \mathbf{C}_d\sum_{i=0}^{k-1} \Phi[k - i - 1]\mathbf{B}_d\mathbf{f}[i] + \mathbf{D}_d\mathbf{f}[k] \tag{8.83}$$

Remark 8.1: If the initial value of the state vector is not $\mathbf{x}[0]$ but $\mathbf{x}[k_0]$, then the solution given in (8.81) must be modified to

$$\mathbf{x}[k_0 + k] = \Phi_d[k]\mathbf{x}[k_0] + \sum_{i=0}^{k-1} \Phi_d[k - i - 1]\mathbf{B}_d\mathbf{f}[k_0 + i] \tag{8.84}$$

Note that for $T \neq 1$, equations (8.81) and (8.83) are modified to

$$\mathbf{x}(kT) = \Phi_d(kT)\mathbf{x}(0) + \sum_{i=0}^{k-1} \Phi_d((k-i-1)T)\mathbf{B}_d\mathbf{f}(iT) \tag{8.85}$$

$$\mathbf{y}(kT) = \mathbf{C}_d\Phi_d(kT)\mathbf{x}(0) + \mathbf{C}_d\sum_{i=0}^{k-1} \Phi((k-i-1)T)\mathbf{B}_d\mathbf{f}(iT) + \mathbf{D}_d\mathbf{f}(kT) \tag{8.86}$$

Remark 8.2: The discrete-time state transition matrix defined by $\mathbf{A}_d^k$ can be evaluated efficiently for large values of k using a method based on the Cayley–Hamilton theorem and described in Section 8.5. It can be also evaluated using the $\mathcal{Z}$-transform method as given in formula (8.91), to be derived in the next subsection.

8.3.4 Solution Using the $\mathcal{Z}$-transform

Applying the $\mathcal{Z}$-transform to the state space equation of a discrete-time linear system,

$$\mathbf{x}[k+1] = \mathbf{A}_d\mathbf{x}[k] + \mathbf{B}_d\mathbf{f}[k] \tag{8.87}$$

we get

$$z\mathbf{X}(z) - z\mathbf{x}[0] = \mathbf{A}_d\mathbf{X}(z) + \mathbf{B}_d\mathbf{F}(z) \tag{8.88}$$

The frequency domain state space vector $\mathbf{X}(z)$ can be expressed as

$$\mathbf{X}(z) = (z\mathbf{I} - \mathbf{A}_d)^{-1}z\mathbf{x}[0] + (z\mathbf{I} - \mathbf{A}_d)^{-1}\mathbf{B}_d\mathbf{F}(z) \tag{8.89}$$

The inverse $\mathcal{Z}$-transform of (8.89) gives $\mathbf{x}[k]$, that is,

$$\mathbf{x}[k] = \mathcal{Z}^{-1}[(z\mathbf{I} - \mathbf{A}_d)^{-1}z]\mathbf{x}[0] + \mathcal{Z}^{-1}[(z\mathbf{I} - \mathbf{A}_d)^{-1}\mathbf{B}_d\mathbf{F}(z)] \tag{8.90}$$

Comparing equations (8.81) and (8.90), we conclude that

$$\Phi_d[k] = \mathcal{Z}^{-1}[(z\mathbf{I} - \mathbf{A}_d)^{-1}z] = \mathbf{A}_d^k, \quad k = 1, 2, 3, \ldots \tag{8.91}$$

and

$$\Phi_d(z) = z(z\mathbf{I} - \mathbf{A}_d)^{-1} \tag{8.91a}$$

The inverse transform of the second term on the right-hand side of (8.90) is obtained directly by the application of the discrete-time convolution, which produces

$$\mathcal{Z}^{-1}\{(z\mathbf{I} - \mathbf{A}_d)^{-1}\mathbf{B}_d\mathbf{F}(z)\} = \sum_{i=0}^{k-1} \Phi_d[k-i-1]\mathbf{B}_d\mathbf{f}[i] \tag{8.92}$$

From (8.90–92), we get the required solution of the discrete-time state space equation as

$$\mathbf{x}[k] = \Phi_d[k]\mathbf{x}[0] + \sum_{i=0}^{k-1} \Phi_d[k-i-1]\mathbf{B}_d\mathbf{f}[i] \tag{8.93}$$

414   Chapter 8   State Space Approach

Using

$$\mathbf{y}[k] = \mathbf{C}_d\mathbf{x}[k] + \mathbf{D}_d\mathbf{f}[k]$$

and (8.93), the system output response is obtained as

$$\mathbf{y}[k] = \mathbf{C}_d\Phi_d[k]\mathbf{x}[0] + \mathbf{C}_d\sum_{i=0}^{k-1}\Phi_d[k-i-1]\mathbf{B}_d\mathbf{f}[i] + \mathbf{D}_d\mathbf{f}[k] \qquad (8.94)$$

The frequency domain form of the output vector $\mathbf{Y}(z)$ is obtained if the $\mathcal{Z}$-transform is applied to the output equation, and $\mathbf{X}(z)$ is substituted from (8.89), leading to

$$\mathbf{Y}(z) = \mathbf{C}_d(z\mathbf{I} - \mathbf{A}_d)^{-1}z\mathbf{x}[0] + [\mathbf{C}_d(z\mathbf{I} - \mathbf{A}_d)^{-1}\mathbf{B}_d + \mathbf{D}_d]\mathbf{F}(z)$$

From this expression, for the zero initial condition (i.e., $\mathbf{x}[0] = 0$), the *discrete matrix transfer function* is defined by

$$\mathbf{H}_d(z) = \mathbf{C}_d(z\mathbf{I} - \mathbf{A}_d)^{-1}\mathbf{B}_d + \mathbf{D}_d \qquad (8.95)$$

EXAMPLE 8.9

Consider the discrete-time system

$$\mathbf{A}_d = \begin{bmatrix} 0 & 1 \\ -\tfrac{1}{6} & -\tfrac{5}{6} \end{bmatrix}, \quad \mathbf{B}_d = \begin{bmatrix} 0 \\ 1 \end{bmatrix}, \quad \mathbf{C}_d = [1 \;\; 0], \quad \mathbf{D}_d = 0$$

The discrete-time state transition matrix in the frequency domain is obtained from (8.91a) as

$$\Phi_d(z) = (z\mathbf{I} - \mathbf{A})^{-1}z = \left(\begin{bmatrix} z & 0 \\ 0 & z \end{bmatrix} - \begin{bmatrix} 0 & 1 \\ -\tfrac{1}{6} & -\tfrac{5}{6} \end{bmatrix}\right)^{-1} z = \begin{bmatrix} z & -1 \\ \tfrac{1}{6} & z + \tfrac{5}{6} \end{bmatrix}^{-1} z$$

$$= \frac{z}{z(z + \tfrac{5}{6}) + \tfrac{1}{6}} \begin{bmatrix} z + \tfrac{5}{6} & 1 \\ -\tfrac{1}{6} & z \end{bmatrix} = \begin{bmatrix} \dfrac{z(z + \tfrac{5}{6})}{(z + \tfrac{1}{2})(z + \tfrac{1}{3})} & \dfrac{z}{(z + \tfrac{1}{2})(z + \tfrac{1}{3})} \\ \dfrac{-\tfrac{1}{6}z}{(z + \tfrac{1}{2})(z + \tfrac{1}{3})} & \dfrac{z^2}{(z + \tfrac{1}{2})(z + \tfrac{1}{3})} \end{bmatrix}$$

The time domain state transition matrix is given by

$$\Phi_d[k] = \mathcal{Z}^{-1}\{\Phi_d(z)\} = \mathcal{Z}^{-1}\left\{\begin{bmatrix} \dfrac{-2z}{z + \tfrac{1}{2}} + \dfrac{3z}{z + \tfrac{1}{3}} & \dfrac{-6z}{z + \tfrac{1}{2}} + \dfrac{6z}{z + \tfrac{1}{3}} \\ \dfrac{z}{z + \tfrac{1}{2}} - \dfrac{z}{z + \tfrac{1}{3}} & \dfrac{3z}{z + \tfrac{1}{2}} - \dfrac{2z}{z + \tfrac{1}{3}} \end{bmatrix}\right\}$$

$$= \begin{bmatrix} -2(-\tfrac{1}{2})^k + 3(-\tfrac{1}{3})^k & -6(-\tfrac{1}{2})^k + 6(-\tfrac{1}{3})^k \\ (-\tfrac{1}{2})^k - (-\tfrac{1}{3})^k & 3(-\tfrac{1}{2})^k - 2(-\tfrac{1}{3})^k \end{bmatrix} u[k]$$

Let us find the response of this system due to

$$f[k] = (-1)^k u[k], \quad \mathbf{x}[0] = \begin{bmatrix} 1 \\ 0 \end{bmatrix}$$

Since the system state transition matrix is already determined, we can use formula (8.93), which produces

$$\mathbf{x}[k] = \Phi_d[k] \begin{bmatrix} 1 \\ 0 \end{bmatrix} + \sum_{i=0}^{k-1} \Phi_d[k-i-1] \begin{bmatrix} 0 \\ 1 \end{bmatrix} (-1)^i$$

$$= \begin{bmatrix} -2(-\tfrac{1}{2})^k + 3(-\tfrac{1}{3})^k \\ (-\tfrac{1}{2})^k - (-\tfrac{1}{3})^k \end{bmatrix} + \sum_{i=0}^{k-1} \begin{bmatrix} -6(-\tfrac{1}{2})^{k-i-1} + 6(-\tfrac{1}{3})^{k-i-1} \\ 3(-\tfrac{1}{2})^{k-i-1} - 2(-\tfrac{1}{3})^{k-i-1} \end{bmatrix} (-1)^i$$

which can be accepted as the final result. Note that using known formulas for series summation (see Appendix B), the preceding formula can be further simplified, and eventually a closed-form solution might be obtained for $\mathbf{x}[k]$. However, if we find in the frequency domain $\mathbf{X}(z)$ using (8.89), then in most cases, the inverse $\mathcal{Z}$-transform will produce a nice closed formula for $\mathbf{x}[k]$. In this example, we have, from (8.89),

$$\mathbf{X}(z) = \frac{z}{(z+\tfrac{1}{2})(z+\tfrac{1}{3})} \begin{bmatrix} z + \tfrac{5}{6} & 1 \\ -\tfrac{1}{6} & z \end{bmatrix} \begin{bmatrix} 1 \\ 0 \end{bmatrix} + \frac{1}{(z+\tfrac{1}{2})(z+\tfrac{1}{3})} \begin{bmatrix} z + \tfrac{5}{6} & 1 \\ -\tfrac{1}{6} & z \end{bmatrix} \begin{bmatrix} 0 \\ 1 \end{bmatrix} \frac{z}{z+1}$$

$$= \frac{z}{(z+\tfrac{1}{2})(z+\tfrac{1}{3})} \begin{bmatrix} z + \tfrac{5}{6} \\ -\tfrac{1}{6} \end{bmatrix} + \frac{z}{(z+\tfrac{1}{2})(z+\tfrac{1}{3})(z+1)} \begin{bmatrix} 1 \\ z \end{bmatrix}$$

$$= \begin{bmatrix} \dfrac{-2z}{z+\tfrac{1}{2}} + \dfrac{3z}{z+\tfrac{1}{3}} \\ \dfrac{z}{z+\tfrac{1}{2}} - \dfrac{z}{z+\tfrac{1}{3}} \end{bmatrix} + \begin{bmatrix} \dfrac{-12z}{z+\tfrac{1}{2}} + \dfrac{9z}{z+\tfrac{1}{3}} + \dfrac{3z}{z+1} \\ \dfrac{6z}{z+\tfrac{1}{2}} + \dfrac{3z}{z+\tfrac{1}{3}} - \dfrac{3z}{z+1} \end{bmatrix}$$

Applying the inverse $\mathcal{Z}$-transform, we obtain the state response

$$\mathbf{x}[k] = \begin{bmatrix} -2(-\tfrac{1}{2})^k + 3(-\tfrac{1}{3})^k \\ (-\tfrac{1}{2})^k - (-\tfrac{1}{3})^k \end{bmatrix} + \begin{bmatrix} -12(-\tfrac{1}{2})^k + 9(-\tfrac{1}{3})^k + 3(-1)^k \\ 6(-\tfrac{1}{2})^k + 3(-\tfrac{1}{3})^k - 3(-1)^k \end{bmatrix}$$

$$= \begin{bmatrix} -14(-\tfrac{1}{2})^k + 12(-\tfrac{1}{3})^k + 3(-1)^k \\ 7(-\tfrac{1}{2})^k + 2(-\tfrac{1}{3})^k - 3(-1)^k \end{bmatrix} = \begin{bmatrix} x_1[k] \\ x_2[k] \end{bmatrix}$$

For the system output response, we have

$$y[k] = \begin{bmatrix} 1 & 0 \end{bmatrix} \mathbf{x}[k] = x_1[k] = -14\left(-\frac{1}{2}\right)^k + 12\left(-\frac{1}{3}\right)^k + 3(-1)^k$$

This problem can be solved using MATLAB as follows:

```
k=0:1:10;  f=(-1).^k;
[y,x]=dlsim(Ad,Bd,Cd,Dd,f,x0)
```

with the system output and state responses being calculated for $k = 0, 1, 2, \ldots, 10$. ∫

8.3.5 Discrete-Time Impulse and Step Responses

Similarly to the continuous-time state space case, the impulse and step responses of multi-input multi-output discrete-time systems are defined for zero initial conditions, and calculated using the formulas

$$\mathbf{x}[k] = \sum_{i=0}^{k-1} \Phi_d[k-i-1]\mathbf{B}_d\mathbf{f}[i], \quad \mathbf{y}[k] = \mathbf{C}_d\mathbf{x}[k] + \mathbf{D}\mathbf{f}[k] \qquad (8.96)$$

Since the input forcing function is a vector of dimension $r \times 1$, we can define the impulse and step responses for every input of the system. We introduce the discrete-time system input function all of whose components are zero except for the jth component, that is,

$$\mathbf{f}^j[k] = [0 \quad \cdots \quad 0 \quad f_j[k] \quad 0 \quad \cdots \quad 0]^T \qquad (8.97)$$

Note that

$$\mathbf{B}_d\mathbf{f}^j[k] = f_j[k]\mathbf{b}_j \qquad (8.98)$$

where $\mathbf{b}_j$ is the jth column of the matrix $\mathbf{B}_d$. The system state and output responses due to the jth component of the input signal are given by

$$\mathbf{x}^j[k] = \sum_{i=0}^{k-1} \Phi_d[k-i-1]\mathbf{b}_j f_j[i], \quad \mathbf{y}^j[k] = \mathbf{C}_d\mathbf{x}^j[k] + \mathbf{d}_j f_j[k] \qquad (8.99)$$

where $\mathbf{d}_j$ is the jth column of the matrix $\mathbf{D}_d$.

If $f_j[k] = \delta[k]$, then formulas (8.99) will produce the *system output impulse response*, $\mathbf{h}^j[k]$, due to the impulse delta function on the jth system input and all other inputs equal to zero, that is,

$$\mathbf{x}^j[k] = \sum_{i=0}^{k-1} \Phi_d[k-i-1]\mathbf{b}_j\delta[i] = \mathbf{A}_d^{k-1}\mathbf{b}_j$$
$$\mathbf{y}^j[k] = \mathbf{C}_d\mathbf{x}^j[k] + \mathbf{d}_j\delta[k] = \mathbf{C}_d\mathbf{A}_d^{k-1}\mathbf{b}_j + \mathbf{d}_j\delta[k] = \mathbf{h}^j[k] \qquad (8.100)$$

Similarly, with $f_j[k] = u[k]$, we can define the *system output step response* due to the unit step function on the jth system input and all other inputs set to zero,

$$\mathbf{x}^j_{\text{step}}[k] = \sum_{i=0}^{k-1} \Phi_d[k-i-1]\mathbf{b}_j u_j[i] = \sum_{i=0}^{k-1} \mathbf{A}_d^{k-i-1}\mathbf{b}_j$$
$$\mathbf{y}^j_{\text{step}}[k] = \mathbf{C}_d\left(\sum_{i=0}^{k-1} \mathbf{A}_d^{k-i-1}\mathbf{b}_j\right) + \mathbf{d}_j u[k] \qquad (8.101)$$

Section 8.3   Discrete-Time Models   417

It is left as an exercise for students (see Problem 8.32) to show that

$$\mathbf{y}_{\text{step}}^j[k] = \sum_{i=1}^{k} \mathbf{h}^j[i], \quad \mathbf{h}^j[k] = \mathbf{y}_{\text{step}}^j[k] - \mathbf{y}_{\text{step}}^j[k-1] \quad (8.102)$$

The last two formulas obtained for multi-input multi-output systems are the vector versions of the corresponding formulas derived in Chapter 5 for single-input single-output systems.

EXAMPLE 8.10

For the discrete-time system given in Example 8.8, we will use MATLAB to find the unit step and impulse responses.

The required time responses can be obtained directly by using MATLAB statements [ys,xs]=dstep(Ad,Bd,Cd,Dd) (for step response) and [yi,xi]=dimpulse(Ad,Bd, Cd,Dd) (for impulse response). The corresponding state and output responses are presented in Figures 8.3 and 8.4.

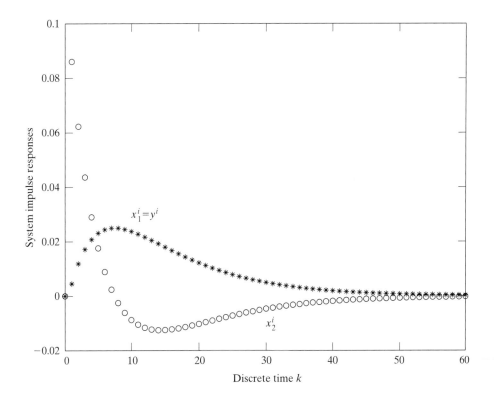

FIGURE 8.3: System state impulse responses

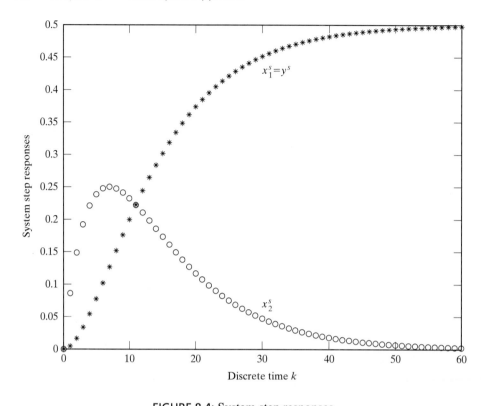

FIGURE 8.4: System step responses

The plot in Figure 8.4 is obtained using the following MATLAB code:

```
xsL=length(xs); k=0:1:xsL-1;
plot(k,xs(1:xsL,1:1),'*',k,xs(1:xsL,2:2),'o')
text(30,0.4,'xs1=ys'); text(30,0.1,'xs2')
```

8.4 SYSTEM CHARACTERISTIC EQUATION AND EIGENVALUES

We saw in Chapter 7 that the characteristic equation and eigenvalues (characteristic values) play important roles in the study of continuous- and discrete-time linear dynamic systems. The system eigenvalues were also tacitly considered in Chapters 4 and 5 within the notion of the system poles. Note that in Chapter 7 we clarified the statement that *all the system poles are the system eigenvalues, but not all the system eigenvalues are the system poles*. In this section, we will first provide a general definition of the system eigenvalues and show that they are identical to the eigenvalues of the system matrix **A**. Then, we will define the eigenvectors of the matrix **A** and indicate their importance for linear dynamic systems.

The *eigenvalues* are defined in linear algebra as scalars, λ, satisfying [4]

$$\mathbf{A}\mathbf{v} = \lambda\mathbf{v}, \quad \mathbf{v} \neq 0 \tag{8.103}$$

where the vectors $\mathbf{v}$ are called the *eigenvectors*. This system of n linear algebraic equations $((\lambda\mathbf{I} - \mathbf{A})\mathbf{v} = 0)$ has a solution $\mathbf{v} \neq 0$ if and only if the corresponding determinant is equal to zero (see Appendix A), that is,

$$|\lambda\mathbf{I} - \mathbf{A}| = 0 \tag{8.104}$$

In the state space technique presented in the previous sections of this chapter, we saw from (8.52) that

$$\mathbf{H}(s) = \mathbf{C}(s\mathbf{I} - \mathbf{A})^{-1}\mathbf{B} + \mathbf{D} = \frac{1}{|s\mathbf{I} - \mathbf{A}|}\mathbf{C}[\text{adj}(s\mathbf{I} - \mathbf{A})]\mathbf{B} + \mathbf{D}$$

$$= \frac{1}{|s\mathbf{I} - \mathbf{A}|}\{\mathbf{C}[\text{adj}(s\mathbf{I} - \mathbf{A})]\mathbf{B} + |s\mathbf{I} - \mathbf{A}|\mathbf{D}\}$$

In this formula, we have used the definition of the matrix inversion given in terms of the matrix determinant and adjoint matrix [4] (see also Appendix A). The characteristic equation of a linear dynamic system represented by the preceding transfer function is defined by

$$|s\mathbf{I} - \mathbf{A}| = 0 \tag{8.105}$$

Obviously, (8.104) and (8.105) are identical algebraic equations. It follows that (8.104) is also the system characteristic equation.

It is left as an exercise (Problem 8.33) to show that for the system matrix in phase variable canonical form, defined by (8.20), the characteristic equation is given by

$$|\lambda\mathbf{I} - \mathbf{A}| = \lambda^n + a_{n-1}\lambda^{n-1} + \cdots + a_1\lambda + a_0 \tag{8.106}$$

Similarity Transformation

A system modeled by the state space technique may have many state space forms. Here, we establish a relationship between two state space forms, using a linear transformation known as the similarity transformation.

For a given system

$$\dot{\mathbf{x}}(t) = \mathbf{A}\mathbf{x}(t) + \mathbf{B}\mathbf{f}(t), \quad \mathbf{x}(0) = \mathbf{x}_0$$

$$\mathbf{y}(t) = \mathbf{C}\mathbf{x}(t) + \mathbf{D}\mathbf{f}(t)$$

we can introduce a new state vector $\hat{\mathbf{x}}(t)$ by a linear coordinate transformation,

$$\mathbf{x}(t) = \mathbf{P}\hat{\mathbf{x}}(t)$$

where $\mathbf{P}$ is some nonsingular $n \times n$ matrix. A new state space model is obtained as

$$\dot{\hat{\mathbf{x}}}(t) = \hat{\mathbf{A}}\hat{\mathbf{x}}(t) + \hat{\mathbf{B}}\mathbf{f}(t), \quad \hat{\mathbf{x}}(0) = \hat{\mathbf{x}}_0$$
$$\mathbf{y}(t) = \hat{\mathbf{C}}\hat{\mathbf{x}}(t) + \hat{\mathbf{D}}\mathbf{f}(t) \tag{8.107}$$

where

$$\hat{\mathbf{A}} = \mathbf{P}^{-1}\mathbf{A}\mathbf{P}, \quad \hat{\mathbf{B}} = \mathbf{P}^{-1}\mathbf{B}, \quad \hat{\mathbf{C}} = \mathbf{C}\mathbf{P}, \quad \hat{\mathbf{D}} = \mathbf{D}, \quad \hat{\mathbf{x}}(0) = \mathbf{P}^{-1}\mathbf{x}(0) \quad (8.108)$$

This transformation, known in the literature as the similarity transformation, plays an important role in linear system theory and practice.

A very important feature of this transformation is that both the system eigenvalues and the system transfer function remain invariant.

Eigenvalue Invariance

A new state space model obtained by the similarity transformation does not change the internal structure of the model, that is, the eigenvalues of the system remain the same. This can be shown as follows:

$$\begin{aligned} |\lambda \mathbf{I} - \hat{\mathbf{A}}| &= |\lambda \mathbf{I} - \mathbf{P}^{-1}\mathbf{A}\mathbf{P}| = |\mathbf{P}^{-1}(\lambda \mathbf{I} - \mathbf{A})\mathbf{P}| \\ &= |\mathbf{P}^{-1}||\lambda \mathbf{I} - \mathbf{A}||\mathbf{P}| = |\lambda \mathbf{I} - \mathbf{A}| \end{aligned} \quad (8.109)$$

Note that in this proof, the following properties of the matrix determinant have been used:

$$\det(\mathbf{M}_1\mathbf{M}_2\mathbf{M}_3) = \det \mathbf{M}_1 \times \det \mathbf{M}_2 \times \det \mathbf{M}_3$$

$$\det \mathbf{M}^{-1} = \frac{1}{\det \mathbf{M}}$$

(see Appendix A).

Transfer Function Invariance

In the similarity transformation, the transfer function remains the same for both models, which can be shown as follows:

$$\begin{aligned} \hat{\mathbf{H}}(s) &= \hat{\mathbf{C}}(s\mathbf{I} - \hat{\mathbf{A}})^{-1}\hat{\mathbf{B}} + \hat{\mathbf{D}} = \mathbf{C}\mathbf{P}(s\mathbf{I} - \mathbf{P}^{-1}\mathbf{A}\mathbf{P})^{-1}\mathbf{P}^{-1}\mathbf{B} + \mathbf{D} \\ &= \mathbf{C}\mathbf{P}[\mathbf{P}^{-1}(s\mathbf{I} - \mathbf{A})\mathbf{P}]^{-1}\mathbf{P}^{-1}\mathbf{B} + \mathbf{D} \\ &= \mathbf{C}\mathbf{P}\mathbf{P}^{-1}(s\mathbf{I} - \mathbf{A})^{-1}\mathbf{P}\mathbf{P}^{-1}\mathbf{B} + \mathbf{D} \\ &= \mathbf{C}(s\mathbf{I} - \mathbf{A})^{-1}\mathbf{B} + \mathbf{D} = \mathbf{H}(s) \end{aligned} \quad (8.110)$$

Note that we have used in (8.110) the matrix inversion property (Appendix A),

$$(\mathbf{M}_1\mathbf{M}_2\mathbf{M}_3)^{-1} = \mathbf{M}_3^{-1}\mathbf{M}_2^{-1}\mathbf{M}_1^{-1}$$

The preceding result is quite logical—the system preserves its input–output behavior no matter how it is mathematically described.

Modal Transformation

One of the most interesting similarity transformations puts matrix $\mathbf{A}$ into diagonal form. Assume that $\mathbf{P} = \mathbf{V} = [\mathbf{v}_1, \mathbf{v}_2, \ldots, \mathbf{v}_n]$, where $\mathbf{v}_i$ are the eigenvectors. We then have

$$\mathbf{V}^{-1}\mathbf{A}\mathbf{V} = \hat{\mathbf{A}} = \mathbf{\Lambda} = \text{diag}(\lambda_1, \lambda_2, \ldots, \lambda_n) \quad (8.111)$$

It is easy to show that the elements $\lambda_i, i = 1, \ldots, n$, on the matrix diagonal of Λ, are the roots of the characteristic equation $|\lambda \mathbf{I} - \Lambda| = |\lambda \mathbf{I} - \mathbf{A}| = 0$, that is, they are the eigenvalues. This can be shown in a straightforward way:

$$|\lambda \mathbf{I} - \Lambda| = \det\{\text{diag}(\lambda - \lambda_1, \lambda - \lambda_2, \ldots, \lambda - \lambda_n)\}$$
$$= (\lambda - \lambda_1)(\lambda - \lambda_2) \cdots (\lambda - \lambda_n)$$

The state transformation (8.111) is known as the *modal transformation*.

Note that the pure diagonal state space form defined in (8.111) is obtained under the assumption that the system matrix has distinct eigenvalues, namely $\lambda_1 \neq \lambda_2 \neq \cdots \neq \lambda_n$, which assures the invertibility of the matrix $\mathbf{V}$ since the corresponding eigenvectors are linearly independent.

There are numerous program packages available to compute both the eigenvalues and eigenvectors of a matrix. In MATLAB this is done using the function `eig`.

Remark 8.3: The presented theory about the system characteristic equation, eigenvalues, eigenvectors, and similarity and modal transformations can be applied directly to discrete-time linear systems with $\mathbf{A}_d$ replacing $\mathbf{A}$.

8.5 CAYLEY–HAMILTON THEOREM

One of the most powerful theorems of linear algebra and matrix analysis is the Cayley–Hamilton theorem. It simply says the following.

THEOREM 8.1. Every square matrix satisfies its characteristic equation. □

The proof of this theorem is also simple. It follows from the definition of the matrix inversion, and is given in Appendix A.

Theorem 8.1 can be interpreted as follows: if the characteristic equation of $\mathbf{A}$ is given by

$$\Delta(\lambda) = \lambda^n + a_{n-1}\lambda^{n-1} + \cdots + a_1\lambda + a_0 = 0 \quad (8.112)$$

then, in addition to that single algebraic equation, the following $n \times n$ algebraic equations are satisfied:

$$\Delta(\mathbf{A}) = \mathbf{A}^n + a_{n-1}\mathbf{A}^{n-1} + \cdots + a_1\mathbf{A} + a_0\mathbf{I} = \mathbf{0}^{n \times n} \quad (8.113)$$

The Cayley–Hamilton theorem can be used to find efficiently the transition matrices for both continuous- and discrete-time systems. From (8.113), we can express $\mathbf{A}^n$ as a linear function of lower powers of $\mathbf{A}$ up to the order $n - 1$. The same can be done for $\mathbf{A}^{n+m}$, that is,

$$\mathbf{A}^{n+m} = \mathcal{F}(\mathbf{A}^{n-1}, \mathbf{A}^{n-2}, \ldots, \mathbf{A}, \mathbf{I})$$
$$= \alpha_0[n+m]\mathbf{I} + \alpha_1[n+m]\mathbf{A} + \cdots + \alpha_{n-1}[n+m]\mathbf{A}^{n-1}, \quad m = 0, 1, 2, 3, \ldots$$
$$(8.114)$$

Using (8.114), the expression for the state transition matrix of the discrete-time systems defined in (8.80) can be found as

$$\Phi_d[k] = \mathbf{A}_d^k = \alpha_0[k]\mathbf{I} + \alpha_1[k]\mathbf{A}_d + \cdots + \alpha_{n-1}[k]\mathbf{A}_d^{n-1} \quad (8.115)$$

Similarly, all powers of n and higher in the infinite series expansion for the continuous-time state transition matrix given in (8.34) can be replaced using a continuous-time formula analogous to discrete-time formula (8.114), which leads to

$$e^{\mathbf{A}t} = \alpha_0(t)\mathbf{I} + \alpha_1(t)\mathbf{A} + \cdots + \alpha_{n-1}(t)\mathbf{A}^{n-1} \quad (8.116)$$

Formulas (8.115) and (8.116) represent the desired expressions for calculating, respectively, the discrete- and continuous-time state transition matrices.

The unknown coefficients in (8.115) can be obtained from

$$\lambda^k = \alpha_0[k] + \alpha_1[k]\lambda + \alpha_2[k]\lambda^2 + \cdots + \alpha_{n-1}[k]\lambda^{n-1}, \quad \lambda = \lambda_1, \lambda_2, \ldots, \lambda_n \quad (8.117)$$

This follows from the fact that we can repeat the same arguments as in (8.113–114) with the corresponding discrete-time scalar characteristic equation. Assuming that the eigenvalues are distinct, we get from (8.117) a system of n linearly independent equations for n unknown coefficients $\alpha_i[k], i = 0, 1, 2, \ldots, n - 1$. Note that in the case of multiple eigenvalues, a set of independent equations can be generated by taking the derivatives of (8.117) *with respect to* λ, as follows. Let q be the multiplicity of λ_i, then the additional independent equations for the required discrete-time coefficients are obtained from

$$\frac{d^j}{d\lambda_i^j}\left(\lambda_i^k = \alpha_0[k] + \alpha_1[k]\lambda_i + \alpha_2[k]\lambda_i^2 + \cdots + \alpha_{n-1}[k]\lambda_i^{n-1}\right), \quad j = 1, 2, \ldots, q - 1$$

$$(8.118)$$

The coefficients $\alpha_i(t), i = 0, 1, 2, \ldots, n - 1$, in (8.116), can be obtained from

$$e^{\lambda t} = \alpha_0(t) + \alpha_1(t)\lambda + \alpha_2(t)\lambda^2 + \cdots + \alpha_{n-1}(t)\lambda^{n-1} \quad (8.119)$$

which follows from the characteristic equation (8.112) and the infinite series expansion of the scalar function $e^{\lambda t} = 1 + \lambda + \lambda^2/2! + \lambda^3/3! + \cdots$. Using different values of λ_i, $i = 1, 2, \ldots, n$, we can generate a system of n linear algebraic equations that determine uniquely the required coefficients $\alpha_i(t), i = 0, 1, 2, \ldots, n - 1$. In the case of multiple eigenvalues, the derivatives must be taken *with respect to* λ. Let q be the multiplicity of λ_i, then the additional independent equations for the required continuous-time coefficients are obtained from

$$\frac{d^j}{d\lambda_i^j}\left(e^{\lambda_i t} = \alpha_0(t) + \alpha_1(t)\lambda_i + \alpha_2(t)\lambda_i^2 + \cdots + \alpha_{n-1}(t)\lambda_i^{n-1}\right), \quad j = 1, 2, \ldots, q - 1$$

$$(8.120)$$

The procedure for finding both continuous- and discrete-time state transition matrices based on the Cayley–Hamilton theorem is demonstrated in the following examples.

EXAMPLE 8.11

By the Cayley–Hamilton method, we have from (8.115) that for a 2×2 discrete-time system matrix, the state transition matrix is given by

$$\mathbf{A}_d^k = \alpha_0[k]\mathbf{I} + \alpha_1[k]\mathbf{A}_d, \quad k = 2, 3, 4, \ldots$$

with $\alpha_0[k]$ and $\alpha_1[k]$ satisfying (8.117), that is,

$$\lambda_1^k = \alpha_0[k] + \alpha_1[k]\lambda_1$$
$$\lambda_2^k = \alpha_0[k] + \alpha_1[k]\lambda_2$$

where λ_1 and λ_2 are distinct eigenvalues of $\mathbf{A}_d$. Let

$$\mathbf{A}_d = \begin{bmatrix} -0.5 & 3 \\ 0 & -1 \end{bmatrix}$$

then $\lambda_1 = -0.5$, $\lambda_2 = -1$. Hence

$$(-0.5)^k = \alpha_0[k] - 0.5\alpha_1[k]$$
$$(-1)^k = \alpha_0[k] - \alpha_1[k]$$

This system of algebraic equations implies

$$\alpha_0[k] = 2(-0.5)^k - (-1)^k, \quad \alpha_1[k] = 2(-0.5)^k - 2(-1)^k$$

and

$$\mathbf{A}_d^k = \alpha_0 \mathbf{I} + \alpha_1 \mathbf{A}_d = \begin{bmatrix} (-0.5)^k & 6(-0.5)^k - 6(-1)^k \\ 0 & (-1)^k \end{bmatrix}, \quad k = 2, 3, \ldots \quad \blacksquare$$

EXAMPLE 8.12

Consider a discrete-time system state matrix that has multiple eigenvalues, that is,

$$\mathbf{A}_d = \begin{bmatrix} -0.5 & 2 \\ 0 & -0.5 \end{bmatrix}$$

with $\lambda_1 = \lambda_2 = -0.5$. In this case, from (8.117), we get only one algebraic equation

$$\lambda_1^k = \alpha_0[k] + \alpha_1[k]\lambda_1$$

We can extract more information from this equation by taking the derivative with respect to λ, which leads to

$$\frac{d}{d\lambda_1}(\lambda_1^k = \alpha_0[k] + \alpha_1[k]\lambda_1) \Rightarrow k\lambda_1^{k-1} = \alpha_1[k] = k(-0.5)^{k-1}$$

Having obtained $\alpha_1[k]$, we get, from the original equation,

$$\alpha_0[k] = \lambda_1^k - \alpha_1[k]\lambda_1 = (-0.5)^k - k(-0.5)^k$$

In this case, the state transition matrix corresponding to $\mathbf{A}_d$ is given by

$$\mathbf{A}_d^k = \alpha_0[k]I + \alpha_1[k]\mathbf{A}_d = \begin{bmatrix} \alpha_0[k] & 0 \\ 0 & \alpha_0[k] \end{bmatrix} + \begin{bmatrix} -0.5\alpha_1[k] & 2\alpha_1[k] \\ 0 & -0.5\alpha_1[k] \end{bmatrix}$$

$$= \begin{bmatrix} \alpha_0[k] - 0.5\alpha_1[k] & 2\alpha_1[k] \\ 0 & \alpha_0[k] - 0.5\alpha_1[k] \end{bmatrix} = \begin{bmatrix} (-0.5)^k & 2k(-0.5)^{k-1} \\ 0 & (-0.5)^k \end{bmatrix} \quad \blacksquare$$

EXAMPLE 8.13

For a 2×2 continuous-time system matrix, the state transition matrix is obtained from (8.116) as

$$e^{\mathbf{A}t} = \alpha_0(t)\mathbf{I} + \alpha_1(t)\mathbf{A}$$

with $\alpha_0(t)$ and $\alpha_1(t)$ satisfying (8.119). Let λ_1 and λ_2 be distinct eigenvalues of $\mathbf{A}$, then according to (8.119), $\alpha_0(t)$ and $\alpha_1(t)$ satisfy

$$e^{\lambda_1 t} = \alpha_0(t) + \alpha_1(t)\lambda_1$$
$$e^{\lambda_2 t} = \alpha_0(t) + \alpha_1(t)\lambda_2$$

Consider

$$\mathbf{A} = \begin{bmatrix} 0 & 1 \\ -3 & -4 \end{bmatrix}$$

then $\lambda_1 = -1$, $\lambda_2 = -3$. Hence

$$e^{-t} = \alpha_0(t) - \alpha_1(t)$$
$$e^{-3t} = \alpha_0(t) - 3\alpha_1(t)$$

This system of algebraic equations has the solution

$$\alpha_0(t) = \frac{1}{2}(3e^{-t} - e^{-3t}), \quad \alpha_1(t) = \frac{1}{2}(e^{-t} - e^{-3t})$$

which implies

$$e^{\mathbf{A}t} = \alpha_0(t)\mathbf{I} + \alpha_1(t)\mathbf{A} = \begin{bmatrix} \alpha_0(t) & \alpha_1(t) \\ -3\alpha_1(t) & \alpha_0(t) - 4\alpha_1(t) \end{bmatrix}$$
$$= \begin{bmatrix} 0.5(3e^{-t} - e^{-3t}) & 0.5(e^{-t} - e^{-3t}) \\ -1.5(e^{-t} - e^{-3t}) & -0.5(e^{-t} - 3e^{-3t}) \end{bmatrix}$$

EXAMPLE 8.14

Consider a continuous-time system whose system matrix has multiple eigenvalues, that is,

$$\mathbf{A} = \begin{bmatrix} 0 & 1 & 0 \\ 0 & 0 & 1 \\ -1 & -3 & -3 \end{bmatrix}$$

The eigenvalues of this matrix are given by $\lambda_1 = \lambda_2 = \lambda_3 = -1$. The state transition matrix $e^{\mathbf{A}t}$, according to (8.116), is given by

$$e^{\mathbf{A}t} = \alpha_0(t)\mathbf{I} + \alpha_1(t)\mathbf{A} + \alpha_2(t)\mathbf{A}^2$$

From (8.119), we have only one equation for three unknowns, that is,

$$e^{\lambda t} = \alpha_0(t) + \alpha_1(t)\lambda + \alpha_2(t)\lambda^2, \quad \lambda = -1$$

By taking the derivatives of this equation, we can generate two additional equations,

$$\frac{d}{d\lambda}(e^{\lambda t} = \alpha_0(t) + \alpha_1(t)\lambda + \alpha_2(t)\lambda^2) \Rightarrow te^{\lambda t} = \alpha_1(t) + 2\alpha_2(t)\lambda$$

and

$$\frac{d^2}{d\lambda^2}(e^{\lambda t} = \alpha_0(t) + \alpha_1(t)\lambda + \alpha_2(t)\lambda^2) \Rightarrow t^2 e^{\lambda t} = 2\alpha_2(t)$$

For $\lambda = -1$, the last equation implies $\alpha_2(t) = 0.5t^2 e^{-t}$. Using the value for $\alpha_2(t)$ in the previous two equations, we obtain $\alpha_1(t) = e^{-t}t(1+t)$ and $\alpha_0(t) = e^{-t}(1+t+0.5t^2)$. Hence, we have

$$e^{\mathbf{A}t} = \alpha_0(t)\mathbf{I} + \alpha_1(t)\mathbf{A} + \alpha_2(t)\mathbf{A}^2$$

$$= \begin{bmatrix} \alpha_0 & 0 & 0 \\ 0 & \alpha_0 & 0 \\ 0 & 0 & \alpha_0 \end{bmatrix} + \begin{bmatrix} 0 & \alpha_1 & 0 \\ 0 & 0 & \alpha_1 \\ -\alpha_1 & -3\alpha_1 & -3\alpha_1 \end{bmatrix} + \begin{bmatrix} 0 & 0 & \alpha_2 \\ -\alpha_2 & -3\alpha_2 & -3\alpha_2 \\ 3\alpha_2 & 8\alpha_2 & 6\alpha_2 \end{bmatrix}$$

$$= \begin{bmatrix} \alpha_0 & \alpha_1 & \alpha_2 \\ -\alpha_2 & \alpha_0 - 3\alpha_2 & \alpha_1 - 3\alpha_2 \\ -\alpha_1 + 3\alpha_2 & -3\alpha_1 + 8\alpha_2 & \alpha_0 - 3\alpha_1 + 6\alpha_2 \end{bmatrix}$$

with the α_i as previously determined.

It is interesting to observe that the procedures for calculating the coefficients $\alpha[k]$ and $\alpha(t)$ from (8.117) and (8.119) always produce real values for these coefficients, even in cases of complex conjugate eigenvalues. However, in such cases, we must deal with complex numbers, but the final results for $\alpha[k]$ and $\alpha(t)$ must be real (see Problems 8.8–9).

Remark 8.4: In addition to the presented methods for finding the transition matrix (matrix exponential) of continuous-time linear systems, there are several other methods that are particularly useful for numerical computations of the transition matrix of higher-order dimensional linear systems [5]. In recent papers [6, 7], it has been shown that the coefficients in (8.116) can be efficiently calculated in terms of fundamental solutions of an nth-order scalar differential equation whose coefficients are equal to the coefficients of the characteristic polynomial of $\mathbf{A}$. Similar results are obtained in [8], where it is shown how to extend the results of [6] to the discrete-time domain, and obtain the coefficients of (8.115) in terms of solutions of an nth-order scalar linear difference equation.

8.6 LINEARIZATION OF NONLINEAR SYSTEMS[†]

We study in this text only time-invariant linear systems. The study of nonlinear systems is rather difficult. However, in some cases it is possible to linearize nonlinear systems and study them as linear. In this section, we show how to perform linearization of systems described by nonlinear differential equations. The procedure introduced is based on the Taylor series expansion and knowledge of nominal system trajectories and nominal system inputs. (Readers particularly interested in the study of nonlinear systems are referred to the very comprehensive book by Khalil [9].)

We will start with a simple scalar first-order nonlinear dynamic system represented by

$$\dot{x}(t) = \mathcal{F}(x(t), f(t)), \quad x(t_0) \text{ given} \qquad (8.121)$$

Assume that under usual working circumstances this system operates along the trajectory $x_n(t)$ while it is driven by the system input $f_n(t)$. We call $x_n(t)$ and $f_n(t)$, respectively, the *nominal system trajectory* and the *nominal system input*. On the nominal trajectory, the following differential equation is satisfied:

$$\dot{x}_n(t) = \mathcal{F}(x_n(t), u_n(t)) \qquad (8.122)$$

Now assume that the motion of the nonlinear system (8.121) is in the neighborhood of the nominal system trajectory, and that the distance from the nominal trajectory is small, that is,

$$x(t) = x_n(t) + \Delta x(t) \qquad (8.123)$$

where $\Delta x(t)$ represents a small quantity. It is natural to assume that the system motion in close proximity to the nominal trajectory will be sustained by a system input that is obtained by adding a small quantity to the nominal system input, that is,

$$f(t) = f_n(t) + \Delta f(t) \qquad (8.124)$$

For system motion in close proximity to the nominal trajectory, from equations (8.121), (8.123), and (8.124), we have

$$\dot{x}_n(t) + \Delta \dot{x}(t) = \mathcal{F}(x_n(t) + \Delta x(t), f_n(t) + \Delta f(t)) \qquad (8.125)$$

Since $\Delta x(t)$ and $\Delta f(t)$ are small quantities, the right-hand side of (8.125) can be expanded into a Taylor series about the nominal system trajectory and input, which produces

$$\dot{x}_n(t) + \Delta \dot{x}(t) = \mathcal{F}(x_n, f_n) + \frac{\partial \mathcal{F}}{\partial x}(x_n, f_n) \Delta x(t) + \frac{\partial \mathcal{F}}{\partial u}(x_n, f_n) \Delta u(t)$$
$$+ \text{ higher-order terms} \qquad (8.126)$$

[†]This section may be skipped without loss of continuity.

Using (8.122) and canceling the higher-order terms (which contain the very small quantities Δx^2, Δf^2, $\Delta x \Delta f$, Δx^3, ...), we obtain the linear differential equation

$$\Delta \dot{x}(t) \approx \frac{\partial \mathcal{F}}{\partial x}(x_n, f_n) \Delta x(t) + \frac{\partial \mathcal{F}}{\partial f}(x_n, f_n) \Delta f(t) \tag{8.127}$$

whose solution represents a valid approximation for $\Delta x(t)$. Note that the *partial derivatives in the linearization procedure are evaluated at the nominal points*. Introducing the notation

$$a_0(t) = -\frac{\partial \mathcal{F}}{\partial x}(x_n, f_n), \quad b_0 = \frac{\partial \mathcal{F}}{\partial f}(x_n, f_n) \tag{8.128}$$

the linear system (8.127) can be approximated as

$$\Delta \dot{x}(t) + a_0(t) \Delta x(t) = b_0(t) \Delta f(t) \tag{8.129}$$

In general, the obtained linear system is time varying. Since in this course we study only time-invariant systems, we will consider only those examples for which the linearization procedure produces time-invariant systems. It remains to find the initial condition for the linearized system, which can be obtained from (8.123) as

$$\Delta x(t_0) = x(t_0) - x_n(t_0) \tag{8.130}$$

Similarly, we can linearize the second-order nonlinear dynamic system

$$\ddot{x} = \mathcal{F}(x, \dot{x}, f, \dot{f}), \quad x(t_0), \dot{x}(t_0) \text{ given} \tag{8.131}$$

by assuming that

$$\begin{aligned} x(t) &= x_n(t) + \Delta x(t), & \dot{x}(t) &= \dot{x}_n(t) + \Delta \dot{x}(t) \\ f(t) &= f_n(t) + \Delta f(t), & \dot{f}(t) &= \dot{f}_n(t) + \Delta \dot{f}(t) \end{aligned} \tag{8.132}$$

and expanding

$$\ddot{x}_n + \Delta \ddot{x} = \mathcal{F}(x_n + \Delta x_n, \dot{x}_n + \Delta \dot{x}, f_n + \Delta f, \dot{f}_n + \Delta \dot{f}) \tag{8.133}$$

into a Taylor series about nominal points x_n, $\dot{x}_n$, f_n, $\dot{f}_n$, which leads to

$$\Delta \ddot{x}(t) + a_1 \Delta \dot{x}(t) + a_0 \Delta x(t) = b_1 \Delta \dot{f}(t) + b_0 \Delta f(t) \tag{8.134}$$

where the corresponding coefficients are evaluated at the nominal points as

$$\begin{aligned} a_1 &= -\frac{\partial \mathcal{F}}{\partial \dot{x}}(x_n, \dot{x}_n, f_n, \dot{f}_n), & a_0 &= -\frac{\partial \mathcal{F}}{\partial x}(x_n, \dot{x}_n, f_n, \dot{f}_n) \\ b_1 &= \frac{\partial \mathcal{F}}{\partial \dot{f}}(x_n, \dot{x}_n, f_n, \dot{f}_n), & b_0 &= \frac{\partial \mathcal{F}}{\partial f}(x_n, \dot{x}_n, f_n, \dot{f}_n) \end{aligned} \tag{8.135}$$

The initial conditions for the second-order linearized system are easily obtained from (8.132) as

$$\Delta x(t_0) = x(t_0) - x_n(t_0), \quad \Delta \dot{x}(t_0) = \dot{x}(t_0) - \dot{x}_n(t_0) \tag{8.136}$$

EXAMPLE 8.15

The mathematical model of a stick-balancing problem is given in [10] by

$$\ddot{\theta}(t) = \sin\theta(t) - f(t)\cos\theta(t) = \mathcal{F}(\theta(t), f(t))$$

where $f(t)$ is the horizontal force of a finger and $\theta(t)$ represents the stick's angular displacement from the vertical. This second-order dynamic system is linearized at the nominal points $(\dot{\theta}_n(t) = \theta_n(t) = 0, f_n(t) = 0)$ using formulas (8.135), which produces

$$a_1 = -\frac{\partial \mathcal{F}}{\partial \dot{\theta}} = 0, \quad a_0 = -\left(\frac{\partial \mathcal{F}}{\partial \theta}\right)_{|n} = -(\cos\theta + f\sin\theta)\Big|_{\substack{\theta_n(t)=0 \\ f_n(t)=0}} = -1$$

$$b_1 = \frac{\partial \mathcal{F}}{\partial \dot{f}} = 0, \quad b_0 = \left(\frac{\partial \mathcal{F}}{\partial f}\right)_{|n} = -(\cos\theta)\Big|_{\theta_n(t)=0} = -1$$

The linearized equation is given by

$$\ddot{\theta}(t) - \theta(t) = -f(t)$$

Note that $\Delta\theta(t) = \theta(t)$, $\Delta f(t) = f(t)$, since $\theta_n(t) = 0$, $f_n(t) = 0$. It is important to point out that the same linearized model could have been obtained by setting $\sin\theta(t) \approx \theta(t)$, $\cos\theta(t) \approx 1$, which is valid for small values of $\theta(t)$.

Of course, we can extend the presented linearization procedure to an nth-order nonlinear dynamic system with one input and one output in a straightforward way. However, for multi-input multi-output systems, this procedure becomes cumbersome. Using the state space model, the linearization procedure for the multi-input multi-output case is quite simple.

Consider now the general nonlinear dynamic control system in matrix form represented by

$$\frac{d}{dt}\mathbf{x}(t) = \mathcal{F}(\mathbf{x}(t), \mathbf{f}(t)), \quad \mathbf{x}(t_0) \text{ given} \tag{8.137}$$

where $\mathbf{x}(t)$, $\mathbf{f}(t)$, and $\mathcal{F}$ are, respectively, the n-dimensional system state space vector, the r-dimensional input vector, and the n-dimensional vector function. Assume that the nominal (operating) system trajectory $\mathbf{x}_n(t)$ is known and that the nominal system input that keeps the system on the nominal trajectory is given by $\mathbf{f}_n(t)$. Using the same logic as for the scalar case, we can assume that the actual system dynamics in the immediate proximity of the system nominal trajectories can be approximated by the first terms of the Taylor series. That is, starting with

$$\mathbf{x}(t) = \mathbf{x}_n(t) + \Delta\mathbf{x}(t), \quad \mathbf{f}(t) = \mathbf{f}_n(t) + \Delta\mathbf{f}(t) \tag{8.138}$$

and
$$\frac{d}{dt}\mathbf{x}_n(t) = \mathcal{F}(\mathbf{x}_n(t), \mathbf{f}_n(t)) \tag{8.139}$$

we expand equation (8.137) as

$$\begin{aligned}\frac{d}{dt}\mathbf{x}_n + \frac{d}{dt}\Delta\mathbf{x} &= \mathcal{F}(\mathbf{x}_n + \Delta\mathbf{x}, \mathbf{f}_n + \Delta\mathbf{f}) \\ &= \mathcal{F}(\mathbf{x}_n, \mathbf{f}_n) + \left(\frac{\partial \mathcal{F}}{\partial \mathbf{x}}\right)_{\big|\substack{\mathbf{x}_n(t)\\ \mathbf{f}_n(t)}} \Delta\mathbf{x} + \left(\frac{\partial \mathcal{F}}{\partial \mathbf{f}}\right)_{\big|\substack{\mathbf{x}_n(t)\\ \mathbf{f}_n(t)}} \Delta\mathbf{f} + \text{higher-order terms}\end{aligned} \tag{8.140}$$

The higher-order terms contain at least quadratic quantities of $\Delta\mathbf{x}$ and $\Delta\mathbf{f}$. Since $\Delta\mathbf{x}$ and $\Delta\mathbf{f}$ are small, their squares are even smaller, and hence the higher-order terms can be neglected. Using (8.139) and neglecting the higher-order terms, an approximation is obtained as

$$\frac{d}{dt}\Delta\mathbf{x}(t) \approx \left(\frac{\partial \mathcal{F}}{\partial \mathbf{x}}\right)_{\big|\substack{\mathbf{x}_n(t)\\ \mathbf{f}_n(t)}} \Delta\mathbf{x}(t) + \left(\frac{\partial \mathcal{F}}{\partial \mathbf{f}}\right)_{\big|\substack{\mathbf{x}_n(t)\\ \mathbf{f}_n(t)}} \Delta\mathbf{f}(t) \tag{8.141}$$

Partial derivatives in (8.141) represent the Jacobian matrices given by

$$\left(\frac{\partial \mathcal{F}}{\partial \mathbf{x}}\right)_{\big|\substack{\mathbf{x}_n(t)\\ \mathbf{f}_n(t)}} = \mathbf{A}^{n \times n} = \begin{bmatrix} \frac{\partial \mathcal{F}_1}{\partial x_1} & \frac{\partial \mathcal{F}_1}{\partial x_2} & \cdots & \cdots & \frac{\partial \mathcal{F}_1}{\partial x_n} \\ \frac{\partial \mathcal{F}_2}{\partial x_1} & \cdots & \cdots & \cdots & \frac{\partial \mathcal{F}_2}{\partial x_n} \\ \cdots & \cdots & \frac{\partial \mathcal{F}_i}{\partial x_j} & \cdots & \cdots \\ \cdots & \cdots & \cdots & \cdots & \cdots \\ \frac{\partial \mathcal{F}_n}{\partial x_1} & \frac{\partial \mathcal{F}_n}{\partial x_2} & \cdots & \cdots & \frac{\partial \mathcal{F}_n}{\partial x_n} \end{bmatrix}_{\big|\substack{\mathbf{x}_n(t)\\ \mathbf{f}_n(t)}} \tag{8.142a}$$

$$\left(\frac{\partial \mathcal{F}}{\partial \mathbf{f}}\right)_{\big|\substack{\mathbf{x}_n(t)\\ \mathbf{f}_n(t)}} = \mathbf{B}^{n \times r} = \begin{bmatrix} \frac{\partial \mathcal{F}_1}{\partial f_1} & \frac{\partial \mathcal{F}_1}{\partial f_2} & \cdots & \cdots & \frac{\partial \mathcal{F}_1}{\partial f_r} \\ \frac{\partial \mathcal{F}_2}{\partial f_1} & \cdots & \cdots & \cdots & \frac{\partial \mathcal{F}_2}{\partial f_r} \\ \cdots & \cdots & \frac{\partial \mathcal{F}_i}{\partial f_j} & \cdots & \cdots \\ \cdots & \cdots & \cdots & \cdots & \cdots \\ \frac{\partial \mathcal{F}_n}{\partial f_1} & \frac{\partial \mathcal{F}_n}{\partial f_2} & \cdots & \cdots & \frac{\partial \mathcal{F}_n}{\partial f_r} \end{bmatrix}_{\big|\substack{\mathbf{x}_n(t)\\ \mathbf{f}_n(t)}} \tag{8.142b}$$

Note that the Jacobian matrices must be evaluated at the nominal points, that is, at $\mathbf{x}_n(t)$ and $\mathbf{f}_n(t)$. With this notation, the linearized system (8.141) can be approximated by

$$\frac{d}{dt}\Delta\mathbf{x}(t) = \mathbf{A}\Delta\mathbf{x}(t) + \mathbf{B}\Delta\mathbf{u}(t), \quad \Delta\mathbf{x}(t_0) = \mathbf{x}(t_0) - \mathbf{x}_n(t_0) \tag{8.143}$$

The output of a nonlinear system, in general, satisfies a nonlinear algebraic equation, that is,

$$\mathbf{y}(t) = \mathcal{G}(\mathbf{x}(t), \mathbf{f}(t)) \tag{8.144}$$

This equation can also be linearized by expanding its right-hand side into a Taylor series about nominal points $\mathbf{x}_n(t)$ and $\mathbf{f}_n(t)$. This leads to

$$\mathbf{y}_n + \Delta\mathbf{y} = \mathcal{G}(\mathbf{x}_n, \mathbf{f}_n) + \left(\frac{\partial \mathcal{G}}{\partial \mathbf{x}}\right)\bigg|_{\substack{\mathbf{x}_n(t)\\ \mathbf{f}_n(t)}} \Delta\mathbf{x} + \left(\frac{\partial \mathcal{G}}{\partial \mathbf{f}}\right)\bigg|_{\substack{\mathbf{x}_n(t)\\ \mathbf{f}_n(t)}} \Delta\mathbf{f} + \text{higher-order terms} \tag{8.145}$$

Note that $\mathbf{y}_n$ cancels the term $\mathcal{G}(\mathbf{x}_n, \mathbf{f}_n)$. By neglecting the higher-order terms in (8.145), the linearized part of the output equation can be approximated by

$$\Delta\mathbf{y}(t) = \mathbf{C}\Delta\mathbf{x}(t) + \mathbf{D}\Delta\mathbf{f}(t) \tag{8.146}$$

where the Jacobian matrices $\mathbf{C}$ and $\mathbf{D}$ satisfy

$$\mathbf{C}^{p\times n} = \left(\frac{\partial \mathcal{G}}{\partial \mathbf{x}}\right)\bigg|_{\substack{\mathbf{x}_n(t)\\ \mathbf{f}_n(t)}} = \begin{bmatrix} \frac{\partial \mathcal{G}_1}{\partial x_1} & \frac{\partial \mathcal{G}_1}{\partial x_2} & \cdots & \cdots & \frac{\partial \mathcal{G}_1}{\partial x_n} \\ \frac{\partial \mathcal{G}_2}{\partial x_1} & & \cdots & \cdots & \frac{\partial \mathcal{G}_2}{\partial x_n} \\ \cdots & \cdots & \frac{\partial \mathcal{G}_i}{\partial x_j} & \cdots & \cdots \\ \cdots & \cdots & \cdots & \cdots & \cdots \\ \frac{\partial \mathcal{G}_p}{\partial x_1} & \frac{\partial \mathcal{G}_p}{\partial x_2} & \cdots & \cdots & \frac{\partial \mathcal{G}_p}{\partial x_n} \end{bmatrix}\bigg|_{\substack{\mathbf{x}_n(t)\\ \mathbf{f}_n(t)}} \tag{8.147a}$$

$$\mathbf{D}^{p\times r} = \left(\frac{\partial \mathcal{G}}{\partial \mathbf{f}}\right)\bigg|_{\substack{\mathbf{x}_n(t)\\ \mathbf{f}_n(t)}} = \begin{bmatrix} \frac{\partial \mathcal{G}_1}{\partial f_1} & \frac{\partial \mathcal{G}_1}{\partial f_2} & \cdots & \cdots & \frac{\partial \mathcal{G}_1}{\partial f_r} \\ \frac{\partial \mathcal{G}_2}{\partial f_1} & & \cdots & \cdots & \frac{\partial \mathcal{G}_2}{\partial f_r} \\ \cdots & \cdots & \frac{\partial \mathcal{G}_i}{\partial f_j} & \cdots & \cdots \\ \cdots & \cdots & \cdots & \cdots & \cdots \\ \frac{\partial \mathcal{G}_p}{\partial f_1} & \frac{\partial \mathcal{G}_p}{\partial f_2} & \cdots & \cdots & \frac{\partial \mathcal{G}_p}{\partial f_r} \end{bmatrix}\bigg|_{\substack{\mathbf{x}_n(t)\\ \mathbf{f}_n(t)}} \tag{8.147b}$$

EXAMPLE 8.16

Let a nonlinear system be represented by

$$\frac{dx_1(t)}{dt} = x_1(t)\sin x_2(t) + x_2(t)f(t)$$

$$\frac{dx_2(t)}{dt} = x_1(t)e^{-x_2(t)} + f^2(t)$$

$$y(t) = 2x_1(t)x_2(t) + x_2^2(t)$$

Assume that the values for the system nominal trajectories and input are known and given by x_{1n}, x_{2n}, and f_n. The linearized state space equation of this nonlinear system is obtained as

$$\begin{bmatrix}\Delta\dot{x}_1(t)\\ \Delta\dot{x}_2(t)\end{bmatrix} = \begin{bmatrix}\sin x_{2n} & x_{1n}\cos x_{2n} + f_n\\ e^{-x_{2n}} & -x_{1n}e^{-x_{2n}}\end{bmatrix}\begin{bmatrix}\Delta x_1(t)\\ \Delta x_2(t)\end{bmatrix} + \begin{bmatrix}x_{2n}\\ 2f_n\end{bmatrix}\Delta f(t)$$

$$\Delta y(t) = \begin{bmatrix}2x_{2n} & 2x_{1n} + 2x_{2n}\end{bmatrix}\begin{bmatrix}\Delta x_1(t)\\ \Delta x_2(t)\end{bmatrix}$$

Having obtained the solution of this linearized system under the given system input $\Delta f(t)$, the corresponding approximation of the nonlinear system trajectories is

$$\mathbf{x}_n(t) + \Delta\mathbf{x}(t) = \begin{bmatrix}x_{1n}(t)\\ x_{2n}(t)\end{bmatrix} + \begin{bmatrix}\Delta x_1(t)\\ \Delta x_2(t)\end{bmatrix}$$

EXAMPLE 8.17

Consider the mathematical model of a single-link robotic manipulator with a flexible joint [11] given by

$$I\ddot{\theta}_1(t) + mgl\sin\theta_1(t) + k(\theta_1(t) - \theta_2(t)) = 0$$

$$J\ddot{\theta}_2(t) - k(\theta_1(t) - \theta_2(t)) = f(t)$$

where $\theta_1(t)$, $\theta_2(t)$ are angular positions, I, J are moments of inertia, m and l are, respectively, link mass and length, and k is the link spring constant. Introducing the change of variables

$$x_1(t) = \theta_1(t), \quad x_2(t) = \dot{\theta}_1(t), \quad x_3(t) = \theta_2(t), \quad x_4(t) = \dot{\theta}_2(t)$$

the manipulator's state space nonlinear model is given by

$$\dot{x}_1(t) = x_2(t)$$

$$\dot{x}_2(t) = -\frac{mgl}{I}\sin x_1(t) - \frac{k}{I}(x_1(t) - x_3(t))$$

$$\dot{x}_3(t) = x_4(t)$$

$$\dot{x}_4(t) = \frac{k}{J}(x_1(t) - x_3(t)) + \frac{1}{J}f(t)$$

Taking the nominal points as $(x_{1n}, x_{2n}, x_{3n}, x_{4n}, f_n)$, the matrices **A** and **B** defined in (8.142) are given by

$$\mathbf{A} = \begin{bmatrix} 0 & 1 & 0 & 0 \\ -\dfrac{k + mgl\cos x_{1n}}{I} & 0 & \dfrac{k}{I} & 0 \\ 0 & 0 & 0 & 1 \\ \dfrac{k}{J} & 0 & -\dfrac{k}{J} & 0 \end{bmatrix}, \quad \mathbf{B} = \begin{bmatrix} 0 \\ 0 \\ 0 \\ \dfrac{1}{J} \end{bmatrix},$$

In [12], the following numerical values are used for system parameters: $mgl = 5$, $I = J = 1$, $k = 0.08$.

Assuming that the output variable is equal to the link's angular position, that is, $y(t) = x_1(t)$, the matrices **C** and **D**, defined in (8.147), are given by

$$\mathbf{C} = [1 \ 0 \ 0 \ 0], \quad \mathbf{D} = 0$$

In the next example, we give state space matrices for two linearized models of an F-15 aircraft obtained by linearizing nonlinear equations for two sets of operating points.

EXAMPLE 8.18

The longitudinal dynamics of an F-15 aircraft can be represented by a fourth-order mathematical model. For two operating conditions (subsonic and supersonic), two linear mathematical models have been derived [13, 14]. The corresponding state space models are given by

$$\begin{bmatrix} \dot{x}_1(t) \\ \dot{x}_2(t) \\ \dot{x}_3(t) \\ \dot{x}_4(t) \end{bmatrix} = \begin{bmatrix} -0.00819 & -25.70839 & 0 & -32.17095 \\ -0.00019 & -1.27626 & 1.0000 & 0 \\ 0.00069 & 1.02176 & -2.40523 & 0 \\ 0 & 0 & 1.0000 & 0 \end{bmatrix} \begin{bmatrix} x_1(t) \\ x_2(t) \\ x_3(t) \\ x_4(t) \end{bmatrix}$$

$$+ \begin{bmatrix} -6.80939 \\ -0.14968 \\ -14.06111 \\ 0 \end{bmatrix} f(t), \quad \mathbf{y}(t) = \mathbf{x}(t)$$

for subsonic flight conditions, and

$$\begin{bmatrix} \dot{x}_1(t) \\ \dot{x}_2(t) \\ \dot{x}_3(t) \\ \dot{x}_4(t) \end{bmatrix} = \begin{bmatrix} -0.01172 & -95.91071 & 0 & -32.11294 \\ -0.00011 & -1.87942 & 1.0000 & 0 \\ 0.00056 & -3.61627 & -3.44478 & 0 \\ 0 & 0 & 1.0000 & 0 \end{bmatrix} \begin{bmatrix} x_1(t) \\ x_2(t) \\ x_3(t) \\ x_4(t) \end{bmatrix}$$

$$+ \begin{bmatrix} -25.40405 \\ -0.22042 \\ -53.42460 \\ 0 \end{bmatrix} f(t), \quad \mathbf{y}(t) = \mathbf{x}(t)$$

for supersonic flight conditions.

Model derivations are beyond the scope of this book. The state space variables represent: $x_1(t)$—velocity in feet per second; $x_2(t)$—angle of attack in radians; $x_3(t)$—pitch rate in radians per second; and $x_4(t)$—pitch attitude in radians: The control input $f(t)$ represents the elevator control in radians.

Finally, we point out that the Simulink package is very convenient for simulation of nonlinear systems. Its function `linmod` can be used to obtain linearized models of nonlinear systems around given operating points (nominal system trajectories and inputs).

8.7 STATE SPACE MATLAB LABORATORY EXPERIMENTS

In this section, we present three MATLAB laboratory experiments on the state space method. These experiments can be used either as supplements for lectures or independently in the corresponding linear system laboratory. Most of the required MATLAB functions have already been introduced in the examples done in this chapter. Students should also consult Appendix C, where a shortened MATLAB manual is given. It is advisable that before using any MATLAB function, students check all its options by typing `help function name`.

8.7.1 Experiment 1—The Inverted Pendulum

Part 1. The linearized equations of the inverted pendulum, obtained by assuming that the pendulum mass is concentrated at its center of gravity [15, 16] are given by

$$(J + mL^2)\ddot{\theta}(t) - mgL\theta(t) + mL\ddot{d}(t) = 0$$
$$(M + m)\ddot{d}(t) + mL\ddot{\theta}(t) = f(t) \tag{8.148}$$

where $\theta(t)$ is the angle of the pendulum from the vertical position, $d(t)$ is the position of the cart, $f(t)$ is the force applied to the cart, M is the mass of the cart, m is the mass of the pendulum, g is the gravitational constant, and J is the moment of inertia about the center of mass. Assuming that normalized values are given by $J = 1$, $L = 1$, $g = 9.81$, $M = 1$, and $m = 0.1$, derive the state space form

$$\dot{\mathbf{x}}(t) = \mathbf{A}\mathbf{x}(t) + \mathbf{B}f(t)$$

where

$$\mathbf{x}(t) = [\theta(t) \quad \dot{\theta}(t) \quad d(t) \quad \dot{d}(t)]^T$$

and $\mathbf{A}^{4 \times 4}$ and $\mathbf{B}^{4 \times 1}$ are the corresponding matrices.

Part 2. Using MATLAB, determine the following:
(a) The eigenvalues, eigenvectors, and characteristic polynomial of matrix $\mathbf{A}$.
(b) The state transition matrix at the time instant $t = 1$.

(c) The unit impulse response (take $\theta(t)$ and $d(t)$ as the output variables) for $0 \le t \le 1$ with $\Delta t = 0.1$. Plot the system output response.
(d) The unit step response for $0 \le t \le 1$ and $\Delta t = 0.1$. Draw the system output response.
(e) The unit ramp response for $0 \le t \le 1$ and $\Delta t = 0.1$. Draw the system output response. Compare the response diagrams obtained in (c), (d), and (e).
(f) The system state response resulting from the initial condition $\mathbf{x}(0) = [-1\ 1\ 1\ 1]^T$ and the input $f(t) = \sin(t)$ for $0 \le t \le 5$ and $\Delta t = 0.1$.
(g) The inverse of the state transition matrix $(e^{\mathbf{A}t})^{-1}$ for $t = 5$.
(h) The state $\mathbf{x}(t)$ at time $t = 5$ assuming that $\mathbf{x}(10) = [10\ 0\ 5\ 2]^T$ and $f(t) = 0$, using the result from (g).
(i) Find the system transfer function.

Part 3. Discretize the continuous-time system defined in (8.148) with $T = 0.02$, and find the discrete-time space model

$$\mathbf{x}[k+1] = \mathbf{A}_d \mathbf{x}[k] + \mathbf{B}_d f[k]$$

Assuming that the output equation of the discrete system is given by

$$\mathbf{y}[k] = \begin{bmatrix} 1 & 0 & 0 & 0 \\ 0 & 0 & 1 & 0 \end{bmatrix} \mathbf{x}[k] = \mathbf{C}_d \mathbf{x}[k]$$

find the system output response for $0 \le k \le 50$ due to initial conditions $\mathbf{x}_0 = [-1\ 1\ -1\ 1]^T$ and unit step input (note that $f[k]$ should be generated as a column vector of 50 elements equal to 1).

Part 4. Consider the continuous-time system given by

$$\frac{d^2 y(t)}{dt^2} + 0.1 \frac{dy(t)}{dt} = f(t) \quad (8.149)$$

(a) Discretize this system with $T = 1$ using the Euler approximation.
(b) Find the system state and output responses of the obtained discrete system for $k = 1, 2, 3, \ldots, 20$, when $f(t) = \sin(0.1\pi t)$ and $y(0) = \dot{y}(0) = 0$.
(c) Find the discrete transfer function, characteristic equation, eigenvalues, and eigenvectors.

Part 5. Discretize the state space form of (8.149) obtained using MATLAB function c2d with $T = 1$. Find the discrete system state and output responses for the initial condition and the input function defined in Part 4(b). Compare the results obtained in Parts 4 and 5. Comment on the results obtained.

8.7.2 Experiment 2—Response of Continuous Systems

Part 1. Consider a continuous-time linear system represented by its transfer function

$$H(s) = \frac{s+5}{s^2 + 5s + 6}$$

(a) Find and plot the impulse response. Use the MATLAB function `impulse`.
(b) Find and plot the step response using the function `step`.
(c) Find the zero-state system output response due to an input given by $f(t) = e^{-3t}, t \geq 0$. Note that you must use the function `lsim` and specify input at every time instant of interest. That can be obtained by t=0:0.1:5 (defines t at $0, 0.1, 0.2, \ldots, 4.9, 5$), f=exp(-3*t), and y=lsim(num,den,f,t). Check that the results obtained agree with analytical results at $t = 1$.
(d) Obtain the state space form for this system using the function `tf2ss`. Repeat Parts (a), (b), and (c) for the corresponding state space representation. Use the following MATLAB instructions

```
[y,x]=impulse(A,B,C,D)
[y,x]=step(A,B,C,D)
[y,x]=lsim(A,B,C,D,f,t)
```

respectively, with f and t as defined in (c). Compare the results obtained.

Part 2. Consider the continuous-time linear system represented by

$$\frac{d^2 y(t)}{dt^2} + 4\frac{dy(t)}{dt} + 4y(t) = \frac{df(t)}{dt} + f(t),$$

$$f(t) = e^{-4t} u(t), \quad t \geq 0, \quad y(0^-) = 2, \quad \dot{y}(0) = 1$$

(a) Find the complete system state and output responses using the MATLAB function `lsim`. Compare the simulation results obtained with analytical results. (*Hint:* Use [y,x]=lsim(A,B,C,D,f,t,x0) with $t = 0:0.1:5$.) Note that the initial condition for the state vector, x0, must be found. This can be obtained by playing algebra with the state and output equations and setting $t = 0$.
(b) Find the zeros and poles of this system using the function `tf2zp`.
(c) Find the system state and output responses due to initial conditions specified in Part 2(a) and the impulse delta function as an input. Since you are unable to specify the system input in time (the delta function has no time structure), you cannot use the `lsim` function. Instead use either the MATLAB function `initial` (zero-input response) or the MATLAB program given at the end of this part. The required response is obtained analytically as

$$\mathbf{x}(t) = e^{\mathbf{A}t}(\mathbf{x}(0) + \mathbf{B})$$

where **A** and **B** stand for the system and input matrices in the state space. Thus, the new initial condition is given by $\mathbf{x}(0) + \mathbf{B}$. (*Hint:* To find and plot the system state response for the given matrix **A** and the corresponding initial conditions, the MATLAB program can be used.)

```
t=0:0.1:5;
for i=0:1:51;
x(:,i)=expm(A*t(i))*(x0+B);
end
plot(t,x(1,:))
plot(t,x(2,:))
```

(d) Justify the answer obtained in Part (c). Solve the same problem analytically using the Laplace transform. Plot results from (c) and compare with these results. Can you draw any conclusion for this "nonstandard" problem from the point of view of the system initial conditions at $t = 0^+$? (The standard problem requires that for the impulse response all initial conditions are set to zero.)

Part 3 Consider the dynamic system [17] represented in the state space form by

$$\mathbf{A} = \begin{bmatrix} -0.01357 & -32.2 & -46.3 & 0 \\ 0.00012 & 0 & 1.214 & 0 \\ -0.0001212 & 0 & -1.214 & 1 \\ 0.00057 & 0 & -9.1 & -0.6696 \end{bmatrix},$$

$$\mathbf{B} = \begin{bmatrix} -0.433 \\ 0.1394 \\ -0.1394 \\ -0.1577 \end{bmatrix}, \quad \mathbf{C} = \begin{bmatrix} 0 & 0 & 0 & 1 \\ 1 & 0 & 0 & 0 \end{bmatrix}, \quad \mathbf{D} = \mathbf{0}^{2 \times 1}$$

This is a real mathematical model of an F-8 aircraft [18]. Using MATLAB, determine the following quantities.

(a) The eigenvalues, eigenvectors, and characteristic polynomial. Take `p=poly(A)` and verify that `roots(p)` also produces the eigenvalues of matrix **A**.
(b) The state transition matrix at the time instant $t = 1$. Use the function `expm`.
(c) The unit impulse response. Plot output variables. (*Hint:* Use `impulse(A,B,C,D)`.)
(d) The unit step response. Plot the corresponding output variables.
(e) Let the initial system condition be $\mathbf{x}(0) = [-1 \ 1 \ 0.5 \ 1]^T$. Find the system state and output responses due to an input given by $f(t) = \sin(t)$, $0 < t < 1000$. (*Hint:* Take `t=0:10:1000` and find the corresponding values for $f(t)$ by using the function `sin` in the form `f=sin(t)`. Then use the `lsim` function.)

(f) Find the system transfer functions. Note that you have one input and two outputs, which implies two transfer functions. (*Hint:* Use the function ss2tf.)

(g) Find the inverse of the state transition matrix $(e^{\mathbf{A}t})^{-1} = e^{-\mathbf{A}t}$ at $t = 2$.

Part 4 Consider a linear continuous-time dynamic system represented by its transfer function

$$H(s) = \frac{(s+1)(s+3)(s+5)(s+7)}{s(s+2)(s+4)(s+6)(s+8)(s+10)}$$

(a) Input the system zeros and poles as column vectors. Note that in this case the static gain $k = 1$. Use the function zp2ss(z,p,k) to get the state space matrices.

(b) Find the eigenvalues and eigenvectors of matrix $\mathbf{A}$.

(c) Verify that the transformation $\mathbf{x} = \mathbf{P}\tilde{\mathbf{x}}$, where $\mathbf{P}$ is the matrix whose columns are the eigenvectors of matrix $\mathbf{A}$, produces in the new coordinates the diagonal system matrix $\mathbf{\Lambda} = \mathbf{P}^{-1}\mathbf{A}\mathbf{P}$ with diagonal elements equal to the eigenvalues of matrix $\mathbf{A}$.

(d) Find the remaining state space matrices in the new coordinates. Find the transfer function in the new coordinates and compare it with the original transfer function.

(e) Compare the unit step responses of the original and transformed systems.

8.7.3 Experiment 3—Response of Discrete Systems

Part 1. Consider a discrete-time linear system represented by its transfer function

$$H(z) = \frac{4z}{z^2 + z + 0.25}$$

(a) Find the impulse response using the MATLAB function dimpulse.

(b) Find the step response using the function dstep, and plot both the state and output responses.

(c) Find the system output response due to a unit step function, $f[k] = u[k]$, and initial conditions specified by $y[-1]=0$, $y[-2]=1$. Use the function dlsim and specify input at every time instant of interest. Take k=0:1:20 (defines k at 0, 1, 2, ..., 19, 20). Check analytically that the results obtained agree with the analytical results for $k = 10$.

(d) Obtain the state space form for this system using the function tf2ss. Repeat Parts (a), (b), and (c). Use the MATLAB statements

```
[y,x]=dimpulse(A,B,C,D)
[y,x]=dstep(A,B,C,D)
[y,x]=dlsim(A,B,C,D,f,x0)
```

respectively, with f and $x[0]$ as defined in (c). Compare the results obtained.

Part 2. Consider the discrete-time linear system represented by

$$y[k+2] + \frac{5}{6}y[k+1] + \frac{1}{6}y[k] = f[k+1],$$

$$f[k] = (0.8)^k u[k], \quad y[-1] = 2, \quad y[-2] = 3$$

(a) Find the system state and output responses using the MATLAB function `dlsim`. (*Hint:* Use `[y,x]=dlsim(A,B,C,D,f,x0)` with `k=0:1:10`.) Note that the initial condition must be found. This can be obtained by playing algebra with the state space and output equations. Compare the obtained simulation results with the analytical results.

(b) Find the zeros and poles of this system using the function `tf2zp`.

(c) Find the system state and output responses due to initial conditions specified in Part 2(a) and with the impulse delta function as an input. Use the `dlsim` function.

(d) Solve the problem in (c) analytically using the $\mathcal{Z}$-transform. Plot results from (c) and compare results.

Part 3. Consider a dynamic system represented in the continuous-time state space form in Section 8.7.2, Experiment 2, Part 3.

(a) Discretize the continuous-time system using the MATLAB function `c2d`. Assume that the sampling period is $T = 1$.

(b) Find the eigenvalues, eigenvectors, and characteristic polynomial of the obtained discrete-time system.

(c) Find the state transition matrix at time instant $k = 5$.

(d) Find the unit impulse response and plot output variables.

(e) Find the unit step response and plot the corresponding output variables.

(f) Assume that the initial system condition is $\mathbf{x}(0) = [-1 \ 0 \ 1 \ -0.5]^T$. Find the system state and output responses due to an input given by $f[k] = \sin[k], 0 \leq k \leq 1000$. Take `k=0:10:1000`. Use the `dlsim` function. Compare the obtained discrete-time results with the continuous-time results for the same system studied in Section 8.7.2, Experiment 2.

(g) Find the system transfer functions. Note that you have one input and two outputs, which implies two transfer functions. The matrices $\mathbf{C}$ and $\mathbf{D}$ are not changed due to discretization. (*Hint:* Use the function `ss2tf`.)

8.8 SUMMARY

Study Guide for Chapter Eight: Students should know how to find state space forms (phase variable canonical forms) of continuous- and discrete-time linear systems represented by differential and difference equations. The central problem is to find the system state and output responses. In that respect, students must known how to find continuous- and discrete-time state transition matrices using either the Cayley–Hamilton method, or frequency domain methods based on the Laplace and $\mathcal{Z}$-transforms. The procedures for discretizing continuous-time systems using integral and Euler's approximations are important. The standard problems are: (1) Find the state transition matrix (continuous and discrete); (2) find the state and output responses of continuous- and discrete-time linear systems; (3) find the system transfer function (continuous- and discrete-time); (4) discretize a continuous-time linear system.

Linear Continuous-Time Dynamic System Representation

(1) Differential Equation:

$$\frac{d^n y(t)}{dt^n} + a_{n-1}\frac{d^{n-1} y(t)}{dt^{n-1}} + \cdots + a_1 \frac{dy(t)}{dt} + a_0 y(t)$$
$$= b_{n-1}\frac{d^{n-1} f(t)}{dt^{n-1}} + \cdots + b_1 \frac{df(t)}{dt} + b_0 f(t)$$

(2) Transfer Function:

$$H(s) = \frac{b_{n-1}s^{n-1} + b_{n-2}s^{n-2} + \cdots + b_1 s + b_0}{s^n + a_{n-1}s^{n-1} + a_{n-2}s^{n-2} + \cdots + a_1 s + a_0}$$

(3) State Space Form:

$$\frac{d}{dt}\mathbf{x}(t) = \dot{\mathbf{x}}(t) = \mathbf{A}\mathbf{x}(t) + \mathbf{B}f(t), \quad \mathbf{x}(0) = \mathbf{x}_0$$
$$y(t) = \mathbf{C}\mathbf{x}(t) + \mathbf{D}f(t)$$

$$\begin{bmatrix} \dot{x}_1(t) \\ \dot{x}_2(t) \\ \vdots \\ \vdots \\ \dot{x}_{n-1}(t) \\ \dot{x}_n(t) \end{bmatrix} = \begin{bmatrix} 0 & 1 & 0 & \cdots & & 0 \\ 0 & 0 & 1 & 0 & \cdots & 0 \\ \vdots & \vdots & \ddots & \ddots & \ddots & \vdots \\ \vdots & \vdots & & \ddots & \ddots & 0 \\ 0 & 0 & \cdots & \cdots & 0 & 1 \\ -a_0 & -a_1 & -a_2 & \cdots & & -a_{n-1} \end{bmatrix} \begin{bmatrix} x_1(t) \\ x_2(t) \\ \vdots \\ \vdots \\ x_{n-1}(t) \\ x_n(t) \end{bmatrix} + \begin{bmatrix} 0 \\ 0 \\ \vdots \\ \vdots \\ 0 \\ 1 \end{bmatrix} f(t)$$

$$y(t) = \begin{bmatrix} b_0 & b_1 & \cdots & b_{n-1} \end{bmatrix} \begin{bmatrix} x_1(t) \\ x_2(t) \\ \vdots \\ x_n(t) \end{bmatrix}$$

Continuous-Time State Transition Matrix (Matrix Exponential) and Its Properties:

$$\Phi(t) = e^{\mathbf{A}t} \triangleq \mathbf{I} + \mathbf{A}t + \frac{1}{2!}\mathbf{A}^2 t^2 + \frac{1}{3!}\mathbf{A}^3 t^3 + \cdots = \sum_{i=0}^{\infty} \frac{1}{i!}\mathbf{A}^i t^i$$

(a) $\Phi(0) = \mathbf{I}$
(b) $\Phi^{-1}(t) = \Phi(-t) \Rightarrow \Phi(t)$ is nonsingular for every t

(c) $\Phi(t_2 - t_0) = \Phi(t_2 - t_1)\Phi(t_1 - t_0)$
(d) $\Phi(t)^i = \Phi(it)$, for $i \in N$
(e) $\frac{d}{dt}e^{\mathbf{A}t} = \mathbf{A}e^{\mathbf{A}t} = e^{\mathbf{A}t}\mathbf{A}$

State Transition Matrix via the Laplace Transform:
$$\Phi(s) = (s\mathbf{I} - \mathbf{A})^{-1} = \frac{1}{\det(s\mathbf{I} - \mathbf{A})}\mathrm{adj}(s\mathbf{I} - \mathbf{A}) = \mathcal{L}\{\Phi(t)\}$$

State Transition Matrix via the Cayley–Hamilton Method:
$$e^{\mathbf{A}t} = \alpha_0(t)\mathbf{I} + \alpha_1(t)\mathbf{A} + \cdots + \alpha_{n-1}(t)\mathbf{A}^{n-1}$$

(distinct eigenvalues)
$$e^{\lambda_i t} = \alpha_0(t) + \alpha_1(t)\lambda_i + \alpha_2(t)\lambda_i^2 + \cdots + \alpha_{n-1}(t)\lambda_i^{n-1}, \quad i = 1, 2, \ldots, n$$

(multiple eigenvalues)
$$\frac{d^j}{d\lambda_i^j}\left(e^{\lambda_i t} = \alpha_0(t) + \alpha_1(t)\lambda_i + \alpha_2(t)\lambda_i^2 + \cdots + \alpha_{n-1}(t)\lambda_i^{n-1}\right), \quad j = 1, 2, \ldots, q - 1$$

Solution of the State Space Equations (System State Response):
$$\mathbf{x}(t) = e^{\mathbf{A}(t-t_0)}\mathbf{x}(t_0) + \int_{t_0}^{t} e^{\mathbf{A}(t-\tau)}\mathbf{B}\mathbf{f}(\tau)\,d\tau$$

$$\mathbf{X}(s) = (s\mathbf{I} - \mathbf{A})^{-1}\mathbf{x}(0^-) + (s\mathbf{I} - \mathbf{A})^{-1}\mathbf{B}\mathbf{F}(s)$$

System Output Response:
$$\mathbf{y}(t) = \mathbf{C}\mathbf{x}(t) + \mathbf{D}\mathbf{f}(t), \quad \mathbf{Y}(s) = \mathbf{C}\mathbf{X}(s) + \mathbf{D}\mathbf{F}(s)$$

System Matrix Transfer Function (multi-input multi-output systems):
$$\mathbf{H}(s) = \mathbf{C}(s\mathbf{I} - \mathbf{A})^{-1}\mathbf{B} + \mathbf{D}$$

Linear Discrete-Time Dynamic System Representation

(1) Difference Equation:
$$y[k+n] + a_{n-1}y[k+n-1] + \cdots + a_1 y[k+1] + a_0 y[k]$$
$$= b_{n-1}f[k+n-1] + \cdots + b_1 f[k+1] + b_0 f[k]$$

(2) Transfer Function:
$$H(z) = \frac{b_{n-1}z^{n-1} + b_{n-2}z^{n-2} + \cdots + b_1 z + b_0}{z^n + a_{n-1}z^{n-1} + a_{n-2}z^{n-2} + \cdots + a_1 z + a_0}$$

(3) State Space Form:

$$\mathbf{x}[k+1] = \mathbf{A}_d\mathbf{x}[k] + \mathbf{B}_d\mathbf{f}[k]$$

$$\mathbf{y}[k] = \mathbf{C}_d\mathbf{x}[k] + \mathbf{D}_d\mathbf{f}[k]$$

$$\begin{bmatrix} x_1[k+1] \\ x_2[k+1] \\ \vdots \\ \vdots \\ x_n[k+1] \end{bmatrix} = \begin{bmatrix} 0 & 1 & 0 & \cdots & 0 \\ 0 & 0 & 1 & \cdots & 0 \\ \vdots & \vdots & \vdots & \ddots & \vdots \\ 0 & 0 & 0 & \cdots & 1 \\ -a_0 & -a_1 & -a_2 & \cdots & -a_{n-1} \end{bmatrix} \begin{bmatrix} x_1[k] \\ x_2[k] \\ \vdots \\ \vdots \\ x_n[k] \end{bmatrix} + \begin{bmatrix} 0 \\ 0 \\ \vdots \\ 0 \\ 1 \end{bmatrix} f[k]$$

$$y[k] = \begin{bmatrix} b_0 & b_1 & \cdots & b_{n-1} \end{bmatrix} \begin{bmatrix} x_1[k] \\ x_2[k] \\ \vdots \\ x_n[k] \end{bmatrix}$$

Discrete-Time State Transition Matrix and Its Properties:

$$\Phi_d[k] = \mathbf{A}_d^k$$

(a) $\Phi_d[0] = \mathbf{A}_d^0 = \mathbf{I} \Leftarrow \mathbf{x}[0] = \Phi_d[0]\mathbf{x}[0]$
(b) $\Phi_d[k_2 - k_0] = \Phi_d[k_2 - k_1]\Phi_d[k_1 - k_0] = \mathbf{A}_d^{k_2-k_1}\mathbf{A}_d^{k_1-k_0} = \mathbf{A}_d^{k_2-k_0}$
(c) $\Phi_d^i[k] = \Phi_d[ik] \Leftarrow (\mathbf{A}_d^k)^i = \mathbf{A}_d^{ik}$
(d) $\Phi_d[k+1] = \mathbf{A}_d\Phi_d[k], \Phi_d[0] = \mathbf{I}$

State Transition Matrix via the $\mathcal{Z}$-Transform:

$$\Phi_d[k] = \mathcal{Z}^{-1}[(z\mathbf{I} - \mathbf{A}_d)^{-1}z] = \mathbf{A}_d^k, \quad k = 1, 2, 3, \ldots$$

State Transition Matrix via the Cayley–Hamilton Method:

$$\Phi_d[k] = \mathbf{A}_d^k = \alpha_0[k]\mathbf{I} + \alpha_1[k]\mathbf{A}_d + \cdots + \alpha_{n-1}[k]\mathbf{A}_d^{n-1}$$

(distinct eigenvalues)

$$\lambda^k = \alpha_0[k] + \alpha_1[k]\lambda + \alpha_2[k]\lambda^2 + \cdots + \alpha_{n-1}[k]\lambda^{n-1}, \quad \lambda = \lambda_1, \lambda_2, \ldots, \lambda_n$$

(multiple eigenvalues)

$$\frac{d^j}{d\lambda_i^j}(\lambda_i^k = \alpha_0[k] + \alpha_1[k]\lambda_i + \alpha_2[k]\lambda_i^2 + \cdots + \alpha_{n-1}[k]\lambda_i^{n-1}), \quad j = 1, 2, \ldots, q-1$$

Solution of the State Space Equations (System State Response):

$$\mathbf{x}[k] = \Phi_d[k]\mathbf{x}[0] + \sum_{i=0}^{k-1} \Phi_d[k - i - 1]\mathbf{B}_d\mathbf{f}[i]$$

$$\mathbf{X}(z) = (z\mathbf{I} - \mathbf{A}_d)^{-1}z\mathbf{x}[0] + (z\mathbf{I} - \mathbf{A}_d)^{-1}\mathbf{B}_d\mathbf{F}(z)$$

System Output Response:

$$\mathbf{y}[k] = \mathbf{Cx}[k] + \mathbf{Df}[k], \quad \mathbf{Y}(s) = \mathbf{CX}(s) + \mathbf{DF}(s)$$
$$\mathbf{Y}(z) = \mathbf{C}_d(z\mathbf{I} - \mathbf{A}_d)^{-1}z\mathbf{x}[0] + [\mathbf{C}_d(z\mathbf{I} - \mathbf{A}_d)^{-1}\mathbf{B}_d + \mathbf{D}_d]\mathbf{F}(z)$$

System Matrix Transfer Function (multi-input multi-output systems):
$$\mathbf{H}_d(z) = \mathbf{C}_d(z\mathbf{I} - \mathbf{A}_d)^{-1}\mathbf{B}_d + \mathbf{D}_d$$

Discretization of Continuous-Time Systems:
Integral Approximation

$$\mathbf{A}_d = e^{\mathbf{A}T}, \quad \mathbf{B}_d = \int_0^T e^{\mathbf{A}\sigma}d\sigma\mathbf{B}, \quad T = \text{sampling period}, \quad \mathbf{C}_d = \mathbf{C}, \quad \mathbf{D}_d = \mathbf{D}$$

Euler Approximation
$$\mathbf{A}_d = \mathbf{I} + T\mathbf{A}, \quad \mathbf{B}_d = T\mathbf{B}, \quad \mathbf{C}_d = \mathbf{C}, \quad \mathbf{D}_d = \mathbf{D}$$

Similarity Transformation

$$\hat{\mathbf{A}} = \mathbf{P}^{-1}\mathbf{AP}, \quad \hat{\mathbf{B}} = \mathbf{P}^{-1}\mathbf{B}, \quad \hat{\mathbf{C}} = \mathbf{CP}, \quad \hat{\mathbf{D}} = \mathbf{D}, \quad \hat{\mathbf{x}}(0) = \mathbf{P}^{-1}\mathbf{x}(0)$$

8.9 REFERENCES

[1] Z. Gajić and M. Lelić, *Modern Control Systems Engineering*, Prentice Hall International, London, 1996.

[2] T. Kailath, *Linear Systems*, Prentice Hall, Englewood Cliffs, NJ, 1980.

[3] L. Qiu and E. Davison, "Performance limitations of non-minimum phase systems in the servo-mechanism problem," *Automatica*, 29:337–49, 1993.

[4] J. Fraleigh and R. Beauregard, *Linear Algebra*, Addison-Wesley, Reading, MA, 1990.

[5] C. Moler and C. Van Loan, "Nineteen dubious ways to compute the exponential of a matrix," *SIAM Review*, 20:801–36, 1978.

[6] I. Leonard, "The matrix exponential," *SIAM Review*, 38:507–12, 1996.

[7] E. Liz, "A note on the matrix exponential," *SIAM Review*, 40:700–02, 1998.

[8] M. Kwapisz, "The power of a matrix," *SIAM Review*, 40:703–05, 1998.

[9] H. Khalil, *Nonlinear Systems*, Prentice Hall, Upper Saddle River, NJ, 2002.

[10] E. Sontag, *Mathematical Control Theory*, Springer-Verlag, New York, 1998.

[11] M. Spong, "Adaptive control of flexible joint manipulators: comments on two papers," *Automatica*, 31:585–90, 1995.

[12] M. Spong and M. Vidyasagar, *Robot Dynamics and Control*, Wiley, New York, 1989.

[13] R. Brumbaugh, "An aircraft model for the AIAA controls design challenge," *Journal of Guidance, Control, and Dynamics,* 17:747–52, 1995.

[14] E. Schomig, M. Sznaier, and U. Ly, "Mixed H_2/H_∞ control of multimodel plants," *Journal of Guidance, Control, and Dynamics,* 18:525–31, 1995.

[15] H. Kwakernaak and R. Sivan, *Linear Optimal Control Systems,* Wiley, New York, 1972.

[16] E. Kamen, *Introduction to Signals and Systems,* Macmillan, New York, 1990.

[17] Z. Gajić and X. Shen, *Parallel Algorithms for Optimal Control of Large Scale Linear Systems,* Springer-Verlag, London, 1993.

[18] D. Teneketzis and N. Sandell, "Linear regulator design for stochastic systems by multiple time-scale method," *IEEE Transactions on Automatic Control,* AC-22:615–21, 1977.

[19] R. Dressler and D. Tabak, "Satellite tracking by combined optimal estimation and control techniques," *IEEE Transactions on Automatic Control,* AC-16:833–40, 1971.

[20] Y. Arkun and S. Ramakrishnan, "Bounds of the optimum quadratic cost of structure constrained regulators," *IEEE Transactions on Automatic Control,* AC-28:924–27, 1983.

[21] D. Petkovski, N. Harkara, and Z. Gajić, "Fast suboptimal solution to the static output control problem of linear singularly perturbed systems," *Automatica,* 27:721–24, 1991.

[22] P. Kokotović, J. Allemong, J. Winkelman, and J. Chow, "Singular perturbation and iterative separation of the time scales," *Automatica,* 16:23–33, 1980.

[23] B. Litkouhi, *Sampled-Data Control Systems with Slow and Fast Modes,* Ph.D. dissertation, Michigan State University, 1983.

[24] Z. Gajić and X. Shen, "Study of the discrete singularly perturbed linear quadratic control problem by a bilinear transformation," *Automatica,* 27:1025–28, 1991.

[25] F. Szidarovszky and A. Bahil, *Linear Systems Theory,* CRC Press, Boca Raton, FL, 1992.

8.10 PROBLEMS

8.1. Find the state space form for the electrical circuit whose mathematical model is given in Example 8.1, by taking for the state space variables the input current and output voltage, that is, by choosing $x_1(t) = i_1(t)$, $x_2(t) = e_o(t)$. In addition, take $y(t) = x_2(t)$ and $f(t) = e_i(t)$.

8.2. Use (8.34) to show that for the diagonal matrices

$$\mathbf{A} = \begin{bmatrix} a_{11} & 0 & \cdots & 0 \\ 0 & a_{22} & \ddots & \vdots \\ \vdots & \ddots & \ddots & 0 \\ 0 & \cdots & 0 & a_{nn} \end{bmatrix}$$

444 Chapter 8 State Space Approach

the matrix exponential is given by

$$e^{At} = \exp\left\{\begin{bmatrix} a_{11} & 0 & \cdots & 0 \\ 0 & a_{22} & \ddots & \vdots \\ \vdots & \ddots & \ddots & 0 \\ 0 & \cdots & 0 & a_{nn} \end{bmatrix} t\right\} = \begin{bmatrix} e^{a_{11}t} & 0 & \cdots & 0 \\ 0 & e^{a_{22}t} & \ddots & \vdots \\ \vdots & \ddots & \ddots & 0 \\ 0 & \cdots & 0 & e^{a_{nn}t} \end{bmatrix}$$

8.3. An antenna model [19] is represented by the transfer function

$$H(s) = \frac{K(s+1)}{s^2(s+6)(s+11.5)(s^2+8s+256)}$$

Find state space matrices for the phase variable canonical form.

8.4. A robotic manipulator called the acrobot [12] has the linearized model

$$\mathbf{A} = \begin{bmatrix} 0 & 0 & 1 & 0 \\ 0 & 0 & 0 & 1 \\ 12.49 & -12.54 & 0 & 0 \\ -14.49 & 29.36 & 0 & 0 \end{bmatrix}, \quad \mathbf{B} = \begin{bmatrix} 0 \\ 0 \\ -2.98 \\ 5.98 \end{bmatrix}$$

Assume that the output matrices are given by

$$\mathbf{C} = [1 \ 0 \ 1 \ 0], \quad \mathbf{D} = 0$$

Use MATLAB to find the following.

(a) Eigenvalues and characteristic polynomial.

(b) Modal canonical form.

(c) Open-loop transfer function.

(d) Phase variable canonical form.

8.5. Consider the harmonic oscillator in the state space form

$$\begin{bmatrix} \dot{x}_1(t) \\ \dot{x}_2(t) \end{bmatrix} = \begin{bmatrix} 0 & 1 \\ -1 & 0 \end{bmatrix} \begin{bmatrix} x_1(t) \\ x_2(t) \end{bmatrix} + \begin{bmatrix} 0 \\ 1 \end{bmatrix} f(t), \quad \begin{bmatrix} x_1(0) \\ x_2(0) \end{bmatrix} = \begin{bmatrix} 0 \\ 1 \end{bmatrix}$$

$$y(t) = [0 \ 1] \begin{bmatrix} x_1(t) \\ x_2(t) \end{bmatrix}$$

(a) Find the state transition matrix.

(b) Find the system state and output responses due to a unit step input.

(c) Verify the obtained answer using the MATLAB functions `ss2zp` and `lsim`.

Answers:

(a) $\Phi(t) = \begin{bmatrix} \cos(t) & \sin(t) \\ -\sin(t) & \cos(t) \end{bmatrix}$

(b) $y(t) = \cos(t) + \sin(t)$

8.6. Given the following continuous-time system matrices, find e^{At} by the Cayley–Hamilton and the Laplace transform methods.

(a) $\mathbf{A} = \begin{bmatrix} 0 & 1 \\ -1 & -2 \end{bmatrix}$

(b) $\mathbf{A} = \begin{bmatrix} -3 & 0 \\ 5 & -1 \end{bmatrix}$

(c) $\mathbf{A} = \begin{bmatrix} 0 & 1 & 0 \\ 0 & 0 & 1 \\ -\frac{1}{12} & -\frac{7}{12} & -\frac{4}{3} \end{bmatrix}$

(d) $\mathbf{A} = \begin{bmatrix} \frac{1}{4} & 1 & 0 \\ 0 & \frac{1}{3} & 1 \\ 0 & 0 & \frac{1}{2} \end{bmatrix}$

Answer:

(a) $\Phi(t) = \begin{bmatrix} (1+t)e^{-t} & te^{-t} \\ -te^{-t} & (1-t)e^{-t} \end{bmatrix}$

8.7. Given the following discrete-time system matrices, find $\mathbf{A}^k$ by the Cayley–Hamilton and the $\mathcal{Z}$-transform methods.

(a) $\mathbf{A} = \begin{bmatrix} 0 & 1 \\ 0 & 0.5 \end{bmatrix}$

(b) $\mathbf{A} = \begin{bmatrix} 0 & 0 \\ 2 & 0 \end{bmatrix}$

(c) $\mathbf{A} = \begin{bmatrix} 0 & 1 & 0 \\ 0 & 0 & 1 \\ 0 & -0.5 & -0.5 \end{bmatrix}$

(d) $\mathbf{A} = \begin{bmatrix} 0 & 1 & 0 \\ 0 & 0 & 1 \\ 0 & 0 & 0.5 \end{bmatrix}$

8.8. Consider a continuous-time system with complex conjugate eigenvalues represented by

$$\mathbf{A} = \begin{bmatrix} 0 & 1 \\ -2 & -2 \end{bmatrix}$$

(a) Use (8.119) to determine the required coefficients for the Cayley–Hamilton method for finding the state transition matrix, and show that they are real.

(b) Find the state transition matrix for this system.

8.9. Repeat Problem 8.8 for the discrete-time system matrix

$$\mathbf{A} = \begin{bmatrix} 0 & 1 \\ -0.5 & 1 \end{bmatrix}$$

8.10. For the system

$$\dddot{y}(t) + 2\dot{y}(t) + y(t) = 6\dot{f}(t) + f(t)$$

find the system state space form. Find the system impulse response, and check the answer obtained using the MATLAB function `impulse`.

8.11. Given a discrete system $\mathbf{x}[k+1] = \mathbf{A}\mathbf{x}[k]$, where

$$\mathbf{A} = \begin{bmatrix} 1 & 2 \\ 0 & 3 \end{bmatrix}$$

find its state response due to the initial condition given by $\mathbf{x}[0] = \begin{bmatrix} 1 & 1 \end{bmatrix}^T$.

Answer:

$$\mathbf{A}^k = \begin{bmatrix} 1 & 3^k - 1 \\ 0 & 3^k \end{bmatrix}, \quad \mathbf{x}[k] = \mathbf{A}^k \mathbf{x}[0] = \begin{bmatrix} 3^k \\ 3^k \end{bmatrix}$$

8.12. Consider a linear time-invariant continuous system

$$\dot{\mathbf{x}}(t) = \mathbf{A}\mathbf{x}(t) + \begin{bmatrix} 0 \\ 1 \end{bmatrix} f(t)$$

with

$$\mathbf{x}(1) = \begin{bmatrix} x_1(1) \\ x_2(1) \end{bmatrix} = \begin{bmatrix} 5 \\ 0 \end{bmatrix}, \quad f(t) = \begin{cases} 1, & t \geq 2 \\ 0, & 0 < t < 2 \end{cases}$$

Assuming that the state transition matrix has a known (given) form as

$$\Phi(t - t_0) = \begin{bmatrix} \phi_{11}(t - t_0) & 0 \\ \phi_{21}(t - t_0) & \phi_{22}(t - t_0) \end{bmatrix}$$

find the system state response for any $t \geq 0$.

8.13. Find the impulse response of the system

$$\dddot{y}(t) + 2\dot{y}(t) + 10y(t) = \ddot{f}(t) - 3\dot{f}(t) + 5f(t)$$

Find the system transfer function and the state space form.

8.14. Find the system output response for $t > 1$, due to initial condition at $t = 1$, for

$$\frac{dy(t)}{dt} + 4y(t) + 3\int y(\tau)d\tau = 0, \quad y(1) = 2$$

8.15. Using the state space method, find the output response of the discrete system

$$y[k+2] + y[k] = (-1)^k, \quad y[0] = 1, \quad y[1] = 0$$

Verify the answer using the MATLAB function `dlsim`.

8.16. A continuous-time system is represented by

$$\ddot{y}(t) + 4\dot{y}(t) + 3y(t) = f(t)$$

(a) Find the transfer function and the impulse response.

(b) Compute the output response $y(t)$ for $y(0^-) = -2$, $\dot{y}(0^-) = 1$ and $f(t)$ equal to a unit step function.

Answers:

(a) $H(s) = \mathbf{C}(s\mathbf{I} - \mathbf{A})^{-1}\mathbf{B} = \dfrac{1}{s^2 + 2s + 3}$, $h(t) = \dfrac{1}{2}(e^{-t} - e^{-3t})u(t)$

(b) $\mathbf{x}^T(t) = \left[\frac{1}{3} - 3e^{-t} + \frac{2}{3}e^{-3t} \quad 3e^{-t} - 2e^{-3t}\right]$, $y(t) = x_1(t)$

8.17. Consider a linear continuous-time system with

$$\mathbf{A} = \begin{bmatrix} 0 & 1 \\ -2 & -3 \end{bmatrix}, \quad \mathbf{B} = \begin{bmatrix} 0 \\ -1 \end{bmatrix}, \quad \mathbf{x}(0) = \begin{bmatrix} -1 \\ 0 \end{bmatrix}, \quad \mathbf{C} = \mathbf{I}, \mathbf{D} = 0$$

(a) Find the state transition matrix.

(b) Find the system transfer function.

(c) Find the system state and output responses due to a unit step input.

(d) Verify the obtained answers using MATLAB.

Answers:

(a) $\Phi(t) = \begin{bmatrix} 2e^{-2t} - e^{-t} & e^{-2t} - e^{-t} \\ -2e^{-2t} + 2e^{-t} & -e^{-2t} + 2e^{-t} \end{bmatrix}$

(b) $\mathbf{H}(s) = \begin{bmatrix} \dfrac{-1}{(s+1)(s+2)} \\ \dfrac{-s}{(s+1)(s+2)} \end{bmatrix}$

(c) $\mathbf{x}(t) = \begin{bmatrix} -e^{-2t} - e^{-t} \\ -1 + e^{-2t} \end{bmatrix}$, $\mathbf{y}(t) = \mathbf{x}(t)$

8.18. A discrete system is given by

$$y[k+1] - 0.5y[k] = 2f[k+1] + f[k]$$

Compute the impulse response. Verify the result using the MATLAB function `dimpulse`.

8.19. Discretize the following system using the Euler approximation.

$$\ddot{y}(t) + 2\dot{y}(t) + 3\sin(y(t)) = f(t), \quad y(0) = 1, \quad \dot{y}(0) = 2$$

Answer:

$y((k+2)T) - 2(1-T)y((k+1)T) + (1-2T)y(kT) + 3T^2 \sin(y(kT)) = T^2 f(kT)$,
$y(0) = 1, \quad y(T) = 1 + 2T$

8.20. Given a time-invariant linear system with the impulse response equal to e^{-t}, find the output response of the system due to an input given by $2\delta(t-1) + 3u(t-2)$, where $\delta(t)$

is the impulse delta function and $u(t)$ is a unit step function. What MATLAB functions can be used to solve this problem?

8.21. A linear discrete system is represented by

$$\mathbf{A} = \begin{bmatrix} 0 & 1 \\ -2 & -3 \end{bmatrix}, \quad \mathbf{B} = \begin{bmatrix} 1 \\ 1 \end{bmatrix}, \quad \mathbf{x}[0] = \begin{bmatrix} -1 \\ 0 \end{bmatrix}, \quad \mathbf{C} = [0 \ 1], \quad \mathbf{D} = 0$$

(a) Find its state transition matrix.

(b) Find the transfer function.

(c) Find the system state and output responses due to $f[k] = k$, assuming that $x_1[0] = 1$ and $x_2[0] = 3$.

Answers:

(a) $\Phi[k] = \begin{bmatrix} 2(-1)^k - (-2)^k & (-1)^k - (-2)^k \\ -2(-1)^k + 2(-2)^k & (-1)^k - (-2)^k \end{bmatrix}$

(b) $H(z) = \dfrac{z-2}{(z+1)(z+2)}$

(c) $Y(z) = \left(\dfrac{23}{12}\right)\dfrac{z}{z+1} + \left(\dfrac{76}{9}\right)\dfrac{z}{z+2} - \left(\dfrac{7}{6}\right)\dfrac{z}{z-1} - \left(\dfrac{1}{6}\right)\dfrac{z}{(z-1)^2}$

8.22. Using the state space method, find the output response of the system

$$\ddot{y}(t) + 2\dot{y}(t) + y(t) = \dot{f}(t) + f(t), \quad f(t) = 2e^{-t}, \quad y(0) = 1, \quad \dot{y}(0) = 1$$

8.23. Given the second-order linear system at rest (initial conditions are zero)

$$\ddot{y}(t) + 2\xi\omega_n \dot{y}(t) + \omega_n^2 y(t) = \omega_n^2 f(t)$$

find its unit step response for $\xi < 1$.

8.24. Consider the discrete system

$$y[k+2] - 6y[k+1] + 8y[k] = 3k + 2, \quad y[0] = 1, \quad y[1] = 1$$

(a) Find the system transition matrix.

(b) Find the output response.

Answers:

(a) $\Phi[k] = \begin{bmatrix} 2(2)^k - (4)^k & -0.5(2)^k + 0.5(4)^k \\ 4(2)^k - 4(4)^k & -(2)^k + 2(4)^k \end{bmatrix}$

(b) $Y(z) = \left(\dfrac{1}{4}\right)\dfrac{z}{(z-1)^2} + \dfrac{3z}{z-1} - \left(\dfrac{7}{2}\right)\dfrac{z}{z-2} + \left(\dfrac{3}{2}\right)\dfrac{z}{z-4}$

8.25. Find the output response of the following continuous system using the state space approach.

$$\ddot{y}(t) + 3\dot{y}(t) - 10y(t) = 2\dot{f}(t) + 5f(t), \quad y(0) = 1, \quad \dot{y}(0) = -1, \quad f(t) = t$$

8.26. Discretize the system $dy(t)/dt = f(t)$ using both the Euler and integral approximations. Compare the discrete systems obtained.

8.27. Given a linear continuous system

$$\dot{\mathbf{x}}(t) = \mathbf{A}\mathbf{x}(t)$$

with

$$\mathbf{A} = \begin{bmatrix} 2 & 1 \\ 2 & 3 \end{bmatrix}$$

find the similarity transformation such that this system has the diagonal form in the new coordinates.

8.28. Find the state transition matrix of a continuous system with

$$\mathbf{A} = \begin{bmatrix} 0 & -1 \\ 1 & 0 \end{bmatrix}$$

Use the Taylor series expansion method.

Answer: It is given in the answer to Problem 8.5 (a).

8.29. Using the state space approach, find the output response of a discrete system represented by

$$y[k+2] + 2y[k] + 1 = (-1)^k, \quad y[0] = y[1] = 1$$

8.30. Find the transition matrix in the complex domain for the system represented by

$$\mathbf{A} = \begin{bmatrix} 1 & 0 & 0 \\ 0 & 0 & 0 \\ 1 & 0 & 0 \end{bmatrix}$$

Answer:

$$\Phi(s) = \begin{bmatrix} \dfrac{1}{s-1} & 0 & 0 \\ 0 & \dfrac{1}{s} & 0 \\ \dfrac{1}{s(s-1)} & 0 & \dfrac{1}{s} \end{bmatrix}$$

8.31. Use the formula for differentiation of integrals whose limits depend on parameters (see the corresponding formula in Appendix B), to derive formula (8.60).

8.32. Justify formulas (8.102).

Hint: At some point you must use formula (2.44).

8.33. Show by induction that the characteristic equation (8.104) of a system in the phase variable canonical form is given by (8.106).

8.34. Linearize a scalar system represented by the first-order differential equation

$$\frac{dx(t)}{dt} = x(t)f(t)e^{-f(t)}, \quad x(0) = 0.9$$

450 Chapter 8 State Space Approach

at a nominal point given by $(x_n(t), f_n(t)) = (1, 0)$.
Answer:
$$\Delta \dot{x}(t) = \Delta f(t), \quad \Delta x(0) = -0.1$$

8.35. Consider a nonlinear continuous-time system given by

$$\frac{d^2 x(t)}{dt^2} = -2 \frac{dx(t)}{dt} \cos f(t) - (1 + f(t))x(t) + 1, \quad x(0) = 1.1, \quad \frac{dx(0)}{dt} = 0.1$$

Derive its linearized equation with respect to a nominal point defined by $(x_n(t), f_n(t)) = (1, 0)$. Find the linearized system response due to $\Delta f(t) = e^{-2t}$.
Answer:

$$\begin{bmatrix} \Delta \dot{x}_1(t) \\ \Delta \dot{x}_2(t) \end{bmatrix} = \begin{bmatrix} 0 & 1 \\ -1 & -2 \end{bmatrix} \begin{bmatrix} \Delta x_1(t) \\ \Delta x_2(t) \end{bmatrix} + \begin{bmatrix} 0 \\ 1 \end{bmatrix} \Delta f(t), \quad \begin{bmatrix} \Delta x_1(0) \\ \Delta x_2(0) \end{bmatrix} = \begin{bmatrix} 0.1 \\ 0.1 \end{bmatrix}, \quad y(t) = \Delta x_1(t)$$

8.36. For a nonlinear system

$$\frac{d^2 x(t)}{dt^2} + 2 \frac{dx(t)}{dt} f(t) + (1 - f(t))x(t) = f^2(t) + 1, \quad x(0) = 0, \quad \frac{dx(0)}{dt} = 1$$

find the nominal system response on the nominal system trajectory defined by $f_n(t) = 1$, subject to $x_n(0) = 0$ and $dx_n(0)/dt = 1.1$. Find the linearized state space equation and its initial conditions.

8.37. The mathematical model of a simple pendulum [16] is given by

$$I \frac{d^2 \theta(t)}{dt^2} + mgl \sin \theta(t) = lf(t), \quad \theta(t_0) = \theta_0, \quad \dot{\theta}(t_0) = \omega_0$$

where I is the moment of inertia, l, m are pendulum length and mass, respectively, and $f(t)$ is an external tangential force, and g is the gravitational constant. Assume that $\theta_n(t) = 0$, $f_n(t) = 0$, $\theta_n(t_0) = 0$, $\dot{\theta}_n(t_0) = 0$, and $\theta_0, \dot{\theta}_0$ are small. Find the linearized equation for this pendulum using formulas (8.135). Determine the initial conditions.
Answer:

$$\Delta \ddot{\theta}(t) - \frac{mgl}{I} \Delta \theta(t) = \frac{l}{I} \Delta f(t), \quad \Delta \theta(t_0) = \theta_0, \quad \Delta \dot{\theta}(t_0) = \omega_0$$

8.38. Linearize the following system at a nominal point $(x_{1n}, x_{2n}, x_{3n}) = (0, 1, 1)$.

$$\dot{x}_1(t) = x_1(t)x_2(t) - \sin x_1(t)$$
$$\dot{x}_2(t) = 1 - 3x_2(t)e^{-x_1}(t)$$
$$\dot{x}_3(t) = x_1(t)x_2(t)x_3(t)$$

Answer:

$$\begin{bmatrix} \Delta \dot{x}_1(t) \\ \Delta \dot{x}_2(t) \\ \Delta \dot{x}_3(t) \end{bmatrix} = \begin{bmatrix} 0 & 0 & 0 \\ 3 & -3 & 0 \\ 1 & 0 & 0 \end{bmatrix} \begin{bmatrix} \Delta x_1(t) \\ \Delta x_2(t) \\ \Delta x_3(t) \end{bmatrix}$$

8.39. Linearize a nonlinear control system represented by

$$\dot{x}_1(t) = f(t)\ln x_1(t) + x_2(t)e^{-f(t)}$$
$$\dot{x}_2(t) = x_1(t)\sin f(t) - \sin x_2(t)$$
$$y(t) = \sin x_1(t)$$

Assume that x_{1n}, x_{2n}, and f_n are known.

8.40. Linearize the Volterra predator–prey mathematical model

$$\dot{x}_1(t) = -x_1(t) + x_1(t)x_2(t)$$
$$\dot{x}_2(t) = x_2(t) - x_1(t)x_2(t)$$

at a nominal point given by $(x_{1n}, x_{2n}) = (0, 0)$.

8.41. A linearized model of a single-link manipulator with a flexible joint [11] is given by

$$J_l\ddot{\theta}_l(t) + B_l\dot{\theta}_l(t) + k(\theta_l(t) - \theta_m(t)) = 0$$
$$J_m\ddot{\theta}_m(t) + B_m\dot{\theta}_m(t) - k(\theta_l(t) - \theta_m(t)) = f(t)$$

where J_l, J_m are moments of inertia, B_l, B_m are damping factors, k is the spring constant, $f(t)$ is the input torque, and $\theta_m(t)$, $\theta_l(t)$ are angular positions. Write the state space form for this manipulator by taking the following change of variables: $x_1(t) = \theta_l(t)$, $x_2(t) = \dot{\theta}_l(t)$, $x_3(t) = \theta_m(t)$, $x_4(t) = \dot{\theta}_m(t)$.

Answer:

$$\begin{bmatrix}\dot{x}_1(t)\\\dot{x}_2(t)\\\dot{x}_3(t)\\\dot{x}_4(t)\end{bmatrix} = \begin{bmatrix}0 & 1 & 0 & 0\\-\dfrac{k}{J_l} & -\dfrac{B_l}{J_l} & \dfrac{k}{J_l} & 0\\0 & 0 & 0 & 1\\\dfrac{k}{J_m} & 0 & -\dfrac{k}{J_m} & -\dfrac{B_m}{J_m}\end{bmatrix}\begin{bmatrix}x_1(t)\\x_2(t)\\x_3(t)\\x_4(t)\end{bmatrix} + \begin{bmatrix}0\\0\\0\\1\end{bmatrix}f(t)$$

8.42. Consider a fifth-order industrial reactor model [20, 21] represented by

$$\mathbf{A} = \begin{bmatrix}-16.11 & -0.39 & 27.2 & 0 & 0\\0.01 & -16.99 & 0 & 0 & 12.47\\15.11 & 0 & -53.6 & -16.57 & 71.78\\-53.36 & 0 & 0 & -107.2 & 232.11\\2.27 & 60.1 & 0 & 2.273 & -102.99\end{bmatrix}$$

$$\mathbf{B} = \begin{bmatrix}11.12 & -2.61 & -21.91 & -53.5 & 69.1\\-12.6 & 3.36 & 0 & 0 & 0\end{bmatrix}^T$$

$$\mathbf{C} = \begin{bmatrix}0 & 0 & 0 & 0 & 1\\0 & 1 & 1 & 0 & 0\end{bmatrix}$$

Chapter 8 State Space Approach

Using MATLAB, find the following.

(a) The system transfer function.

(b) The impulse response.

(c) The output response due to inputs $f_1(t) = e^{-t} + \sin(t)$ and $f_2(t) = 0$.

8.43. Discretize the system given in Problem 8.42 using the MATLAB function c2d with $T = 0.1$, and repeat steps (a), (b), and (c).

8.44. The model of a synchronous machine connected to an infinite bus [22] has the system matrix

$$\mathbf{A} = \begin{bmatrix} -0.58 & 0 & 0 & -0.27 & 0 & 0.2 & 0 \\ 0 & -1 & 0 & 0 & 0 & 1 & 0 \\ 0 & 0 & -5 & 2.1 & 0 & 0 & 0 \\ 0 & 0 & 0 & 0 & 337 & 0 & 0 \\ -0.14 & 0 & 0.14 & -0.2 & -0.28 & 0 & 0 \\ 0 & 0 & 0 & 0 & 0 & 0.08 & 2 \\ -17.2 & 66.7 & -11.6 & 40.9 & 0 & -66.7 & -16.7 \end{bmatrix}$$

(a) Find the eigenvalues, eigenvectors, and similarity transformation that puts this system into a diagonal form.

(b) Discretize this system with $T = 1$.

(c) Find the system response of the discrete system obtained in part (b) due to the initial condition $\mathbf{x}[0] = [1 \ 1 \ 1 \ 1 \ 1 \ 1 \ 1]^T$, and draw the corresponding response for the time interval $0 \le k \le 10$. Use MATLAB.

8.45. A linearized mathematical model of an aircraft considered in [23] and [24] has the form

$$\mathbf{A} = \begin{bmatrix} -0.015 & -0.0805 & -0.0011666 & 0 \\ 0 & 0 & 0 & 0.03333 \\ -2.28 & 0 & -0.84 & 1 \\ 0.6 & 0 & -4.8 & -0.49 \end{bmatrix}$$

$$\mathbf{B} = \begin{bmatrix} -0.0000916 & 0.0007416 \\ 0 & 0 \\ -0.11 & 0 \\ -8.7 & 0 \end{bmatrix}$$

Obtain the following (using MATLAB).

(a) Discretize this model with $T = 1$.

(b) Find its response due to a unit ramp input.

(c) Find the system transfer function and the system poles.

8.46. Dynamics of eye movement can be modeled by a sixth-order linear continuous-time system whose state space form [25] is represented by the matrices

$$\mathbf{A} = \begin{bmatrix} 0 & 0 & 0 & 1 & 0 & 0 \\ 35.79 & -52.97 & 0 & 0 & 0.29 & 0 \\ 75.41 & 0 & -111.61 & 0 & 0 & -0.60 \\ -120000 & 56818 & 56811 & -1409 & 0 & 0 \\ 0 & 0 & 0 & 0 & -103 & 0 \\ 0 & 0 & 0 & 0 & 0 & -526 \end{bmatrix}$$

$$\mathbf{B} = \begin{bmatrix} 0 & 0 & 0 & 0 & 103 & 0 \\ 0 & 0 & 0 & 0 & 0 & 526 \end{bmatrix}^T$$

In this model, the state variables $x_1(t)$ and $x_4(t)$ represent, respectively, the position of the eye and eye velocity. (Physical meanings of the remaining state space variables and input signals are discussed in [25].) The initial conditions defined in [25] are $x_1(0) = x_4(0) = 0$, $x_2(0) = x_3(0) = 1.1$, and $x_5(0) = x_6(0) = 0.2$. We are particularly interested in the state variables $x_1(t)$ and $x_4(t)$, so that the system output can be defined by the relation

$$y(t) = [1\ 0\ 0\ 1\ 0\ 0]\mathbf{x}(t) = \mathbf{Cx}(t)$$

Using MATLAB, evaluate the following quantities.

(a) The eigenvalues, eigenvectors, and similarity transformation that puts this system into a diagonal form. Write down the system state space diagonal form.

(b) The impulse response.

(c) The step response for the state variables $x_1(t)$ and $x_4(t)$.

(d) The zero-input output response.

Hint: You may use either the MATLAB function `initial` or the appropriately modified MATLAB program from Section 8.7.2.

(e) The complete response for $x_1(t)$ and $x_4(t)$ due to the given initial conditions, and inputs $f_1(t) = u(t) + 2u(t-1) - u(t-2)$ and $f_2(t) = u(t) - 0.5u(t-1) + 0.25u(t-2)$.

8.47. Consider the ATM computer communication network switch model introduced in Section 1.3. The queue length, $q[k]$, and the packet arrival rate, $y[k]$, of the available bit rate traffic, can be modeled by the following set of difference equations,

$$q[k+1] = q[k] + q[k+1-d] - f[k]$$

$$y[k+1] = y[k] - \sum_{j=0}^{1} \alpha_j(q[k-j] - q^0) - \sum_{i=0}^{d} \beta_i y[k-i]$$

where $f[k]$ is the service rate of the buffer, q^0 is the desired buffer queue length at steady state, d is the round trip delay between the source and the switch, and α_j and β_i are the gains to be chosen by a control (network) engineer (designer). In addition, the following relationships exist:

$$\sum_{i=0}^{d} \beta_i = 0, \text{ and } \sum_{j=0}^{1} \alpha_j > 0$$

(a) Introduce the change of variables

$$\mathbf{x}[k] = \begin{bmatrix} q[k] - q^0 \\ q[k-1] - q^0 \\ y[k] - y[k-d] \\ y[k-1] - y[k-d] \\ \vdots \\ y[k-d+1] - y[k-d] \end{bmatrix}$$

and show that the discrete-time system can be put in the standard discrete-time state space form

$$\mathbf{x}[k+1] = \mathbf{A}\mathbf{x}[k] + \mathbf{B}f[k]$$

where

$$f[k] = -\alpha_0 x_1[k] - \alpha_1 x_2[k] - \beta_0 x_3[k] - \beta_1 x_4[k] - \cdots - \beta_{d-1} x_{d+2}[k]$$
$$= -\mathbf{F}\mathbf{x}(k)$$

$$\mathbf{F}^{1 \times (d+2)} = [\alpha_0 \quad \alpha_1 \quad \beta_0 \quad \beta_1 \quad \cdots \quad \beta_{d-1}]$$

with the system state space matrices given by

$$\mathbf{A}^{(d+2) \times (d+2)} = \begin{bmatrix} 2 & -1 & 0 & \cdots & & 0 & 1 \\ 1 & 0 & 0 & \cdots & & \cdots & 0 \\ 0 & 0 & 1 & 0 & \cdots & 0 & -1 \\ \vdots & \vdots & 1 & 0 & \ddots & \vdots & -1 \\ \vdots & \vdots & 0 & 1 & \ddots & 0 & \vdots \\ \vdots & \vdots & \vdots & \ddots & \ddots & 0 & -1 \\ 0 & 0 & 0 & \cdots & 0 & 1 & -1 \end{bmatrix}$$

$$\mathbf{B}^{(d+2) \times 1} = [0 \quad 0 \quad 1 \quad 0 \quad \cdots \quad \cdots \quad 0]^T$$

(b) Using the preceding relations, establish a simple formula for the dynamics of the available bit rate as

$$y[k+1] = y[k] + f[k]$$

(c) In Section 5.3.4, using realistic data, we showed that the round-trip time delay is equal to $d = 12$. Assume that the initial conditions for this state space model are given by $x_1[0] = 10$, $x_2[0] = 10$, and $x_3[0] = x_4[0] = \cdots = x_{14}[0] = 0$. Guess some values for the gains α_j and β_i and simulate the response of this state space model using MATLAB and/or Simulink.

Hint: The buffer queue length $q[k]$ will go to the desired queue length q^0 (equivalently, $x_1[k]$ will tend to zero) only if the matrix $\mathbf{A} - \mathbf{BF}$ is asymptotically stable (has all eigenvalues within the unit circle). Choosing the gains α_j and β_i such that this condition is satisfied ($|\rho(\mathbf{A} - \mathbf{BF})| < 1$) is, in general, a difficult problem for higher-order matrices. However, it can be solved using the MATLAB function place as F=place(A,B,p), where **p** is the column vector of dimension 14 that specifies the desired locations of the eigenvalues of the matrix $\mathbf{A} - \mathbf{BF}$ within the unit circle. Hence, choose $p_i, i = 1, 2, \ldots, 14$ such that $|p_i| < 1$, and use the preceding MATLAB statement to find the vector **F**. Note that the eigenvalues are either real or complex conjugate pairs.

PART THREE

LINEAR SYSTEMS IN ELECTRICAL ENGINEERING

CHAPTER 9

Signals in Digital Signal Processing

Some digital signal processing concepts, notions, and elements have already been introduced in this book: discrete-time signals (Chapter 2), one-sided $\mathcal{Z}$-transform and its application to discrete-time linear systems (Chapter 5), discrete-time convolution (Sections 6.3 and 6.4), discrete-time linear systems described by linear constant coefficient difference equations (Chapter 7), and discrete-time state space form (Section 8.3).

In this chapter, we begin by complementing these topics with rigorous derivations of the famous *sampling theorem* and by justifying the expression for the Nyquist sampling frequency, in Section 9.1. The Fourier transform of discrete-time signals, known as the *discrete-time Fourier transform* (DTFT) is derived in Section 9.1.1 as the *Fourier transform of a continuous-time signal sampled by an ideal sampler*.

In Section 9.2, we present the main properties of the DTFT. The discrete-time Fourier transform is a special form of the more general *two-sided $\mathcal{Z}$-transform*, whose specific features (similar to those considered for the one-sided $\mathcal{Z}$-transform) are considered in detail in Section 9.3. The properties and common pairs for the double-sided $\mathcal{Z}$-transform are presented similarly to those of the one-sided $\mathcal{Z}$-transform. In Section 9.4, we present the discrete Fourier transform (DFT) as a technique for numerical evaluation of the DTFT. The most efficient numerical algorithm for computation of the DFT, known as the *fast Fourier transform* (FFT), is discussed in Section 9.4.1. In Section 9.5, we introduce the *discrete-time Fourier series* (DFS) from the DFT and indicate how to use the DFS for finding the zero-state response of discrete-time linear systems due to periodic inputs.

Section 9.6 presents fundamentals of a signal operation known as discrete-time signal correlation, an operation that has some similarities to discrete-time convolution. Signal correlation is used to measure the power (energy) of received signals, and for signal estimation from noisy measurements. Signal filtering is one of the most fundamental operations of digital signal processing. Two types of digital filters—*finite impulse response* (FIR) and *infinite impulse response* (IIR)—are introduced in Section 9.7. Section 9.8 represents a laboratory experiment on digital signal processing.

9.1 SAMPLING THEOREM

In the introductory section of this book, Section 1.1.1, we introduced the concept of continuous-time signal sampling, which is graphically represented in Figure 1.3. In this section, we will further clarify the sampling operation, and establish conditions under which a continuous-time signal represented in terms of its sample values can be recovered uniquely from its discrete-time samples, which basically constitutes the sampling theorem.

In the sampling procedure, a continuous-time signal $x(t)$ is represented by a sequence of numbers (samples) that represent the signal values at particular time instants. The sampling process is performed at equidistant time instants, so that the sampling period T_s (the time between two adjacent samples) is constant. The samples of the continuous-time signal $x(t)$ are defined by $x(kT_s)$, $k = 0, \pm1, \pm2, \ldots$. It should be pointed out that "the sampling process represents a very drastic chopping operation on the original signal $x(t)$, and therefore, it will introduce a lot of spurious *high-frequency* components into the frequency spectrum" [1].

Sampling is a mandatory step in preparing signal data for digital computer processing. However, since real physical systems operate in continuous time, at some point we must recover the continuous-time signal from its discrete-time version (from its sample values). We would like to perform this operation as accurately as possible and without generating redundant data that will cause unnecessary computational burden. To that end, a natural question to be asked is: *What is the minimal value for the sampling period T_s such that the original signal can be uniquely (at all time instants) reproduced from its discrete-time values?* Another more general question would be: *Can all continuous-time signals be discretized such that a meaningful (with tolerable errors) recovery procedure can be performed?* The answers to these two fundamental questions are given in the sampling theorem.

Before we state the sampling theorem, we must introduce the formal definition of bandlimited signals. (Note that we have tacitly used this definition in previous chapters of this book.)

> **DEFINITION 9.1:** A signal is *bandlimited* if its magnitude spectrum is equal to zero for all frequencies greater than $\omega_{\max}$. The frequency $f_{\max} = \omega_{\max}/2\pi$ is called the signal bandwidth frequency.

Similarly we have time-limited signals. We dealt with time-limited signals in Chapter 6, where we gave a formal definition. For completeness, we restate the definition of time-limited signals here.

> **DEFINITION 9.2:** A signal is *time-limited* if it is different from zero only in a finite time interval, say $[t_{\min}, t_{\max}]$.

We observed in Chapter 3, while finding the spectrum (Fourier transform) of continuous-time signals, that some time-limited signals (like the impulse delta signal, and rectangular and triangular pulses) have infinite spectrum; and that signals with bandlimited spectrum have infinite duration in the time domain (which in fact follows from the duality property of the Fourier transform). It can be shown in general that *a signal cannot be both time-limited and bandlimited*. The proof of this interesting signal property is outside of the scope of this book.

Now we state the sampling theorem.

Section 9.1 Sampling Theorem

THEOREM 9.1. SAMPLING THEOREM A continuous-time bandlimited signal $x(t)$ with bandwidth frequency $f_{\max}$ can be uniquely reconstructed from its sample values $x(kT_s)$, $k = 0, \pm 1, \pm 2, \ldots$, if the sampling frequency $f_s = 1/T_s$ satisfies

$$f_s = \frac{1}{T_s} \geq 2 f_{\max} \qquad (9.1)$$

The frequency $2f_{\max}$ is called the Nyquist frequency, and the frequency interval $[-f_{\max} \quad f_{\max}]$ is called the Nyquist interval. $\square$

Note that the sampling theorem is often called Shannon's sampling theorem in honor of his celebrated paper published in 1949 [2]. However, the main results stated in the sampling theorem were known in mathematics for many years before Shannon's paper was published. In fact, the main result of the sampling theorem can be deduced from the 1928 paper published by Nyquist [3]. For that reason, the frequency $2f_{\max}$ is called the Nyquist frequency.

In the following, we derive the relationship between a continuous-time signal $x(t)$ and its sample values (which constitutes the proof of the sampling theorem), and draw the conclusions stated in the sampling theorem. The proof is easy and does not go beyond basic Fourier analysis. *Since the proof is somewhat lengthy, in the interest of time, students and/or instructors can skip the proof and go to the proof comments.*

Proof of the Sampling Theorem

Consider a bandlimited signal $x(t) \leftrightarrow X(j\omega)$ with bandwidth frequency $f_{\max} = \omega_{\max}/2\pi$, whose spectrum is presented in Figure 9.1, and form a periodic frequency domain signal by replicating the magnitude spectrum of the original bandlimited signal as presented in Figure 9.2. (For simplicity of the spectrum plots only, we assume that the time domain signal is real and even, so that its spectrum is also real and even; see Section 3.3.2).

Denote the obtained signal by $X_p(j\omega)$. It follows that $X(j\omega) = X_p(j\omega)$ for $|\omega| \leq \omega_{\max}$. For the signal $x(t)$, we have the basic Fourier transform pair relations, that is,

$$x(t) = \frac{1}{2\pi} \int_{-\infty}^{\infty} X(j\omega) e^{j\omega t} \, d\omega, \quad X(j\omega) = \int_{-\infty}^{\infty} x(t) e^{-j\omega t} \, dt \qquad (9.2)$$

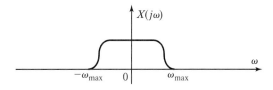

FIGURE 9.1: Magnitude spectrum for a bandlimited signal

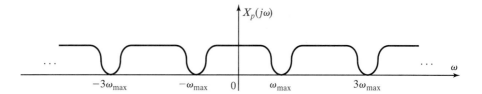

FIGURE 9.2: A periodic frequency domain signal magnitude spectrum, obtained by replicating the original bandlimited signal magnitude spectrum

The frequency domain $2\omega_{max}$-periodic signal $X_p(j\omega)$ can be represented via the Fourier series as follows

$$X_p(j\omega) = X_p(j(\omega + 2\omega_{max})) = \sum_{k=-\infty}^{\infty} X_k(jkT_0)e^{jkT_0\omega}, \quad (9.3)$$

$$T_0 = \frac{2\pi}{2\omega_{max}} = \frac{1}{2f_{max}}, \quad k = 0, \pm 1, \pm 2, \ldots$$

with

$$X_k(jkT_0) = \frac{1}{2\omega_{max}} \int_{-\omega_{max}}^{\omega_{max}} X_p(j\omega)e^{-jkT_0\omega}\, d\omega$$

$$= X_k\left(j\frac{k}{2f_{max}}\right) = \frac{1}{2\omega_{max}} \int_{-\omega_{max}}^{\omega_{max}} X(j\omega)e^{-jk\left(\frac{1}{2f_{max}}\right)\omega}\, d\omega \quad (9.4)$$

Note that the frequency domain Fourier series representation defined in (9.3–4) is completely analogous to the time domain Fourier series representation considered in Section 3.1 and its formulas (3.1), (3.11), and (3.14). (To see this, we have only to replace t by ω and ω_0 by T_0.)

From (9.2) for $t = -k/(2f_{max})$, $k = 0, \pm 1, \pm 2, \ldots$, we have an interesting relation between the original signal sample values and the Fourier series coefficient defined in (9.4):

$$x\left(-\frac{k}{2f_m}\right) = \frac{1}{2\pi}\int_{-\infty}^{\infty} X(j\omega)e^{-jk\left(\frac{1}{2f_{max}}\right)\omega}\, d\omega = 2f_{max}X_k\left(j\frac{k}{2f_{max}}\right)$$

Hence, the Fourier series coefficients for the frequency domain periodic function obtained by replicating the frequency spectrum of the original signal are related to the signal samples by a very simple formula:

$$X_k\left(j\frac{k}{2f_{max}}\right) = \frac{1}{2f_{max}}x\left(-\frac{k}{2f_{max}}\right), \quad k = 0, \pm 1, \pm 2, \ldots \quad (9.5)$$

Using (9.5) in (9.3), we obtain

$$X_p(j\omega) = \sum_{k=-\infty}^{\infty} \frac{1}{2f_{max}} x\left(-\frac{k}{2f_{max}}\right) e^{jk\left(\frac{1}{2f_{max}}\right)\omega} \quad (9.6)$$

Now we are ready to establish a relation between the original signal $x(t)$ and its sample values. From (9.2), (9.6), and the fact that $X(j\omega) = X_p(j\omega)$ for $|\omega| \leq \omega_{max}$, we have

$$x(t) = \frac{1}{2\pi} \int_{-\omega_{max}}^{\omega_{max}} X(j\omega) e^{j\omega t} d\omega$$

$$= \frac{1}{2\pi} \int_{-\omega_{max}}^{\omega_{max}} \sum_{k=-\infty}^{\infty} \frac{1}{2f_{max}} x\left(-\frac{k}{2f_{max}}\right) e^{jk\left(\frac{1}{2f_{max}}\right)\omega} e^{j\omega t} d\omega$$

Interchanging the order of summation and integration, and factoring out the constant terms in front of the integral and the terms that are not functions of ω, the last formula produces

$$x(t) = \frac{1}{4\pi f_{max}} \sum_{k=-\infty}^{\infty} x\left(-\frac{k}{2f_{max}}\right) \int_{-\omega_{max}}^{\omega_{max}} e^{j\omega\left(t+\left(\frac{k}{2f_{max}}\right)\right)} d\omega$$

Evaluating this integral, the following relationship is obtained:

$$x(t) = \sum_{k=-\infty}^{\infty} x\left(\frac{k}{2f_{max}}\right) \frac{\sin\left(\omega_{max}\left(t - \frac{k}{2f_{max}}\right)\right)}{\omega_{max}\left(t - \frac{k}{2f_{max}}\right)}$$

$$= \sum_{k=-\infty}^{\infty} x\left(\frac{k}{2f_{max}}\right) \text{sinc}(2f_{max}t - k) \quad (9.7)$$

This relation establishes the fact that the bandlimited signal $x(t)$ can be uniquely represented in terms of its sample values, and provides the proof of the sampling theorem.

Proof Comments

Formula (9.7) shows that it is possible to uniquely determine the continuous-time signal values at any continuous-time instant in terms of the sample values, which represents the essence of the sampling theorem. The derivations have been done for a bandlimited signal. The proof requires that the replicated frequency domain signal $X_p(j\omega)$ presented in Figure 9.2 be such that $X(j\omega) = X_p(j\omega)$ for $|\omega| \leq \omega_{max}$, that is,

$$X(j\omega) = X_p(j\omega) = \sum_{k=-\infty}^{\infty} \frac{1}{2f_{max}} x\left(\frac{k}{2f_{max}}\right) e^{-jk\omega\left(\frac{1}{2f_{max}}\right)}, \quad |\omega| \leq \omega_{max} \quad (9.8)$$

Since the replicated $X_p(j\omega)$ must be periodic, this is possible only under the assumption that replicated signals do not overlap; that is, when $T_0 \geq 1/2f_{max}$. Using the fact that the signal samples are defined as $x(-k/(2f_{max}))$, we conclude that $T_s = T_0 \geq 1/(2f_{max})$. In practice, the sampling frequency is slightly increased, leading to $f_s = 1/T_s > 2f_{max}$.

Important Observations

It can be concluded from the sampling theorem that *signal sampling makes the signal frequency spectrum periodic*. This means that due to sampling every frequency component of the original bandlimited signal, say ω, is replicated infinitely many times at high frequencies as $\omega \pm k\omega_s, k = \pm 1, \pm 2, \ldots$. Note that we observed the corresponding phenomenon in Section 3.1 regarding the Fourier series: *signal periodicity makes the signal frequency spectrum discrete (line spectra)*, which implies that sampling of the frequency spectrum corresponds to periodicity in the time domain. A nice story about an interplay of sampling and periodicity in both the time and frequency domains can be found in a popular article by Deller [4].

Aliasing

In the case when the conditions of the sampling theorem are not satisfied (bandlimited signal sampled with the sampling frequency $f_s = 1/T_s \geq 2f_{max}$) high-frequency components of the replicated baseband spectrum can overlap with the original baseband spectrum and cause problems in recovery of the original signal. The aliasing phenomenon of a bandlimited signal, for which the condition $f_s \geq 2f_{max}$ is not satisfied, is demonstrated in Figure 9.3. The sampling theorem condition $f_s > 2f_{max}$ ($\omega_s > 2\omega_{max}$) apparently eliminates aliasing of frequencies by providing a frequency guard band between the original frequency spectrum and its replicas introduced by the sampling process.

In practice, many signals are not bandlimited and contain high-frequency components that do not carry important information. Such signals can be made bandlimited by using an analog prefilter, as demonstrated in Figure 9.4.

Such filters are called *antialiasing filters,* since they either drastically reduce or completely eliminate the aliasing phenomenon (when $\omega_s \geq \omega_{max}$, and assuming ideal low-pass filtering).

Analog signal prefiltering also reduces the requirement on the sampling frequency imposed by the sampling theorem, which is important when the hardware used in the sampling process imposes limitations on the upper value of the sampling frequency.

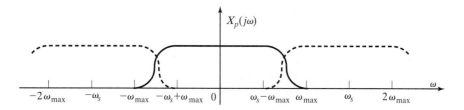

FIGURE 9.3: Aliasing phenomenon: baseband signal spectrum (solid line) and spectrum replicas (dashed lines)

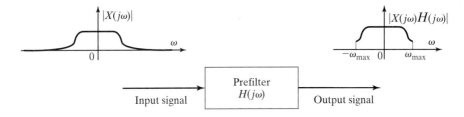

FIGURE 9.4: Analog signal prefiltering to avoid signal aliasing

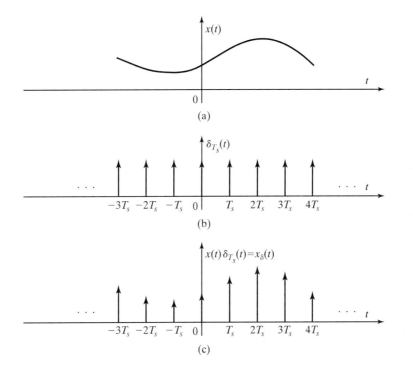

FIGURE 9.5: A continuous-time signal sampled by an ideal sampler

In the following, we consider ideal and practical sampling techniques. The ideal method is primarily used to derive the discrete-time Fourier transform and draw some conclusions. The practical sampling procedure indicates what actually can be done in practice.

9.1.1 Sampling with an Ideal Sampler and the DTFT

In this section we present the sampling operation using an ideal sampler and derive the discrete-time Fourier transform (DTFT). Properties of the DTFT are discussed in Section 9.2.

Consider a continuous-time signal as presented in Figure 9.5(a). This signal is sampled by an ideal sampler, as shown in Figure 9.5(b). The ideal sampler can be mathematically represented by a periodic train of impulse delta signals,

$$\delta_{T_s}(t) = \sum_{k=-\infty}^{\infty} \delta(t - kT_s) \tag{9.9}$$

where T_s is the sampling period. Hence, the signal sampled by an ideal sampler is given by

$$x_\delta(t) = x(t)\delta_{T_s}(t) \tag{9.10}$$

The signal $x_\delta(t)$ is called the *ideal sampled signal,* and the signal $\delta_{T_s}(t)$ is called the *Dirac comb.*

Since $\delta_{T_s}(t)$ is a periodic signal, it can be represented by the Fourier series. It was shown in Example 3.11 that its complex Fourier series coefficients are equal to $1/T_s$ for every k, so that

$$\delta_{T_s}(t) = \sum_{k=-\infty}^{\infty} \delta(t - kT_s) = \frac{1}{T_s}\sum_{k=-\infty}^{\infty} e^{-jk\omega_s t}, \quad \omega_s = \frac{2\pi}{T_s} \tag{9.11}$$

Using the relationship between the complex Fourier series coefficients and the trigonometric Fourier series coefficients, as given in (3.9), we have

$$a_0 = \frac{2}{T_s}, \quad a_k = \frac{2}{T_s}, \quad b_k = 0, \quad k = 1, 2, \ldots \tag{9.12}$$

so that the trigonometric form of the Fourier series for the train of impulse delta signals is given by

$$\delta_{T_s}(t) = \frac{1}{T_s} + \frac{2}{T_s}\sum_{k=1}^{\infty} \cos(n\omega_s t) \tag{9.13}$$

The ideal sampled signal (9.10) can be represented as

$$\begin{aligned}x_\delta(t) &= x(t)\delta_{T_s}(t) \\ &= \cdots + x(-2T_s)\delta(t + 2T_s) + x(-T_s)\delta(t + T_s) + x(0)\delta(t) \\ &\quad + x(T_s)\delta(t - T_s) + x(2T_s)\delta(t - 2T_s) + \cdots \\ &= \sum_{k=-\infty}^{\infty} x(kT_s)\delta(t - kT_s)\end{aligned} \tag{9.14}$$

Since the Fourier transform of the shifted impulse delta signal is given by $e^{-jkT_s\omega} \leftrightarrow \delta(t - kT_s)$, it follows that the Fourier transform of the ideal sampled signal is

$$\mathcal{F}\{x_\delta(t)\} = X_\delta(j\omega) = \sum_{k=-\infty}^{\infty} x(kT_s)e^{-jkT_s\omega} \tag{9.15}$$

This formula, in fact, defines the *discrete-time Fourier transform*. We will study the DTFT in more detail in the next section.

On the other hand, by using the trigonometric form of the train of impulse delta signals as given in (9.13), it can be easily shown that

$$\mathcal{F}\{x_\delta(t)\} = \mathcal{F}\{x(t)\delta_{T_s}(t)\}$$

$$= \frac{1}{T_s}\mathcal{F}\{x(t)[1 + 2\cos(\omega_s t) + 2\cos(2\omega_s t) + 2\cos(3\omega_s t) + \cdots]\} \quad (9.16)$$

$$= X_\delta(j\omega) = \frac{1}{T_s} \sum_{k=-\infty}^{\infty} X(j(\omega - k\omega_s)), \quad \omega_s = \frac{2\pi}{T_s}$$

where the Fourier transform modulation property has been used in deriving the last expression. Also, the frequency shift property of the Fourier transform applied to (9.15) directly implies (9.16). Formula (9.16) indicates the expected result that the *frequency spectrum of the ideal sampled signal is periodic*. In addition, it relates the frequency spectrum of the original continuous-time signal to the frequency spectrum of the ideal sampled (discrete-time) signal, that is,

$$X(j\omega) = T_s X_\delta(j\omega), \quad |\omega| \leq \omega_{\max} \quad (9.17)$$

9.1.2 Sampling with a Physically Realizable Sampler

Using the train of impulse delta signals defined in (9.9) to perform sampling as defined in (9.10) is not practically realizable. Instead of the train of impulse delta signals, we can practically use any periodic train of narrow pulses—such as a train of rectangular (or triangular) pulses with a very narrow width, whose waveform is presented in Figure 9.6.

For simplicity, we assume that the rectangular pulse height is equal to 1, but any value can be used for the pulse height. For example, we can use the pulse height equal to $1/\tau$ so that the pulse area is equal to 1 (as in the case of the impulse delta signal). Also, for simplicity, we assume that the train of pulses is an even function, so that its Fourier series expansion contains only harmonics corresponding to the cosine terms. Due to the periodicity of a rectangular pulse train, it can be represented by the Fourier series, whose

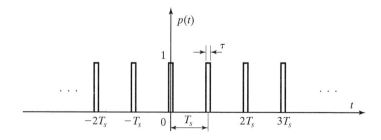

FIGURE 9.6: A train of rectangular pulses with a narrow width, used for practical sampling

trigonometric form with complex coefficients, (3.10), is given by

$$p(t) = \sum_{k=-\infty}^{\infty} p_\tau(t - kT_s)$$

$$= P_0 + 2\sum_{k=1}^{\infty} |P_k(k\omega_s)| \cos(\omega_s t + \angle P_k(k\omega_s)), \quad \omega_s = \frac{2\pi}{T_s} \tag{9.18}$$

(Note that at this point and in the following derivations we can also use the trigonometric form of the Fourier series with real coefficients, due to a nice form of the train of rectangular pulses.) The complex Fourier series coefficients introduced in (9.18) are obtained from

$$P_0 = \frac{1}{2T_s} \int_{-T_s/2}^{T_s/2} p_\tau(t)\, dt = \frac{a_0}{2} = \frac{\tau}{2T_s} = \text{const}$$

$$P_k(k\omega_s) = \frac{1}{T_s} \int_{-T_s/2}^{T_s/2} p_\tau(t)e^{-jk\omega_s t}\, dt = \frac{1}{2}(a_k - jb_k), \quad k = 1, 2, \ldots \tag{9.19}$$

where a_k, b_k are trigonometric Fourier series coefficients as defined in (3.3). The assumption that the pulse is an even function implies that $b_k = 0$, $\forall k$, so that $P_k(k\omega_s)$ are real for every k, which further implies that $\angle P_k(k\omega_s) = 0$, $\forall k$. It should be emphasized that this condition is not crucial for the considered sampling procedure. It simplifies the derivations, but the main conclusions will remain the same if we use trains of pulses that are neither even nor odd functions.

The sampled signal is now given by

$$x_s(t) = x(t)p(t) = x(t) \times \text{const} + x(t) \times 2\sum_{k=1}^{\infty} |P_k(k\omega_s)| \cos(\omega_s t) \tag{9.20}$$

Note that this constant can be made equal to 1 by choosing the pulse height as $2T_s/\tau$. The Fourier transform of this sampled signal is given by

$$\mathcal{F}\{x_s(t)\} = X(j\omega) \times \text{const}$$

$$+ \sum_{k \neq 0,\ k=-\infty}^{\infty} |P_k(k\omega_s)|\{X(j(\omega - k\omega_s)) + X(j(\omega + k\omega_s))\} \tag{9.21}$$

Assuming that $x(t)$ is a bandlimited signal—that is, assuming that its frequency spectrum is different from zero only in the interval $[-\omega_{\max}\ \omega_{\max}]$—it can be noticed from (9.21) that the *sampling theorem condition $f_s \geq 2f_{\max}$ ($\omega_s \geq 2\omega_{\max}$) will eliminate aliasing of frequencies in this case also*. Namely, all frequency components in the infinite sum in (9.21) are outside the frequency range of the original signal, $[-\omega_{\max}\ \omega_{\max}]$. Simply, a low-pass filter will be able to extract the original signal in the frequency domain.

9.2 DISCRETE-TIME FOURIER TRANSFORM (DTFT)

The discrete-time Fourier transform is derived in Section 9.1.2, formula (9.15), as the *Fourier transform of a continuous-time signal sampled by an ideal sampler*, $x_\delta(t)$. The signal $x_\delta(t)$ is given in (9.14).

In the study of the DTFT, it is customary to use the concept of the *digital frequency* defined by

$$\Omega = T_s \omega \tag{9.22}$$

Employing the notation that we have used in this text for representation of discrete-time signals, that is, $x[k] \triangleq x(kT_s)$, the DTFT can be represented by the infinite summation

$$X_\delta(j\omega) = \mathcal{F}\{x_\delta(t)\} = \sum_{k=-\infty}^{\infty} x(kT_s) e^{-jkT_s\omega} \triangleq X(j\Omega) = \sum_{k=-\infty}^{\infty} x[k] e^{-jk\Omega} \tag{9.23}$$

Using the definition formula (9.23), we can find the DTFT of some simple signals, as demonstrated in the following examples.

EXAMPLE 9.1

The DTFT of the discrete-time delta impulse signal $\delta[k]$ is given by

$$X(j\Omega) = \sum_{k=-\infty}^{\infty} x[k] e^{-jk\Omega} = \sum_{k=-\infty}^{\infty} \delta[k] e^{-jk\Omega} = 1 \leftrightarrow \delta[k]$$

The shifted impulse delta signal $\delta[k-m]$ has the DTFT

$$X(j\Omega) = \sum_{k=-\infty}^{\infty} \delta[k-m] e^{-jk\Omega} = e^{-jm\Omega} \leftrightarrow \delta[k-m] \qquad \clubsuit$$

EXAMPLE 9.2

Consider the discrete-time signal

$$x[k] = \begin{cases} 1, & k = -1 \\ 2, & k = 0 \\ 3, & k = 2 \\ 0, & \text{otherwise} \end{cases}$$

Since this signal has only three nonzero terms, the infinite sum simplifies into a sum of three terms, and the corresponding DTFT is given by

$$X(j\Omega) = \sum_{k=-\infty}^{\infty} x[k] e^{-jk\Omega} = e^{j\Omega} + 2 + 3e^{-j2\Omega} \qquad \clubsuit$$

We have observed that $X_\delta(j\omega)$ is a periodic function in the frequency domain, with period $\omega_s = 2\pi/T_s$ (see (9.16)). The corresponding digital frequency period is $\Omega_s = T_s\omega_s = 2\pi$. Hence, the *frequency Ω-domain signal $X(j\Omega) = X_\delta(j\omega)$ is periodic with period equal to 2π*. The 2π-periodicity of the DTFT can be also confirmed from the definition formula, since

$$X(j\Omega) = X(j(\Omega + 2\pi)) = \sum_{k=-\infty}^{\infty} x[k]e^{-jk(\Omega+2\pi)} = \sum_{k=-\infty}^{\infty} x[k]e^{-jk\Omega} \quad (9.24)$$

where we have used the fact that $e^{-jk2\pi} = 1$ for any integer k.

In addition, we can observe from the definition formula (9.23) that

$$X(-j\Omega) = X^*(j\Omega) \quad (9.25)$$

which indicates that the *magnitude spectrum of the DTFT is an even function and the phase spectrum of the DTFT is an odd function,* that is,

$$|X(-j\Omega)| = |X(j\Omega)|, \quad \angle X(-j\Omega) = -\angle X(j\Omega) \quad (9.26)$$

Due to these facts, we indeed need plot the spectrum of the DTFT only in the frequency interval $[0, \pi]$. It follows that low-frequency signal components are around zero frequency (and, due to periodicity, around $n2\pi$, for some integer n), and high-frequency signal components are around π frequency (and $(2n+1)2\pi$, for some integer n).

Formula (9.23) can be viewed as the frequency domain Fourier series expansion of the 2π-periodic signal $X(j\Omega)$, with $x[k]$ playing the role of the corresponding Fourier series coefficients, $X(j\Omega) \leftrightarrow x[k]$. This leads to the conclusion that the signal sample values $x[k]$ can be obtained in terms of $X(j\Omega)$, using the formula for the Fourier series coefficients, that is,

$$x[k] = \frac{1}{2\pi} \int_{-\pi}^{\pi} X(j\Omega) e^{jk\Omega} \, d\Omega \quad (9.27)$$

This formula defines the *inverse* DTFT. The discrete-time domain signal $x[k]$ and the Ω-domain frequency signal $X(j\Omega)$ form the corresponding pair, which we denote by $x[k] \leftrightarrow X(j\Omega)$.

Comment: We have introduced and derived the discrete-time Fourier transform, as the Fourier transform of a discrete-time signal, obtained by sampling a continuous-time signal by an ideal sampler. It is interesting to point out that the DTFT as defined in (9.23), and its inverse (9.27), can be derived using the discrete-time Fourier series (as was done in [5]). By analogy to the continuous-time Fourier series, the discrete-time Fourier series are applicable to discrete-time periodic signals. A discrete-time aperiodic signal can be considered in the limit as a discrete-time periodic signal whose period tends to infinity. In derivations such as those done in [5], one avoids the use of an ideal sampler, but performs a limiting operation similar to what we did in Section 3.1.1 while deriving the Fourier transform from the Fourier series. The approach taken here directly relates the DTFT to the concept of signal sampling.

Existence Condition for DTFT

The DTFT exists if the infinite sum defined in (9.23) exists. This leads to the following condition:

$$\left|\sum_{k=-\infty}^{\infty} x[k]e^{-jk\Omega}\right| \leq \sum_{k=-\infty}^{\infty} |x[k]e^{-jk\Omega}| \leq \sum_{k=-\infty}^{\infty} |x[k]||e^{-jk\Omega}|$$
$$\leq \sum_{k=-\infty}^{\infty} |x[k]| < \infty \qquad (9.28)$$

It follows that if the signal $x[k]$ is absolutely summable, then the corresponding DTFT will exist. Note that (9.28) is only a sufficient condition, which means that if condition (9.28) is satisfied, then the DTFT exists, but the reverse is not generally true.

EXAMPLE 9.3

Consider the signal defined by

$$x[k] = a^k u[k], \quad |a| < 1, \quad u[k] = \text{unit step}$$

It follows from (9.23) that

$$X(j\Omega) = \sum_{k=-\infty}^{\infty} x[k]e^{-jk\Omega} = \sum_{k=0}^{\infty} a^k e^{-jk\Omega} = \sum_{k=0}^{\infty} \left(\frac{a}{e^{j\Omega}}\right)^k$$
$$= \frac{1}{1 - (a/e^{j\Omega})} = \frac{e^{j\Omega}}{e^{j\Omega} - a}$$

Note that the condition $|a/e^{j\Omega}| < 1$, which is the consequence of $|a| < 1$, has been used to sum the obtained geometric series. Hence, the convergence condition (9.28) is satisfied and the corresponding DTFT exists.

In the following, we state the main properties of the DTFT and prove most of them.

Property 1: Linearity

The DTFT is defined by an infinite sum (9.23). The linearity property of summation says that for pairs $x_i[k] \leftrightarrow X_i(j\Omega)$, $i = 1, 2, \ldots, n$, and an arbitrary set of constants α_i, $i = 1, 2, \ldots, n$, the following holds:

$$\alpha_1 x_1[k] \pm \alpha_2 x_2[k] \pm \cdots \pm \alpha_n x_n[k] \leftrightarrow \alpha_1 X_1(j\Omega) \pm \alpha_2 X_2(j\Omega) \pm \cdots \pm \alpha_n X_n(j\Omega) \quad (9.29)$$

The proof of this property is a direct consequence of the linearity property of summation.

Property 2: Time Shifting

Let the pair $x[k] \leftrightarrow X(j\Omega)$ exist. Then, signal time shifting implies the pair

$$x[k-m] \leftrightarrow e^{-j\Omega m} X(j\Omega) \tag{9.30}$$

It should be emphasized that this formula is *valid for both positive and negative values of* m.

Proof. In order to establish the proof, we use the definition formula (9.23) and a change of variables $n = k - m$, that is,

$$\sum_{k=-\infty}^{\infty} x[k-m]e^{-jk\Omega} = \sum_{n=-\infty}^{\infty} x[n]e^{-j(n+m)\Omega}$$

$$e^{-jm\Omega} \sum_{n=-\infty}^{\infty} x[n]e^{-jn\Omega} = e^{-jm\Omega} X(j\Omega) \qquad \square$$

Property 3: Frequency Shifting

Let the pair $x[k] \leftrightarrow X(j\Omega)$ exist; then the frequency shifted signal produces the pair

$$x[k]e^{jk\Omega_0} \leftrightarrow X(j(\Omega - \Omega_0)) \tag{9.31}$$

Proof. This property can be proved as follows:

$$X(j(\Omega - \Omega_0)) = \sum_{k=-\infty}^{\infty} x[k]e^{-jk(\Omega-\Omega_0)} = \sum_{k=-\infty}^{\infty} \{x[k]e^{jk\Omega_0}\}e^{-jk\Omega} \leftrightarrow x[k]e^{jk\Omega_0}$$

$\square$

Property 4: Time Reversal

Let $x[k] \leftrightarrow X(j\Omega)$. It follows directly from the definition formula (9.23) that time reversal produces the pair

$$x[-k] \leftrightarrow X(-j\Omega) \tag{9.32}$$

Proof. To prove this property, we introduce a change of variables in the definition formula as follows:

$$\sum_{k=-\infty}^{\infty} x[-k]e^{-jk\Omega} = \sum_{n=\infty}^{-\infty} x[n]e^{jn\Omega} = \sum_{n=-\infty}^{\infty} x[n]e^{jn\Omega}$$

$$= \sum_{n=-\infty}^{\infty} x[n]e^{-jn(-\Omega)} = X(-j\Omega)$$

Note that in this proof we have replaced $-k$ by n. $\square$

Property 5: Conjugation

Let $x[k] \leftrightarrow X(j\Omega)$. It follows from the definition formula (9.23) that conjugation produces the pair

$$x^*[k] \leftrightarrow X^*(-j\Omega) \tag{9.33}$$

Hence, signal conjugation in the time domain implies both conjugation and frequency reversal in the frequency domain.

Proof. The proof is as follows:

$$\sum_{k=-\infty}^{\infty} x^*[k]e^{-jk\Omega} = \left(\sum_{k=-\infty}^{\infty} x[k]e^{jk\Omega}\right)^*$$

$$= \left(\sum_{k=-\infty}^{\infty} x[k]e^{-jk(-\Omega)}\right)^* = X^*(-j\Omega) \qquad \square$$

Property 6: Frequency Differentiation

Let $x[k] \leftrightarrow X(j\Omega)$; then the differentiation of $X(j\Omega)$ with respect to Ω produces

$$kx[k] \leftrightarrow j\frac{dX(j\Omega)}{d\Omega} \tag{9.34}$$

Proof. The proof follows simply from the definition formula (9.23):

$$\frac{dX(j\Omega)}{d\Omega} = \sum_{k=-\infty}^{\infty} (-j)kx[k]e^{-jk\Omega}$$

which implies

$$\frac{dX(j\Omega)}{d\Omega} \leftrightarrow (-j)kx[k] \quad \text{or} \quad j\frac{dX(j\Omega)}{d\Omega} \leftrightarrow kx[k] \qquad \square$$

Property 7: Modulation

The modulation property is a generalization of the frequency shifting property. Let $x[k] \leftrightarrow X(j\Omega)$; then the modulation property states the results

$$x[k]\cos(\Omega_0 k) \leftrightarrow \frac{1}{2}[X(j(\Omega+\Omega_0)) + X(j(\Omega-\Omega_0))] \tag{9.35}$$

and

$$x[k]\sin(\Omega_0 k) \leftrightarrow \frac{j}{2}[X(j(\Omega+\Omega_0)) - X(j(\Omega-\Omega_0))] \tag{9.36}$$

The modulation property has applications in digital signal processing and digital communication systems. The proof of this property requires first representation of sine and cosine

474   Chapter 9   Signals in Digital Signal Processing

functions via Euler's formulas, that is,

$$\cos(\Omega_0 k) = \frac{1}{2}(e^{j\Omega_0 k} + e^{-j\Omega_0 k}), \quad \sin(\Omega_0 k) = \frac{1}{2j}(e^{j\Omega_0 k} - e^{-j\Omega_0 k}) \qquad (9.37)$$

and then follows the proof of the frequency shifting property.

Property 8: Time Convolution

Let $x_1[k] \leftrightarrow X_1(j\Omega)$ and $x_2[k] \leftrightarrow X_2(j\Omega)$. Then, the time domain convolution of these two signals corresponds to a product in the frequency domain, that is,

$$x_1[k] * x_2[k] \leftrightarrow X_1(j\Omega)X_2(j\Omega) \qquad (9.38)$$

This property is useful for deriving theoretical results about the response of linear time-invariant systems. Note that for this purpose we have the analogous result obtained using the one-sided $\mathcal{Z}$-transform, as demonstrated in Chapter 5. The proof of (9.38) is similar to the corresponding convolution property proof from Chapter 5.

Property 9: Periodic Frequency Domain Convolution

Let $x_1[k] \leftrightarrow X_1(j\Omega)$ and $x_2[k] \leftrightarrow X_2(j\Omega)$. Then the time domain product of these two signals corresponds to the frequency domain convolution, called the *periodic convolution,* defined by

$$x_1[k]x_2[k] \leftrightarrow \frac{1}{2\pi} \int_{-\pi}^{\pi} X_1(j\theta)X_2(j(\Omega-\theta))\, d\theta \qquad (9.39)$$

Proof. The proof of this property uses the definitions of both the DTFT and inverse DTFT, as follows.

$$\text{DTFT}\{x_1[k]x_2[k]\} = \sum_{k=-\infty}^{\infty} x_1[k]x_2[k]e^{-jk\Omega}$$

$$= \sum_{k=-\infty}^{\infty} \left(\frac{1}{2\pi} \int_{-\pi}^{\pi} X_1(j\theta)e^{jk\theta}\, d\theta \right) x_2[k]e^{-jk\Omega}$$

$$= \frac{1}{2\pi} \int_{-\pi}^{\pi} X_1(j\theta) \left[\sum_{k=-\infty}^{\infty} x_2[k]e^{-jk(\Omega-\theta)} \right] d\theta$$

$$= \frac{1}{2\pi} \int_{-\pi}^{\pi} X_1(j\theta)X_2(j(\Omega-\theta))\, d\theta \qquad \square$$

Section 9.2 Discrete-Time Fourier Transform (DTFT)

Note that for $\Omega = 0$, formula (9.39) gives the expression for the discrete-time signal total energy:

$$E_\infty = \sum_{k=-\infty}^{\infty} |x[k]|^2 = \frac{1}{2\pi} \int_{-\pi}^{\pi} |X(j\theta)|^2 \, d\theta \qquad (9.40)$$

which in fact represents the *Parseval theorem* for discrete-time signals.

Generalized DTFT

We observed in Chapter 3 that signals that do not satisfy the existence condition of the Fourier transform, or more precisely, signals for which the Fourier integral does not exist in terms of regular functions can still be transformed into the frequency domain using generalized functions (continuous-time impulse delta function, $\delta(t)$). Similarly, we can define the generalized DTFT by using the continuous-time impulse delta signal (the signal that accounts for infinitely large values). This procedure is demonstrated in the following examples.

EXAMPLE 9.4

Consider now a discrete-time signal that does not satisfy the DTFT existence condition (9.28), given by

$$x[k] = 1, k = 0, \pm 1, \pm 2, \ldots$$

It is obvious that for this signal, the sum in (9.28) is equal to infinity. It follows from (9.23) that

$$X(j\Omega) = \sum_{k=-\infty}^{\infty} x[k] e^{-jk\Omega} = \sum_{k=-\infty}^{\infty} e^{-jk\Omega}$$

The infinite sum of complex exponential functions can be evaluated using the results established in Example 3.11, where we showed that

$$\sum_{k=-\infty}^{\infty} e^{-jnT_0\omega} = \omega_0 \sum_{k=-\infty}^{\infty} \delta(\omega - k\omega_0), \quad \omega_0 = \frac{2\pi}{T_0}$$

Using the digital frequency as $\Omega = T_0\omega$ and the property of the continuous-time impulse delta function ($\delta(at) = \delta(t)/|a|$), the preceding expression implies

$$\sum_{k=-\infty}^{\infty} e^{-jn\Omega} = 2\pi \sum_{k=-\infty}^{\infty} \delta(\Omega - k2\pi), \quad \omega_0 = \frac{2\pi}{T_0} \qquad (9.41)$$

This establishes the generalized DTFT pair

$$x[k] = 1, \ k = 0, \pm 1, \pm 2, \ldots \leftrightarrow 2\pi \sum_{k=-\infty}^{\infty} \delta(\Omega - k2\pi) \qquad (9.42)$$

EXAMPLE 9.5

Using the results established in the previous example, we can find the DTFT of sine and cosine signals as follows. By modulating the signal

$$x[k] = 1, \ k = 0, \pm 1, \pm 2, \ldots$$

using the cosine function and evoking the modulation property, we have

$$x[k]\cos(\Omega_0 k) \leftrightarrow \frac{1}{2}\left[2\pi \sum_{k=-\infty}^{\infty} \delta(\Omega + \Omega_0 - k2\pi) + 2\pi \sum_{k=-\infty}^{\infty} \delta(\Omega - \Omega_0 - k2\pi)\right]$$

or

$$\cos(\Omega_0 k) \leftrightarrow \pi \sum_{k=-\infty}^{\infty} \delta(\Omega + \Omega_0 - k2\pi) + \pi \sum_{k=-\infty}^{\infty} \delta(\Omega - \Omega_0 - k2\pi) \qquad (9.43)$$

Similarly, by modulating the signal $x[k]$ by the sine function, we obtain

$$x[k]\sin(\Omega_0 k) \leftrightarrow j\pi \sum_{k=-\infty}^{\infty} \delta(\Omega + \Omega_0 - k2\pi) - j\pi \sum_{k=-\infty}^{\infty} \delta(\Omega - \Omega_0 - k2\pi) \qquad (9.44)$$

9.2.1 DTFT in Linear Systems

The DTFT can be used in the analysis and design of discrete-time linear systems with constant coefficients in the same way that the Fourier transform is used in continuous-time linear systems with constant coefficients. Hence, the DTFT can be used to find the zero-state response of discrete-time linear time-invariant systems.

Such a discrete-time system is defined in Section 5.3, in formula (5.40), as

$$\begin{aligned} y[k+n] + a_{n-1}y[k+n-1] + \cdots + a_1 y[k+1] + a_0 y[k] \\ = b_m x[k+m] + b_{m-1}x[k+m-1] + \cdots + b_1 x[k+1] + b_0 x[k] \end{aligned} \qquad (9.45)$$

where a_i, b_i are constant coefficients and n is the system order. Note that it is more common in signal processing to use the system formulation defined in (5.41), which we have called the integral formulation. However, both formulations give the same system transfer function, which is relevant in this case.

Let $y[k] \leftrightarrow Y(j\Omega)$ and $x[k] \leftrightarrow X(j\Omega)$. Applying the DTFT to (9.45) and using the time shifting property, we have

$$\begin{aligned} \left[e^{jn\Omega} + a_{n-1}e^{j(n-1)\Omega} + \cdots + a_1 e^{j\Omega} + a_0\right]Y(j\Omega) \\ = \left[b_m e^{jm\Omega} + b_{m-1}e^{j(m-1)\Omega} + \cdots + b_1 e^{j\Omega} + b_0\right]X(j\Omega) \end{aligned}$$

from which we obtain

$$Y(j\Omega) = \frac{b_m e^{jm\Omega} + b_{m-1} e^{j(m-1)\Omega} + \cdots + b_1 e^{j\Omega} + b_0}{e^{jn\Omega} + a_{n-1} e^{j(n-1)\Omega} + \cdots + a_1 e^{j\Omega} + a_0} X(j\Omega) \triangleq H(j\Omega) X(j\Omega) \quad (9.46)$$

where the quantity

$$H(j\Omega) \triangleq \frac{b_m e^{jm\Omega} + b_{m-1} e^{j(m-1)\Omega} + \cdots + b_1 e^{j\Omega} + b_0}{e^{jn\Omega} + a_{n-1} e^{j(n-1)\Omega} + \cdots + a_1 e^{j\Omega} + a_0} \quad (9.47)$$

defines the *discrete-time system digital frequency transfer function*.

Assuming that the input signal is equal to the discrete-time impulse delta function $\delta[k]$, whose DTFT is equal to 1, we see that the system impulse response in the Ω-frequency domain is equal to $Y(j\Omega) = H(j\Omega) \times 1 = H(j\Omega)$. It can be concluded that $H(j\Omega)$ is the DTFT of the *discrete-time system impulse response,* so that we have the following pair:

$$\text{impulse response} = h[k] \leftrightarrow H(j\Omega) = \text{transfer function} \quad (9.48)$$

From the time convolution property of the DTFT and (9.46), it follows that the *linear discrete-time system response to any excitation $x[k]$ is given by the convolution of that input signal and the system impulse response,* that is,

$$y_{zs}[k] = x[k] * h[k] \quad (9.49)$$

Note that the convolution result obtained represents only the *zero-state* component of the system response, since the DTFT (like the Fourier transform) is unable to take into account the system initial conditions and provide the zero-input component of the system response.

It should be pointed out that finding the response of a linear discrete-time, time-invariant system using the DTFT can be computationally involved. In addition, the existence condition (9.28) of the DTFT is very strong, and many real signals do not satisfy it. In many cases, it is impossible to find analytically the DTFT, so that we must resort to numerical techniques. Hence, *for finding exclusively the time domain response of time-invariant linear discrete-time systems, we have better ways than to use the DTFT approach, as discussed in Chapters 5, 7, and 8*. However, for digital signal processing and digital communications, especially for noncausal signals and corresponding noncausal systems, the DTFT is very important, first of all in the form of its numerical equivalents, the DFT and the FFT, which open doors for numerous applications. Furthermore, in the next section, we will extend the DTFT to the double-sided $\mathcal{Z}$-transform, which is applicable to much wider classes of discrete-time signals than the DTFT.

Linear System Response to Sinusoidal Inputs

Linear discrete-time system responses due to sinusoidal inputs can be found employing the same technique as in Section 5.5.1, which leads to the same result as in (5.92), with $H(e^{j\omega T})$ being replaced by $H(j\Omega)$. Let the system input be given by $x[k] = \cos(k\Omega_0 + \varphi)$; then, by replacing $H(e^{j\omega T})$ with $H(j\Omega_0)$ in (5.92), the system output is

$$y_{ss}[k] = |H(j\Omega_0)| \cos(k\Omega_0 + \varphi + \angle H(j\Omega_0)) \quad (9.50)$$

where $H(j\Omega_0)$ is the system transfer function (9.47) evaluated at the digital frequency Ω_0. It should be emphasized that the result in (9.50) is valid at steady state.

9.2.2 From DTFT to the Double-Sided $\mathcal{Z}$-Transform

The double-sided $\mathcal{Z}$-transform can be obtained in a straightforward manner from the discrete-time Fourier transform. Namely, starting with

$$X_\delta(j\omega) = \sum_{k=-\infty}^{\infty} x(kT_s)e^{-jkT_s\omega} \triangleq X(j\Omega) = \sum_{k=-\infty}^{\infty} x[k]e^{-jk\Omega}$$

and replacing $e^{jT_s\omega} = e^{j\Omega}$ by z, we obtain the definition of the double-sided $\mathcal{Z}$-transform,

$$X(z) \triangleq \sum_{k=-\infty}^{\infty} x[k]z^{-k} \tag{9.51}$$

This transformation is also called the *bilateral $\mathcal{Z}$-transform*. Note that $e^{jT_s\omega} = e^{j\Omega}$ represents the unit circle in the complex plane, and that the convergence condition for the corresponding DTFT is

$$\sum_{k=-\infty}^{\infty} |x[k]| < \infty$$

It is known that many real discrete-time signals do not satisfy this convergence condition. The definition of the double-sided $\mathcal{Z}$-transform can be made even more general by defining the complex variable z as $z = \rho e^{j\Omega}$, $0 < \rho < \infty$; that is, by considering the whole complex plane (recall at this point the procedure used in the passage from the Fourier transform to the Laplace transform). Now, the convergence (existence) condition of the double-sided $\mathcal{Z}$-transform becomes parameter dependent, that is,

$$\sum_{k=-\infty}^{\infty} |\rho^{-k}x[k]| < \infty \tag{9.52}$$

It is obvious that the parameter ρ may help the convergence of this infinite summation. Note that different values of ρ may be required for positive and negative discrete-time instants. The values for the parameter ρ must be chosen such that condition (9.52) is satisfied. Such values of ρ define the region of existence (convergence) of the double-sided $\mathcal{Z}$-transform. Note that the *region of convergence* (ROC) is particularly important for the double-sided $\mathcal{Z}$-transform, since two different signals can have the same double-sided $\mathcal{Z}$-transform, but different regions of convergence (see Example 9.8, in the next section). Hence, the region of convergence can be used to distinguish between two signals that have the same double-sided $\mathcal{Z}$-transform. A detailed study of the ROC for the double-sided $\mathcal{Z}$-transform, including its properties in various combinations of discrete-time signals, can be found in [5].

Section 9.2 Discrete-Time Fourier Transform (DTFT)

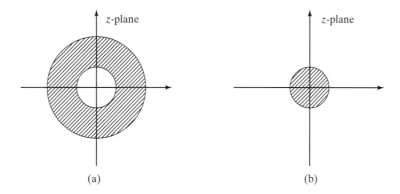

(a) (b)

FIGURE 9.7: Regions of convergence denoted by shaded areas of the double-sided $\mathcal{Z}$-transform for (a) noncausal signals and (b) anticausal signals

It should be pointed out that the *one-sided (unilateral) $\mathcal{Z}$-transform* (presented in Chapter 5 as one of the main tools for finding responses of linear discrete-time dynamic systems with constant coefficients) can be obtained from (9.51–52) by specializing these formulas to causal signals (whose values are equal to zero for negative times). Consequently, in the case of the one-sided $\mathcal{Z}$-transform, the lower summation limit starts at $k = 0$.

It should emphasized that in the case of causal signals the region of convergence of the corresponding one-sided $\mathcal{Z}$-transform is outside of a circle of a given radius centered at the origin of the z-plane (see Figure 5.2). We will see from examples to be solved in the next section that the region of convergence of noncausal signals in general is represented by a ring centered at the origin of the z-plane (see Figure 9.7(a)). Also, the region of convergence of *anticausal signals (equal to zero for positive times)* is inside of a circle of a given radius (see Figure 9.7(b)).

EXAMPLE 9.6

Consider the discrete-time signal represented by the expression

$$x[k] = \left(\frac{1}{3}\right)^{-k} u[-k] + \left(\frac{1}{2}\right)^{k} u[k]$$

Its double-sided $\mathcal{Z}$-transform is given by

$$\sum_{k=-\infty}^{\infty} x[k] z^{-k} = \sum_{k=-\infty}^{0} \left(\frac{1}{3}\right)^{-k} z^{-k} + \sum_{k=0}^{\infty} \left(\frac{1}{2}\right)^{k} z^{-k}$$

$$= \left[1 + \left(\frac{z}{3}\right) + \left(\frac{z}{3}\right)^{2} + \cdots \right] + \left[1 + \left(\frac{1}{2z}\right) + \left(\frac{1}{2z}\right)^{2} + \cdots \right]$$

The causal part (the second bracketed term on the right-hand side) is convergent for $|1/2z| < 1 \Rightarrow |z| > 1/2$. The anticausal part (the first bracketed term) is convergent for $|z/3| < 1 \Rightarrow |z| < 3$. Hence, for $1/2 < |z| < 3$, the double-sided $\mathcal{Z}$-transform of the considered signal is

$$\frac{1}{1-\frac{1}{3}z} + \frac{1}{1-\frac{1}{2z}} = \frac{1}{1-\frac{1}{3}z} + \frac{z}{z-\frac{1}{2}}$$

However, if we consider the signal

$$x[k] = (3)^{-k}u[-k] + \left(\frac{1}{2}\right)^k u[k]$$

we will find, by repeating the procedure, that the region of convergence must satisfy $1/2 < |z| < 1/3$; such a region of convergence does not exist, and consequently the corresponding double-sided $\mathcal{Z}$-transform does not exist.

The double-sided $\mathcal{Z}$-transform has applications in digital signal processing, where noncausal signals play an important role. It is also used in digital communications and the study of discrete-time random processes. In the next section, we present the main properties of the double-sided $\mathcal{Z}$-transform.

9.3 DOUBLE-SIDED $\mathcal{Z}$-TRANSFORM

The double-sided $\mathcal{Z}$-transform is defined in (9.51). The inverse of the double-sided $\mathcal{Z}$-transform has a complex form (as does the inverse of the one-sided $\mathcal{Z}$-transform; see formulas (5.31–32)), which involves contour integration. Dealing directly with the definition formula of the inverse of the double-sided $\mathcal{Z}$-transform can be avoided in many cases, in the same manner as for the one-sided $\mathcal{Z}$-transform.

The properties of the double-sided $\mathcal{Z}$-transform are very similar to the properties of the discrete-time Fourier transform. These properties and outlines of their proofs are given below. The proofs that are similar to those for the one-sided $\mathcal{Z}$-transform are omitted.

Property 1: Linearity

The double-sided $\mathcal{Z}$-transform is defined by an infinite sum (9.51). The linearity property of summation says that for pairs $x_i[k] \leftrightarrow X_i(z)$, $i = 1, 2, \ldots, n$, and an arbitrary set of constants α_i, $i = 1, 2, \ldots, n$, the following holds:

$$\alpha_1 x_1[k] \pm \alpha_2 x_2[k] \pm \cdots \pm \alpha_n x_n[k] \leftrightarrow \alpha_1 X_1(z) \pm \alpha_2 X_2(z) \pm \cdots \pm \alpha_n X_n(z) \quad (9.53)$$

Let $\mathfrak{R}_i$ define respectively the ROCs of signals $x_i[k]$. Then the ROC of the signal represented by a weighted sum of $x_i[k]$ signals, denoted by $\mathfrak{R}$, is the intersection of the ROCs of all other signals, that is,

$$\mathfrak{R} = \mathfrak{R}_1 \cap \mathfrak{R}_2 \cap \cdots \cap \mathfrak{R}_n \quad (9.53a)$$

Property 2: Time Shifting

Let the pair

$$x[k] \leftrightarrow X(z) = \sum_{k=-\infty}^{\infty} x[k]z^{-k}$$

exist. Then, signal time shifting implies the pair

$$x[k+m] \leftrightarrow z^m X(z) \tag{9.54}$$

Proof. The proof of this property requires the following change of variables:

$$\mathcal{Z}\{x[k+m]\} \triangleq \sum_{k=-\infty}^{\infty} x[k+m]z^{-k}$$

$$= \sum_{i=-\infty}^{\infty} x[i]z^{-(i-m)} = z^m \sum_{i=-\infty}^{\infty} x[i]z^{-i} = z^m X(z) \qquad \square$$

EXAMPLE 9.7

The double-sided $\mathcal{Z}$-transform of the discrete-time impulse delta signal is simply obtained as

$$\sum_{k=-\infty}^{\infty} \delta[k]z^{-k} = 1 \leftrightarrow \delta[k] \tag{9.55}$$

The shifted discrete-time impulse delta signal has the double-sided $\mathcal{Z}$-transform given by

$$\sum_{k=-\infty}^{\infty} \delta[k+m]z^{-k} = z^m \leftrightarrow \delta[k+m] \tag{9.56}$$

EXAMPLE 9.8

The double-sided $\mathcal{Z}$-transform of the discrete-time unit step signal is simply obtained from the definition formula,

$$\sum_{k=-\infty}^{\infty} u[k]z^{-k} = \sum_{k=0}^{\infty} u[k]z^{-k} = \frac{z}{z-1} \leftrightarrow u[k] \tag{9.57}$$

Note that in this case the double-sided $\mathcal{Z}$-transform is identical to the single-sided $\mathcal{Z}$-transform, since the signal $u[k]$ is causal. It is shown in Example 5.3 that the region of convergence for the unit step signal is $|z| > 1$. Consider now the signal given by $-u[-k-1]$. Its double-sided $\mathcal{Z}$-transform is given by

$$\sum_{k=-\infty}^{\infty}(-1)u[-k-1]z^{-k} = (-1)\sum_{k=-\infty}^{-1} z^{-k} = (-1)[z+z^2+z^3+\cdots]$$

$$= -z(1+z+z^2+\cdots) = -z\frac{1}{1-z} = \frac{z}{z-1} \tag{9.58}$$

Note that the geometric series obtained is convergent if $|z| < 1$. Hence, we have obtained identical double-sided $\mathcal{Z}$-transforms, but in the first case the convergence region is $|z| > 1$ and in the second case the convergence region is $|z| < 1$.

Property 3: Frequency Scaling

Let the pair $x[k] \leftrightarrow X(z)$ exist, with ROC equal to $\mathfrak{R}$; then the frequency scaled signal produces the pair

$$x[k]a^{-k} \leftrightarrow X(az), \quad \text{ROC} = |a|\mathfrak{R} \tag{9.59}$$

Proof. The proof is as follows:

$$\mathcal{Z}\{a^{-k}x[k]\} = \sum_{k=-\infty}^{\infty} x[k]a^{-k}z^{-k} = \sum_{k=-\infty}^{\infty} x[k](az)^{-k} \stackrel{\triangle}{=} X(az) \tag{9.60}$$

Property 4: Time Reversal

Let $x[k] \leftrightarrow X(z)$. Then the double-sided $\mathcal{Z}$-transform of the time reversed signal $x[-k]$ is given by $X(1/z)$.

Proof.

$$\begin{aligned}\mathcal{Z}\{x[-k]\} &= \sum_{k=-\infty}^{\infty} x[-k]z^{-k} = \sum_{n=\infty}^{-\infty} x[n]z^{-(-n)} \\ &= \sum_{n=-\infty}^{\infty} x[n]\left(\frac{1}{z}\right)^{-n} \stackrel{\triangle}{=} X\left(z^{-1}\right) = X\left(\frac{1}{z}\right) \leftrightarrow x[-k]\end{aligned} \tag{9.61}$$

where we have used a change of variables $n = -k$. Note that the region of convergence also changes. Assuming that $X(z)$ has the region of convergence given by $\rho_{\min} < |z| < \rho_{\max}$, then the region of convergence of $X(1/z)$ is $1/\rho_{\min} > |z| > 1/\rho_{\max}$.

EXAMPLE 9.9

The double-sided $\mathcal{Z}$-transform of $a^k u[k]$ is equal to $z/(z-a)$ with the region of convergence $|z| > |a|$ (we know this result from Example 5.4). Using the time reversal property, it follows that

$$\mathcal{Z}\{a^{-k}u[-k]\} = \frac{\frac{1}{z}}{\frac{1}{z}-a} = \frac{1}{1-az} \tag{9.62}$$

The region of convergence of the time-reversed signal is $|z| < 1/|a|$.

Property 5: Complex Conjugation

The double-sided $\mathcal{Z}$-transform can be applied to complex signals as well. In that case, we have the property

$$x^*[k] \leftrightarrow X^*(z^*) \tag{9.63}$$

Proof. The proof of this property is as follows:

$$\mathcal{Z}\{x^*[k]\} = \sum_{k=-\infty}^{\infty} x^*[k] z^{-k} = \left(\sum_{k=-\infty}^{\infty} x[k](z^*)^{-k}\right)^* \triangleq X^*(z^*) \tag{9.64}$$

$\square$

Property 6: Frequency Differentiation (Time Multiplication)

For the pair $x[k] \leftrightarrow X(z)$, the following holds:

$$kx[k] \leftrightarrow -z\frac{dX(z)}{dz} \tag{9.65}$$

with the ROC remaining unchanged.

The proof of this property can be obtained by starting with the definition of the double-sided $\mathcal{Z}$-transform and taking the derivative with respect to z, which is identical to the corresponding one-sided $\mathcal{Z}$-transform proof.

Property 7: Time Convolution

Let $x_1[k] \leftrightarrow X_1(z)$ and $x_2[k] \leftrightarrow X_2(z)$, with ROCs respectively equal to $\Re_1$ and $\Re_2$. Then, the time domain convolution of these two signals corresponds to a product in the frequency domain, that is,

$$x_1[k] * x_2[k] \leftrightarrow X_1(z)X_2(z), \quad \text{ROC} = \Re_1 \cap \Re_2 \tag{9.66}$$

The proof of this property follows the same steps as the corresponding proof for the one-sided $\mathcal{Z}$-transform.

Inversion of the Double-Sided $\mathcal{Z}$-transform

The inverse of the double-sided $\mathcal{Z}$-transform can be found using the partial fraction expansion technique, as was done in Chapter 5 in finding the inverse of the one-sided $\mathcal{Z}$-transform. As pointed out, the inverse double-sided $\mathcal{Z}$-transform is not unique, unless the region of convergence is specified. The partial fraction expansion can be performed using the MATLAB `residuez`.

EXAMPLE 9.10

We want to find the inverse double-sided $\mathcal{Z}$-transform of

$$X(z) = \frac{2z}{(z-5)(z+1)} = \frac{\frac{4}{3}z}{z-0.5} - \frac{\frac{4}{3}z}{z+1}$$

484   Chapter 9   Signals in Digital Signal Processing

It can be shown that the signals $a^k u[k]$ and $-a^k u[-k-1]$ have the same double-sided $\mathcal{Z}$-transform, but different ROCs (see Problem 9.13). The first term in this frequency domain signal corresponds either to $(4/3)(0.5)^k u[k]$ or $-(4/3)(0.5)^k u[-k-1]$. Similarly, the second part of the signal corresponds either to $-(4/3)(-1)^k u[k]$ or $(4/3)(-1)^k u[-k-1]$. Hence, we have four possibilities, but we must also check whether the corresponding ROCs exist.

The first possibility is that both signals are causal, in which case

$$x[k] = \frac{4}{3}(0.5)^k u[k] - \frac{4}{3}(-1)^k[k]$$

$$\text{ROC} = \{|z| > 0.5\} \cap \{|z| > 1\} = \{|z| > 0.5\}$$

Since the corresponding ROC exists (is a nonempty set), this signal is a valid inverse double-sided $\mathcal{Z}$-transform.

The second possibility is that the first signal is causal and the second signal is anticausal. In this case, we have

$$x[k] = \frac{4}{3}(0.5)^k u[k] + \frac{4}{3}(-1)^k u[-k-1]$$

$$\text{ROC} = \{|z| > 0.5\} \cap \{|z| < 1\} = \{0.5 < |z| < 1\}$$

The third possibility is that the first signal is anticausal and the second signal is causal, which implies

$$x[k] = -\frac{4}{3}(0.5)^k u[-k-1] - \frac{4}{3}(-1)^k u[k]$$

$$\text{ROC} = \{|z| < 0.5\} \cap \{|z| > 1\} = \{\emptyset\}$$

Such a time domain signal does not exist, since its ROC is an empty set.

The fourth possibility is that both signals are anticausal, that is,

$$x[k] = -\frac{4}{3}(0.5)^k u[-k-1] + \frac{4}{3}(-1)^k u[-k-1]$$

$$\text{ROC} = \{|z| < 0.5\} \cap \{|z| < 1\} = \{|z| < 0.5\}$$

which is also a valid inversion.

In summary, we have identified three different time domain signals that have the same double-sided $\mathcal{Z}$-transform, but different ROCs. In Problem 9.18, the student is asked to plot and compare the discrete-time signals obtained.

9.3.1   Double-Sided $\mathcal{Z}$-Transform in Linear Systems

The double-sided $\mathcal{Z}$-transform can be used in the analysis of discrete-time linear systems with constant coefficients in the same manner that the one-sided $\mathcal{Z}$-transform was used in Chapter 5. In addition, the double sided $\mathcal{Z}$-transform can handle noncausal signals, so that it is used for the analysis and design of digital filters. Since the DTFT can be used for the same purpose, this section parallels Section 9.2.1 on the use of the DTFT in linear systems.

Consider the time-invariant linear discrete-time system in (9.45). Let $y[k] \leftrightarrow Y(z)$ and $x[k] \leftrightarrow X(z)$. Applying the double-sided $\mathcal{Z}$-transform to this system and using its time shifting property, we obtain

$$[z^n + a_{n-1}z^{n-1} + \cdots + a_1 z + a_0]Y(z) = [b_m z^m + b_{m-1}z^{m-1} + \cdots + b_1 z + b_0]X(z)$$

from which we have

$$Y(z) = \frac{b_m z^m + b_{m-1}z^{m-1} + \cdots + b_1 z + b_0}{z^n + a_{n-1}z^{n-1} + \cdots + a_1 z + a_0} X(z) \triangleq H(z)X(z) \qquad (9.67)$$

where the quantity

$$H(z) \triangleq \frac{b_m z^m + b_{m-1}z^{m-1} + \cdots + b_1 z + b_0}{z^n + a_{n-1}z^{n-1} + \cdots + a_1 z + a_0} \qquad (9.68)$$

defines the *discrete-time system transfer function*. Assuming that the input signal is equal to the discrete-time impulse delta function $\delta[k]$, whose double-sided $\mathcal{Z}$-transform is equal to 1, we conclude that $H(z)$ is the double-sided $\mathcal{Z}$-transform of the *discrete-time system impulse response,* and we have the pair

$$h[k] \leftrightarrow H(z) \qquad (9.69)$$

From the time convolution property of the double-sided $\mathcal{Z}$-transform and (9.67), it can be concluded that the *linear discrete-time system response to any excitation $x[k]$ is given by the convolution of that input signal and the system impulse response,* that is,

$$y[k] = x[k] * h[k] \qquad (9.70)$$

Note that the convolution result obtained represents only the *zero-state* component of the system response, since double-sided $\mathcal{Z}$-transform (like the DTFT and the Fourier transform) is unable to take into account the system initial conditions and provide the zero-input component of the system response.

9.4 DISCRETE FOURIER TRANSFORM (DFT)

The discrete-time Fourier transform (DTFT) can also be derived in the process of the numerical evaluation of the integral that defines the Fourier transform. Consider the basic definition of the Fourier transform,

$$X(j\omega) = \int_{-\infty}^{\infty} x(t) e^{-j\omega t} dt$$

and approximate it by an infinite sum obtained by performing sampling (discretization of the time axis), with sampling period T_s. In such a case, we have

$$X(j\omega) \approx T_s \left\{ \sum_{k=-\infty}^{\infty} x(kT_s)e^{-j\omega kT_s} \right\} = T_s X_\delta(j\omega) = T_s X(j\Omega)$$

$$= T_s \sum_{k=-\infty}^{\infty} x[k] e^{-jk\Omega}, \quad \Omega = \omega T_s$$

(9.71)

and

$$X(j\omega) = \lim_{T_s \to 0} \{T_s X_\delta(j\omega)\} = \lim_{T_s \to 0} \{T_s X(j\Omega)\} \qquad (9.72)$$

In order to calculate this sum numerically (using a computer), we must approximate the infinite duration signal $x[k]$ by its finite duration approximation. To that end, we first define the DTFT *of length L* as

$$\sum_{k=0}^{L-1} x[k] e^{-jk\Omega} \qquad (9.73)$$

This approximation is meaningful if the signal values are concentrated in the time interval $[0, L-1]$, and those outside of this interval have negligible values. Since L can be arbitrarily chosen, and noncausal signals can be shifted to the right (in the direction of positive time instants), in most cases we can find L such that the signal DTFT of length L can be used to approximate well the signal DTFT (and the signal Fourier transform, assuming that T_s is small enough—see (9.72)).

Our goal is to numerically compute the DTFT (and the Fourier transform), which is a function of the digital frequency Ω. Since $X(j\Omega)$ is a 2π-periodic function, in numerical computations of the corresponding sum we need only consider the interval $\Omega \in [0, 2\pi]$. For the purpose of such computations, this frequency interval must be divided into N subintervals, so that the digital frequencies are evaluated at

$$\Omega_n = \frac{2\pi n}{N}, \quad n = 0, 1, 2, \ldots, N-1 \qquad (9.74)$$

(Note that these frequencies are equidistantly distributed on the unit circle $e^{j\Omega}$.) Now, we are ready to define the DFT, more precisely the *N-point DFT of a length L signal*, as

$$\text{DFT}\{x[k]\} \triangleq X(j\Omega_n) = \sum_{k=0}^{L-1} x[k] e^{-jk\Omega_n} = \sum_{k=0}^{L-1} x[k] e^{-jk(2\pi n/N)},$$

$$n = 0, 1, 2, \ldots, N-1$$

(9.75)

The number of time samples L and the number of frequency samples N can be independently chosen. It is convenient to make $L = N$. If $L < N$, we can add zeros at the end of the signal time samples to increase its length to N. This procedure is called *zero* padding, and is demonstrated in the next example. Note that zero padding does not affect the DFT result.

EXAMPLE 9.11

Consider the following signal samples, and assume that $N = 8$:

$$x = [x_0, x_1, x_2, x_3, x_4], \quad L = 5$$

This signal is zero padded to the length $L = N = 8$ as follows:

$$x = [x_0, x_1, x_2, x_3, x_4, 0, 0, 0], \quad L = 8$$

Matrix Form of the DFT

The N-point DFT of the length L signal (formula (9.75)) can be easily and conveniently recorded in matrix form. Let us first form the vectors of signal time samples and the signal frequency domain DFT samples as

$$\mathbf{x} = \begin{bmatrix} x[0] \\ x[1] \\ \vdots \\ x[L-1] \end{bmatrix}, \quad \mathbf{X} = \begin{bmatrix} X[0] \\ X[1] \\ \vdots \\ X[N-1] \end{bmatrix} = \begin{bmatrix} X(j\Omega_0) \\ X(j\Omega_1) \\ \vdots \\ X(j\Omega_{N-1}) \end{bmatrix} \quad (9.76)$$

Then, (9.75) has the matrix form

$$\mathbf{X} = \mathbf{A}\mathbf{x} = \text{DFT}\{\mathbf{x}\} \quad (9.77)$$

where the matrix $\mathbf{A}$ is defined by

$$\mathbf{A}^{N \times L} = \{A_{nk}\} = \{e^{-jk\Omega_n}\} = \{e^{-jk(2\pi n/N)}\} \triangleq \{W_N^{nk}\} \quad (9.78)$$

All elements of the matrix $\mathbf{A}$ can be easily determined. Observe first that $A_{0k} = e^0 = 1$, $\forall k$, and $A_{n0} = e^0 = 1$, $\forall n$, which implies that all elements in the first row and in the first column of the matrix $\mathbf{A}$ are equal to 1. Simple algebra (evaluation of $e^{-jk(2\pi n/N)}$ for given values of k, n, N) produces other entries in matrix $\mathbf{A}$. For example,

$$\mathbf{A}^{2\times 2} = \begin{bmatrix} 1 & 1 \\ 1 & -1 \end{bmatrix}, \quad \mathbf{A}^{4\times 4} = \begin{bmatrix} 1 & 1 & 1 & 1 \\ 1 & -j & -1 & j \\ 1 & -1 & 1 & -1 \\ 1 & j & -1 & -j \end{bmatrix} \quad (9.79)$$

It is left as an exercise for the student to find matrix $\mathbf{A}^{3\times 3}$ (see Problem 9.21).

Matrix Form for the Inverse DFT (IDFT)

Assuming that matrix $\mathbf{A}^{N \times L}$ is square ($N = L$) and invertible, then (9.77) provides a simple formula for the recovery of the signal time domain samples from the frequency domain DFT samples. From (9.77), under the preceding assumptions, we have the following definition of the *inverse* DFT, denoted by IDFT:

$$\mathbf{x} = \mathbf{A}^{-1}\mathbf{X} = \text{IDFT}\{\mathbf{X}\} \tag{9.80}$$

where the vectors $\mathbf{x}$ and $\mathbf{X}$ are defined as in (9.76). Hence, the crucial step is to show that the square matrix $\mathbf{A}$ is invertible. This can be done by multiplying the matrix $\mathbf{A}$ by its complex conjugate, $\mathbf{A}^*$, obtained by taking the complex conjugate of each element in the matrix. It can be easily shown that

$$\frac{1}{N}\mathbf{A}\mathbf{A}^* = \mathbf{I}_N \tag{9.81}$$

where $\mathbf{I}_N$ is an identity matrix of dimension N. This implies that the inverse of matrix $\mathbf{A}$ has a very simple form, given by

$$\mathbf{A}^{-1} = \frac{1}{N}\mathbf{A}^* \tag{9.82}$$

Hence, the *IDFT is given by the very simple formula*

$$\mathbf{x} = \mathbf{A}^{-1}\mathbf{X} = \frac{1}{N}\mathbf{A}^*\mathbf{X} = \text{IDFT}\{\mathbf{X}\} \tag{9.83}$$

Scalar Form of the IDFT

From (9.83) we can recover every component of the discrete-time signal, by observing from (9.78) and (9.82) the fact that

$$\mathbf{A}^{-1} = \frac{1}{N}\mathbf{A}^* = \left\{\frac{1}{N}A_{nk}^*\right\} = \left\{\frac{1}{N}e^{jk(2\pi n/N)}\right\} \triangleq \left\{\frac{1}{N}W_N^{-nk}\right\} \tag{9.84}$$

From (9.80), for each particular sample value of $\mathbf{x}$, we have

$$x[k] = \frac{1}{N}\sum_{n=0}^{N-1} W_N^{-nk} X(j\Omega_n) = \frac{1}{N}\sum_{n=0}^{N-1} e^{jk\Omega_n} X(j\Omega_n) \tag{9.85}$$

Hence, the *scalar form* of the IDFT is given by

$$x[k] = \frac{1}{N}\sum_{n=0}^{N-1} X(j\Omega_n)e^{jk\Omega_n}, \quad k = 0, 1, 2, \ldots, N-1 \tag{9.86}$$

where Ω_n is defined as in (9.74). Note that (9.86) was derived under the assumption that $N = L$. If that were not the case, then L time domain samples of the vector $\mathbf{x}$ would not be uniquely recovered from N frequency domain DFT samples of the vector $\mathbf{X}$.

Section 9.4 Discrete Fourier Transform (DFT) 489

Discrete-Time Signal Wrapping (Modulo-N Reduction)
In the case when $L > N$, a very simple procedure called signal wrapping can reduce the original signal length to N. Note that the value for L is dictated by the condition of the sampling theorem, which sometimes produces large values for L. The wrapping procedure is demonstrated in the next example.

EXAMPLE 9.12

Let $L = 8$ and $N = 4$, with the signal values being represented by

$$x = [x_0, x_1, x_2, x_3, x_4, x_5, x_6, x_7], \quad L = 8$$

The modulo-N reduced (wrapped) signal is given by

$$x = [x_0 + x_4, x_1 + x_5, x_2 + x_6, x_3 + x_7], \quad N = 4$$

When L is not an integer multiple of N, the signal of length L can first be zero padded and then wrapped:

Let $L = 10$ and $N = 4$. We first form a signal of length $12 = 3N$ by zero padding, that is,

$$x = [x_0, x_1, x_2, x_3, x_4, x_5, x_6, x_7, x_8, x_9, 0, 0], \quad L = 12$$

and then wrap this signal into a signal of length $N = 4$:

$$x = [x_0 + x_4 + x_8, x_1 + x_5 + x_9, x_2 + x_6 + 0, x_3 + x_7 + 0], \quad N = 4$$

The wrapping procedure is a consequence of the following simple fact. Let us denote the wrapped signal by $\tilde{\mathbf{x}}$, and let $L = rN$. Then the N-point DFT of a length-L signal is given by

$$\mathbf{X}^{N \times 1} = \mathbf{A}^{N \times L} \mathbf{x}^{L \times 1} = \underbrace{[\tilde{\mathbf{A}}^{N \times N} \tilde{\mathbf{A}}^{N \times N} \cdots \tilde{\mathbf{A}}^{N \times N}]}_{r \text{ times}} \mathbf{x}^{L \times 1} \Leftrightarrow X^{N \times 1} = \tilde{\mathbf{A}}^{N \times N} \tilde{\mathbf{x}}^{N \times 1} \quad (9.87)$$

where the N columns of matrix $\tilde{\mathbf{A}}^{N \times N}$ are identical to the first N columns of matrix $\mathbf{A}$, that is,

$$\{\tilde{A}_{nk}\} = \{A_{nk}\} = \{e^{-jk(2\pi n/N)}\} = \{W_N^{nk}\}, \quad 0 \leq n, k \leq N - 1 \quad (9.88)$$

It should be emphasized that the *inverse DFT for the wrapped signal produces the samples of the wrapped signal;* that is, from (9.87) we have

$$\tilde{\mathbf{x}} = (\tilde{\mathbf{A}})^{-1} \mathbf{X} = \frac{1}{N}(\tilde{\mathbf{A}})^* \mathbf{X} = \text{IDFT}\{\mathbf{X}\} \quad (9.89)$$

IDFT in Terms of DFT

An interesting relation can be obtained from (9.77) and (9.83). It follows from (9.83) that

$$\text{IDFT}(\mathbf{X}) = \frac{1}{N}\mathbf{A}^*\mathbf{X} = \frac{1}{N}(\mathbf{A}\mathbf{X}^*)^* = \frac{1}{N}(\text{DFT}(\mathbf{X}^*))^*$$

so that we have

$$\text{IDFT}\{\mathbf{X}\} = \frac{1}{N}(\text{DFT}\{\mathbf{X}^*\})^* \qquad (9.90)$$

This property of the DFT and IDFT is used in the fast Fourier transform, to be introduced in the next section.

9.4.1 Fast Fourier Transform (FFT)

In order to calculate the DFT from (9.87), we must multiply an $N \times N$ dimensional matrix by a vector of length N. In general, multiplying a square matrix of order N by a corresponding vector requires N^2 scalar multiplications. In a celebrated paper that marks the beginning of the scientific discipline of digital signal processing [6], it was shown how to exploit the special structure of the matrix $\tilde{\mathbf{A}}$ to evaluate the required product more efficiently and to reduce the number of required scalar multiplications. During the last thirty-five years, many FFT algorithms have been developed. The main idea of these algorithms is to evaluate the N-point DFT in terms of two $N/2$-point DFTs, and then to evaluate the $N/2$-point DFT in terms of two $N/4$-point DFTs, and so on. Those algorithms have a common feature, in that the number of scalar multiplications required to evaluate the DFT is given by

$$\frac{1}{2}N\log_2(N) \ll N^2, \quad N \text{ large} \qquad (9.91)$$

which, for large values of N, brings significant savings. Note that the function $0.5 \times \log_2(N)$ grows much more slowly than a linear function N as N increases.

Detailed study of FFT algorithms is beyond the scope of this book. We refer the interested reader to [1] for a discussion of the standard FFT algorithms. Furthermore, since digital computers now perform scalar multiplications and scalar additions with equal speed, we will certainly see in the near future many new FFT algorithms that are efficient in view of the number of both scalar multiplications and scalar additions needed to evaluate the DFT (9.87).

For the purpose of this course, it is sufficient to evaluate the FFT using the MATLAB function fft. This can be done in a very simple manner by the expression X=fft(x,N), where **x** is the N-data point vector and **X** is the sought DFT. Note that in evaluating fft(x,N) the data vector **x** is padded with zeros if it has less than N samples and truncated if it has more than N samples. The inverse FFT is computed via MATLAB using the function ifft as x=ifft(X,N), which basically evaluates the DFT as indicated in formula (9.90). These MATLAB functions will be used in the MATLAB laboratory experiment in Section 9.8.

9.5 DISCRETE-TIME FOURIER SERIES (DFS)

In Section 9.1 we introduced the DTFT through the sampling operation of a continuous-time signal, and in Section 9.4 we introduced the DFT from the DTFT. As indicated in Section 9.2, the DTFT could have been derived from the discrete-time Fourier series (DFS), as the Fourier transform was derived in Chapter 3 from the continuous-time Fourier series. This approach to deriving the DFT has been taken by several authors (see, for example, [5]). We have already introduced the DFT; now the DFS comes as a by-product of the DFT.

Discrete-time Fourier series can be easily deduced from the N-point DFT and its IDFT. Namely, from (9.75) and (9.86), we have the pair

$$x[k] = \frac{1}{N} \sum_{n=0}^{N-1} X(jn\Omega_0) e^{jkn\Omega_0}, \quad \Omega_0 = \frac{2\pi}{N}, \quad k = 0, 1, 2, \ldots, N-1 \quad (9.92)$$

and

$$X(jn\Omega_0) = \sum_{k=0}^{N-1} x[k] e^{-jkn\Omega_0}, \quad \Omega_0 = \frac{2\pi}{N}, \quad n = 0, 1, 2, \ldots, N-1 \quad (9.93)$$

Knowing that $x[k]$ is a periodic signal with period N, that is, $x[k] = x[k+N]$, we conclude that (9.92) represents the periodic signal expansion in terms of N harmonics of its *fundamental digital harmonic* $\Omega_0 = 2\pi/N$ (see (9.22)). The fact that we need only N harmonics to represent a periodic signal follows from the result

$$e^{j(k+rN)n\Omega_0} = e^{jkn\Omega_0} e^{jrn2\pi} = e^{jkn\Omega_0}, \quad r = 0, 1, 2, \ldots$$

Hence, the *discrete-time Fourier series* are defined by formula (9.92). Consequently, formula (9.93) defines the *discrete-time Fourier series coefficients*. Since a periodic signal has an infinite duration, the *signal obtained in (9.92) must be periodically extended along the entire time axis* so that $x[N] = x[0]$, $x[N+1] = x[1], \ldots$. The same extension should be done for the negative values of the discrete-time instants.

It can be observed from (9.92) that, in contrast to the continuous-time Fourier series (which represents a continuous-time periodic signal in terms of an infinite sum of its harmonics, (3.14)), the DFS *is a finite sum* that contains N terms, where N is the period of the discrete-time periodic signal. Being represented by a finite sum, the DFS is always convergent, in contrast to the continuous-time Fourier series, whose convergence requires that the Dirichlet conditions be satisfied (see (3.12)).

EXAMPLE 9.13

Consider the periodic discrete-time signal

$$x[k] = \begin{cases} 1, & k = 0 \pm rN \\ 1, & k = 1 \pm rN \\ 0, & k = 2 \pm rN \\ 0, & k = 3 \pm rN \end{cases}, \quad N = 4, r = 0, 1, 2, \ldots$$

with period $N = 4$. The signal fundamental digital harmonic is equal to $\Omega_0 = 2\pi/N = 2\pi/4 = \pi/2$. The DFS coefficients from (9.93) are given by

$$X(jn\Omega_0) = \sum_{k=0}^{N-1} x[k]e^{-jkn\Omega_0} = X\left(jn\frac{\pi}{2}\right) = \sum_{k=0}^{3} x[k]e^{-jkn(\pi/2)}, \quad n = 0, 1, 2, 3$$

Evaluating each of the four Fourier series coefficients, we obtain

$$X(j0) = x[0] + x[1] + x[2] + x[3] = 2$$

$$X\left(j\frac{\pi}{2}\right) = x[0] + x[1]e^{-j(\pi/2)} + x[2]e^{-j\pi} + x[3]e^{-j3(\pi/2)} = 1 - j$$

$$X(j\pi) = x[0] + x[1]e^{-j\pi} + x[2]e^{-j2\pi} + x[3]e^{-j3\pi} = 0$$

$$X\left(j3\frac{\pi}{2}\right) = x[0] + x[1]e^{-j3(\pi/2)} + x[2]e^{-j3\pi} + x[3]e^{-j9(\pi/2)} = 1 + j$$

Hence, the DFS representation of the given periodic signal is

$$x[k] = \frac{1}{4} \sum_{n=0}^{3} X(jn\Omega_0) e^{jkn(\pi/2)}, \quad k = 0, 1, 2, 3$$

Of course, this must be periodically extended to any discrete-time instant k, since $x[k] = x[k+4]$. We can easily check that the obtained DFS represents the original signal, that is,

$$x[0] = \frac{1}{4}\left(X(j0) + X\left(j\frac{\pi}{2}\right) + X(j\pi) + X\left(j3\frac{\pi}{2}\right)\right)$$

$$= \frac{1}{4}(2 + 1 - j + 0 + 1 + j) = 1$$

$$x[1] = \frac{1}{4}\left(X(j0) + X\left(j\frac{\pi}{2}\right)e^{j(\pi/2)} + X(j\pi)e^{j\pi} + X\left(j3\frac{\pi}{2}\right)e^{j3(\pi/2)}\right)$$

$$= \frac{1}{4}(2 + (1-j)j + 0 - j(1+j)) = 1$$

$$x[2] = \frac{1}{4}\left(X(j0) + X\left(j\frac{\pi}{2}\right)e^{j\pi} + X(j\pi)e^{j2\pi} + X\left(j3\frac{\pi}{2}\right)e^{j3\pi}\right)$$

$$= \frac{1}{4}(2 - (1-j) + 0 - (1+j)) = 0$$

$$x[3] = \frac{1}{4}\left(X(j0) + X\left(j\frac{\pi}{2}\right)e^{j3(\pi/2)} + X(j\pi)e^{j3\pi} + X\left(j3\frac{\pi}{2}\right)e^{j9(\pi/2)}\right)$$

$$= \frac{1}{4}(2 - j(1-j) + 0 + j(1+j)) = 0$$

Discrete-Time Linear System Response to Periodic Inputs

Using the convolution formula that determines the discrete-time system zero-state response due to the given input $x[k]$, we can find the discrete-time steady-state response due to a periodic input. Similarly, as discussed in Section 5.5.1, we conclude that a hypothetical system input $(e^{j n \Omega_0})^k$ produces the discrete-time system output $(e^{j n \Omega_0})^k H(j n \Omega_0)$, where $H(j n \Omega_0)$ is the discrete-time system digital frequency transfer function, as defined in (9.47). Using the linearity principle, the discrete-time system periodic input of the period N and the fundamental digital harmonic $\Omega_0 = 2\pi/N$, represented by its DFS as defined in (9.92), produces the steady-state response

$$y_{ss}[k] = \frac{1}{N} \sum_{n=0}^{N-1} X(j n \Omega_0) H(j n \Omega_0) e^{j k n \Omega_0} = \frac{1}{N} \sum_{n=0}^{N-1} Y(j n \Omega_0) e^{j k n \Omega_0} \qquad (9.94)$$

where the DFS coefficients for the system output are given by

$$Y(j n \Omega_0) = X(j n \Omega_0) H(j n \Omega_0) \qquad (9.95)$$

It can be concluded that the *system output is also periodic with the same period (the same fundamental digital frequency) as the input signal.*

9.6 CORRELATION OF DISCRETE-TIME SIGNALS

Signal correlation is an operation similar to signal convolution, but with a completely different physical meaning. Recall that signal convolution physically represents the response of a linear system with one signal playing the role of the system impulse response and another representing the system input signal. Signal correlation can be performed with one signal (autocorrelation) or between two signals (crosscorrelation). Physically, signal autocorrelation indicates how the signal energy (power) is distributed within the signal, and as such is used to measure signal power. Typical applications of signal autocorrelation are in radar, sonar, satellite, and wireless communications systems. Devices that measure signal power using signal correlation are known as signal correlators. There are also many applications of signal crosscorrelation in signal processing systems, especially when a signal of interest is corrupted by another undesirable signal (noise), so that signal estimation (detection) from a noisy signal must be performed. Signal crosscorrelation can also be considered a measure of the similarity between two signals. In the following, we give definitions of autocorrelation and crosscorrelation.

DEFINITION 9.3: DISCRETE-TIME AUTOCORRELATION AND CROSSCORRELATION Given two discrete-time real signals (sequences) $x[k]$ and $y[k]$, the *autocorrelation* and *croosscorrelation* functions are respectively defined by

$$R_{xx}[k] = \sum_{m=-\infty}^{\infty} x[m]x[m-k], \quad R_{yy}[k] = \sum_{m=-\infty}^{\infty} y[m]y[m-k] \qquad (9.96)$$

$$R_{xy}[k] = \sum_{m=-\infty}^{\infty} x[m]y[m-k], \quad R_{yx}[k] = \sum_{m=-\infty}^{\infty} y[m]x[m-k] \qquad (9.97)$$

where the parameter k is any integer, $-\infty \leq k \leq \infty$.

Using the definition for the total discrete-time signal energy given in (2.55), we see that for $k = 0$, the autocorrelation function represents the total signal energy, that is,

$$R_{xx}[0] = E_\infty^x, \quad R_{yy}[0] = E_\infty^y \tag{9.98}$$

Naturally, the autocorrelation and crosscorrelation sums, (9.96) and (9.97), are convergent under assumptions that the signals $x[k]$ and $y[k]$ have finite total energy. It can be observed that $R_{xx}[k] \leq R_{xx}[0] = E_\infty^x$. In addition, it is easy to show that the autocorrelation function is an even function, that is,

$$R_{xx}[k] = R_{xx}[-k] \tag{9.99}$$

Hence, the autocorrelation function is symmetric with respect to the vertical axis. Also, it can be shown that

$$R_{xy}[k] = R_{xy}[-k] \tag{9.100}$$

(see Problem 9.29).

Note that the autocorrelation and crosscorrelation functions are applicable to energy signals (which have finite total energy); they can also be defined for power signals (which have infinite energy, but finite power). In that case, we must redefine the sums in (9.96–97) in keeping with the definition formula of the discrete-time signal average power as given in (2.56). For more information on the correlation of power signals (periodic signals), the interested reader is referred to [7].

It is interesting to observe that the autocorrelation and crosscorrelation functions can be evaluated using the discrete-time convolution as follows:

$$R_{xx}[k] = x[k] * x[-k], \quad R_{xy}[k] = x[k] * y[-k] \tag{9.101}$$

It is left to the student as an exercise to establish these results (Problem 9.30).

As a measure of the similarity between two signals, we can use the *correlation coefficient* defined by

$$c_{xy} = \frac{R_{xy}[0]}{\sqrt{R_{xx}[0]R_{yy}[0]}} \tag{9.102}$$

Note that the correlation coefficient satisfies $-1 \leq c_{xy} \leq 1$. This can be established by observing that $R_{xy}[0] = \mathbf{x} \cdot \mathbf{y}$ is the inner product of the vectors that contain, respectively, the samples of $x[k]$ and $y[k]$. Similarly, the relations $R_{xx}[0] = \mathbf{x} \cdot \mathbf{x} = |\mathbf{x}|^2$ and $R_{yy}[0] = \mathbf{y} \cdot \mathbf{y} = |\mathbf{y}|^2$ represent the squares of the corresponding vector Euclidean norms, so that the correlation coefficient geometrically represents the angle between the vectors $\mathbf{x}$ and $\mathbf{y}$, that is,

$$-1 \leq c_{xy} = \frac{R_{xy}[0]}{\sqrt{R_{xx}[0]R_{yy}[0]}} = \frac{\mathbf{x} \cdot \mathbf{y}}{\sqrt{|\mathbf{x}|^2 |\mathbf{y}|^2}} = \frac{\mathbf{x} \cdot \mathbf{y}}{|\mathbf{x}| |\mathbf{y}|} \triangleq \cos(\mathbf{x}, \mathbf{y}) \leq 1 \tag{9.103}$$

When the correlation coefficient is close to 1, the signals are similar (they almost overlap), and when the correlation coefficient is close to 0, the signals are very different (orthogonal, as a matter of fact). When the correlation coefficient is close to -1, the signals are *asimilar* (opposite direction, but almost the same sample values). Note that the correlation coefficient can also be defined in terms of parameter k, that is,

$$-1 \le c_{xy}[k] = \frac{R_{xy}[k]}{\sqrt{R_{xx}[0]R_{yy}[0]}} \le 1 \qquad (9.104)$$

in which case the same lower and upper bounds hold, due to the fact that $|R_{xy}[k]| \le \sqrt{|R_{xx}[0]||R_{yy}[0]|}$ (see Problem 9.31).

The crosscorrelation operation is used for the detection (estimation) of signals from measured signals that contain the original signal corrupted by an additive noise; that is, $x[k] + w[k]$, where $x[k]$ is the original signal and $w[k]$ is noise. The signal that has the highest correlation with the signal $x[k] + w[k]$ is considered the best estimate of the signal $x[k]$, and denoted by $\hat{x}[k]$.

Note that the discrete-time autocorrelation and crosscorrelation functions can be evaluated using the MATLAB function xcorr, respectively, as Rxx=xcorr(x) and Rxy=xcorr(x,y).

DTFT of Autocorrelation and Crosscorrelation Functions

The DTFT of the auto- and crosscorrelation functions can be found similarly to the DTFT of the convolution function. Define the DTFT pairs $x_1[k] \leftrightarrow X_1(j\Omega)$ and $x_2[k] \leftrightarrow X_2(j\Omega)$. Then, the auto- and crosscorrelation functions of these two signals satisfy

$$R_{x_1 x_1}[k] \leftrightarrow X_1(j\Omega)X_1^*(j\Omega) = |X_1(j\Omega)|^2, \quad R_{x_1 x_2}[k] \leftrightarrow X_1(j\Omega)X_2^*(j\Omega) \qquad (9.105)$$

The proof of this property follows the convolution property proof (see Problem 9.9). The quantity $|X_1(j\Omega)|^2$ is called the *energy spectral density* of the signal $x_1[k]$. Hence, the discrete-time signal energy spectral density is the DTFT of the signal autocorrelation function.

9.7 IIR AND FIR FILTERS

Filtering is one of major operations performed on digital signals. For that purpose there are two main classes of digital filters: *infinite-duration impulse response* (IIR) and *finite-duration impulse response* (FIR) filters.

IIR filters are also known as *recursive filters,* since they are represented by linear difference equations with constant coefficients. In general, they behave like linear discrete-time systems with constant coefficients.

The simplest form of the IIR filter is known as the *autoregressive* (AR) filter. It is given by

$$y[k-n] + a_{n-1}y[k-(n-1)] + \cdots + a_1 y[k-1] + a_0 y[k] = x[k] \qquad (9.106)$$

The transfer function of the autoregressive IIR filter is

$$H(z) = \frac{Y(z)}{X(z)} = \frac{1}{a_0 + a_1 z^{-1} + \cdots + z^{-n}} \qquad (9.107)$$

Note that in digital signal processing it is customary to express the transfer function in terms of z^{-1}, since z^{-i} indicates a time delay of i discrete-time units (steps).

A more complex IIR filter than that defined in (9.106) is the *autoregressive moving average* (ARMA) filter, whose difference equation is given by

$$y[k-n] + a_{n-1}y[k-(n-1)] + \cdots + a_1 y[k-1] + a_0 y[k]$$
$$= b_m x[k-m] + b_{m-1} x[k-(m-1)] + \cdots + b_1 x[k-1] + b_0 x[k] \quad (9.108)$$

The transfer function of the autoregressive moving average IIR filter can be easily obtained from (9.108) as

$$H(z) = \frac{Y(z)}{X(z)} = \frac{b_0 + b_1 z^{-1} + \cdots + b_m z^{-m}}{a_0 + a_1 z^{-1} + \cdots + z^{-n}} \quad (9.109)$$

It is obvious that the impulse response of IIR filters has infinite duration, since it is obtained as $h[k] = \mathcal{Z}^{-1}\{H(z)\}$, where $H(z)$ is a rational function that has n poles.

The FIR filters are represented by the difference equation

$$y[k] = b_0 x[k] + b_1 x[k-1] + \cdots + b_m x[k-m] \quad (9.110)$$

whose transfer function is

$$H(z) = \frac{Y(z)}{X(z)} = b_0 + b_1 z^{-1} + \cdots + b_m z^{-m} \quad (9.111)$$

The impulse response of an FIR filter is given by

$$h[k] = \mathcal{Z}^{-1}\{H(z)\} = \begin{cases} b_k, & k = 0, 1, 2, \ldots, m \\ 0, & \text{otherwise} \end{cases} \quad (9.112)$$

Hence, FIR filters have only a finite number of nonzero samples in their impulse response.

Assuming that the filter coefficients are known, the filter response can be obtained using the MATLAB function `filter`, specifying the vector of the coefficients $\mathbf{a} = [a_0 \; a_1 \; \cdots \; a_{n-1} \; 1]^T$, $\mathbf{b} = [b_0 \; b_1 \; \cdots \; b_m]^T$, and the input data sequence x, as `y=filter(b,a,x)`. Note that the output sequence y will have the same length as the input sequence.

There are many techniques for the design of IIR and FIR digital filters. The general digital filter design problem can be formulated as follows: choose the parameter vectors $\mathbf{a}$ and $\mathbf{b}$ (including the orders n and m) such that the filter frequency characteristics (magnitude and phase spectra with given frequency bandwidth, frequency pick, attenuation outside of the frequency bandwidth, phase linearity, and so on) have the desired properties. FIR digital filters are designed using the same principles as analog filters studied in electrical circuits (notch, comb, Butterworth, Chebyshev, and other analog filters), and mapping those filters into the z-domain via the bilinear transformation, defined in Chapter 5. IIR filters are designed directly in the z-domain using various "windowing" techniques (among which Hamming and Kaiser windows are the most popular). Digital filter design techniques are outside of the scope of this book. The interested reader is referred to [1] and [8] for coverage of this very broad digital signal processing area.

9.8 LABORATORY EXPERIMENT ON SIGNAL PROCESSING

Purpose: By performing this experiment, the student will receive a better understanding of the use and power of the FFT algorithm in evaluating the corresponding discrete (time) Fourier transforms, continuous-time Fourier transform, and discrete-time convolution. As the computational tool, we will use the MATLAB functions `fft` and `ifft`.

Part 1. In this part we use the FFT algorithm, as implemented in MATLAB, to find the DFT of some discrete-time signals. In addition, we demonstrate the use of the IFFT in recovering original discrete-time signals.

(a) Consider the discrete-time signal

$$x[k] = \begin{cases} 1, & k = 0, 1, 2, 3 \\ 0, & \text{otherwise} \end{cases}$$

and find analytically its DTFT.

(b) Use the MATLAB function `X=fft(x,N)` to find the DFT of the preceding signal for $N = 4, 8, 12, 16, 24, 32$. Use the MATLAB function `x=ifft(X,N)` to recover the original discrete-time signal. Plot the DFTs and IDFTs, and comment on the results obtained.

(c) Consider the signal

$$x[k] = \begin{cases} 1, & k = 0, 1, 2, \ldots, 11 \\ 0, & \text{otherwise} \end{cases}$$

and repeat Parts (a) and (b).

(d) Consider the signal whose nonzero values are between $k=0$ and $k=8$, respectively defined by $x = [1, 2, 3, 4, 5, 4, 3, 2, 1]$, $L = 9$, and repeat Parts (a) and (b). Comment on the results obtained.

Part 2. Formula (9.71), $X(j\omega) \approx T_s X_\delta(j\Omega)$, can be used for an approximate evaluation of the continuous-time Fourier transform. In this formula, T_s is the sampling interval used for sampling the continuous-time signal $x(t)$ into $x(kT_s) \stackrel{\Delta}{=} x[k]$, and $X_\delta(j\Omega)$ is the corresponding DFT.

(a) Consider the continuous-time signals presented in Figures 3.22 and 3.23. Sample these signals with $T_s = 0.1$ and find the DFTs of the obtained discrete-time signals. Calculate and plot the corresponding magnitude spectra for the approximate Fourier transforms, and compare them to the results obtained analytically.

(b) Repeat Part (a) with $T_s = 0.01$.

Part 3. Discrete-time signal convolution can be efficiently evaluated via the DTFT and its convolution property. The relation $y[k] = x[k] * h[k]$ implies $Y(j\Omega) = X(j\Omega)H(j\Omega)$. Hence, discrete-time convolution via DFT can be evaluated as $y[k] = \text{IDFT}\{\text{DFT}(x[k])\text{DFT}(h[k])\}$. Note that such an obtained signal $y[k]$ is, in general, the wrapped signal, so that the corresponding convolution is called mod-N *circular convolution* [1].

Use the preceding formula to find the convolution of the discrete-time signals defined in Problems 6.15 and 6.16. Do the results obtained represent unwrapped or wrapped signals?

9.9 SUMMARY

Study Guide for Chapter Nine: The sampling theorem states the very important result that for the exact reconstruction of a signal from its samples, the sampling period must be at least $1/(2f_{\max})$. It is important to remember that *sampling in the time domain implies periodicity in the frequency domain and, conversely, sampling in the frequency domain implies periodicity in the time domain*. Also, if time domain samples are transformed into frequency domain samples, then both the time and frequency domain signals are periodic. Students should be familiar with the definitions and master the properties of both the DTFT and the double-sided $\mathcal{Z}$-transform (see Tables 9.1 and 9.2). Note that these properties are very similar to the corresponding properties of the one-sided $\mathcal{Z}$-transform studied in Chapter 5. Numerical evaluation of the DTFT leads to the DFT. The DFT is evaluated using an extraordinarily simple matrix-vector product, as is the IDFT, due to a special structure of the given matrix. The most efficient numerical method for evaluating the DFT is the FFT. Due to a nice property of the matrix involved, the inverse FFT, called the IFFT, can be evaluated as another FFT. The DFS is deduced from the DTF. Remember that signal correlation is used in signal receivers to measure signal power and to estimate signals from noisy measurements (this topic is studied in more advanced signal processing courses). IIR and FIR filtering is the central theme of signal processing courses.

Sampling Theorem:

$$x(t) = \sum_{k=-\infty}^{\infty} x\left(\frac{k}{2f_{\max}}\right) \operatorname{sinc}(2f_{\max}t - k), \quad T_s = \frac{1}{2f_{\max}}$$

In general,

$$\frac{1}{T_s} = f_s \geq 2f_{\max}, \quad T_s = \text{sampling period}$$

Discrete-Time Fourier Transform (DTFT):

$$\text{DTFT}\{x(t)\} = \mathcal{F}\{x_\delta(t)\} = \mathcal{F}\{x(t) \sum_{k=-\infty}^{\infty} \delta(t - kT_s)\}$$

$$= \sum_{k=-\infty}^{\infty} x(kT_s)e^{-jkT_s\omega} \triangleq \sum_{k=-\infty}^{\infty} x[k]e^{-jk\Omega} = \text{DTFT}\{x[k]\}$$

$$\Omega = T_s\omega = \text{digital frequency}$$

Double-Sided $\mathcal{Z}$-Transform:

$$X(z) = \sum_{k=-\infty}^{\infty} x[k]z^{-k}, \quad \sum_{k=-\infty}^{\infty} |\rho^{-k}x[k]| < \infty \Rightarrow \text{ROC}$$

DTFT in Linear System Analysis:

$$Y(j\Omega) = H(j\Omega)X(j\Omega), \quad y[k] = h[k] * x[k]$$

$$H(j\Omega) = \frac{b_m e^{jm\Omega} + b_{m-1} e^{j(m-1)\Omega} + \cdots + b_1 e^{j\Omega} + b_0}{e^{jn\Omega} + a_{n-1} e^{j(n-1)\Omega} + \cdots + a_1 e^{j\Omega} + a_0}$$

Discrete-Time System Response due to $\cos(\Omega_0 k + \varphi)$:

$$y_{ss}[k] = |H(j\Omega_0)| \cos(k\Omega_0 + \varphi + \angle H(j\Omega_0))$$

Double-Sided $\mathcal{Z}$-Transform in Linear System Analysis:

$$Y(z) = H(z)X(z), \quad y[k] = h[k] * x[k]$$

$$H(z) = \frac{b_m z^m + b_{m-1} z^{m-1} + \cdots + b_1 z + b_0}{z^n + a_{n-1} z^{n-1} + \cdots + a_1 z + a_0}$$

Discrete Fourier Transform (DFT):

$$\text{DFT}\{x[k]\} \triangleq X(j\Omega_n) = \sum_{k=0}^{L-1} x[k] e^{-jk\Omega_n} = \sum_{k=0}^{L-1} x[k] e^{-jk(2\pi n/N)}$$

$$n = 0, 1, 2, \ldots, N$$

Inverse Discrete Fourier Transform (IDFT):

$$x[k] = \frac{1}{N} \sum_{n=0}^{N-1} X(j\Omega_n) e^{jk\Omega_n}, \quad k = 0, 1, 2, \ldots, N-1$$

Matrix Form of the DFT:

$$\mathbf{X} = \mathbf{A}\mathbf{x} = \text{DFT}\{\mathbf{x}\}$$

$$\mathbf{A}^{N \times L} = \{A_{nk}\} = \{e^{-jk\Omega_n}\} = \{e^{-jk(2\pi n/N)}\} \triangleq \{W_N^{nk}\}$$

Matrix Form of the Inverse DFT (for $N = L$):

$$\mathbf{x} = \mathbf{A}^{-1}\mathbf{X} = \frac{1}{N}\mathbf{A}^*\mathbf{X} = \text{IDFT}\{\mathbf{X}\}$$

IDFT in Terms of DFT:

$$\text{IDFT}\{\mathbf{X}\} = \frac{1}{N}(\text{DFT}\{\mathbf{X}^*\})^*$$

Discrete-Time Fourier Series:

$$x[k] = \frac{1}{N} \sum_{n=0}^{N-1} X(jn\Omega_0) e^{jkn\Omega_0}, \quad \Omega_0 = \frac{2\pi}{N}, \quad k = 0, 1, 2, \ldots, N-1$$

Chapter 9 Signals in Digital Signal Processing

Discrete-Time Fourier Series Coefficients:

$$X(jn\Omega_0) = \sum_{k=0}^{N-1} x[k]e^{-jkn\Omega_0}, \qquad \Omega_0 = \frac{2\pi}{N}, \quad n = 0, 1, 2, \ldots, N-1$$

Discrete-Time System Response to Periodic Inputs:

$$y_{ss}[k] = \frac{1}{N}\sum_{n=0}^{N-1} Y(jn\Omega_0)e^{jkn\Omega_0}, \qquad Y(jn\Omega_0) = X(jn\Omega_0)H(jn\Omega_0)$$

Discrete-Time Signal Autocorrelation:

$$R_{xx}[k] = \sum_{m=-\infty}^{\infty} x[m]x[m-k] = x[k] * x[-k]$$

$$R_{xx}[k] \le R_{xx}[0], \quad R_{xx}[k] = R_{xx}[-k], \quad -\infty \le k \le \infty$$

Discrete-Time Signal Crosscorrelation:

$$R_{xy}[k] = \sum_{m=-\infty}^{\infty} x[m]y[m-k] = x[k] * y[-k]$$

$$R_{xy}[k] = R_{xy}[-k], \quad -\infty \le k \le \infty$$

Discrete-Time Total Signal Energy:

$$R_{xx}[0] = E_\infty^x, \qquad R_{yy}[0] = E_\infty^y$$

Discrete-Time Signal Correlation Coefficients:

$$-1 \le c_{xy} = \frac{R_{xy}[0]}{\sqrt{R_{xx}[0]R_{yy}[0]}} = \frac{\mathbf{x} \cdot \mathbf{y}}{\sqrt{|\mathbf{x}|^2|\mathbf{y}|^2}} = \frac{\mathbf{x} \cdot \mathbf{y}}{|\mathbf{x}||\mathbf{y}|} \stackrel{\triangle}{=} \cos(\mathbf{x}, \mathbf{y}) \le 1$$

$$-1 \le c_{xy}[k] = \frac{R_{xy}[k]}{\sqrt{R_{xx}[0]R_{yy}[0]}} \le 1$$

Autoregressive (AR) IIR Filter:

$$y[k-n] + a_{n-1}y[k-(n-1)] + \cdots + a_1 y[k-1] + a_0 y[k] = x[k]$$

Autoregressive Moving Average (ARMA) IIR Filter:

$$y[k-n] + a_{n-1}y[k-(n-1)] + \cdots + a_1 y[k-1] + a_0 y[k]$$
$$= b_m x[k-m] + b_{m-1}x[k-(m-1)] + \cdots + b_1 x[k-1] + b_0 x[k]$$

FIR Filter:

$$y[k] = b_0 x[k] + b_1 x[k-1] + \cdots + b_m x[k-m]$$

TABLE 9.1 Properties of the DTFT

Discrete-time domain	$\Omega = T_s \omega$-domain
$a_1 x_1[k] \pm a_2 x_2[k]$	$a_1 X_1(j\Omega) \pm a_2 X_2(j\Omega)$
$x[k-m]$	$e^{-j\Omega m} X(j\Omega)$
$x[k] e^{j\Omega_0}$	$X(j(\Omega - \Omega_0))$
$x[-k]$	$X(-j\Omega)$
$x^*[k]$	$X^*(-j\Omega)$
$kx[k]$	$j \dfrac{dX(j\Omega)}{d\Omega}$
$x[k]\cos(\omega kT)$	$\dfrac{1}{2}[X(j(\Omega+\Omega_0)) + X(j(\Omega-\Omega_0))]$
$x[k]\sin(\omega kT)$	$\dfrac{j}{2}[X(j(\Omega+\Omega_0)) - X(j(\Omega-\Omega_0))]$
$x_1[k] * x_2[k]$	$X_1(j\Omega) X_2(j\Omega)$
$x_1[k] x_2[k]$	$\dfrac{1}{2\pi} \displaystyle\int_{-\pi}^{\pi} X_1(j\theta) X_2(j(\Omega - \theta))\, d\theta$
$R_{x_1 x_2}[k]$	$X_1(j\Omega) X_2^*(j\Omega)$

TABLE 9.2 Properties of the Double-Sided $\mathcal{Z}$-Transform

Discrete-time domain	z-domain
$a_1 x_1[k] \pm a_2 x_2[k]$	$a_1 X_1(z) \pm a_2 X_2(z)$
$x[k+m]$	$z^m X(z)$
$a^{-k} x[k]$	$X(az)$
$x[-k]$	$X\left(\dfrac{1}{z}\right)$
$x^*[k]$	$X^*(z^*)$
$kx[k]$	$-z \dfrac{d}{dz} X(z)$
$x_1[k] * x_2[k]$	$X_1(z) X_2(z)$

9.10 REFERENCES

[1] S. Orfanidis, *Introduction to Signal Processing*, Prentice Hall, Upper Saddle River, NJ, 1996.

[2] C. Shannon, "Communication in the presence of noise," *Proceedings of IRE*, 37:10–21, 1949.

[3] H. Nyquist, "Certain topics in telegraph transmission theory," *AIEE Transactions*, 47:617–44, 1928.

[4] J. Deller, "Tom, Dick, and Mary discover the DFT," *IEEE Signal Processing Magazine*, 11:36–50, 1994.

[5] A. Oppenheim and A. Willsky with H. Nawab, *Signals & Systems*, Prentice Hall, Upper Saddle River, NJ, 1997.

[6] J. Cooley and J. Tukey, "An algorithm for the machine computation of complex Fourier series," *Mathematical Computations*, 19:297–301, 1965.

[7] M. O'Flynn and E. Moriarty, *Linear Systems—Time Domain and Tranform Analysis*, Wiley, New York, 1987.

[8] V. Ingle and J. Proakis, *Digital Signal Processing using MATLAB*, Brooks/Cole, Pacific Grove, CA, 2000.

9.11 PROBLEMS

9.1. Find the DTFT of the signal represented by

$$x[k] = \begin{cases} -2, & k = -5 \\ 3, & k = 2 \\ -1, & k = 10 \\ 0, & \text{otherwise} \end{cases}$$

9.2. Find the DTFT of the signal $x[k] = (-0.5)^k u[k]$, where $u[k]$ denotes the discrete-time unit step signal.

9.3. Find the DTFT of the discrete-time rectangular pulse $p_5[k]$.
Answer: $\sin(3\Omega)/\sin(\Omega/2)$

9.4. Find the DTFT and sketch the magnitude spectrum for the signal $x[k] = \left(\frac{1}{3}\right)^k u[k]$.

9.5. Find the DTFT and plot the magnitude spectrum for the signal

$$x[k] = \left(\frac{1}{2}\right)^{|k|}, \quad k = 0, \pm 1, \pm 2, \ldots$$

9.6. Verify the result obtained in (9.43) using the inverse DTFT formula.

9.7. Find the DTFT of the discrete-time unit step signal $u[k]$.
Answer:
$$\pi \sum_{k=-\infty}^{\infty} \delta(\Omega - k2\pi) + \frac{e^{j\Omega}}{e^{j\Omega} - 1}$$

9.8. Find the DTFT of the following signals

(a) $\cos\left(3k + \dfrac{\pi}{3}\right)$

(b) $\sin\left(5k - \dfrac{\pi}{4}\right)$

Answer:

(a) $\pi \displaystyle\sum_{k=-\infty}^{\infty} [e^{-j\pi/3}\delta(\Omega + 3 - k2\pi) + e^{j\pi/3}\delta(\Omega - 3 - k2\pi)]$

9.9. Using the definitions of the auto- and crosscorrelation functions given in (9.96) and (9.97), find their DTFT; that is, verify formula (9.105).

Hint: The proof is similar to the convolution property proof. Hence, the convolution property proof should be provided first.

9.10. Find the double-sided $\mathcal{Z}$-transform of the signal presented in Problem 9.1.

9.11. Find double-sided $\mathcal{Z}$-transform of the signal presented in Problem 9.5 and determine the ROC (region of convergence).

9.12. Find the double-sided $\mathcal{Z}$-transform of the signal

$$x[k] = \left(\dfrac{1}{2}\right)^k u[k] + \left(-\dfrac{1}{3}\right)^k u[k] + (-2)^{-k} u[-k]$$

and determine the ROC.

9.13. Show that the signals $a^k u[k]$ and $-a^k u[-k-1]$ have the same double-sided $\mathcal{Z}$-transform. Find the corresponding regions of convergence.

9.14. Find the double-sided $\mathcal{Z}$-transform of the signal

$$x[k] = x_1[k]u[k] + x_2[k]u[-k]$$

9.15. Solve Problem 9.14 assuming that $x_1[k] = (-0.5)^k$ and $x_2[k] = (-1)^k$. Determine the corresponding region of convergence.

9.16. Let $X_1(z) = 1 + z^{-1}$ and $X_2(z) = 2 + 3z^{-2}$. Find $X_1(z)X_2(z)$.

Hint: Use the convolution formula.

9.17. Find the inverse double-sided $\mathcal{Z}$-transforms of the z-domain signal

$$X(z) = \dfrac{5z}{(z - \tfrac{1}{4})(z + \tfrac{1}{2})}$$

Associate the results obtained with the corresponding ROCs.

Hint: See Example 9.10.

9.18. Sketch the three time domain signals obtained in Example 9.10 by performing the inverse double-sided $\mathcal{Z}$-transform using different regions of convergence.

9.19. Find the inverse double-sided $\mathcal{Z}$-transforms of the z-domain signal

$$X(z) = \frac{5z(z+1)}{(z-1)(z+\frac{1}{2})^2}$$

Associate the results obtained with the corresponding ROCs.

9.20. Repeat Problem 9.17 for the signal

$$X(z) = \frac{10(z-1)}{(z-\frac{1}{5})(z+\frac{1}{2})(z+\frac{1}{3})}$$

9.21. Use formula (9.78) to evaluate the elements of the matrix $\mathbf{A}^{3\times 3}$.

9.22. Show that the matrix $\mathbf{A}^{4\times 4}$ given in (9.79) satisfies (9.81).

9.23. Assume for the DFT that the sample lengths are $L=6$ and $N=10$. Consider the discrete-time signal

$$x = [2, 3, 0, -2, -1, 6], \quad L = 6$$

Use the zero padding technique to form the discrete-time signal x with $L = 10$.
Answer: $x = [2, 3, 0, -2, -1, 6, 0, 0, 0, 0], \quad L = 10$

9.24. Repeat Problem 9.23 for the signal

$$x = [1, 2, 3, -2, -3, -1, 1, 5], \quad L = 8$$

and $N = 12$.

9.25. Let $N = 3$ and $L = 7$. Perform modulo-N reduction of the signal

$$x = [1, 2, -1, -2, 3, -3, 4], \quad L = 7$$

Answer: $x = [3, 5, -4], \quad N = 3$

9.26. Repeat Problem 9.25 for the signal

$$x = [2, -2, 3, -3, 4, -4, 1, 1, 1], \quad L = 9$$

and $N = 4$.

9.27. Find the DFS coefficients for the discrete-time periodic signal represented by

$$x[k] = \begin{cases} 1, & k = 0 \pm rN \\ 2, & k = 1 \pm rN \\ 1, & k = 2 \pm rN \\ 0, & k = 3 \pm rN \end{cases}, \quad N = 4, \quad r = 0, 1, 2, \ldots$$

9.28. Find the DFS for a periodic discrete-time signal whose values are equal to 2 at the even discrete-time instants and -1 at the odd discrete-time instants.

9.29. Starting with the definition formulas of the auto- and crosscorrelation functions, and using appropriate changes of variables, establish formulas (9.99) and (9.100).

9.30. Establish the proof for the formulas given in (9.101), by introducing appropriate change of variables in the corresponding sums.

9.31. Prove that $|R_{xy}[k]| \leq \sqrt{|R_{xx}[0]||R_{yy}[0]|}$, and hence establish the fact that the correlation coefficient defined in (9.104) satisfies $-1 \leq c_{xy}[k] \leq 1$.
Hint: Use the Cauchy–Schwartz inequality given in Appendix B.

9.32. Find and plot the autocorrelation functions for the signals defined in Problems 9.1 and 9.3.

9.33. Let $y[k] = x[k] * h[k]$ be the output of a linear time-invariant system whose impulse response is $h[k]$ and whose input signal is $x[k]$. Find the autocorrelation function for $y[k]$.

9.34. Using MATLAB and formulas given in (9.101) find and plot auto- and crosscorrelation functions for the signals defined in Problems 9.2 and 9.3. Find and plot the correlation coefficients as defined in (9.104).

9.35. Repeat Problem 9.34 for the signal defined in Problem 9.5.

9.36. Use MATLAB to find DFTs of the discrete-time signals presented in Problems 9.23 and 9.24. Apply IFFT to the results obtained and recover discrete-time signals.

9.37. Use MATLAB to find DFTs of the discrete-time signals presented in Problems 9.25 and 9.26. Apply IFFT to the results obtained and find the corresponding discrete-time signals. Are they wrapped or unwrapped signals? Comment on the results obtained.

9.38. Use FFT and MATLAB to convolve the discrete-time signals defined in Problems 9.23 and 9.24.

9.39. Use FFT and MATLAB to convolve the discrete-time signals defined in Problems 9.25 and 9.26. Consider two cases, $N = 5$ and $N = 10$.

9.40. A student intends to take a course in linear control systems for which a linear systems and signals course is a prerequisite. The student polls their colleagues, who have taken the control course, about the quality of the course (questions 1–4) and the teaching effectiveness of the instructor (questions 5–7). The student collects answers to seven questions from five of their colleagues, each answer marked from 1 (bad) to 5 (excellent). The answers obtained can be considered as five vector signals of length seven. Assume that such an obtained set of signals has the samples

$$\mathbf{x}_1 = \begin{bmatrix} 5 \\ 4 \\ 4 \\ 5 \\ 3 \\ 4 \\ 3 \end{bmatrix}, \quad \mathbf{x}_2 = \begin{bmatrix} 2 \\ 3 \\ 2 \\ 4 \\ 1 \\ 2 \\ 2 \end{bmatrix}, \quad \mathbf{x}_3 = \begin{bmatrix} 4 \\ 4 \\ 4 \\ 5 \\ 5 \\ 4 \\ 4 \end{bmatrix}, \quad \mathbf{x}_4 = \begin{bmatrix} 3 \\ 4 \\ 4 \\ 3 \\ 3 \\ 4 \\ 3 \end{bmatrix}, \quad \mathbf{x}_5 = \begin{bmatrix} 2 \\ 1 \\ 1 \\ 2 \\ 3 \\ 2 \\ 1 \end{bmatrix}$$

A simple way to make a positive or negative decision is to find the mean values with respect to each component of the vector signals, and if either all means (or particular means, or groups of means) are above a determined threshold the decision is positive, otherwise the decision is negative.

(a) Use MATLAB to find the mean value of the vector signals, that is, evaluate the vector

$$\mathbf{x}_{mean} = (\mathbf{x}_1 + \mathbf{x}_2 + \mathbf{x}_3 + \mathbf{x}_4 + \mathbf{x}_5)/5$$

Assume that the student, based on this information, might take the course if all means are above 3.5, or if the average of the means for either the course quality or for the teaching effectiveness is above 3.5. Find the means.

(b) Another more sophisticated procedure would be for the student to evaluate the correlation of each signal with respect to the signal mean value $\mathbf{x}_{mean}$, and to remove weakly correlated signals (signals with small correlation coefficients as defined in (9.103)). To that end, the student finds the correlation coefficients $c_{x_{mean}x_i}[0]$, removes two signals that have relatively small correlation coefficients, calculates the means of the remaining three signals, and makes the decision according to the criteria set in part (a).

(c) A third procedure would be to find the correlation coefficients among all signals and remove those that are weakly correlated (with small correlation coefficients) to the remaining signals (some data might be unreliable, inaccurate, or even biased). Remove the signal that is the least correlated with the remaining signals and make the decision according to the criteria set in part (a). Will the student take the course based on the data and the chosen criteria?

9.41. Use MATLAB to find the zero-state response of the discrete-time linear system represented by

$$y[k+2] - y[k+1] + \frac{1}{4}y[k] = x[k]$$

due to the periodic input defined in Example 9.13.

CHAPTER 10

Signals in Communication Systems

In this chapter, we consider fundamentals of communication systems, their basic principles and concepts, using the linear system theory presented in the first eight chapters of this book. Traditionally, continuous-time linear and time-invariant communication systems are studied in the frequency domain using the Fourier transform. In addition to deterministic inputs, communication systems are very often driven by random external and internal signals (noise, random arrival of signals (calls) in a telephone network, and so on) so that they should be studied also as stochastic systems (see Section 1.4 on system classification). This fact implies that in addition to Fourier analysis, probability and stochastic processes are fundamental mathematical tools for studying communication systems. Since it is outside of the scope of this chapter (and this book) to study linear systems driven by random processes (stochastic systems), we will present only deterministic communication system fundamentals that can be derived and understood using Fourier analysis and other material presented in this book.

The main principles of communication systems have historically been formulated in continuous time. Nowadays, most communication systems operate in the discrete-time domain—they are *digital communication systems*. Our presentation will focus on continuous-time communication system principles that establish fundamentals of signal transmission. Presentation of digital communication systems would require additional study of digital signal processing techniques, information theory, and computer/communication networking topics that we will not encompass in this book. We will indicate only the essence of digital communication systems. Hence, our presentation is limited to the fundamentals of *deterministic continuous-time communication systems*.

We first cover some basic concepts of signal transmission in communication systems and present the signal energy and power spectra in Sections 10.1 and 10.2. In Section 10.3, we introduce the Hilbert transform using standard Fourier analysis from Chapter 3. There are two forms of the Hilbert transform, which can be derived independently for causal and real signals. For completeness, both forms of the Hilbert transform are presented in this chapter. For causal signals, the Hilbert transform relates the real and imaginary parts of the signal Fourier transform. As such, it is used in linear electrical circuits and electric power systems. For real signals, the Hilbert transform introduces the signal phase shift of $-90°$ for positive frequencies and $90°$ for negative frequencies. Applications of this form of the

508 Chapter 10 Signals in Communication Systems

Hilbert transform in communication systems is discussed in Section 10.5. Ideal filters play an important theoretical role in the study of communication systems. The low-pass ideal filter is presented in Section 10.4. The modulation operation, one of the most fundamental operations in communication systems, was introduced in Chapter 3. In this chapter the modulation concept is thoroughly studied, and its inverse operation, demodulation, is introduced and explained in Section 10.5. Section 10.6 presents the main principles of digital signal transmission, which is now extensively used for transmission of voice, video, and data in computer networks and digital communication systems.

10.1 SIGNAL TRANSMISSION IN COMMUNICATION SYSTEMS

The main role of a communication system is to transmit signals (information) from the source of information (system input) to the user or destination (system output). The transmission is done over a *communication channel* using a *transmitter* and a *receiver*. A simplified basic communication system is presented in Figure 10.1.

The original signal, usually called the *baseband* signal (this name will be justified after we explain the modulation concept), is first transformed into a signal convenient for transmission (called the *transmitted signal*) by the transmitter. The transmitter sends the signal which may be electrical or optical (electromagnetic), over a communication channel, which represents a physical medium convenient for propagation of electromagnetic waves (low signal attenuation and distortion). Communication channels can be guided media (such as copper wire or optical fiber cable), or free-space channels (such as satellite or wireless (radio)). The role of the receiver is to convert received signals, theoretically, into baseband signals, and pass them to the user. Due to channel attenuation, distortion, and noise, the receiver produces a signal that is similar but not identical to the baseband channel. Such a signal is called an estimated or reconstructed signal. The *estimated signal* can be slightly different than the original (baseband) signal, especially for voice and video transmissions, since the human eye and ear are unable to detect small errors. However, in the case of data, signal transmission must be error free.

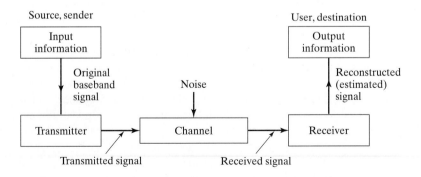

FIGURE 10.1: Basic communication system

Modulation

In a standard communication system, the transmitter is a modulator, and the receiver is a demodulator. The modulator and demodulator taken together are called a *modem*. We have introduced the modulation concept within the properties of the Fourier transform (see Section 3.2, Property 6). The modulation property of the Fourier transform says: Let the signal $x(t)$ have the Fourier transform $X(j\omega)$, where $\omega = 2\pi f$. Then, the Fourier transform of the *modulated signal*, defined as $x(t)\cos(\omega_c t)$, $\omega_c = 2\pi f_c$, is given by

$$\mathcal{F}\{x(t)\cos(\omega_c t)\} = \frac{1}{2}X(j(\omega+\omega_c)) + \frac{1}{2}X(j(\omega-\omega_c)) \tag{10.1}$$

Since $|X(j\omega)|$ denotes the signal spectrum, it can be seen that the *spectrum of the modulated signal is shifted left and right by* ω_c, as represented in Figure 10.2. The frequency ω_c is called the *carrier frequency*. The original signal spectrum is the baseband signal spectrum, and the other two spectra are the modulated signal spectra. (This justifies the name *baseband signal*.)

Note that due to magnitude spectrum symmetry, the positive frequencies carry all the information contained in a given signal. We can make two observations from Figure 10.2:

1. The spectrum of the modulated signal (including negative frequencies) is double that of the baseband signal (including negative frequencies). It contains the *upper frequency sidebands* in the frequency ranges $\omega \in [\omega_c, \omega_c + \omega_{\max}]$ and $\omega \in [-\omega_c - \omega_{\max}, -\omega_c]$ and *the lower frequency sidebands* in the frequency ranges $\omega \in [\omega_c - \omega_{\max}, \omega_c]$ and $\omega \in [-\omega_c, -\omega_c + \omega_{\max}]$.
2. Negative frequencies of the original baseband signal come into the picture due to frequency translation, and form the lower frequency sideband for positive frequencies and the upper frequency sideband for negative frequencies.

Hence, in its original definition, amplitude modulation requires doubling in the spectrum requirements (with waste of the frequency band). In Section 10.5, we will study a technique that remedies this problem.

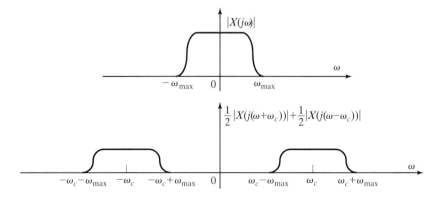

FIGURE 10.2: Spectra of original and modulated signals

The modulation concept as presented in Figure 10.2 indicates the extraordinary possibility that the same channel can be used to simultaneously transmit several signals by appropriately shifting their spectra such that they do not overlap in the frequency domain. (Note that the signals may overlap in the time domain.) In Example 3.8 and Figure 3.8 we demonstrated this possibility by considering a telephone network that must transmit many telephone signals (calls) simultaneously.

It can be observed from Figure 3.8 that users share the frequency band. If we assume that the channel frequency bandwidth is equal to ω_{BW} and that the channel must serve N users (baseband signals), then we see that each user has reserved at all times a part of the channel frequency band equal to ω_{BW}/N. Such channel sharing is called *frequency division multiplexing* (FDM). Note that FDM is presently used in communication systems in a much broader sense than presented here within the basic modulation concept, but the main idea of frequency band sharing remains the same in all cases.

Another channel sharing technique used in communication system practice is *time division multiplexing* (TDM), a technique in which each user gets the whole frequency band of the channel, but only during a limited period of time. In such a case the users are switched on and off according to the given time schedule. For example, each user uses the whole channel frequency band during the time period of Δt, and they rotate so that each gets a turn after $N\Delta t$ time units (fair sharing of the channel). Note that there is no single criteria by which to judge that one of the channel sharing techniques is better, due to the very simple fact that a *channel with a larger frequency bandwidth has a higher capacity* (it can transmit more units of information per unit of time; it is a faster channel). Hence, there is an interplay between transmitting at high speeds during short periods of time (TDM) and transmitting constantly at low speeds (FDM). Sometimes these two techniques are equivalent, but for particular applications one of them may have an advantage over the other. It is obvious that it is desirable to work with channels that have large capacity (high speed). Of course, the larger the channel bandwidth (faster channel), the more expensive the channel.

Demodulation

The demodulation process is reciprocal to the modulation process. Demodulation is an operation that reconstructs the original baseband signal from its modulated signal. Technically speaking, the demodulator must cut out (filter out) the frequency band that corresponds to the given baseband signal. Theoretically, demodulation can be performed by remodulating the modulated signal, that is,

$$\mathcal{F}\{[x(t)\cos(\omega_c t)]\cos(\omega_c t)\} = \mathcal{F}\left\{\frac{1}{2}x(t) + \frac{1}{2}x(t)\cos(2\omega_c t)\right\}$$
$$= \frac{1}{2}X(j\omega) + \frac{1}{4}X(j(\omega+2\omega_c)) + \frac{1}{4}X(j(\omega-2\omega_c)) \tag{10.2}$$

By passing this signal through a low-pass filter, we can recover the original signal multiplied by 0.5, that is $0.5x(t)$. Note that since the second modulation is done at another location, which might be very far from the sender, both modulators (the transmitter and the receiver) must be very accurately synchronized. In Section 10.5, we say more about the modulation and demodulation processes.

Section 10.1 Signal Transmission in Communication Systems

In the remaining part of this section, we introduce some notions frequently used in signal transmission in communication systems.

Signal-to-Noise Ratio

As mentioned, channel noise is most often random in nature. Despite the fact that we will not study channels from the stochastic point of view, we can define a simple quantity that tells us how noisy, on average, a given channel is. Let P_s denote the average signal power (as defined in (2.53)) and let P_n denote the average noise power. The *signal-to-noise ratio (SNR) in decibels* [dB] is defined by

$$\text{SNR [dB]} = 10 \log_{10}\left(\frac{P_s}{P_n}\right) \tag{10.3}$$

Apparently, according to (10.3), the higher the SNR, the better the channel.

Channel Capacity

It can be experimentally observed that channel capacity is directly proportional to frequency bandwidth (note that frequency bandwidth is defined in (3.74)). It has also been observed that higher SNR implies higher channel capacity. An exact formula that relates channel capacity in bits per second, channel frequency bandwidth in Hz, and the channel signal power to noise power ratio was derived by Shannon (and is also known as the Shannon–Harteley formula). It is given by

$$C\left[\frac{b}{s}\right] = w_{BW} \log_2\left(1 + \frac{P_s}{P_n}\right) \tag{10.4}$$

Note that this formula is valid for channels with Guassian noise (where the noise statistics are completely described by the first- and second-order moments, mean, and variance). In the case of non-Gaussian noise, formula (10.4) gives only an approximate lower bound.

Optical Fiber Cable

The "waveguide medium of the future" [1], optical fiber cable has a huge frequency bandwidth that theoretically can reach several hundred THz (1 terahertz is equal to 10^{12} Hz). It also has very low signal attenuation of only 0.2 dB/km, which means that

$$10 \log_{10}\left(\frac{P^{\text{in}}}{P^{\text{out}}}\right) = 0.2 \text{ [dB]} \tag{10.5}$$

where P^{in} and P^{out} represent, respectively, the input and output signal powers. In addition, optical fiber cable has very low signal distortion. It is made of silica glass (dielectric) that transports light signals, also called optical signals. Similarly to frequency division multiplexing, in optical communication systems *wavelength division multiplexing* (WDM) is used to transmit simultaneously many signals (eighty or more) over the same optical fiber channel. The optical wavelength is defined by $\lambda = v/f$, where v is the speed of light (equal to $c = 3 \times 10^8$ m in a vacuum and $c/\sqrt{\epsilon_r \mu_r} \approx 0.6c$ in a guided media, where ϵ_r

and μ_r are, respectively, the medium permittivity and permeability constants) and f is the frequency of the corresponding light signal. *Dense wavelength frequency division multiplexing* (DWDM) has optical wavelength channels densely spaced every 10^{-9} m = 1 nm.

It is interesting to point out that a channel represents a dynamic system that can be linear or nonlinear, time-invariant or time varying, deterministic or stochastic (see system classification in Section 1.4). For example, telephone channels are linear systems in most cases, wireless channels can be considered time-varying linear systems, optical fiber channels are nonlinear time-invariant systems that are often linearized (see section on linearization of nonlinear systems, Section 8.6), and satellite channels are nonlinear. Another classification of channels distinguishes between *bandlimited channels* such as telephone networks and *power-limited channels* such as optical fiber and satellite channels. These channel classifications will not be further considered in this brief chapter about communication systems.

10.2 SIGNAL CORRELATION, ENERGY, AND POWER SPECTRA

In addition to system frequency bandwidth, signal power represents another important quantity that engineers are particularly concerned with while transmitting signals. We have already defined signal energy and power in the time domain in Section 2.3. Here, we present their representations in the frequency domain, and relate them to the quantity known as the signal correlation function.

Devices called signal correlators are used to measure the power of incoming signals in many communication (and signal processing) systems. For example, in wireless communication systems, correlators at the base station measure at all times the signal power of all mobiles in the base station area (cell). These signal powers are periodically adjusted such that each mobile has sufficient signal power for a good quality transmission but not so much signal power as to cause unnecessary interference to the other mobiles that use the same frequency band. Correlation receivers are also used in radar and sonar systems. The interested reader can find more details about correlation receivers and their application to a radar system in [2].

Continuous-Time Signal Correlation

The analytical expression for signal correlation is very similar to the convolution integral, even though signal correlation and signal convolution have completely different physical meanings.

The correlation of two continuous-time signals $x_1(t)$ and $x_2(t)$ is defined by the integral

$$R_{12}(\tau) = \int_{-\infty}^{\infty} x_1(t) x_2(t + \tau) \, dt \qquad (10.6)$$

where τ is a parameter, $-\infty \leq \tau \leq \infty$. More precisely, $R_{12}(\tau)$ is called the *cross-correlation function*. Assuming that the signals $x_1(t)$ and $x_2(t)$ have Fourier transforms respectively given by $X_1(j\omega)$ and $X_2(j\omega)$, that is,

$$x_1(t) = \frac{1}{2\pi} \int_{-\infty}^{\infty} X_1(j\omega)e^{j\omega t}\, d\omega, \quad x_2(t) = \frac{1}{2\pi} \int_{-\infty}^{\infty} X_2(j\omega)e^{j\omega t}\, d\omega \qquad (10.7)$$

then we have

$$R_{12}(\tau) = \int_{-\infty}^{\infty} x_1(t) \left[\frac{1}{2\pi} \int_{-\infty}^{\infty} X_2(j\omega)e^{j\omega(t+\tau)}\, d\omega \right] dt$$

$$= \frac{1}{2\pi} \int_{-\infty}^{\infty} \left[\int_{-\infty}^{\infty} x_1(t)e^{j\omega t}\, dt \right] X_2(j\omega)e^{j\omega\tau}\, d\omega \qquad (10.8)$$

$$= \frac{1}{2\pi} \int_{-\infty}^{\infty} X_1^*(j\omega)X_2(j\omega)e^{j\omega\tau}\, d\omega$$

Note that $X_1^*(j\omega) = X_1(-j\omega)$. The last formula indicates that $R_{12}(\tau)$ and $X_1^*(j\omega)X_2(j\omega)$ form the Fourier transform pair, that is,

$$R_{12}(\tau) \leftrightarrow X_1^*(j\omega)X_2(j\omega) \qquad (10.9)$$

In the case when $x_1(t) = x_2(t) = x(t)$, we have the definition of the *autocorrelation function* as

$$R(\tau) = \int_{-\infty}^{\infty} x(t)x(t+\tau)\, dt \qquad (10.10)$$

In this case, in view of formula (10.8), we have

$$R(\tau) = \frac{1}{2\pi} \int_{-\infty}^{\infty} X^*(j\omega)X(j\omega)e^{j\omega\tau}\, d\omega = \frac{1}{2\pi} \int_{-\infty}^{\infty} |X(j\omega)|^2 e^{j\omega\tau}\, d\omega \qquad (10.11)$$

that is, the autocorrelation function and $|X(j\omega)|^2$ form the Fourier transform pair

$$R(\tau) \leftrightarrow |X(j\omega)|^2 \qquad (10.12)$$

It can be shown that the autocorrelation function has the following properties:

1. The autocorrelation function is even; that is, $R(\tau) = R(-\tau)$.
2. $R(0) = E_\infty$, where E_∞ stands for the total signal energy. (This follows from (3.39) and (10.10).)

3. $R(0) \geq R(\tau)$, $\forall \tau$.
4. $R(\tau)$ is continuous in time (like convolution).

The quantity $|X(j\omega)|^2$ defines the signal power at the given frequency ω, so that $|X(j\omega)|^2$ is called the *power spectrum*. $|X(j\omega)|^2$ is also called the *energy density spectrum*, for reasons soon to be clear. Introducing the notation

$$|X(j\omega)|^2 = S(\omega) \tag{10.13}$$

we have

$$S(\omega) = \int_{-\infty}^{\infty} R(t)e^{-j\omega t} dt, \quad R(\tau) = \frac{1}{2\pi} \int_{-\infty}^{\infty} S(\omega)e^{-j\omega\tau} d\omega \tag{10.14}$$

Note that $S(\omega)$ is a real positive and even function, that is $S(\omega) = S(-\omega) \geq 0$.
It follows from $R(0) = E_\infty$ and the second formula in (10.14) that

$$E_\infty = \frac{1}{2\pi} \int_{\omega=-\infty}^{\omega=\infty} S(\omega) d\omega \tag{10.15}$$

It is clear that $S(\omega)$ represents the *energy density in the frequency domain*, which justifies the use of the name "energy density spectrum" for $|X(j\omega)|^2 = S(\omega)$.

If one intends to find the signal energy in the frequency domain in any frequency range, say (ω_1, ω_2), then the knowledge of the signal energy density $S(\omega)$ gives the following formula:

$$W(\omega_1, \omega_2) = \frac{1}{2\pi} \int_{\omega_1}^{\omega_2} S(\omega) d\omega + \frac{1}{2\pi} \int_{-\omega_2}^{-\omega_1} S(\omega) d\omega = \frac{1}{\pi} \int_{\omega_1}^{\omega_2} S(\omega) d\omega \tag{10.16}$$

The last expression follows from the fact that $S(\omega)$ is a positive and symmetric function of frequency. Formula (10.16) determines the distribution of the signal energy in the frequency domain.

Note that while deriving the Fourier series and later the Fourier transform, for reasons of mathematical convenience (compactness of obtained expressions; compare formulas (3.10) and (3.14)), we introduced "negative" frequencies. Here, we see how those "negative" frequencies come into the picture, and how the signal energy can be completely expressed in terms of positive frequencies, which reflects physical reality.

EXAMPLE 10.1

In this example, we find the frequency range that contains a given percentage (50%) of the signal energy. Consider the signal frequency spectrum presented in Figure 10.3.

Section 10.2 Signal Correlation, Energy, and Power Spectra

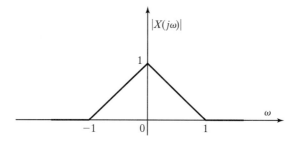

FIGURE 10.3: The frequency spectrum of a signal

According to the problem formulation, we are looking for the frequency ω_1 such that

$$W(0,\omega_1) = \frac{1}{2}W = \frac{1}{2}\frac{1}{2\pi}\int_{-\infty}^{\infty} S(\omega)\,d\omega = \frac{1}{4\pi}\int_{-1}^{1} |X(j\omega)|^2\,d\omega$$

$$= \frac{1}{2\pi}\int_{0}^{1}(1-\omega)^2\,d\omega = \frac{1}{6\pi}$$

Hence, we have the equality

$$W(0,\omega_1) = \frac{1}{\pi}\int_{0}^{\omega_1}(1-\omega^2)\,d\omega = \frac{1}{2}W = \frac{1}{6\pi}$$

Using the change of variables $1 - \omega = \mu$, the preceding integral can be easily calculated, which leads to $\omega_1 = 1 - 1/\sqrt[3]{2}$. Note that in this problem we have tacitly taken into account the contribution of "negative" frequencies to the signal energy. ∫

As a measure of the similarity of two signals, the correlation coefficient can be defined as

$$-1 \le c_{12}(\tau) = \frac{R_{12}(\tau)}{\sqrt{R_{11}(0)R_{22}(0)}} \le 1 \qquad (10.17)$$

When the correlation coefficient is close to 1, then the signals are similar.

Correlation of Periodic Signals

When signals are periodic, the correlation function can be obtained using the Fourier series. Let $x_1(t) = x_1(t+T)$ and $x_2(t) = x_2(t+T)$, $T < \infty$, then the *crosscorrelation function for periodic signals* is defined by

$$R_{12}(\tau) = \frac{1}{T}\int_{-T/2}^{T/2} x_1(t)x_2(t+\tau)\,dt \qquad (10.18)$$

Using the fact that periodic functions can be expressed using the Fourier series, that is,

$$x_1(t) = \sum_{n=-\infty}^{n=\infty} X_1(jn\omega_0)e^{jn\omega_0 t}, \quad x_2(t) = \sum_{n=-\infty}^{n=\infty} X_2(jn\omega_0)e^{jn\omega_0 t}, \quad \omega_0 = \frac{2\pi}{T} \quad (10.19)$$

we have

$$R_{12}(\tau) = \frac{1}{T}\int_{-T/2}^{T/2} x_1(t)\left[\sum_{n=-\infty}^{n=\infty} X_2(jn\omega_0)e^{jn\omega_0(t+\tau)}\right]dt$$

$$= \sum_{n=-\infty}^{n=\infty} X_2(jn\omega_0)\left[\frac{1}{T}\int_{-T/2}^{T/2} x_1(t)e^{jn\omega_0 t}\,dt\right]e^{jn\omega_0\tau} \quad (10.20)$$

$$= \sum_{n=-\infty}^{n=\infty} X_1^*(jn\omega_0)X_2(jn\omega_0)e^{jn\omega_0\tau}$$

Note that $R_{12}(\tau)$ and $X_1^*(jn\omega_0)X_2(jn\omega_0)$ are the corresponding Fourier series pair, which implies that

$$X_1^*(jn\omega_0)X_2(jn\omega_0) = \frac{1}{T}\int_{-T/2}^{T/2} R_{12}(\tau)e^{-jn\omega_0\tau}\,d\tau \quad (10.21)$$

When $x_1(t) = x_2(t) = x(t)$, we can define the *autocorrelation function for periodic signals* as

$$R(\tau) = \frac{1}{T}\int_{-T/2}^{T/2} x(t)x(t+\tau)\,dt \quad (10.22)$$

Formula (10.20) in this case becomes

$$R(\tau) = \sum_{n=-\infty}^{n=\infty} |X(jn\omega_0)|^2 e^{jn\omega_0\tau} \quad (10.23)$$

Introducing the notion of the power spectrum $S(n\omega_0) = |X(jn\omega_0)|^2$ of a periodic signal (or line power spectrum, to be consistent with terminology from Section 3.1), we have the corresponding Fourier series pair

$$R(\tau) = \sum_{n=-\infty}^{n=\infty} S(n\omega_0)e^{jn\omega_0\tau}, \quad S(n\omega_0) = \frac{1}{T}\int_{-T/2}^{T/2} R(\tau)e^{-jn\omega_0\tau}\,d\tau \quad (10.24)$$

Note that for periodic signals the autocorrelation function is also an even function. The corresponding spectrum is an even and positive function. In addition, $R(0)$ defines the signal energy during one time period, that is,

$$R(0) = \sum_{n=-\infty}^{n=\infty} S(n\omega_0) = \sum_{n=-\infty}^{n=\infty} |X(jn\omega_0)|^2 = \frac{1}{T} \int_{-T/2}^{T/2} x^2(t)\,dt = W_T \qquad (10.25)$$

This relation also represents *Parseval's theorem for periodic signals*.

10.3 HILBERT TRANSFORM

The Hilbert transform plays an important role in communication systems. It can be easily derived using knowledge from Chapter 3 about the Fourier transform. There are two forms of the Hilbert transform; the first is valid for causal signals and the second holds for real signals. The first form of the Hilbert transform has applications in linear electrical circuits and electric power systems, and the second form is used in communication systems.

Hilbert Transform for Causal Signals

In the following, we show that in the case of causal signals, the Hilbert transform, in fact, relates the real and imaginary parts of the corresponding Fourier transform. Such a relationship holds for any *causal* real or complex signal (function) $x(t)$. Recall that causal signals are equal to zero for $t < 0$.

Due to causality, we have

$$\mathcal{F}(x(t)) = X(j\omega) = X_{\text{Re}}(\omega) + jX_{\text{Im}}(\omega) = \int_0^\infty x(t)e^{-j\omega t}\,dt \qquad (10.26)$$

Also, causality implies $x(t) = x(t)u(t)$, where $u(t)$ is the unit step function. The application of the Fourier transform to the left- and right-hand sides of the last expression produces

$$\mathcal{F}(x(t)) = X(j\omega) = \mathcal{F}(x(t)u(t)) = \frac{1}{2\pi}X(j\omega) * U(j\omega) \qquad (10.27)$$

Recall from Chapter 3 that the product in the time domain corresponds to the convolution in the frequency domain (Property 9, frequency convolution). Using the expression for the Fourier transform of the unit step function found in Example 3.12, we obtain

$$X(j\omega) = X_{\text{Re}}(\omega) + jX_{\text{Im}}(\omega) = \frac{1}{2\pi}[X_{\text{Re}}(\omega) + jX_{\text{Im}}(\omega)] * \left[\pi\delta(\omega) + \frac{1}{j\omega}\right] \qquad (10.28)$$

It is known from Chapter 6 that the convolution of any signal with the impulse delta signal produces that signal. Using this fact, equation (10.28) is simplified into

$$X_{\text{Re}}(\omega) + jX_{\text{Im}}(\omega) = \frac{1}{2}X_{\text{Re}}(\omega) + j\frac{1}{2}X_{\text{Im}}(\omega) - \frac{1}{2\pi}X_{\text{Re}}(\omega) * \frac{j}{\omega} + \frac{1}{2\pi}X_{\text{Im}}(\omega) * \frac{1}{\omega} \quad (10.29)$$

Equating the real and imaginary parts in the last equation, we obtain two relationships,

$$X_{\text{Re}}(\omega) = X_{\text{Im}}(\omega) * \frac{1}{\pi\omega} \quad (10.30)$$

and

$$X_{\text{Im}}(\omega) = -X_{\text{Re}}(\omega) * \frac{1}{\pi\omega} \quad (10.31)$$

Formulas (10.30) and (10.31) relate the real and imaginary parts of the Fourier transform of the causal signal $x(t)$, and define the Hilbert transform. Using the definition of the frequency domain convolution from Chapter 3, the last two formulas can be written in the following form:

$$X_{\text{Re}}(\omega) = \frac{1}{\pi}\int_{-\infty}^{\infty} X_{\text{Im}}(\nu)\frac{1}{\omega-\nu}\,d\nu \quad (10.32)$$

and

$$X_{\text{Im}}(\omega) = -\frac{1}{\pi}\int_{-\infty}^{\infty} X_{\text{Re}}(\nu)\frac{1}{\omega-\nu}\,d\nu \quad (10.33)$$

EXAMPLE 10.2

The unit step signal $u_h(t)$ is a causal signal whose Fourier transform has both the real and imaginary parts, that is,

$$\mathcal{F}\{u_h(t)\} = \pi\delta(\omega) + \frac{1}{j\omega} = \pi\delta(\omega) - j\frac{1}{\omega}$$

Using (10.33), it can be confirmed that the imaginary part of this Fourier transform is related to its real part through the Hilbert transform, that is,

$$-\frac{1}{\pi}\int_{-\infty}^{\infty} \pi\delta(\nu)\frac{1}{\omega-\nu}\,d\nu = -\frac{1}{\omega} \qquad \qquad \textit{f}$$

This form of the Hilbert transform is used in linear electrical circuits and electric power systems to find the imaginary part of the Fourier transform when its real part is known (obtained experimentally), and vice versa, to find the real part from the imaginary part. For completeness, it is presented in this section, together with the second form of the Hilbert transform, which has a particular importance for the modulation process in communication systems.

Hilbert Transform for Real Signals

The second form of the Hilbert transform is derived for real Fourier transformable signals. Let $x(t) \leftrightarrow X(j\omega)$, that is,

$$x(t) = \frac{1}{2\pi} \int_{-\infty}^{\infty} X(j\omega) e^{j\omega t} \, d\omega \tag{10.34}$$

Consider the signal $x_+(t)$ whose spectrum is zero for negative frequencies and $2X(j\omega)$ for positive frequencies, that is,

$$x_+(t) = \frac{1}{\pi} \int_{0}^{\infty} X(j\omega) e^{j\omega t} \, d\omega \tag{10.35}$$

Then the following relationship exists between the signals $x(t)$ and $x_+(t)$:

$$x_+(t) = x(t) + j\hat{x}(t) \tag{10.36}$$

where $\hat{x}(t)$ is the Hilbert transform of $x(t)$, defined by

$$\hat{x}(t) = x(t) * \frac{1}{\pi t} = \frac{1}{\pi} \int_{-\infty}^{\infty} x(\tau) \frac{1}{t-\tau} \, d\tau \tag{10.37}$$

The result stated in (10.36–37) can be established as follows. From (10.34) and (10.35), we have

$$X_+(j\omega) = 2U_h(j\omega) X(j\omega) \tag{10.38}$$

where $U_h(j\omega)$ represents the unit step function in the frequency domain. Note that $x_+(t)$ is the signal whose frequency spectrum is zero for negative frequencies and identical to the frequency spectrum of the signal $x(t)$ for positive frequencies (up to a multiplication factor of two). It can be shown using the duality property of the Fourier transform that $\mathcal{F}^{-1}\{U_h(j\omega)\} = \delta(t)/2 + j/2\pi t$ (see also Problem 3.28). Since the product in the frequency domain corresponds to the convolution in the time domain, we have, from (10.38),

$$x_+(t) = x(t) * \left[\delta(t) + \frac{j}{\pi t}\right] = x(t) + j\hat{x}(t) \tag{10.39}$$

which justifies the claim made in (10.35).

Note that the convolution integral (10.37) corresponds to a product in the frequency domain, that is,

$$\hat{X}(j\omega) = \mathcal{F}\left(\frac{1}{\pi t}\right) X(j\omega) \tag{10.40}$$

Problem 3.11(c) and its answer indicate that the preceding Fourier transform is equal to $-j\,\text{sgn}(\omega)$, which implies that

$$\hat{X}(j\omega) = -j\,\text{sgn}(\omega)X(j\omega) \tag{10.41}$$

Finding analytically this form of the Hilbert transform requires that the signal Fourier transform is multiplied by $-j\,\text{sgn}(\omega)$, and that the inverse Fourier transform is applied to the result obtained. This procedure is demonstrated in the next example.

EXAMPLE 10.3

The Hilbert transform of the sine function is obtained from (10.41) as

$$-j\,\text{sgn}(\omega)\mathcal{F}\{\sin(\omega_0 t)\} = j\,\text{sgn}(\omega) \times j\pi[\delta(\omega+\omega_0) - \delta(\omega-\omega_0)]$$
$$= -\pi[\delta(\omega+\omega_0) + \delta(\omega-\omega_0)] = -\mathcal{F}\{\cos(\omega_0 t)\}$$

Hence, the Hilbert transform of the signal $\sin(\omega_0 t)$ is equal to the Fourier transform of $-\cos(\omega_0 t)$. ∫

The Hilbert transform can be also found by using the MATLAB function `hilbert`. This is clarified in the MATLAB laboratory experiment in Section 10.7.

The sgn function is defined in (2.3). Using that definition, we have

$$\hat{X}(j\omega) = \begin{cases} -jX(j\omega) = e^{-j(\pi/2)}X(j\omega), & \omega > 0 \\ jX(j\omega) = e^{j(\pi/2)}X(j\omega), & \omega < 0 \end{cases} \tag{10.42}$$

It can be seen that *for real signals the Hilbert transform introduces a phase shift of* $-90° = -\pi/2$ *for positive frequencies and a phase shift of* $90° = \pi/2$ *for negative frequencies*.

The signal defined in (10.36) is called the *positive frequency pre-envelope signal* of $x(t)$. Its main feature is given by its spectrum formula,

$$X_+(j\omega) = \begin{cases} 2X(j\omega), & \omega > 0 \\ X(0), & \omega = 0 \\ 0, & \omega < 0 \end{cases} \tag{10.43}$$

The middle relation in (10.43) follows from (10.38) and the property of the frequency domain unit step function, that is, $X_+(0) = 2U_h(0)X(0) = X(0)$. Similarly, we can define the *negative frequency pre-envelope signal* of $x(t)$ by $x_-(t) = x(t) - j\hat{x}(t)$. Consequently, its spectrum is

$$X_-(j\omega) = \begin{cases} 0, & \omega > 0 \\ X(0), & \omega = 0 \\ 2X(j\omega), & \omega < 0 \end{cases} \tag{10.44}$$

Applications of this form of the Hilbert transform in communication systems will be discussed in Section 10.5, within the single sideband modulation technique which uses only the upper (or lower) frequency sideband for signal transmission.

10.4 IDEAL FILTERS

Signal filtering plays a very important role in communication systems. Filters can extract from a given frequency spectrum low-frequency components (*low-pass filtering*), high-frequency components (*high-pass filtering*), or signal components that belong to a certain frequency range (*band-pass filtering*). In addition, a filter can eliminate certain components from the signal frequency spectrum (*band-stop filtering*).

It is important to know that an ideal filter that exactly passes a given range of frequency components and exactly suppresses the frequency components outside of that range is not physically realizable. However, the ideal filter has theoretical importance in understanding the interplay between the time and frequency domains. Moreover, with slight modifications we can construct realizable filters starting with the frequency characteristics of ideal filters. The frequency characteristics of an ideal low-pass filter are presented in Figure 10.4. The frequency ω_0 is called the *filter cut-off frequency*.

According to Figure 10.4, the ideal low-pass filter transfer function is given by

$$H(j\omega) = \begin{cases} 1 \times e^{-j\omega t_d}, & |\omega| \leq \omega_0 \\ 0, & \text{otherwise} \end{cases} \qquad (10.45)$$

Note that we have assumed that the phase of the ideal filter changes linearly in frequency, which corresponds to the time shift (see the time shift property of the Fourier transform) of the filter input signals by t_d, known as the time delay.

In the following, we derive the impulse response of the ideal low-pass filter, and show that this impulse response does not correspond to that of a causal (real physical) system. We know from Example 3.16 that a rectangular frequency domain pulse has the time domain Fourier equivalent (note that in this case $\tau = 2\omega_0$) given by

$$p_{2\omega_0}(\omega) \leftrightarrow \frac{2\omega_0}{2\pi} \text{sinc}\left(\frac{2\omega_0}{2\pi}t\right) \qquad (10.46)$$

Using the time shift property of the Fourier transform, we have

$$H(j\omega) = p_{\omega_0}(\omega)e^{-j\omega t_d} \leftrightarrow \frac{\omega_0}{\pi}\text{sinc}\left(\frac{\omega_0}{\pi}(t-t_d)\right) = h(t) \qquad (10.47)$$

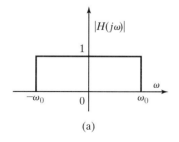 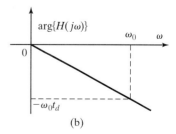

FIGURE 10.4: The frequency spectra of an ideal low-pass filter: (a) magnitude, (b) phase

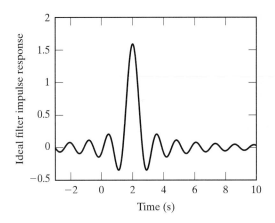

FIGURE 10.5: MATLAB plot of the impulse response of an ideal low-pass filter

The basic waveform of the sinc signal is presented in Figure 2.6. In (10.47), we have the shifted sinc signal. Its maximum, at $t = t_d$, is equal to ω_0/t_d. The waveform is present both left and right from the point $t = t_d$, having infinite duration in both directions (see Figure 10.5, where we use MATLAB to plot the corresponding impulse response for $\omega_0 = 5$ rad/s and $t_d = 2$ s).

It can be concluded that the ideal low-pass filter impulse response produces a nonzero waveform even for the times ($t < 0$) before the impulse delta input is applied to the filter. That violates the causality of the filter, that is, its physical realizability.

10.5 MODULATION AND DEMODULATION

Signal modulation formula (10.1) has been the cornerstone formula for the development of modern communication theory and its applications. More precisely, the formula

$$\mathcal{F}\{x(t)A_c \cos(\omega_c t)\} = 0.5 A_c X(j(\omega + \omega_c)) + 0.5 A_c X(j(\omega - \omega_c))$$

defines *amplitude modulation*. The signal $A_c \cos(\omega_c t)$ is the carrier signal, with carrier frequency ω_c and carrier amplitude A_c. The modulated signal is $x(t) A_c \cos(\omega_c t)$. The baseband signal $x(t)$ is also called the *message signal, modulating signal,* or *original signal*. The spectra of the original (baseband) and modulated signals are presented in Figure 10.2.

There are several types of modulation techniques. In addition to amplitude modulation, we have *frequency modulation* and *phase modulation* techniques, in which, respectively, the carrier signal frequency and the carrier signal phase are affected by the baseband signal. Hence, in those cases the carrier frequency and the carrier phase bear information about the original signal $x(t)$. Frequency and phase modulation are outside the scope of this introductory chapter on communication systems.

In addition to a sinusoidal carrier, a *train of pulses* can be used as the carrier signal in communication systems. In that case, we have again *amplitude modulation* (the pulse magnitude is proportional to the original signal magnitude at the given time instant), *pulse duration modulation* (the pulse width is proportional to the magnitude of the original signal), and *pulse position modulation* (the pulse position with respect to the reference position is determined by the magnitude of the original signal). These pulse modulation techniques are specific to continuous-time (analog) signals. Due to space limitations and the introductory nature of this chapter, continuous-time pulse modulation techniques will not be discussed in any detail.

For digital signals, we have the *pulse code modulation* technique, in which signals are binary encoded, and bits carrying information about signal magnitude are transmitted. We will say a little more about this modulation technique in the next section, where we present the essence of digital communication systems.

Amplitude Modulation

It can be seen from the preceding analysis that the carrier signal amplitude is equal to A_c. Being multiplied by $x(t)$, the carrier signal changes its magnitude according to $x(t)A_c$, which is how the carrier signal carries information about the signal $x(t)$. There are several variants of the amplitude modulation technique. In this section, we will say more about some of them. Let us first demonstrate in a simple example that the envelope of a modulated signal can carry information about the original signal. At the same time, we will establish conditions required for such a signal transmission.

EXAMPLE 10.4

Consider a simple signal $x(t) = te^{-t}u_h(t)$. Its Fourier transform is given by $\mathcal{F}(x(t)) = 1/(1 + j\omega)^2 = X(j\omega)$. The modulated signal $x(t)A_c \cos(\omega_c t)$, $A_c = 1$, and the original signal $x(t)$ are presented in Figures 10.6 and 10.7, for $\omega_c = 20$ rad/s and $\omega_c = 2$ rad/s, respectively.

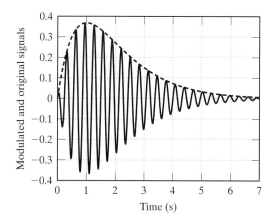

FIGURE 10.6: Modulated (solid line) and original (dashed line) signals for $\omega_c = 20$ rad/s

524   Chapter 10   Signals in Communication Systems

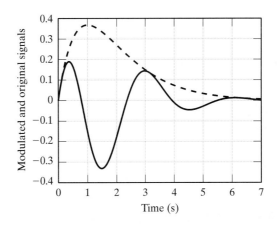

FIGURE 10.7: Modulated (solid line) and original (dashed line) signals for $\omega_c = 2$ rad/s

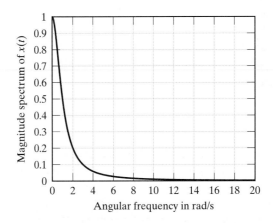

FIGURE 10.8: Magnitude spectrum of the original signal

It can be seen from Figure 10.6 that the modulated signal in its envelope basically carries information about the original signal. It is natural to expect that such information is sufficient for recovery of the original signal. However, it follows from Figure 10.7 that, in this case, the recovery of the original signal from the modulated signal is very difficult if not impossible.

In Figure 10.8, we present the magnitude spectrum of the original signal. It can be seen from this figure that the signal has a significant frequency component at $\omega_c = 2$ rad/s and almost negligible frequency component at $\omega_c = 20$ rad/s. Using these facts and the information about the shapes of the modulated signals presented in Figures 10.6 and 10.7, we can draw the conclusion that for easy and accurate signal recovery from the modulated signal the *carrier frequency must be much higher than the frequency of any significant spectral component* of the signal. Figures 10.7 and 10.8 have been generated using the following MATLAB code.

```
t=0:0.01:7;
x=t.*exp(-t);
% modulated signal and its plot
wc=2; % wc=20 for Figure 10.6
xmod=x.*cos(wc*t);
figure (1); % Figure 10.7
plot(t,xmod,t,x,'- -')
xlabel('Time (s)')
ylabel('Modulated and original signals'); grid
% spectrum of the original signal
w=0:0.01:20;
num=1; den=[1 2 1];
X=freqs(num,den,w); magX=abs(X);
figure (2); % Figure 10.8
plot(w,magX)
xlabel('Angular frequency in rad/s')
ylabel('Magnitude spectrum of x(t)'); grid
```

Note that in Example 10.4 we used our knowledge from Section 3.3.2 about frequency spectra to plot the magnitude spectrum, by finding first the Fourier transform of the signal, and then calculating and plotting the magnitude of the Fourier transform obtained. It was shown in Chapter 9 that this spectrum can be also found using the MATLAB function fft, finding the signal discrete Fourier transform via the fast Fourier transform algorithm. In such a case, one must also specify the sampling interval T_s. The Fourier transform is obtained from the discrete-time Fourier transform via a simple formula (9.72), which for small values of T_s implies

$$X(j\omega) \approx T_s X(j\Omega) \tag{10.48}$$

with $X(j\Omega)$ evaluated using the FFT.

Amplitude Modulation with a Transmitted Carrier

Note that in Example 10.4 the original signal $x(t) = te^{-t}u_h(t)$ is positive for all t. If $x(t)$ changes its sign for some t, then the modulated signal $x(t)A_c \cos(\omega_c t)$ will change phase at that time, such that its envelope will be distorted and it will no longer preserve the shape of the original signal. To prevent this problem, we can define the modulated signal using a slightly different modulation formula,

$$(1 + k_a x(t)) A_c \cos(\omega_c t) \leftrightarrow A_c \pi \delta(\omega + \omega_c) + A_c \pi \delta(\omega - \omega_c)$$
$$+ 0.5 k_a A_c X(j(\omega + \omega_c)) + 0.5 k_a A_c X(j(\omega - \omega_c)) \quad (10.49)$$

where k_a is an arbitrary constant called either the *amplitude sensitivity* or *index of modulation*. By choosing this constant such that

$$1 + k_a x(t) > 0 \Rightarrow |k_a x(t)| < 1, \quad \forall t \quad (10.50)$$

the envelope of the modulated signal will have the shape of the original signal and hence carry information about the original signal at all times. This property will facilitate the use of simple modulators for signal modulation and simple demodulators (envelope detectors) for signal reconstruction.

The frequency domain price for this time domain convenience is the presence of two additional delta impulses in the frequency spectrum of the modulated signal (see formula (10.49)). Since A_c may have a large value (as in the case of the switching modulator), a considerable amount of power is wasted in this kind of modulation, known as *double sideband with transmitted carrier* (DSB-TC) modulation. The originally considered modulation technique $(x(t) A_c \cos(\omega_c t))$ does not require independent carrier transmission. It is known as *double sideband with suppressed carrier* (DSB-SC) modulation. However, in both DSB-SC and DSB-TC the lower and upper signal frequency sidebands are transmitted. Since the signal information is completely contained in either the upper or lower frequency sideband, we conclude that these two modulation techniques waste a significant amount of the channel's frequency band. Exactly half of the frequency band can be saved by transmitting only the lower or upper signal frequency sideband. This can be facilitated by *single sideband* (SSB) modulation. Theoretical foundations for SSB modulation lie in the Hilbert transform considered in Section 10.3.

Switching Modulator and Envelope Detector (Demodulator)

Amplitude modulation with transmitted carrier and corresponding demodulation are easily performed by simple electrical devices known as the *switching modulator* and *envelope detector (demodulator)*. They are presented in Figures 10.9 and 10.10.

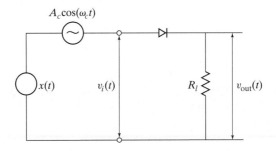

FIGURE 10.9: Switching modulator

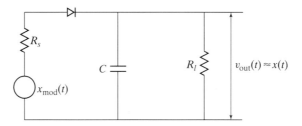

FIGURE 10.10: Envelope detector

In the MATLAB laboratory experiment in Section 10.7, we will say more about the actual operations of these devices. In the same experiment, the students will compare waveforms produced by these devices to the corresponding analytical results, and draw conclusions about the choice of relative values for the elements of the switching modulator and envelope detector.

Amplitude modulation for DSB-SC signals requires the use of more complex modulators, the most common of which is the ring modulator. As the corresponding demodulator, the Costas receiver is generally recommended. It is beyond the scope of this text to go into detail about these devices. For more information about the ring modulator and Costas receiver (demodulator), the interested reader is referred to any undergraduate communication systems textbook, for example [3].

Note that MATLAB has the modulation function `modulate`, which can be used for any of the preceding three modulation techniques. Its general form is `xmod=modulate(x,fc,fs,'method',parameter)`, where x represents samples of the original continuous-time signal sampled with the frequency `fs`; `fc` is the carrier frequency ($f_c = \omega_c/2\pi$); method is `amdsb-sc`, `amdsb-tc`, or `amssb`, denoting, respectively, the modulation method used: DSB-SC, DSB-TC, or SSB. The choice of parameter should be such that the modulating signal is positive with the minimum equal to zero. (parameter is set to zero for DSB-SC and SSB; it can be also omitted since its default value is zero.) Similarly, the MATLAB function `demod` performs demodulation, which can be achieved using the MATLAB statement `x=demod(xmod,fc,fs,'method')`.

Single Sideband Amplitude Modulation

As indicated in Section 10.3, theoretical foundations for the development of single sideband amplitude modulation lie in the Hilbert transform. The SSB amplitude modulated signal can be obtained by using the Hilbert transform as follows. Consider the cosine modulated original signal, that is,

$$s_{\text{mod}}^{\cos}(t) = x(t)\cos(\omega_c t) \leftrightarrow \frac{1}{2}X(j(\omega - \omega_c)) + \frac{1}{2}X(j(\omega + \omega_c)) \tag{10.51}$$

and the Hilbert transform of $x(t)$, denoted by $\hat{x}(t)$, modulated by the sine signal, that is,

$$\hat{s}_{\text{mod}}^{\sin}(t) = \hat{x}(t)\sin(\omega_c t) \leftrightarrow \frac{j}{2}\hat{X}(j(\omega + \omega_c)) - \frac{j}{2}\hat{X}(j(\omega - \omega_c)) \tag{10.52}$$

Since the signals $x(t)$ and $\hat{x}(t)$ are related through the Hilbert transform, it follows from (10.41) that
$$\hat{X}(j\omega) = -j\mathrm{sgn}(\omega)X(j\omega)$$
The use of this relation in (10.52) produces
$$\hat{s}_{\mathrm{mod}}^{\sin}(t) = \hat{x}(t)\sin(\omega_c t) \leftrightarrow \frac{1}{2}\mathrm{sgn}(\omega+\omega_c)X(j(\omega+\omega_c)) \qquad (10.53)$$
$$-\frac{1}{2}\mathrm{sgn}(\omega-\omega_c)X(j(\omega-\omega_c))$$

If we form the new modulated signal as
$$s_{\mathrm{mod}}(t) = \frac{1}{2}s_{\mathrm{mod}}^{\cos}(t) - \frac{1}{2}\hat{s}_{\mathrm{mod}}^{\sin}(t) \qquad (10.54)$$

its frequency spectrum, from (10.51) and (10.53), will be given by
$$\frac{1}{4}[1+\mathrm{sgn}(\omega-\omega_c)]X(j(\omega-\omega_c)) + \frac{1}{4}[1+\mathrm{sgn}(\omega-\omega_c)]X(j(\omega+\omega_c)) \qquad (10.55)$$

Bearing in mind the expression for the signum function, we see from (10.55) that the spectrum of this modulated signal has the form
$$\mathcal{F}(s_{\mathrm{mod}}(t)) = \mathcal{F}\left(\frac{1}{2}s_{\mathrm{mod}}^{\cos}(t) - \frac{1}{2}\hat{s}_{\mathrm{mod}}^{\sin}(t)\right) = \begin{cases} 0.5X(j(\omega-\omega_c)), & \omega \geq \omega_c \\ 0.5X(j(\omega+\omega_c)), & \omega \leq -\omega_c \\ 0, & \text{otherwise} \end{cases} \qquad (10.56)$$

This spectrum is presented in Figure 10.11.

Similarly, it can be shown that the frequency magnitude spectrum of the signal
$$\frac{1}{2}s_{\mathrm{mod}}^{\cos}(t) + \frac{1}{2}\hat{s}_{\mathrm{mod}}^{\sin}(t) \qquad (10.57)$$
contains only the lower frequency sidebands, that is,
$$\mathcal{F}\left(\frac{1}{2}s_{\mathrm{mod}}^{\cos}(t) + \frac{1}{2}\hat{s}_{\mathrm{mod}}^{\sin}(t)\right) = \begin{cases} 0.5X(j(\omega-\omega_c)), & 0 \leq \omega \leq \omega_c \\ 0.5X(j(\omega+\omega_c)), & 0 \geq \omega \geq -\omega_c \\ 0, & \text{otherwise} \end{cases} \qquad (10.58)$$

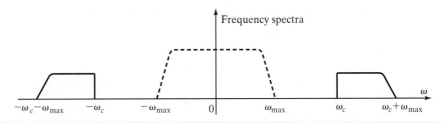

FIGURE 10.11: The frequency spectra of the original (dashed line) and single sideband modulated signal (solid lines)

Demodulation of SSB Signals

The original signal can be extracted from an SSB amplitude modulated signal using the idea presented in (10.2). Namely, the modulated signal is remodulated using a signal of the same frequency and phase (careful synchronization is required, so that the transmitter and the receiver use exactly the same frequency and phase for the carrier signal). Then, the original signal can be easily extracted using a low-pass filter, since

$$\mathcal{F}\{[x(t)\cos(\omega_c t) - \hat{x}(t)\sin(\omega_c t)]\cos(\omega_c t)\}$$
$$= \mathcal{F}\left\{\frac{1}{2}x(t) + \frac{1}{2}x(t)\cos(2\omega_c t) + \frac{1}{2}\hat{x}(t)\sin(2\omega_c t)\right\}$$
$$= \frac{1}{2}X(j\omega) + \frac{1}{4}X(j(\omega+2\omega_c)) + \frac{1}{4}X(j(\omega-2\omega_c))$$
$$+ \frac{j}{4}\hat{X}(j(\omega+2\omega_c)) - \frac{j}{4}\hat{X}(j(\omega-2\omega_c))$$
(10.59)

Due to the need for synchronization, this demodulation technique is called *coherent demodulation*, in contrast to the envelope detector technique, which does not require synchronization. This demodulation technique can also be used for DSB-SC signals.

10.6 DIGITAL COMMUNICATION SYSTEMS

Signal transmission in current communication systems is primarily digital. The main advantage of digital signal transmission techniques is their improved tolerance to noise. Noise is unavoidably present in all communication channels. The rapid development of digital computer networks, digital signal processing, and fast electronic and photonic switching devices during the last ten years has facilitated powerful signal transmission techniques that can make digital communication systems more efficient than corresponding analog systems. We have witnessed the merger of computer networks, used primarily for data transmission, with digital communication networks, used primarily for voice and video transmission. At this point, it is difficult to distinguish between digital computer networks and digital communication networks (systems) [4].

In Section 1.1.1, we introduced the concept of discretization of continuous-time signals with a given sampling period, which leads to the formation of discrete-time signals. Here, we just point out that, according to the Shannon sampling theorem, the sampling period must be taken as

$$T_s \leq \frac{\pi}{\omega_{\max}} = \frac{1}{2f_{\max}}$$
(10.60)

such that the original continuous-time signal can be reconstructed without error from its discrete-time equivalent. Signal reconstruction can be easily performed using a low-pass filter with cut-off frequency equal to half of the sampling frequency. (Note that in Section 9.1 we rigorously proved the Shannon theorem.)

530 Chapter 10 Signals in Communication Systems

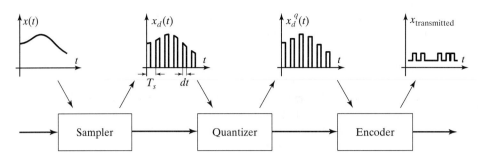

FIGURE 10.12: Pulse code modulation

The device that performs signal discretization (sampling) is called the *sampler*. In addition to being discretized, in digital communication systems, signals are also quantized (discretized with respect to magnitude). The device that performs magnitude quantization is called the *quantizer*. The discretized and quantized signal is called the digital signal. Finally, the digital signal is encoded into a stream of bits. This process of sampling, quantization, and encoding is known as *pulse code modulation* (PCM). It is symbolically presented in Figure 10.12.

The transmitter in a digital communication system performs pulse code modulation on an incoming signal and forms the encoded binary signal. The encoded binary signal is then sent over a communication channel as a stream of bits.

In this section, we have presented only the essential idea of digital communications. Further study of digital communications is beyond the scope of this chapter. For detailed studies of the PCM technique, and digital communication systems in general, the reader is referred to standard textbooks on digital communications, for example [3].

EXAMPLE 10.5   PCM for Speech Signals

Speech (telephone) signals are sampled every $125\,\mu$s, which generates 8000 samples per second. Quantization of speech signals is performed at $128 = 2^7$ levels, with each quantized sample encoded using 8 bits (one bit for the sign). This generates $8000 \times 8 = 64000$ bits per second, commonly denoted as 64 kbps (kilobits per second). Hence, while talking on the telephone, each user (speaker) generates 64 kb every second. Owing to recent advances in digital communication networks that use optical fiber channels, such a heavy bit stream can be easily handled. A synchronous optical network (SONET), is predominantly used today for the transmission of speech (telephone) signals [1].

10.7 COMMUNICATION SYSTEMS LABORATORY EXPERIMENT

Purpose: MATLAB and Simulink ([5], see also Appendix D) are convenient tools for studying flows of signals in any system, including communication systems. This is particularly important when analytical expressions for signals either become cumbersome or are not available due to the complexity of operations performed on signals. In this

Section 10.7   Communication Systems Laboratory Experiment

laboratory experiment, we will use MATLAB to study some basic problems related to the material presented in this chapter. In addition, we will use Simulink to simulate the switching modulator and the envelope detector, that is, to simulate the entire modulation-demodulation process.

Part 1. Consider the signal $x_1(t) = x(t)$ given in Figure 3.22. Find analytically its Fourier transform and plot its magnitude frequency spectrum. Present this signal in MATLAB using the signal representation considered in Example 2.11. Take `t=-2:0.01:3`. This signal is basically sampled with $T_s = 0.01$. Compare the obtained results with the frequency magnitude spectrum obtained using `X=Ts*fft(x)` and `plot(abs(X))`, a method used in MATLAB for numerical calculation of the Fourier transform.

Part 2. Find the Hilbert transform of the signal considered in Part 1. Use the MATLAB statement `hx=imag(hilbert(x))`. Form the amplitude modulated signal using `xmod=x.*cos(wc*t)` with `wc=100`. Find and plot the frequency magnitude spectrum of xmod. Form the amplitude modulated SSB signals defined in (10.54) and (10.57) and find their frequency magnitude spectra. Compare these frequency magnitude spectra with the frequency magnitude spectrum of the double sideband modulated signal.

Part 3. Use Simulink to make block diagrams for the switching modulator and the envelope detector (demodulator) whose schemes are presented in Figures 10.9 and 10.10. Connect them into a communication system. Note that you must use a relatively large value for the carrier signal amplitude A_c (as a matter of fact, it is required that $A_c \gg |x(t)|$). Under such a condition, the output voltage of the switching modulator is equal to

$$v_{\text{out}}(t) = [x(t) + A_c \cos(\omega_c t)] p_{T_c}^{\text{train}}(t), \quad T_c = \frac{2\pi}{\omega_c} \qquad (10.61)$$

where $p_{T_c}^{\text{train}}(t)$ is a periodic pulse train of duty cycle $0.5T_c$ and magnitude 1 (it is presented in Figure 3.18 with $T_c = T$ and $E = 1$). The Fourier series of such a pulse train is given by

$$p_{T_c}^{\text{train}}(t) = \frac{1}{2} + \frac{2}{\pi} \sum_{n=1}^{\infty} \frac{(-1)^{n-1}}{2n-1} \cos(\omega_c(2n-1)t) \qquad (10.62)$$

Show from (10.61) and (10.62) that the output component of the first harmonic is given by

$$\frac{A_c}{2}\left[1 + \frac{4}{\pi A_c} x(t)\right] \cos(\omega_c t) \qquad (10.63)$$

which represents the DSB-TC signal.

Part 4. Pass the signal defined in Part 1 through the modulator-demodulator configuration developed in Part 3. Observe on an oscilloscope the signal at the output of the modulator (modulated signal) and the signal at the output of the envelope detector (demodulated signal). Note that the parameters of the envelope detector must satisfy

$$(r_d + R_s)C \ll \frac{2\pi}{\omega_c} \ll R_l C \ll \frac{1}{f_{x\max}} \qquad (10.64)$$

where r_d is the diode resistance in the forward-biased region and $f_{x\max}$ approximately denotes the largest frequency in the original signal frequency magnitude spectrum (the capacitor charging time constant must be short and the discharging time constant must be long). Plot the waveform of the modulated signal. Plot the waveform of the demodulated signal and compare it with the original signal.

Part 5. Perform modulation-demodulation operations defined in Part 4 using the MATLAB functions `modulate` and `demod`. Compare the results obtained in Parts 4 and 5.

Comment: Reference [6] represents a very comprehensive guide to the use of MATLAB in communication systems.

10.8 SUMMARY

Study Guide for Chapter Ten: The essence of this chapter revolves around the use of the Fourier transform modulation property in the design of practical signal transmission techniques for communication systems. In that respect, students should master the basic amplitude modulation and demodulation techniques. To make amplitude modulation more efficient from the frequency band point of view, single sideband amplitude modulation was developed. Its theoretical foundations lie in the Hilbert transform, which is derived using the properties of the Fourier transform. Since frequency bandwidth and signal power are the most important resources in signal transmission, in this chapter, we give independent coverage of the signal power and energy. Standard problems: (1) find the signal power, energy, and correlation; (2) find the Hilbert transform (in many cases we have to use MATLAB), (3) find modulated and demodulated signals.

Aperiodic Signal Correlation Function:

$$R_{12}(\tau) = \int_{-\infty}^{\infty} x_1(t) x_2(t+\tau)\, dt, \quad R_{12}(\tau) \leftrightarrow X_1^*(j\omega) X_2(j\omega)$$

Aperiodic Signal Autocorrelation Function and Signal Power Spectrum:

$$R(\tau) = \int_{-\infty}^{\infty} x(t) x(t+\tau)\, dt, \quad R(\tau) \leftrightarrow |X(j\omega)|^2 = S(\omega)$$

Aperiodic Signal Total Energy:

$$E_\infty = \frac{1}{2\pi} \int_{\omega=-\infty}^{\omega=\infty} S(\omega)\,d\omega = R(0)$$

Aperiodic Signal Energy in the Given Frequency Band:

$$W(\omega_1,\omega_2) = \frac{1}{2\pi}\int_{\omega_1}^{\omega_2} S(\omega)\,d\omega + \frac{1}{2\pi}\int_{-\omega_2}^{-\omega_1} S(\omega)\,d\omega = \frac{1}{\pi}\int_{\omega_1}^{\omega_2} S(\omega)\,d\omega$$

Periodic Signal Correlation Function:

$$R_{12}(\tau) = \frac{1}{T}\int_{-T/2}^{T/2} x_1(t)x_2(t+\tau)\,dt, \quad R_{12}(\tau) \leftrightarrow X_1^*(jn\omega_0)X_2(jn\omega_0), \quad \omega_0 = \frac{2\pi}{T}$$

Periodic Signal Autocorrelation Function and Line Power Spectrum:

$$R(\tau) = \frac{1}{T}\int_{-T/2}^{T/2} x(t)x(t+\tau)\,dt = \sum_{n=-\infty}^{n=\infty} S(n\omega_0)e^{jn\omega_0\tau}, \quad R(\tau) \leftrightarrow S(n\omega_0)$$

Periodic Signal Energy During One Time Period:

$$R(0) = \sum_{n=-\infty}^{n=\infty} S(n\omega_0) = \sum_{n=-\infty}^{n=\infty} |X(jn\omega_0)|^2 = \frac{1}{T}\int_{-T/2}^{T/2} x^2(t)\,dt = W_T$$

Hilbert Transform for Causal Signals:

$$x(t) \leftrightarrow X(j\omega) = X_{\text{Re}}(\omega) + jX_{\text{Im}}(\omega)$$

$$X_{\text{Re}}(\omega) = X_{\text{Im}}(\omega) * \frac{1}{\pi\omega}, \quad X_{\text{Im}}(\omega) = -X_{\text{Re}}(\omega) * \frac{1}{\pi\omega}$$

Hilbert Transform for Real Signals:

$$\hat{x}(t) = x(t) * \frac{1}{\pi t} = \frac{1}{\pi}\int_{-\infty}^{\infty} x(\tau)\frac{1}{t-\tau}\,d\tau, \quad \hat{x}(t) \leftrightarrow \hat{X}(j\omega)$$

$$\hat{X}(j\omega) = -j\operatorname{sgn}(\omega)X(j\omega) = \begin{cases} -jX(j\omega) = e^{-j(\pi/2)}X(j\omega), & \omega > 0 \\ jX(j\omega) = e^{j(\pi/2)}X(j\omega), & \omega < 0 \end{cases}$$

Chapter 10 Signals in Communication Systems

Ideal Filter:

$$H(j\omega) = \begin{cases} 1 \times e^{-j\omega t_d}, & |\omega| \leq \omega_0 \\ 0, & \text{otherwise} \end{cases} \leftrightarrow \frac{\omega_0}{\pi}\text{sinc}\left(\frac{\omega_0}{\pi}(t - t_d)\right) = h(t)$$

Double Sideband Amplitude Modulation with Carrier Suppressed:

$$x(t)A_c \cos(\omega_c t) \leftrightarrow 0.5A_c X(j(\omega + \omega_c)) + 0.5A_c X(j(\omega - \omega_c))$$

Double Sideband Amplitude Modulation with Carrier Transmitted:

$$(1 + k_a x(t))A_c \cos(\omega_c t) \leftrightarrow A_c \pi \delta(\omega + \omega_c) + A_c \pi \delta(\omega - \omega_c)$$
$$+ 0.5 k_a A_c X(j(\omega + \omega_c)) + 0.5 k_a A_c X(j(\omega - \omega_c))$$

Single Upper Sideband Amplitude Modulation:

$$0.5[x(t)\cos(\omega_c t) - \hat{x}(t)\sin(\omega_c t)] \leftrightarrow \begin{cases} 0.5X(j(\omega - \omega_c)), & \omega \geq \omega_c \\ 0.5X(j(\omega + \omega_c)), & \omega \leq -\omega_c \\ 0, & \text{otherwise} \end{cases}$$

Single Lower Sideband Amplitude Modulation:

$$0.5[x(t)\cos(\omega_c t) + \hat{x}(t)\sin(\omega_c t)] \leftrightarrow \begin{cases} 0.5X(j(\omega - \omega_c)), & 0 \leq \omega \leq \omega_c \\ 0.5X(j(\omega + \omega_c)), & 0 \geq \omega \geq -\omega_c \\ 0, & \text{otherwise} \end{cases}$$

Double Sideband Amplitude Demodulation with Low-Pass Filtering:

$$\mathcal{F}\{[x(t)\cos(\omega_c t)]\cos(\omega_c t)\} = \mathcal{F}\left\{\frac{1}{2}x(t) + \frac{1}{2}x(t)\cos(2\omega_c t)\right\}$$

$$= \frac{1}{2}X(j\omega) + \frac{1}{4}X(j(\omega + 2\omega_c)) + \frac{1}{4}X(j(\omega - 2\omega_c))$$

Single Lower Sideband Amplitude Demodulation with Low-Pass Filtering:

$$\mathcal{F}\{[x(t)\cos(\omega_c t) - \hat{x}(t)\sin(\omega_c t)]\cos(\omega_c t)\} = \frac{1}{2}X(j\omega) + \text{HFS}$$

HFS = high frequency spectrum

10.9 REFERENCES

[1] R. Ramaswami and K. Sivarajan, *Optical Networks: A Practical Perspective,* Morgan and Kaufmann, San Francisco, 1998.

[2] F. Taylor, *Principles of Signals and Systems,* McGraw-Hill, New York, 1994.

[3] S. Haykin, *Communication Systems,* Wiley, New York, 1994.

[4] J. Walrand and P. Varaiya, *High-Performance Communication Networks,* Morgan and Kaufmann, San Francisco, 2000.

[5] Simulink 3.0, Student Version, The MathWorks Inc., Natick, MA, 1999.

[6] J. Proakis and M. Salehi, *Contemporary Communication Systems: Using MATLAB,* PWS Publishing Company, Boston, 1998.

10.10 PROBLEMS

10.1. Find the input signal power attenuation in dB for an optical fiber of length 25 km.

10.2. Assume that an optical fiber communication channel has frequency bandwidth equal to $\omega_{BW} = 50 \times 10^{12}$ Hz. Find the channel capacity, assuming that the ratio of the signal power to the noise power is 1000.

10.3. Find the frequency range that contains 70% of the signal energy for the signal whose frequency spectrum is presented in Figure 3.25.

10.4. Assuming that the signal $x(t)$ has the autocorrelation function $R_x(\tau)$, find the autocorrelation function of the signal $dx(t)/dt$.

Answer:
$$-\frac{d^2 R_x(\tau)}{d\tau^2}$$

10.5. Assuming that the signal $x(t)$ has the spectrum given by $S_x(\omega)$, find the spectrum of the signal $dx(t)/dt$.

Answer: $\omega^2 S_x(\omega)$

10.6. Find the autocorrelation function for the signal given by $u(t) - u(t - 2T)$.

10.7. Find the autocorrelation function for the exponential signal defined in Example 3.4.

10.8. Find the power spectrum and the autocorrelation function for the sinc signal defined in (2.16).

10.9. Establish the validity of the limits for the correlation coefficient as given in (10.17).

Hint: Use the Cauchy–Schwartz inequality given in Appendix B.

10.10. Use MATLAB to find the power spectrum of the signal defined by

$$x(t) = \begin{cases} 2\sin(10t) + 5\cos(5t), & 0 \leq t \leq 2 \\ 0, & \text{otherwise} \end{cases}$$

10.11. Find the Hilbert transform of the impulse delta signal.

Answer: $1/\pi t$.

10.12. Find the Hilbert transform of the following signals.

(a) $\dfrac{1}{t}$

(b) $\cos(\omega_0 t)$

(c) $\dfrac{\sin(2t)}{t}$

Answer:

(a) $-\pi \delta(t)$

(b) $\sin(\omega_0 t)$

10.13. Use MATLAB to find the Hilbert transform of the signal presented in Problem 2.8. Form its positive frequency pre-envelope signal and draw the corresponding frequency magnitude spectrum.

Hint: Use `imag(hilbert(x))`.

10.14. Use MATLAB to find the Hilbert transform of the signal defined in Problem 10.10.

10.15. Derive formula (10.47) using $h(t) = \mathcal{F}^{-1}\{H(j\omega)\} = \mathcal{F}^{-1}\{1 \times e^{-j\omega t_d}\}$ for the frequency range $-\omega_0 \leq \omega \leq \omega_0$.

10.16. Find the step response of an ideal filter.

Hint: Use the fact that the step response is the integral of the impulse response.

10.17. The modulating signal $x(t) \leftrightarrow X(j\omega)$ has a bandlimited frequency magnitude spectrum with maximal frequency $\omega_{\max}$. This signal is added to the carrier signal $A_c \cos(\omega_c t)$ and passed through a nonlinear element that has the characteristic

$$v_{\text{out}}(t) = a_1 v_{\text{in}}(t) + a_2 v_{\text{in}}^2(t)$$

Find the time domain and frequency domain forms of the output signal. Choose the carrier frequency such that the magnitude spectrum of the original signal $x(t)$ can be extracted from the output signal using a band-pass filter.

10.18. Find the transmitted power in DSB-TC and DSB-SC signals, and compare their values.

10.19. Derive formula (10.62).

10.20. Use MATLAB to evaluate and plot autocorrelation functions of the signals presented in Figure 2.17.

10.21. Use MATLAB to evaluate and plot crosscorrelation functions of the signals presented in Figures 2.17.

10.22. Use MATLAB to form DSB-SC and DSB-TC amplitude modulated signals from the signal presented in Problem 2.8. Draw the corresponding frequency magnitude spectrum.

Hint: Use `X=Ts*fft(x)` to find the Fourier transform numerically, where `Ts` is the sampling interval used for representation of the original signal.

10.23. Repeat Problem 10.22 for the signals presented in Figure 3.22.

10.24. Use MATLAB to form the SSB amplitude modulated signal of the signal presented in Problem 2.8. Obtain both the single upper sideband and single lower sideband amplitude spectra.

10.25. Use MATLAB and the demodulation technique presented in formula (10.59) to demodulate the signal that is modulated in Problem 10.24.

CHAPTER 11

Linear Electrical Circuits

In this chapter we study electrical circuits, which in most cases represent typical examples of linear time-invariant dynamic systems. However, electrical circuits can also be nonlinear. It is well known that an electrical circuit is composed of active and passive elements. Active elements are voltage and current sources; passive elements are resistors, capacitors, and inductors. Electrical circuit time invariance is contributed by constant values of circuit passive elements. If any circuit passive element varies in time, the circuit does not obey the time-invariance principle, and it represents a linear time-varying system. Linear time-varying systems are much more difficult to analyze than linear time-invariant systems, and are not treated in this textbook. Electrical circuit inputs are voltage and/or current sources that are, in general, time functions. In some applications, voltage and current sources have constant values. Electrical circuit outputs are voltages and currents in any set of branches of the given electrical circuit.

Linear electrical circuits composed only of constant voltage (current) sources and resistors represent linear static systems. They are described by systems of linear algebraic equations, which can be easily set using Ohm's and Kirchhoff's laws. The solution of systems of linear algebraic equations is also an easy problem, well known to all undergraduate engineering students. Such electrical circuits are studied in elementary physics and engineering courses, and will not be covered in this chapter.

The goal of this chapter is to show how the theory presented in previous chapters can be used to analyze linear electrical circuits that contain at least one capacitor or inductor, so that they are described by linear differential equations with constant coefficients (linear time-invariant dynamic systems). The Laplace transform presented in Chapter 4 will be our main tool for analyzing and studying the dynamics of simple linear electrical circuits. For more complex electrical circuits described by higher-order linear differential equations, we will use the state space technique. We will start with simple circuits that contain an active source, a resistor, and either a capacitor or an inductor. It will be clear that such circuits are described by first-order linear differential equations, and we typically call them first-order linear electrical circuits. An electrical circuit that has two capacitors, two inductors, or a capacitor and an inductor, is described by second-order differential equations; such circuits are called second-order electrical circuits. In general, *the total number of circuit capacitors and inductors determines the order of the circuit, that is, the order of the corresponding differential equation that describes the circuit dynamics.*

Note that *capacitors and inductors are energy storing elements*. A capacitor is able to store electrostatic energy; the presence of such energy in a capacitor is demonstrated by the corresponding capacitor voltage. Once connected in an electrical circuit, such a voltage plays the role of the capacitor's initial voltage. An inductor stores electromagnetic energy, which is demonstrated by the presence of an initial current through the inductor. *The initial voltages on capacitors and initial currents in inductors determine the electrical circuit initial conditions.* Recall from the previous chapters of this book that the initial conditions determine the system zero-input response. In the case of asymptotically stable linear systems, the zero-input response decays to zero in the long run. It is well known that *linear electrical circuits with constant parameters are asymptotically stable* (unless the circuit has no resistors when it is only stable, that is, the motion of its variables is bounded). Hence, their zero-input responses decay to zero (rather quickly), so that electrical circuit responses in the long run (steady state) are determined by the forcing functions. More precisely, linear electrical circuit responses in the long run have the form of the forcing functions. If the forcing function is sinusoidal, the steady state circuit response will be sinusoidal; if the forcing function has the form of the step function, the circuit response will have the form of the step function, and so on. It is interesting to point out that the steady state response of linear electrical circuits (composed of resistors, capacitors, and inductors) due to sinusoidal forcing functions (either voltage or current sources) can be studied using phasor algebra, which leads to simple algebra with complex numbers. Such simplified electrical circuit problems will not be covered in this chapter. However, electrical circuit responses subject to both sinusoidal inputs and initial conditions should be considered using the methodology presented here.

This chapter is organized as follows. In Section 11.1, we review some basic relationships between voltages and currents for resistors, inductors, and capacitors; introduce the notion of the electrical power; and define the electrostatic and electromagnetic energy. Section 11.2 presents first-order linear electrical circuits that are described by first-order linear differential equations. Second-order linear electrical circuits, whose models are represented by second-order linear differential equations, are studied in Section 11.3. Electrical circuits whose dynamics are described by higher-order differential equations are considered in Section 11.4, using the state space technique. The main theme in Sections 11.2–4 is finding electrical circuit responses due to given voltage and/or current sources and circuit initial conditions. A MATLAB laboratory experiment on electrical circuits is designed in Section 11.5.

11.1 BASIC RELATIONS

We first review some basic formulas from an elementary electrical circuit (or elementary physics) course, which relate voltage, current, electrical charge, and electromagnetic flux, for resistors, capacitors, and inductors. In addition, we review formulas for the power and energy for these three basic electrical circuit passive elements.

Resistor Voltage, Current, Power, and Energy

It is well known, from Ohm's law [1–4], that the voltage on the resistor, denoted by $v_R(t)$, is given by

$$v_R(t) = Ri_R(t) \tag{11.1}$$

where $i_R(t)$ is the current through the resistor R. The resistor power, $p_R(t)$, is defined by

$$p_R(t) = v_R(t)i_R(t) = Ri_R^2(t) = \frac{1}{R}v_R^2(t) \tag{11.2}$$

The expression for the resistor energy (energy dissipated by the resistor—transformed into heat—during a certain time interval) can be obtained from the well-known relation between the power and the energy (the power is the energy time rate of change, $p(t) = dw(t)/dt$); that is,

$$w_R(t) = \int_{t_0}^{t} p_R(\sigma)\,d\sigma = R\int_{t_0}^{t} i_R^2(\sigma)\,d\sigma = \frac{1}{R}\int_{t_0}^{t} v_R^2(\sigma)\,d\sigma \tag{11.3}$$

Capacitor Voltage, Current, Power, and Energy
Ohm's law for the capacitor [1–4] is given by

$$v_C(t) = \frac{1}{C}\int_{-\infty}^{t} i_C(\sigma)\,d\sigma = \frac{1}{C}\int_{-\infty}^{0^-} i_C(\sigma)\,d\sigma + \frac{1}{C}\int_{0^-}^{t} i_C(\sigma)\,d\sigma$$

$$= v_C(0^-) + \frac{1}{C}\int_{0^-}^{t} i_C(\sigma)\,d\sigma \tag{11.4}$$

where $i_C(t)$ is the current through the capacitor C and $v_C(0^-)$ is the initial voltage on the capacitor. From (11.4), by taking the derivative, we can easily obtain the expression for the capacitor current in terms of the capacitor voltage as

$$i_C(t) = C\frac{d(v_C(t) - v_C(0^-))}{dt} \tag{11.5}$$

The capacitor electric charge is given by

$$q(t) = q(0^-) + \int_{0^-}^{t} i_C(\sigma)\,d\sigma \tag{11.6}$$

where $q(0^-)$ is the initial charge on the capacitor. From this expression, the current through the capacitor, in terms of capacitor charge, is given by

$$i_C(t) = \frac{d(q(t) - q(0^-))}{dt} \tag{11.7}$$

If we divide (11.6) by C, we obtain formula (11.4), with

$$q(t) = Cv_C(t), \quad q(0^-) = Cv_C(0^-) \tag{11.8}$$

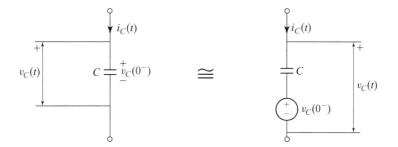

FIGURE 11.1: Equivalent representations of a charged capacitor

The capacitor power, $p_C(t)$, is defined by

$$p_C(t) = v_C(t) i_C(t) \tag{11.9}$$

The capacitor electrostatic energy, $w_C(t)$, during the time interval $[t_0, t]$, is given by

$$w_C(t) = \int_{t_0}^{t} p_C(\sigma)\,d\sigma = \int_{t_0}^{t} v_C(\sigma) i_C(\sigma)\,d\sigma = \frac{1}{C} \int_{q(t_0)}^{q(t)} q\,dq$$

$$= \frac{q^2(t)}{2C} - \frac{q^2(t_0)}{2C} = \frac{1}{2} C v_C^2(t) - \frac{1}{2} C v_C^2(t_0) \tag{11.10}$$

In our consideration, the initial time instant will almost always be taken as $t_0 = 0^-$.

It follows from formula (11.4) that a charged capacitor (with the initial energy stored) can be represented as indicated in Figure 11.1.

This figure helps to better understand the charged capacitor.

Current Direction

In this chapter, we will use a standard convention regarding current direction. The direction of the current through a passive electrical element is assumed to be from the element node with positive voltage to the node with negative voltage. This convention has been used in Figure 11.1.

Inductor Voltage, Current, Power, and Energy

It is well known [1–4] that the inductor current, $i_L(t)$, is given by

$$i_L(t) = \frac{1}{L} \int_{-\infty}^{t} v_L(\sigma)\,d\sigma = \frac{1}{L} \int_{-\infty}^{0^-} v_L(\sigma)\,d\sigma + \frac{1}{L} \int_{0^-}^{t} v_L(\sigma)\,d\sigma$$

$$= i_L(0^-) + \frac{1}{L} \int_{0^-}^{t} v_L(\sigma)\,d\sigma \tag{11.11}$$

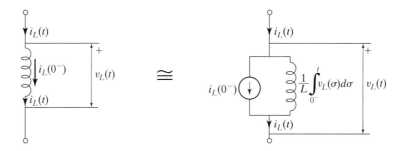

FIGURE 11.2: Equivalent representations of an inductor with initial current

where $i_L(0^-)$ denotes the inductor initial current and L stands for the inductor inductance. Differentiating (11.11), we obtain the expression for the inductor voltage in terms of its current as

$$v_L(t) = L \frac{d(i_L(t) - i_L(0^-))}{dt} \tag{11.12}$$

The electromagnetic flux, $\phi(t)$, of an inductor is related to its current by a very simple relation,

$$\phi(t) = L i_L(t) \tag{11.13}$$

The inductor power is defined by

$$p_L(t) = v_L(t) i_L(t) \tag{11.14}$$

and its electromagnetic energy is given by

$$w_L(t) = \int_{t_0}^{t} p_L(\sigma) \, d\sigma = \int_{t_0}^{t} u_L(\sigma) i_L(\sigma) \, d\sigma = L \int_{i_L(t_0)}^{i_L(t)} i_L \, di_L$$

$$= \frac{1}{2} L i_L^2(t) - \frac{1}{2} L i_L^2(t_0) = \frac{\phi^2(t)}{2L} - \frac{\phi^2(t_0)}{2L} \tag{11.15}$$

It follows from formula (11.11) that an inductor with initial current (or more precisely, *initial electromagnetic energy stored*) can be represented using an equivalent circuit, as indicated in Figure 11.2.

11.1.1 Equivalence Between Voltage and Current Sources

A simple equivalence relation that exists between voltage and current sources can facilitate the simplification of electrical circuits, and the establishment of equivalent electrical circuits. Using this equivalence, the number of different basic electrical circuit configurations that must be studied is reduced. The corresponding equivalence is presented in Figure 11.3.

This equivalence says that the cascade connection of a voltage source and a resistor can be replaced by an equivalent parallel connection of the same resistor and a current source, and vice versa. Assuming that the voltage source has the value $v_s(t)$, the equivalent

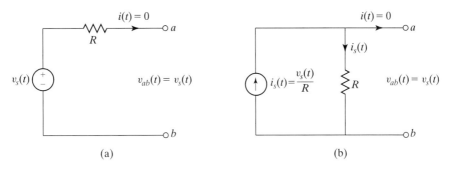

FIGURE 11.3: Equivalent voltage and current sources

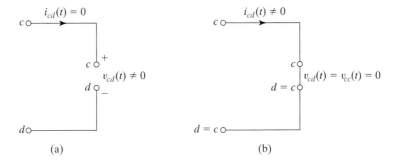

FIGURE 11.4: (a) Open and (b) short electrical circuits

current source is equal to $i_s(t) = v_s(t)/R$. Indeed, it can be observed that both circuits have the same voltage between the points (nodes) a and b, in both cases equal to $v_s(t)$. Also, for both circuits presented in the figure, between the nodes a and b there is an *open circuit*, so that the corresponding current $i_{ab}(t)$ is zero in both cases. In addition, due to the facts that an ideal voltage source has zero resistance and an ideal current source has infinite resistance [1–4], it can be concluded that the resistances between the nodes a and b are identical in both cases and equal to R.

Open and Short Circuits

By definition [1–4], an *open circuit has zero current and arbitrary voltage,* and a *short circuit has zero voltage and arbitrary current* (see Figure 11.4). The concepts of open and short circuits will be used in subsequent sections of this chapter.

11.2 FIRST-ORDER LINEAR ELECTRICAL CIRCUITS

First-order linear electrical circuits are described by first-order linear differential equations. Their passive electrical elements consist of one inductor or capacitor, and an arbitrary number of resistors. In the simplest cases, only one resistor is present; such first-order linear electrical circuits are divided into two categories, RC and RL circuits—containing either one resistor and one capacitor (RC circuits) or one resistor and one inductor (RL circuits).

In addition, first-order linear electrical circuits can be driven by at least one active element, a voltage or current source. In both RC and RL circuits, the corresponding elements can be put in either parallel or series (cascade) connection. Hence, we can form four types of basic RC and four types of basic RL electrical circuits. Using the equivalence between voltage and current sources, the actual number of basic RC and RL circuits that must be analyzed can be reduced.

First-order linear electrical circuits can be analyzed using either the Laplace transform from Chapter 4 or techniques for solving first-order differential equations presented in Chapter 7. In this chapter, we will use the Laplace transform, since it is simpler.

11.2.1 RC Electrical Circuits

In the following we analyze cascade (series) and parallel connections in first-order RC circuits. Using the Ohm and Kirchhoff laws, we derive RC circuit mathematical models and find corresponding responses due to both input sources and initial conditions.

Resistor and Capacitor in Series Driven by a Voltage Source

An RC electrical circuit composed of a series (cascade) connection of voltage source, resistor, and capacitor is presented in Figure 11.5.

The voltage source is connected to the circuit at time $t = 0$, which is indicated by the switch being closed at $t = 0$. In general, the capacitor can be initially charged, so that we assume that its initial voltage is $v_C(0^-)$.

Note that in this circuit, the resistor and capacitor currents are the same, that is, $i_R(t) = i_C(t) = i(t)$. Using the Kirchhoff law for voltage (the sum of all voltages around a closed loop is zero), we have

$$v_s(t) - v_R(t) - v_C(t) = 0 \qquad (11.16)$$

or

$$v_s(t) = Ri(t) + \frac{1}{C}\int_{0^-}^{t} i(\sigma)\,d\sigma + v_C(0^-) \qquad (11.17)$$

For given $v_s(t)$, equation (11.17) represents the integral equation with respect to $i(t)$ that is equivalent to the first-order differential equation (by taking the time derivative). We can use

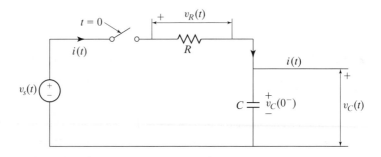

FIGURE 11.5: A cascade RC circuit driven by a voltage source

the Laplace transform to solve this integral equation. By applying the Laplace transform to (11.17), we obtain

$$V(s) = RI(s) + \frac{1}{sC}I(s) + \frac{v_C(0^-)}{s} \qquad (11.18)$$

where $V(s) = \mathcal{L}\{v_s(t)\}$ and $I(s) = \mathcal{L}\{i(t)\}$. Note that $v_C(0^-)$ is constant, hence its Laplace transform is given by $v_C(0^-)/s$. From this algebraic equation, we obtain easily the expression for the circuit current as

$$I(s) = \frac{V(s)}{R + \dfrac{1}{sC}} - \frac{v_C(0^-)}{s\left(R + \dfrac{1}{sC}\right)} = \frac{sCV(s)}{sRC + 1} - \frac{Cv_C(0^-)}{sRC + 1}$$

$$= \frac{s\dfrac{1}{R}}{s + \dfrac{1}{RC}} V(s) - \frac{\dfrac{1}{R}}{s + \dfrac{1}{RC}} v_C(0^-) \qquad (11.19)$$

The first part of the right-hand side of (11.19) determines the zero-state circuit response (obtained by assuming that $v_C(0^-) = 0$), and the second part on the right-hand side of (11.19) determines the zero-input response (obtained assuming that $v_s(t) = 0$). We can see that this electrical circuit has an asymptotically stable pole located at $-1/RC$. The *reciprocal value of the system pole is called the system time constant.* Hence, the time constant for the RC electrical circuit is simply given by the product $\tau = RC$.

For the given $v_s(t)$, we can find its Laplace transform $V(s)$. Applying the inverse Laplace transform to (11.19) produces the sought time domain solution for the RC circuit current, $i(t) = \mathcal{L}^{-1}\{I(s)\}$. Once the current $i(t)$ is determined, all voltages for the RC circuit can be easily calculated.

EXAMPLE 11.1

Assume that the input voltage to the electrical circuit represented in Figure 11.5 is constant, that is, $v_s(t) = E$. Then, the current in the RC circuit, obtained from (11.19), is given by

$$i(t) = \mathcal{L}^{-1}\{I(s)\} = \mathcal{L}^{-1}\left\{\frac{s\dfrac{1}{R}}{\left(s + \dfrac{1}{RC}\right)} \frac{E}{s} - \frac{\dfrac{1}{R}}{s + \dfrac{1}{RC}} v_C(0^-)\right\}$$

$$= \mathcal{L}^{-1}\left\{\frac{\dfrac{E}{R}}{s + \dfrac{1}{RC}} - \frac{\dfrac{1}{R}}{s + \dfrac{1}{RC}} v_C(0^-)\right\} = \frac{E}{R} e^{-(1/RC)t} - \frac{v_C(0^-)}{R} e^{-(1/RC)t}$$

or

$$i(t) = \left(\frac{E}{R} - \frac{v_C(0^-)}{R}\right) e^{-(1/RC)t}, \qquad t \geq 0$$

Hence, the current in this circuit decays exponentially to zero. As a matter of fact, since resistance is usually measured in kΩ (kiloohms) and capacitance in μF (microfarads), it follows that RC is of the order of 10^{-3}, which implies that the current $i(t)$ rapidly decays to zero. The resistor voltage, defined by (11.1), also decays rapidly to zero. The energy dissipated by the resistor as heat is given by the expression

$$w_R(\infty) = R \int_{0^-}^{\infty} i^2(\sigma)\, d\sigma = 2C(E - v_C(0^-))^2$$

On the other hand, the capacitor is charged during the same short period of time to the new value E. This can be analytically verified using (11.4), as follows:

$$v_C(t) = v_C(0^-) + \frac{1}{C}\int_{0^-}^{t} i_C(\sigma)\, d\sigma$$

$$= v_C(0^-) + \frac{1}{C}\left(\frac{E}{R} - \frac{v_C(0^-)}{R}\right)\int_{0^-}^{t} e^{-(1/RC)\sigma}\, d\sigma$$

$$= v_C(0^-) + (E - v_C(0^-))\left(1 - e^{-(1/RC)t}\right)$$

The last expression indicates that $v_C(0^+) = v_C(0^-)$. Consequently, the formula $i_C(t) = C dv_C(t)/dt$ produces no impulse delta function at $t = 0$ for capacitor current. Hence, no instant jump in capacitor voltage takes place at $t = 0$. Note that an instant jump in capacitor voltage is possible only when the capacitor is exposed to current that contains an impulse delta signal (see formula (11.4)).

It can be seen from the preceding formula that the steady state value for capacitor voltage is given by $v_C(\infty) = v_C(0^-) + (E - v_C(0^-)) = E$. Since the circuit current (the current through the capacitor) is equal to zero, *the capacitor at steady state (driven by a constant input) behaves like an open circuit* (zero current and nonzero voltage; in this

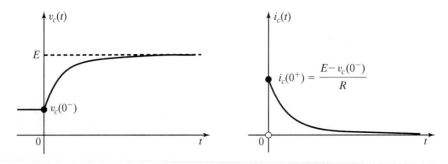

FIGURE 11.6: Capacitor voltage and current waveforms in an RC circuit driven by a constant voltage source

case, the voltage is equal to E). It follows also that $v_C(0^+) = v_C(0^-)$ and $i_C(0^+) = (E - v_C(0^-))/R$. Note that by imposing the assumption that $v_C(0^-) = 0$, it can be seen that the capacitor at $t = 0$ behaves like a short circuit.

Waveforms for capacitor voltage and current are presented in Figure 11.6. It can be observed that the capacitor is charged from its initial voltage $v_C(0^-)$ to the new value E. On the other hand, Figure 11.6 shows that the capacitor current decays to zero exponentially.

The capacitor is charged, in this case using a constant voltage source. Though the problem of charging a capacitor looks very simple, it becomes challenging when we want to achieve the same goal with the minimum dissipation of energy by the resistor. It was recently shown [5] that in such a case the optimal input source is given by a sum of step and ramp inputs.

Resistor and Capacitor in Parallel Driven by a Voltage Source

The corresponding circuit configuration is presented in Figure 11.7. Analysis of this electrical circuit is fairly simple, since it follows directly from formulas (11.1) and (11.5). The resistor and capacitor voltages are the same and equal to the source voltage, that is, $v_R(t) = v_C(t) = v_s(t)$.

The current through the resistor is simply obtained from (11.1) as $i_R(t) = v_s(t)/R$. The capacitor current is obtained from (11.5). It is interesting to observe that the capacitor current may contain an impulse delta function. Assuming that $v_s(t) = E = \text{const}$, the current through the capacitor is given by

$$i_C(t) = C(E - v_C(0^-))\delta(t) \qquad (11.20)$$

Furthermore, if the source voltage has no jump discontinuity at $t = 0^-$, the delta impulse in the capacitor current can come from the capacitor initial condition (note that an ideal voltage source has zero resistance, and that the capacitor and the voltage source are placed in parallel). Since the duration of the impulse delta function is infinitely short, the delta impulse current signal is harmless to the circuit. Note that in this case, the capacitor is

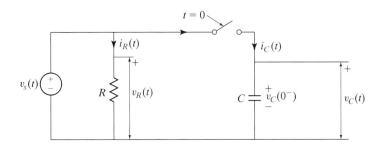

FIGURE 11.7: A parallel RC circuit driven by a voltage source

instantaneously charged to the new value $v_C(0^+) = E$, and that $v_C(t) = E, \forall t \geq 0$. This can also be analytically justified using (11.4), as follows:

$$v_C(t) = v_C(0^-) + \int_{0^-}^{t} (E - v_C(0^-))\delta(\sigma)\,d\sigma = v_C(0^-) + E - v_C(0^-) = E$$

The current $i(t)$ through the voltage source is equal to the sum of the resistor and capacitor currents, that is, $i(t) = i_R(t) + i_C(t)$. It can be noticed that the capacitor current after an initial delta impulse at $t = 0$ is equal to zero for all other positive times (this capacitor behaves like an open circuit), and that $i(t) = i_R(t) = E/R, t > 0$.

Resistor and Capacitor in Parallel Driven by a Current Source

Using the equivalence between voltage and current sources as presented in Figure 11.3, we can conclude that an electrical circuit represented by a serial connection of a capacitor and a resistor driven by a voltage source, as given in Figure 11.5, is *equivalent* to an electrical circuit represented by a parallel connection of a resistor and a capacitor driven by a current source whose current is equal to $i_s(t) = v_s(t)/R$. Such an equivalent electrical circuit, driven by the current source, is presented in Figure 11.8.

It is left as an exercise for the student to solve equations describing the dynamics of this electrical circuit, and to find the corresponding circuit currents and voltages (see Problems 11.2 and 11.3).

If one intends to solve this electrical circuit without using the equivalence relation, one must first set the corresponding differential equation, which for this electrical circuit is given by

$$i_C(t) + i_R(t) = i_s(t) \Rightarrow C\frac{dv_C(t)}{dt} + \frac{v_C(t)}{R} = i_s(t), \qquad v_C(0^-) \text{ given} \qquad (11.21)$$

Assuming that the value for $i_s(t)$ is known, the solution of this differential equation can be easily found.

Resistor and Capacitor in Series Driven by a Current Source

This electrical circuit is presented in Figure 11.9. Its analysis is straightforward. The current through the resistor and capacitor is equal to the source current, $i_s(t)$. The source

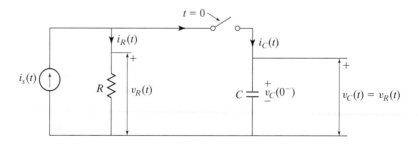

FIGURE 11.8: A parallel RC circuit driven by a current source

Section 11.2 First-Order Linear Electrical Circuits

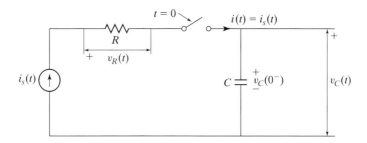

FIGURE 11.9: A series RC circuit driven by a current source

current produces the voltage on the resistor given by formula (11.1), that is $v_R(t) = Ri_s(t)$. The capacitor voltage is obtained from formula (11.4) as

$$v_C(t) = v_C(0^-) + \frac{1}{C}\int_{0^-}^{t} i_s(\sigma)\,d\sigma \tag{11.22}$$

Note that no current can flow due to capacitor initial voltage (or capacitor initial charge $q(0^-) = Cv_C(0^-)$), since the current source has an infinite resistance, which implies an open circuit.

Assume that in this electrical circuit the source current is constant. This can be represented by $i_s(t) = I_s u(t)$, where $u(t)$ is the unit step function and $I_s = \text{const}$. The voltage on the capacitor is given by the expression

$$v_C(t) = v_C(0^-) + \frac{I_s}{C}\int_{0^-}^{t} u(\sigma)\,d\sigma = v_C(0^-) + \frac{I_s}{C}r(t)$$

where $r(t)$ is the unit ramp function. It can be concluded that in such a case, the capacitor voltage grows unbounded in time, which will destroy the capacitor in the long run. Hence, we should not expose a capacitor to a constant current source for a long period of time.

11.2.2 RL Electrical Circuits

In this section we consider first-order linear electrical circuits obtained by combining a resistor and an inductor either in parallel or cascade and driven by either voltage or current sources. All four types of such electrical circuits will be presented in this section. For generality, we will assume that the inductor has some initial energy stored.

Note that in order to have an inductor with initial current flow, say $i_L(0^-)$, in an electrical circuit, the inductor should be a part of another electrical circuit in which the current through the branch containing the inductor is $i_L(0^-)$ (note that the electromagnetic energy, and not the current, is stored in the inductor). At the initial time $t = 0$, the inductor should be disconnected from that branch circuit and connected to the circuit under consideration. This situation is presented in Figure 11.10 using dashed lines and a switch that opens at $t = 0$. In this chapter, for simplicity of figures, such branch circuits that set up inductor initial current are henceforth omitted.

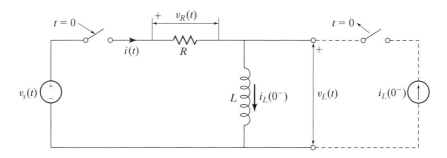

FIGURE 11.10: A cascade RL electrical circuit driven by a voltage source

Resistor and Inductor in Series Driven by a Voltage Source

An RL circuit composed of a series (cascade) connection of voltage source, resistor, and inductor is represented in Figure 11.10.

Due to the cascade connection, both the resistor and inductor have the same current, that is, $i_R(t) = i_L(t) = i(t)$. The Kirchhoff voltage law applied to the circuit in Figure 11.10 implies

$$v_s(t) = v_R(t) + v_L(t) = Ri(t) + L\frac{di(t)}{dt} \qquad (11.23)$$

Using the Laplace transform, we have

$$V(s) = RI(s) + sLI(s) - Li_L(0^-) \qquad (11.24)$$

which produces

$$I(s) = \frac{1}{sL+R}V(s) + \frac{L}{sL+R}i_L(0^-) = \frac{\frac{1}{L}}{s+\frac{R}{L}}V(s) + \frac{1}{s+\frac{R}{L}}i_L(0^-) \qquad (11.25)$$

Applying the inverse Laplace transform, we obtain

$$i(t) = i_{zs}(t) + i_{zi}(t) = \mathcal{L}^{-1}\{I(s)\}$$

$$i_{zs}(t) = \mathcal{L}^{-1}\left\{\frac{\frac{1}{L}}{s+\frac{R}{L}}V(s)\right\}$$

$$i_{zi}(t) = \mathcal{L}^{-1}\left\{\frac{1}{s+\frac{R}{L}}i_L(0^-)\right\} = e^{-(R/L)t}i_L(0^-) \qquad (11.26)$$

Note that in this electrical circuit the time constant is given by $\tau = L/R$. Since resistance is usually measured in kΩ (kiloohms) and inductance in mH (millihenrys), it follows that

L/R is of the order of 10^{-6}, which implies that the current $i_{zi}(t)$ rapidly decays to zero. In order to find the expression for $i_{zs}(t)$, we must know the form of the input function. In the next example, we find $i_{zs}(t)$ assuming that $v_s(t) = E = \text{const}$.

EXAMPLE 11.2

From (11.25), we have

$$i_{zs}(t) = \mathcal{L}^{-1}\{I_{zs}(s)\} = \mathcal{L}^{-1}\left\{\frac{\frac{1}{L}}{\left(s+\frac{R}{L}\right)}\frac{E}{s}\right\}$$

$$= \mathcal{L}^{-1}\left\{\frac{\frac{E}{R}}{s} - \frac{\frac{E}{R}}{s+\frac{R}{L}}\right\} = \frac{E}{R} - \frac{E}{R}e^{-(R/L)t}, \qquad t \geq 0$$

so that

$$i(t) = i_{zs}(t) + i_{zi}(t) = \frac{E}{R} + \left(i_L(0^-) - \frac{E}{R}\right)e^{-(R/L)t}$$

It follows that $i(0^+) = i_L(0^+) = i_L(0^-)$, which indicates that the *inductor cannot change its current instantaneously* (unless it is driven by the impulse delta function). This observation also follows analytically from formula (11.11).

It can be seen from the last formula that the circuit current tends to a constant value of E/R as t increases. Since the current through the inductor tends to a constant, it follows from (11.12) that the voltage on the inductor tends to zero. This can also be seen from the expression for the inductor voltage,

$$v_L(t) = L\frac{d(i_{zs}(t) + i_{zi}(t))}{dt} = (E - Ri_L(0^-))e^{-(R/L)t}$$

which implies that $v_L(\infty) = 0$. Note that because the inductor current has no jump discontinuity at $t = 0$, the corresponding inductor voltage cannot contain an impulse delta signal at $t = 0$. The voltage on the resistor is given by

$$v_R(t) = Ri(t) = R(i_{zs}(t) + i_{zi}(t)) = E + (Ri_L(0^-) - E)e^{-(R/L)t}$$

During the initial short time period, electromagnetic energy initially stored in the inductor dissipates in the form of heat produced by the resistor, and the inductor thereafter (in the long run, at steady state) *behaves like a short circuit* (its current is nonzero, but its voltage is zero).

It can be noticed also that for this electrical circuit, $v_L(0) = E - Ri_L(0^-)$. Assuming that $i_L(0^-)$ is equal to zero, we see that the inductor at $t = 0$ behaves like an open circuit.

The waveforms for the inductor voltage and current are presented in Figure 11.11. It can be seen from this figure that the inductor current increases (when $E/R > i_L(0^-)$) and

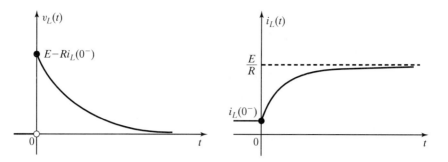

FIGURE 11.11: Waveforms for inductor voltage and current in a cascade RL electrical circuit driven by a constant voltage source

decreases (when $E/R < i_L(0^-)$) exponentially from its initial value $i_L(0^-)$ to its steady state value E/R. It can be seen also that the inductor voltage starting from its initial value $v_L(0) = E - Ri_L(0^-)$ decays exponentially to zero.

Resistor and Inductor in Parallel Driven by a Voltage Source

An RL circuit composed of a parallel connection of voltage source, resistor, and inductor is represented in Figure 11.12. Analysis of this electrical circuit follows directly from the basic relations established in Section 11.1. Since the resistor and inductor voltages are identical and equal to the value of the voltage source $v_s(t)$, that is $v_R(t) = v_L(t) = v_s(t)$, we have

$$i_R(t) = \frac{v_s(t)}{R}, \quad i_L(t) = i_L(0^-) + \frac{1}{L}\int_{0^-}^{t} v_s(\sigma)\, d\sigma \tag{11.27}$$

Assuming that the voltage source has constant value equal to E, $t \geq 0$, it follows from the preceding formula that the inductor current increases linearly in time, that is,

$$i_L(t) = i_L(0^-) + \frac{E}{L}r(t) = i_L(0^-) + \frac{E}{L}t, \quad t \geq 0 \tag{11.28}$$

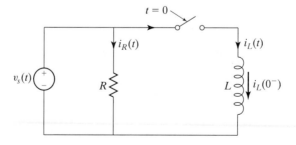

FIGURE 11.12: A parallel RL electrical circuit driven by a voltage source

where $r(t)$ is the ramp function (signal). Of course, as the current increases to large values, the inductor will be damaged and eventually destroyed. Hence, we should not expose an inductor to a constant voltage source over a long period of time.

Resistor and Inductor in Series Driven by a Current Source

This electrical circuit is presented in Figure 11.13. The resistor voltage is simply given by $v_R(t) = Ri_s(t)$. The inductor voltage can be obtained from formula (11.12) (see also Figure 11.2),

$$v_L(t) = L \frac{d(i_s(t) - i_L(0^-))}{dt} \tag{11.29}$$

Assuming that the source current is constant and equal to $i_s(t) = I_s u(t)$, we have

$$v_R(t) = RI_s, \quad v_L(t) = L(I_s - i_L(0^-))\delta(t) \tag{11.30}$$

which implies that $v_L(t) = 0$ for $t > 0$. It can be concluded that in this electrical circuit, for a constant current source, the inductor voltage after the initial delta impulse at $t = 0$ drops instantaneously to zero and stays zero for all $t > 0$. On the other hand, the resistor voltage instantaneously becomes equal to the constant value of RI_s and remains constant at all times. For different waveforms of the current source, formula (11.29) must be used appropriately.

Resistor and Inductor in Parallel Driven by a Current Source

This electrical circuit, presented in Figure 11.14, is equivalent to the electrical circuit presented in Figure 11.10. The correspondence follows from the equivalence of current and voltage sources as given in Figure 11.3. (Students are required in Problem 11.12 to perform complete analysis of this electrical circuit using this correspondence.)

One can also independently analyze this electrical circuit starting with its basic equations,

$$i_s(t) = i_R(t) + i_L(t)$$
$$Ri_R(t) = v_R(t) = v_L(t) = L\frac{di_L(t)}{dt} \tag{11.31}$$

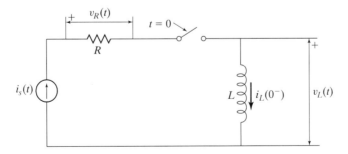

FIGURE 11.13: A series RL electrical circuit driven by a current source

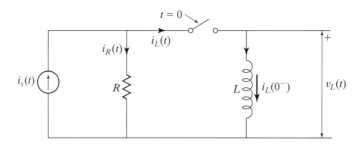

FIGURE 11.14: A parallel RL electrical circuit driven by a current source

which lead to

$$L\frac{di_L(t)}{dt} + Ri_L(t) = Ri_s(t), \quad i_L(0^-) = \text{given} \quad (11.32)$$

11.3 SECOND-ORDER LINEAR ELECTRICAL CIRCUITS

There are several types of second-order linear electrical circuits. Such circuits contain one capacitor and one inductor, two capacitors, or two inductors. In addition, they are driven by voltage and/or current sources and may have an arbitrary number of resistors. In the following we present two main types of second-order electrical circuits: cascade and parallel connections of resistor, capacitor, and inductor driven by either voltage or current sources. For simplicity, we call such configurations RLC cascade and parallel second-order electrical circuits. Other configurations of second-order electrical circuits can be studied using techniques similar to those presented here for RLC cascade and parallel electrical circuits.

As a special case of a cascade RLC circuit driven by a voltage source, we have for $R=0$ a cascade LC circuit driven by a voltage source. Such a circuit displays oscillations, an interesting phenomenon characterized by a periodic sinusoidal circuit response that can be caused either by the LC circuit initial conditions or by its forcing function. In the following, we first study the LC circuit.

11.3.1 Cascade LC Circuit Driven by a Voltage Source

An LC electrical circuit obtained by connecting inductor and capacitor in series, and driving them by a voltage source, is presented in Figure 11.15. A specific feature of this electrical circuit is that it displays oscillations, in which the electrostatic energy is transformed into electromagnetic energy and vice versa. Due to the fact that the energy cannot dissipate into heat (since no resistive element is present), the oscillations are infinite in duration. It will be seen that such oscillations can be caused by the circuit initial conditions or excited by a forcing function (such as impulse or step inputs).

Note that for this second-order electrical circuit the initial conditions are specified by given values for $i_L(0^-)$ and $v_C(0^-)$. It will be clear from Section 11.4 (where we use the state space technique, with the state variables defining inductor currents and capacitor voltages) that all we need know to specify the circuit initial conditions are the initial currents through inductors and initial voltages on capacitors. In this section, we derive the second-order differential equation for $v_C(t)$. Hence, we must know both $v_C(0^-)$ and $v_C^{(1)}(0^-)$.

Section 11.3 Second-Order Linear Electrical Circuits

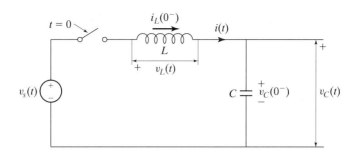

FIGURE 11.15: A cascade LC electrical circuit driven by a voltage source

The expression for $v_C^{(1)}(0^-)$ can be obtained using the fact that the capacitor initial energy cannot change instantaneously, so that we have

$$\frac{dw_C(0^-)}{dt} = p_C(0^-) = v_C(0^-)i_C(0^-) = v_C(0^-)C\frac{dv_C(0^-)}{dt}$$

$$= \frac{dw_C(0^+)}{dt} = p_C(0^+) = v_C(0^+)i_C(0^+) = v_C(0^+)C\frac{dv_C(0^+)}{dt} \quad (11.33)$$

We have also observed that $v_C(0^-) = v_C(0^+)$ and $i_L(0^-) = i_L(0^+)$. Formula (11.33) now implies

$$\frac{dv_C(0^-)}{dt} = v_C^{(1)}(0^-) = v_C^{(1)}(0^+) = \frac{dv_C(0^+)}{dt}$$

$$= \frac{i_C(0^+)}{C} = \frac{i_L(0^+)}{C} = \frac{i_L(0^-)}{C} \quad (11.34)$$

The balance of voltages for the electrical circuit in Figure 11.15 produces

$$v_s(t) = v_L(t) + v_C(t) = L\frac{di(t)}{dt} + v_C(t) \quad (11.35)$$

Since

$$i(t) = i_L(t) = i_C(t) = C\frac{dv_C(t)}{dt} \Rightarrow \frac{di(t)}{dt} = C\frac{d^2v_C(t)}{dt^2} \quad (11.36)$$

we have

$$v_s(t) = LC\frac{d^2v_C(t)}{dt^2} + v_C(t) \quad (11.37)$$

Hence, the capacitor voltage can be determined from

$$\frac{d^2v_C(t)}{dt^2} + \frac{1}{LC}v_C(t) = \frac{1}{LC}v_s(t),$$

$$v_C(0^-) = \text{given}, \quad \frac{dv_C(0^-)}{dt} = \frac{i_L(0^-)}{C} = \text{given} \quad (11.38)$$

Using the Laplace transform, we obtain the algebraic equation

$$s^2 V_C(s) - sv_C(0^-) - v_C^{(1)}(0^-) + \frac{1}{LC}V_C(s) = \frac{1}{LC}V_s(s) \quad (11.39)$$

whose solution is given by

$$V_C(s) = \frac{sv_C(0^-) + i_L(0^-)/C}{s^2 + \frac{1}{LC}} + \frac{1}{LC}\frac{V_s(s)}{s^2 + \frac{1}{LC}} \quad (11.40)$$

Applying the Laplace inverse, we obtain two components of the circuit response: the zero-input response (or circuit natural response) due to its initial conditions $v_{Czi}(t)$, and the zero-state response (or circuit forced response) due to its forcing function $v_{Czs}(t)$. The zero-input response is given by

$$v_{Czi}(t) = \mathcal{L}^{-1}\left\{\frac{sv_C(0^-) + \frac{i_L(0^-)}{C}}{s^2 + \frac{1}{LC}}\right\} = \mathcal{L}^{-1}\left\{\frac{sv_C(0^-)}{s^2 + \frac{1}{LC}}\right\} + \mathcal{L}^{-1}\left\{\frac{\frac{i_L(0^-)}{C}}{s^2 + \frac{1}{LC}}\right\}$$

$$= v_C(0^-)\cos\left(\frac{1}{\sqrt{LC}}t\right) + \sqrt{\frac{L}{C}}i_L(0^-)\sin\left(\frac{1}{\sqrt{LC}}t\right) \quad (11.41)$$

Using simple trigonometric transformations, the preceding expression can be written in the form

$$v_{Czi}(t) = \sqrt{v_C^2(0^-) + \frac{L}{C}i_L^2(0^-)}\cos\left(\frac{1}{\sqrt{LC}}t - \tan^{-1}\left(\sqrt{\frac{L}{C}}\frac{i_L(0^-)}{v_C(0^-)}\right)\right) \quad (11.42)$$

It can be observed from the last two formulas that the zero-input capacitor voltage response is periodic and sinusoidal with *frequency of oscillations*

$$\omega = \frac{1}{\sqrt{LC}} \quad (11.43)$$

The zero-input component of the circuit current is obtained in a straightforward manner by differentiating the corresponding component of the capacitor voltage, that is, $i_{zi}(t) = i_{Czi}(t) = Cdv_{Czi}(t)/dt$.

The forced capacitor zero-state response $v_{Czs}(t)$ can be similarly found for the known voltage input. In the next two examples, we consider two cases: first, when $v_s(t) = \delta(t)$, and second, when $v_s(t) = Eu(t)$, $E = \text{const}$, and we find the corresponding impulse and step responses.

EXAMPLE 11.3

The impulse response of the LC circuit presented in Figure 11.15 can be obtained from (11.40), with $V_s(s) = 1$ and assuming that the initial conditions are zero. This leads to

$$v_{Czs}(t) = h_C(t) = \mathcal{L}^{-1}\left\{\frac{\frac{1}{LC}}{s^2 + \frac{1}{LC}}\right\} = \frac{1}{\sqrt{LC}}\sin\left(\frac{1}{\sqrt{LC}}t\right) \quad (11.44)$$

Section 11.3 Second-Order Linear Electrical Circuits

Hence, the forced response excited by the impulse delta function is also sinusoidal. The corresponding capacitor current is given by

$$i_{Czs}(t) = C\frac{dv_C(t)}{dt} = \frac{1}{L}\cos\left(\frac{1}{\sqrt{LC}}t\right) \qquad (11.45)$$

EXAMPLE 11.4

The step response is obtained by using the Laplace transform of the unit step function in (11.40), that is, $V_s(s) = E/s$, assuming that the initial conditions are zero, and finding the Laplace inverse of the obtained expression, that is,

$$v_{Czs}(t) = \mathcal{L}^{-1}\left\{\frac{1}{LC}\frac{E}{\left(s^2 + \frac{1}{LC}\right)s}\right\} \qquad (11.46)$$

Performing partial fraction expansion using the technique from Chapter 4, we have

$$v_{Czs}(t) = \mathcal{L}^{-1}\left\{\frac{E}{s} + \frac{-E/2}{s - j\frac{1}{\sqrt{LC}}} + \frac{-E/2}{s + j\frac{1}{\sqrt{LC}}}\right\}$$

$$= E + E\cos\left(\frac{1}{\sqrt{LC}}t - \pi\right) = E - E\cos\left(\frac{1}{\sqrt{LC}}t\right) \qquad (11.47)$$

Note that the capacitor voltage oscillates between 0 and $2E$. The corresponding capacitor (circuit) current is given by

$$i_{Czs}(t) = C\frac{dv_{Czs}(t)}{dt} = E\sqrt{\frac{C}{L}}\sin\left(\frac{1}{\sqrt{LC}}t\right) \qquad (11.48)$$

Analysis of a cascade LC electrical circuit driven by a current source is rather simple. It is left to the student to find the impulse and step response of this circuit (see Problems 11.14 and 11.15).

11.3.2 Series Connection of R, L, and C Elements

An RLC electrical circuit obtained by connecting in series a resistor, inductor, and capacitor and driving them by a voltage source, is presented in Figure 11.16. This circuit is a little more complex than the LC circuit considered in the previous section.

The balance of voltages for this electrical circuit gives

$$v_s(t) = v_R(t) + v_L(t) + v_C(t) = Ri(t) + L\frac{di(t)}{dt} + v_C(t) \qquad (11.49)$$

Since the circuit current satisfies

$$i(t) = i_C(t) = C\frac{dv_C(t)}{dt} \quad \Rightarrow \quad \frac{di(t)}{dt} = C\frac{d^2v_C(t)}{dt^2} \qquad (11.50)$$

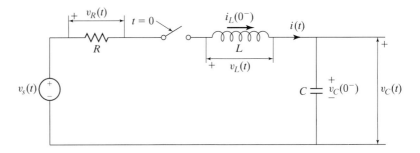

FIGURE 11.16: A cascade RLC electrical circuit driven by a voltage source

we have

$$v_s(t) = LC\frac{d^2v_C(t)}{dt^2} + RC\frac{dv_C(t)}{dt} + v_C(t) \tag{11.51}$$

Hence, the capacitor voltage can be determined from

$$\frac{d^2v_C(t)}{dt^2} + \frac{R}{L}\frac{dv_C(t)}{dt} + \frac{1}{LC}v_C(t) = \frac{1}{LC}v_s(t),$$

$$v_C(0^-) = \text{given}, \quad \frac{dv_C(0^-)}{dt} = \frac{i_L(0^-)}{C} = \text{given} \tag{11.52}$$

This second-order differential equation can be solved using the Laplace transform, assuming that the Laplace transform of $v_s(t)$ can be found analytically. In such a case, we have

$$s^2 V_C(s) - sv_C(0^-) - v_C^{(1)}(0^-) + s\frac{R}{L}V_C(s) - \frac{R}{L}v_C(0^-) + \frac{1}{LC}V_C(s) = \frac{1}{LC}V_s(s) \tag{11.53}$$

Solving this algebraic equation with respect to $V_C(s)$, we obtain

$$V_C(s) = \frac{sv_C(0^-) + v_C^{(1)}(0^-) + \frac{R}{L}v_C(0^-)}{s^2 + \frac{R}{L}s + \frac{1}{LC}} + \frac{1}{LC}\frac{1}{s^2 + \frac{R}{L}s + \frac{1}{LC}}V_s(s) \tag{11.54}$$

Given the value for $V_s(s)$, the capacitor voltage can be obtained by applying the inverse Laplace transform, producing

$$v_C(t) = v_{Czi}(t) + v_{Czs}(t) \tag{11.55}$$

It is left to the student in the form of exercises to find the expression for the capacitor voltage for standard voltage sources.

Note that in Section 12.2 we study a general second-order system. Many conclusions drawn there are applicable here. For example, in the case of the step input, the circuit response (capacitor voltage) can display damped oscillations (in which electrostatic and electromagnetic energies are transformed into heat by the resistor).

Section 11.4 Higher-Order Linear Electrical Circuits 559

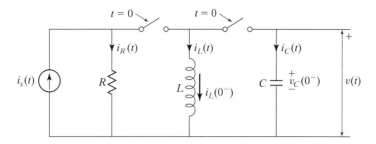

FIGURE 11.17: A parallel RLC electrical circuit driven by a current source

11.3.3 Parallel Connection of R, L, and C Elements

An RLC electrical circuit obtained by connecting in parallel a resistor, inductor, and capacitor, and driving them by a current source, is presented in Figure 11.17.

In the following, we will show that this electrical circuit is equivalent to the electrical circuit studied in the previous section. *This means that the two electrical circuits are described by differential equations of the same order that have the same form* (only the coefficients may differ).

From the first Kirchhoff law, we have

$$i_s(t) = i_L(t) + \frac{v(t)}{R} + C\frac{dv(t)}{dt} \tag{11.56}$$

Since $v(t) = v_L(t) = Ldi_L(t)/dt$, we obtain

$$i_s(t) = i_L(t) + \frac{L}{R}\frac{di_L(t)}{dt} + LC\frac{d^2 i_L(t)}{dt^2} \tag{11.57}$$

The corresponding initial conditions are $i_L(0^-)$ and $i_L^{(1)}(0^-) = v_C(0^-)/L$, since the accumulated inductor energy cannot change instantaneously (see (11.33) and (11.34)). Second-order differential equation (11.57) can be rewritten in the form

$$\frac{d^2 i_L(t)}{dt^2} + \frac{1}{RC}\frac{di_L(t)}{dt} + \frac{1}{LC}i_L(t) = \frac{1}{LC}i_s(t),$$

$$i_L(0^-) = \text{given}, \quad i_L^{(1)}(0^-) = \frac{v_C(0^-)}{L} = \text{given} \tag{11.58}$$

It can be seen that differential equations (11.52) and (11.58) have the same form and differ only in one coefficient. Hence, by appropriately changing the corresponding coefficient in the obtained solution, all results for the cascade RLC circuit driven by a voltage source can be extended to the parallel RLC circuit driven by a current source, and vice versa.

11.4 HIGHER-ORDER LINEAR ELECTRICAL CIRCUITS

More complex electrical circuits are obtained by connecting multiple passive and active electrical elements: resistors, inductors, capacitors, and voltage and current sources. These circuits are described by differential equations of higher order. The state space technique

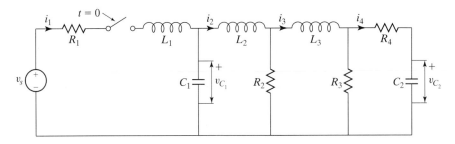

FIGURE 11.18: A complex electrical circuit

is well suited for analysis of such electrical circuits. Further more, by choosing *inductor currents and capacitor voltages for the state variables*, the circuit initial conditions are directly determined from the initial inductor currents and capacitor voltages.

A complex electrical circuit is represented in Figure 11.18. For simplicity, we assume that the voltage source is connected at $t = 0$ and that no initial energy is present in the inductors and capacitors.

As pointed out, the total number of inductors and capacitors determines the order of the differential equation that describes the given electrical circuit. In this particular example, we have three inductors and two capacitors, thus the order of this electrical circuit is $n = 5$. Having a dynamic system of order $n = 5$ indicates that the required number of first-order differential equations to be set up is five. If one sets up more than n differential equations, for a system of order n, some of them are redundant. Redundant equations must be eliminated since they do not carry any new information; they are just linear combinations of the remaining equations.

Using basic laws for currents and voltages, we can easily set up five first-order differential equations; for voltages around loops we have

$$v_s(t) - L_1 \frac{di_1(t)}{dt} - R_1 i_1(t) - v_{C_1}(t) = 0$$
$$L_2 \frac{di_2(t)}{dt} - v_{C_1}(t) + R_2(i_2(t) - i_3(t)) = 0 \quad (11.59)$$
$$L_3 \frac{di_3(t)}{dt} + R_3(i_3(t) - i_4(t)) - R_2(i_2(t) - i_3(t)) = 0$$

and for currents,

$$i_1(t) - i_2(t) - C_1 \frac{dv_{C_1}(t)}{dt} = 0$$
$$i_4(t) - C_2 \frac{dv_{C_2}(t)}{dt} = 0 \quad (11.60)$$

We have set up five equations for four currents and two voltages. Current $i_4(t)$ can be eliminated by using the following algebraic voltage balance equation, which is valid for the last loop:

$$R_3(i_3(t) - i_4(t)) = R_4 i_4(t) + v_{C_2}(t) \quad (11.61)$$

from which the current $i_4(t)$ can be expressed as

$$i_4(t) = \frac{R_3}{R_3 + R_4} i_3(t) - \frac{1}{R_3 + R_4} v_{C_2}(t) \tag{11.62}$$

Replacing current $i_4(t)$ in (11.59) and (11.60) by the expression obtained in (11.62), the following five first-order differential equations are obtained:

$$\begin{aligned}
\frac{di_1(t)}{dt} &= -\frac{R_1}{L_1} i_1(t) - \frac{1}{L_1} v_{C_1}(t) + \frac{1}{L_1} v_s(t) \\
\frac{di_2(t)}{dt} &= -\frac{R_2}{L_2} i_2(t) + \frac{R_2}{L_3} i_3(t) + \frac{1}{L_2} v_{C_1}(t) \\
\frac{di_3(t)}{dt} &= \frac{R_2}{L_3} i_2(t) - \frac{R_2 R_3 + R_2 R_4 + R_3 R_4}{L_3 (R_3 + R_4)} i_3(t) - \frac{R_3}{L_3 (R_3 + R_4)} v_{C_2}(t) \quad (11.63) \\
\frac{dv_{C_1}}{dt} &= \frac{1}{C_1} i_1(t) - \frac{1}{C_1} i_2(t) \\
\frac{dv_{C_2}(t)}{dt} &= \frac{R_3}{C_2 (R_3 + R_4)} i_3(t) - \frac{1}{C_2 (R_3 + R_4)} v_{C_2}(t)
\end{aligned}$$

The matrix form of this system of first-order differential equations represents the system state space form. Take $x_1(t) = i_1(t)$, $x_2(t) = i_2(t)$, $x_3(t) = i_3(t)$, $x_4(t) = v_{C_1}(t)$, and $x_5(t) = v_{C_2}(t)$, where $f(t) = v_s(t)$; then the state space form is given by

$$\begin{bmatrix} \dot{x}_1 \\ \dot{x}_2 \\ \dot{x}_3 \\ \dot{x}_4 \\ \dot{x}_5 \end{bmatrix} = \begin{bmatrix} -\frac{R_1}{L_1} & 0 & 0 & -\frac{1}{L_1} & 0 \\ 0 & -\frac{R_2}{L_2} & \frac{R_2}{L_2} & \frac{1}{L_2} & 0 \\ 0 & \frac{R_2}{L_3} & a_{33} & 0 & -\frac{R_3}{L_3(R_3+R_4)} \\ \frac{1}{C_1} & -\frac{1}{C_1} & 0 & 0 & 0 \\ 0 & 0 & \frac{R_3}{C_2(R_3+R_4)} & 0 & -\frac{1}{C_2(R_3+R_4)} \end{bmatrix} \begin{bmatrix} x_1 \\ x_2 \\ x_3 \\ x_4 \\ x_5 \end{bmatrix} + \begin{bmatrix} \frac{1}{L_1} \\ 0 \\ 0 \\ 0 \\ 0 \end{bmatrix} f$$

(11.64)

with $a_{33} = -(R_2 R_3 + R_2 R_4 + R_3 R_4)/(L_3(R_3 + R_4))$.

Remark on Electrical Circuit Initial Conditions

In general, electrical circuit initial conditions must be consistent with the electrical circuit configuration, unless we use a switch on every energy storing element and assume that all switches can be simultaneously turned on and off at the initial time instant. The consistent initial conditions can be determined from a given electrical circuit—assuming that it has been under given constant inputs long enough that the circuit currents and voltages have reached their steady state values. In doing so, *at the steady state, all inductors should be replaced by short circuits and all capacitors should be replaced by open circuits.* Such

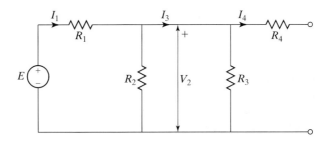

FIGURE 11.19: Simplified electrical circuit that determines consistent initial conditions for the circuit in Figure 11.18

an obtained circuit of resistors driven by constant voltage and/or current sources determines inductor currents (equal to the branch currents where inductors are present) and capacitor voltages (equal to the voltages between the nodes where capacitors are connected).

For example, for the electrical circuit presented in Figure 11.18, assuming that the circuit is driven by the constant voltage source E, by setting short circuits instead of inductors and open circuits instead of capacitors, we get a pure resistive electrical circuit, presented in Figure 11.19.

Analyzing this simple circuit, we find that the currents are

$$i_{L_1}(0^-) = i_{L_2}(0^-) = I_1 = \frac{E}{R_1 + \dfrac{R_2 R_3}{R_2 + R_3}} = \frac{E(R_2 + R_3)}{R_1 R_2 + R_1 R_3 + R_2 R_3}$$

and

$$i_{L_3}(0^-) = I_3 = \frac{R_2}{R_2 + R_3} I_1 = \frac{E R_2}{R_1 R_2 + R_1 R_3 + R_2 R_3}, \quad i_{L_4}(0^-) = I_4 = 0$$

For the voltages, we obtain

$$v_{C_1}(0^-) = v_{C_2}(0^-) = V_2 = \frac{\dfrac{R_2 R_3}{R_2 + R_3}}{R_1 + \dfrac{R_2 R_3}{R_2 + R_3}} E = \frac{R_2 R_3}{R_1 R_2 + R_1 R_3 + R_2 R_3} E$$

Remark on Mesh Currents and Nodal Analysis Techniques

These very well-known techniques, used in the analysis of static resistive electrical circuits, are also applicable in the analysis of higher-order dynamic linear electrical circuits that contain resistors, inductors, and capacitors. These technique can be useful in setting up circuit equations, from which we can obtain the circuit state space form. In such a case, one also must figure out the corresponding circuit initial conditions and relate them to the original circuit initial conditions. The state space equations for an electrical circuit can be set up rather directly by choosing the state space variables to represent inductor currents and capacitor voltages. This will give directly the initial conditions for the state space form. However, in that case we might be faced with the problem of eliminating redundant state variables in order to represent the electrical circuit of order n by the state space form of order n.

Remark on the Stability of Passive Electrical Circuits

We would like to indicate an important feature of passive electrical circuits (composed of resistors, inductors, and capacitors), known as the *passivity property*. This property says that passive electrical circuits can only dissipate and not generate energy. This means that as time passes the passive electrical circuit loses energy (except in the case of ideal LC circuits, where that energy stays constant), so that its natural response (response due to the initial energy stored) cannot grow in time—it is bounded, and in most cases decays, in time. This is equivalent to the stability (asymptotic stability) definition(s) of linear time-invariant systems (see Definition 7.1). We can conclude this section with an important statement: *passive electrical circuits are either stable or asymptotically stable*, which implies that all their poles are in the closed left half complex plane (including the imaginary axis). Furthermore, this stability is robust, in the sense that any perturbation of passive electrical circuit elements will preserve stability (asymptotic stability), so that there is no danger that oscillatory modes will become unstable.

11.5 MATLAB LABORATORY EXPERIMENT

Consider the electrical circuit presented in Figure 11.18. Assume that all resistors in the circuit are $10\,\text{k}\Omega$, the inductors $1\,\text{mH}$, and the capacitors $20\,\mu\text{F}$. Assume that the circuit has been under the constant voltage input $v_s = E = 6\,\text{V}$ for a long period of time.

Part 1. Use the resistive circuit presented in Figure 11.19 and the corresponding formulas to find the currents through the inductors and the voltages on the capacitors. The values obtained will represent the initial conditions for the problem defined in Part 2.

Part 2. Assume that the input voltage source is changed to the new value $v_s = E = 12\,\text{V}$. Using state space analysis, form the matrices **A** and **B** and the vector of initial conditions. Use MATLAB to find the currents through the inductors and voltages on the capacitors as functions of time. Observe that such state space variables settle down at steady state to some constant values. Plot all state space variables during the initial time interval until they reach their steady state values. Check the steady state values obtained via MATLAB, using the electrical circuit given in Figure 11.19 and the corresponding formulas. The current and voltage steady state values obtained will serve as the initial conditions in Part 3.

Part 3. Assume that this electrical circuit is exposed to the input voltage $v_s(t) = r(t) - r(t-1)$, where $r(t)$ stands for the ramp signal and the time is measured in seconds. Using MATLAB, find and plot the currents through the inductors and the voltages on the capacitors during the time interval of 2 s.

11.6 SUMMARY

Study Guide for Chapter Eleven: The student should be familiar with the basic electrical laws (such as Ohm's and Kirchhoff's laws) applicable to time-invariant resistors, capacitors, and inductors, as well as to linear electrical circuits composed of active and passive elements. Direct applications of these laws to given circuits produce corresponding differential equations that describe circuit dynamics. The differential equations obtained are commonly solved via either the Laplace transform (Chapter 4) or the state space approach (Chapter 8), which is convenient for higher-order electrical circuits. One can also use time domain analysis from Chapter 7, which requires the direct solution in time of differential equations that represent electrical circuit dynamics, but that approach may be computationally involved. Hence, there is no need for the student to study specialized methods for solving specialized linear electrical circuits. All of them can be analyzed in the same way: *set up and solve differential equations that describe circuit dynamics.* Some useful facts are summarized in the following list.

1. It can be concluded from Sections 11.2.1 and 11.2.2 that a *capacitor cannot change its voltage instantaneously* (unless it is driven by the impulse delta function), and that the *inductor cannot change its current instantaneously* (unless it is driven by the impulse delta function). These observations follow analytically also, from formulas (11.4) and (11.11), respectively, and are a consequence of the fact that the initial energy cannot change instantaneously.
2. In an electrical circuit driven by constant input sources, a *capacitor at steady state behaves like an open circuit* (it has zero current and nonzero voltage).
3. In an electrical circuit driven by constant input sources, *an inductor at steady state behaves like a short circuit* (its current is nonzero, but its voltage is zero).
4. In electrical circuits composed of inductors, capacitors, and resistors, *the total number of inductors and capacitors indicates the order of the linear dynamic system.*
5. When using the state space approach to analyze electrical circuits, *take for the state space variables inductor currents and capacitor voltages.*

11.7 REFERENCES

[1] J. Nilsson and S. Riedel, *Electrical Circuits,* Addison Wesley, Reading, MA, 1996.

[2] R. Thomas and A. Rosa, *Analysis and Design of Linear Circuits,* Prentice Hall, Englewood Cliffs, NJ, 1994.

[3] W. Siebert, *Circuits, Signals, and Systems,* MIT Press, Cambridge, MA, 1986.

[4] V. Del Toro, *Engineering Circuits,* Prentice Hall, Englewood Cliffs, NJ, 1987.

[5] S. Paul, A. Schlaffer, and J. Nossek, "Optimal charging of a capacitor," *IEEE Transactions on Circuits and Systems I: Fundamental Theory and Applications,* 47: 1009–16, 2000.

11.8 PROBLEMS

11.1. Use the Laplace transform and draw the frequency domain equivalents for the inductor and the capacitor with initial conditions whose time domain equivalents are presented in Figures 11.1 and 11.2.

11.2. For a parallel RC electrical circuit driven by a current source, find the currents through the resistor and capacitor, and the capacitor (resistor) voltage. Assume $v_C(0^-) = 0$. Use

11.3. Repeat Problem 11.2, assuming that the capacitor has the initial voltage $v_C(0^-) = 2$ V. Assuming $R = 10\,\text{k}\Omega$, $C = 100\,\mu\text{F}$, and the constant current source given by $i_s(t) = I_s u(t)$, $I_s = 100\,\text{mA}$, find the capacitor current.

11.4. Find the impulse response (with initial condition zero) of the electrical circuit given in Figure 11.5. Consider the capacitor voltage as the circuit output.

Answer:
$$v_C^{\text{imp}}(t) = \frac{1}{RC} e^{-(1/RC)t} u(t)$$

11.5. Find the impulse response (with initial condition zero) of the electrical circuit presented in Figure 11.10. Consider the inductor current as the circuit output.

Answer:
$$i_L^{\text{imp}}(t) = \frac{1}{L} e^{-(R/L)t} u(t)$$

11.6. What is the impulse response of a simple electrical circuit that contains a capacitor and a voltage source?

Answer: The unit doublet.

11.7. What is the impulse response of a simple electrical circuit that contains an inductor and a current source?

Answer: The unit doublet.

11.8. Find the voltage on the capacitor of the electrical circuit presented in Figure 11.8, subject to the constant current source $i_s(t) = I_s u(t)$, $I_s = \text{const}$, and the initial capacitor voltage $v_C(0^-) = V_{C0}$. Use the following values: $R = 10\,\text{k}\Omega$, $C = 100\,\mu\text{F}$, $I_s = 100\,\text{mA}$, and $v_C(0^-) = 1$ V.

11.9. Find the step response (with initial condition zero) of the electrical circuit presented in Figure 11.20. Consider the capacitor voltage as the circuit output.

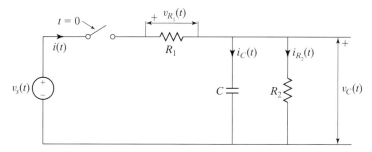

FIGURE 11.20

11.10. Assume that the electrical circuit in Figure 11.20 is driven by a sinusoidal voltage source of the form $v_s(t) = E \cos(\omega_0 t) u(t)$, $E = \text{const}$, and that the capacitor in the circuit was initially charged to $v_C(0^-) = E/2$ with its positive terminal connected to the node between R_1 and R_2. Find all currents and all voltages in this electrical circuit.

11.11. Find the step response (with initial condition zero) of the electrical circuit presented in Figure 11.21. Consider the inductor current as the circuit output.

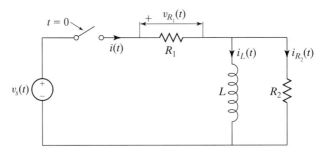

FIGURE 11.21

11.12. Using the results obtained for the electrical circuit in Figure 11.3 (the equivalence between voltage and current sources), find the inductor voltage for the electrical circuit presented in Figure 11.14.

11.13. For the LC circuit represented in Figure 11.15, find the zero-input component of the capacitor current for the following.

(a) $v_C(0^-) = 0, i_L(0^-) \neq 0$

(b) $v_C(0^-) \neq 0, i_L(0^-) = 0$

(c) $v_C(0^-) \neq 0, i_L(0^-) \neq 0$

11.14. Find the impulse response (inductor and capacitor voltages) for the electrical circuit formed of a cascade connection of current source, inductor, and capacitor. Consider all three cases (real and distinct, real and double, and complex conjugate poles).

11.15. Find the step response (inductor and capacitor voltages) for the electrical circuit formed of a cascade connection of current source, inductor, and capacitor. Consider all three cases (real and distinct, real and double, and complex conjugate poles).

11.16. Assuming that all initial conditions are zero, find the capacitor voltage for the circuit presented in Figure 11.22. Take for the input voltage source $v_s(t) = 6$ V, with $R = 10$ kΩ, $L = 50$ mH, and $C = 100$ μF.

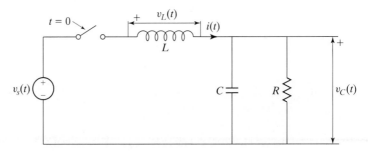

FIGURE 11.22

11.17. Assume that the switch in the electrical circuit presented in Figure 11.23 has been in position a for a long time, with $v_s(t) = Eu(t)$, $E = $ const. Let the switch move to position b at $t = 0$. Find the circuit current and voltages for $t > 0$. Assume the following numerical values: $v_s(t) = 2\,\text{V}$, $R = R_1 = R_2 = 200\,\text{k}\Omega$, $L = 50\,\text{mH}$, and $C = 100\,\mu\text{F}$.

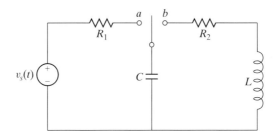

FIGURE 11.23

11.18. Set up a system of linear differential equations for the electrical circuit presented in Figure 11.24, and obtain the corresponding state space form.

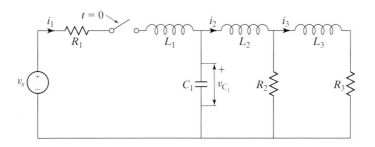

FIGURE 11.24

11.19. Consider an electrical circuit with no initial energy stored, presented in Figure 11.25.

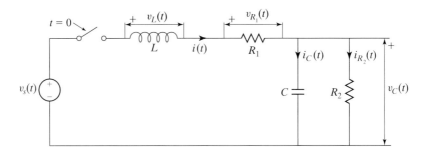

FIGURE 11.25

Assume that the switch is closed at $t = 0$ and that $v_s(t) = Eu(t)$, $E = \text{const}$. Find all currents and voltages in this electrical circuit. Find the steady state values for the currents and voltages. Assume the following numerical values: $v_s(t) = 6\,\text{V}$, $R_1 = 100\,\text{k}\Omega$, $R_2 = 500\,\text{k}\Omega$, $L = 20\,\text{mH}$, and $C = 100\,\mu\text{F}$. Using MATLAB, plot the results obtained.

11.20. Using MATLAB, find and plot the zero-input response for the electrical circuit in Figure 11.25, assuming the initial conditions $i_L(0^-) = 20\,\text{mA}$ in the direction of R_1, and $v_C(0^-) = 2\,\text{V}$ with the positive terminal at R_1.

11.21. Repeat Problem 11.19 for the electrical circuit given in Figure 11.25, assuming the initial conditions $i_L(0^-)$ and $v_C(0^-)$ and sinusoidal input voltage, that is, $v_s(t) = E\cos(\omega_0 t)u(t)$, $E = \text{const}$. Take $i_L(0^-) = 50\,\text{mA}$ and $v_C(0^-) = 1\,\text{V}$. Using MATLAB, plot the results obtained.

11.22. Set up a system of linear differential equations for the electrical circuit presented in Figure 11.26 and obtain the corresponding state space form.

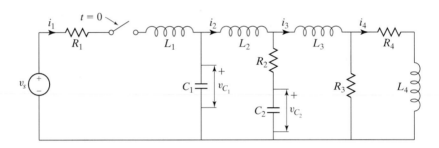

FIGURE 11.26

Assume $R_1 = R_2 = R_3 = R_4 = 100\,\text{k}\Omega$, $L_1 = L_2 = L_3 = L_4 = 50\,\text{mH}$, $C_1 = C_2 = 10\,\mu\text{F}$, and $v_s = 12\,\text{V}$. Using MATLAB, find and plot the waveforms for currents and voltages that correspond to state variables.

11.23. Can we build a Simulink block diagram for the linear electrical circuit presented in Figure 11.26?

CHAPTER 12

Linear Control Systems

In the previous chapters of this book, we were primarily concerned with the problem of finding the response of linear time-invariant systems in the time or frequency domains. In this chapter, our goal is to shape the response of linear systems using special input signals, called control signals, such that the system response has the desired shape. To that end, we utilize the concept of feedback, and study the transient response of continuous-time linear feedback systems and derive linear feedback system response steady state errors. We also present Bode diagrams as a tool frequently used in control system design, and use them to define the system stability margins. At the end, we briefly present the classic PID controllers, still commonly used in industry. The presentation of this chapter is restricted to the continuous-time domain. Similar conclusions and results can be obtained in the discrete-time domain. Also, for simplicity, we consider only single-input single-output systems.

The *concept of system feedback,* though extraordinary simple (the feeding back of output signals and use of them as input signals), is one of the most fundamental engineering discoveries of the twentieth century. The power of feedback is demonstrated by its ability to reject system disturbances, reduce system sensitivity to inaccuracies in system parameters (modeling errors) and system measurements, improve system transient response, reduce system steady state errors, and above all stabilize unstable systems. The concept of feedback was introduced in Section 4.4, where we defined closed-loop (feedback) linear systems. The main features of linear feedback systems will be discussed in detail in Section 12.1.

In Section 12.2, we consider the *transient response* of second-order linear systems and define the transient response parameters: maximum percent overshoot, pick time, settling time, and rise time. In this section we also introduce the concept of the dominant system poles, and discuss their relation to the transient response of higher-order linear dynamic systems.

System output steady state errors are defined and studied in Section 12.3 for common input signals (unit step, unit ramp, and unit parabolic signals). The analytical expressions for corresponding steady state errors are derived in terms of the steady state constants (position, velocity, and acceleration constants). Requirements are imposed on the feedback loop such that the steady state errors can be reduced to zero.

Frequency response characteristics (and specifications) are summarized in Section 12.4, where we define the concept of the open- and closed-loop frequency transfer functions, and define system bandwidth, resonant frequency, and peak resonance. We indicate the importance of some frequency domain parameters that can be used to design controllers for linear time-invariant systems. To that end, we show how to read the frequency domain characteristics from the magnitude diagram of the closed-loop frequency transfer function, and how to relate them to some of the transient response and steady state parameters.

In Section 12.5, we present a technique for drawing magnitude and phase *Bode diagrams* using Bode diagrams of elementary transfer functions. In addition, we show how to read from the Bode diagrams the gain and phase stability margins, and how to relate the peak of the magnitude Bode diagram to the transient response overshoot. Note that Bode diagrams are used in almost all areas of electrical engineering.

Section 12.6 presents the basic dynamic controllers—proportional and derivative (PD), proportional and integral (PI), and proportional, integral, and derivative (PID) controllers—and briefly discusses the use of more advanced controllers. We draw a general conclusion that the PD controller introduces a positive phase in the feedback loop and improves the system output transient response. On the other hand, the PI controller introduces a negative phase in the feedback loop and improves the system output steady state errors. The PID controller, which represents a combination of the PD and PI controllers, can be used to improve both the system transient response and the system output steady state errors. Of course, the primary task of any controller is system stabilization. Despite tremendous developments in control theory during the last twenty years, more than 90% of the controllers presently used in industry are the classic PD, PI, and PID controllers.

12.1 THE ESSENCE OF FEEDBACK

We can find the response of an open-loop linear system due to given system input using any of the methods presented in this book. The response obtained is strictly determined by information about the system mathematical model, input, and initial conditions. In control theory and its applications, given a system input and initial conditions, the goal is very often for the system output to have a desired response (shape). The system output can be shaped by closing the feedback loop around the system, and using particular static and/or dynamic elements in the feedback path. By closing the feedback loop around the system, several interesting and important features can be achieved:

1. feedback can stabilize the system (the most important feature);
2. feedback reduces system output sensitivity to system parameter variations (due to aging of system components or internal system disturbances);
3. feedback reduces the impact of external system disturbances on the system output;
4. feedback can improve system output transient response;
5. feedback can improve system output steady state response;
6. feedback can reduce system sensitivity to inaccuracies in system measurements (measurement noise).

Within this section we will discuss features 1–3. Feedback features 4 and 5 will become evident in Sections 12.3, 12.4, and 12.6. In Section 12.6 we will present simple controllers that can improve transient and steady state responses. Complete clarification of feature 6 is outside of the scope of this text.

Feedback Stabilization of Linear Time-Invariant Systems

Consider first the system with unity feedback represented in Figure 4.7(a). Its closed-loop transfer function is derived in (4.59). Assume that the open-loop transfer function $H(s)$ is represented by the ratio of two polynomials, that is, $H(s) = N(s)/D(s)$. The zeros of $D(s)$ represent the open-loop system poles, and we say that the open-loop transfer function is asymptotically stable and BIBO stable (see Section 7.7) if all the poles of $H(s)$ are strictly in the left half complex plane (none of them is on the imaginary axis). The closed-loop transfer function of the feedback system in Figure 4.7(a) is given by

$$M(s) = \frac{H(s)}{1 + H(s)} = \frac{N(s)/D(s)}{1 + N(s)/D(s)} = \frac{N(s)}{D(s) + N(s)}$$

The algebraic equation

$$1 + H(s) = 0 \Rightarrow D(s) + N(s) = 0 \qquad (12.1)$$

defines the *characteristic equation of the closed-loop system* with unity feedback. The zeros of the closed-loop characteristic equation represent the closed-loop system poles. Note that the *open-loop characteristic equation* is simply $D(s) = 0$. Hence, the feedback changes the system characteristic equation and the location of the system poles.

It can happen that the open-loop system is unstable, but the closed-loop system is stable, as demonstrated in the next example.

EXAMPLE 12.1

Consider an open-loop unstable linear system represented by

$$H(s) = \frac{1}{(s - 0.5)(s + 1)}$$

Its open-loop poles are located at -1 and 0.5. The closed-loop characteristic equation is

$$(s - 0.5)(s + 1) + 1 = s^2 + 0.5s + 0.5 = 0$$

The solutions of this algebraic equation, given by $s_{1,2} = -0.25 \pm j0.25\sqrt{7}$, represent the closed-loop system poles. Obviously, both poles are in the open left half complex plane and the system is asymptotically stable.

Sometimes unity feedback is not sufficient to stabilize an open-loop unstable system, as demonstrated in the following.

EXAMPLE 12.2

Consider the following open-loop unstable system

$$H(s) = \frac{1}{(s - 1)(s + 2)}$$

Its closed-loop characteristic equation is

$$(s - 1)(s + 2) + 1 = s^2 + s - 1 = 0$$

The solutions of this algebraic equation are $s_{1,2} = -1 \pm \sqrt{2}$, hence the closed-loop system is unstable, since the pole $\sqrt{2} - 1$ is in the right half complex plane. ♦

In order to remedy the problem identified in Example 12.2, we use nonunity feedback. The simplest form of nonunity feedback is when the element in the feedback path is static, that is, $G(s) = g = \text{const}$. In such a case, it follows from formula (4.60) that the closed-loop transfer function is

$$M(s) = \frac{H(s)}{1 + gH(s)} = \frac{N(s)}{D(s) + gN(s)} \qquad (12.2)$$

so that the closed-loop system characteristic equation is given by

$$1 + gH(s) = 0 \Rightarrow D(s) + gN(s) = 0 \qquad (12.3)$$

EXAMPLE 12.3

Consider the open-loop transfer function from Example 12.2 in configuration with nonunity static feedback elements $G(s) = g$. The closed-loop characteristic equation obtained from formula (12.3) is given by

$$(s-1)(s+2) + g = s^2 + s + (g-2) = 0$$

It can be easily confirmed that for all static feedback elements with $g > 2$, both closed-loop poles are in the open left half complex plane, which implies that the closed-loop system is asymptotically stable for any $g > 2$. ♦

For more complex linear systems, systematic procedures are developed that find the range of values for the static element g that provides stability of the closed-loop linear systems. The Roth-Hurwitz stability test, presented in Section 7.7.2, can be used for that purpose, as demonstrated in Example 7.28.

In some cases, it can happen that for all choices of the static element g the closed-loop system remains unstable. In such cases, we should try to stabilize the system using a dynamic feedback element $G(s)$. It follows from (4.60) that the corresponding closed-loop characteristic equation is given by

$$1 + H(s)G(s) = 0 \Rightarrow D(s) + N(s)G(s) = 0 \qquad (12.4)$$

Note that $G(s)$ can also be chosen as a ratio of two polynomials. It should be pointed out that system stabilization with a dynamic feedback element is more complicated than stabilization with a static feedback element. Detailed consideration of the stabilization problem with dynamic feedback is outside the scope of this textbook.

Feedback Reduction of System Output Sensitivity

A unity feedback system is presented in Figure 4.7(a). The perturbed form of this system is given in Figure 12.1, where $\Delta H(s)$ denotes perturbations of the system open-loop

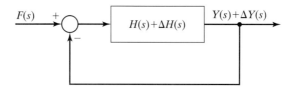

FIGURE 12.1: A unity feedback system and its perturbation

transfer function. The perturbations in the system open-loop transfer function are primarily due to aging of the system components and system internal disturbances (this also includes the case when some of the system parameters are not perfectly known—parameter uncertainties). For the closed-loop unity feedback system without perturbations, the system output is given by

$$Y(s) = M(s)F(s) = \frac{H(s)}{1 + H(s)} F(s)$$

In the case of perturbations, we obtain from Figure 12.1 the following expression for the perturbed value of the system output

$$Y(s) + \Delta Y(s) = \frac{H(s) + \Delta H(s)}{1 + H(s) + \Delta H(s)} F(s)$$

which implies

$$\Delta Y(s) = \frac{H(s) + \Delta H(s)}{1 + H(s) + \Delta H(s)} F(s) - Y(s)$$

$$= \frac{H(s) + \Delta H(s)}{1 + H(s) + \Delta H(s)} F(s) - \frac{H(s)}{1 + H(s)} F(s)$$

$$= \frac{\Delta H(s)}{(1 + H(s) + \Delta H(s))(1 + H(s))} F(s)$$

Since, in general, $|\Delta H(s)| \ll |H(s)|$, we can approximate the last expression as

$$\Delta Y(s) \approx \frac{\Delta H(s)}{(1 + H(s))(1 + H(s))} F(s)$$

$$= \frac{\Delta H(s) H(s)}{H(s)(1 + H(s))(1 + H(s))} F(s) = \frac{\Delta H(s)}{H(s)} \frac{1}{(1 + H(s))} Y(s)$$

In summary, we have derived the following relationship:

$$\frac{\Delta Y(s)}{Y(s)} = \frac{1}{(1 + H(s))} \frac{\Delta H(s)}{H(s)} \stackrel{\Delta}{=} S(s) \frac{\Delta H(s)}{H(s)} \qquad (12.5)$$

which indicates that the relative change in the system output is proportional to the relative change in the system transfer function. The proportionality factor $S(s)$ is called the system output *sensitivity function*. Due to the fact that $|S(s)| = |1/(1 + H(s))| < 1$, we can conclude that the impact of system parameter variations on the system output is relatively reduced by the factor of $|S(s)| < 1$.

In the case of systems with nonunity feedback, $G(s) \neq 1$, we can follow the same derivations, and obtain the following form for the system output sensitivity function:

$$S(s) = \frac{1}{1 + G(s)H(s)} \quad (12.6)$$

In Problem 12.2, contributed by an anonymous reviewer of this textbook, it will be demonstrated how the feedback loop can drastically reduce the system DC gain error tolerance caused by a large DC gain uncertainty in the open-loop transfer function.

Feedback Disturbance Rejection

Consider a feedback system under the influence of an external disturbance, as demonstrated in the block diagram presented in Figure 12.2.

Using the basic transfer function rules, we obtain

$$Y(s) = D(s) + H(s)(F(s) - Y(s))$$

which implies

$$Y(s) = \frac{H(s)}{1 + H(s)} F(s) + \frac{1}{1 + H(s)} D(s) = \frac{H(s)}{1 + H(s)} F(s) + S(s)D(s) \quad (12.7)$$

It follows from the last formula that the sensitivity function causes the disturbance to be attenuated, due to the fact that $|S(s)| < 1$. In general, $S(s)$ is a function of frequency, hence we need $S(s)$ *to be very small, at least in the frequency range of the system disturbance*. We can conclude from (12.7) that the feedback configuration has a nice property of rejecting the system disturbance by reducing its impact on the system output. As in the previous cases, we can perform the same derivations for the case of a nonunity feedback control system, that is, with $G(s) \neq 1$, which implies

$$Y(s) = \frac{H(s)}{1 + G(s)H(s)} F(s) + S(s)D(s)$$

with the sensitivity function $S(s)$ as defined in (12.6).

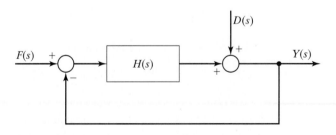

FIGURE 12.2: A feedback system with an external disturbance

12.2 TRANSIENT RESPONSE OF SECOND-ORDER SYSTEMS

The system transient response is dominant during the initial time interval. For asymptotically stable systems, the transient response decays to zero rather quickly, so that after the initial time interval, the system response is determined by its steady state component. In unstable systems, some of the system variables can take on huge values during the initial time interval, and the system can be damaged. Even in the case of asymptotically stable systems, we would like to have the capability to shape the system transient response.

In this section, we study the system transient response and define the most important transient response parameters. The transient response parameters can be explicitly defined only for second-order systems. For higher-order systems, analytical derivations are in general impossible. In such cases, the transient response parameters can be estimated using simulation.

Consider a general second-order feedback system represented by the block diagram in Figure 12.3.

In this transfer function, K is the system static gain and T represents the system time constant. The closed-loop transfer function of this system is given by

$$M(s) = \frac{Y(s)}{F(s)} = \frac{\frac{K}{T}}{s^2 + \frac{1}{T}s + \frac{K}{T}} \tag{12.8}$$

which can be rewritten in the form

$$\frac{Y(s)}{F(s)} = \frac{\omega_n^2}{s^2 + 2\zeta\omega_n s + \omega_n^2}, \quad \zeta = \frac{1}{2\omega_n T}, \quad \omega_n^2 = \frac{K}{T} \tag{12.9}$$

The parameter ζ is called the *system damping ratio,* and ω_n is known as the *system natural frequency.* The system poles are obtained from (12.9) as

$$p_{1,2} = -\zeta\omega_n \pm j\omega_n\sqrt{1-\zeta^2} = -\zeta\omega_n \pm j\omega_d \tag{12.10}$$

where ω_d is called the *system damped frequency.* The locations of the system poles for an asymptotically stable system and $\zeta < 1$ are presented in Figure 12.4.

Depending on the value of ζ, the system poles may be complex conjugate ($\zeta < 1$), multiple and real ($\zeta = 1$), or real and distinct ($\zeta > 1$). These three cases are respectively called the under-damped, critically damped, and over-damped cases. The step responses for all three cases are plotted in Figure 12.5 using MATLAB.

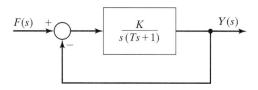

FIGURE 12.3: Block diagram of a second-order system

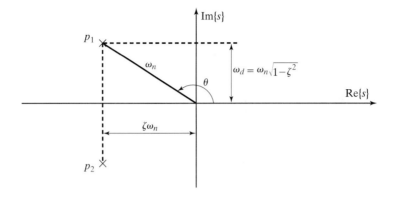

FIGURE 12.4: Second-order system poles for $\zeta < 1$

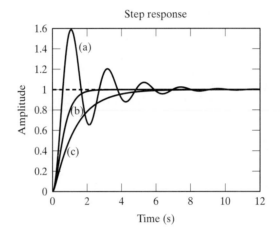

FIGURE 12.5: Responses of second-order systems: (a) under-damped system; (b) critically damped system; (c) over-damped system

The step responses presented in Figure 12.5 correspond to the following second-order systems:

$$(a)\ M(s) = \frac{9}{s^2 + s + 9} \Rightarrow \omega_n = 3, \quad \zeta = 1/6$$

$$(b)\ M(s) = \frac{9}{s^2 + 6s + 9} \Rightarrow \omega_n = 3, \quad \zeta = 1$$

$$(c)\ M(s) = \frac{9}{s^2 + 12s + 9} \Rightarrow \omega_n = 3, \quad \zeta = 2$$

The under-damped case is the most important from the application point of view, since almost all real physical dynamic control systems go through oscillations before settling at the desired steady state value.

The step response of the considered second-order closed-loop system is obtained from

$$Y(s) = M(s)F(s), \quad F(s) = \frac{1}{s} \Rightarrow Y(s) = \frac{\omega_n^2}{s\left(s^2 + 2\zeta\omega_n s + \omega_n^2\right)} \quad (12.11)$$

In the under-damped case, the system has a pair of complex conjugate poles, which leads to the partial fraction expansion

$$Y(s) = \frac{k_1}{s} + \frac{k_2}{s + \zeta\omega_n + j\omega_d} + \frac{k_2^*}{s + \zeta\omega_n - j\omega_d} \quad (12.12)$$

Using the inverse Laplace transform, we obtain the system output in the time domain as

$$y(t) = 1 + \frac{e^{-\zeta\omega_n t}}{\sqrt{1 - \zeta^2}} \sin\left[\left(\omega_n\sqrt{1 - \zeta^2}\right)t - \theta\right] \quad (12.13)$$

where the angle θ is as defined in Figure 12.4. It is left as an exercise for the student to derive formula (12.13) (see Problem 12.5).

The general under-damped step response obtained from (12.13) is plotted in Figure 12.6, where we define the most important transient response parameters: response overshoot (OS), settling time (t_s), peak time (t_p), and rise time (t_r).

The response overshoot is obtained by finding the maximum of $y(t)$ with respect to time. Taking the derivative of (12.13), we have

$$\frac{dy(t)}{dt} = -\frac{\zeta\omega_n}{\sqrt{1 - \zeta^2}} e^{-\zeta\omega_n t} \sin(\omega_d t - \theta) + \frac{\omega_d}{\sqrt{1 - \zeta^2}} e^{-\zeta\omega_n t} \cos(\omega_d t - \theta) = 0$$

which leads to the algebraic equation

$$\zeta\omega_n \sin(\omega_d t - \theta) - \omega_d \cos(\omega_d t - \theta) = 0$$

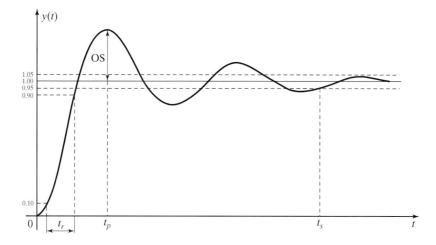

FIGURE 12.6: Response of an under-damped second-order system

From Figure 12.4, we have $\zeta = -\cos\theta$ and $\omega_d = \omega_n \sin\theta$. Using these formulas in the last expression and applying the well-known trigonometric identity, $\sin(\alpha)\cos(\beta) + \sin(\beta)\cos(\alpha) = \sin(\alpha + \beta)$, we obtain

$$\sin \omega_d t = 0 \qquad (12.14)$$

The solutions of this simple equation are

$$\omega_d t = i\pi, \quad i = 0, 1, 2, \ldots \qquad (12.15)$$

The *peak time* is obtained from (12.15) for $i = 1$, that is,

$$t_p = \frac{\pi}{\omega_d} = \frac{\pi}{\omega_n\sqrt{1-\zeta^2}} \qquad (12.16)$$

The steady state value of $y(t)$ can be obtained using the final value theorem of the Laplace transform, which leads to

$$y_{ss} = \lim_{t \to \infty}\{y(t)\} = \lim_{s \to 0}\{sY(s)\} = \lim_{s \to 0}\left\{\frac{\omega_n^2}{s^2 + 2\zeta\omega_n s + \omega_n^2}\right\} = 1 \qquad (12.17)$$

The *response overshoot* is defined by

$$\begin{aligned} \text{OS} = y(t_p) - y_{ss} &= 1 + \frac{e^{-\zeta\omega_n t_p}}{\sqrt{1-\zeta^2}}\sin\left(\omega_d\frac{\pi}{\omega_d} - \theta\right) - 1 \\ &= e^{-\zeta\omega_n t_p}\frac{\sin(\pi - \theta)}{\sin(\theta)} = e^{-(\zeta\pi/\sqrt{1-\zeta^2})} \end{aligned} \qquad (12.18)$$

The response overshoot is very often expressed as a percentage, and the corresponding quantity is called the *maximum percent overshoot*,

$$\text{MPOS} = \text{OS}(\%) = e^{-(\zeta\pi/\sqrt{1-\zeta^2})} \times 100(\%) \qquad (12.19)$$

The expression for the response 5% *settling time* is evaluated from the equality

$$y(t_s) = 1 + \frac{e^{-\zeta\omega_n t_s^{5\%}}}{\sqrt{1-\zeta^2}} = 1.05 \qquad (12.20)$$

which leads to

$$t_s^{5\%} = -\frac{1}{\zeta\omega_n}\ln\left(0.05\sqrt{1-\zeta^2}\right) \approx \frac{3}{\zeta\omega_n} \qquad (12.21)$$

The last approximation is obtained assuming that ζ is small. We can also define the 2% settling time. By repeating the same procedure as in (12.20–21), we obtain

$$t_s^{2\%} \approx \frac{4}{\zeta\omega_n} \qquad (12.22)$$

The validity of approximations (12.21–22) can be tested using MATLAB and Simulink. Note that in some cases, (12.21) and (12.22) give rather rough estimates for the settling times.

The response *rise time* is defined as the time needed for the unit step response to change from 10% to 90% of its steady state value. It is not easy to derive an analytical expression for the system rise time. In general, the smaller the peak time, the smaller the rise time and the faster the system response. Fast system response is very often desirable for control systems. It is known from general electrical circuit courses that fast systems have large frequency bandwidth. Hence, one way to decrease the response rise time is to increase the system bandwidth.

EXAMPLE 12.4

Consider the second-order closed-loop system represented by

$$\frac{Y(s)}{F(s)} = \frac{9}{s^2 + 3s + 9}$$

Its natural frequency is $\omega_n^2 = 9 \Rightarrow \omega_n = 3$ rad/s. The damping ratio is obtained from $2\zeta\omega_n = 3 \Rightarrow \zeta = 0.5$, and the damped frequency is given by

$$\omega_d = \omega_n\sqrt{1 - \zeta^2} = 1.5\sqrt{3} \text{ rad/s}$$

The peak time is evaluated from formula (12.16) as

$$t_p = \pi/\omega_d = (2\pi)/(3\sqrt{3}) = 1.21 \text{ s}$$

The 5% settling time is obtained by using formula (12.21),

$$t_s^{5\%} \approx 3/\zeta\omega_n = 3/1.5 = 2 \text{ s}$$

The maximum percent overshoot, obtained from (12.19), is given by

$$\text{MPOS} = e^{-(\zeta\pi/\sqrt{1-\zeta^2})}100(\%) = 100e^{-(\pi/\sqrt{3})} = 16.3\%$$

12.2.1 Transient Response of Higher-Order Systems

The transient response parameters of higher-order linear time-invariant systems cannot be obtained analytically. However, they can be determined via computer simulation, for example using MATLAB and Simulink. In the special case of an asymptotically stable linear time-invariant system with a pair of complex conjugate dominant poles (see Figure 12.7), the system response can be approximated by the response of a second-order system. The pair of dominant poles is located in the stable half complex plane close to the imaginary axis, with the remaining system poles located much farther to the left in the stable half complex plane. The transient response contributed by the nondominant system poles decays rather quickly to zero, since they have large negative real parts.

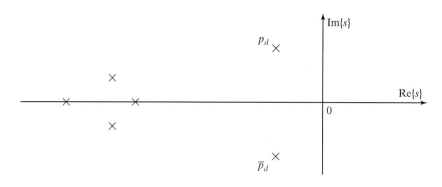

FIGURE 12.7: Complex conjugate dominant system poles

The procedure for performing order reduction of a higher-order system and approximating it by a second-order system is presented in the next example.

EXAMPLE 12.5

Consider the closed-loop transfer function of a fourth-order system given by

$$M(s) = \frac{Y(s)}{F(s)} = \frac{200(s+3)}{(s^2+4s+5)(s+10)(s+20)}$$

This system has the complex conjugate dominant poles $p_d = -2 \pm j1$, and simple distinct poles at $p_1 = -10$ and $p_2 = -20$. The asymptotically stable poles p_1 and p_2 provide negligible contributions to the system transient response since e^{-10t} and e^{-20t} decay very rapidly to zero. The approximation of the original system by as second-order system is obtained as follows:

$$M(s) = \frac{200(s+3)}{(s^2+4s+5)(s+10)(s+20)}$$

$$= \frac{200(s+3)}{(s^2+4s+5)10\left(\frac{s}{10}+1\right)20\left(\frac{s}{20}+1\right)} \approx \frac{s+3}{s^2+4s+5} = M_r(s)$$

In the last expression, we have neglected $s/10$ and $s/20$ with respect to 1, which is a valid approximation, at least at low frequencies. If we are interested in the step response, the corresponding approximation is certainly valid since in this case the system response must settle down to a constant value (DC component, zero-frequency component). It can be concluded that in the order-reduction procedure we have eliminated appropriately the dynamics of real poles p_1 and p_2. Certainly, the system impulse response has a much higher content of high-frequency components than the corresponding system step response, so that a comparison of the impulse responses of the original and reduced-order systems will not produce as good an agreement as the step responses (see Problem 12.22).

The step responses of the original and reduced second-order systems are presented in Figure 12.8, from which we see a very good agreement between these two responses.

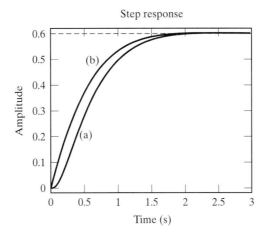

FIGURE 12.8: System step responses for (a) original and (b) second-order approximate systems

Note that the steady state response values are equal to 3/5 is both cases, which can be easily verified by applying the final value theorem of the Laplace transform. Namely, $y_{ss} = M(0) = 3/5$, and $y_{rss} = M_r(0) = 3/5$.

12.3 FEEDBACK SYSTEM STEADY STATE ERRORS

In the design of linear control systems, in addition to obtaining the desired system output transient response, we are also interested in achieving desired steady state values for the system output. In this section we will define and analyze three types of system output steady state errors, and suggest a methodology to reduce these errors to zero.

Consider the unity feedback configuration presented in Figure 4.7(a). Let us assume that the system input signal is equal to the desired system output signal, say $f(t) = y_d(t)$. Such an input signal is often called the reference input. It can be seen from Figure 4.7(a) that the actual system output $y(t)$ is fed back and compared with the desired system output $y_d(t)$. The corresponding difference represents the input signal applied to the system. The difference between the actual and desired system outputs defines the error signal, that is, $y(t) - y_d(t) = e(t)$. If the actual output is equal to the desired output then the error signal is zero. In such a case, the actual input to the system is zero, and hence there is no need to control the system. However, if $e(t) \neq 0$, the error signal will be used to control the system and achieve the goal of reducing the steady state error to zero. The methodology that achieves this goal for three different types of desired system output signals (step, ramp, and parabolic signals) is described in the following.

Consider first the unity feedback system given in Figure 4.7(a) with $F(s) = Y_d(s)$. The error signal is given by

$$E(s) = Y_d(s) - Y(s) = Y_d(s) - H(s)E(s) \qquad (12.23a)$$

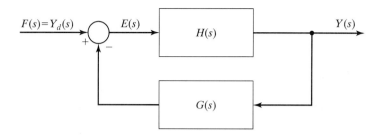

FIGURE 12.9: Feedback system with a dynamic feedback element

which implies

$$E(s) = \frac{Y_d(s)}{1 + H(s)}$$

Our goal is to make $e_{ss} = \lim_{t \to \infty}\{e(t)\} = 0$, where $e(t) = \mathcal{L}^{-1}\{E(s)\}$. As given by formula (12.23a), the error signal depends on the desired output signal and the system dynamics represented by the system transfer function $H(s)$. For the given $Y_d(s)$ and $H(s)$, we can easily find $E(s)$ and then apply the final value theorem of the Laplace transform to obtain e_{ss}. Unless the value obtained for e_{ss} is zero, we will have no possibility to reduce this error to zero. In order to be able to reduce the steady state errors to zero for different types of reference inputs (step, ramp, parabolic), in all cases, we will need to use a dynamic element in the feedback loop, as demonstrated in Figure 12.9.

For generality, we will proceed with the feedback configuration presented in Figure 12.9. The definition of the feedback system error given in (12.23a) must be slightly redefined, and the dynamic feedback element $G(s)$ must be chosen such the newly defined error signal is reduced to zero at steady state. The new definition of the feedback system error is given by

$$E(s) = Y_d(s) - H(s)Y(s) \qquad (12.23b)$$

From Figure 12.9, we have

$$E(s) = Y_d(s) - G(s)H(s)E(s)$$

which implies

$$E(s) = \frac{Y_d(s)}{1 + G(s)H(s)} \qquad (12.24)$$

Using the final value theorem of the Laplace transform, we obtain the steady state value for the error as

$$e_{ss} = \lim_{t \to \infty} e(t) = \lim_{s \to 0}\{sE(s)\} = \lim_{s \to 0}\left\{\frac{sY_d(s)}{1 + G(s)H(s)}\right\} \qquad (12.25)$$

In the following, we consider the problem of choosing $G(s)$ such that the steady state error is reduced to zero for different types of reference inputs: unit step, unit ramp, and unit parabolic. Note that the same derivations hold for nonunit step, ramp, and parabolic reference inputs. Before we present the procedure for reducing the steady state errors to zero, we must introduce the following definition.

DEFINITION 12.1: The *type of feedback control system* represents the number of poles of the loop transfer function $G(s)H(s)$ located at the origin.

In general, the loop transfer function (see Definition 4.2) is given by

$$G(s)H(s) = \frac{K(s+z_1)\cdots(s+z_m)}{s^j(s+p_{j+1})(s+p_{j+2})\cdots(s+p_n)} \quad (12.26)$$

The exponent j defines the type of the feedback system. For $j = 0$, we have feedback system type 0, $j = 1$ indicates feedback system type 1, and so on.

The pole at the origin is commonly called the pure integrator.

Unit Step Reference Input

For $f(t) = y_d(t) = u(t)$, that is $Y_d(s) = 1/s$, we have, from (12.25),

$$e_{ss}^{step} = \lim_{s \to 0} \left\{ \frac{s}{1+G(s)H(s)} \frac{1}{s} \right\} = \frac{1}{1+\lim_{s \to 0}\{G(s)H(s)\}} = \frac{1}{1+K_p} \quad (12.27)$$

The constant K_p, called the *position constant*, is defined by

$$K_p = \lim_{s \to 0}\{G(s)H(s)\} \quad (12.28)$$

It can be concluded that the steady state error due to the unit step reference input is equal to zero for $K_p = \infty$. That will be the case if the system feedback type is at least equal to 1, that is $j \geq 1$.

Unit Ramp Reference Input

For $f(t) = y_d(t) = tu(t)$, we have $Y_d(s) = 1/s^2$. Formula (12.25) produces

$$e_{ss}^{ramp} = \lim_{s \to 0}\{sE(s)\} = \lim_{s \to 0}\left\{\frac{s}{1+G(s)H(s)}\frac{1}{s^2}\right\} = \frac{1}{\lim_{s \to 0}\{sG(s)H(s)\}} = \frac{1}{K_v} \quad (12.29)$$

where the *velocity constant* K_v is defined by

$$K_v = \lim_{s \to 0}\{sG(s)H(s)\} \quad (12.30)$$

It follows that for $K_v = \infty$ we have $e_{ss}^{ramp} = 0$. From the expression for K_v and (12.29), we conclude that the steady error due to the unit ramp input is zero for $j \geq 2$. Hence, the feedback system must have at least two pure integrators (poles at the origin) to provide zero steady state error due to the ramp reference input. In practice, we prefer to keep the number of required pure integrators at a minimum, since pure integrators represent open-loop system poles at the origin, which can impair system stability. Hence, we take $j = 2$ for $e_{ss}^{ramp} = 0$.

Unit Parabolic Reference Input
In this case, $f(t) = y_d(t) = t^2 u(t)$ and $Y_d(s) = 2/s^3$, which implies

$$e_{ss}^{par} = \lim_{s \to 0} \left\{ \frac{s}{1 + G(s)H(s)} \frac{2}{s^3} \right\} = \frac{2}{\lim_{s \to 0} \{s^2 G(s)H(s)\}} = \frac{2}{K_a} \quad (12.31)$$

The constant K_a, given by

$$K_a = \lim_{s \to 0} \{s^2 G(s) H(s)\} \quad (12.32)$$

is called the *acceleration constant*. We can conclude that $K_a = \infty$ for $j \geq 3$, which produces $e_{ss}^{par} = 0$. Hence, we need three pure integrators to reduce the steady state parabolic error to zero.

EXAMPLE 12.6

Consider the feedback type 1 system represented by

$$G(s)H(s) = \frac{20(s+1)(s+3)}{s(s+4)(s+10)}$$

The corresponding steady state constants and steady state errors are given by

$$K_p = \infty \Rightarrow e_{ss} = 0 \quad \text{(step)}$$
$$K_v = 1.5 \Rightarrow e_{ss} = 2/3 \quad \text{(ramp)}$$
$$K_a = 0 \Rightarrow e_{ss} = \infty \quad \text{(parabolic)}$$

EXAMPLE 12.7

Let a system open-loop transfer function be given by

$$H(s) = \frac{s+4}{(s+1)(s+2)}$$

To reduce the system steady state step error to zero, we need a feedback element that has at least one pure integrator, that is $G(s) = 1/s$. Reduction of the steady state ramp error to zero requires $G(s) = 1/s^2$. This dynamic feedback element will produce

$$K_a = \lim_{s \to 0} \{s^2 G(s) H(s)\} = \lim_{s \to 0} \left\{ s^2 \frac{1}{s^2} H(s) \right\} = H(0) = 2$$

The corresponding steady state parabolic error, evaluated from (12.31), is

$$e_{ss}^{par} = \frac{2}{K_a} = 1$$

If we want to reduce the parabolic steady state error to zero, we will need three integrators in the feedback loop, that is $G(s) = 1/s^3$. Note that in some cases integrators are present in the system open-loop transfer function $H(s)$. Such systems, in general, have low steady state errors. For example, let

$$H(s) = \frac{s+2}{s^2(s+1)}$$

This system has $e_{ss}^{step} = 0$ and $e_{ss}^{ramp} = 0$ even for unity feedback, but $K_a = 2 \Rightarrow e_{ss}^{par} = 1$. In order to reduce e_{ss}^{par} to zero, we only need a dynamic feedback element with one pure integrator, that is, $G(s) = 1/s$.

12.4 FEEDBACK SYSTEM FREQUENCY CHARACTERISTICS

Frequency transfer functions of continuous-time, time-invariant, linear systems were introduced in Sections 3.3.1. and 3.3.2. In this section, we present a similar study within the context of linear feedback systems. The *open control loop frequency transfer function* is defined by

$$G(j\omega)H(j\omega) \quad (12.33)$$

where ω takes all possible frequencies, that is $\omega \in [0, \infty)$. Similarly, the *closed-loop frequency transfer function* is defined by

$$\frac{H(j\omega)}{1 + G(j\omega)H(j\omega)} = M(j\omega) \quad (12.34)$$

Since these frequency transfer functions represent complex numbers for any fixed value of ω, we can draw the corresponding magnitude and phase diagrams in terms of ω. We can read from these diagrams important information about control system characteristics. In Figure 12.10, we present such diagrams for the open-loop frequency transfer function.

In Figure 12.10, we have denoted (defined) the phase (Pm) and gain (Gm) stability margins, and the corresponding phase (ω_{cp}) and gain (ω_{cg}) crossover frequencies. It is known from engineering experience, which can also be analytically confirmed using the Nyquist stability criterion (see for example, [1–2]), that systems with phase margins between $30°$ and $60°$ and gain margins $Gm > 2$ (note that $Gm > 2$ corresponds to $Gm[\text{dB}] > 6\,\text{dB}$, since $20\log_{10}(2) = 6$) have very robust stability, which means that they remain stable despite disturbances.

Other frequency response parameters can be obtained from the magnitude diagram of the closed-loop frequency transfer function: closed-loop system bandwidth (ω_{BW}), closed-loop peak resonance (M_r), and closed-loop resonant frequency (ω_r). These are defined in Figure 12.11, and mathematically obtained as explained in the following.

The *closed-loop system bandwidth* represents the frequency range in which the magnitude of the closed-loop frequency transfer function is no more than 3 dB lower than its

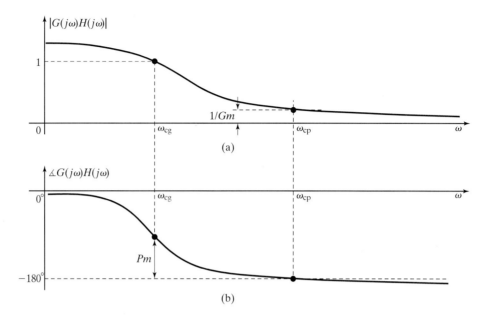

FIGURE 12.10: (a) Magnitude and (b) phase diagrams of the open control loop frequency transfer function

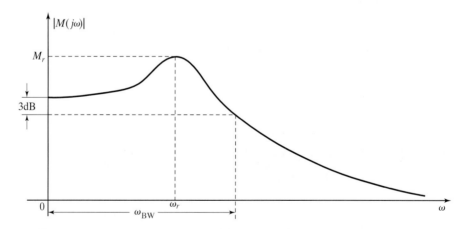

FIGURE 12.11: Magnitude diagram of the closed-loop frequency transfer function

zero-frequency value. The system closed-loop bandwidth is analytically defined by

$$|M(j\omega_{\text{BW}})| = \frac{1}{\sqrt{2}}|M(0)| \Rightarrow \omega_{\text{BW}} \qquad (12.35)$$

It is well known from the study of elementary electrical circuits that the wider the system bandwidth the faster the system response, hence the step response rise time can be shortened by increasing the system bandwidth.

The *closed-loop peak resonance* is obtained by finding the maximum of the function $|M(j\omega)|$ with respect to frequency ω. Systems that have large maximum percent overshoot have also large peak resonance.

It follows from the definition of the closed-loop peak resonance that the *closed-loop resonant frequency* is obtained from

$$\frac{d}{d\omega}|M(j\omega)| = 0 \Rightarrow \omega_r \qquad (12.36)$$

12.5 BODE DIAGRAMS

Bode diagrams represent the magnitude and phase plots of the system frequency transfer function with respect to the frequency ω. In control theory and its applications, Bode diagrams are plotted for the loop frequency transfer function $G(j\omega)H(j\omega)$. The magnitude Bode diagram is plotted in dB, that is $|G(j\omega)H(j\omega)|_{\text{dB}} = 20 \log_{10} |G(j\omega)H(j\omega)|$, with respect to $\log_{10}(\omega)$. The phase Bode diagram is plotted in degrees with respect to $\log_{10}(\omega)$. Note that $\log_{10}(\omega)$ is taken for convenience, since in practice the frequency ω takes a broad range of values. The use of the log-scale makes the points $\ldots 10^{-2}, 10^{-1}, 10^0, 10^1, 10^2, \ldots$ equidistantly distributed along the frequency axis.

Since the loop frequency transfer function contains elementary frequency transfer functions (such as constant term, real poles and real zeros, and complex conjugate poles and zeros), in the following we will first show how to plot Bode diagrams for elementary frequency transfer functions. Because the loop frequency transfer function $G(j\omega)H(j\omega)$ is given by products and ratios of elementary loop frequency transfer functions, both the magnitude and phase Bode diagrams are obtained by summing, respectively, magnitude and phase Bode diagrams of elementary loop frequency transfer functions. For the magnitude we have

$$\begin{aligned}
|G(j\omega)H(j\omega)|_{\text{dB}} &= 20 \log_{10} \left| \frac{K(j\omega+z_1)(j\omega+z_2)\cdots(j\omega+z_m)}{(j\omega)(j\omega+p_2)(j\omega+p_3)\cdots(j\omega+p_n)} \right| \\
&= 20\{\log_{10}|K| + \log_{10}|j\omega+z_1| + \log_{10}|j\omega+z_2| + \cdots + \log_{10}|j\omega+z_m|\} \\
&\quad + 20\left\{\log_{10}\left|\frac{1}{j\omega}\right| + \log_{10}\left|\frac{1}{j\omega+p_2}\right| + \cdots + \log_{10}\left|\frac{1}{j\omega+p_n}\right|\right\} \qquad (12.37)
\end{aligned}$$

This nice additivity property of elementary frequency transfer functions is the main reason why the Bode magnitude diagram is plotted in dB.

For the phase Bode diagrams, we have

$$\begin{aligned}
\angle\{G(j\omega)H(j\omega)\} &= \angle\{K\} + \angle\{j\omega+z_1\} + \angle\{j\omega+z_2\} + \cdots + \angle\{j\omega+z_m\} \\
&\quad - \angle\{j\omega\} - \angle\{j\omega+p_2\} - \cdots - \angle\{j\omega+p_n\} \qquad (12.38)
\end{aligned}$$

Note that within the context of Bode diagrams, we have assumed that $z_i \geq 0, i = 1, 2, \ldots, m$ and $p_i \geq 0, i = 1, 2, \ldots, n$, hence all zeros and poles of the defined frequency transfer functions are in the left half complex plane (stable half plane).

588 Chapter 12 Linear Control Systems

In the following, we show how to plot Bode diagrams of elementary transfer functions. *The magnitude and phase Bode diagrams of more complex transfer functions are obtained simply by adding Bode diagrams of elementary transfer functions.* For convenience, all Bode diagram plots in this section are plotted using the MATLAB function bode. In general, this function has the form bode(sys,{wmin,wmax}), where sys defines the linear system. (For example, if the coefficients of the system numerator and denominator are known, then sys=tf(num,den) defines the system using the transfer function representation, where num and den contain, in descending order, respectively the coefficient of the system transfer function numerator and denominator.) Scalar variables wmin and wmax define the lower and upper bounds for the desired frequency range in which the Bode diagrams are plotted. Note that we can also use simply bode(num,den,wmin,wmax) or the even simpler bode(num,den). In the latter case, MATLAB will automatically choose the appropriate frequency range for the given system transfer function.

Bode Diagrams for a Constant

It is well known that for a constant term the following holds:

$$K_{dB} = 20 \log_{10} |K| = \begin{cases} \text{positive number,} & |K| > 1 \\ \text{negative number,} & |K| < 1 \end{cases}$$

$$\angle K = \begin{cases} 0°, & K > 0 \\ 180° = -180°, & K < 0 \end{cases}$$

(12.39)

The corresponding magnitude and phase Bode diagrams are presented in Figure 12.12 for a positive constant, and in Figure 12.13 for a negative constant whose magnitude is smaller than 1, so that its magnitude Bode diagram is also negative (due to the $\log_{10}(K)$ operation). Note that since $-180° = 180°$, MATLAB presents the phase diagram as being equal to 180°. In plotting these figures, we used the MATLAB instructions num=1.1; den=1; bode(num,den).

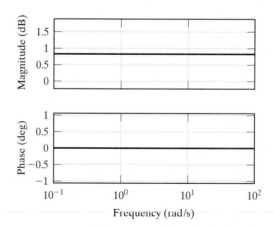

FIGURE 12.12: Magnitude and phase Bode diagrams for a positive constant, $K = 1.1$

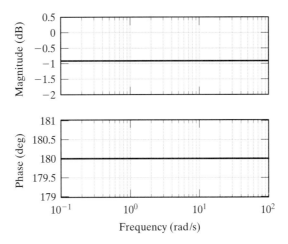

FIGURE 12.13: Magnitude and phase Bode diagrams for a negative constant, $K = -0.9$.

Bode Diagrams for the Pure Integrator

The pure integrator is represented by the simple pole located at the origin. The frequency transfer function of the pure integrator is given by

$$H(j\omega) = \frac{1}{j\omega} \tag{12.40}$$

(We use the name "pure integrator" to distinguish it from "integrators," the name often used for real system poles.) The magnitude and phase of the transfer function in (12.40) are

$$|H(j\omega)|_{dB} = 20\log_{10}\frac{1}{\omega} = -20\log_{10}\omega, \quad \angle H(j\omega) = -90° \tag{12.41}$$

The magnitude of the pure integrator in the $\log_{10}(\omega)$ scale is a straight line that intersects the frequency axis at $\omega = 1$ and has negative slope of -20 dB/decade. The phase of the pure integrator is constant and equal to $-90°$. The corresponding magnitude and phase Bode diagrams are presented in Figure 12.14. Note that in the case of multiple pure integrators (multiple poles at the origin), the corresponding slopes and phase angles add up; that is, the double pure integrator has negative slope of -40 dB/decade and phase angle of $-180°$, the triple pure integrator has negative slope of -60 dB/decade and phase angle of $-270°$, and so on.

Bode Diagrams for the Pure Differentiator

The frequency transfer function of the pure differentiator is

$$H(j\omega) = j\omega \tag{12.42}$$

Its magnitude and phase are

$$|H(j\omega)|_{dB} = 20\log_{10}\omega, \quad \angle H(j\omega) = 90° \tag{12.43}$$

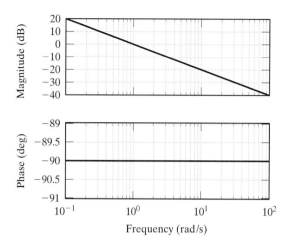

FIGURE 12.14: Magnitude and phase Bode diagrams for the pure integrator

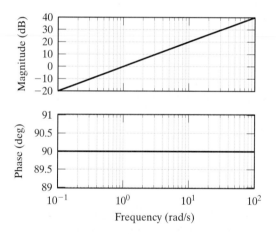

FIGURE 12.15: Magnitude and phase Bode diagrams for the pure differentiator

(We use the name "pure differentiator" to distinguish it from "differentiator," the name often used for real system zeros.) The magnitude of the pure differentiator in the $\log_{10}(\omega)$ scale is a straight line that intersects the frequency axis at $\omega = 1$ and has positive slope of 20 dB/decade. The phase of the pure differentiator is constant and equal to $90°$. The corresponding magnitude and phase Bode diagrams are presented in Figure 12.15. We plotted this figure using the MATLAB statements num=[1 0]; den=1; bode(num,den). A multiple pure differentiator (multiple zeros at the origin), say of order q, has positive slope of $q \times 20$ dB/decade and phase angle of $q \times 90°$.

Bode Diagrams for a Real Pole

The frequency transfer function of a real pole is given by

$$H(j\omega) = \frac{p}{p + j\omega} = \frac{1}{1 + j\dfrac{\omega}{p}} \qquad (12.44)$$

Its magnitude and phase are

$$|H(j\omega)|_{dB} = -20\log_{10}\left[1 + \left(\frac{\omega}{p}\right)^2\right]^{1/2}, \quad \angle H(j\omega) = -\tan^{-1}\left(\frac{\omega}{p}\right) \qquad (12.45)$$

The frequency at which $\omega = p$ is called the *corner frequency*, ω_c. At the corner frequency, $|H(j\omega_c)|_{dB} = -20\log_{10}\sqrt{1+1} = -3$ dB, $-\tan^{-1}(1) = -45°$. At low frequencies, $\omega \ll p$, the magnitude is equal to a very small negative quantity, practically equal to zero, since $|H(j\omega)|_{dB} = -20\log_{10} 1 = 0$ dB; and the phase also takes very small negative values close to zero, since $-\tan^{-1}(0) = 0°$. At high frequencies, $\omega \gg p$, the phase contribution is $-90°$, since $-\tan^{-1}(\infty) = -90°$. In the same frequency range, for $\omega \gg p$, the magnitude can be approximated by $|H(j\omega)|_{dB} = -20\log_{10}(p/\omega) = -20\log_{10}(p) - 20\log_{10}(\omega)$, which is a straight line with negative slope of -20 dB/decade that intersects the frequency axis at the corner frequency. Based on this information, we can easily draw the corresponding magnitude and phase Bode diagrams for a real pole, as presented in Figure 12.16, where the corner frequency is taken to be $\omega_c = p = 1$. The corresponding MATLAB code used to plot Figure 12.16 is num=1; den=[1 1]; bode(num,den,{0.1,10}). In the same figure, we have plotted the asymptote for the magnitude Bode diagram of -20 dB/decade that starts at the corner frequency $\omega_c = p = 1$, and the asymptote of $-90°$ for the phase Bode diagram.

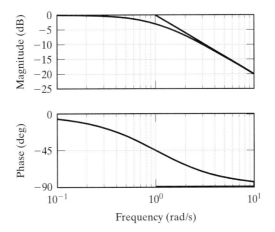

FIGURE 12.16: Magnitude and phase Bode diagrams for a real pole, $H(j\omega) = 1/(1+j\omega)$

Bode Diagrams for a Real Zero

The frequency transfer function of a real zero is given by

$$H(j\omega) = 1 + j\left(\frac{\omega}{z}\right) \tag{12.46}$$

Its magnitude and phase are

$$|H(j\omega)|_{\mathrm{dB}} = 20\log_{10}\left[1 + \left(\frac{\omega}{z}\right)^2\right]^{1/2}, \quad \angle H(j\omega) = \tan^{-1}\left(\frac{\omega}{z}\right) \tag{12.47}$$

Since the zero is the mirror image of the pole, we repeat the same analysis as for the real pole. For low frequencies, an asymptote for the magnitude is equal to zero, and for high frequencies the magnitude asymptote has a positive slope of 20 dB/decade and intersects the real axis at the corner frequency $\omega_c = z$. The phase diagram for low frequencies has an asymptote equal to zero, and for high frequencies an asymptote of $90°$. The corresponding magnitude and phase Bode diagrams are represented in Figure 12.17, where the corner frequency is taken as $\omega_c = z = 1$. In Figure 12.17, we have also plotted the asymptote for the magnitude Bode diagram of 20 dB/decade that starts at the corner frequency $\omega_c = 1$, and the asymptote of $90°$ for the phase Bode diagram.

Bode Diagrams for Complex Conjugate Poles

The frequency transfer function corresponding to a pair of complex conjugate poles is, in general, given by

$$H(j\omega) = \frac{\omega_n^2}{(j\omega)^2 + 2\zeta\omega_n(j\omega) + \omega_n^2} = \frac{1}{\left(1 - \dfrac{\omega^2}{\omega_n^2}\right) + j2\zeta\dfrac{\omega}{\omega_n}} \tag{12.48}$$

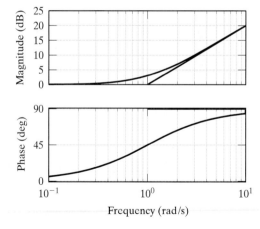

FIGURE 12.17: Magnitude and phase Bode diagrams for a real zero, $H(j\omega) = 1 + j\omega$

Its magnitude and phase are

$$|H(j\omega)|_{\text{dB}} = -20\log_{10}\left[\left(\frac{2\zeta\omega}{\omega_n}\right)^2 + \left(1 - \frac{\omega^2}{\omega_n^2}\right)^2\right]^{1/2}$$
(12.49)
$$\angle H(j\omega) = -\tan^{-1}\left(\frac{2\zeta\omega_n\omega}{\omega_n^2 - \omega^2}\right)$$

For high frequencies, we have

$$|H(j\omega)|_{\text{dB}} \approx -20\log_{10}\left(\frac{\omega^2}{\omega_n^2}\right) = -40\log_{10}\left(\frac{\omega}{\omega_n}\right)$$

$$\angle\{H(j\omega)\} \approx -\tan^{-1}\left(\frac{2\zeta\omega_n}{-\omega}\right) \to -\tan^{-1}(0^-) = -180°$$

For low frequencies, we obtain

$$H(j\omega) \approx \frac{\omega_n^2}{\omega_n^2} = 1 \Rightarrow |H(j\omega)|_{\text{dB}} = 0, \quad \angle\{H(j\omega)\} = 0°$$

Note that $\omega = \omega_n$ represents the corner frequency for a pair of complex conjugate poles. It can be concluded that the asymptotes for low and high frequencies are, respectively, zero and -40 dB/decade for the magnitude, and zero and $-180°$ for the phase. At the corner frequency $\omega_c = \omega_n$, the magnitude is equal to $|H(j\omega_c)|_{\text{dB}} \approx -20\log_{10}(2\zeta)$, and the phase is equal to $-90°$. The corresponding Bode diagrams are presented in Figure 12.18. It is clear from this figure that the magnitude Bode diagram asymptote is -40 dB/decade, and that the phase Bode diagram asymptote is $-180°$.

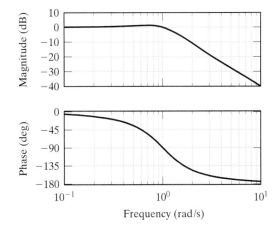

FIGURE 12.18: Magnitude and phase Bode diagrams for a pair of complex conjugate poles, $H(j\omega) = 1/((j\omega)^2 + j\omega + 1)$.

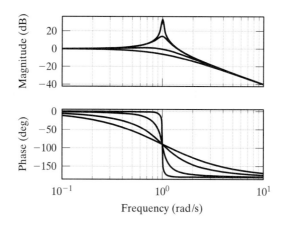

FIGURE 12.19: Magnitude and phase Bode diagrams in terms of ζ for the transfer function $H(j\omega) = 1/((j\omega)^2 + 2\zeta(j\omega) + 1)$, with $\zeta = 0.01, 0.1, 0.5, 1$

Note that the actual plot in the neighborhood of the corner frequency depends on the values of the damping ratio ζ. Magnitude and phase Bode diagrams for the second-order transfer function (12.48), for different values of the damping ratio ($0.01 \leq \zeta \leq 1$), are presented in Figure 12.19. Since the bottom curve in the magnitude diagram corresponds to $\zeta = 1$ and the top curve in the same diagram corresponds to $\zeta = 0.01$, it can be concluded that the smaller the ζ (the higher the overshoot), the higher the peak of the magnitude Bode diagram. The impact of ζ on the phase diagram is demonstrated by the fact that smaller ζ implies larger slope of the phase curve around the corner frequency.

Bode Diagrams for Complex Conjugate Zeros

The frequency transfer function for a pair of complex conjugate zeros is given by

$$H(j\omega) = 1 + 2\zeta j\left(\frac{\omega}{\omega_n}\right) + \left(\frac{j\omega}{\omega_n}\right)^2 = 1 - \left(\frac{\omega}{\omega_n}\right)^2 + j2\zeta\left(\frac{\omega}{\omega_n}\right) \quad (12.50)$$

The corresponding Bode diagrams are the mirror images of the Bode diagrams obtained for the complex conjugate poles. The complex conjugate zeros have asymptotes for low frequencies equal to zero for both the magnitude and phase; for high frequencies, the magnitude asymptote has positive slope of 40 dB/decade and starts at the corner frequency of $\omega_c = \omega_n$, and the phase plot has an asymptote of $180°$. The corresponding magnitude and phase Bode diagrams for a pair of complex conjugate zeros with the corner frequency $\omega_c = \omega_n = 1$ are presented in Figure 12.20.

The *additivity property* for both the magnitude and phase Bode diagrams, established in (12.37–38), can be used for plotting general Bode diagrams by first plotting Bode diagrams of elementary transfer functions and then adding them. The additivity property of Bode diagrams can be easily observed in the following example.

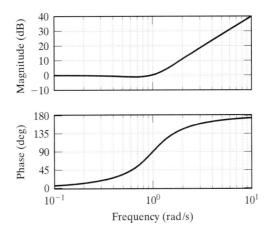

FIGURE 12.20: Magnitude and phase Bode diagrams for a pair of complex conjugate zeros, $H(j\omega) = 1 + (j\omega) + (j\omega)^2$

EXAMPLE 12.8

Consider the transfer function

$$G(s)H(s) = \frac{s+1}{s(s+10)}$$

This transfer function is composed of the pure integrator, a real zero element with corner frequency at 1 rad/s, and a real pole element with corner frequency at 10 rad/s. Bode diagrams of the pure integrator, real pole element, and real zero element are presented, respectively, in Figures 12.14, 12.16, and 12.17. (Note that in this case the corner frequency for the real pole is at 10 rad/s.) By adding the three magnitude Bode diagrams of the elementary transfer functions involved, we can observe that the magnitude Bode diagram of $G(s)H(s)$ at low frequency has a slope of -20 dB/decade, since in that region only the pure integrator is active. The real zero element that contributes a slope of 20 dB/decade becomes active at its corner frequency of 1 rad/s. It can be seen that in the frequency range from 1 rad/s to 10 rad/s, the slope of the magnitude Bode diagram is close to 0 dB/decade. At 10 rad/s, the real pole element becomes active and contributes another -20 dB/decade so that the magnitude Bode diagram for frequencies greater than 10 rad/s has slope close to -20 dB/decade.

The corresponding Bode diagrams for this transfer function are presented in Figure 12.21.

The additivity property can be similarly observed for the phase Bode diagram, by looking at the corresponding elementary phase Bode diagrams presented in Figures 12.14, 12.16, and 12.17. Note that the pure integrator contributes $-90°$ at all frequencies, and that the real pole and real zero elements contribute, respectively, $-90°$ and $90°$ at high

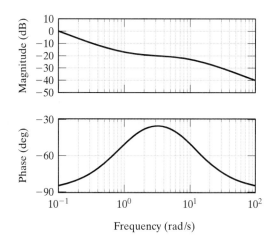

FIGURE 12.21: Magnitude and phase Bode diagrams for the system in Example 12.8

frequencies, so that the overall phase Bode diagram at high frequencies is equal to $-90°$. At low frequencies, only the pure integrator is active, hence, the total phase in this range is equal to $-90°$. In the range from 1 rad/s to 10 rad/s, the phase Bode diagram first increases since the zero element becomes active at 1 rad/s, and then decreases due to activity of the real pole at 10 rad/s.

In general, we can plot Bode diagrams of more complex systems using the MATLAB function bode(num,den), where num and den represent, respectively, the transfer function coefficients of the numerator and denominator, in descending order.

EXAMPLE 12.9

Consider a control system whose loop transfer function is given by

$$G(s)H(s) = \frac{s+1}{s(s+2)(s^2+10s+25)}$$

The magnitude and phase Bode diagrams of this control system are represented in Figure 12.22.

The Bode diagrams in this example are obtained using the following MATLAB script:

```
num=[1 1];
d1=[1 0]; d2=[1 2]; d3=[1 10 25]; d23=conv(d2,d3);
den=conv(d1,d23);
bode(num,den,{0.1, 100})
```

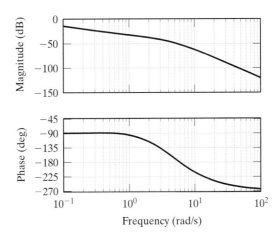

FIGURE 12.22: Magnitude and phase Bode diagrams for the system in Example 12.9

Bode Diagrams and Phase and Gain Stability Margins

The phase and gain stability margins were introduced in Figure 12.10 of Section 12.4. It can be seen from Figure 12.10 that the gain crossover frequency is found at the frequency where $|G(j\omega_{cg})H(j\omega_{cg})| = 1$. Since the magnitude Bode diagram is plotted in dB, it follows that $|G(j\omega_{cg})H(j\omega_{cg})|_{dB} = 0$. The gain crossover frequency is obtained at the frequency where the magnitude Bode diagram intersects the horizontal (frequency) axis. Hence, the phase stability margin can be defined analytically as follows.

DEFINITION 12.2: The phase stability margin is defined by

$$Pm = -180° - \angle G(j\omega_{cg})H(j\omega_{cg}) \tag{12.51}$$

where ω_{cg} is the gain crossover frequency at which $|G(j\omega_{cg})H(j\omega_{cg})| = 1$.

The phase crossover frequency, as denoted in Figure 12.10, is the frequency at which the phase Bode diagram intersects the $-180°$ horizontal line. At that frequency, we read the value for the gain stability margin. The gain stability margin can be analytically defined as follows.

DEFINITION 12.3: The gain stability margin is defined by

$$Gm[dB] = -|G(j\omega_{cp})H(j\omega_{cp})|_{dB} \tag{12.52}$$

ω_{cp} is the phase crossover frequency at which $\angle G(j\omega_{cp})H(j\omega_{cp}) = -180°$.

It is well known that the system steady state errors are improved by increasing the Bode gain. On the other hand, the phase and gain margins are improved by reducing the Bode gain. For the general linear system transfer function (12.37), the Bode gain is defined as follows.

DEFINITION 12.4: Let the loop transfer function (12.26) be of feedback type j (it has j poles at the origin). The Bode gain for this transfer function is defined by

$$K_B = \frac{K z_1 z_2 \cdots z_m}{p_{j+1} p_{j+2} \cdots p_n} \tag{12.53}$$

We can easily obtain the phase and gain stability margins using the MATLAB function `margin(num,den)`, where `num` and `den` represent, respectively, the coefficients of the numerator and denominator of the loop transfer function $G(s)H(s)$, specified in descending order. In the next example, we use MATLAB to plot Bode diagrams and find the phase and gain stability margins, which are clearly denoted in the associated figure.

EXAMPLE 12.10

Consider a control system whose loop transfer function is given by

$$G(s)H(s) = \frac{K(s+3)}{(s+2)(s+5)(s^2+6s+25)}$$

The phase and gain stability margins for $K = 100$ are presented in Figure 12.23, which is obtained by using the following MATLAB script.

```
num=[100 300];
d1=[1 2]; d2=[1 5]; d3=[1 6 25];
d23=conv(d2,d3);
den=conv(d1,d23);
margin(num,den)
```

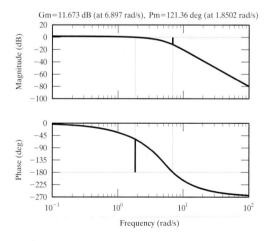

FIGURE 12.23: Phase and gain stability margins for the system in Example 12.10

Section 12.6   Common Dynamic Controllers: PD, PI, and PID   599

Students are asked in Problem 12.25 to vary K in the range from 1 to 500 and observe dependence of the phase and gain stability margins on the Bode gain.

In addition to providing information about the phase and gain stability margins, Bode diagrams can be used as efficient tools for the design of controllers. These controllers are inserted in either the forward path or feedback loop to improve some system characteristics so that they take desired values. Controller design techniques based on Bode diagrams can be found in all undergraduate control theory and applications textbooks. In the next section, we will indicate how we can improve feedback system characteristics by using very simple controllers, known as PI, PD, and PID controllers. More than 90% of the controllers used in industry today are of the PID type [3–4]. The main reason for their popularity is their simplicity; they can be used even by nonexperts.

12.6 COMMON DYNAMIC CONTROLLERS: PD, PI, AND PID

Control systems must satisfy certain specifications so that the systems under consideration have the desired behavior (such as required stability margins, transient, and steady state responses). If the desired specifications are not intrinsically met, controllers are designed and placed in either the forward path or the feedback loop, such that the desired specifications are obtained. A dynamic controller is inserted in the feedback loop, usually in front of the system, as demonstrated in Figure 12.24.

The desired specifications are assigned in terms of required values (or upper and/or lower limits) for phase and gain stability margins, settling time, rise time, peak time, maximum percent overshoot, and steady state errors. Additional specifications can be defined in the frequency domain, such as system frequency bandwidth, resonant frequency, and resonance peak.

It is impossible to meet all these specifications at once. Sometimes in some cases, requirements are conflicting, and in others, they are not feasible. Thus, control engineers must compromise while trying to satisfy all imposed control system requirements. Fortunately, we can identify the most important requirements. First of all, *systems must be stable*. Hence, the main goal of controller design is to stabilize the system under consideration. Furthermore, to preserve system stability under internal and external disturbances (to

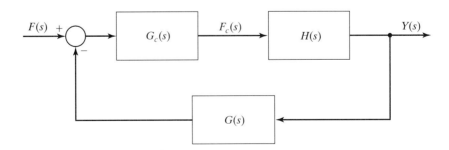

FIGURE 12.24: Feedback system controlled by a dynamic controller

provide stability robustness), it is required that the system phase and gain stability margins be maintained at specific values (usually $Pm \in [30°, 60°]$ and $Gm > 2 \Leftrightarrow Gm[\text{dB}] > 6$ dB). Second, *systems should have limited overshoot and settling time,* and *steady state errors should be kept within admissible bounds.*

PID controllers can successfully cope with most of the specified requirements. In rare cases, we can achieve the desired control system specifications using only a static controller, that is $G_c(s) = K = $ const. In general, by increasing the static gain, we improve the steady state errors, but degrade the system transient response—the response overshoot is increased. Furthermore, large values of K can impair system stability.

In the following, we present the basic structures of the classic PD, PI, and PID controllers and indicate their main properties. It is interesting to point out that these controllers originated in the forties [5], and became very popular again in the nineties. A recent survey paper [6] lists more than three hundred journal papers, published in the nineties, about the use, design, tuning, and applications of PID controllers.

PD Controller

The *proportional and derivative* (PD) controller is used to improve system transient response, while providing system stability. Its transfer function is given by

$$G_c(s) = K_p + K_d s \tag{12.54}$$

where the constants K_p and K_d must be determined in order to obtain the desired transient response characteristics (desired overshoot, settling time, rise time). This controller introduces the positive phase in the loop transfer function, which for the feedback control systems represented in Figure 12.24 is given by $G_c(s)G(s)H(s)$. The introduced positive phase comes from the fact that $\angle G_c(j\omega) = \angle(K_p + jK_d\omega) > 0$ for $K_p > 0$ and $K_d > 0$. There are algorithms in the control literature for the choice of the constants (gains) K_p and K_d, such that the desired system response specifications are met [3]. This controller is often used in practice for control of robots and communication networks.

PI Controller

The *proportional and integral* (PI) controller is primarily used to improve steady state errors. It has a pole at the origin that increases the system feedback type by 1. Its transfer function is given by

$$G_c(s) = K_p + K_i \frac{1}{s} = \frac{1}{s}(K_p s + K_i) \tag{12.55}$$

In addition, the constants K_p and K_i can be found such that the controlled system has desired phase and gain stability margins [7]. This controller is very often used in chemical engineering and process industries.

PID Controller

The *proportional, integral, and derivative controller* (PID) is used to improve both the transient response and the steady state errors, and of course to provide a stable closed-loop system with good stability margins. It has the transfer function

$$G_{\text{PID}}(s) = K_p + K_i \frac{1}{s} + K_d s = \frac{1}{s}\left(K_i + K_p s + K_d s^2\right) \qquad (12.56)$$

This controller is very often called the *industrial controller* due to its broad popularity in industry. The controller parameters K_p, K_i, K_d are tuned until desired specifications for the transient response and the steady state errors are achieved. In some simple cases, tuning can be simply done manually. For efficient tuning techniques for PID controllers, the interested reader is referred to [3–4]. It should be pointed out that there are many techniques for tuning, PID controllers, and that none of them is universal (all of them are problem and system dependent).

In practical applications, the PID controller defined in (12.56) is very often replaced by the PID controller with approximate derivative whose transfer function is given by

$$G_{\text{PID}}^{\text{app}}(s) = K_p + K_i \frac{1}{s} + K_d \frac{s}{\tau s + 1} \qquad (12.57)$$

where τ takes small values. Note that both the PID controller and the PID controller with approximate derivative are implemented in Simulink.

Other variants of PID controllers are also available for practical implementations. PID controller that has recently become very popular [3–6] provides the following input signal to the system:

$$F_c(s) = K_p E(s) + K_i \frac{1}{s} E(s) + K_d \frac{s}{\frac{K_d}{n_f}s + 1} Y(s), \quad E(s) = F(s) - Y(s), \quad G(s) = 1$$

$$(12.58)$$

with n_f usually taken as 10.

Other Controllers

There are many other types of controllers, starting with classic controllers (such as phase-lag, phase-lead, phase-lag-lead), then state-space based controllers, up to modern optimal controllers that optimize a certain performance criterion [8]. Modern optimal controllers culminated in the nineties in robust optimal controllers [9]. These controllers are studied in specialized courses on control theory and its applications.

12.7 LABORATORY EXPERIMENT ON CONTROL SYSTEMS

Problem: A Pitch Controller for a Commercial Aircraft The linearized equations governing the motion of a commercial aircraft, obtained by using the linearization methodology presented in Section 8.6, are given [10] by

$$\frac{d\alpha(t)}{dt} = -0.313\alpha(t) + 56.7q(t) + 0.232\delta_e(t)$$

$$\frac{dq(t)}{dt} = -0.0139\alpha(t) - 0.426q(t) + 0.0203\delta_e(t) \qquad (12.59)$$

$$\frac{d\theta(t)}{dt} = 56.7q(t)$$

where $\theta(t)$ represents the pitch angle. The corresponding open-loop transfer function obtained from (12.59) is given by

$$\frac{\Theta(s)}{\Delta(s)} = \frac{1.151s + 0.1774}{s^3 + 0.739s^2 + 0.921s} = \frac{1.151(s + 0.1541)}{s(s^2 + 0.739s + 0.921)} = 1.151H(s)$$

In this experiment we design an autopilot that controls the pitch angle $\theta(t)$ of this aircraft. The autopilot is obtained by forming a closed-loop system with unity feedback and a controller of the form $KG_c(s)$. For simplicity, we assume that $K = 1.151K'$, so that the loop transfer function is $KG_c(s)H(s)$.

Part 1. Find the steady state unit step and unit ramp errors of the original closed-loop system ($K = 1, G_c(s) = 1, M(s) = H(s)/(1 + H(s))$). The unity feedback closed-loop transfer function can also be obtained using the MATLAB statement [cnum,cden]=feedback(num,den,1,1,-1). Plot the closed-loop system step and ramp responses and observe (check) the corresponding steady state errors. (*Hint:* In order to find the ramp response, use the MATLAB function y=lsim(cnum,cden,t,t) with t=0:0.1:30.

Part 2. Find the value for the static gain K such that the steady state error due to the unit ramp is reduced to $e_{ss}^{ramp} \leq 0.1$. For the obtained value of K, plot the corresponding closed-loop system step and ramp responses and notice the steady state unit ramp error improvement. Also observe the transient step response worsen due to an increased value of the static gain K. (*Hint:* Use the same time interval as in Part 1.)

Part 3. Plot Bode diagrams with the value for the static gain K obtained in Part 2. Use bode(K*num,den); and margin(K*num,den) to find the phase and gain stability margins, and observe that the phase margin is quite pure. Design a PD controller of the form $G_c(s) = G_{PD}(s) = 1 + K_d s$ to improve the phase stability margin, such that the controlled system has a phase stability margin close to $50°$ (see Figure 12.24, and note that unity feedback is used by the design requirement). Find the closed-loop step response of the system controlled by the PD controller, and compare it to the closed-loop step response of the uncontrolled system whose static gain K is found in Part 2. Comment on the transient response improvement of the controlled system. (*Hint:* Try several values for the derivative constant in the range $0 < K_d < 10$.)

Part 4. Use a PI controller of the form $G_c(s) = G_{PI}(s) = 1 + K_i/s$. Though the PI controller introduces negative phase shift, you still can improve the system phase stability margin by placing the new gain crossover frequency in the range of low frequencies. Find the closed-loop step response of the system controlled by the PI controller and compare it to the closed-loop step response of the system controlled by the PD controller. Observe that the PI controller reduces the system frequency bandwidth (ω_{cg} gets smaller), so that the system step response with the PI controller is slower, that is, it has longer rise time. Which controller do you prefer? Explain.

Part 5. Use a PID controller of the form $G_c(s) = G_{\text{PID}}(s) = 1 + K_d s + K_i/s$ and tune the values for K_d and K_i such that satisfactory results are obtained for both the transient response parameters and the steady state errors. Try also to achieve satisfactory stability margins, that is, $Pm \in [30°, 60°]$. Compare the results obtained with the corresponding results obtained in Parts 3 and 4. Which among the three controllers considered for this particular aircraft control problem do you prefer?

12.8 SUMMARY

Study Guide for Chapter Twelve: Control of systems is facilitated using feedback signals. Fundamentally important features of feedback are its ability to stabilize unstable systems, and the insensitivity (robustness) of feedback systems to system parameter changes (disturbances). In addition, in control systems we intend to shape both the system transient and steady state responses such that they have desired properties, which can also be done using feedback. The student must be familiar with the basic transient response parameters and expressions for steady state errors. An important rule to remember: one integrator is needed in the feedback loop to eliminate the step response steady state error, and two integrators to eliminate the ramp response steady state error. Bode diagrams are useful for the design of controllers and for finding system stability margins. The most widely used controller in industry is the PI controller, whose integral part improves the steady state errors. The derivative part of the PD controller improves the system transient response, and the PID controller can improve both the transient response and the steady state errors, but is more difficult to design (tune).

Transient Response Parameters of a Second-Order System

Second-order Closed-loop System:

$$\frac{Y(s)}{F(s)} = \frac{\omega_n^2}{s^2 + 2\zeta\omega_n s + \omega_n^2}, \quad \zeta = \frac{1}{2\omega_n T}, \quad \omega_n^2 = \frac{K}{T}$$

Maximum Percent Overshoot:

$$\text{MPOS} = \text{OS}(\%) = e^{-(\zeta\pi/\sqrt{1-\zeta^2})} \times 100(\%)$$

5% and 2% Settling Time:

$$t_s^{5\%} = -\frac{1}{\zeta\omega_n} \ln\left(0.05\sqrt{1-\zeta^2}\right) \approx \frac{3}{\zeta\omega_n}, \quad t_s^{2\%} \approx \frac{4}{\zeta\omega_n}$$

Peak Time:

$$t_p = \frac{\pi}{\omega_d} = \frac{\pi}{\omega_n\sqrt{1-\zeta^2}}$$

Steady State Constants and Steady State Errors:

Position Constant and Steady State Unit Step Error:

$$K_p = \lim_{s \to 0}\{G(s)H(s)\}, \quad e_{ss}^{step} = \frac{1}{1+K_p}$$

Velocity Constant and Steady State Unit Ramp Error:

$$K_v = \lim_{s \to 0}\{sG(s)H(s)\}, \quad e_{ss}^{ramp} = \frac{1}{K_v}$$

Acceleration Constant and Steady State Unit Parabolic Error:

$$K_a = \lim_{s \to 0}\{s^2 G(s)H(s)\}, \quad e_{ss}^{par} = \frac{2}{K_a}$$

Bode Diagrams and Stability Margins

Magnitude Bode Diagrams:

$$|G(j\omega)H(j\omega)|_{dB}$$
$$= 20\log_{10}\left|\frac{K(j\omega+z_1)(j\omega+z_2)\cdots(j\omega+z_m)}{(j\omega)(j\omega+p_2)(j\omega+p_3)\cdots(j\omega+p_n)}\right|$$
$$= 20\{\log_{10}|K| + \log_{10}|j\omega+z_1| + \log_{10}|j\omega+z_2| + \cdots + \log_{10}|j\omega+z_m|\}$$
$$+ 20\left\{\log_{10}\left|\frac{1}{j\omega}\right| + \log_{10}\left|\frac{1}{j\omega+p_2}\right| + \cdots + \log_{10}\left|\frac{1}{j\omega+p_n}\right|\right\}$$

Phase Bode Diagrams:

$$\angle\{G(j\omega)H(j\omega)\} = \angle\frac{K(j\omega+z_1)(j\omega+z_2)\cdots(j\omega+z_m)}{(j\omega)(j\omega+p_2)(j\omega+p_3)\cdots(j\omega+p_n)}$$
$$= \angle\{K\} + \angle\{j\omega+z_1\} + \angle\{j\omega+z_2\} + \cdots + \angle\{j\omega+z_m\}$$
$$- \angle\{j\omega\} - \angle\{j\omega+p_2\} - \cdots - \angle\{j\omega+p_n\}$$

Phase Stability Margin:

$$Pm = -180° - \angle G(j\omega_{cg})H(j\omega_{cg}), \quad |G(j\omega_{cg})H(j\omega_{cg})| = 1$$

Gain Stability Margin:

$$Gm[dB] = -|G(j\omega_{cp})H(j\omega_{cp})|_{dB}, \quad \angle G(j\omega_{cp})H(j\omega_{cp}) = -180°$$

Bode Gain for j Type Feedback System (j poles at the origin):

$$K_B = \frac{Kz_1 z_2 \cdots z_m}{p_{j+1}p_{j+2}\cdots p_n}$$

PD, PI, and PID Controllers

PD Controller:

$$G_c(s) = G_{\text{PD}}(s) = K_p + K_d s$$

PI Controller:

$$G_c(s) = G_{\text{PI}}(s) = K_p + K_i \frac{1}{s} = \frac{1}{s}(K_p s + K_i)$$

PID Controller:

$$G_c(s) = G_{\text{PID}}(s) = K_p + K_i \frac{1}{s} + K_d s = \frac{1}{s}(K_i + K_p s + K_d s^2)$$

PID Controller with Approximative Derivative:

$$G_c(s) = G_{\text{PID}}^{\text{app}}(s) = K_p + K_i \frac{1}{s} + K_d \frac{s}{\tau s + 1}$$

12.9 REFERENCES

[1] Z. Gajić and M. Lelić, *Modern Control System Engineering,* Prentice Hall, London, 1996.

[2] B. Kuo, *Automatic Control Systems,* Prentice Hall, Englewood Cliffs, NJ, 1995.

[3] K. Astrom and T. Hagglund, *PID Controllers: Theory, Design, and Tuning,* Instrument Society of America, Research Triangle Park, NC, 1995.

[4] C. Yu, *Autotuning of PID Controllers: Relay Feedback Approach,* Springer Verlag, London, 1999.

[5] J. Ziegler and N. Nichols, "Optimum settings for automatic controllers," *Transactions of ASME,* 65:433–44, 1943.

[6] M. Lelić and Z. Gajić, "A reference guide to PID controllers in the nineties," *Proceedings of the IFAC Workshop on Past, Present, and Future of PID Control,* Terrassa, Spain, April 2000, 73–82.

[7] K. Astrom and T. Hagglund, "Automatic tuning of simple regulators with specifications on phase and amplitude margins," *Automatica,* 20:645–51, 1984.

[8] F. Lewis, *Applied Optimal Control & Estimation,* Prentice Hall, Englewood Cliffs, NJ, 1992.

[9] K. Zhou and J. Doyle, *Essentials of Robust Control,* Prentice Hall, Upper Saddle River, NJ, 1998.

[10] W. Messner and D. Tilbury, *Control Tutorials for MATLAB and Simulink: A Web-Based Approach,* Addison Wesley, Menlo Park, CA, 1998.

12.10 PROBLEMS

12.1. Find the sensitivity function $S^p(s)$ for the feedback system presented in Figure 12.25.

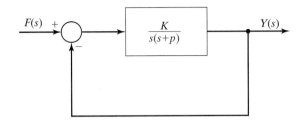

FIGURE 12.25

Answer:

$$S^p(s) = \frac{-ps}{s^2 + ps + K}$$

12.2. Consider the unity feedback system represented in Figure 12.1, with

$$H(s) + \Delta H(s) = \frac{100(10 \pm 5)}{s + 10} = \frac{1000}{s + 10} \pm \frac{500}{s + 10} = H(s) \pm \Delta H(s)$$

The static gain of the open-loop system is fairly uncertain, since

$$H(0) + \Delta H(0) = \frac{100(10 \pm 5)}{10} = 100 \pm 50$$

causing 50% DC gain error tolerance. Show that the unity feedback can reduce the DC gain error tolerance to less than 2%.

12.3. Find the sensitivity function for the unity feedback system defined in Problem 12.2. Use MATLAB to plot its magnitude frequency characteristics. What is the frequency range of disturbances that can be harmful to this system?

12.4. For the feedback system presented in Figure 12.26, find the system output in terms of the system input and the system disturbance. Suggest possible choices for the controller and the feedback element that attenuate the disturbance in the system output.

12.5. Derive formula (12.13).

12.6. Find the step response of the second-order system given in (12.11), under the assumption that the system poles are real and equal (critically damped case).

12.7. Find the step response of the second-order system given in (12.11), under the assumption that the system poles are real and distinct (over-damped case).

12.8. Derive the formula for the 2% settling time given in (12.22).

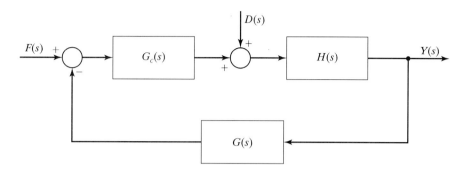

FIGURE 12.26

12.9. Consider the following second-order systems represented by their closed-loop transfer functions.

(a) $M_1(s) = \dfrac{25}{s^2 + 8s + 25}$

(b) $M_2(s) = \dfrac{6}{s^2 + 4s + 16}$

Find in both cases the maximum percent overshoot and the 5% settling time.

Answer:

(a) MPOS $= 1.52\,\%$, $t_s^{5\%} = 0.75$ s

12.10. Find the transient response parameter and the steady state errors for the eye movement dynamic model presented in formula (1.39). Assume unity feedback.

12.11. Following the procedure of Example 12.5, approximate the higher-order system

$$H(s) = \dfrac{500(s+6)}{(s^2+10s+100)(s+25)(s+50)}$$

by its second-order reduced system. Find the maximum percent overshoot and 2% settling time for the second-order approximate system.

12.12. Assuming unity feedback, find the steady state errors for the following systems.

(a) $H_1(s) = \dfrac{s}{(s+1)(s^2+7s+10)(s+14)}$

(b) $H_2(s) = \dfrac{5(s+5)}{s(s+1)^2(s+2)(s^2+s+1)}$

(c) $H_3(s) = \dfrac{100(s+2)(s+4)}{s^2(s+3)(s+5)(s+10)}$

Answer:

(c) $K_p = \infty$, $K_v = \infty$, $K_a = 16/3$

12.13. Assuming unity feedback, find the steady state errors for the vehicle lateral error dynamics model given in formula (1.40).

12.14. Repeat Problem 12.13 for the commercial aircraft pitch angle dynamics whose model is given in (1.43).

12.15. Repeat Problem 12.13 for the ship positioning system defined in Problem 4.64.

12.16. Repeat Problem 12.13 for the voltage regulator system defined in Problem 4.65.
Answer: $e_{ss}^{step} = 0.0023$, $e_{ss}^{ramp} = \infty$, $e_{ss}^{par} = \infty$

12.17. Find the resonant frequency, peak resonance, and frequency bandwidth for the second-order closed-loop system whose transfer function is defined by

$$M(j\omega) = \frac{\omega_n^2}{(j\omega)^2 + 2\zeta\omega_n(j\omega) + \omega_n^2}$$

Answer:
$$\omega_r = \omega_n\sqrt{1-\zeta^2}, \quad M_p = \frac{1}{2\zeta\sqrt{1-\zeta^2}}$$

12.18. Sketch Bode diagrams of the following transfer functions.

(a) $H_1(s) = \dfrac{s}{(s+10)(s+100)}$

(b) $H_2(s) = \dfrac{100}{s(s^2 + 10s + 100)}$

(c) $H_3(s) = \dfrac{10(s+10)}{s(s+2)(s+50)}$

(d) $H_4(s) = \dfrac{200(s+10)}{s^2(s+100)}$

12.19. Draw the block diagram for the PID controller defined in (12.58).

12.20. Write a MATLAB program to find the transient response parameters for a general linear continuous-time time-invariant system. Using that program, find the transient response parameters for the real physical systems defined in Problems 12.13–16.

12.21. Using MATLAB, plot the step response of the second-order system considered in Example 12.4, and check the values for the corresponding transient response parameters and steady state errors.

12.22. Applying the initial value theorem of the Laplace transform, check the initial values for the impulse responses of the full- and reduced-order systems defined in Example 12.5, and conclude that they are far apart. Using MATLAB plot the impulse responses of the full- and reduced-order systems. Comment on the proximity of the impulse responses obtained.

12.23. Using MATLAB, plot Bode diagrams of the systems whose transfer functions are given in Problem 12.12, and find the corresponding phase and gain stability margins.

12.24. Using MATLAB, plot Bode diagrams of the real physical systems defined in Problems 12.13–16.

12.25. Consider the loop transfer function defined in Example 12.10. Vary the static gain K in the range from 1 to 500 with increments of 50. Using MATLAB, find the corresponding phase and gain margins. Plot the phase and gain margins obtained in terms of the Bode gain.

12.26. Consider the ship position system defined in Problem 4.64. Find (guess) the parameters for the PD controller defined in (12.54), such that the closed loop system with unity feedback (see Figure 12.24) is stable with as high as possible stability margins. Use MATLAB and Simulink to solve this problem.

12.27. Consider the voltage regulator system defined in Problem 4.65. Find (guess) the parameters for the PID controller defined in (12.56), such that the closed loop system with unity feedback (see Figure 12.24) provides overshoot less than 15%, 5% settling time less than 4s, and steady state unit step error less than 0.1. Use MATLAB and Simulink to solve this problem.

APPENDIX A

Linear Algebra

Linear algebra plays an important role in the study of linear dynamic systems and signals and their applications. The main objects in linear algebra are matrices and vectors. In this appendix, we review some basic linear algebra results used in this book. To that end, we define matrices and basic operations with matrices. Here, we present mostly the results and methodology. For details, derivations, and proofs, the interested reader is referred to standard books on linear algebra and matrices, for example [1–5].

A matrix is an array of elements that in general has m rows and n columns. It is defined as follows.

$$\mathbf{A} = \begin{bmatrix} a_{11} & a_{12} & \cdots & a_{1n} \\ a_{21} & a_{22} & \cdots & a_{2n} \\ \vdots & \vdots & \vdots & \vdots \\ a_{m1} & a_{m2} & \cdots & a_{mn} \end{bmatrix} \triangleq \{a_{ij}\}, i = 1, 2, \ldots, m; j = 1, 2, \ldots, n \quad (A.1)$$

The elements of matrix $\mathbf{A}$, denoted by a_{ij}, are most commonly represented as real and complex numbers, but in general, they can be mathematical formulas and functions. (Note that in this book we use boldfaced letters to denote matrices and vectors.) It follows that a vector is a special form of a matrix, which has either only one row ($m = 1$, row vector) or only one column ($n = 1$, column vector).

SQUARE MATRICES AND MATRIX TRACE

A square matrix has the same number of rows and columns, $m = n$. For a square matrix, we define the *matrix trace as the sum of the matrix diagonal elements,* that is,

$$\text{tr}\{\mathbf{A}\} = \sum_{i=1}^{m} a_{ii} \quad (A.2)$$

IDENTITY AND DIAGONAL MATRICES

A square matrix in which all elements are equal to zero except for the elements on the diagonal, which are equal to 1, is called an identity matrix. The identity matrix is very often used in linear algebra. Its structure is given in the following:

$$\mathbf{I}_m = \begin{bmatrix} 1 & 0 & \cdots & & \cdots & 0 \\ 0 & 1 & 0 & & \cdots & 0 \\ \vdots & 0 & \ddots & & \vdots & \vdots \\ \vdots & \vdots & \vdots & & \ddots & 0 \\ 0 & 0 & \cdots & & 0 & 1 \end{bmatrix} \quad (A.3)$$

Note that this matrix has the same number of rows and columns, which in the case of formula (A.3) is denoted by the subscript m. A matrix in which all elements are equal to zero except for those on the diagonal is called a *diagonal matrix*. Such a matrix can be represented in a simplified form as $\mathbf{D} = \text{diag}\{d_{11}, d_{22}, \ldots, d_{mm}\}$. Apparently, $\mathbf{I}_m = \text{diag}\{1, 1, \ldots, 1\}$.

Several operations can be defined with matrices.

MATRIX ADDITION (SUBTRACTION)

We can add or subtract two or more matrices that have the *same dimensions*, the same number of rows and columns. In that process, the corresponding elements of the given matrices are added (subtracted), that is,

$$\begin{aligned} \mathbf{A} \pm \mathbf{B} &= \begin{bmatrix} a_{11} & a_{12} & \cdots & a_{1n} \\ a_{21} & a_{22} & \cdots & a_{2n} \\ \vdots & \vdots & \vdots & \vdots \\ a_{m1} & a_{m2} & \cdots & a_{mn} \end{bmatrix} \pm \begin{bmatrix} b_{11} & b_{12} & \cdots & b_{1n} \\ b_{21} & b_{22} & \cdots & b_{2n} \\ \vdots & \vdots & \vdots & \vdots \\ b_{m1} & b_{m2} & \cdots & b_{mn} \end{bmatrix} \\ &= \begin{bmatrix} a_{11} \pm b_{11} & a_{12} \pm b_{12} & \cdots & a_{1n} \pm b_{1n} \\ a_{21} \pm b_{21} & a_{22} \pm b_{22} & \cdots & a_{2n} \pm b_{2n} \\ \vdots & \vdots & \vdots & \vdots \\ a_{m1} \pm b_{m1} & a_{m2} \pm b_{m2} & \cdots & a_{mn} \pm b_{mn} \end{bmatrix} \end{aligned} \quad (A.4)$$

MATRIX PRODUCT

In order to multiply two matrices, the number of rows in the first matrix must be equal to the number of columns in the second matrix. Such matrices are called *product compatible matrices*. The matrix multiplication procedure goes as follows: the first row of the first matrix is multiplied by the first column of the second matrix (first element by first element, second element by second element, and so on), and the results of products are added. The result obtained represents the one-one element of the matrix product. Next, the first row of the first matrix is multiplied by the second column of the second matrix, and the result constitutes the one-two element of the matrix product. This procedure is repeated with the first row until all elements of the first row of the matrix product are obtained. Then, the procedure is repeated with the second row of the first matrix, and the results form the second row in the matrix product. This procedure is repeated until all rows of the first matrix are multiplied by the columns of the second matrix. This matrix product mechanism can be recorded analytically as well. Let $\mathbf{A}^{m \times n} = \{a_{ij}\}, i = 1, 2, \ldots, m; j = 1, 2, \ldots, n$ and $\mathbf{B}^{n \times q} = \{b_{ij}\}, i = 1, 2, \ldots, n; j = 1, 2, \ldots, q$ be two product compatible matrices. The product of these two matrices is analytically defined by

$$\mathbf{A}^{m \times n} \mathbf{B}^{n \times q} = \mathbf{C}^{m \times q} = \{c_{ij}\}, \quad i = 1, 2, \ldots, m; \quad j = 1, 2, \ldots, q$$

$$c_{ij} = \sum_{k=1}^{n} a_{ik} b_{kj} \tag{A.5}$$

This matrix product procedure can be generalized to several matrices when all of them are mutually compatible (the number of rows of the matrix that multiplies must be equal to the number of columns of the matrix being multiplied). An interesting property of the matrix product is that it does not represent a commutative operation; that is, in general, for matrices that are product compatible in both directions, for example, $\mathbf{A}^{m \times n}$ and $\mathbf{B}^{n \times m}$, we have

$$\mathbf{A}^{m \times n} \mathbf{B}^{n \times m} \neq \mathbf{B}^{n \times m} \mathbf{A}^{m \times n} \tag{A.6}$$

This result is obvious for $m \neq n$, since the results obtained have different dimensions. Note that two matrices are identical only if they have the same dimensions and all corresponding elements are identical. The result stated in (A.6) holds in general—even in the case $m = n$. Hence, the commutativity of scalar multiplication ($ab = ba$, a, b are scalars) does not extend to matrices.

MULTIPLICATION OF A MATRIX BY A SCALAR

A matrix is multiplied by a scalar such that all matrix elements are multiplied by the given scalar. This operation can be analytically presented as

$$\alpha \mathbf{A}^{m \times n} = \{\alpha a_{ij}\}, \quad i = 1, 2, \ldots, m; j = 1, 2, \ldots, n \tag{A.7}$$

MATRIX TRANSPOSE

Matrix transposition is achieved by interchanging rows and columns, so that the matrix with m rows and n columns becomes a matrix with n rows and m columns. This operation is denoted by $\mathbf{A}^T$, and demonstrated as follows:

$$\mathbf{A}^T = \begin{bmatrix} a_{11} & a_{12} & \cdots & a_{1n} \\ a_{21} & a_{22} & \cdots & a_{2n} \\ \vdots & \vdots & \vdots & \vdots \\ a_{m1} & a_{m2} & \cdots & a_{mn} \end{bmatrix}^T = \begin{bmatrix} a_{11} & a_{21} & \cdots & a_{m1} \\ a_{12} & a_{22} & \cdots & a_{m2} \\ \vdots & \vdots & \vdots & \vdots \\ a_{1n} & a_{2n} & \cdots & a_{mn} \end{bmatrix} \tag{A.8}$$

A matrix that satisfies $\mathbf{A} = \mathbf{A}^T$ is called a *symmetric matrix*.

The following interesting property holds for the transpose of the matrix product:

$$(\mathbf{A}^{m \times n} \mathbf{B}^{n \times m})^T = (\mathbf{B}^{n \times m})^T (\mathbf{A}^{m \times n})^T \tag{A.9}$$

MATRIX DETERMINANT

The determinant is only defined for *square matrices* (which have the same number of rows and columns). The determinant of a 2×2 matrix is given by

$$\det\{\mathbf{A}^{2\times 2}\} = \det\left\{\begin{bmatrix} a_{11} & a_{12} \\ a_{21} & a_{22} \end{bmatrix}\right\} \triangleq \begin{vmatrix} a_{11} & a_{12} \\ a_{21} & a_{22} \end{vmatrix} = a_{11}a_{22} - a_{12}a_{21} \quad (A.10)$$

The determinant of a 3×3 matrix is obtained as follows:

$$\det\{\mathbf{A}^{3\times 3}\} = \det\left\{\begin{bmatrix} a_{11} & a_{12} & a_{13} \\ a_{21} & a_{22} & a_{23} \\ a_{31} & a_{32} & a_{33} \end{bmatrix}\right\} \triangleq \begin{vmatrix} a_{11} & a_{12} & a_{13} \\ a_{21} & a_{22} & a_{23} \\ a_{31} & a_{32} & a_{33} \end{vmatrix}$$

$$= (-1)^{1+1} a_{11} \begin{vmatrix} a_{22} & a_{23} \\ a_{32} & a_{33} \end{vmatrix} + (-1)^{1+2} a_{12} \begin{vmatrix} a_{21} & a_{23} \\ a_{31} & a_{33} \end{vmatrix}$$

$$+ (-1)^{1+3} a_{13} \begin{vmatrix} a_{21} & a_{22} \\ a_{31} & a_{32} \end{vmatrix} \triangleq a_{11}|\mathbf{C}_{11}| - a_{12}|\mathbf{C}_{12}| + a_{13}|\mathbf{C}_{13}|$$

$$= a_{11}(a_{22}a_{33} - a_{23}a_{32}) - a_{12}(a_{21}a_{33} - a_{23}a_{31}) + a_{13}(a_{21}a_{32} - a_{22}a_{31})$$
$$(A.11)$$

The submatrices $\mathbf{C}_{11}, \mathbf{C}_{12}, \mathbf{C}_{13}$ are obtained from matrix $\mathbf{A}$ by omitting the first row and the corresponding column. Similarly, we can expand the determinant of a fourth-order square matrix into four determinants of third-order matrices, and evaluate determinants of the third-order matrices as before in terms of second-order matrices.

The procedure of expanding the determinant of higher-order matrices can be generalized to a square matrix of an arbitrary order, say $m \times m$. In that process, we define the so-called *cofactors* of the matrix element a_{ij} by $(-1)^{i+j}|\mathbf{C}_{ij}|$, where the submatrix $\mathbf{C}_{ij}$ of order $(m-1) \times (m-1)$ is obtained from matrix $\mathbf{A}$ by omitting the ith row and the jth column. Using the definition of cofactors, the determinant of a matrix of order $m \times m$ can be expressed in terms of its cofactors with respect to the first row [4] as follows:

$$|\mathbf{A}| = a_{11}(-1)^{1+1}|\mathbf{C}_{11}| + a_{12}(-1)^{1+2}|\mathbf{C}_{12}| + \cdots$$
$$+ a_{1m-1}(-1)^m|\mathbf{C}_{1m-1}| + a_{1m}(-1)^{m+1}|\mathbf{C}_{1m}| \quad (A.12)$$

Similarly, the determinant of an $m \times m$ matrix can be expressed in terms of its cofactors with respect to any row or any column [4], which in the case of a row expansion can be written as

$$|\mathbf{A}| = \sum_{j=1}^{m} a_{ij}(-1)^{i+j}|\mathbf{C}_{ij}|, \quad i = 1, 2, \ldots, m \quad (A.13)$$

Note that the cofactors are obtained by evaluating determinants of m matrices of order $(m-1) \times (m-1)$.

Determinant of Matrix Product

The following results hold for the determinant of a matrix product

$$\det\{\mathbf{A}_1\mathbf{A}_2\} = \det\{\mathbf{A}_1\}\det\{\mathbf{A}_2\} \quad (A.14)$$

For the proof of this statement the reader is referred to [2]. This result can be generalized to the product of a finite number of matrices.

MATRIX INVERSION

Matrix inversion is defined only for square matrices. It comes originally from the problem of solving systems of linear algebraic equations. Along those lines, the unique solution of the algebraic equation $\mathbf{A}^{m \times m}\mathbf{x} = \mathbf{b}$ is denoted by $\mathbf{x} = \mathbf{A}^{-1}\mathbf{b}$, where $\mathbf{A}^{-1}$ represents the inversion of the matrix $\mathbf{A}$. Hence, $\mathbf{A}^{-1}$ is obtained in the process of solving the system of linear algebraic equations, which can be done using the Gaussian elimination method or any other method for solving systems of linear algebraic equations.

The matrix inversion can also be analytically defined in terms of the *adjoint matrix* (also called the adjugate matrix) of $\mathbf{A}$, defined by

$$\text{adj}\{\mathbf{A}\} = \begin{bmatrix} (-1)^{1+1}|\mathbf{C}_{11}| & (-1)^{1+2}|\mathbf{C}_{12}| & \cdots & (-1)^{1+m}|\mathbf{C}_{1m}| \\ (-1)^{2+1}|\mathbf{C}_{21}| & (-1)^{2+2}|\mathbf{C}_{22}| & \cdots & (-1)^{2+m}|\mathbf{C}_{2m}| \\ \vdots & \vdots & \vdots & \vdots \\ (-1)^{m+1}|\mathbf{C}_{m1}| & (-1)^{m+2}|\mathbf{C}_{m2}| & \cdots & (-1)^{m+m}|\mathbf{C}_{mm}| \end{bmatrix}^T \qquad (A.15)$$

It follows from (A.15) that the *adjoint matrix of $\mathbf{A}$ is obtained by replacing every element in $\mathbf{A}$ by its cofactor and transposing*. Using the notions of the matrix determinant and the adjoint matrix, it can be shown using (A.13) that the matrix inversion is given by

$$\mathbf{A}^{-1} = \frac{1}{|\mathbf{A}|}\text{adj}\{\mathbf{A}\} \qquad (A.16)$$

The matrix inverse has the following property:

$$\mathbf{A}^{-1}\mathbf{A} = \mathbf{A}\mathbf{A}^{-1} = \mathbf{I}_m \qquad (A.17)$$

Determinant of Matrix Inversion

By using the rule for the determinant of a product, we can establish the following formula:

$$\det\{\mathbf{A}^{-1}\} = \frac{1}{\det\{\mathbf{A}\}} \qquad (A.18)$$

This can be proved as follows.

$$\det\{\mathbf{A}\mathbf{A}^{-1}\} = \det\{\mathbf{I}\} = 1 = \det\{\mathbf{A}\}\det\{\mathbf{A}^{-1}\} \Rightarrow \det\{\mathbf{A}^{-1}\} = \frac{1}{\det\{\mathbf{A}\}} \qquad (A.19)$$

Matrix Product Inversion

Consider the problem of finding the matrix inversion of $\mathbf{A}_1\mathbf{A}_2$. The inversion is a matrix whose product with $\mathbf{A}_1\mathbf{A}_2$ produces an identity matrix, that is

$$\mathbf{A}_1\mathbf{A}_2\,[\text{Inverse}] = \mathbf{I} \qquad (A.20)$$

It can be checked that the inverse of the form $\mathbf{A}_2^{-1}\mathbf{A}_1^{-1}$ satisfies (A.20). This inversion of a product formula can be easily generalized to the product of a finite number of matrices.

SYSTEMS OF LINEAR ALGEBRAIC EQUATIONS

We first introduce some terminology used often in linear algebra, especially in systems of linear algebraic equations.

Independent Vectors

A set of vectors $\mathbf{x}_1, \mathbf{x}_2, \ldots, \mathbf{x}_n$ is *linearly independent* if no scalars $\alpha_1, \alpha_2, \ldots, \alpha_n$ (not all of them equal to zero) can be found such that

$$\alpha_1 \mathbf{x}_1 + \alpha_2 \mathbf{x}_2 + \cdots + \alpha_n \mathbf{x}_n = \mathbf{0} \tag{A.21}$$

If such a set of α_i exists, then one of the vectors $\mathbf{x}_1, \mathbf{x}_2, \ldots, \mathbf{x}_n$ can be expressed in terms of the remaining $n - 1$ vectors and the vectors $\mathbf{x}_1, \mathbf{x}_2, \ldots, \mathbf{x}_n$ will be *linearly dependent*.

Matrix Rank

The rank of a matrix is defined by the *number of linearly independent rows or the number of linearly independent columns*. That *the matrix row rank is equal to its column rank* is one of the fundamental theorems of linear algebra.

Consider a solvable system of linear algebraic equations in n unknowns,

$$\mathbf{A}\mathbf{x} = \mathbf{b} \tag{A.22}$$

with dim$\{\mathbf{A}\} = m \times n$. Equation (A.22) has a solution if and only if

$$\text{rank}\{[\mathbf{A} \quad \mathbf{b}]\} = \text{rank}\{\mathbf{A}\} \tag{A.23}$$

In addition, if rank$\{\mathbf{A}\} = m$, then (A.22) always has a solution. For $n = m$ and rank$\{\mathbf{A}\} = m$ the solution obtained is unique, and $\mathbf{A}$ is a square invertible matrix (nonsingular matrix whose det$(\mathbf{A}) \neq 0$). In such a case, $\mathbf{x} = \mathbf{A}^{-1}\mathbf{b}$. It follows also that $\mathbf{A}^{m \times m}\mathbf{x} = 0$ has a solution $\mathbf{x} \neq 0$ only if the matrix $\mathbf{A}$ has no full rank ($\mathbf{A}$ is singular (det$(\mathbf{A}) = 0$) and noninvertible), otherwise $\mathbf{x} = 0$.

Matrix Null Space

The null space of a matrix $\mathbf{A}$ of dimensions $m \times n$ is the space spanned by vectors $\mathbf{v}$ that satisfy $\mathbf{A}\mathbf{v} = \mathbf{0}$. If the matrix $\mathbf{A}$ has full rank, then the null space is empty.

Matrix Range Space

The range space of a matrix $\mathbf{A}$ of dimensions $m \times n$ is the space spanned by vectors $\mathbf{y}$ that satisfy $\mathbf{A}\mathbf{v} = \mathbf{y}$ for all n-dimensional vectors $\mathbf{v}$. The range space of $\mathbf{A}$ is in fact the space spanned by the columns of matrix $\mathbf{A}$.

EIGENVALUES AND EIGENVECTORS

Due to their importance for the state space representation of linear systems, these are treated in Section 8.4. Here, we state an important theorem about the eigenvalues.

SPECTRAL THEOREM If $\mathbf{A}$ is a symmetric matrix, then its eigenvalues are real and $\mathbf{A}$ is diagonalizable, that is, there exists a similarity transformation $\mathbf{P}$ such that $\mathbf{P}^{-1}\mathbf{A}\mathbf{P}$ is diagonal. Furthermore, the transformation is unitary, that is $\mathbf{P}^{-1} = \mathbf{P}^T$.
$\square$

Proof of this theorem can be found in many standard books on linear algebra and matrices (see, for example, [4]).

MATRIX INTEGRATION AND DIFFERENTIATION

If the integral is to be applied to a matrix it must be applied to all of its elements. Similarly, the derivative of a matrix requires that the derivative be applied to each matrix element. Hence, we have

$$\int \mathbf{A}(t)dt = \left\{ \int a_{ij}(t)dt \right\}, i = 1, 2, \ldots, m; j = 1, 2, \ldots, n \quad (A.24)$$

$$\frac{d}{dt}\mathbf{A}(t) = \left\{ \frac{d}{dt}a_{ij}(t) \right\}, i = 1, 2, \ldots, m; j = 1, 2, \ldots, n \quad (A.25)$$

MATRIX EXPONENTIAL

The matrix exponential $e^{\mathbf{A}t}$ is defined and studied in detail in Section 8.2, due to its importance to linear system state space formulation. Note that the matrix exponential $e^{\mathbf{A}t}$ represents the transition matrix of a linear continuous-time system.

INTEGRAL OF A MATRIX EXPONENTIAL

The following matrix integral formula is useful in some applications:

$$\int_0^T e^{\mathbf{A}t}dt = (e^{\mathbf{A}T} - I)\mathbf{A}^{-1} \quad (A.26)$$

provided that the matrix $\mathbf{A}$ is nonsingular. In addition, if all eigenvalues of matrix $\mathbf{A}$ are asymptotically stable, then

$$\int_0^\infty e^{\mathbf{A}t}dt = -\mathbf{A}^{-1} \quad (A.27)$$

PROOF OF THE CAYLEY-HAMILTON THEOREM

The proof of this famous theorem is fairly simple. It does not go beyond the material presented in this appendix. Consider a square matrix $\mathbf{A}$ of dimensions $n \times n$. Its characteristic equation is defined by

$$\Delta_\mathbf{A}(\lambda) = \lambda^n + a_{n-1}\lambda^{n-1} + \cdots + \lambda a_1 + a_0 = 0 \quad (A.28)$$

The Cayley-Hamilton theorem states that

$$\Delta_\mathbf{A}(\mathbf{A}) = \mathbf{A}^n + a_{n-1}\mathbf{A}^{n-1} + \cdots + a_1\mathbf{A} + a_0\mathbf{I} = \mathbf{0} \qquad (A.29)$$

Let us start with the definition of the inverse of the matrix $\lambda\mathbf{I} - \mathbf{A}$, that is,

$$(\lambda\mathbf{I} - \mathbf{A})^{-1} = \frac{1}{\det(\lambda\mathbf{I} - \mathbf{A})} \mathrm{adj}\{(\lambda\mathbf{I} - \mathbf{A})\} = \frac{1}{\Delta_\mathbf{A}(\lambda)} \mathbf{B}(\lambda)$$

or

$$\Delta_\mathbf{A}(\lambda)\mathbf{I} = (\lambda\mathbf{I} - \mathbf{A})\mathbf{B}(\lambda)$$

Note that since the adjoint matrix is obtained by omitting a row and a column in matrix $\mathbf{A}$ and finding the determinant of the corresponding $(n-1) \times (n-1)$ submatrix, the highest power of λ in $\mathbf{B}(\lambda)$ is $n-1$. The matrix $\mathbf{B}(\lambda)$ can be represented by

$$\mathbf{B}(\lambda) = \mathbf{B}_{n-1}\lambda^{n-1} + \mathbf{B}_{n-2}\lambda^{n-2} + \cdots + \mathbf{B}_1\lambda + \mathbf{B}_0$$

Hence, we have

$$(\lambda^n + a_{n-1}\lambda^{n-1} + \cdots + a_1\lambda + a_0)\mathbf{I}$$
$$= (\lambda\mathbf{I} - \mathbf{A})(\mathbf{B}_{n-1}\lambda^{n-1} + \mathbf{B}_{n-2}\lambda^{n-2} + \cdots + \mathbf{B}_1\lambda + \mathbf{B}_0)$$

Equating coefficients with respect to the same power of λ, we obtain

$$\mathbf{B}_{n-1} = \mathbf{I}$$
$$\mathbf{B}_{n-2} - \mathbf{A}\mathbf{B}_{n-1} = a_{n-1}\mathbf{I} \Rightarrow \mathbf{B}_{n-2} = \mathbf{A}\mathbf{B}_{n-1} + a_{n-1}\mathbf{I} = \mathbf{A} + a_{n-1}\mathbf{I}$$
$$\mathbf{B}_{n-3} - \mathbf{A}\mathbf{B}_{n-2} = a_{n-2}\mathbf{I} \Rightarrow \mathbf{B}_{n-1} = \mathbf{A}\mathbf{B}_{n-2} + a_{n-2}\mathbf{I} = \mathbf{A}^2 + a_{n-1}\mathbf{A} + a_{n-2}\mathbf{I}$$
$$\vdots$$
$$\mathbf{B}_1 - \mathbf{A}\mathbf{B}_2 = a_1\mathbf{I} \Rightarrow \mathbf{B}_1 = \mathbf{A}\mathbf{B}_2 + a_1\mathbf{I}$$
$$-\mathbf{A}\mathbf{B}_1 = a_0\mathbf{I} \Rightarrow \mathbf{0} = \mathbf{A}\mathbf{B}_1 + a_0\mathbf{I}$$

Eliminating $\mathbf{B}_i$ starting from the top equation, as demonstrated in the top three equations, the last equation will produce

$$\mathbf{0} = \mathbf{A}^n + a_{n-1}\mathbf{A}^{n-1} + \cdots + a_1\mathbf{A} + a_0\mathbf{I}$$

which is the statement of the Cayley-Hamilton theorem.

REFERENCES

[1] J. Fraleigh and R. Beauregard, *Linear Algebra*, Addison-Wesley, Reading, MA, 1990.

[2] G. Stewart, *Introduction to Matrix Computations*, Academic Press, New York, 1973.

[3] G. Strang, *Linear Algebra and Its Applications*, Saunders College Publishing, Orlando, FL, 1988.

[4] P. Lancaster and M. Tismenetsky, *Theory of Matrices*, Academic Press, Orlando, FL, 1985.

[5] G. Golub and C. Van Loan, *Matrix Computations*, John Hopkins University Press, Baltimore, MD, 1996.

APPENDIX B

Some Results from Calculus

In this appendix, we present some results from calculus that are used in the text, or that must be used while solving problems defined in the text.

STANDARD TRIGONOMETRIC FORMULAS

$$\begin{aligned}
\sin(\alpha \pm \beta) &= \sin(\alpha)\cos(\beta) \pm \cos(\alpha)\sin(\beta) \\
\cos(\alpha \pm \beta) &= \cos(\alpha)\cos(\beta) \mp \sin(\alpha)\sin(\beta) \\
\sin(2\alpha) &= 2\sin(\alpha)\cos(\alpha) \\
\cos(2\alpha) &= \cos^2(\alpha) - \sin^2(\alpha)
\end{aligned} \qquad (B.1)$$

BASICS OF COMPLEX NUMBERS

A complex number, its complex conjugate pair, and the complex number magnitude (modulo) and phase (angle) are defined as follows:

$$\begin{aligned}
z &= a + jb, \quad z^* = a - jb, \quad j = \sqrt{-1}, \quad a, b \text{ real numbers} \\
|z| &= zz^* = \sqrt{a^2 + b^2}, \quad \angle z = \tan^{-1}\left(\frac{b}{a}\right), \quad z = |z|\angle z
\end{aligned} \qquad (B.2)$$

Sum, product, and ratio of two complex numbers are defined by

$$\begin{aligned}
z_1 &= a_1 + jb_1 = |z_1|\angle z_1, \quad z_2 = a_2 + jb_2 = |z_2|\angle z_2 \\
z_1 \pm z_2 &= (a_1 \pm a_2) + j(b_1 \pm b_2) \\
z_1 z_2 &= |z_1||z_2|(\angle z_1 + \angle z_2) \\
\frac{z_1}{z_2} &= \frac{|z_1|}{|z_2|}(\angle z_1 - \angle z_2)
\end{aligned} \qquad (B.3)$$

SUMMATION FORMULAS

$$\sum_{m=0}^{\infty} a^m = \frac{1}{1-a}, \quad |a| < 1; \quad \sum_{m=0}^{k} a^m = \frac{1 - a^{k+1}}{1 - a} \qquad (B.4)$$

Appendix B Some Results from Calculus

$$\sum_{m=0}^{k} ma^m = \frac{a}{(1-a)^2}[1 - (k+1)a^k + ka^{k+1}], \quad a \neq 1 \tag{B.5}$$

$$\Rightarrow \sum_{m=0}^{\infty} ma^m = \frac{a}{(1-a)^2}, \quad |a| < 1$$

$$\sum_{m=0}^{\infty} m^2 a^m = \frac{a(1+a)}{(1-a)^3}, \quad |a| < 1 \tag{B.6}$$

RESULTS ABOUT INTEGRALS

Integration by parts formula:

$$\int_a^b u\,dv = uv\Big|_a^b - \int_a^b v\,du \tag{B.7}$$

Some indefinite integrals often encountered in linear systems and signals:

$$\int te^{at}\,dt = \frac{1}{a}e^{at}\left(t - \frac{1}{a}\right) \tag{B.8}$$

$$\int t^2 e^{at}\,dt = \frac{1}{a}e^{at}\left(t^2 - \frac{2}{a}t + \frac{2}{a^2}\right) \tag{B.9}$$

$$\int t\sin(bt)\,dt = \frac{1}{b^2}\sin(bt) - \frac{1}{b}t\cos(bt) \tag{B.10}$$

$$\int t\cos(bt)\,dt = \frac{1}{b^2}\cos(bt) + \frac{1}{b}t\sin(bt) \tag{B.11}$$

$$\int e^{at}\sin(bt)\,dt = \frac{1}{a^2+b^2}e^{at}(a\sin(bt) - b\cos(bt)) \tag{B.12}$$

$$\int e^{at}\cos(bt)\,dt = \frac{1}{a^2+b^2}e^{at}(a\cos(bt) + b\sin(bt)) \tag{B.13}$$

Leibniz Formula

This formula presents the rule for differentiation of an integral with respect to a parameter α, where the parameter α appears in both the integration limits and in the integrand [1].

$$\frac{d}{d\alpha}\left\{\int_{L(\alpha)}^{U(\alpha)} f(t,\alpha)\,dt\right\} = \int_{L(\alpha)}^{U(\alpha)} \frac{\partial}{\partial \alpha}\{f(t,\alpha)\}\,dt \tag{B.14}$$

$$+ f(U(\alpha),\alpha)\frac{dU(\alpha)}{d\alpha} - f(L(\alpha),\alpha)\frac{dL(\alpha)}{d\alpha}$$

This formula is valid even in the case when the integrand is a matrix function, thus it can be used in derivations of some state space results.

DERIVATIVE RESULT

Differentiation of a ratio of two functions:

$$\frac{d}{dt}\left(\frac{u(t)}{v(t)}\right) = \frac{1}{v^2(t)}\left(v(t)\frac{du(t)}{dt} - u(t)\frac{dv(t)}{dt}\right) \qquad (B.15)$$

CAUCHY-SCHWARZ INEQUALITIES

Continuous-time formula [2]:

$$\left|\int_a^b x(t)y(t)\,dt\right|^2 \leq \left(\int_a^b |x(t)|^2\,dt\right)\left(\int_a^b |y(t)|^2\,dt\right) \qquad (B.16)$$

Discrete-time formula [2, 3]:

$$\left|\sum_{k=1}^n x_k y_k\right|^2 \leq \left(\sum_{k=1}^n |x_k|^2\right)\left(\sum_{k=1}^n |y_k|^2\right) \qquad (B.17)$$

Vector formula:

$$|\mathbf{x}(\mathbf{y}^*)^T|^2 \leq \|\mathbf{x}\|^2 \|\mathbf{y}\|^2, \qquad \|\mathbf{x}\|^2 = \mathbf{x}(\mathbf{x}^*)^T, \qquad \|\mathbf{y}\|^2 = \mathbf{y}(\mathbf{y}^*)^T \qquad (B.18)$$

where $\mathbf{x}(\mathbf{y}^*)^T$ defines the inner product of the vectors $\mathbf{x}$ and $\mathbf{y}$, and $\|\ \|$ stands for the vector Euclidean norm.

The proofs of (B.16) and (B.17) can be found in [4].

REFERENCES

[1] M. Greenberg, *Fundamentals of Applied Mathematics,* Prentice Hall, Englewood Cliffs, NJ, 1978.

[2] A. Papoulis, *Signal Analysis,* McGraw-Hill, New York, 1977.

[3] C. Wylie, *Advanced Engineering Mathematics,* McGraw-Hill, New York, 1995.

[4] M. O'Flynn and E. Moriarty, *Linear Systems: Time Domain and Transform Analysis,* Wiley, New York, 1987.

APPENDIX C

Introduction to MATLAB[†]

In this appendix, we present MATLAB functions that are related to the basic mathematical operations and to procedures for the input of data.

BASIC MATLAB FUNCTIONS

The Help Facility

MATLAB has an extensive built-in help system. If help is required for any *function*, simply type `help` followed by the name of the function in question. MATLAB displays a brief text indicating the use of the function and examples of its usage. For example, `help eig` provides information on the use of the eigenvalue function. A Help facility is also available, providing on-line information on most MATLAB *topics*. To get a list of Help topics type `help`. To get Help on a specific topic, type `help` followed by the topic.

MATLAB was written so that the appearance of the notation is much the same as if you were writing on paper. For example, to find the solution to 21/3, you would type

```
>>21/3
```

and MATLAB would return

```
ans =
     7
```

Note that the double right caret, >>, indicates that MATLAB is ready for your input. (Also note that comments in MATLAB are preceded by a percent sign.) It is important to point out that no variable is either defined or used in the preceding operation. To assign a value

---

[†]This appendix is adopted from Z. Gajić and M. Lelić, *Modern Control Systems Engineering,* Prentice Hall International, London, 1996.

to the variable x, you use the assignment operator "=". For example,

```
>>x=21/3
x=
        7
```

Now the variable x has been assigned a value.

There are two things to remember in this context. The first is that x will hold its assigned value until it is changed or cleared, or the MATLAB session is terminated. Typing the command who causes MATLAB to show all of the variables that have been declared thus far. The command whos lists all variables that have been declared plus additional information (such as the type and size of the variables, and the total amount of RAM consumed due to all of the variables that have been assigned). Issuing the `clear` command purges all declared variables from memory. The second thing to remember is that MATLAB is by default case sensitive; that is, x does not equal X.

To see the present value of a variable, simply type the variable name and press return.

```
>>x
x=
        7
```

Basic Mathematics Operators

+ addition
− subtraction
* multiplication
/ division
^ power

Brackets are reserved for the identification of vectors and matrices. One of the special operators is the semicolon. Issuing the semicolon at the end of a line prevents the result from being displayed. (In the preceding examples, without the semicolon, the results were immediately displayed.) The advantages of this will become obvious when writing and debugging scripts.

```
>>x^2;
>>% does not display the result
>>x
x=
        49
```

The concatenation function, for continuing equations that will not fit on one line, is represented by an ellipsis (three dots):

```
>>x=1+2+3+4...
+5+6
```

returns

```
x=
    21
```

Numbers and Arithmetic Expressions

Conventional decimal notation, with optional decimal point and leading minus sign, is used for numbers. A power-of-ten scale factor can be included as a suffix. For example,

```
3          -99         0.0001
9.64595    1.606E-20   6.066e23
```

MATLAB includes several predefined variables, including `i`, `j`, `pi`, `inf`, and `NaN`. The `i` and `j` variables are the square root of –1, `pi` is π, `inf` is ∞, and `NaN` is not a number (e.g., division by zero). MATLAB will return an `NaN` error message if a division by zero occurs.

VECTORS AND MATRICES

Entering a vector or matrix is a painless operation. Simply use brackets to delimit the elements of the vector, separated by at least one space. Do not use a comma to separate the elements. For example

```
>>x=[1 2 3 4]
```

results in the output

```
x =
    1   2   3   4
```

Entering matrices is basically the same, except that each row is delimited by a semicolon.

```
>>y=[1 2;3 4]
y =
    1   2
    3   4
```

A matrix may also be entered using carriage returns rather than semicolons:

```
>>y=[1 2
     3 4]
y =
     1   2
     3   4
```

The expression

```
>>y(2,2)
```

results in

```
ans=
     4
```

Using the parenthesis in this fashion allows the user to access any element in the matrix—in this case, the element in the second row and second column.

MATRIX OPERATIONS

Transpose

The special character ' (apostrophe) denotes the transpose of a matrix. The statements

```
>>A=[1 2 3;4 5 6;7 8 0]
>>B=A'
```

result in

```
A =
     1   2   3
     4   5   6
     7   8   0
B =
     1   4   7
     2   5   8
     3   6   0
```

Matrix Addition and Subtraction

Addition and subtraction of matrices are denoted by $+$ and $-$. For example, with the preceding matrices, the statement

```
>>C=A+B
```

results in

```
C =
        2    6   10
        6   10   14
       10   14    0
```

Matrix Multiplication

Multiplication is denoted by *. For example, the statement

```
>>C=A*B
```

multiplies matrices **A** and **B** and stores the obtained result in matrix **C**.

Matrix Powers

The expression A^p raises **A** to the pth power, and is defined if **A** is a square matrix and p is a scalar.

Eigenvalues and Eigenvectors

If **A** is an $n \times n$ matrix, the n scalars λ that satisfy $\mathbf{Av} = \lambda \mathbf{v}$ are the eigenvalues of **A**. They are found by using the function eig. For example,

```
>>A=[2 1;0 3]
>>eig(A)
```

produces

```
ans =
        2
        3
```

Eigenvectors are obtained with the statement

```
>>[V,D]=eig(A)
```

in which case the diagonal elements of **D** are the eigenvalues and the columns of **V** are the corresponding eigenvectors.

Characteristic Polynomial

The coefficients of the characteristic polynomial of the matrix **A** are obtained by using the function `poly`. The characteristic equation, a polynomial equation, is solved by using the function `roots`. For example, the statement

```
>>poly(A)
```

for the matrix **A** given by

```
>>A=[1 2 3;4 5 6;7 8 0]
```

produces

```
p =
     1  -6  -72  -27
```

The characteristic polynomial is given by $s^3 - 6s^2 - 72s - 27$. The roots of the characteristic equations are obtained using the function `roots` as

```
>>r=roots(p)
```

producing

```
r =
    12.1229
    -5.7345
    -0.3884
```

POLYNOMIALS

Entering polynomials is as simple as entering a vector. The polynomial $h(s) = 5s^4 + 10s^2 + 18s + 23$ would be entered starting with the highest order first, as follows:

```
>>h=[5 0 10 18 23];
```

We have already seen that the function `roots(h)` finds all solutions of the polynomial equation $h(s) = 0$. The other useful MATLAB functions dealing with polynomials follow.

1. `polyval(h,10)`
   evaluates the polynomial $h(s)$ at $s = 10$
2. `[r,p,k]=residue(a,b)`
   performs partial fraction expansion, where **a** = numerator, **b** = denominator, **r** = residues, **p** = poles, and k = direct (constant) term
3. `c=conv(a,b)`
   multiplies polynomials $a(s)$ and $b(s)$—note that the vector **c** contains coefficients of the polynomial $c(s)$ in descending order
4. `[q,r]=deconv(c,a)`
   divides polynomial $c(s)$ by $a(s)$, with the quotient given by $q(s)$ and remainder by $r(s)$

For example,

```
>>p=[1 5 6];
>>polyval(p,1)
```

evaluates the polynomial $s^2 + 5s + 6$ at $s = 1$, producing

```
ans=
      12
```

and

```
>>r=roots(p)
```

finds roots of $p(s) = 0$ as

```
r=
    -3
    -2
```

In the following, two polynomials are multiplied using the convolution function `conv`:

```
>>a=[1 2 3];
>>b=[4 5 6];
>>c=conv(a,b)
c=
      4 13 28 27 18
```

PLOTS

The `plot` command creates linear *x-y* plots. If **y** is a vector, `plot(y)` produces a linear plot of the elements of **y**. Notice that the data are autoscaled and that *x-y* axes are drawn.

LOOPS

MATLAB provides several methods of looping. The commands `for`, `while`, and `elseif` are the most useful. Each requires an `end` statement at the end of the loop. They are used much the same as in any programming language, with the exception of incrementing, which is done using the colon. Two examples of these methods follow.

```
>>for k=1:2:10
for m=1:5
if k==m
A(k,m)=2;
elseif abs(k-m)==1
A(k,m)=-1;
else
A(k,m)=0;
end
end
end

>>count=0
>>n=10
>>while (n-1)>=2
count=count+1;
n=n-1;
end
```

SCRIPT FILES

Script files (subsequently referred to as m-files) are simply text files containing all of the code necessary to perform some function. The name m-files comes from the fact that the extension for all scripts must be "m," that is, "script.m" is an acceptable filename. An m-file can also act as a function, where arguments are passed from MATLAB to the m-file and processed. Also, m-files may be written to execute interactively, prompting the user for information or data. MATLAB was written in C and uses many of the I/O functions found in C, increasing its flexibility even more.

(Normally, when writing an m-file, several lines of comments are placed at the top of the file to indicate the purpose and use of the file. The help function can access these lines and display them as it would any of the other functions in the system. So, long after you have forgotten how to use a particular m-file, simply type `help filename` (no extension is required) and those comments at the top of the file will be displayed.)

Take as an example an m-file that finds the poles and zeros, and evaluates $H(s)$ at $s = -10$, where $H(s)$ is given by

$$H(s) = \frac{s^3 + 6.4s^2 + 11.29s + 6.76}{s^4 + 14s^3 + 46s^2 + 64s + 40}$$

The filename is sample1.m, and its contents read as follows.

```
num=[1 6.4 11.29 6.76];
den=[1 14 46 64 40];
pole=roots(den);
zero=roots(num);
value=(polyval(num,-10))/(polyval(den,-10));
```

To execute this m-file, type the filename without the extension,

```
>>sample1
```

and the file will run by itself.

SOME MATLAB FUNCTIONS USED IN LINEAR SYSTEMS

Some of the system/signal MATLAB functions are described in the following list.

1. `[num,den]=ss2tf(A,B,C,D)`
   finds the system transfer function from its state space form
2. `[A,B,C,D]=tf2ss(num,den)`
   state space form of the transfer function

3. [z,p,k]=ss2zp(A,B,C,D,in)
gives the factored expression for the transfer function

$$H(s) = \frac{k(s+z_1)(s+z_2)\cdots(s+z_m)}{(s+p_1)(s+p_2)\cdots(s+p_n)}$$

4. [A,B,C,D]=zp2ss(z,p,k)
5. [z,p,k]=tf2zp(num,den)
6. [num,den]=zp2tf(z,p,k)
(Note: functions 4–6 are self-explanatory.)
7. [y,x]=impulse(A,B,C,D)
the impulse response; also,
[y,x]=impulse(num,den)
8. [y,x]=step(A,B,C,D)
the step response; also,
[y,x]=step(num,den)
9. [y,x]=lsim(A,B,C,D,f,t)
the system response during the time interval up to $t = l \times \Delta t$, due to arbitrary inputs whose values at the discrete time instants l are defined in the matrix $\mathbf{F} \in R^{l \times r}$; also,
[y,x]=lsim(A,B,C,D,f,t,x0)
where $\mathbf{x}_0$ stands for the system initial condition
10. [Ad,Bd]=c2d(A,B,T)
from the continuous-time linear system to the discrete-time linear system, where T stands for the sampling period
11. plot(r,'-')
plotting function
12. [mag,phase]=bode(num,den,w)
Bode diagram, where mag, phase, and w represent magnitude, phase, and frequency, respectively
13. [Gm,Ph,wcp,wcg]=margin(num,den)
the phase and gain stability margins, and the corresponding crossover frequencies

(Note that the functions impulse, step, and lsim have the prefix d when applied to discrete-time domain systems.)

Detailed coverage of MATLAB functions can be found in [1]. The use of MATLAB in linear systems and signals has been explained throughout this text. Also, [2] and [3] give a broad coverage of the use of MATLAB in linear systems and signals.

A summary of the main linear algebra MATLAB functions used in linear systems is presented in Table C.1.

TABLE C.1 Linear Algebra Functions

`B=A'`	Matrix transpose
`C=A+B`	Matrix addition (subtraction)
`C=A*B`	Matrix multiplication
`expm(A)`	Matrix exponent of **A** (i.e., $e^{\mathbf{A}}$)
`inv(A)`	Matrix inversion of **A**
`eig(A)`	Eigenvalues of **A**
`[X,D]=eig(A)`	Eigenvectors and eigenvalues of **A**
`rank(A)`	Rank of **A**
`p=poly(A)`	Characteristic polynomial of **A**
`r=roots(p)`	Root of the polynomial equation
`det(A)`	Determinant of **A**

REFERENCES

[1] D. Hanselman and B. Littlefield, *Matering MATLAB 6: A Comprehensive Tutorial and Reference,* Prentice Hall, Upper Saddle River, NJ, 2001.

[2] R. Strum and D. Kirk, *Contemporary Linear Systems: Using MATLAB,* Brooks/Cole, Pacific Grove, CA, 2000.

[3] T. ElAli and M. Karim, *Continuous Signals and Systems with MATLAB,* CRC Press, Boca Raton, FL, 2001.

APPENDIX D

Introduction to Simulink

We have shown that Simulink can be used for the analysis of linear continuous- and discrete-time dynamic systems, and demonstrated its use in several examples. Here, we give a brief introduction to Simulink and the blocks that it uses to built dynamic system block diagrams. Simulink operates in the MATLAB environment and can be accessed from MATLAB by typing `simulink`.

Simulink has several libraries, from which we take blocks (icons) to build a block diagram. In addition to the libraries described in the following paragraphs, Simulink contains other libraries not important for the purpose of this course.

The *Continuous* library contains the integrator, derivative, state space, transfer function, and zero-pole (transfer function) blocks, alongwith others not very important for us.

The *Discrete* library contains the discrete transfer function, discrete zero-pole (transfer function), unit delay ($1/z$ element), discrete state space, zero-order hold, discrete filter blocks, among others.

The *Math* (mathematics) library contains the sum, min, max, product, dot product, gain, math function, trigonometric function, and absolute value blocks, among many others.

The *Source* library contains the clock, constant, step, ramp, sine wave, pulse generator, signal generator, from file, and from workspace blocks, among other blocks for sources of input signals.

The *Sink* library contains the scope, display, to file, and to workspace blocks.

Drawing a block diagram using blocks from the libraries is simple, and is based on several very simple rules given below.

BASIC RULES FOR BUILDING BLOCK DIAGRAMS

A library is opened by clicking the *left mouse button once* on the given library. All icons in the given library will be displayed. We can drag any icon (block) to the working window by pointing the arrow at the icon, and *holding the left mouse button while dragging* the icon to the desired position in the working window. After the mouse button is released, the icon will stay stationary at the given position. After collecting all blocks and positioning them in the working window according to the structure of the block diagram, we must connect them by lines (to indicate the flow of signals). The lines are drawn using the left mouse button: the arrow is positioned at the output of a block, and (holding the left mouse botton) the line is drawn to the input of the next block. To start a new line at a point on an existing

634 Appendix D   Introduction to Simulink

line, the arrow is brought to that point and the *right mouse button is used to start the new line* (the only operation involving the right mouse button). By releasing the right mouse button, the new line will be terminated, and it can be continued to any other block using the left mouse button as just explained.

Every block in the working window can be opened by *double clicking it with the left mouse button*. When the block is opened, new parameters can be set. Also, any block can be enlarged by pointing the arrow at the corner and moving the mouse while holding the left mouse button.

When all block parameters are set, simulation can be started from the simulation menu by clicking on "start" (again with the left mouse button). The simulation parameter can be redefined as necessary.

EXAMPLE D.1

We will demonstrate the procedure for building the block diagram presented in Figure 4.10 (repeated in Figure D.1).

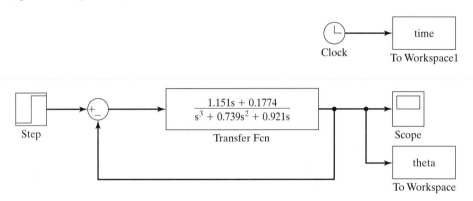

FIGURE D.1: Simulink block diagram from Example 4.26.

In order to build this block diagram, we must get the transfer function block from the Continuous library, step and clock blocks from the Sources library, sum block from the Math library, and scope and to workspace blocks from the Sources library. By clicking on the "page" icon, a new working window is opened, in which we create a new model. First, we drag all of the required blocks from the given libraries into the working window and position them as desired. Then we connect them by drawing the lines among the blocks, as explained in the Basic Rules.

The next step is to set parameters. For example, we double click on the step block with the left mouse button, and change the step start time from 1 (the default value) to 0, and confirm the change by clicking on "OK." The sum block signs should be set to +-, and the transfer function block parameters should be changed to the system transfer function parameters, equal to [1.151 0.1774] and [1 0.739 0.921 0], respectively, for the numerator and denominator.

With the block diagram drawn and parameters set, we go to the simulation menu and click on "start." The result of simulation, the system output, will be displayed on the scope by double clicking the left mouse button on the scope. In addition, the block "To Workspace" passes the simulation results to MATLAB. Similarly, the variable `time` will be passed to MATLAB. If necessary, the simulation parameters (for example, the simulation time interval, the accuracy of numerical techniques used in simulation, and others) can be adjusted. Note that for running discrete-time simulation, the step size should be set to "fixed."

Similarly, we can build much more complex block diagrams using Simulink. For complete coverage of Simulink and its features, the reader is referred to [1].

REFERENCE

[1] J. Dabney and T. Harman, *Mastering SIMULINK 4,* Prentice Hall, Upper Saddle River, NJ, 2001.

Index

Acceleration constant, 584
aliasing, 464
amplitude modulation, 522
amplitude sensitivity, 526
analog signal prefiltering, 465
antialiasing filters, 464
anticausal signal, 58
ARMA filter, 496
asymptotic stability, 353, 365, 371
auxiliary polynomial, 360

Bandlimited signal, 460, 464
bandwidth, 113, 115, 585
baseband signal, 508
BIBO stability, 352, 363–364, 369, 371
bilinear transformation, 368
block diagrams, 183
Block diagram rules
    product, 186, 256
    sum, 186, 256
Bode diagrams, 115, 259
    complex conjugate poles, 592
    complex conjugate zeros, 595
    constant, 587
    corner frequency, 592
    gain, 598
    magnitude, 587
    phase, 587
    pure differentiator, 589
    pure integrator, 589
    real pole, 591
    real zero, 581

Capacitor
    current, 540
    electric charge, 540
    energy, 541
    power, 541
    voltage, 540
carrier frequency, 525
cascade connection, 186, 256
Cauchy–Schwarz inequalities, 621
causal signal, 58
causal system, 26
Cayley–Hamilton theorem, 421, 617
channel capacity, 511
characteristic equation, 320, 329, 571
characteristic polynomial, 176, 419
characteristic values, 320, 329
continuous signal, 3
communication channel, 508
communication system, 507
Closed-loop
    bandwidth, 585
    characteristic equation, 571
    frequency transfer function, 585
    peak resonance, 585
    resonant frequency, 585
complete response, 175, 238, 246, 348–349
Convolution
    continuous-time, 51, 332, 335
    discrete-time, 52, 270, 334
    graphical, 283
    numerical, 305

Convolution (*Continued*)
    sliding tape method, 298
    theorem, 279, 294, 305
Convolution properties
    associativity, 280, 296
    commutativity, 280, 296
    continuity, 280
    distributivity, 280, 296
    duration, 280, 296
    time shifting, 280, 296
corner frequency, 592
Correlation
    autocorrelation, 493, 513
    coefficient, 494
    continuous-time, 52, 512
    crosscorrelation, 493, 512
    discrete-time, 52
    periodic signals, 515
correlators, 512
current source, 543
cut-off frequency, 521

Damped frequency, 575
damping ratio, 575
DC gain, 168
demodulation, 510, 529
demodulator, 509
DFS, 470, 491
DFT, 485
    inverse, 488
    matrix form, 487
    scalar form, 488
digital communication
        system, 507
Difference equation, 6, 318, 327
    characteristic equation, 329
    characteristic values, 329
    homogeneous solution, 329
    initial conditions, 328
    particular solution, 331, 334

Differential equation, 7, 318
    characteristic equation, 320
    characteristic values, 320
    homogeneous solution, 7, 319
    initial conditions, 319, 323
    particular solution, 7, 319, 325
digital frequency, 469
digital signal, 4
digital transfer function, 477
Dirac comb, 466
Dirac delta impulse signal, 43
Dirichlet's conditions, 80
discrete signal, 4
Discretization, 408
    Euler's method, 411
    integral approximation, 408
disturbance, 574
dominant poles, 579
DTFT, 467, 469
    existence condition, 471
    generalized, 475
    inverse, 470
DTFT properties
    conjugation, 473
    frequency differentiation, 473
    frequency shifting, 472
    linearity, 471
    modulation, 473
    periodic convolution, 474
    time convolution, 474
    time reversal, 472
    time shifting, 472

Eigenvalue invariance, 420
eigenvalues, 320, 329, 363, 419
eigenvectors, 419
Electrical circuits
    first-order, 543
    frequency of oscillations, 556
    higher-order, 559

initial conditions, 561
LC, 554
RC, 544
RL, 549
RLC, 557, 559
second-order, 554
state variables, 560
energy signals, 58
energy spectral density, 495
energy spectrum, 97
envelope detector, 526
Euler's method, 411
estimated signal, 508

Feedback
configuration, 185, 255
disturbance rejection, 574
essence, 570
feedback path, 185, 254
sensitivity function, 574
sensitivity reduction, 572
stabilization, 571
type, 583
forward difference, 52
forward path, 184, 254
FFT, 490
FIR filters, 495
Fourier series, 74
complex coefficient trigonometric, 79–80
convergence conditions, 80
exponential, 80
real coefficient trigonometric, 76
Fourier transform, 85
existence condition, 85
generalized, 98, 100
inverse, 85, 104
periodic signals, 100
Fourier transform properties, 86
conjugate symmetry, 97
duality, 98–99

frequency convolution, 96
frequency shifting, 92
integration, 102
linearity, 89
modulation, 93
symmetry, 98
time convolution, 95
time derivative, 94
time multiplication, 91
time reversal, 97
time scaling, 90
time shifting, 89
frequency division multiplexing, 510
frequency domain, 2, 73, 165
frequency magnitude spectra, 111, 258
frequency modulation, 522
frequency phase spectra, 111, 258
frequency spectra, 110, 258
Function
Dirac impulse, 43
distribution, 33, 42
even, 76
even periodic, 76
generalized, 43, 88
Kronecker, 49
odd, 76
odd periodic, 76
ordinary, 33
poles, 156
proper rational, 157
sign, 34
singular, 33, 42
signum, 34
strictly proper rational, 157
fundamental angular frequency, 74
fundamental harmonic, 75
fundamental digital harmonic, 491

Gain crossover frequency, 585, 597
gain stability margin, 585, 597

640 Index

generalized derivative, 49–50, 165
geometric series, 211
Gibbs phenomenon, 79
graphical convolution, 283

Hurwitz polynomial, 355
Heaviside's unit step, 34, 101, 104, 138
Hilbert transform, 517, 519

Ideal filter, 521
ideal sampler, 465, 469
IIR filter, 495
Impulse delta signal, 42, 86, 146, 165, 170, 172, 293, 304, 319
    discrete-time, 48, 211, 469, 481
    discrete-time shifted, 49, 211
    derivative property, 47
    frequency domain, 117
    shifted, 44, 466
    sifting property, 44, 49, 293
    time scaling property, 46
    train of impulse delta signals, 100–101, 466–467
    unit doublet, 46
    unit triplet, 46
    unit-n-tuplet, 46
impulse response, 108, 169, 235, 279, 303, 335, 338, 343, 370, 485
independent vectors, 616
Inductor
    current, 541
    electromagnetic flux, 542
    energy, 542
    power, 542
    voltage, 541
index of modulation, 526
industrial controller, 601
integral approximation method, 408
internal stability, 352, 365
instability theorem, 356

Jacobian matrices, 429
Jury's table, 367
Jury's test, 367

Kronecker function, 49

Laplace transform, 125, 144, 401
    convergence region, 125
    double-sided, 125
    existence condition, 125, 145
    inverse, 145, 157
    one-sided, 125, 144
    starred, 190
    unilateral, 144
Laplace transform properties, 146
    final value theorem, 155
    frequency shifting, 150
    initial value theorem, 154
    integral, 154
    linearity, 147
    modulation, 151
    time convolution, 153
    time derivative, 152
    time multiplication, 150
    time scaling, 149
    time shifting, 147
line spectra, 81
linearity principle, 14
linearization, 426
loop transfer function, 185
lower frequency sidebands, 509

Magnitude line spectra, 81
marginal stability, 353
mathematical modeling, 13
Mathematical model
    acrobot, 444
    aircraft, 206, 452
    amortization, 21
    car suspension system, 142

Index   641

commercial aircraft, 24, 182, 188
communication network switch, 23, 249, 453, 601
dynamics of vehicle lateral error, 23
electrical circuit, 18, 120, 389
eye movement, 23, 453
F-8 aircraft, 436
F-15 aircraft, 432
flexible beam, 395
heartbeat dynamics, 22, 180
industrial reactor, 451
inverted pendulum, 433
manipulator with a flexible joint, 206, 431, 451
mechanical system, 124
national income, 20
neuromuscular system, 206
robotic manipulator, 386
ship positioning system, 206
simple pendulum, 450
steam power system, 386
stick balancing, 428
synchronous machine, 452
translational mechanical system, 19, 390
voltage regulator, 206
MATLAB, 2, 26, 241
    basic mathematical operators, 623
    functions, 630
    help, 622
    introduction, 622
    loops, 629
    matrix operations, 625
    numbers and arithmetic expressions, 624
    plots, 629
    polynomials, 628
    script files, 630
    vectors and matrices, 624

Matrix
    addition, 612
    adjoint, 615
    Cayley–Hamilton theorem, 617
    cofactors, 614
    determinant, 614
    determinant of a product, 614
    determinant of inversion, 615
    diagonal, 400, 612
    differentiation, 617
    exponential, 396, 425
    identity, 611
    integral, 617
    integral of matrix exponential, 617
    inversion, 615
    multiplication by a scalar, 613
    null space, 616
    product, 612
    product compatible matrices, 612
    product inversion, 615
    range space, 616
    rank, 616
    spectral theorem, 617
    subtraction, 612
    symmetric, 613
    trace, 611
    transfer function, 404
    transpose, 613
maximum percent overshoot, 578
measurement noise, 570
message signal, 522
modal transformation, 420
modem, 509
modulated signal, 509
modulating signal, 522
modulation, 509
modulator, 509
modulo-N-reduction, 487
multi-input multi-output system, 13
multivariable system, 13

Natural frequency, 575
natural mode, 109, 117
negative frequencies, 509
nominal system trajectory, 426
nonlinear systems, 426
Nyquist frequency, 461
Nyquist interval, 461

Open circuit, 543, 546, 561
Open loop
    characteristic equation, 571
    frequency transfer function, 585
optical fiber amplifier, 138
optical fiber cable, 511
over-damped system, 575–576
overshoot, 578

Parallel connection, 186, 256
Parseval formula
    aperiodic continuous-signals,
        96–97
    discrete-time signals, 475
    periodic continuous signals,
        84, 517
Partial fraction expansion, 157
    complex conjugate distinct poles,
        161, 225
    distinct real poles, 158, 223
    multiple complex conjugate
        poles, 164
    multiple real poles, 159, 224
PD controller, 600
peak resonance, 113, 585, 587
peak time, 578
periodic inputs, 116, 118, 492
periodic outputs, 118
periodic spectrum, 464
phase crossover frequency, 585, 597
phase line spectra, 81
phase modulation, 522

phase stability margin, 585, 597
phase variable canonical form, 393
PI controller, 600
PID controller, 600
    practical implementation, 601
    with approximate derivative, 601
Poles, 117
    complex conjugate distinct, 161
    distinct real, 158
    multiple complex conjugate, 164
    multiple real, 159
position constant, 583
power signals, 58
pre-envelope signal, 520
pulse code modulation, 523, 530
pulse duration modulation, 523
pulse position modulation, 523

Quantizer, 530

Ramp signal, 35–36
rectangular pulse, 37, 38, 87
reference input, 583
Resistor
    energy, 540
    power, 540
    voltage, 539
resonant frequency, 113, 585, 587
rise time, 579
Routh criterion, 357
Routh–Hurwitz stability
    criterion, 355
Routh table, 357
    singular cases, 359
Routh theorem, 358

Sampling
    operation, 4, 459
    period, 4
    theorem, 4, 461

Sampler, 530
    ideal, 465
    physically realizable, 467
sensitivity function, 574
settling time, 578
short circuit, 543, 551, 561
sign signal, 34
Signal
    aliasing, 464
    anticausal, 58
    bandlimited, 460
    carrier, 93
    causal, 58
    cosine, 39
    damped sinusoidal, 69
    discrete rectangular pulse, 38
    discrete sinc, 41
    discrete unit ramp, 36
    discrete unit step, 35
    energy, 56, 513
    energy density spectrum, 115
    energy spectrum, 115
    even, 56
    exponential, 87, 146, 212
    frequency bandwidth, 115
    Heaviside's unit step, 34, 101, 104, 138
    ideal sampled, 466–467
    impulse delta, 42, 86
    magnitude spectrum, 114
    modulo-N-reduction, 469
    odd, 56
    parabolic, 36
    periodic, 56, 464
    phase spectrum, 115
    power, 56
    prefiltering, 465
    rectangular pulse, 37, 87
    sampling, 459
    sawtooth, 82
    sinc, 39
    sine, 39
    spectra, 114
    time duration, 115
    time limited, 460
    triangular pulse, 38, 102
    unit ramp, 35
    unit step, 34, 146, 212
    with jump discontinuities, 49–50, 165
    wrapped, 489
Signal average power
    continuous-time, 57
    discrete-time, 57
signal correlators, 493
Signal energy
    continuous-time, 57
    discrete-time, 57
signal-to-noise ratio, 511
signum signal, 34
signal transmission, 508
similarity transformation, 419
Simulink, 26, 253, 633
    printing, 253
    rules for building diagrams, 633
    setting parameters, 634
    simulation parameters, 635
sinc signal, 39, 41
single-input single-output system, 13
sinusoidal inputs, 116, 121, 260, 477
sliding tape method, 298
s–plane, 209
SONET, 530
Stability
    asymptotic, 353, 365, 371
    BIBO, 352, 363–364, 369, 371
    gain crossover frequency, 585, 597
    gain margin, 585, 597
    internal, 352, 365
    instability, 356
    marginal, 353

Stability (*Continued*)
    phase crossover frequency, 585, 597
    phase margin, 585, 597
    stable (system), 352, 365
static gain, 167, 234
steady state errors, 581
steady state gain, 167, 234
steady state response, 9–10, 116, 122, 260, 349, 493
step response, 173, 240
step signal, 34, 35, 146, 212, 481, 518
summation formulas, 619
summation of infinite series, 227
superposition principle, 14
System
    analog, 26
    at rest, 13, 108, 170, 279
    bandwidth, 113, 585
    causal, 26
    characteristic polynomial, 176, 419
    complete response, 175, 238, 246, 348–349
    continuous-time, 5
    derivative formulation, 228, 242
    deterministic, 26
    digital, 26
    discrete-time, 5
    disturbance, 574
    dynamic, 24
    elementary impulse response, 339, 344
    eigenfunction, 110
    eigenvalues, 418–419
    finite dimensional, 25
    forced response, 7, 176
    linear dynamic, 25, 113
    linearity, 13
    identification, 20
    impulse response, 108, 169, 235, 279, 303, 335, 338, 343, 370, 405, 416
    infinite dimensional, 25
    initial conditions, 6, 172
    input, 6
    input-in transient, 231, 232, 243
    input signal differentiation, 172, 323
    integral formulation, 228
    instantaneous, 26
    memoryless, 26
    modes, 173
    multi-input multi-output, 13
    multivariable, 13
    natural response, 176
    nonanticipatory, 26
    nonlinear dynamic, 25
    order, 6
    output, 6
    poles, 117, 234, 320, 363, 414
    ramp response, 173
    sensitivity, 574
    single-input single-output, 13
    static, 24
    step response, 173, 240
    steady state errors, 581
    steady state response, 9–10, 116, 122, 260, 349
    stochastic, 26, 507
    time delayed formulation, 228
    time forwarded formulation, 228
    time invariance, 16–17, 293
    time-invariant, 11–12, 25, 165, 318
    time-varying, 25
    transfer function, 2, 107, 167, 233
    transient response, 9–10, 127
    with concentrated parameters, 24
    with distributed parameters, 24
    with lumped parameters, 24

with memory, 26
zero-input response, 7, 108, 228, 237, 245, 334
zero-state response, 7, 106–107, 165, 170, 228, 236, 244, 292, 304, 322, 349
zeros, 234
system analysis, 106, 127, 165, 228, 476, 484
systems of linear difference equations, 249, 616
System state
    impulse response, 405, 416
    phase variable canonical form, 393
    response, 399
    space approach, 2, 388
    space form, 388–389, 407
    step response, 405, 416
    transition matrix, 397, 412, 421–422, 425
    transfer function, 404
    zero-input response, 402
    zero-state response, 402
switching modulator, 526

Telephone signal, 93, 530
time constant, 545, 575
time delay, 164
time division multiplexing, 510
time domain, 2, 279
time invariance, 16–17, 293
time limited signal, 460
Transfer function, 107, 167, 185, 233, 404, 485, 585
    closed-loop, 184–185, 255
    DC gain, 168
    invariance, 420
    open-loop, 183–184, 254
    poles, 167, 234
    static gain, 167
    steady state gain, 167, 234
    zeros, 167, 234
train of impulse delta signals, 100, 466
train of rectangular pulses, 467
transient response, 9–10, 127
transition matrix, 397, 412, 421–422, 425
transmitted signal, 508
transmitter, 508
triangular pulse, 38, 102
type of feedback control system, 583

Under-damped system, 575–576
underdetermined coefficients method, 325, 331
unit circle, 209, 367
upper frequency sidebands, 509

Variation of parameters method, 325, 331
velocity constant, 583
voltage source, 543

Wavelength division multiplexing, 511
wrapping, 489

Zero-input response, 7, 108, 228, 237, 245, 334
zero-order hold, 189–190
zero padding, 486
zero-state response, 7, 106–107, 165, 170, 228, 236, 244, 292, 304, 322, 349
z–plane, 209
$\mathcal{Z}$-transform, 189, 191, 208, 413, 478
    bilateral, 191, 478
    convergence condition, 210
    inverse, 222
    one-sided, 191, 209

$\mathcal{Z}$-transform (*Continued*)
   region of convergence, 210, 478
   two-sided, 191, 210, 478
   unilateral, 191
$\mathcal{Z}$-transform properties, 209
   complex conjugation, 483
   convolution, 219, 483
   final value theorem, 220
   frequency differentiation, 483
   frequency scaling, 482
   frequency shifting, 217
   initial value theorem, 220
   left shift in time, 215,
   linearity, 212, 480
   modulation, 217
   right shift in time, 213
   time shifting, 481
   time multiplication, 216, 483
   time reversal, 482

TABLE 4.1 Properties of the Laplace Transform

$\mathcal{L}\{\alpha_1 f_1(t) \pm \alpha_2 f_2(t)\}$	$\alpha_1 F_1(s) \pm \alpha_2 F_2(s)$
$\mathcal{L}\{f(t - t_0) u(t - t_0)\}$	$e^{-st_0} F(s), \ t_0 > 0$
$\mathcal{L}\{f(at)\}$	$\dfrac{1}{a} F\left(\dfrac{s}{a}\right), \ a > 0$
$\mathcal{L}\{t^n f(t)\}$	$(-1)^n \dfrac{d^n}{ds^n} F(s)$
$e^{\lambda t} f(t)$	$F(s - \lambda)$
$f(t)\cos(\omega_0 t)$	$\dfrac{1}{2}[F(s + j\omega_0) + F(s - j\omega_0)]$
$f(t)\sin(\omega_0 t)$	$\dfrac{j}{2}[F(s + j\omega_0) - F(s - j\omega_0)]$
$\mathcal{L}\left\{\dfrac{d}{dt} f(t)\right\}$	$sF(s) - f(0^-)$
$\mathcal{L}\left\{\dfrac{d^2}{dt^2} f(t)\right\}$	$s^2 F(s) - sf(0^-) - f^{(1)}(0^-)$
$\mathcal{L}\left\{\dfrac{d^n}{dt^n} f(t)\right\}$	$s^n F(s) - s^{n-1} f(0^-) - s^{n-2} f^{(1)}(0^-) - \cdots - f^{(n-1)}(0^-)$
$\mathcal{L}\{f_1(t) * f_2(t)\}$	$F_1(s) F_2(s)$
$\mathcal{L}\left\{\int_0^t f(\tau)\, d\tau\right\}$	$\dfrac{1}{s} F(s)$
$\lim\limits_{t \to 0^+} \{f(t)\}$	$\lim\limits_{s \to \infty} \{sF(s)\}$
$\lim\limits_{t \to \infty} \{f(t)\}$	$\lim\limits_{s \to 0} \{sF(s)\}$